Crop Wild Relative Conservation and Use

Crop Wild Relative Conservation and Use

Edited by

N. Maxted,

School of Biosciences, University of Birmingham, Edgbaston, Birmingham, B15 2TT, UK.

B.V. Ford-Lloyd,

School of Biosciences, University of Birmingham, Edgbaston, Birmingham, B15 2TT, UK.

S.P. Kell,

School of Biosciences, University of Birmingham, Edgbaston, Birmingham, B15 2TT, UK.

J.M. Iriondo,

Area de Biodiversidad y Conservación, ESCET, Universidad Rey Juan Carlos, c/Tulipán s/n, E-28933 Móstoles, Madrid, Spain.

E. Dulloo,

Bioversity International, Maccarese 00057, Rome, Italy.

and

J. Turok,

Bioversity International, Maccarese 00057, Rome, Italy.

www.cabi.org

CABI is a trading name of CAB International

CABI Head Office
Nosworthy Way
Wallingford
Oxfordshire OX10 8DE
UK

Tel: +44 (0)1491 832111
Fax: +44 (0)1491 833508
E-mail: cabi@cabi.org
Website: www.cabi.org

CABI North American Office
875 Massachusetts Avenue
7th Floor
Cambridge, MA 02139
USA

Tel: +1 617 395 4056
Fax: +1 617 354 6875
E-mail: cabi-nao@cabi.org

A catalogue record for this book is available from the British Library, London, UK.

Library of Congress Cataloging-in-Publication Data

Crop wild relative conservation and use / edited by N. Maxted ... [et al.].
p. cm.
ISBN 978-1-84593-099-8 (alk. paper) -- ISBN 978-1-84593-307-4 (ebook)
1. Crops--Germplasm resources. 2. Germplasm resources, Plant. 3. Genetic resources conservation. I. Maxted, Nigel. II. Title.
SB123. 3. C768 2007
333. 95' 34--dc22

2007017714

ISBN 978 1 84593 099 8

Typeset by SPi, Pondicherry, India.
Printed and bound in the UK by Biddles Ltd., King's Lynn.

Contents

Part II Establishing Inventories and Conservation Priorities

Part III Threat and Conservation Assessment

Contributors

S. Abbo, *The Levi Eshkol School of Agriculture, The Hebrew University of Jerusalem, Rehovot 76100, Israel.*

J.M. Abreu, *Universidade do Porto, Centro de Estudos de Ciencia Animal, ICETA, Campus Agrário de Vairão, R. Monte-Crasto 4485-661, Vairão, Portugal.*

M. Agami, *Institute for Cereal Crops Improvement, Tel Aviv University, Tel Aviv 69978, Israel.*

C. Agbangla, *Laboratoire de Génétique, FAST-Université d'Abomey-Calavi, BP 526 Cotonou, Bénin.*

N. Al-Atawneh, *Dryland Agrobiodiversity Project, Ministry of Agriculture/ United Nations Development Programme, P.O. Box 524, Hebron, Palestine.*

S. Allahham, *Rangeland specialist, Conservation and Sustainable Use of Dry Land Agrobiodiversity Project, Palestinian Authority.*

A. Alves, *Instituto de Investigação Científica Tropical, Centro de Florestas e dos Produtos Florestais, Tapada da Ajuda 1349-017 Lisboa, Portugal.*

A. Amri, *International Centre for Agriculture Research in the Dry Areas (ICARDA), West Asia office, Amman, Jordan.*

R. Assi, *Lebanon Agricultural Research Institute, Tel Amara, Lebanon.*

G. Augusto, *European Topic Centre on Biological Diversity, Muséum National d'Histoire Naturelle, 57 rue Cuvier, 75231 Paris cedex 05, France.*

A. Avagyan, *EC Food Security Programme in Armenia, Ministry of Agriculture, Government Building 3, Republic Square, Yerevan 375010, Armenia.*

N. Azzu, *Seed and Plant Genetic Resources Service, Agriculture Department, Food and Agriculture Organization of the UN, Via delle Terme di Caracalla 00100, Rome, Italy.*

M.N. Baco, *Institut National des Recherches Agricoles du Bénin, Station d'Ina, BP 526 Parakou, Bénin.*

J.-P. Baudoin, *Unité de Phytotechnie tropicale et d'Horticulture, Faculté Universitaire des Sciences Agronomiques de Gembloux, BE-5030, Gembloux, Belgium.*

G. Bedini, *Botanico di Pisa, Dipartimento di Scienze Botaniche, Università degli Studi di Pisa, 56100 Pisa, Italy.*

M. Benchacho, *Université Ibn Tofaïl, Faculté des Sciences, Departement de Biologie, 1400 Kénitra, Morocco.*

P. Ben-Yehuda, *Institute for Cereal Crops Improvement, Tel Aviv University, Tel Aviv 69978, Israel.*

E. Bettencourt, *Department of Genetic Resources and Breeding, Estação Agronómica Nacional – INIAP, 2784-505 Oeiras, Portugal.*

C. Bonomi, *Giardino Botanico Alpino 'Viotte', Museo Tridentino di Scienze Naturali, 38100 Trento, Italy.*

P. Bordoni, *Global Facilitation Unit for Underutilized Species, Bioversity International, Maccarese 00057, Rome, Italy.*

F. Branco, *Dipartimento di Biologia Vegetale e Biotecnologie Agro-ambientali e Zootecniche, Università degli Studi di Perugia, Borgo XX Giugno 74, 06121 Perugia, Italy.*

C. Can, *Department of Biology, University of Gaziantep, 27310 Gaziantep, Turkey.*

G. Castellini, *Dipartimento di Biologia Vegetale e Biotecnologie Agro-ambientali e Zootecniche, Università degli Studi di Perugia, Borgo XX Giugno 74, 06121 Perugia, Italy.*

L. Collette, *Seed and Plant Genetic Resources Service, Agriculture Department, Food and Agriculture Organization of the UN, Via delle Terme di Caracalla 00100, Rome, Italy.*

E. Collin, *Cemagref, Unité de Recherches Ecosystèmes Forestiers, Nogents/Vernisson, France.*

T.L.P. Couvreur, *IRD (Institut de Recherche pour le Développement), UMR 1097 DGPC/DYNADIV, 911 Avenue Agropolis BP 64501, 34394 Montpellier cedex 5, France/Present address: Wageningen University, Biosystematics Group, National Herbarium Nertherland, General Foulkesweg 37, 6703 BL Wageningen, The Netherlands.*

S. Curtis, *Neal's Yard Remedies, Peacemarsh, Gillingham, Dorset SP8 4EU, UK.*

O. Daïnou, *Laboratoire de Génétique, FAST-Université d'Abomey-Calavi, BP 526 Cotonou, Bénin.*

B. De Cuyper, *Institute for Forestry and Game Management (IBW), Hoeilaart, Belgium.*

L. De Hond, *Depto. Biología Vegetal, EUIT Agrícola, Universidad Politécnica de Madrid, E-28040 Madrid, Spain.*

J. Degreef, *National Botanical Garden of Belgium, Domein van Bouchout, BE-1860 Meise, Belgium.*

K.J. Dehmer, *Leibniz Institute of Plant Genetics and Crop Plant Research (IPK), Corrensstr. 3, 06466 Gatersleben, Germany.*

S. Dias, *Bioversity International, Maccarese 00057, Rome, Italy.*

D. Draper, *Universidade de Lisboa, Museu Nacional de História Natural, Jardim Botánico, Rua da Escola Politécnica no 58, 1200-102 Lisboa, Portugal.*

M.E. Dulloo, *Bioversity International, Maccarese 00057, Rome, Italy.*

K. Durah, *Bioversity International (CWANA), c/o* ICARDA, P.O. Box 5466, Aleppo, Syria.

D. Evans, *Scottish Natural Heritage, seconded to the European Topic Centre on Biological Diversity, Edinburgh, UK.*

B. Fady, *Institut National de la Recherche Agronomique (INRA), URFM Unité de Recherches Forestières Méditerranéennes (UR629), Domaine Saint Paul, Site Agroparc 84914 Avignon Cedex 9, France.*

S. Ferreira, *Dias Instituto Superior de Agronomia, Centro de Estudos Agro-Alimentares, Tapada da Ajuda, 1349-017 Lisboa, Portugal.*

K. Flath, *Federal Biological Research Centre for Agriculture and Forestry, Institute for Plant Protection in Field Crops and Grassland, 14532 Kleinmachnow, Germany.*

C.S. Ford, *Plant Science Laboratories, School of Biological Sciences, The University of Reading, RG6 6AS, UK.*

B.V. Ford-Lloyd, *School of Biosciences, University of Birmingham, Edgbaston, Birmingham, B15 2TT, UK.*

L. Frese, *Federal Centre for Breeding Research on Cultivated Plants, Gene Bank, Braunschweig 38116, Germany.*

S. Gaiji, *Bioversity International, Maccarese 00057, Rome, Italy.*

P. García, *Área de Genética, Facultad de Ciencias Biológicas y Ambientales, Universidad de León, E-24071 León, Spain.*

C.U. Germeier, *Federal Centre for Breeding Research on Cultivated Plants, Gene Bank, Braunschweig 38116, Germany.*

E. Große, *Federal Biological Research Centre for Agriculture and Forestry, Institute for Nematology and Vertebrate Research, 14532 Kleinmachnow, Germany.*

L. Guarino, *Secretariat of the Pacific Community (SPC), Private Mail Bag, Suva, Fiji Islands.*

A. Habekuß, *Federal Centre for Breeding Research on Cultivated Plants, Institute of Epidemiology and Resistance Resources, 06449 Aschersleben, Germany.*

R. Hadas, *Israeli Gene Bank, ARO, The Volcani Center, P.O. Box 6, Bet Dagan 50250, Israel.*

R. Hajjar, *Bioversity International, Maccarese 00057, Rome, Italy.*

M. Harutyunyan, *Armenian State Agrarian University, PGR Laboratory, Teryan 74, Yerevan 375009, Armenia.*

H. Hasasneh, *Crop breeder, Conservation and Sustainable Use of Dry Land Agrobiodiversity Project, Palestinian Authority.*

J. Helenius, *Department of Applied Biology, P.O. Box 27, FI-00014, University of Helsinki, Finland.*

V.H. Heywood, *Plant Science Laboratories, School of Biological Sciences, The University of Reading, Reading, RG6 2AS, UK.*

T. Hodgkin, *Bioversity International, Maccarese 00057, Rome, Italy.*

R. Hoekstra, *Centre for Genetic Resources, The Netherlands (CGN), Wageningen University and Research Centre, P.O. Box 16, 6700 AA Wageningen, The Netherlands.*

I. Hoeschle-Zeledon, *Global Facilitation Unit for Underutilized Species, Bioversity International, Maccarese 00057, Rome, Italy.*

A. Horovitz, *Department of Genetics, The Hebrew University of Jerusalem, Jerusalem 91904, Israel.*

M. Hovhannisyan, *Armenian State Agrarian University, PGR Laboratory, Teryan 74, Yerevan 375009, Armenia.*

J.M. Iriondo, *EUIT Agrícolas, Departamento de Biología Vegetal, Universidad Politécnica de Madrid, 28040 Madrid, Spain.*

S.L. Jury, *Centre for Plant Diversity and Systematics, Plant Science Laboratories, The University of Reading, Whiteknights, P.O. Box 221, Reading, Berkshire, RG6 6AS, UK.*

D. Kaplan, *Israel Nature and Parks Authority, Megido Post, 19230, Israel.*

C.S. Kariyawasam, *Biodiversity Secretariat, Ministry of Environment and Natural Resources, "Sampathpaya", Battaramulla, Sri Lanka.*

S.P. Kell, *School of Biosciences, University of Birmingham, Edgbaston, Birmingham, B15 2TT, UK.*

H. Korpelainen, *Department of Applied Biology, P.O. Box 27, FI-00014, University of Helsinki, Finland.*

J. Koskela, *Bioversity International, Maccarese 00057, Rome, Italy.*

H. Knüpffer, *Leibniz Institute of Plant Genetics and Crop Plant Research (IPK), Corrensstrasse 3, D-06466 Gatersleben, Germany.*

C. Lara, *PUCE (Pontificia Universidad Católica del Ecuador), Laboratorio de Genética Molecular de Eucariotes, Av. 12 de Octubre y Roca, Quito, Ecuador.*

F. Lefèvre, *Institut National de la Recherche Agronomique (INRA), URFM Unité de Recherches Forestières Méditerranéennes (UR629), Domaine Saint Paul, Site Agroparc 84914 Avignon Cedex 9, France.*

J. Labokas, *Institute of Botany, Žaliųjų Ežerų 49, LT-08406 Vilnius, Lithuania.*

A. Lane, *Bioversity International, Maccarese 00057, Rome, Italy.*

S. Lev-Yadun, *Department of Biology, University of Haifa-Oranim, Tivon 36006, Israel.*

G. Loïs, *European Centre for Nature Conservation, seconded to the European Topic Centre on Biological Diversity, UK.*

B. Ludeña, *PUCE (Pontificia Universidad Católica del Ecuador), Laboratorio de Genética Molecular de Eucariotes, Av. 12 de Octubre y Roca, Quito, Ecuador.*

J. Magos Brehm, *School of Biosciences, University of Birmingham, Edgbaston, Birmingham, B15 2TT, UK/Universidade de Lisboa, Museu Nacional de História Natural, Jardim Botânico, R. Escola Politécnica 58, 1269-102 Lisboa, Portugal.*

J. Manisterski, *Institute for Cereal Crops Improvement, Tel Aviv University, Tel Aviv 69978, Israel.*

M.A. Martins-Loução, *Universidade de Lisboa, Museu Nacional de História Natural, Jardim Botânico, R. Escola Politécnica 58, 1269-102 Lisboa, Portugal/Universidade de Lisboa, Faculdade de Ciências, Centro de Ecologia e Biologia Vegetal, Campo Grande C2, Piso 4, 1749-016 Lisboa, Portugal.*

N. Maxted, *School of Biosciences, University of Birmingham, Edgbaston, Birmingham, B15 2TT, UK.*

E. Millet, *Institute for Cereal Crops Improvement, Tel Aviv University, Tel Aviv 69978, Israel.*

M. Mitchell, *School of Biosciences, University of Birmingham, Edgbaston, Birmingham, B15 2TT, UK.*

J.D. Moore, *Plantkind Consulting, 9 Guys Cliffe Terrace, Warwick, CV34 4LP, UK.*

V. Negri, *Dipartimento di Biologia Vegetale e Biotecnologie Agro-ambientali e Zootecniche, Università degli Studi di Perugia, Borgo XX Giugno 74, 06121 Perugia, Italy.*

B. Odé, *Stichting Floristisch Onderzoek Nederland (Floron), P.O. Box 9514, 2300 RA Leiden, The Netherlands.*

M. Ouédraogo, *Institut de l'Environnement et de Recherches Agricoles, 04 BP 8645, Ouagadougou 04, Burkina Faso.*

M. Ozaslan, *Department of Biology, University of Gaziantep, 27310 Gaziantep, Turkey.*

S. Padulosi, *Global Facilitation Unit for Underutilized Species, Bioversity International, Maccarese 00057, Rome, Italy.*

M. Parra-Quijano, *Facultad de Agronomía, Universidad Nacional de Colombia sede Bogotá, Ciudad Universitaria, Avenida Carrera 30 No 45-03, A.A. 14490, Bogotá, Colombia.*

G. Pendinen, *Federal Centre for Breeding Research on Cultivated Plants, Institute of Horticultural Crops, 06484 Quedlinburg, Germany.*

M. Pérez de la Vega, *Área de Genética, Facultad de Ciencias Biológicas y Ambientales, Universidad de León, E-24071 León, Spain.*

D. Pignone, *CNR – Institute of Plant Genetics, Via Amendola 165/A, 70126 Bari, Italy.*

J.-C. Pintaud, *IRD (Institut de Recherche pour le Développement), UMR 1097 DGPC/DYNADIV, 911 Avenue Agropolis BP 64501, 34394 Montpellier cedex 5, France / 2 IRD, Whimper 442 y Coruña, A. P. 17-12-857, Quito, Ecuador.*

J.-L. Pham, *IRD (Institut de Recherche pour le Développement), UMR 1097 DGPC/DYNADIV, 911 Avenue Agropolis BP 64501, 34394 Montpellier cedex 5, France.*

J. Radušienė, *Institute of Botany, Žaliujų Ežerų 49, LT-08406 Vilnius, Lithuania.*

T.S. Rajicic, *Leibniz Institute of Plant Genetics and Crop Plant Research (IPK), Corrensstr. 3, 06466 Gatersleben, Germany.*

R.S.S. Ratnayake, *Biodiversity Secretariat, Ministry of Environment and Natural Resources, 'Sampathpaya', Battaramulla, Sri Lanka.*

D. Richard, *European Topic Centre on Biological Diversity, Muséum National d'Histoire Naturelle, 57 rue Cuvier, 75231 Paris cedex 05, France.*

O.J. Rocha, *Escuela de Biología, Universidad de Costa Rica, Ciudad Universitaria R. Facio, San José, Costa Rica.*

J.C. Rodrigues, *Instituto de Investigação Científica Tropical, Centro de Florestas e dos Produtos Florestais, Tapada da Ajuda 1349-017 Lisboa, Portugal.*

G. Rossi, *Dipartimento di Ecologia del Territorio, Università degli Studi di Pavia, 27100 Pavia, Italy.*

B. Ruge-Wehling, *Federal Centre for Breeding Research on Cultivated Plants, Institute of Agricultural Crops, 18190 Groß Lüsewitz, Germany.*

L.E. Sáenz de Miera, *Área de Genética, Facultad de Ciencias Biológicas y Ambientales, Universidad de León, E-24071 León, Spain.*

N. Scarcelli, *Equipe DYNADIV, UMR 1097 Diversité et Génomes des Plantes Cultivées, Institut de Recherche pour le Développement (IRD), BP 64501, 34394 Montpellier cedex 5, France.*

M.A. Scholten, *School of Biosciences, University of Birmingham, Edgbaston, Birmingham, B15 2TT, UK.*

O. Schrader, *Department of Biotechnology, University of St. Petersburg, St. Petersburg, Russia.*

M. Scholz, *Federal Centre for Breeding Research on Cultivated Plants, Institute of Agricultural Crops, 18190 Groß Lüsewitz, Germany.*

S. Sharrock, *Botanic Gardens Conservation International, Descanso House, 199 Kew Road, Richmond, Surrey TW9 3BW, UK.*

A. Sirota, *Israeli Gene Bank, ARO, The Volcani Center, P.O. Box 6, Bet Dagan 50250, Israel.*

T.N. Smekalova, *N.I. Vavilov Institute of Plant Industry (VIR), 42–44 Bolshaya Morskaya Street, St. Petersburg 190000, Russia.*

G. Sonnante, *CNR – Institute of Plant Genetics, Via Amendola 165/A, 70126 Bari, Italy.*

C.M. Sousa-Correia, *University of Birmingham, School of Biosciences, Birmingham, Edgbaston, B15 2TT, UK.*

P.P. Smith, *Millennium Seed Bank Project, Royal Botanic Gardens Kew, Wakehurst Place, Ardingly, RH17 6TN, UK.*

S. Takaluoma, *Department of Applied Biology, P.O. Box 27, FI-00014 University of Helsinki, Finland.*

I. Thormann, *Bioversity International, Maccarese 00057, Rome, Italy.*

E. Torres, *EUIT Agrícolas, Departamento de Biología Vegetal, Universidad Politécnica de Madrid, 28040 Madrid, Spain.*

S. Tostain, *Equipe DYNADIV, UMR 1097 Diversité et Génomes des Plantes Cultivées, Institut de Recherche pour le Développement (IRD), BP 64501, 34394 Montpellier cedex 5, France.*

A. Toussaint, *Unité de Phytotechnie tropicale et d'Horticulture, Faculté Universitaire des Sciences Agronomiques de Gembloux, BE-5030, Gembloux, Belgium.*

J. Turok, *Bioversity International, Maccarese 00057, Rome, Italy.*

M.G.P. van Veller, *Centre for Genetic Resources, The Netherlands (CGN), Wageningen University and Research Centre, P.O. Box 16, 6700 AA Wageningen, The Netherlands.*

F.J. Vences, *Área de Genética, Facultad de Ciencias Biológicas y Ambientales, Universidad de León, E-24071 León, Spain.*

Y. Vigouroux, *Equipe DYNADIV, UMR 1097 Diversité et Génomes des Plantes Cultivées, Institut de Recherche pour le Développement (IRD), BP 64501, 34394 Montpellier cedex 5, France.*

P. Wehling, *Federal Centre for Breeding Research on Cultivated Plants, Institute of Agricultural Crops, 18190 Groß Lüsewitz, Germany.*

M.J. Wilkinson, *Plant Science Laboratories, School of Biological Sciences, The University of Reading, RG6 6AS, UK.*

G. von Wühlisch, *Institute for Forest Genetics and Forest Tree Breeding (BFH), Grosshansdorf, Germany.*

D. Wyse-Jackson, *Botanic Gardens Conservation International, Descanso House, 199 Kew Road, Richmond, Surrey TW9 3BW, UK.*

D. Zohary, *Department of Evolution, Systematics and Ecology, The Hebrew University, Jerusalem 91904, Israel.*

I. Zoro Bi, *Unité de Formation et de Recherche des Sciences de la Nature, Université d'Abobo-Adjamé, 02 BP 801, 02 Abidjan, Côte d'Ivoire.*

Jack Hawkes: Plant Collector, Researcher, Mentor and Visionary

Professor John Gregory Hawkes, known to everyone simply as Jack, died peacefully on the evening of 6 September 2007; he was 92 years old. Jack graduated with first class honours from the University of Cambridge in 1937, and received his PhD in 1941 from the same university. His thesis was one of the first studies on the diversity and taxonomy of potatoes, based on the materials he had collected during his 1938–1939 expeditions to South America. After joining the Imperial Bureau of Plant Breeding and Genetics in Cambridge, UK, Jack travelled to Leningrad to meet with Russian potato experts, including N.I. Vavilov, and thus began a life-long interest in genetic resources and their use.

In 1948 he moved to Colombia to help establish a national potato programme, working with Nelson Estrada to broaden the genetic foundation of potato breeding using crop wild relatives. In 1952, he returned to the United Kingdom to accept a Lectureship at the University of Birmingham. In 1961 he was awarded a Personal Chair in Taxonomic Botany and subsequently was appointed Mason Professor of Botany and Head of Department in 1967, remaining in the department until his retirement in 1982.

Jack's contribution to the taxonomy and biosystematics of wild and cultivated potatoes was enormous and he was awarded the ScD degree from the University of Cambridge in 1957 for this research. He returned on several occasions to Central and South America to collect potatoes and their crop wild relatives. He was also involved in establishing the genetic resources programme at the International Potato Centre in Lima, Peru. He established Birmingham University as a centre of excellence in crop plant evolution and taxonomic studies, and led the first computer-mapped flora project for the English county of Warwickshire.

Jack's lifelong passion and major contribution was to genetic resources conservation and use, a subject in which he worked with Sir Otto Frankel, Jack Harlan and Erna Bennett, among others, to establish the science. He recognized the skills shortage in this field and in 1969 established the Master's course in Conservation and Utilization of Plant Genetic Resources in Birmingham. In

Professor Jack Hawkes

1915–2007

this sense, Jack not only passed his enthusiasm on to many cohorts of PGR students but mentored the emerging discipline of genetic resources conservation itself. About 1,500 students have passed through this course since its inception, most coming from developing countries.

Throughout his career Jack was a close collaborator with several centres of the CGIAR, in particular with CIP (in Lima) and Bioversity (in Rome) and was a personal advisor to a number of national PGR programmes. Among the many awards that Jack received were the Frank N Meyer Memorial Award of the American Genetic Association (1973), the Congress Medal from the XII International Botanical Congress in Leningrad (1975) and the Linnean Society Gold Medal (1984). He was President of the Linnean Society of London from 1991–1994. In 1994 Her Majesty Queen Elizabeth invested Jack with the OBE (Officer of the Order of the British Empire).

Adapted from text provided by
M.T. Jackson

Preface

Crop wild relatives (CWR), which are wild species closely related to crops, are a neglected global natural resource; yet they make a concrete contribution to global wealth creation and food security (estimated at US$350 million per year in 1986 in the USA alone). CWR genetic diversity is severely threatened by habitat loss, fragmentation and simplification, the impact of invasive species, overexploitation and global change, but the diversity of CWR is not adequately conserved, either *in situ* or *ex situ*. In recent years, slowly but steadily, the science of CWR conservation has been developed, applying both *in situ* and *ex situ* techniques, in parallel with their growing exploitation as gene donors for the broad range of crops. Especially the EC-funded PGR Forum project, which this volume grew out of, has made significant methodological advances. The project culminated in the First International Conference on Crop Wild Relative Conservation and Use, which was held in Sicily, Italy, in September 2005. As the Conference title suggests, the scope was broader than simply reporting the products of a European project – for the first time, it provided a platform for CWR scientists, potential users and other stakeholders to debate the broad range of issues relating to CWR conservation and use. It aimed to provide a comprehensive review of CWR conservation and use methods, highlight exemplar case studies and draw attention to new initiatives. The specific objectives of the Conference were to:

- Promote the importance of wild plant species of socio-economic value to the international community;
- Review the establishment of CWR inventories and establish a baseline for their conservation assessment;
- Assess procedures for establishing conservation priorities for CWR;
- Review the current status of information access and management for CWR;
- Evaluate methodologies for *in situ* and *ex situ* CWR conservation;

- Explore ways of strengthening CWR conservation and use through international and interagency collaboration;
- Disseminate PGR Forum products to the European and global PGR community, and discuss their wider application and continued use.

This volume presents the major products of the PGR Forum project and draws on the wide range of expertise and geographic coverage represented by delegates to the CWR Conference. Some of the chapters are directly related to the presentations given, while others have been written independently of the Conference. Like the Conference itself, this volume aims to address all aspects of CWR conservation and use.

We hope that this volume will further act as a stimulus to improved global CWR conservation and use, to raise the profile of CWR both within the PGR community and among CWR users, but most importantly to ensure that they remain available to meet the demands of future generations. The chapters, and particularly the Global Strategy for CWR Conservation and Use, offer detailed protocols to ensure that these goals are met – now we need to implement them.

The Editors

Foreword

This comprehensive volume marks an important landmark in the conservation of crop genetic resources. It arose out of the First International Conference on Crop Wild Relative Conservation and Use, which was organized within the framework of the EC-funded PGR Forum project, in which Bioversity was a partner. The Conference – which was jointly organized by the University of Birmingham, United Kingdom, Istituto Sperimentale per la Frutticoltura di Roma, Italy, and the International Plant Genetic Resources Institute (now Bioversity International), with local assistance from the Agricultural Extension Service of the Regional Administration of Agrigento, Sicily, Italy – came at a very opportune and strategic time. This is a subject of fundamental importance, not only to agriculture, but also to health and the environment and the livelihood of people. Mechanisms for sharing knowledge in relation to this area of work, and for strengthening partnerships and strategic alliances for future collaboration are therefore critically needed. That is what the Conference was set up to do.

Wild species related to crops are valuable sources of genes for crop improvement and adaptation to changing environmental conditions. Crop wild relatives (CWR) have been used for crop improvement in the formal breeding sector for more than 100 years, especially for resistance to insect pests and diseases. Their genes provide resistance against pests such as the wheat curl mite and diseases such as the potato virus and grassy stunt in rice. Wild relative genes are being used to improve tolerance of stressful abiotic conditions such as drought in wheat and acid sulphate soils in rice. Breeders have also used genes from CWR to boost the nutritional value of foods, such as enhancing the protein content in durum wheat and increasing provitamin A in tomato. The high anticancer properties found in some varieties of broccoli originated in a Sicilian wild relative: by crossing cultivated broccoli with the wild relative, scientists are breeding a variety that contains higher levels of the cancer-fighting antioxidant that destroys compounds damaging to DNA.

The increasing genetic uniformity of some crops as a result of the adoption of modern high-yielding varieties increases their susceptibility to biotic and abiotic stresses. Large-scale crop failures resulting from genetic vulnerability have occurred as a result of the late potato blight epidemic of the 1840s, when a large proportion of the mostly susceptible potato varieties grown at that time were eliminated as the blight spread across Ireland, Europe and North America. Devastating losses caused by the Southern corn blight outbreak in the USA in the 1970s highlighted the real risk of relying on a few high-yielding varieties. Genetic vulnerability also caused large-scale losses in rice in the Philippines and Indonesia in the 1970s. CWR contain enormous genetic diversity, and their use in breeding resistance and tolerance can safeguard crops against devastation from a range of threats.

The economic benefits of improved crop production and quality through breeding with CWR are therefore highly significant. For example, the desirable traits of wild sunflower (*Helianthus* spp.), when bred into cultivated sunflower, are worth an estimated US$267–384 million annually to the sunflower industry in the USA. One wild tomato has contributed to a 2.4% increase in solid contents, worth US$250 million. Three different wild groundnuts have provided resistance to root knot nematodes that cost groundnut growers around the world approximately US$1 billion each year. This reduced risk of crop losses for subsistence farmers improves food security and sustainable livelihood, and has far-reaching and invaluable social and health benefits.

CWR are becoming increasingly recognized at the global level as critical for the sustainability of agriculture and adaptation to global change. The Science Council of the Consultative Group on International Agricultural Research (CGIAR) includes CWR under CGIAR research priority 1 (2005–2015), recognizing that most of the diversity critical to improve agricultural performance and production lies in understudied wild relatives. The Council also recognizes that wild relatives are under-represented in most CGIAR *ex situ* collections and that evaluation of the collections for traits related to resistance and tolerance is needed in order to utilize their enormous potential. *In situ* wild relatives are at risk, however, and the Council advocates a concerted effort to study their distribution and ensure that accessions are adequately collected and represent the broad range of genetic diversity.

Target 9 of the Global Strategy for Plant Conservation, adopted by the Parties to the Convention on Biological Diversity in April 2002, recognizes the central role that within-species genetic diversity plays in improving production and use of crops and wild species. The need to conserve this genetic diversity has long been recognized by national and international activities, particularly the Global Plan of Action. In another international initiative, and in recognition of the critical need to conserve CWR *in situ*, the International Union for the Conservation of Nature (IUCN) Species Survival Commission Plant Conservation Committee recently established a Crop Wild Relative Specialist Group (CWRSG). Together with 36 other plant Specialist Groups, it forms the plant conservation network that feeds into IUCN's Plant Conservation Strategy 2000–2005, as well as the Global Strategy for Plant Conservation. The CWRSG will pay special attention to effective conservation and use of wild plant species of socio-economic and conservation value by promoting integrated conservation and by developing strategies for gathering, documenting and disseminating information on these species. It will also maintain a global inventory and threat assessment of wild plant species of socio-economic value.

Despite this growing recognition of their high importance, CWR remain underconserved. Many countries have conservation initiatives in place (such as gene banks and protected areas), but few of these target CWR. An assessment of *in situ* conservation of *Lupinus* spp. in Spain, for example, showed that protected areas do not consider CWR unless they are an endangered species. A number of CWR species are known to be threatened, primarily by loss of habitat. For example, more than 1 in 20 of the species of *Poaceae*, which also includes cereal crops such as wheat, rice, maize, barley and millet and their wild relatives, are under threat of extinction, and wild soybean, which once grew over much of China's Yellow River Delta and Sanjiang Plain, can now be found only at a few sites. The acute lack of awareness of the importance of CWR and poor capacity to incorporate information are also major constraints to effective CWR conservation and use.

In view of the incalculable value of CWR to people's livelihood and the paucity of targeted action and knowledge for their conservation and use, Bioversity International is leading a UNEP/GEF-supported project in five countries to address major constraints and threats to the conservation and use of CWR. Each of the project countries – Armenia, Bolivia, Madagascar, Sri Lanka and Uzbekistan – has a significant number of important and threatened taxa of CWR. All but Uzbekistan are among the world's biodiversity hot spots – places that have the highest concentrations of unique biodiversity on the planet, but that are also at the greatest risk of diversity loss. In each country, the project will work towards policies that support plant genetic resources conservation and use; public awareness campaigns will raise understanding and appreciation of CWR; capacity of national institutions will be strengthened to implement conservation actions; the distribution of wild relatives will be mapped; protected areas that are important for conservation of priority CWR will be identified and management plans will be implemented; and information on CWR will be made available for use at the national and international levels. The results of these pilot activities in each country will be used to broaden CWR conservation efforts nationally and internationally, to raise global awareness and to stimulate further research on these important species.

Bioversity International has recently embarked on a new strategy, in which it has chosen to place 'people' at the centre of its mission. In this new strategy, CWR cut through all areas of work. Bioversity International will continue to remain engaged on this subject and will strive to remain a strategic partner to other organizations and to countries in the work on agricultural biodiversity and CWR. CWR conservation and use require continued engagement and scientific action with strong links to development.

Emile Frison
Director General

Kwesi Atta-Krah
Deputy Director General

Acknowledgements

This volume grew out of the EC-funded project, PGR Forum (the European Crop Wild Relative Diversity Assessment and Conservation Forum – EVK2-2001-00192 – available at: http://www.pgrforum.org/) and the First International Conference on Crop Wild Relative Conservation and Use, which incorporated the PGR Forum Final Dissemination Conference. As such, many of the concepts presented in this volume were stimulated by PGR Forum discussions. PGR Forum was funded by the EC Fifth Framework Programme for Energy, Environment and Sustainable Development and the editors wish to acknowledge the support of the European Community in providing the forum for discussion and publication of this volume.

I

Crop Wild Relative Conservation and Use: an Overview

1 Crop Wild Relative Conservation and Use: Establishing the Context

N. Maxted, S.P. Kell and B.V. Ford-Lloyd

1.1 Introduction

Crop wild relatives (CWR) are species closely related to crops that include crop progenitors, and that may contribute beneficial traits to crops, such as pest or disease resistance, and yield improvement or stability. They are identified as critical resources that are vital for wealth creation and food security in the 21st century, as well as contributing to environmental sustainability (Prescott-Allen and Prescott Allen, 1983; Hoyt, 1988; Maxted *et al.*, 1997a; Meilleur and Hodgkin, 2004; Stolton *et al.*, 2006). However, CWR, like any other group of wild species, are subject to an increasing range of threats in their host habitats.

Historically, conservation of plant genetic resources (PGR) has focused almost explicitly on cultivated plants. However, recently, CWR and wild-harvested species have been acknowledged as being equally important. Therefore, PGR may be defined as the taxonomic and genetic diversity of plants that is of value as a resource for present and future generations of people. They present a tangible resource of actual or potential economic benefit for humankind at national, regional and global levels. Exploitation of plant genetic diversity has existed for millennia, with farmers using variation within species to improve their crops. Subsistence farmers in Mexico, for example, would annually grow cultivated maize near its wild relatives to facilitate introgression between the CWR and the crop as a means of crop enhancement (Hoyt, 1988). These species and this process are as important to humankind today as they were to the earliest farmers.

Development in the biotechnology industries has also allowed the transfer of genes from more distantly related species, further enhancing the value of CWR species, both closely and more distantly related (see Hodgkin and Hajjar, Chapter 38, this volume). These species have contributed significantly to improving food production; for example, Prescott-Allen and Prescott Allen (1986) calculated that the yield and quality contribution to US-grown or imported crops was over US$350 million/year in 1986. The contribution of CWR is growing and has largely been through the donation of useful genes coding for pest and disease resistance, higher salt tolerance, abiotic stress tolerance and higher nutritional value (Hajjar and Hodgkin, 2007). Primarily, single-gene-controlled traits have been introduced from CWR into crops providing virus resistance in rice (*Oryza sativa*), blight resistance in potato (*Solanum tuberosum*), powdery mildew resistance in wheat (*Triticum aestivum*) and *Fusarium* and nematode resistance in tomato (*Lycopersicon esculentum*). Increased nutritional value of the crops has been fulfilled through the introduction of genes for higher protein content in wheat and vitamin C content in tomato. Genes from wild *Brassica oleracea* plants have created domestic broccoli with high levels of anti-cancer compounds (Hodgkin and Hajjar, Chapter 38, this volume).

Gene introductions have tended to be most effective when the wild species are close relatives of the crop, or are even direct ancestors to it. However, transfer of more distantly related species expands the value of CWR by increasing their usefulness into the secondary and tertiary crop gene pools (Meilleur and Hodgkin, 2004). Tanksely and McCouch (1997) argued that breeders were not fully exploiting the potential of CWR; historically, they relied on searching for specific beneficial traits associated with certain CWR rather than searching more generally for beneficial genes. Hajjar and Hodgkin (2007) comment that although quantitative trait loci have been identified in many CWR groups, the potential to exploit them as a breeding resource using new molecular technologies has yet to be fully realized. This position is likely to improve with time, underpinning the need for the availability of broad CWR diversity, and emphasizing the conservation–use linkage and the need for the conservation community to meet the evolving needs of the users.

The Convention on Biological Diversity (UNEP, 1992) attempted to address these issues through promotion of biodiversity conservation, sustainable use of its components and the equitable sharing of the benefits arising from the use of biodiversity. Specifically, in relation to plants, the Global Strategy for Plant Conservation (GSPC) (CBD, 2002a) was adopted by the CBD at its sixth conference of the parties. It includes global targets that are to be achieved by 2010, such as: '60% of the world's threatened species conserved *in situ*; 60 % of threatened plant species in accessible *ex situ* collections . . . and 10 % of them included in recovery and restoration programmes', and specifically in relation to PGR, '70 % of the genetic diversity of crops and other major socio-economically valuable plant species conserved' (CBD, 2002a). Following on from the GSPC, in Europe, the European Plant Conservation Strategy (EPCS) was proposed and submitted to the CBD Subsidiary Body on Scientific, Technical and Technological Advice (SBSTTA) by Planta Europa and the Council of Europe (Anonymous, 2002). Its vision was 'a world in which wild plants are valued – now and in the future', and its goal was 'to halt the loss of wild plants diversity in Europe'. These were to be achieved by 2007, using 43 targets including target 17: 'Management plan for wild crop relatives initiated in at least one protected area in each of 5 or more European countries'; target 24: '30% of wild crop relatives and other socio-economically and ethnobotanically important species stored in genebanks'; and target 27: 'Manual with guidelines and case studies of best practice for integrated (*in situ* and *ex situ*) plant conservation programmes made available on the web' (Anonymous, 2002).

In the broader context of biodiversity and landscape, the Pan-European Biological and Landscape Diversity Strategy (PEBLDS) (ECNC, 1998–2006), when implemented at the regional and national levels, will undoubtedly reduce genetic erosion and promote agrobiodiversity conservation. The International Treaty on Plant Genetic Resources for Food and Agriculture (ITPGRFA) specifically focuses on agrobiodiversity (FAO, 2001), its objectives being the 'conservation and sustainable use of plant genetic resources for food and agriculture and the fair and equitable sharing of the benefits arising out of their use'. Article 5 states that each contracting party shall: 'Survey and inventory plant genetic resources for food and agriculture, taking into account the status and degree of variation in existing populations, including those that are of potential use and, as feasible, assess any threats to them. . . . Promote *in situ* conservation of wild crop relatives and wild plants for food production, including in protected areas.'

The current threats to genetic diversity from genetic erosion and extinction were further recognized by the CBD 2010 Biodiversity Target (CBD, 2002b), which committed the parties 'to achieve by 2010 a significant reduction of the current rate of biodiversity loss at the global, regional and national level as a contribution to poverty alleviation and to the benefit of all life on earth'. To address this target, along with the requirements of other relevant international, regional and national strategies and legislation, we need to be able to assess biodiversity change and threats, which requires precise knowledge of what biodiversity exists.

It is generally acknowledged that PGR are a finite world resource that we know are being eroded or lost, in part because of careless, unsustainable human practices. This loss of botanical diversity can occur at each biodiversity level – genes, species and communities. If the threat to species is taken as an example, it is estimated that, of the 20,590 European vascular plant species (Kell *et al.*, Chapter 5, this volume), 21% were classified as threatened by the 1994 International Union for the Conservation of Nature and Natural Resources (IUCN) Red List Categories and Criteria, 50% of Europe's 8624 vascular plant endemics are considered to be threatened to some degree and 64 taxa are already extinct (Walters and Gillett, 1998). The Gran Canaria Declaration called for a Global Program for Plant Conservation (Anonymous, 2000), stating that 'as many as two-thirds of the world's plant species are in danger of extinction in nature during the course of the 21st century...genetic erosion and narrowing of the genetic basis of many species'. The same declaration recognized that plants are vital for the planet in maintaining ecosystem stability and providing food, fibre, fuel, clothing and medicine for humankind.

It is even more difficult, if not impossible, to estimate the precise rates of loss of genetic diversity within species. However, it must be faster than the loss of species, because there will be some genetic erosion (loss of genetic diversity) from the species that remain extant and complete loss of genetic diversity from those species that become extinct (Maxted *et al.*, 1997b). It therefore seems likely that virtually all species are currently suffering loss of genetic diversity to varying degrees and it is likely that 25–35% of plant genetic diversity will be lost between the ratification of the CBD in 1993 and the 2010 Biodiversity Target date (Maxted *et al.*, 1997b). Loss of any genetic diversity means that plants may not be able to adapt to changing conditions quite so readily in the future – at a time of ecosystem instability this is a serious concern, since these species are the basis of our future food security.

The CBD and ITPGRFA marked an important watershed in PGR conservation. These treaties highlighted not only the need to conserve the breadth of agrobiodiversity, but also the gaps in our knowledge of biodiversity. Furthermore, they refocused their activities on to *in situ* PGR conservation, while revealing the lack of appropriate genetic conservation techniques for socio-economic species. This drew fresh attention to the poorly conserved and underutilized wild species of socio-economic importance that are related to crops – the CWR.

It seems self-evident to note that CWR are those wild plant species related to a crop, but what actually constitutes a CWR and how closely does a taxon have to be related to a crop to be considered a CWR? In the light of contemporary biotechnological advances, most, if not all, species are potential gene donors to a crop. However, within the utilitarian sense of conservation for food and agriculture, it seems likely that exploitation will remain primarily focused on CWR; therefore, an accurate definition of the relationship between a crop and its close wild relatives is required, so that conservationists competing for limited conservation resources may objectively prioritize taxa for study (Kell and Maxted, 2003; Meilleur and Hodgkin, 2004; Maxted *et al.*, 2006). In addition, if CWR diversity is to be exploited by the user community, it must be available to them, so there is a need for effective identification, systematic conservation and preuse characterization.

It was specifically to discuss and address these issues that the First International Conference on Crop Wild Relative Conservation and Use was held and, within the European context, that the European Crop Wild Relative Diversity Assessment and Conservation Forum (PGR Forum), an EC Framework 5-funded Thematic Network, was initiated (see PGR Forum 2003–2005) to raise awareness and to improve the scientific basis for CWR conservation and use throughout Europe and globally.

1.2 PGRFA and CWR Conservation

PGR are the 'genetic material of plants which is of value as a resource for the present and future generations of people' (IPGRI, 1993), and plant genetic resources for food and agriculture (PGRFA) are the PGR most directly associated with human food production and agriculture. PGRFA may be partitioned into six components: modern cultivars; breeding lines and genetic stocks; obsolete cultivars; primitive forms of cultivated plants and landraces; weedy races; and CWR. Modern cultivars, breeding lines, genetic stocks and obsolete cultivars are directly associated with modern breeding activities and constitute the bulk of gene bank holdings. Landraces, weedy races and CWR accessions are of less immediate breeding potential and are therefore less well represented in gene banks.

CWR may be defined by their characteristics: they are species more or less closely related to crops, the possible progenitors or direct ancestors of crops and can possibly act as gene donors to crops. Their potential as gene donors for crop improvement was recognized by N.I. Vavilov in the 1920s and they have been formally used in breeding programmes since the 1940s (Loskutov, 1999). However, farmers and plant breeders have been using wild species to improve the quality and yield of crops for thousands of years. Recent attention to CWR as a critical resource for future human well-being has been underpinned by Prescott-Allen and Prescott Allen (1983), Hoyt (1988), Maxted *et al.* (1997a), Meilleur and Hodgkin (2004) and Stolton *et al.* (2006), who each identified CWR as being a neglected natural resource and a deserving target for systematic conservation. This new conservation focus is associated with our improved knowledge of their taxonomy and genetic diversity, and the need for greater production of food (Valdés *et al.*, 1997).

In recent times, much European-scale PGR conservation activity has been associated with the work of the Bioversity International (formerly the International Plant Genetic Resources Institute – IPGRI) and two of its associated networking programmes: the European Crop Genetic Resources Network (formerly the European Cooperative Programme for Crop Genetic Resources Network – ECP/GR, available at: www.ecpgr.cgiar.org) and European Forest Genetic Resources Programme (EUFORGEN). Both networks are collaborative programmes involving European countries and are aimed at facilitating long-term conservation on a cooperative basis and the increased sustainable utilization of PGR in Europe. Within the context of CWR conservation it should be noted that the ECP/GR 'Task Force on wild species conservation in genetic reserves' has as its prime objective the establishment and development of a CWR inventory for Europe and the raising of public awareness of the importance of, and need for, CWR diversity, as well as the assessment of taxonomic and genetic diversity of European CWR and the development of methods to conserve CWR diversity.

Initially, PGR (and specifically CWR) conservation efforts concentrated on *ex situ* conservation, but in the late 1980s, the trend started to turn towards *in situ* and the application of complementary methods of conservation (Maxted *et al.*, 1997c). Practically, CWR maintenance is often associated with conservation in natural habitats, primarily due to the large numbers of species included as CWR and the difficulty of collecting and conserving this vast array of species and their entire genetic diversity *ex situ*. The maintenance of CWR diversity is considered a public good, normally associated with public institutional conservation, but often can be seen to fall between the priorities of different communities – the agricultural community gives CWR a low action priority because they are wild species and the ecological community also gives them a low action priority because they are associated with crop plants. Recognition of the fact that CWR conservation is falling through the cracks between existing conservation agencies should encourage action to rectify the situation and promote active conservation.

1.3 European and Mediterranean CWR Diversity and Threats

The combined European and Mediterranean region has an estimated extant flora of 30,983 vascular plant species and around 80% of these species are of current or potential socio-economic use (Kell *et al.*, Chapter 5, this volume). However, within this region, both taxonomic and genetic diversity of CWR species are being eroded and lost at an ever-increasing rate (Maxted, 2003). The EPCS (Council of Europe and Planta Europa, 2001) acknowledges that 'although Europe was one of the first regions to address the conservation of plants, the Council of Europe commissioned and published the first ever regional list of threatened plants in the 1970s, Europe's plant life continues to decline and its conservation is not yet receiving the attention it deserves'. Meanwhile, the European Community Biodiversity Strategy (European Commission, 2000) states that 24% of certain groups are threatened and 3456 European plant species are on the 1997 IUCN Red List (Walters and Gillett, 1998).

The EPCS recognizes that Europe is rich in diversity of domesticated and economically important plant resources and their wild relatives. Europe has significant endemic genetic diversity of global importance in major crops and their wild relatives (Heywood, 1999), such as oats (*Avena sativa* L.), sugarbeet (*Beta vulgaris* L.), carrot (*Daucus carota* L.), apple (*Malus domestica* Borkh.), annual meadow grass (*Festuca pratensis* Huds.), perennial rye grass (*Lolium perenne* L.) and white clover (*Trifolium repens* L.). Gene pools of many minor crop species and their wild relatives are also present in the region, such as arnica (*Arnica montana* L.), asparagus (*Asparagus officinalis* L.), lettuce (*Lactuca sativa* L.), sage (*Salvia officinalis* L.) and raspberries and blackberries (*Rubus* spp.), as well as herbs and aromatic plants such as mints (*Mentha* spp.) and chives (*Allium* spp.). Europe is also an important region for forest genetic resources, such as pine (*Pinus* spp.), and ornamental plants, such as sweet pinks (*Dianthus* spp.) and violets (*Viola* spp.).

The Mediterranean region is a particularly rich centre of CWR diversity as it comprises or borders three important Vavilov centres of crop diversity (Vavilov, 1926). It is also the major centre of CWR diversity for important crops such as wheat (*T. aestivum* L.), barley (*Hordeum vulgare* L.), oats (*A. sativa* L.), chickpea (*Cicer arietinum* L.), lentil (*Lens culinaris* Medik.), pea (*Pisum sativum* L.), faba bean (*Vicia faba* L.), lucerne (*Medicago sativa* L.), white clover (*T. repens* L.), grape (*Vitis vinifera* L.), fig (*Ficus carica* L.), olive (*Olea europaea* L.) and pistachio (*Pistacia vera* L.), as well as the minor crops flax (*Linum usitatissimum* L.), melon (*Cucumis melo* L.), lettuce (*L. sativa* L.) and sage (*S. officinalis* L.). For each of these groups, the crop species and its wild relatives are found within the Euro-Mediterranean region.

However, the diversity of Euro-Mediterranean CWR is increasingly under threat from a number of factors including habitat loss through deforestation, destruction and fragmentation, introduction of exotic alien species outcompeting native species or changing the habitat characteristics, changes in agricultural practices and land use, and overexploitation by humans (Maxted *et al.*, 1997c; Maxted, 2003). Each of these threats is associated with detrimental human activities. Intensification of agricultural systems has led to the loss of prime habitats for a number of CWR species that favour disturbed habitats and are now restricted to field margins.

Threats to genetic diversity are as important as those to taxonomic (species) diversity, but perhaps genetic erosion is more insidious as it is less easy to quantify. Estimating the loss of genetic diversity within CWR species is difficult as it is both costly and time-consuming (Maxted and Guarino, 2006; Ford-Lloyd *et al.*, Chapter 6, this volume), and the loss of genetic diversity in itself leads to species becoming more vulnerable as they are less able to adapt to any changing conditions (Maxted *et al.*, 1997d). Climate change is likely to be an important threat to genetic diversity worldwide. Changes in temperature and more importantly in water availability are likely to result in alterations in species distribution and assemblages and in strong selection pressure for more adaptive genotypes. For example, in the United Kingdom, southerly distributed species may increase their range due to warmer weather conditions further north, allowing them to exploit previously uninhabitable regions, provided they are capable of migrating; but this cannot be the case for those species located in the north. Such changes are likely to radically alter the distribution and diversity of CWR species (United Kingdom CIP, 2001), and are likely to question current *in situ* conservation paradigms.

In the face of such threats, it is salutary to realize that an analysis of data contained in the European Internet Search Catalogue of *Ex Situ* PGR Accessions, (EURISCO, available at: http://eurisco.ecpgr.org) revealed that both native and exotic CWR are represented by only approximately 4% of germplasm holdings in European *ex situ* collections (Dias and Gaiji, 2005). The EURISCO analysis found that 37,528 accessions of 2629 species in 613 genera are CWR out of a total of 925,000 accessions of 7950 species in 1280 genera (Dias and Gaiji, 2005). While the number of wild species included in collections is estimated to be around 33% in total, less than 50% of genera containing crops (Kell *et al.*, Chapter 5, this volume) have wild relatives conserved *ex situ*, indicating that the breadth of CWR coverage is seriously limited. Further, the ratio of the number of accessions of cultivated species to wild species is striking, with an average of 167 for each cultivated species and 14 for each wild species, giving a ratio of 12:1, which is particularly surprising, given that most diversity is located in wild species. It is obvious from the growing threats, this preliminary review of the conservation status of CWR in European *ex situ* collections and the fact that there are no protected areas in Europe where CWR conservation is an explicit focus that CWR genetic diversity is far from secure and more concerted conservation action is required.

To address the threats to CWR diversity and meet the CBD 2010 Biodiversity Target, there is a clear requirement for a baseline assessment against which to assess change, and as a first step, there is a need to establish regional and national inventories of this diversity. Although lists of CWR exist, especially those proposed for Europe by Zeven and Zhukovsky (1975) and Heywood and Zohary (1995), and for individual countries by Schlosser *et al.* (1991) for the former German Democratic Republic and by Mitteau and Soupizet (2000) for France, none is complete and the definitions used for what constitutes a CWR, although formulated on the basis of expert knowledge, are subjective.

Despite the vital natural resource provided by CWR to sustain food security in every country, no comprehensive inventories of species exist, their diversity is threatened with extinction or genetic erosion and their utilization is hampered by less effective conservation efforts. PGR Forum took on the challenge to address these issues by creating a comprehensive web-enabled CWR Catalogue for Europe and the Mediterranean region, as well as formulating conservation methodologies to specifically aid CWR conservation and use. To publicize these issues in a global forum and to disseminate the products of a series of PGR Forum workshops attended by PGR Forum partners and invited international experts, the First International Conference on Crop Wild Relative Conservation and Use was held and the results will, we hope, prove timely for CWR conservation and use.

1.4 PGR Forum: a European Initiative for CWR Conservation

1.4.1 Overview

PGR Forum was a Thematic Network funded under the EC Framework 5 Programme for Research, Key action 2 'Global change, climate and biodiversity', 2.2.3 'Assessing and conserving biodiversity' (see PGR Forum, 2003–2005). It provided a European forum for the assessment of taxonomic and genetic diversity of European CWR and the development of appropriate methodologies for their conservation. To achieve this broad aim, PGR Forum had five subordinate objectives:

1. Debate the assessment and conservation of European CWR at both the species and component population levels.
2. Produce a European inventory of baseline biodiversity data, threat and conservation status for CWR.
3. Debate data structures and documentation methodologies, formulate management and monitoring regimes, establish a means of assessing genetic erosion and genetic pollution as an aid to their *in situ* conservation.
4. Communicate project results to European stakeholders, policy makers and user groups.
5. Establish a dialogue between European national and regional CWR conservationists and user stakeholders, and also with policy makers, end-users and the broader international stakeholder communities.

The Forum brought together 23 partners from 21 countries throughout Europe, with the addition of partners representing IUCN – The World Conservation Union and the IPGRI (now Bioversity International). The PGR Forum network included a broad cross section of the professional European PGR community, including conservationists, taxonomists, plant breeders, information managers, policy makers and other end-users. The Forum facilitated a dialogue both within the European plant conservation community (protected area managers, Natura 2000, gene bank managers and academics), and with the germplasm user community (plant breeders and other genetic resource users, policy makers, CBD Clearing-House Mechanism, European Community and the European public). The products of the PGR Forum project have made a valuable contribution to improved CWR conservation and exploitation within the professional community, and have also increased national and European CWR capacities and raised public awareness. The major achievements of PGR Forum are summarized below.

1.4.2 Catalogue of crop wild relatives for Europe and the Mediterranean

One of the primary objectives of the PGR Forum project was to create a web-enabled European CWR database, incorporating baseline biodiversity data with current conservation and threat status for all the CWR taxa present in Europe. The CWR Catalogue for Europe and the Mediterranean (Kell *et al.*, 2005 and Chapter 5, this volume), available through the web-enabled information system, is one of the major outputs of the project. The methodology used for creating the initial list of European and Mediterranean CWR was produced through a process of data harmonization and cross-checking between a number of databases, primarily Euro+Med PlantBase (available at: http://www.euromed.org.uk/) and Mansfeld's World Database of Agricultural and Horticultural Crops (Hanelt and IPK 2001; available at: http://Mansfeld.ipk-gatersleben.de/Mansfeld/) (Kell *et al.*, Chapter 5, this volume). The initial list was created by selecting the accepted taxa contained in Euro+Med PlantBase within genera matching those contained in Mansfeld's World Database of Agricultural and Horticultural Crops. The same process was then repeated for forestry, ornamental, medicinal and aromatic plant species, using different data sources. The Catalogue contains more than 25,000 crop and CWR species and more than 273,000 records of taxon occurrences in 130 geographical units across the Euro-Mediterranean region (Kell *et al.*, Chapter 5, this volume). More than 17,000 of these species occur in Europe alone.

1.4.3 CWR data structures

Prior to this project, there were no widely agreed standards for the management of CWR data, which are essential for effective management and exchange of CWR information and urgently needed for CWR conservation and sustainable use. Although standards for exchange of *ex situ* data already existed (see the Food and Agriculture Organization, FAO/IPGRI multi-crop passport descriptors: FAO/IPGRI, 2001), standards for the collation, analysis and exchange of *in situ* data were not yet available.

The minimum data types needed to document CWR *in situ* population data were debated and identified, and a set of CWR descriptors were generated (see Kell *et al.*, Chapter 33, this volume). A user acceptance testing panel was established to test and provide feedback on their development. It was agreed that, where feasible, internationally accepted standards such as those of the International Working Group on Taxonomic Databases (TWDG) (available at: http://www.tdwg.org/) and those recommended by the Global Biodiversity Information Facility (GBIF) (available at: http://www.gbif.org/) should be used.

Together with the definition of data types, the requirements for data modelling tools were reviewed, i.e. those associated with filling the gaps in the knowledge of species distribution, providing the potential distribution of species, projected distribution under altered climatic conditions and to identify species 'hot spots' and conservation priorities. Several programs were discussed and recommended: DIVA (available at: http://www.diva-gis.org/), which comes with basic data sets and provides specific functions like extraction of climate data for localities, spatial autocorrelation, histograms and scattergrams of gridfiles, prediction of crop adoption (Ecocrop); GARP (available at: http://www.lifemapper.org/desktopgarp/), which is a genetic algorithm that creates an ecological niche model for a species that represents the environmental conditions where this species would be able to maintain populations; and BIOCLIM (available at: http://www.andra.fr/bioclim/), which is a prediction system using bioclimatic parameters derived from mean monthly climate estimates to approximate energy and water balances at a given location.

1.4.4 Crop Wild Relative Information System

Using the data models generated by PGR Forum, the Crop Wild Relative Information System (CWRIS) was developed (PGR Forum, 2005). CWRIS provides access to CWR inventories and taxon information, and assists in the process of planning and implementing conservation and sustainable use strategies (Kell *et al.*, Chapter 33, this volume). CWRIS comprises two interlinked elements: (i) an information management model (CWR descriptors) for CWR conservation and sustainable use, with an emphasis on site and population data (and corresponding XML schema); and (ii) an online information management system and portal providing access to crop and CWR inventories, as well as to information on taxonomic status and nomenclature, distribution, uses and other types of information which are relevant to CWR conservation and sustainable use. A snapshot of the CWRIS database (incorporating the CWR Catalogue for Europe and the Mediterranean) has also been published on CD-ROM for fast access to the data without an Internet connection (see Moore and Kell, 2005). The CWR descriptors provide a comprehensive set of data standards that can be used to effectively manage genetic conservation of CWR taxa and their component populations.

Figure 1.1 is a conceptual model of CWRIS. The CWR inventory (currently the CWR Catalogue for Europe and the Mediterranean, but any taxonomic information can be used) is at the core of the system and ancillary information on the taxa contained in the database is enabled through external web links. Many more ancillary information sources have been linked to CWRIS than indicated in the model – the system has also been developed to be highly flexible and it is possible to add further links at any stage during further development. In particular, the ability to incorporate links to national data sets and/or to provide a loading mechanism for national data sets will be an important part of the future development of the system (Kell *et al.*, Chapter 33, this volume). CWRIS uses open standards for data storage and retrieval – the open database, MySQL, has been used to support the CWRIS data schema, which is a generic 'data warehouse' design that could equally be managed in other database management systems, if necessary. The CWRIS data retrieval modules have been

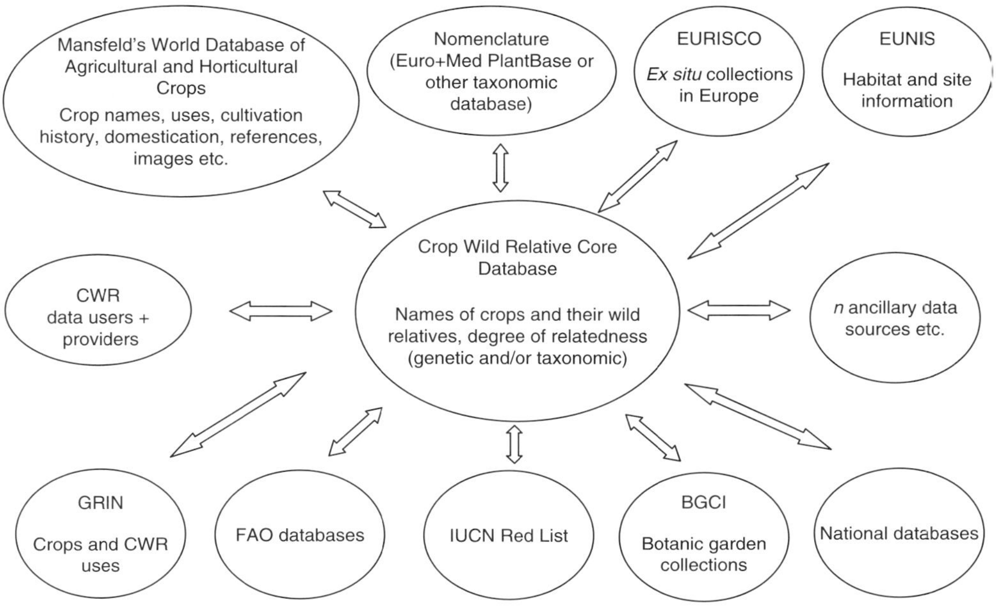

Fig. 1.1. Crop Wild Relative Information System (CWRIS): conceptual model.

written in the open language Perl, and also, in parallel, in the proprietary Microsoft language VB Script. Because of this, CWRIS can be hosted on any web server, either open or proprietary. A corresponding XML schema has been written as part of PGR Forum's commitment to enabling access and sharing of CWR data (see Moore and Kell, 2005).

To ensure its continued development and use, CWRIS has been transferred to Bioversity International (formerly IPGRI), where it will be hosted and maintained on behalf of the Secretariat of the European Cooperative Programme for Plant Genetic Resources (ECPGR, available at: http://www.ecpgr.cgiar.org/Introduction/AboutECPGR.htm). It has been agreed by ECPGR in principle that both CWRIS and the European *ex situ* PGR search catalogue, EURISCO (available at: http://eurisco.ecpgr.org/), will be developed to make a single entry point for all European PGR information.

1.4.5 Population *in situ* management and monitoring methodologies

Much of the development in the science of plant genetic conservation has been devoted to *ex situ* conservation techniques (see Frankel and Bennett, 1970; Frankel, 1973; Frankel and Hawkes, 1975; Hawkes, 1980; Brown *et al*., 1989; Guarino *et al*., 1995); in fact Hawkes (1991) went so far as to comment that *in situ* techniques by comparison are still very much in their infancy. With the increased interest in the application of *in situ* conservation techniques post CBD, there was a need to develop practical *in situ* conservation methodologies.

This was recognized and first addressed by Horovitz and Feldman (1991) and Maxted *et al.* (1997a); but there remained a requirement to further develop population management and monitoring techniques specifically appropriate for the *in situ* genetic conservation of CWR.

While recognizing that CWR can be found both within and outside of existing protected areas, such as field borders and roadsides, which are often overlooked in biodiversity conservation, the limited experience available shows that newly established genetic reserves (protected areas where CWR genetic diversity conservation is a priority) are most often associated with existing protected areas (Maxted *et al.*, 1997e, 2007). The pragmatic reasons for this are that: (i) existing protected areas already have an associated long-term conservation ethos and are less prone to hasty management changes where conservation value and sustainability will always be vital management consideration; (ii) it is relatively easy to amend the existing site management to facilitate genetic conservation of CWR species; and (iii) creating novel conservation sites can be avoided, thus avoiding the prohibitive cost of acquiring land. Therefore, the simplest way forward in economic and political terms has proven to be for countries to locate genetic reserves in existing protected areas, e.g. national parks or heritage sites. Here, the reserves can provide benefits to local people, and their establishment and ongoing management are therefore more likely to gain their support.

Existing protected areas are already likely to include some CWR species, but they may not be associated with the species of conservation concern and are therefore not actively monitored and managed, or their status as CWR may be unrecognized. In genetic reserve conservation, the profile of the CWR taxa is raised so that the site is actively managed to promote the conservation of the genetic diversity within CWR populations. The methodology for generating protected area management plans and protected area population monitoring is one of the major outputs from PGR Forum. The subject generated much discussion within the Forum and the guidelines produced (Iriondo *et al.*, 2007; Iriondo and De Hond, Chapter 20, this volume) aim to provide the PGR community and, in particular, protected area managers with practical protocols and recommendations for CWR management and monitoring. The Genetic Reserve Management Guidelines will soon be published (Iriondo *et al.*, 2007) and will address the integration of CWR conservation with protected area management, location and design of genetic reserve, structure and writing of a management plan, methodologies used for population monitoring, techniques for population and habitat recovery and the necessity and practical application of an *ex situ* back-up strategy.

1.4.6 Threat and conservation assessment

Alongside *in situ* techniques, the assessment of threat and conservation status in the establishment of CWR conservation priorities has too often been ignored. With limited conservation resources available, how can they be employed efficiently without prior systematic threat and conservation assessment? Within PGR Forum, this subject was discussed under four related topics.

Threat assessment

In recent years, there has been extensive development of threat assessment techniques, at least at the taxon level, through the application of the IUCN Red List Categories and Criteria (IUCN, 2001). The PGR community as a whole has historically had little contact with these initiatives – perhaps because of the focus on genetic, rather than species conservation; however, the Red List assessment provides a powerful tool for the PGR conservation community, simply because if for no other reason if a species goes extinct then all genetic diversity within that species is also necessarily lost. Red List assessment data can be used to monitor the status of CWR taxa and populations over time and the published Red List itself can help raise awareness of the need for conservation action. A Red List training course was run at a PGR Forum workshop with the aim of training project participants in applying the IUCN Red List Categories and Criteria and the associated Regional Application Guidelines (IUCN, 2003) to CWR taxa. As a result, several European CWR taxa from Germany, Hungary, the Netherlands, Poland, Portugal, Spain and the United Kingdom were assessed (see Magos Brehm *et al.*, Chapter 13, this volume).

If the 2004 IUCN Red List of Threatened Species is queried, only 184 CWR taxa have been assessed and are considered threatened, and these are nearly all trees (Kell *et al.*, Chapter 5, this volume). We know that at least 488 European CWR species were categorized as globally threatened (Kell *et al.*, Chapter 5, this volume) in the 1997 Red List (Walters and Gillett, 1998), but these were assessed using the 1994 Red List Categories and Criteria (IUCN, 1994) in which the assessment was much more subjective. Many more CWR taxa are also likely to be listed in national Red Lists. There is an urgent need to review national Red Lists and begin a systematic process of regional and global Red Listing for CWR (Kell *et al.*, Chapter 5, this volume).

PGR Forum also recommended that research be undertaken on developing novel means of assessing threat for PGR at two additional diversity levels: gene pool and crop landrace. The former could possibly be associated with genetic conservation gap analysis discussed below, while the latter is likely to require more detailed genetic diversity studies.

Conservation priority-setting

The CWR Catalogue for Europe and the Mediterranean (Kell *et al.*, 2005) lists over 25,000 CWR species. Some critics would say that this list is too large and that the Catalogue should have been limited to only the crops generally considered to be of major economic importance. However, limiting the Catalogue in this way, would not only discount the value of other crop groups, such as underutilized species, but also hamper access to information on the full range of crop resources. Of course, with limited conservation resources available, the question of how to conserve such an extensive list immediately arises. It is obviously not possible to direct conservation resources into all 25,000 species. The answer to this question is that an objective means of establishing conservation priorities for CWR is needed. Furthermore, analysis of which species in the Catalogue have already been afforded conservation efforts can be undertaken (Kell *et al.*, Chapter 5, this volume). PGR Forum reviewed the options for priority-setting (see Ford-Lloyd *et al.*, Chapter 6, this volume). Among others, these include the economic value of the associated crop, level of threat and the relative endemicity of each CWR.

Systematic conservation planning

The need for systematic conservation planning is well established within the PGR community – ecogeographic analysis is now routinely applied before taking conservation action (Maxted *et al.*, 1995). Perhaps one conservation initiative that the PGR conservation community has been slow to implement is the use of species hot spot analysis to identify geographic conservation priorities. Within the PGR community there has been little significant advance since Vavilov's identification of the Centres of Crop Diversity (1926), with the exception of recent applications of geographical information system (GIS) applications (Jones *et al.*, 1997; Guarino *et al.*, 2002). The ecologically based conservation community is systematically identifying important plant areas (Anderson, 2002; Palmer and Smart, 2004) where conservation action may be focused. For example, 122 important plant areas have been identified in Turkey (Özhatay *et al.*, 2005), covering 11 million ha (equivalent to 13% of the land surface of Turkey), to conserve 10,000 native species (34.4% of which are endemic).

The GIS applications that are currently being applied in PGR conservation are largely associated with identifying specific sites where closely related taxa are concentrated, rather than general locations that have a high concentration of all CWR taxa. We need to locate areas of high CWR diversity (which may also be areas of high landrace diversity). As a first step towards identifying the important CWR locations in Britain, Maxted *et al.* (in press) established 17 'best' locations for the establishment of genetic reserves based on the 226 rarest and most threatened CWR species (Fig. 1.2). They found that these 17 protected areas contain 152 (67%) of the priority British CWR species.

Genetic conservation gap analysis

In prioritizing species for conservation and in formulating efficient conservation strategies, an assessment of current conservation action is a necessary consideration. It would waste limited resources to conserve a species that is already effectively conserved – we need to identify the gaps in current conservation actions. Following debate within PGR Forum, a methodology was proposed for plant genetic conservation gap analysis.

The identification of disparities or 'gaps' in current conservation action is generally referred to as 'gap analysis' and is used as a means of prioritizing future conservation action. This concept was put forward as a conservation evaluation technique that identifies areas in which selected elements of biodiversity are represented (Margules, 1989) and was largely applied to indigenous forests, particularly on small islands rich in endemic species. However, the concept of gap analysis can equally be applied to document taxonomic and genetic diversity and its distribution in existing wild populations, and to develop strategies for their

Fig. 1.2. Seventeen United Kingdom CWR hot spots using the iterative method (total numbers of CWR taxa present in each shown, as well as additional CWR taxa in brackets) (Maxted *et al.*, in press).

genetic conservation. Burley (1988) identified four steps in gap analysis: (i) identify and classify biodiversity; (ii) locate areas managed primarily for biodiversity; (iii) identify biodiversity that is under-represented in the managed areas; and (iv) set priorities for conservation action. This application is clearly associated with ecosystem conservation, but the basic methodologies can equally be applied to PGR conservation as was illustrated in the recent application for cowpea (*Vigna unguiculata* (L.) Walp.) and its wild relatives from Africa (Maxted *et al.*, 2004). Maxted *et al.* (in preparation a), in proposing the methodology for plant genetic conservation gap analysis, state that the procedure involves comparing natural *in situ* plant diversity for the target taxon with the sample of the diversity that is actively conserved. The methodology may be broken down into a series of steps, as follows: (i) circumscription of target taxon and target area; (ii) assessment of natural diversity; (iii) assessment of current *in situ* and *ex situ* conservation strategies; and (iv) setting priorities for conservation action to ensure the comprehensive conservation of the target taxon's gene pool.

Establishing national conservation priorities for CWR conservation

Genetic conservation gap analysis may be associated with establishing national CWR conservation priorities, but it would be difficult to apply the methodology for an entire country's CWR diversity. There is a need to provide guidelines on how individual countries might generate national CWR conservation strategies. There is unlikely to be one method for establishing a national CWR strategy; each will be unique due to the conservation resources available, the amount and availability of baseline biodiversity data and the type of agency writing the strategy (e.g. agricultural or environmental, formal or NGO sector) (Maxted *et al.*, in preparation b). However, the foundation for any national strategy for CWR conservation is likely to be the national CWR checklist (Fig. 1.3).

The CWR Catalogue for Europe and the Mediterranean (Kell *et al.*, 2005) and the associated national CWR Catalogues that may be extracted from it means that in Europe (and the neighbouring Mediterranean countries) it is relatively straightforward to generate a national CWR inventory. The national list is likely to be extensive in terms of the number of taxa included; therefore, the second step will be to establish conservation priorities, particularly if the number of taxa exceeds the number that can be conserved using the resources available. The third step is to collate the available baseline ecogeographic data, and undertake threat assessment and gap analysis culminating in a clear national CWR strategy. The strategy can then be implemented by individual protected-area managers or germplasm collectors. These steps require implementation at two distinct levels: (i) national agencies are responsible for producing the inventory, establishing the taxon and site priorities, and ensuring the conserved diversity is used; and (ii) individual protected-area and gene bank managers are responsible for conserving actual populations *in situ* or *ex situ*. These complementary responsibilities are reflected in the proposed model (Fig. 1.3).

Although the two levels are interconnected, they can also be seen as distinct and with quite separate goals. The national CWR strategy developed for an individual country aims to ensure the conservation of the maximum taxonomic and genetic diversity of the country's CWR. However, for individual CWR-protected area or gene bank managers, the aim is to ensure the conservation of the maximum CWR taxonomic and genetic diversity – whether in the protected area or in the gene bank accessions – while at the same time promoting use of the conserved diversity. The national CWR strategy is more extensive and has policy implications for national conservation and exploitation agencies. It leads to the conservation of priority CWR taxa in key protected area sites and possibly the establishment of important plant areas with a specific focus on CWR genetic diversity conservation. The individual CWR-protected area or gene bank's conservation activities may be seen as being more focused and practical in terms of conserving CWR, and may involve the identification of CWR found in a single, existing protected area, possible refocusing of the protected area management plan, or filling gaps identified in the gene bank's CWR coverage.

Thus, the national phase is composed of various steps that lead to the selection of key protected area sites and identification of diversity under-represented in *ex situ* collections, but must also be linked to multiple applications of individual protected areas or targeted collecting, to ensure that maximum taxonomic and

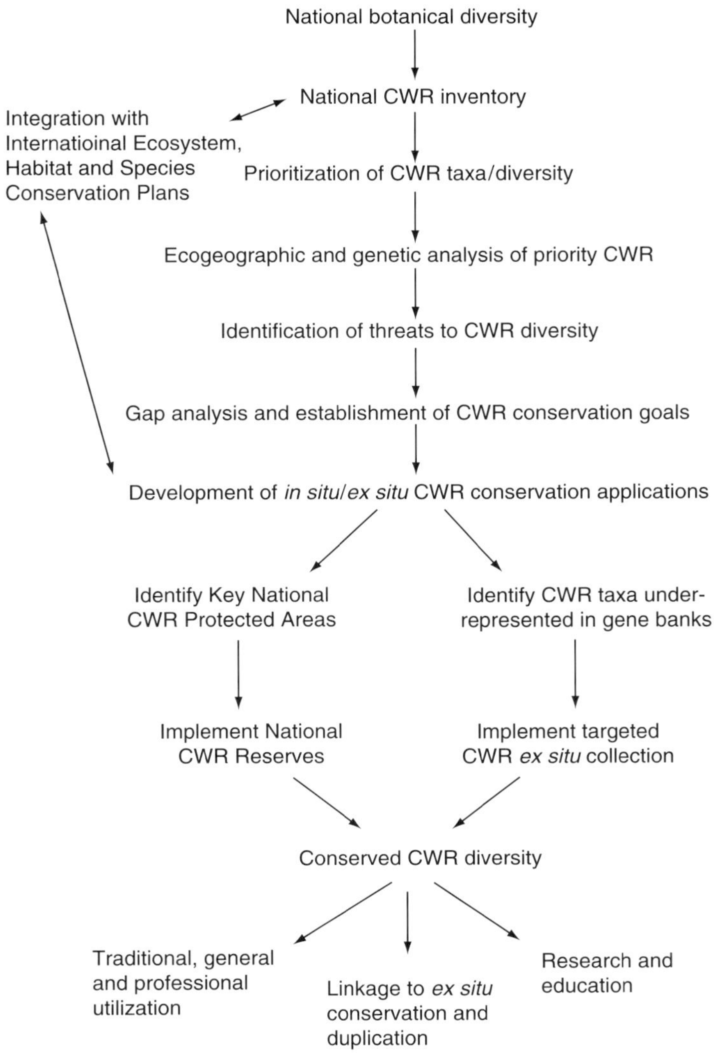

Fig. 1.3. Model for development of a national CWR strategy (Maxted *et al.*, in preparation b).

genetic diversity of the country's CWR are conserved. The two levels of conservation activity must work together to ensure a practical national CWR strategy.

It is undoubtedly the case that numerous, existing protected areas contain a wealth of CWR. However, these areas are likely to have been established to conserve habitats or megafauna, rather than CWR species. Therefore, the number of CWR species monitored within the protected areas is unlikely to be large, unless they are coincidentally keystone or indicator species, as well as being CWR. In such cases, the management of the CWR species is passive, and individual CWR populations may possibly decline or even be lost without changes to the management plan being triggered. An important function of PGR Forum was to raise the profile of CWR as a vital national resource and to stimulate active protected area conservation.

1.4.7 Genetic erosion and genetic pollution

Within the PGR literature, incidents of genetic erosion are often alluded to but rarely quantified. The recent debate concerning genetically modified organisms (GMOs) has raised the issue of genetic pollution, but again it is rarely quantified and it is likely that non-GMO introgression has been polluting CWR taxa ever since the emergence of plant breeding and distribution of cultivars. It is logical that CWR are the wild species most threatened by genetic pollution as they are often found growing sympatrically with the crop and are most likely to naturally introgress with crops. Both genetic erosion and genetic pollution threaten CWR diversity and need to be monitored and assessed, as they will directly impinge on CWR diversity and the COP 2010 Biodiversity Target.

Discussion of assessment and monitoring methodologies for CWR species has highlighted existing protocols and analytical methods (see Wilkinson and Ford, Chapter 17, this volume), which can be applied broadly in the assessment of change. The ease with which they can be used is clearly dependent upon whether they are direct or indirect. Direct measures of genetic erosion and genetic pollution are associated with the array of available molecular techniques, but these remain impractical on the broadest scale because of the expense of application. Indirect measurements are associated with assessing a reliable 'proxy' of change that can be appropriately used for the large number of CWR species that exists. Such proxies might include relative threat according to position in gene pools 1b, 2 and 3 of a crop, coverage of related crop number, relative sophistication of the crop in terms of breeding, forestry and botanic garden conservation facilities, numbers of genetic reserves, on-farm and home garden projects, the number of CWR protected by legislation and those included within national CWR conservation and use strategies.

1.4.8 Raising professional and public awareness

Perhaps one of the longest-lasting legacies of the PGR Forum project is the creation of a greater understanding among the professional and public sectors of the importance of CWR and the threats to CWR populations. There would have been little point in producing the CWR Catalogue for Europe and the Mediterranean, developing CWRIS and the associated conservation methodologies, if they were not to be used. At the outset, the project partners agreed that CWR awareness raising was required both among the general public and within the professional PGR conservation and user communities. The latter might perhaps surprisingly be evidenced by the fact that only 4% of European *ex situ* germplasm holdings comprise CWR (Dias and Gaiji, 2005) and there are no protected areas in Europe where CWR conservation is focused on explicitly. PGR Forums' efforts to raise awareness included:

- *Project web site* (PGR Forum, 2003–2005);
- *Web-enabled database* of CWR Catalogue for Europe and the Mediterranean available through CWRIS (PGR Forum, 2005);
- *Project CD-ROM* containing the various PGR Forum products;
- *Final Dissemination Conference* – the First International Conference on Crop Wild Relative Conservation and Use;
- *Crop Wild Relative* – the five issues of the newsletter *Crop Wild Relative* are available online through the project web site. With the completion of PGR Forum, the newsletter will, in future, be associated with the newly established IUCN Species Survival Commission (SSC) CWR Specialist Group (CWRSG).

1.5 A Closer Look at the Definition of CWR

CWR are commonly defined in terms of wild species related to agricultural and horticultural crops; therefore, a broad definition of a CWR would be any wild taxon belonging to the same genus as a crop. This definition is intuitively accurate and can be simply applied, and was therefore adopted by PGR Forum. However, application of this broad definition results in the possible inclusion of a wide range of species that may be either closely or remotely related to the crop itself. If the European and the Mediterranean floras are taken as examples, approximately 80% of species in the Euro-Mediterranean region (Kell *et al.*, Chapter 5, this volume) can be regarded as being CWR and other socio-economically important species, including the crops themselves. Therefore, there is a need to estimate the degree of CWR relatedness to enable limited conservation resources to be focused on priority species – those most closely related to the crop.

To establish the degree of crop relatedness, one method which could be applied is the Harlan and de Wet (1971) gene pool concept – close relatives being found in the primary gene pool (GP1), more remote ones in the secondary gene pool (GP2) and very remote ones in the tertiary gene pool (GP3). Harlan and de Wet (1971) commented that GP2 may be seen as encompassing the whole genus of the crop. This simple application of the gene pool concept remains functional for the crop complexes where hybridization experiments have been performed and the pattern of genetic diversity within the gene pool is well understood. However, for the majority of crop complexes, particularly those in the tropics, the wild species related to crops have been described and classified using a combination of morphological characteristics; therefore, the degree of reproductive differentiation among species remains unknown, making application of the gene pool concept impossible.

As a pragmatic solution to the lack of crossing and genetic diversity data for the majority of crops and related taxa, an alternative solution using the existing taxonomic hierarchy was proposed (Maxted *et al.*, 2006). This can be applied to define a ranking for CWR as follows: taxon group 1a – crop; taxon group 1b – same species as crop; taxon group 2 – same series or section as crop; taxon group 3 – same subgenus as crop; taxon group 4 – same genus; and taxon group 5 – same tribe but different genus from crop. Therefore, for

CWR taxa where we have little or no information about reproductive isolation or compatibility, the taxon group concept can be used to establish the degree of CWR relatedness of a taxon. Application of the taxon group concept assumes that taxonomic distance is positively related to genetic distance. Flint (1991), Heywood (1994), Johnson (1995) and Maxted *et al.* (2006) among others draw attention to the fact that this relationship may not always hold because of inconsistencies among taxonomists when describing species; species are not all separated by the same amount of genetic isolation. Nevertheless, the taxonomic hierarchy is likely to be a reasonable approximation; therefore, for practical purposes, classical taxonomy remains an extremely useful means of estimating genetic relationships. The taxon group concept can be applied to all crop and CWR taxa, and can be used to define relative CWR relatedness for 78% of crop and CWR taxa where the gene pool concept is not understood (Maxted *et al.*, 2006), as long as the existing classification of the genus contains an infrageneric structure.

Application of the gene pool and the taxon group concepts to a crop and its wild relatives would ideally be expected to be congruent, but as discussed earlier and acknowledged by Harlan (1992), inconsistencies among taxonomists will mean that where both taxonomic and genetic information are available, the two concepts may not match perfectly. However, Maxted *et al.* (2006) provided the example of gene pool and taxon group concepts applied to narbon vetch (*Vicia narbonensis* L.) and its wild relatives (Table 1.1). It is interesting to note the close correlation between the applications of the two concepts for this crop. Thus, the combined use of the gene pool and taxon group concepts provides the best pragmatic means available to determine whether a species is a CWR and how closely a wild relative is related to the associated crop.

A working definition of a CWR was also provided by Maxted *et al.* (2006):

> A crop wild relative is a wild plant taxon that has an indirect use derived from its relatively close genetic relationship to a crop; this relationship is defined in terms of the CWR belonging to gene pools 1 or 2, or taxon groups 1 to 4 of the crop.

Therefore, taxa which belong to GP1B, or TG1b and TG2 may be considered close to CWR and demand higher conservation priority, while those in GP2 or TG3 and TG4 are more remote and may be afforded lower priority. Those in GP3 and TG5 would be excluded from being considered wild relatives of that particular crop. Therefore, application of the gene pool and taxon group concepts provides a pragmatic way of establishing the degree of CWR relatedness and thus assists in establishing conservation priorities.

1.6 CWR Specialist Group of the IUCN Species Survival Commission

Stimulated by the interest in CWR raised by PGR Forum, the SSC of IUCN – the World Conservation Union – established a specialist group to focus activities on CWR conservation. IUCN was established in 1948 and is now a global science-based network of 1000 volunteer experts. The goal of the SSC is: 'The

Table 1.1. Application of gene pool and taxon group concepts for the crops *Vicia narbonensis* and their wild relatives.

	Gene pool concept					
Crop	GP1A[a]	GP1B	GP2	GP3		
Narbon bean	*Vicia narbonensis* L. var. *narbonensis*	*V. narbonensis* L. var. *salmonea* (Mout.) H. Schäfer var. *jordanica* H. Schäfer var. *affinis* Kornhuber ex Asch. & Schweinf. var. *aegyptiaca* Kornhuber ex Asch. & Schweinf.	*V. kalakhensis* Khattab, Maxted & Bisby *V. johannis* Tamamschjan in Karyagin *V. galilaea* Plitm. & Zoh. in Plitm. *V. serratifolia* Jacq. *V. hyaeniscyamus* Mout.	All other *Vicia* spp.		
	Taxon group concept					
	TG1A[b]	TG1B	TG2	TG3	TG4	TG5
Narbon bean	*V. narbonensis* L. var. *narbonensis*	*V. narbonensis* L. var. *salmonea* (Mout.) H. Schäfer var. *jordanica* H. Schäfer var. *affinis* Kornhuber ex Asch. & Schweinf. var. *aegyptiaca* Kornhuber ex Asch. & Schweinf.	Sect. *Narbonensis* (Radzhi) Maxted *V. kalakhensis* Khattab, Maxted & Bisby *V. johannis* Tamamschjan in Karyagin *V. galilaea* Plitm. & Zoh. in Plitm. *V. serratifolia* Jacq. *V. hyaeniscyamus* Mout. *V. eristalioides* Maxted	All non-section *Narbonensis* (Radzhi) Maxted *Vicia* L. subgenus *Vicia* spp.	All *Vicia* subgenus *Vicilla* Schur spp.	*Lens* Miller *Lathyrus* L. *Pisum* L. *Vavilovia* A. Fed.

[a]Gene pool concept for *Vicia narbonensis* is taken from Enneking and Maxted (1995).
[b]Taxon group concept for *Vicia narbonensis* is derived from the classification provided in Maxted (1995).

extinction crisis and massive loss in biodiversity are universally adopted as a shared responsibility, resulting in action to reduce this loss of diversity within species, among species and of ecosystems.' The main role of the SSC is to provide information to IUCN on the conservation of species, their inherent values and their roles in ecosystem health and functioning, the provision of ecosystem services and the provision of support to human livelihood. The SSC Specialist Groups provide the breadth of expertise across the wide diversity of species on the planet.

The CWRSG was established in 2006. The major focus of the group is to facilitate national, regional and global CWR conservation and use. The Specialist Group will continue to raise the profile of CWR and stimulate their conservation and use, by improving awareness among the scientific community and general public. Clarification of the importance to agriculture and the environment of wild plant species of socio-economic value among governments, institutions, decision makers and the general public cannot be overemphasized. A key element of the policies that result in adequately conserved and sustainably utilized CWR and world heritage site (WHS) diversity is the need for systematic analysis of the gaps in current conservation and utilization strategies, and where gaps are identified, implementation of remedial actions.

Experts within the CWRSG, as well as engaged in the IUCN Red List assessment of threat to CWR and WHS species, will develop and promote *in situ* and *ex situ* techniques for conservation that are applicable at national, regional and global scales, using diverse case studies. The group will help establish effective strategies for CWR data gathering, analysis and dissemination of information.

Finally, experts from the CWRSG will provide an information source and 'help desk' for those seeking advice, expertise and access to appropriate contacts to enhance the actions of individuals or organizations working on the conservation of all wild plant species of socio-economic value. Thus, the impact of the CWRSG will be far-reaching, helping to ensure our utilization options are maintained for the future in a changing environment. More information on the CWRSG is provided by Dulloo and Maxted (Chapter 48, this volume).

1.7 Global Strategy for CWR Conservation and Use

The First International Conference on CWR Conservation and Use provided a platform for the development of a 'Global Strategy for CWR Conservation and Use', which is currently undergoing a process of review and development (see Heywood *et al.*, Chapter 49, this volume). The Strategy was reviewed by Conference delegates both at the Conference and in working groups. Subsequently, its adoption is being led by the FAO to be taken forward in the context of the ITPGRFA.

The Strategy essentially provides an action plan for nations and regions to refer to in addressing the critical issues of effective CWR conservation and use. Practical steps that can be taken are included, based on existing experience and knowledge such as the identification (internationally, and within each region and country) of a small number of priority sites (international – 100, regional – 25 and national – 5) for the establishment of active CWR genetic reserves. These reserves should form an interrelated network of internationally, regionally and nationally important CWR genetic reserve sites for *in situ* conservation. It is hoped that those charged with the task of taking forward the action plan will view the Strategy in the context of existing policy, legislation and conservation initiatives where possible. The Strategy can also provide the backdrop for the development of specific national and regional policy and legislative instruments.

1.8 Conclusions

There has been a growing interest among genetic conservationists in *in situ* conservation techniques, because of the urgent need to protect ecosystems threatened with imminent change and also because of the realization of the difficulty or impossibility of conserving sufficient genetic diversity of crops and wild relatives *ex situ* to underpin our sustainable agricultural future. Even though both *in situ* and *ex situ* conservation strategies have advantages and disadvantages (Maxted *et al*., 1997c) and they should not be seen as alternatives or in opposition to one another (Ford-Lloyd and Maxted, 1993), for CWR, *in situ* conservation provides the most important and practical option for securing adequate conservation. Although the majority of plant genetic conservation research and activity has thus far focused on *ex situ* techniques, the emphasis has now shifted to *in situ* genetic conservation of CWR diversity. This has necessitated the development of novel *in situ* methodologies, which are now available. These techniques have been practically tested, presenting no barrier to CWR conservation in natural habitats – now they must be implemented nationally, regionally and globally.

When implementing *in situ* genetic conservation of CWR diversity, the approach may be divided between what might be termed monographic and floristic. The former selects certain crop complexes and focuses on the establishment of genetic reserves for their conservation, while the latter takes a broader hot spot approach to identify locations where CWR taxa overlap and where genetic reserves should be established. It can be argued that the monographic approach will inevitably focus resources on a subset of crop complexes, possibly related to the most economically important crops, while other minor crops will be excluded. With severely limited resources, the monographic approach will always be adopted, but it is important to stress that if the aim is to conserve a country's overall CWR diversity and to retain maximum food security potential for the future, the floristic approach is equally appropriate. Ideally, within any individual country or region, the monographic and floristic approaches will proceed in tandem to capitalize on potential CWR exploitation in the short and longer term.

Although PGR Forum is now complete the legacy is likely to be:

- PGR Forum CWR Catalogue for Europe and the Mediterranean, which contains more than 25,000 species records in 130 geographical units across the region;

- PGR Forum CWRIS, the first information management system designed specifically to facilitate CWR conservation and use;
- Individual methodologies and practical guidelines for the *in situ* conservation of CWR (*in situ* data management, threat and conservation assessment, gap analysis, genetic reserve location and design, population monitoring and management, and genetic erosion and pollution assessment).

Apart from these practical scientific achievements, equally important has been the goal of raising the profile of CWR conservation and use within both the public and professional communities. Many publications associated with PGR Forum and this volume itself will contribute significantly to this goal.

It is important to underline that the achievements have only been possible due to the collaborative efforts of a network of committed individuals who have the common aim of conserving these vital resources. But in many ways, both within Europe and globally, the work is only just beginning. There is now a need to act on the recommendations that have arisen from the project, those present at the conference and the collaborative efforts of all involved in CWR conservation and use globally.

Acknowledgements

The concepts discussed in this chapter were stimulated by PGR Forum (the European Crop Wild Relative Diversity Assessment and Conservation Forum – EVK2-2001-00192, available at: http://www.pgrforum.org/), funded by the EC Fifth Framework Programme for Energy, Environment and Sustainable Development.

References

Anderson, S. (2002) *Identifying Important Plant Areas*. Plantlife International, Salisbury, UK.
Anonymous (2000) *Gran Canaria Declaration: Calling for a Global Programme of Plant Conservation*. Botanic Gardens Conservation International, Surrey, UK.
Anonymous (2002) *European Plant Conservation Strategy*. Council of Europe and Planta Europa, London.
Brown, A.H.D., Marshall, D.R., Frankel, O.H. and Williams, J.T. (1989) *The Use of Plant Genetic Resources*. Cambridge University Press, Cambridge.
Burley, F.W. (1988) Monitoring biological diversity for setting priorities in conservation. In: Wilson, E.O. and Peter, F.M. (eds) *Biodiversity*. National Academy Press, Washington, DC, pp. 227–230.
CBD (2002a) *Global Strategy for Plant Conservation*. Secretariat of the Convention on Biological Diversity, Montreal. Available at: http://www.biodiv.org/decisions/?lg=0&dec=VI/9
CBD (2002b) *2010 Biodiversity Target*. Secretariat of the Convention on Biological Diversity, Montreal. Available at: http://www.biodiv.org/2010-target/default.aspx
Council of Europe and Planta Europa (2001) *European Plant Conservation Strategy*. Plantlife International, Salisbury, UK.
Dias, S. and Gaiji, S. (2005) *EURISCO a Window on Europe's Plant Genetic Diversity: Overview of European CWR Collections*. Presentation given at the First International Conference on Crop Wild Relative Conservation and Use, Agrigento, Italy, September 2005.

ECNC (1998–2006) *The Strategy Guide for the Pan-European Biological and Landscape Diversity Strategy*. European Centre for Nature Conservation. Available at: http://www.strategyguide.org/

Enneking, D. and Maxted, N. (1995) Narbon bean: *Vicia narbonensis* L. (Leguminosae). In: Smartt, J. and Simmonds, N.W. (eds) *Evolution of Crop Plants*, 2nd edn. Longman Group, Harlow, UK, pp. 316–321.

European Commission (2000) *European Community Biodiversity Strategy*. European Community, Brussels.

FAO (2001) *International Treaty on Plant Genetic Resources for Food and Agriculture*. Food and Agriculture Organization of the United Nations. Available at: http://www.fao.org/ag/cgrfa/itpgr.htm

FAO/IPGRI (2001) *Multi-Crop Passport Descriptors*. FAO/IPGRI, Rome, Italy.

Flint, M. (1991) *Biological Diversity and Developing Countries: Issues and Options*. Overseas Development Administration, London.

Ford-Lloyd, B.V. and Maxted, N. (1993) Preserving diversity. *Nature* 361, 579.

Ford-Lloyd, B.V. and Maxted, N. (1997) Genetic Conservation Information Management. In: Maxted, N., Ford-Lloyd, B.V. and Hawkes, J.G. (eds) *Plant Genetic Conservation: the* In Situ *Approach*. Chapman & Hall, London, pp. 284–309.

Frankel, O.H. (1973) *Survey of Crop Genetic Resources in Their Centre of Diversity*. FAO/IBP, Rome, Italy.

Frankel, O.H. and Bennett, E. (1970) *Genetic Resources in Plants – Their Exploration and Conservation*. Blackwell Scientific Publications, Oxford, UK.

Frankel, O.H. and Hawkes, J.G. (1975) *Crop Genetic Resources for Today and Tomorrow*. Cambridge University Press, Cambridge.

Guarino, L., Ramanatha Rao, V. and Reid, R. (1995) *Collecting Plant Genetic Diversity: Technical Guidelines*. CAB International, Wallingford, UK.

Guarino, L., Jarvis, A., Hijmans, R.J. and Maxted, N. (2002) Geographical information systems (GIS) and the conservation and use of plant genetic resources. In: Engels, J.M.M., Ramanatha Rao, V.R., Brown, A.H.D. and Jackson, M.T. (eds) *Managing Plant Genetic Diversity*. CAB International, Wallingford, UK, pp. 387–404.

Hajjar, R. and Hodgkin, T. (2007) The use of wild relatives in crop improvement: a survey of developments over the last 20 years. *Euphytica* 10.1007/s10681-007-9363-0.

Hanelt, P. and IPK (2001) *Mansfeld's Encyclopaedia of Agricultural and Horticultural Crops*. Springer, Berlin.

Harlan, J.R. (1992) *Crops and Man*. American Society of Agronomy, Madison, Wisconsin.

Harlan, J.R. and de Wet, J.M.J. (1971) Towards a rational classification of cultivated plants. *Taxon* 20, 509–517.

Hawkes, J.G. (1980) *Crop Genetic Resources Field Collection Manual*. IBPGR/EUCARPIA, Rome, Italy.

Hawkes, J.G. (1991) International workshop on dynamic *in situ* conservation of wild relatives of major cultivated plants: summary of final discussion and recommendations. *Israel Journal of Botany* 40, 529–536.

Heywood, V.H. (1994) The measurement of biodiversity and the politics of implementation. In: Forey, P.L., Humphries, C.J. and Vane-Wright, R.I. (eds) *Systematics and Conservation Evaluation*. Systematic Association Special Volume 50. Oxford University Press, Oxford, UK, pp. 15–22.

Heywood, V.H. (1999) The role of botanic gardens in *ex situ* conservation of agrobiodiversity. In: Gass, T., Frese, L., Begemann, F. and Lipman, L. (compilers) *Implementation of the Global Plan of Action in Europe. Conservation and Sustainable Utilization of Plant Genetic Resources for Food and Agriculture. Proceedings of the European Symposium on Plant Genetic Resources for Food and Agriculture, Braunschweig, Germany, 30 June–4 July 1998*. International Plant Genetic Resources Institute, Rome, Italy, pp. 102–107.

Heywood, V.H. and Zohary, D. (1995) A catalogue of the wild relatives of cultivated plants native to Europe. *Flora Mediterranea* 5, 375–415.

Horovitz, A. and Feldman, M. (1991) Population dynamics of the wheat progenitor, *Triticum turgidum* var. *dioccoides*, in a natural habitat in Eastern Galilee. *Israel Journal of Botany* 40(5–6), 349–536.

Hoyt, E. (1988) *Conserving the wild relatives of crops*. IBPGR/IUCN/WWF, Rome, Italy.

IPGRI (1993) *Diversity for Development*. International Plant Genetic Resources Institute, Rome, Italy.

Iriondo, J.M., Dulloo, E. and Maxted, N. (eds) (2007) *Plant Genetic Population Management*. CAB International, Wallingford, UK.

IUCN (1994) *The IUCN Red List of Threatened Species: 1994 Categories and Criteria (v.2.3)*. Available at: http://www.iucnredlist.org/info/categories_criteria1994

IUCN (2001) *The IUCN Red List of Threatened Species: 2001 Categories and Criteria (v. 3.1)*. Available at: http://www.iucnredlist.org/info/categories_criteria2001

IUCN (2003) *Guidelines for Application of IUCN Red List Criteria at Regional Levels: Version 3.0*. IUCN Species Survival Commission. IUCN, Gland, Switzerland/Cambridge. Available at: http://www.iucn.org/themes/ssc/redlists/regionalguidelines.htm

Johnson, N. (1995) *Biodiversity in the Balance: Approaches to Setting Geographic Conservation Priorities*. Biodiversity Support Program, Washington, DC.

Jones, P.G., Beebe, S.E. Tohme, J. and Galwey, N.W. (1997) The use of geographical information systems in biodiversity exploration and conservation. *Biodiversity and Conservation* 6, 947–958.

Kell, S.P. and Maxted, N. (compilers) (2003) *European Crop Wild Relative Diversity Assessment and Conservation Forum. Report of Workshop 1: European Crop Wild Relative Assessment, 5–7 February, 2003*. University of Birmingham, Birmingham, UK. Available at: http://www.pgrforum.org/Documents/WS%20reports/WS1/WS1%20Report.pdf

Kell, S.P., Knüpffer, H., Jury, S.L., Maxted, N. and Ford-Lloyd, B.V. (2005) *Catalogue of Crop Wild Relatives for Europe and the Mediterranean*. Available online via the Crop Wild Relative Information System (CWRIS – http://cwris.ecpgr.org/) and on CD-ROM. University of Birmingham, Birmingham, UK.

Loskutov, I.G. (1999) *Vavilov and his Institute: a History of the World Collection of Plant Genetic Resources in Russia*. International Plant Genetic Resources Institute, Rome, Italy.

Margules, C.R. (1989) Introduction to some Australian developments in conservation evaluation. *Biological Conservation* 50, 1–11.

Maxted, N. (1995) *An Ecogeographic Study of* Vicia *subgenus* Vicia. Systematic and Ecogeographic Studies in Crop Genepools 8. IBPGR, Rome, Italy.

Maxted, N. (2003) Conserving the genetic resources of crop wild relatives in European protected areas. *Biological Conservation* 113(3), 411–417.

Maxted, N. and Guarino, L. (2006) Genetic erosion and genetic pollution of crop wild relatives. In: Ford-Lloyd, B.V., Dias, S.R. and Bettencourt, E. (eds) *Genetic Erosion and Pollution Assessment Methodologies*. IPGRI, Rome, Italy, pp. 35–46.

Maxted, N., van Slageren, M.W. and Rihan, J. (1995) Ecogeographic surveys. In: Guarino, L., Ramanatha Rao, V. and Reid, R. (eds) *Collecting Plant Genetic Diversity: Technical Guidelines*. CAB International, Wallingford, UK, pp. 255–286.

Maxted, N., Ford-Lloyd, B.V. and Hawkes, J.G. (1997a) *Plant Genetic Conservation: the* In Situ *Approach*. Chapman & Hall, London.

Maxted, N., Hawkes, J.G., Guarino, L. and Sawkins, M. (1997b) The selection of taxa for plant genetic conservation. *Genetic Resources and Crop Evolution* 44, 337–348.

Maxted, N., Ford-Lloyd, B.V. and Hawkes, J.G. (1997c) Complementary Conservation Strategies. In: Maxted, N., Ford-Lloyd, B.V. and Hawkes, J.G. (eds) *Plant Genetic Conservation: the* In Situ *Approach* Chapman & Hall, London, pp. 20–55.

Maxted, N., Hawkes, J.G., Ford-Lloyd, B.V. and Williams, J.T. (1997d) A practical model for *in situ* genetic conservation. In: Maxted, N., Ford-Lloyd, B.V. and Hawkes, J.G. (eds) *Plant Genetic Conservation: the* In Situ *Approach*. Chapman & Hall, London, pp. 545–592.

Maxted, N., Guarino, L. and Dulloo, M.E. (1997e) Management and monitoring. In: Maxted, N., Ford-Lloyd, B.V. and Hawkes, J.G. (eds) *Plant Genetic Conservation: the* In Situ *Approach*. Chapman & Hall, London, pp. 231–258.

Maxted, N., Mabuza-Dlamini, P., Moss, H., Padulosi, S., Jarvis, A. and Guarino, L. (2004) *An Ecogeographic Survey: African* Vigna. Systematic and Ecogeographic Studies of Crop Genepools 10. IPGRI, Rome, Italy.

Maxted, N., Ford-Lloyd, B.V., Jury, S.L., Kell, S.P. and Scholten, M.A. (2006) Towards a definition of a crop wild relative. *Biodiversity and Conservation* 15(8), 2673–2685.

Maxted, N., Iriondo, J., Dulloo, E. and Lane, A. (2007) Introduction: the integration of PGR conservation with protected area management. In: Iriondo, J.M., Dulloo, E. and Maxted, N. (eds) *Genetic Reserve Management Guidelines*. CAB International, Wallingford, UK.

Maxted, N., Scholten, M.A., Codd, R. and Ford-Lloyd, B.V. (in press) Creation and use of a national inventory of crop wild relatives. *Biological Conservation*.

Maxted, N., Dulloo, E, Ford-Lloyd, B.V., Iriondo, J. and Jarvis, A. *Gap analysis: a tool for effective genetic conservation assessment of agrobiodiversity* (in preparation a).

Maxted, N., Scholten, M., Kell, S.P. and Ford-Lloyd, B.V. *Developing a national plant genetic resource strategy: crop wild relatives* (in preparation b).

Meilleur, B.A. and Hodgkin, T. (2004) *In situ* conservation of crop wild relatives. *Biodiversity and Conservation* 13, 663–684

Mitteau, M. and Soupizet, F. (2000) Preparation of a preliminary list of priority target species for in situ conservation in Europe. In: Laliberté, B., Maggioni, L., Maxted, N. and Negri, V. (compilers) *ECP/GR In situ and On-farm Conservation Network Report of a Task Force on Wild Species Conservation in Genetic Reserves and a Task Force on On-farm Conservation and Management Joint meeting, 18–20 May 2000, Isola Polvese, Italy*. IPGRI, Rome, Italy, pp. 32–42.

Moore, J.D. and Kell, S.P. (eds) (2005) *PGR Forum CD-ROM*. University of Birmingham, Birmingham, UK.

Moore, J., Kell, S.P., Maxted, N., Ford-Lloyd, B.V. Development of an XML schema for crop wild relative conservation and use (in preparation).

Özhatay, N., Byfield, A. and Atay, S. (2005) *Türkiye'nin 122 Önemli Bitki Alani. (Important plant areas of Turkey)*. WWF Türkiye (Dogal Hayati Koruma Vakfi), Istanbul, Turkey.

Palmer, M. and Smart, J. (2004) *Guidelines to the selection of Important Plant Areas in Europe*. Planta Europa, London. Available at: http://www.plantlife.org.uk/international/plantlife-ipas.html

PGR Forum (2003–2005) *European Crop Wild Relative Diversity Assessment and Conservation Forum*. University of Birmingham, Birmingham, UK. Available at: http://www.pgrforum.org/

PGR Forum (2005) *Crop Wild Relative Information System (CWRIS)*. University of Birmingham, Birmingham, UK. Available at: http://cwris.ecpgr.org/

Prescott-Allen, R. and Prescott Allen, C. (1983) *Genes from the Wild: Using Wild Genetic Resources for Food and Raw materials*. Earthscan Publications, London.

Prescott-Allen, R. and Prescott Allen, C. (1986) *The First Resource: Wild Species in the North American economy*. Yale University Press, New Haven, Connecticut.

Schlosser, S., Reichhoff, L. and Hanelt, P. (1991) *Wildpflanzen Mitteleuropas*. Nutzung und Schutz. Deutscher Landwirtschaftsverlag Berlin GmbH, Berlin.

Stolton, S., Maxted, N., Ford-Lloyd, B., Kell, S.P. and Dudley, N. (2006). *Food Stores: Using Protected Areas to Secure Crop Genetic Diversity*. WWF arguments for protection series. Gland, Switzerland.

Tanksely, S. and McCouch, S. (1997) Seed banks and molecular maps: unlocking genetic potential from the wild. *Science* 277, 1063–1066.

UNEP (1992) *Convention on Biological Diversity: Text and Annexes*. United Nations Environment Programme, Nairobi, Kenya.

UK CIP (2001) *Newsletter of the U.K. Department of the Environment, Transport and the Region's Climate Impacts*. LINK Project Report 18.

Valdés, B., Heywood, V.H., Raimondo, F.M. and Zohary, D. (1997 Proceedings of the Workshops on 'Conservation of the Wild Relatives of European Cultivated Plants'. *Bocconea* 7, 1–479.

Vavilov, N.I. (1926) Studies on the origin of cultivated plants. *Bulletin of Applied Botany, Genetics and Plant Breeding* 16, 1–248.

Walters, K.S. and Gillett, H.J. (1998) *1997 IUCN Red List of Threatened Plants*. Compiled by the World Conservation Monitoring Centre. IUCN – The World Conservation Union, Gland, Switzerland/Cambridge.

Zeven, A.C. and Zhukovsky, P.M. (1975) Dictionary of cultivated plants and their centres of diversity. PUDOC, Wageningen, The Netherlands.

2 Addressing the Conservation and Sustainable Utilization of Crop Wild Relatives: the International Policy Context*

N. Azzu and L. Collette

2.1 Introduction

Crop wild relatives (CWR) are an important component of plant genetic resources for food and agriculture (PGRFA). At the international policy level, they are addressed within the context of PGRFA. When speaking of CWR, this also includes wild food or crop species, minor crops and underutilized crops – crop species which are important not only for breeding and environmental health, but also as additional sources of income and nutrition for rural households.

The conservation and sustainable utilization of CWR are addressed in international policy in both the agriculture and the environmental sectors. While specifically addressing CWR, this chapter highlights their contribution not only to agriculture, but also as part of the wider environment. Section 2 introduces the importance of CWR in the context of agroecosystems, and their socio-economic and cultural bearing. Section 3 describes the international policy under which CWR are addressed, both in agriculture and environment. It also describes how FAO's Commission on Genetic Resources for Food and Agriculture and its various instruments, and the Convention on Biological Diversity (CBD) and its relevant programmes take into account the conservation and sustainable utilization of CWR. Finally, Section 4 briefly presents further considerations to strengthen the importance of conservation and sustainable utilization of CWR in international policy.

* Since the writing of this chapter in 2005, changes in the international policy arena have occurred, and important events have taken place. Of note are the 11th Regular Session of the Commission on Genetic Resources for Food and Agriculture (11–15 June 2007) – and the subsequent adoption of the Multi-year Programme of Work of the Commission, the first session of the Governing Body of the International Treaty on Plant Genetic Resources for Food and Agriculture (the second session is 29 October–2 November 2008), and updated status of preparation of the Country Reports.

2.2 The Importance of Crop Wild Relatives

CWR play a very important role in agricultural innovation, because they are the raw materials on which breeders depend to develop improved plant varieties and to respond to unexpected shocks such as climate change, and evolving human needs. CWR are also important both nutritionally and culturally, as well as having a socio-economic value, to many people. During the reporting process for the preparation of the first *State of the World's Plant Genetic Resources for Food and Agriculture*, a number of countries reported the use of wild foods during periods of famine and especially during the hunger season that precedes crop harvests. Such foods form an integral part of the daily diets of many poor rural households. Wild foods are a source of important vitamins, minerals and other nutrients that complement the staple crops eaten by many of the most vulnerable people, including children and the elderly. Wild crop species (including roots and tubers, leafy vegetables and fruits) also represent ready sources of income for cash-poor households and may provide a significant portion of total household income, particularly where farming is marginal. Lastly, wild crop species are also important for medicinal purposes. Indeed, with the increased realization that some wild species are being overexploited (for medicinal purposes), a number of agencies are recommending that wild species be brought into cultivation systems (Schippmann *et al*., 2002).

In the last 20 years, since the need for conserving CWR in their natural habitat and ecosystems has been recognized, the concept of *in situ* conservation of plant genetic resources (PGRs) has been broadened to include the maintenance of varieties and cultivars of crop plants in agroecosystems. While conservation efforts of major crops have been undertaken *ex situ*, in the past years much effort has been placed by international organizations on traditional agroecosystems and their sustainable management. Awareness for the need to conserve CWR within the framework of ecosystems and natural habitat conservation was raised since the early Technical Conferences of the Food and Agriculture Organization (FAO) of the United Nations (the 1960s and the 1970s), but achievements were slow. Today, however, national programmes increasingly consider the conservation and sustainable use of natural resources, including PGR, over the long term and incorporate them into national programmes and strengthen their work on *in situ* conservation (Ng, 2005). The conservation and sustainable utilization of PGR is critical to improving agricultural productivity and sustainability, thereby contributing to national development, food security and poverty alleviation.

The CBD, at its second meeting of the Conference of Parties (decision II/15) recognized the special nature of agricultural biodiversity, its distinctive features and problems demanding distinctive solutions. The CBD (CBD, 1992) defines agricultural biodiversity as a broad term that includes all components of biological diversity of relevance to food and agriculture and that constitute the agroecosystem: the variety and variability of animals, plants and microorganisms at the genetic, species and ecosystem levels, which are necessary to sustain key functions of the agroecosystem, its structure and processes (decision V/5 of the Conference of Parties). One of the four dimensions of agricultural biodiversity identified by the CBD (decision V/5) that is of relevance to CWR refers to genetic resources for food and agriculture – these constitute the main units of production in agriculture, including cultivated species, domesticated species and managed wild plants and animals, as well as wild relatives of cultivated and domesticated species.

2.3 International Policy Context: CWR Conservation and Sustainable Use

The importance of the conservation and sustainable utilization of CWR is recognized in a number of international instruments, agreements and fora. The main international policy context in which CWR are addressed is within FAO, through its Commission on Genetic Resources for Food and Agriculture (the Commission) and its Global System for the Conservation and Sustainable Utilization of Plant Genetic Resources for Food and Agriculture (FAO, 2006a,b), as well as legally binding instruments including the *International Treaty on Plant Genetic Resources for Food and Agriculture* (Box 2.1). CWR are also addressed in the CBD.

The importance of agriculture is recognized in the environmental forum of the CBD, which 'recognizes the special nature of agricultural biodiversity, its distinctive features and problems needing distinctive solutions' (decision II/15).

2.3.1 International agricultural policy

FAO's Commission on Genetic Resources for Food and Agriculture

At the beginning of the 1950s, FAO served as a forum for the development of common international action on PGRFA. In 1983, the Commission on Plant

Box 2.1. Main international policy environment

AGRICULTURE
- FAO
  - Global System on Plant Genetic Resources for Food and Agriculture
- Report on the State of the World's Plant Genetic Resources for Food and Agriculture
- Global Plan of Action for the Conservation and Sustainable Utilization of Plant Genetic Resources for Food and Agriculture (GPA)
  - International Treaty on Plant Genetic Resources for Food and Agriculture
  - International Plant Protection Convention

ENVIRONMENT
- Convention on Biological Diversity
  - Ecosystem approach
  - Programme of Work on Agricultural Biodiversity
  - Global Strategy for Plant Conservation
  - 2010 Biodiversity Target

Genetic Resources for Food and Agriculture was established by the FAO Conference (Resolution 9/83), to deal with issues related to PGRFA. In 1995, its mandate was broadened (Resolution 3/95) to cover *all components of agro-biodiversity of relevance to food and agriculture*. It was then renamed the Commission on Genetic Resources for Food and Agriculture. It is a permanent forum where governments discuss and negotiate matters relevant to genetic resources for food and agriculture. The main objectives of the Commission are to ensure the conservation and sustainable utilization of genetic resources for food and agriculture, as well the fair and equitable sharing of benefits derived from their use, for present and future generations. The Commission aims to reach international consensus on areas of global interest through negotiations.

From its establishment by the 1983 FAO Conference, the Commission on Plant Genetic Resources for Food and Agriculture (now the Commission on Genetic Resources for Food and Agriculture) has coordinated, overseen and monitored the development of an FAO Global System for the Conservation and Sustainable Utilization of Plant Genetic Resources for Food and Agriculture. Its objectives are to 'ensure the safe conservation, and promote the availability and sustainable use of plant genetic resources by providing a flexible framework for sharing the benefits and burdens'. The Commission also monitors, reviews and updates the rolling Global Plan of Action for the Conservation and Sustainable Utilization of Plant Genetic Resources for Food and Agriculture (GPA).

The Global System comprises international agreements, a variety of codes of conduct, scientific standards, technical mechanisms and global instruments for PGRFA, which are outlined in Table 2.1.

Table 2.1. Components of the FAO Global System.

International agreements
International Undertaking on Plant Genetic Resources (1983)[a]
Global instruments
State of the World's Plant Genetic Resources (First Report 1998)
Global Plan of Action for the Conservation and Sustainable Utilization of Plant Genetic Resources for Food and Agriculture (June 1996)
Global mechanisms
International Network of *Ex Situ* Collections under the Auspices of FAO
International Networks on Plant Genetic Resources for Food and Agriculture
World Information and Early Warning System
Codes of conduct and international standards
Code of Conduct for Germplasm Collecting and Transfer (November 1993)
Gene Bank Standards and Guidelines (April 1993)
Code of Conduct on Biotechnology (draft)

[a]The International Undertaking was revised (beginning in 1994, through the FAO Commission on Plant Genetic Resources for Food and Agriculture – which is now the Commission on Genetic Resources for Food and Agriculture) and underwent a series of negotiations, culminating in the adoption of the International Treaty on Plant Genetic Resources for Food and Agriculture (November 2001, and entered into force on 29 June 2004).

The Report on the State of the World's Plant Genetic Resources and the GPA are the two key elements of the Global System, through which CWR are addressed.

First State of the World's Plant Genetic Resources for Food and Agriculture

PROCESS. At its 26th session (1991), the FAO Conference agreed that a first *Report on the State of the World's Plant Genetic Resources for Food and Agriculture* (the Report) should be developed as part of the Global System for the Conservation and Sustainable Utilization of Plant Genetic Resources for Food and Agriculture. At its 27th session (1993), the Conference agreed that this should be done through a country-driven process under the guidance of the Commission, in preparation for the International Technical Conference on Plant Genetic Resources, held in Leipzig, Germany, in June 1996. The preparation of a *Report on the State of the World's Plant Genetic Resources* and its adoption at the International Technical Conference were also recommended by the United Nations Conference on Environment and Development, in its Agenda 21,[1] and supported by the Conference of the Parties to the CBD.[2]

OUTCOMES. Developed through a participatory, country-driven process, and the result of 154 country reports submitted by governments, the Report emphasized the contribution of PGRFA to world food security. The Report furthermore described the current situation of PGRFA, at the global level, and identified the gaps and needs for their conservation and sustainable utilization, as well as for emergency situations, and laid the foundation for the GPA to be adopted by the International Technical Conference in 1996.

A main conclusion of the first Report (FAO, 1998) was the need for an integrated approach to the conservation and utilization of PGRFA. In particular, with respect to minor crops and underutilized species, it was noted that there was a need for greater investment in their conservation and sustainable utilization, in order to broaden the base of agriculture and meet the needs of nation states and local people who are dependent on these species. With respect to *in situ* management of PGRFA, whether in protected areas, on farms or through ecosystem management outside protected areas and farms, it was observed that there should be a greater focus on the need for further work with regard to: ecological research, the development of protocols for conservation of wild relatives in protected areas, ethnobotanical research and studies of farmer management of PGRFA, conservation networks including protected areas for the conservation of CWR and other wild PGRFA, sustainable harvesting of wild food plants and underutilized species with commercial potential, participation of local communities in ecosystem management and support for on-farm management.

The utilization aspect of PGR was also highlighted as a prerequisite to meeting the challenges of development, food security and poverty alleviation

[1] Agenda 21, paragraph 14.60 (c).

[2] Decision 11/15 of the Second Session of the Conference of the Parties to the CBD, Jakarta, Indonesia, 6–17 November 1995.

especially as PGRFA would be required in order to produce varieties adapted to the extreme and highly variable environments of low-productivity areas.

Second State of the World's Plant Genetic Resources for Food and Agriculture

PROCESS. At its tenth regular session (FAO, 2004), the Commission decided that the second Report should provide objective information and analysis and identify priorities, as a basis for updating the rolling GPA. The Commission encouraged members, and other countries and relevant organizations, such as the International Plant Genetic Resources Institute (IPGRI),[3] to participate in the preparatory process. It adopted the steps for preparing the second Report recommended in the report of the Second Session of the Commission's Intergovernmental Working Group on Plant Genetic Resources, and requested FAO to revise the time line, on the basis that the second Report would be completed in 2008. It confirmed that the country-driven preparatory process for the second Report should be fully integrated into the process of monitoring the implementation of the GPA, in order to minimize the reporting burden on members (CGRFA-10/4/REP).

COUNTRY REPORTS. Guidelines for the preparation of country reports have been prepared for country contributions to the Second State of the World's Plant Genetic Resources for Food and Agriculture. These guidelines are available in CGRFA/WG-PGR-3/05/Inf.5.[4] Of particular relevance to CWR, it is strongly suggested that countries consider Chapter 1 of the Report (the State of Diversity) to assist in determining the scope and focus of their country report. The main headings used in this chapter are: major crops; minor crops; underutilized species (species that are utilized at a local level, either through cultivation or harvesting; multi-purpose plants; and crops that contribute to agricultural diversification); wild species; and crop varieties (modern varieties and landraces/farmers' varieties). As the second Report is intended to update the first Report as much as possible, it would be extremely helpful if country reports address the same genetic resources addressed in Chapter 1 of the first Report, and in doing so, use the same subject headings. Countries are also encouraged to review Annex I of the International Treaty, the List of Crops Covered under the Multilateral System (CGRFA-10/04/Inf.8).

It is also recommended that countries in assessing the state of their PGRFA, and their roles and values, attempt to describe the related aspects of agricultural biodiversity, the production systems and the environments in which these resources are being used, the range of products and services which they provide, the consumption patterns and sociocultural practices associated with them, the ecosystem functions which they sustain and their roles in agricultural production and in achieving food security (CGRFA-10/04/Inf.8).

[3] The International Plant Genetic Resources Institute (IPGRI) is now 'Bioversity International'.

[4] Available at: http://www.fao.org/waicent/FaoInfo/Agricult/AGP/AGPS/pgr/ITWG3rd/docsp1.htm; http://www.fao.org/waicent/FaoInfo/Agricult/AGP/AGPS/pgr/ITWG3rd/pdf/p3i5E.pdf (English version).

The deadline for completion of country reports, decided on the basis of the Commission's request (10th Session), is 30 June 2007. May 2008 is the deadline for completion of the first draft of the second Report, in order for it to be considered by the 12th Regular Session of the Commission (CGRFA/WG-PGR-3/05/Inf.5).

Global Plan of Action for the Conservation and Sustainable Utilization of Plant Genetic Resources for Food and Agriculture

PROCESS. In 1991, the Commission on Genetic Resources for Food and Agriculture requested the development of a rolling *Global Plan of Action for the Conservation and Sustainable Utilization of Plant Genetic Resources for Food and Agriculture*, with programmes and activities aimed at filling gaps, overcoming constraints and facing emergency situations. The periodically updated GPA would permit the Commission to recommend priorities and promote the rationalization and coordination of efforts. The first GPA was developed under the guidance of the Commission, through a country-driven preparatory process. It was adopted in 1996 by 150 countries, at the Fourth International Technical Conference in Leipzig, Germany.

At the FAO International Conference on Plant Genetic Resources (1996), 50 countries agreed that 'overall progress in the implementation of the Global Plan of Action and of the related follow-up processes would be monitored and guided by the national governments and other Members of FAO, through the Commission on Genetic Resources for Food and Agriculture'. To this end, the Commission was asked to set the formats for receiving progress reports from all the parties concerned and establish criteria and indicators to assess progress in the implementation of the GPA.

SCOPE. The Global Plan of Action for the Conservation and Sustainable Utilization of Plant Genetic Resources for Food and Agriculture (GPA) is a framework providing recommendations and activities which grows logically out of the first State of the World report. The first global instrument recognizing the importance of PGRFA for food security, the GPA formalized the concern and initiatives related to PGRFA, and also contributes to achieving the objectives of the CBD and Agenda 21. It promotes a new and rational approach for the conservation of genetic material and represents the first general recognition and support for on-farm management and for an integrated improvement of PGR, including CWR. The GPA makes a priority to restore locally adapted PGR and their improvement, and establishes a new initiative to rebuild agricultural systems devastated by natural disaster, war and civil strife.

The GPA is intended as a framework, guide and catalyst for action at community, national, regional and international levels. It seeks to create an efficient system for the conservation and sustainable use of PGR, through better cooperation, coordination and planning, and through the strengthening of capacities. The GPA is a supporting component of the *International Treaty on Plant Genetic Resources for Food and Agriculture* (Article 14 of the International Treaty). The main objectives of the GPA are to:

- Ensure the conservation of PGRFA as the basis of food security;
- Promote sustainable use of PGR to foster development and reduce hunger and poverty;
- Promote the fair and equitable sharing of the benefits arising from the use of PGR;
- Assist countries and institutions to identify priorities for action;
- Strengthen existing programmes and enhance institutional capacity.

The first GPA comprises 20 priority activity areas in four main groups covering:

- *In situ* conservation and development;
- *Ex situ* conservation;
- PGR utilization;
- Institutions and capacity building.

These groups, and their priority activity areas, are presented in Table 2.2. The GPA is also an essential contribution to successful implementation of the CBD.

CROP WILD RELATIVES IN THE GLOBAL PLAN OF ACTION. CWR are covered by the definition of PGR, as defined by FAO and the International Treaty. For the

Table 2.2. GPA four areas and 20 related priority activities.

In situ *conservation and development*
Surveying and inventorying PGRFA
Supporting on-farm management and improvement of PGRFA
Assisting farmers in disaster areas to restore agricultural systems
Promoting *in situ* conservation of CWR and wild plants for food production
Ex situ *conservation*
Sustaining existing *ex situ* collections
Regenerating threatened *ex situ* accessions
Supporting planned and targeted collecting of PGRFA
Expanding *ex situ* conservation activities
Utilization of plant genetic resources
Expanding the characterization, evaluation and number of core collections to facilitate use
Increasing genetic enhancement and base-broadening efforts
Promoting sustainable agriculture through diversification of crop production and broader diversity in crops
Promoting development and commercialization of underutilized crops and species
Supporting seed production and distribution
Developing new markets for local varieties and 'diversity-rich' products
Institutions and capacity building
Building strong national programmes
Promoting networks for PGRFA
Constructing comprehensive information system of PGRFA
Developing monitoring and early warning systems for loss of PGRFA
Expanding and improving education and training
Promoting public awareness of the value of PGRFA conservation and use

purpose of this chapter, Priority Activity 4 is of particular relevance: 'Promoting *in situ* conservation of CWR and wild plants for food production.' Priority Activity 4 recognized the need for complementing *ex situ* conservation efforts, including the conservation of CWR genetic material, with *in situ* conservation. It defines both long- and intermediate-term objectives, policy or strategy issues, capacity, and coordination or administration. Priority Activity 4 is closely linked with other activities and aspects of the GPA, including:

- Surveying and inventorying PGRFA;
- Building strong national programmes;
- Constructing comprehensive information systems for PGRFA;
- Supporting on-farm management and improvement of PGRFA;
- Promoting development and commercialization of underutilized crops and species;
- Supporting planned and targeted collections of PGRFA;
- Promoting public awareness of the value of PGRFA conservation and use.

MONITORING THE GLOBAL PLAN OF ACTION: A DISCUSSION OF AREAS SPECIFIC TO CROP WILD RELATIVES. Surveys for monitoring GPA implementation were carried out in 1998 and, following a methodological refinement, in 2000 and 2003. Meanwhile, a new approach for monitoring the implementation of the GPA was proposed to the First Session of the Intergovernmental Technical Working Group on Plant Genetic Resources for Food and Agriculture (July 2001) and to the Ninth Regular Session of the Commission (2002). During 2003–2004, the new monitoring approach was tested in a number of countries. In 2004, the Commission's Tenth Regular Session adopted the indicators and reporting format for monitoring the implementation of the GPA, revised after the pilot testing, and supported the application of the new monitoring approach to all countries, in view of the integration of these monitoring activities with the preparation of the second *Report of the State of the World's Plant Genetic Resources for Food and Agriculture*. The new approach for monitoring the implementation of the GPA, as described in a number of documents before the Commission, was tested and successfully implemented during 2003 and 2004 in seven countries: Cuba, the Czech Republic, Ecuador, Ghana, Kenya, Papua New Guinea and Fiji. The approach was also evaluated in Germany. In view of the positive early results observed in these countries, the new monitoring approach was also initiated in Mali, the Philippines and Uzbekistan, in 2004.

As of June 2005, 18 countries had completed or were in the course of monitoring implementation of the GPA using the new approach. In addition, preliminary discussions were held and funding secured for undertaking the new monitoring approach in an additional ten countries. Three international agricultural research centres, namely ICRISAT, IPGRI and IRRI, are monitoring their implementation of the GPA using the new approach. Remaining centres of the CGIAR system agreed to make available relevant information on the implementation of the GPA, based on the experience built up by the three centres.

In brief, the surveys are based on a simple questionnaire reflecting the 20 priority activity areas, clustered into the four thematic groups: *in situ* conservation and development; *ex situ* conservation; utilization of PGRFA; and institutions and capacity building. For each of these, information was requested on:

- Actions undertaken since mid-1995, and funding sources;
- Countries' prioritized main needs and the main constraints;
- Opportunities for further action, in the near future, at national and sub-regional level;
- Support required from regional or international organizations.

The analysis conducted on the first thematic group, *in situ* conservation and development found that there was a general improvement in the implementation of all four activities (including Priority Activity 4) with the exception of the African region. Countries are giving high priority to inventory activities, funded almost solely by national budgets, while international support tends to concentrate on activities related to on-farm management covering mainly crop improvement in all regions, and improvement of on-farm seed supply in African countries. Promoting conservation of CWR is receiving increasing attention at the national level in the European, African, Asian and Pacific regions, with stronger support from donors.

More specifically, the analysis of Priority Activity 4 is based on the changes that occurred since the last monitoring exercise in the 72 countries that participated in the survey in 2004 covering the period of 2001–2003, and integrates the information from five countries that have completed pilot testing of the comprehensive monitoring system: Cuba, Czech Republic, Ecuador, Ghana and Kenya.

Overall, 15 countries, most of them in the European and Asian regions, reported 23 internationally supported projects, which had a major focus on the promotion of the conservation of CWR – for example, the project on the conservation and sustainable utilization of genetic diversity of local plants in Senegal (Préservation et exploitation durable de la diversité génétique des plantes locales cultivées au Sénégal). The number of countries that have included planning and implementation activities in their national programmes to promote the conservation of CWR and wild plants for food continues to grow. Increases were observed in Canada, and in more than 80% of reporting countries in the Asian, Pacific, African and European regions. The percentage of countries with *in situ* conservation of CWR and wild plants for food production is still low in the Latin American and the Caribbean regions, with only 60% of countries reporting such projects. The involvement of local communities in national activities for the conservation of CWR and wild plants for food production according to the survey has only increased in Europe and Latin America, but even in these regions, 40% of the countries reported no participation of local communities in these activities (CGRFA-10/04/Inf.6).

Regarding the thematic group on the utilization of PGR, the survey results showed that despite significant efforts in characterization in the Asian, Pacific and European regions, overall investment in characterization of *ex situ* collection remains rather low. It appears that more emphasis is currently given to the establishment of collections, rather than to active utilization. Genetic enhancement and base-broadening activities increased since 2001, resulting from both national and external support. In spite of the potential to enhance the use of locally underutilized or biodiversity-rich products reported by countries, there appear to be inadequate incentives and flexible policy frameworks for commercialization of local varieties and diversity-rich products (CGRFA-10/04/Inf. 6). Specific to Priority Activity 12 (promoting development and commercialization of underutilized crops and species), almost 70% of countries reported some activities to promote commercialization of underutilized crops, although countries reported a low number of projects focused on Activity Area 12. Some countries indicated a lack of incentives for quality seed production of local varieties and underutilized crops, thus, appropriate policies and marketing incentives may be needed to improve the trend. Almost 60% of reporting countries indicated that they did not provide training during the reporting period for this activity area (CGRFA-10/04/Inf.6)

THE INTERNATIONAL TREATY ON PLANT GENETIC RESOURCES FOR FOOD AND AGRICULTURE. A major issue regarding the conservation and sustainable utilization of CWR, which has been and is being considered in international fora, is access and benefit sharing of PGR. Concerns of developing countries over the imbalance in the sharing of the benefits of the diversity built up by their farmers were addressed in 1961, with the adoption of the International Convention for the Protection of New Varieties of Plants. In order to address the conservation and sustainable utilization of these resources, and access and benefit sharing, the International Undertaking on Plant Genetic Resources was adopted as an international instrument, in 1983, by the FAO Conference (Stannard *et al.*, 2004). In 1993, the FAO Conference made the FAO Commission on Genetic Resources for Food and Agriculture the forum for negotiations among governments for the revision of the International Undertaking, in harmony with the CBD; and for considerations of the issue of access on mutually agreed terms to PGRFA, including *ex situ* collections not addressed by the CBD. The International Undertaking is the precursor of the International Treaty.

After 7 years of negotiations, the FAO Conference (Resolution 3/2001) adopted the (multilateral) International Treaty on Plant Genetic Resources for Food and Agriculture,[5] in November 2001. This is a legally binding treaty which covers all plant genetic resources relevant for food and agriculture, and is in harmony with the CBD. The International Treaty came into force on 29 June 2004.

The objectives of the International Treaty are the conservation and sustainable use of PGRFA and the fair and equitable sharing of benefits derived from their use, for sustainable agriculture and food security. The International Treaty recognizes that PGRFA are crucial in feeding the world's population, that PGRs are the raw material that farmers and plant breeders use to improve the quality and productivity of crops, and that the future of agriculture depends on international cooperation and on the open exchange of the crops and their genes that farmers all over the world have developed and exchanged over 10,000 years. No country is sufficient in itself – all depend on crops and the genetic diversity within these crops from other countries and regions.

[5] Information on the International Treaty is available at: http://www.fao.org/ag/cgrfa/itpgr.htm

The International Treaty also affirms that the rights recognized in the International Treaty to save, use, exchange and sell farm-saved seed and other propagating material, and to participate in decision making regarding, and in the fair and equitable sharing of the benefits arising from, the use of PGRFA are fundamental to the realization of farmers' rights, as well as the promotion of farmers' rights at national and international levels. The International Treaty recognizes that, in the exercise of their sovereign rights over their PGRFA, states may mutually benefit from the creation of an effective multilateral system for facilitated access to a negotiated selection of these resources and for the fair and equitable sharing of the benefits arising from their use. Article 11 of the International Treaty states that 'the multilateral system shall cover the PGRFA listed in Annex I, established according to the criteria of food security and interdependence'.

The International Treaty defines *in situ* conservation as the conservation of ecosystems and natural habitats and the maintenance and recovery of viable populations of species in their natural surroundings and, in the case of domesticated or cultivated plant species, in the surroundings where they have developed their distinctive properties (Article 2). Thus, *in situ* conservation of PGRFA should take into consideration other biological diversity coexisting in the same ecosystem.

Of particular relevance to this chapter are Articles 5 and 6 of the International Treaty, and especially Article 5. Article 6 on 'Sustainable Use of Plant Genetic Resources' states that contracting parties shall develop and maintain appropriate legal and policy measures that promote the sustainable use of PGR, including through a number of measures described in the article. Article 5 deals with the 'conservation, exploration, collection, characterization, evaluation and documentation of Plant Genetic Resources for Food and Agriculture'. Article 5 is presented in its entirety in Box 2.2. Article 5.1 (d) specifically states to 'promote *in situ* conservation of crop wild relatives and wild plants for food production, including in protected areas, by supporting, *inter alia*, the efforts of indigenous and local communities'.

The importance of the International Treaty as a legally binding instrument for PGRFA is also acknowledged by the CBD, more specifically through decision VI/6 of the Conference of Parties, which recognizes the important role that the International Treaty on Plant Genetic Resources for Food and Agriculture will have, in harmony with the CBD, for the conservation and sustainable utilization of this important component of agricultural biological diversity, for facilitated access to PGRFA, and for the fair and equitable sharing of the benefits arising out of their utilization. In the same decision, the Conference of Parties decided to establish and maintain cooperation with the Commission on Genetic Resources for Food and Agriculture acting as the Interim Committee for the International Treaty and, upon the entry into force of the Treaty, with the governing body. The CBD further supports the International Treaty by urging parties and other governments to ratify the International Treaty since 'the Treaty will be an important instrument for the conservation and sustainable use of genetic resources leading to hunger reduction and poverty alleviation' (decision VII/3).

OTHER INTERNATIONAL AGRICULTURAL POLICY. CWR are also considered within the context of the International Plant Protection Convention (IPPC), which originally

Box 2.2. Article 5: Conservation, exploration, collection, characterization, evaluation and documentation of plant genetic resources for food and agriculture

5.1 Each contracting party shall, subject to national legislation, and in cooperation with other contracting parties where appropriate, promote an integrated approach to the exploration, conservation and sustainable use of plant genetic resources for food and agriculture and shall in particular, as appropriate:

(a) Survey and inventory plant genetic resources for food and agriculture, taking into account the status and degree of variation in existing populations, including those that are of potential use and, as feasible, assess any threats to them;

(b) Promote the collection of plant genetic resources for food and agriculture and relevant associated information on those plant genetic resources that are under threat or are of potential use;

(c) Promote or support, as appropriate, farmers and local communities' efforts to manage and conserve on-farm their plant genetic resources for food and agriculture;

(d) Promote *in situ* conservation of wild crop relatives and wild plants for food production, including in protected areas, by supporting, *inter alia*, the efforts of indigenous and local communities;

(e) Cooperate to promote the development of an efficient and sustainable system of *ex situ* conservation, giving due attention to the need for adequate documentation, characterization, regeneration and evaluation, and promote the development and transfer of appropriate technologies for this purpose with a view to improving the sustainable use of plant genetic resources for food and agriculture;

(f) Monitor the maintenance of the viability, degree of variation and the genetic integrity of collections of plant genetic resources for food and agriculture.

5.2 The contracting parties shall, as appropriate, take steps to minimize or, if possible, eliminate threats to plant genetic resources for food and agriculture.

came into force in April 1952. Invasive alien species, addressed in this international convention, are considered to represent one of the primary threats to biodiversity, and risks may be increasing due to increased global trade, transport, tourism and climate change. For example, Pakistan, in reporting to the CBD on invasive and alien species, states that 'many primitive land races/cultivars and wild relatives of agricultural crops have suffered from genetic erosion due to introduction of high yielding varieties of these [alien] crops and habitat degradation'. Among those risks, risks to plant life are covered by the IPPC; the Seventh Conference of Parties of the CBD (February 2004) recommended that parties to the Convention and other governments ratify the revised IPPC and work activities to enhance its implementation. The new revised text of the IPPC came into force on 2 October 2005.

2.3.2 International environmental policy

As illustrated so far, the agricultural fora recognizes the importance of CWR for agriculture. CWR are also addressed through the CBD's Programme of Work on Agricultural Biodiversity, and bearing in mind the ecosystem approach (which is the primary framework for action under the CBD).

The Convention on Biological Diversity

One of the key international agreements adopted at the United Nations Conference on Environment and Development (1992), the CBD, has as main objectives the conservation of biological diversity; sustainable use of its components; and fair and equitable sharing of the benefits arising out of the utilization of genetic resources, including by appropriate access to genetic resources. The CBD describes the scope of agricultural biodiversity as a broad term that includes all components of biological diversity of relevance to food and agriculture and that constitute the agroecosystem: the variety and variability of animals, plants and microorganisms, at the genetic, species and ecosystem levels, which are necessary to sustain key functions of the agroecosystem, its structure and processes.

The CBD furthermore recognizes the work undertaken by international fora on agriculture such as FAO (in particular, the Convention Executive Secretary requested FAO to support the development of the Programme of Work on Agricultural Biodiversity), and also recognized the contribution of global agricultural instruments, policy or legislation such as the International Treaty and the GPA, for achieving the objectives of the Convention on Biodiversity.

CROP WILD RELATIVES AND THE ECOSYSTEM APPROACH. The ecosystem approach principles and operational guidance were developed during several workshops and expert meetings, and endorsed at COP V (COP decision V/6). In 2004, parties to the CBD in decision VII/11 agreed 'that the priority at this time lies in facilitating the implementation of the EA and welcomed additional guidelines to this effect'. Based on the assessment of parties' experience in implementing EA, these guidelines contribute to the further refinement and elaboration of the EA (COP decision VII/11). The EA defines 12 principles related to the holistic 'management of land, water and living resources', and provides five points of operational guidance. It is important to conserve CWR within the context of the ecosystem as a whole, given their important contribution not only to ecosystem health and resilience, but in the provision of essential ecosystem services.

PROGRAMME OF WORK ON AGRICULTURAL BIODIVERSITY. At its third meeting of the Conference of Parties, the Convention on Biodiversity decided to establish a multi-year Programme of Work on Agricultural Biodiversity (decision III/11), which was endorsed at the fifth meeting of the Conference of Parties (decision V/5) in 2000. The proposed elements of the Programme of Work on Agricultural Biodiversity were developed bearing in mind the need to 'build upon existing international plans of action, programmes and strategies that have been agreed by countries, in particular, the Global Plan of Action for the Conservation and Sustainable Utilization of Plant Genetic Resources for Food and Agriculture . . . and the International Plant Protection Convention (IPPC)'.

The Programme of Work on Agricultural Biodiversity is divided into four elements. In particular, element 2, 'adaptive management', calls for a series of activities to be undertaken by parties, including carrying out a series of case studies, in a range of environments and production systems, and in each region, including to 'monitor and assess the actual and potential impacts of existing and new agricultural technologies. This activity would address the multiple goods and services provided by the different levels and functions of agricultural biodiversity and the interaction between its various components, with a focus on certain specific and cross-cutting issues, such as . . . the role and potential of wild, underutilized and neglected species, varieties and breeds, and products'.

The objective of element 4 'mainstreaming' is to support the development of national plans or strategies for the conservation and sustainable use of agricultural biodiversity and to promote their mainstreaming and integration in sectoral and cross-sectoral plans and programmes. In order to achieve this objective, element 4 outlines an activity to 'promote ongoing and planned activities for the conservation, on farm, *in situ*, and *ex situ*, in particular, in the countries of origin, of the variability of genetic resources for food and agriculture, including their wild relatives'.

In addition to the Programme of Work on Agricultural Biodiversity, CWR are highlighted as an important component of this international agreement, and which need to be conserved and managed sustainably, in a number of different ways, including through the Global Strategy on Plant Conservation and the 2010 Biodiversity Target.

GLOBAL STRATEGY FOR PLANT CONSERVATION. The Global Strategy for Plant Conservation (the Global Strategy) in particular recognizes that in addition to the small number of crop plants used for basic food and fibres, many thousands of wild plants have great economic and cultural importance and potential, providing food, medicine, fuel, clothing and shelter for vast numbers of people throughout the world. Plants also play a key role in maintaining the planet's basic environmental balance and ecosystem stability, and provide an important component of the habitats for the world's animal life. The Global Strategy, adopted at the sixth meeting of the Conference of Parties (decision VI/9), has the ultimate and long-term objective to halt the current and continuing loss of plant diversity. To do so, the Global Strategy sets out 16 global targets for 2010, and provides a framework to facilitate harmony between existing initiatives aimed at plant conservation and identify gaps where new initiatives are required. National and/or regional targets for plant conservation may be developed within this flexible framework.

The Global Strategy, which applies to plant genetic diversity, plant species and communities and their associated habitats and ecosystems, is also comprised of sub-objectives which consider different aspects of plant biodiversity, including plants in the wild:

1. Understanding and documenting plant diversity:
    - Document the plant diversity of the world, including its use and distribution in the wild, in protected areas and in *ex situ* collections.
    - Monitor the status and trends in global plant diversity and its conservation, and threats to plant diversity, and identify plant species, plant communities and associated habitats and ecosystems, at risk, including consideration of 'red lists'.
    - Develop an integrated, distributed, interactive information system to manage and make accessible information on plant diversity.
    - Promote research on the genetic diversity, systematics, taxonomy, ecology and conservation biology of plants and plant communities, and associated habitats and ecosystems, and on social, cultural and economic

factors that impact biodiversity, so that plant diversity, both in the wild and in the context of human activities, can be well understood and utilized to support conservation action.

2. Conserving plant diversity:
   - Improve long-term conservation, management and restoration of plant diversity, plant communities and the associated habitats and ecosystems, *in situ* (both in more natural and in managed environments), and, where necessary, to complement *in situ* measures, *ex situ*, preferably in the country of origin. The Global Strategy will pay special attention to the conservation of the world's important areas of plant diversity, and to the conservation of plant species of direct importance to human societies.
3. Using plant diversity sustainably:
   - Strengthen measures to control unsustainable utilization of plant resources.
   - Support the development of livelihoods based on sustainable use of plants, and promote the fair and equitable sharing of benefits arising from the use of plant diversity.
4. Promoting education and awareness about plant diversity:
   - Articulate and emphasize the importance of plant diversity, the goods and services that it provides and the need for its conservation and sustainable use, in order to mobilize necessary popular and political support for its conservation and sustainable use.
5. Building capacity for the conservation of plant diversity:
   - Enhance human resources, necessary physical and technological infrastructure and necessary financial support for plant conservation.
   - Link and integrate actors to maximize action and potential synergies in support of plant conservation.

At its seventh meeting, the Conference of the Parties, by decision VII/10, encouraged parties to identify focal points for the Global Strategy in order, *inter alia*, to promote and facilitate implementation and monitoring of the Strategy. The Conference of Parties also decided to integrate the targets of the Global Strategy into all relevant programmes of work of the Convention, as these programmes become due for review. The progress made in reaching the global targets will be reviewed at the eighth and tenth meetings of the Conference of the Parties, in 2006 and 2010, respectively. In order to facilitate the implementation of the Global Strategy, countries are strongly encouraged to use existing instruments (such as the Global Plan of Action for the Conservation and Sustainable Utilization of Plant Genetic Resources for Food and Agriculture).

2010 BIODIVERSITY TARGET. In decision VI/26, the sixth meeting of the Conference of the Parties adopted the Strategic Plan for the CBD. In its mission statement, parties committed themselves to a more effective and coherent implementation of the three objectives of the Convention, to achieve by 2010, a significant reduction in the current rate of biodiversity loss at the global, regional and national level as a contribution to poverty alleviation and to the benefit of all life on earth. This target was subsequently endorsed by the World Summit on Sustainable Development (Johannesburg, 2002).

In decision VII/30, the seventh meeting of the Conference of the Parties adopted a framework to facilitate the assessment of progress towards 2010 and communication of this assessment, to promote coherence among the programmes of work of the Convention and to provide a flexible framework within which national and regional targets may be set, and indicators identified. The framework includes seven focal areas. The Conference of the Parties identified indicators for assessing progress towards, and communicating the 2010 target at the global level, and goals and sub-targets for each of the focal areas, as well as a general approach for the integration of goals and sub-targets into the programmes of work of the Convention.

The seven focal areas of the 2010 Biodiversity Target are:

1. Reducing the rate of loss of the components of biodiversity, including:
   - Biomes, habitats and ecosystems;
   - Species and populations;
   - Genetic diversity.
2. Promoting sustainable use of biodiversity;
3. Addressing the major threats to biodiversity, including those arising from invasive alien species, climate change, pollution and habitat change;
4. Maintaining ecosystem integrity, and the provision of goods and services provided by biodiversity in ecosystems, in support of human well-being;
5. Protecting traditional knowledge, innovations and practices;
6. Ensuring the fair and equitable sharing of benefits arising out of the use of genetic resources;
7. Mobilizing financial and technical resources, especially for developing countries, in particular, least developed countries and small island developing states among them, and countries with economies in transition, for implementing the Convention and the Strategic Plan.

2.4 Further Considerations

CWR are addressed in international policy and fora – both in the agricultural and the environmental arenas. They are distinctly recognized as being of crucial importance to food security, through agriculture. In particular, it is recognized that CWR are important not only for their intrinsic value, but also for:

- Plant breeding;
- Crop productivity;
- Nutritional value of food;
- Providing traits such as disease resistance, tolerance to extreme temperatures, tolerance to salinity and resistance to drought.

CWR are also included in discussions pertaining to the conservation of PGRFA, protected areas, *ex situ* conservation, ecosystem management, traditional knowledge and intellectual property rights (as addressed in the work on the Intergovernmental Committee on Genetic Resources, Traditional Knowledge and Folklore of the World Intellectual Property Organization [WIPO] and in national policy considerations (e.g. National Biodiversity Strategy and Action Plans), prepared by national governments. In these national plans, considerations of CWR are also addressed, for example, in the National Biodiversity Strategy and Action Plan of Armenia, which discusses the unsustainable harvesting by local populations of such plants). Another example of national policy considerations is the development of national agricultural biodiversity programmes, such as the one developed in Lao PDR, which addresses the issue of CWR.

The issue of the conservation and use of CWR has an opportunity to advance considerably in the agriculture fora, where the use of genetic resources for food and agriculture is considered by the Commission on Plant Genetic Resources for Food and Agriculture. In particular, the preparation of the Second State of the World's Plant Genetic Resources affords an opportunity for countries to provide concrete input into the Report, through the National Focal Points for the GPA. The National Focal Points for the GPA are responsible for coordinating the preparation of the country reports, which are used as valuable inputs for the preparation of the second Report. Concerns, achievements and technical inputs have an opportunity, at the national level, to be considered for the preparation of the Second State of the World's Plant Genetic Resources.

Hence, countries are encouraged to participate actively in preparing country reports for the preparation of the Second State of the World's Plant Genetic Resources. Results of the second Report will have an impact on future considerations in agriculture, at the FAO Commission on Genetic Resources for Food and Agriculture. Furthermore, the second Report will provide input to the rolling GPA.

References

CBD (1992) *Convention on Biological Diversity: Text and Annexes*. Secretariat of the Convention on Biological Diversity, Montreal. Available at: http://www.biodiv.org/convention/convention.shtml

FAO (1998) *State of the World's Plant Genetic Resources for Food and Agriculture*. FAO, Rome, Italy. Available at: http://www.fao.org/ag/aGp/AGPS/Pgrfa/pdf/swrfull.pdf

FAO (2004) *Commission on Genetic Resources for Food and Agriculture 10th Session Final Report (CGRFA-10/4/REP)*, FAO, Rome, Italy. Available at: ftp://ftp.fao.org/ag/cgrfa/cgrfa10/r10repe.pdf

FAO (2006a) *Global System on Plant Genetic Resources*. FAO, Rome, Italy. Available at: http://www.fao.org/ag/cgrfa/PGR.htm#diagram

FAO (2006b) *Monitoring the Implementation of the Global Plan of Action for the Conservation and Sustainable Use of PGRFA*. FAO, Rome, Italy. Available at: http://apps3.fao.org/wiews/wiewspage.jsp?i_l=EN&show=GPAMonitor

Ng, N.Q. (2005) *Views and Perspectives of* in situ *Conservation and Development of Plant Genetic Resources for Food and Agriculture: Reference to Some Initiatives of the CBD and FAO*. Presentation given at the FAO *in situ* conservation workshop and GIS training course, 29 August to 2 September 2005. Bangkok, Thailand.

Schippmann, U., Leaman, D.J. and Cunningham, A.B. (2002) Impact of cultivation and gathering of medicinal plants on biodiversity: trends and issues. In: *Biodiversity and the Ecosystem Approach in Agriculture, Forestry and Fisheries*. FAO, Rome, Italy. Available at: http://www.fao.org/DOCREP/005/Y4586E/Y4586E00.HTM

Stannard, C., van der Graaff, N., Randell, A., Lallas, P. and Kenmore, P. (2004) Agricultural biological diversity for food security: shaping international initiatives to help agriculture and the environment. *Howard Law Journal* 48, 397–430.

3 Crop Wild Relatives: Putting Information in a European Policy Context

D. Richard, G. Augusto, D. Evans and G. Loïs

3.1 Introduction

Over the last half decade, biodiversity has gained increased political visibility and support, being recognized as critical for both sustainable development and poverty eradication. Commitments have been taken by heads of states and governments on halting or significantly reducing the current rate of loss of biodiversity by 2010 (the 2010 targets). This was addressed at:

- The European Union (EU) level, within the 6th Environmental Action Programme 'Environment 2010. Our future, our choice' (2001–2010) and subsequently in the EU Sustainable Development Strategy (2001);
- Global level, within the Strategic Plan for the Convention on Biological Diversity (CBD, 2002), supported by the World Summit for Sustainable Development Plan of Implementation (Johannesburg, 2002);
- Pan-European level, in the Kyiv Resolution on Biodiversity (2003), under the United Nations Economic Commission for Europe process 'Environment for Europe' and the Pan-European Biological and Landscape Diversity Strategy (PEBLDS).

More recently, the United Nations has proclaimed 22 May the International Day for Biological Diversity, focusing, in May 2005, on the topic: 'Biodiversity, Life Insurance for Our Changing World'.

In this context, raising the profile of crop wild relatives (CWR) is important since it links nature conservation with agronomic, forestry and medical, as well as cultural concerns. In a situation where attempts are being made to put economic value on biodiversity, highlighting the need to conserve species diversity, not only for its own sake, but also as a potential resource for human well-being in an increasingly uncertain environment, is crucial.

As a specialized branch of the European Environment Agency (EEA), the European Topic Centre on Biological Diversity (ETC/BD) has the mandate – in the specific field of biodiversity – 'to support sustainable development and to help achieve significant and measurable improvement in Europe's environment, through the provision of timely, targeted, relevant and reliable information to policy making agents and the public'. Based on a consortium of nine biodiversity-related institutions from various European countries, coordinated by the French 'Museum National d'Histoire Naturelle' the ETC/BD collects, assesses and synthesizes information and reports on nature and biodiversity at the European scale, as a basis for policy implementation. Identifying needs and opportunities to report on CWR in Europe in a policy context is thus part of the work of the EEA–ETC/BD.

The purpose of this chapter is not to assess the influence of European policies on the status of CWR. In particular, it is not in our scope to assess the current and future impacts of the Common Agricultural Policy on CWR or how the new EC Regulation on genetic resources for agriculture (EC Regulation 870/2004) will apply to CWR. Instead, the aim is to show how information collected in the framework of various policy processes can be used in the evaluation of CWR occurrence and status, and therefore help in defining conservation priorities.

3.2 European Nature Conservation Policy of Relevance for Crop Wild Relatives

Seen from a nature protection perspective, there are three main instruments of relevance for the CWR conservation in Europe.

3.2.1 Convention on the Conservation of European Wildlife and Natural Habitats (Bern Convention)

The Bern Convention is a binding international legal instrument in the field of nature conservation, which covers the whole of the natural heritage of the European continent and extends to some states of Africa (40 member states of the Council of Europe, as well as to Burkina Faso, Morocco, Senegal, Tunisia and the European Community). Its aims are to conserve wild flora and fauna and their natural habitats, and to promote European cooperation in that field. Two aspects are of particular relevance to the protection of plant species, and therefore potentially of CWR:

- Appendix I of the Bern Convention provides a list of 656 plant species (excluding algae, mosses and liverworts) which should be strictly protected. It means that 'appropriate and necessary legislative and administrative measures' should be taken by Contracting Parties including prohibition of deliberate picking, collecting, cutting, uprooting and, as appropriate, possession or sale. An assessment of the Crop Wild Relative Catalogue for Europe and the Mediterranean (Kell *et al.*, 2005) showed that 415 plant species considered to be CWR are included in Appendix I of the Bern

Convention, which represents 63% of the species protected under this legal instrument (Kell *et al.*, 2005a).

- Recommendation No. 16 (1989) and Resolution No. 3 (1996) of the Standing Committee of the Bern Convention call for the conservation of natural habitats of endangered species through 'areas of special conservation interest' within the 'Emerald network'. The plant species whose habitats have to be protected are those listed in Appendix I of the Bern Convention, as well as those listed in Annex II of the Habitats Directive (see below). Appendix VIII – Resolution No. 4 (1996) of the Standing Committee lists endangered natural habitat requiring specific conservation measures. Launched in 1999, the Emerald Network consists of a programme of national pilot projects set up with a view to developing a pilot database on selected 'areas of special conservation interest'. Twenty-two pilot projects have been organized in Europe to date (excluding the previous 15 EU countries), and two pilot projects are under way in African states (Senegal and Burkina Faso). Other European and African states may launch pilot projects in the coming months. The information collected through this process includes, when relevant, data on occurrence of plant species (Appendix I of the Bern Convention as well as those listed in Annex II of the Habitats Directive) as well as habitats. An analysis of the database to assess occurrence of CWR remains to be done.

3.2.2 European Council Directive 92/43/EEC of 21 May 1992 on the Conservation of Natural Habitats and of Wild Fauna and Flora (Habitats Directive)

Together with the Council Directive 79/409/EEC of 2 April 1979 (Bird Directive), the so-called Habitats Directive is effectively a transposition of the Bern Convention in the EU regulation. The Habitats Directive, which now applies to 27 EU Member States, is intended to help maintain biodiversity in the Member States by defining a common framework for the conservation of wild plants and animals and habitats of community interest. Although the following assessments only relate to the previous 25 EU Member States.

Relevant annexes of the Habitats Directive

Three annexes of the Directive, related to species, are of relevance for CWR:

- Annex II – provides a list of species which should be maintained in a 'favourable conservation status', through the designation and adequate management by Member States of Special Areas of Conservation within the so-called Natura 2000 network: 540 vascular plant species are included in this annex.
- Annex IV – provides a list of species which are strictly protected throughout the EU (with some local exemptions). As many as 599 vascular plant species are included in this annex. All species listed in Annex II of the Habitats Directive are also listed in Annex IV but there are other additional species.

- Annex V – provides a list of species of Community interest whose wild harvesting and exploitation may be subject to management measures. As many as 26 angiosperms and all *Lycopodium* spp. are included in this annex. Two species of algae, one species of lichen and all sphagnum except *Sphagnum pylaisii* Brid. (which is included in Annexes II and IV) are also included.

An assessment of the Catalogue of CWR for Europe and the Mediterranean showed that out of the 628 vascular plant species listed in the Habitats Directive annexes, 395 are CWR, which represents 63% of the species protected under the Habitats Directive (Kell *et al.*, 2005a). They are distributed as follows:

- 334 CWR plant species are listed in Annex II of the Habitats Directive (53%);
- 377 CWR plant species are listed in Annex IV of the Habitats Directive (60%);
- 18 CWR plant species are listed in Annex V of the Habitats Directive (60%).

Apart from these three annexes related to species, another annex related to habitats is relevant to CWR. Annex I provides a list of 225 habitats of Community importance. As in the case of Annex II, Member States have to ensure their 'favourable conservation status' through the designation and adequate management of Special Areas of Conservation within the Natura 2000 network. A description of these habitats is provided in the Interpretation Manual of European Union Habitats (EC, 2003). Obvious examples of habitats related to CWR are '*Quercus suber* forests'. For other habitat types, the relation to CWR is not as straightforward. However, as the descriptions of Annex I habitat types are often supported by lists of indicator or characteristic species, this information can be derived (see Kell *et al.*, Chapter 5, this volume). For example, the description of habitat 1230 'vegetated sea cliffs of the Atlantic and Baltic coasts' refers to the presence of *Brassica oleracea* and *Beta vulgaris*, among others; habitat 40A0, 'subcontinental peri-Pannonic scrub', refers to the presence of *Asparagus officinalis*', among others.

The Natura 2000 network

The Natura 2000 network is a network of areas designated as Special Protection Areas under the Birds Directive and as mentioned earlier as Special Areas of Conservation under the Habitats Directive, specifically for habitats listed in Annex I and species listed in Annex II. So far, the Habitats Directive component of the Natura 2000 network includes 20,587 sites across the EU, which represent about 12% of the EU territory.

Each site is described according to a standard data form and the information is handled and managed by the European Topic Centre on Biological Diversity on behalf of the European Commission. This includes information on occurrence on the site of Annex I habitats and/or Annex II species. Thus, out of 20,587 sites, 4519 sites have records of at least one Annex II plant species. In addition, some countries report the occurrence of non-Annex II species which may however have a particular value to characterize the site, such as IUCN Red List species; 6769 sites have such information on non-Annex II plant species.

As Member States have the duty to report every 6 years on the conservation status of species and habitats for which the sites have been designated, information on both ecological and socio-economic aspects, as well as management plans, will also be available in a standardized format from 2007 onwards. An analysis of the Natura 2000 database to assess occurrence of CWR remains to be done.

3.2.3 EC Regulation 338/97 (amended 1497/2003 of 18 August 2003) of the European Council of 9 December 1996 related to the protection of wild fauna and flora species by trade control

In order to protect endangered wildlife from unsustainable trade exploitation, the Convention on International Trade in Endangered Species of Wild Fauna and Flora (CITES) was signed in Washington in 1973. CITES distinguishes three levels of threat to species in relation to trade, as reflected in three annexes:

- Appendix I – includes species threatened with extinction, for which trade must be subject to stricter regulation and can only be authorized in exceptional circumstances for specimens of wild origin. Commercial trade in wild specimens of Appendix-I listed species is generally not allowed.
- Appendix II – includes species that are not necessarily threatened now with extinction, but may become so unless trade is strictly regulated. Appendix II further contains the so-called look-alike species (see Article II, paragraph 2(b) of CITES), which are controlled because of their similarity in appearance to the other regulated species, thereby facilitating a more effective control thereof.
- Appendix III – includes species that are subject to regulation within the jurisdiction of a Party and for which the cooperation of other Parties is needed to prevent or restrict their exploitation.

The EU has adopted wildlife trade regulations, aiming at a sound implementation of the CITES. Thus, EC Regulation 338/97 amended 1497/2003 deals with imports and exports of wild animals and plants, and their products to and from the EU, as well as commerce between Member States. It is supported by four annexes (Table 3.1). A cross-analysis with European CWR should only consider plant species for which export is regulated by EC Regulation 338/97.

3.3 Other Information of Relevance for CWR in Europe

3.3.1 Information on Red List species

Database on the most threatened endemic and subendemic species

In 2003, in collaboration with the Council of Europe, the European Topic Centre on Biological Diversity has asked the Conservatoire Botanique National

Table 3.1. The four annexes of EC Regulation 338/97.

Annex A Commercial trade from, to and within the Community is, as a general rule, prohibited for wild specimens of Annex A species. External trade is governed by provisions comparable to those applicable to Appendix I species under CITES.	• All CITES Appendix I species • Some CITES Appendixes II and III species, for which the EU has adopted stricter domestic measures • Some non-CITES species This concerns 313 wild plant species
Annex B Annex B contains species for which trade into and from the Community requires the issuance of import permits, export permits and re-export certificates along the lines of the provisions applicable to CITES Appendix II species.	• All other CITES Appendix II species • Some CITES Appendix III species • Some non-CITES species This concerns 26,005 plant species
Annex C Species in Annex C are not subject to the stricter Community requirement of an import permit. Imports can take place on the basis of a CITES (re-)export certificate or a certificate of origin and an import notification.	• All other CITES Appendix III species This concerns 46 plant species
Annex D Annex D lists species that do not have a CITES equivalent. Imports of Annex D specimens require an import notification. As the purpose of Annex B is to ensure sustainable trade in species and thus prevent them from becoming Annex A candidates, the Annex D monitoring system is intended to allow an early detection of possible conservation concerns to the species listed.	• Some CITES Appendix III species for which the EU holds a reservation • Some non-CITES species This includes 52 plant species

de Brest (France) to survey the most threatened endemic and subendemic plants in Europe. The survey was implemented in five steps:

1. Analytical review of the 1997 IUCN global Red List of plants;
2. Literature review – some 48 red books/lists from 36 countries have been reviewed;
3. Preliminary national lists of threatened endemic and subendemic species – one of the most innovative approaches of the survey is the application, in a harmonized way, of the IUCN 1996 criteria – modified as IUCN 2000 – to assign a global status (EX, EW or CR) to species assessed within different national contexts;
4. Expert consultation and validation – this phase is still ongoing;
5. Information incorporated into a European database which contains the following information for each assessed plant species:

- IUCN categories and criteria;
- Geographical distribution;
- Cultivation frequency: *ex situ* measures;
- Location of cultivation sites;

- Protection status under Bern Convention (1979) and Habitats Directive (1992);
- Threat categories;
- Recovery programmes;
- Comments;
- Bibliographical references.

The assessment shows that 763 European plant taxa can be considered as extinct or close to extinction, out of which 75 are extinct in the wild (5 totally extinct (EX) and 20 extinct in the wild but remaining in collections (EW)). Final validation of the database through the IUCN network of plant specialists is foreseen before making it available to the public.

Information on national Red Lists

Information on national Red Lists of species has been compiled by the European Topic Centre on Biological Diversity for some European countries. This information has been incorporated into the 'European Nature Information System' (EUNIS). However, there is a need to organize a more systematic and comprehensive provision of such information from national to European level. This is why the European Topic Centre on Biological Diversity will consider how a 'priority data flow' on Red Lists of species can be set up within the official network of member countries of the European Environment Agency, the so-called European Information and Observation Network (EIONET).

3.3.2 Information on nationally designated areas

As part of a collaborative process among the European Environment Agency, the Council of Europe and the UNEP–World Conservation Monitoring Centre, a 'Common database on Designated Areas' (CDDA) in Europe has been set up. This database includes information on more than 75,000 nationally designated sites across 48 European countries. However, no information on occurrence of species in these sites is available yet in the CDDA, although it may be available at national level. The CDDA is therefore of limited value at the moment to identify possible occurrence of CWR in Europe. On the other hand, for CWR species for which distribution is well known, it is possible to assess in which protected areas it may occur.

3.3.3 Information related to 'high nature value farming' areas

As part of the above-mentioned Kiev Resolution on Biodiversity (2003), the European Environment Ministers and Heads of Delegations of the States participating in the process of the Pan-European Biological and Landscape Diversity have called for:

- By 2006, the identification, using agreed common criteria, of all high nature value (HNV) areas in agricultural ecosystems in the pan-European region will be complete. By 2008, a substantial proportion of these areas will be under biodiversity-sensitive management by using appropriate mechanisms such as rural development instruments, agrienvironmental programmes and organic agriculture, to *inter alia* support their economic and ecological viability.

Within the EU context, the 'Message from Malahide', endorsed by the Council of Ministers in June 2004 called for:

- High nature value areas (including the Natura 2000 network) threatened with loss of biodiversity and abandonment identified, and measures to address those threats provided.

A preliminary survey to identify such HNV has been undertaken by the European Environment Agency (EEA, 2004) largely based on land-cover data. Further work is foreseen in 2006 on the basis of CORINE Land Cover 2000, biodiversity and socio-economic data. This exercise will probably not bring additional information on occurrence of CWR, as compared to other sources of data mentioned earlier. However, it is very important that the CWR issue is addressed within this HNV framework and those relevant experts are associated in the designing of the map at European level.

3.4 Conclusions

As one of the three components of biodiversity recognized by the Convention on Biological Diversity, genetic diversity is a case for policy action. Specific regulations and treaties, as well as programmes related to agricultural genetic resources are set up at international and European level. In the specific case of CWR, it is also important to consider which other policy contexts mainly targeted to conservation of wild plant diversity can be applied. Information collected at the European level in support of the implementation of the Bern Convention, the EC Habitats Directive as well as the EC Regulation in application to CITES, provides opportunities for assessing potential distribution, threat status and protection status of European CWR. In addition, other information related to Red List species, nationally protected areas and sites recognized as 'high nature value farming areas' can be useful. Such information is partly available from the EUNIS set up by the European Topic Centre on Biological Diversity on behalf of the European Environment Agency. It is important to initiate proper communication between managers of genetic resources and wild biodiversity, including protected areas managers, in order to ensure synergy between *ex situ* and *in situ* conservation of CWR wherever appropriate and possible.

References

CBD (2002) *Strategic Plan for the Convention on Biological Diversity*. Secretariat of the Convention on Biological Diversity, Montreal, Canada. Available at: http://www.biodiv.org/sp/default.shtml

EC (2003) *Interpretation Manual of European Union Habitats*. European Commission Available at: http://ec.europa.eu/environment/nature/nature_conservation/eu_enlargement/2004/pdf/habitats_im_en.pdf

European Environment Agency (2004) *High Nature Value Farmland. Characteristics, Trends and Policy Challenges*. Office for Official Publications of the European Communities, Luxembourg.

Kell, S.P., Knüpffer, H., Jury, S.L. Maxted, N. and Ford-Lloyd, B.V. (2005) *Catalogue of Crop Wild Relatives for Europe and the Mediterranean*. Available online via the Crop Wild Relative Information System (CWRIS – http://cwris.ecpgr.org/) and on CD-ROM. University of Birmingham, Birmingham, UK.

Kell, S.P., Knüpffer, H., Jury, S.L. Maxted, N. and Ford-Lloyd, B.V. (2005a) *Creating a regional catalogue of crop taxa and their wild relatives: a methodology illustrated for Europe and the Mediterranean*. Presentation given at the First International Conference on Crop Wild Relative Conservation and Use, incorporating the PGR Forum Final Dissemination Conference. Agrigento, Sicily, Italy, 14–17 September 2005.

Web References

2010 Biodiversity targets: http://biodiversity-chm.eea.eu.int/convention/F1117799202

Convention on the Conservation of European Wildlife and Natural Habitats: http://www.coe.int/T/E/Cultural%5FCo%2Doperation/Environment/Nature%5Fand%5Fbiological%5Fdiversity/Nature%5Fprotection/

Emerald network: http://www.coe.int/TIE/Cultural_Co-operation/Environment/Nature_and_biological_diversity/Ecological_networks/The_Emerald_Network/02General_information.asp#TopOfPage

EU implantation of CITES for protection of Wild Fauna and Flora Species by Trade Control: http://ec.europa.eu/environment/cites/legislation_en.htm

EUNIS: http://eunis.eea.europa.eu/

European Council Directive 92/43/EEC of 21 May 1992 on the Conservation of Natural Habitats and of Wild Fauna and Flora: http://europa.eu.int/comm/environment/nature/nature_conservation/eu_nature_legislation/habitats_directive/index_en.htm

The European Environment Agency: http://www.eea.eu.int/main_html

The European Topic Centre on Biological Diversity: http://biodiversity.eionet.europa.eu/

NATURA 2000 network: http://europa.eu.int/comm/environment/nature/home.htm

4 Crop Wild Relatives in Armenia: Diversity, Legislation and Conservation Issues

A. Avagyan

Being one of the oldest known sites of agriculture, Armenia has preserved the traces of residence of prehistoric man in its territory and there is much archaeological evidence of various household goods from the Stone Age. As is well known, primitive man mainly populated the territories rich in specific composition and abundance of edible plants. Armenia, where many species, varieties and forms of wild wheat, rye, barley, pea, lentil, flax, beet, spinach, lettuce, etc., as well as a large diversity of wild berry and nectar plants are concentrated, belongs to this type of territories. Due to abundance of wild relatives of cultivated plants Armenia was selected as a centre of cultivated plants diversity by N.I. Vavilov; as part of the Western Asia centre of origin of cultivated plants being rich in soft and durum wheat, pea, lentil and grape. High concentration of wild progenitors of cultivated plants represents a very rich gene pool for creation of new crop varieties providing resistance to diseases, drought and cold among other adaptive characteristics.

Primarily, the Western Asian gene centre is distinguished internationally by the diversity of wheat species and ecotypes. Out of the four known wild wheat species, three occur in Armenia – *Triticum boeoticum* Boiss., *T. urartu* Thum. ex Gandil. and *T. araraticum* Jakubz. (Ghandilyan, 1972). Diploid self-pollinated wild wheat *T. urartu* provided the A genome for the tetraploid hard wheat *T. turgidum* and hexaploid bread wheat *T. aestivum*. This wild wheat species grows on tertiary red clays and on basalts, at altitudes of 1300–1400 m, as a component of the semi-arid, herbaceous vegetation. Wild einkorn wheat *T. boeoticum* is fully interfertile and shares homology chromosomes, and is considered the direct progenitor of *T. monococcum*. This species is often found in mixed populations with other wild wheat species (*T. urartu* and *T. araraticum*) and several species of *Aegilops*. Tetraploid self-pollinated wild wheat *T. araraticum* grown in semi-desert and mountain steppe conditions has been identified as a progenitor of the cultivated *T. timofeevii* Zhuk. wheat.

Along with other wild wheat species, it is protected in the Erebuni state reserve (Gabrielian and Zohary, 2004).

Nine *Aegilops* species, with wide interspecific diversity, have been discovered in the Republic of Armenia. Greatest breeding interest has been focused on *Aegilops tauschii* Cosson (*A. squarrosa* L.) as an annual diploid self-pollinated species and the donor of D genome of modern hexaploid wheat. It is widely distributed in semi-desert and steppe habitats, at altitudes of 700–1300 m. The other species of goat grass (*A. cylindrica* Host, *A. triuncialis* L., *A. triaristata* Willd., *A. crassa* Boiss, *A. biuncialis* Vis., *A. columnaris* Zhuk., *A. mutica* (Boiss.) Eig. and *Amblyopyrum muticum* Boiss. and *A. umbellulata* Zhuk.) provide a rich reservoir of genes for drought resistance, poor soil tolerance and pest and disease resistance (Harutyunyan, 1991). As for other cereals, two species of wild rye, annual *Secale vavilovii* Grosch. and perennial *S. montanum* Guss. (Matevosyan, 1987), as well as eight species of wild barley, including two-rowed *Hordeum spontaneum* C. Koch and *H. bulbosum* L., are found and are of special interest for breeding (Avagyan, 1992).

Numerous indigenous forms of cultivated legumes have also been identified, though there are also wild forms, such as:

- Lentil species – annual diploid wild lentil *Lens orientalis* (Boiss.) Hand., which is closely related to crop and rare and distant from the crop lentil species *L. ervoides* Grande.
- Two widely distributed wild forms of pea *Pisum sativum* L. – *P. elatius* M. Bieb. [*P. sativum* L. subsp. *elatius* (M. Bieb.) Aschers. & Graebn.] and *P. sativum* L. subsp. *humile* (Holmb.) Greut., Matthäs & Risse [*P. sativum* L. subsp. *syriacum* Berger] – and one more distant form of crop alpine perennial pea *Vavilovia formosa* (Steven) Fed., which is rare and insufficiently studied.
- Wild forms of bitter vetch *Vicia ervilia* (L.) Willd. are widely distributed in six floristic regions of Armenia at altitudes ranging from 1300 to 2000 m.
- Two species of liquorices – widely distributed *Glycyrrhiza glabra* L. occupied mainly swampy, sometimes saline places, and the comparatively rare species *G. echinata* L. (Gabrielian and Zohary, 2004).

Armenia is also a primary and secondary centre of origin for many vegetable plants, 280 species of which are found in the territory of the republic, such as beet (widely distributed wild and weedy forms of *Beta vulgaris* subsp. *maritima* (L.) Arcang [*B. perennis* (L.) Freyn.], diploid *B. lomatogona* Fisch. et C.A. Mey and *B. macrorrhiza* Steven, and tetraploid wild species *B. corolliflora* Zoss. ex Battler), carrot, purslane, watermelon, melon, species of lettuce, asparagus and sorrel (Melikyan, 2001; Gabrielian and Zohary, 2004). Some wild species of oil-bearing plants that occur in Armenia include:

- Different wild and weedy forms of flax (*Linum* L.) usually referred to as *L. bienne* Mill. [*Linum usitatissimum* L. subsp. *angustifolium* (Huds) Thell.];
- Wild and weedy forms of hemp (*Cannabis sativa* L.) found in eight floristic regions at altitudes of 700–2000 m;

- Wild and weedy forms of gold of pleasure (*Camelina sativa* L.) locally grown in few places;
- Weedy forms of turnip (*Brassica rapa* L. [*B. campestris* L.]);
- Species of safflower – found in five floristic regions and infesting crop sowings *Carthamus oxyacanthus* M. Bieb. and *C. gypsicola* Iljin occupying clays rich in gypsum, as well as saline places and dry stony slopes;
- Wild-growing weedy rape forms (*Brassica napus* L.) etc. (Takhtajan, 1992).

As many as 18 species of condiments are distributed in Armenia. Most of them are used for aromatic leaves or seeds that serve to flavour foods and drinks (caraway, summer savory, tarragon, sumac, brown mustard, hop and coriander), as well as for medicinal purposes (thymus, mints and lemon balm) or extraction of essential oils (oregano and wormwood).

Western Asia has a wide diversity of native fruit species, such as grape, pear, cherry plum, sweet cherry, pomegranate, walnut, almond and fig, and *Pyrus* spp. are significant for their remarkable diversity in drought, cold and poor soil tolerance (Gabrielian, 1991). The genus *Sorbus* is represented by about 13 polymorphic species with a great diversity of forms. A special emphasis should be given also to the genus *Crataegus* with its extremely polymorphic species with breeding, ornamental and medicinal importance. The principal CWR, defined in terms of economic value and threat, for Armenia are presented in Table 4.1.

The natural populations of many crop wild relatives (CWR) are increasingly at risk in Armenia. The greatest threats to biodiversity result directly and indirectly from human activities. As a result of multi-year blockade, fuel and energy crisis and disafforestation on large forest areas in some regions, the situation with some CWR species has become catastrophic. A combination of poor forest management and illegal felling creates a threat to forest gene resources (particularly to native pine stands). Anthropogenic effects have demonstratively led to a drastic reduction in the genetic diversity of Armenian CWR species. The very species diversity that is being lost could serve as an indispensable basis for enrichment of crop gene pools and wealth creation in Armenia.

Many wild relatives of crops are now under the threat of extinction, reduction in population number, narrowing of distributional range and genetic erosion. Hence, immediate action is required to conserve CWR in Armenia. There is a need to prevent the degradation of natural resources, loss of biodiversity and desertification, as well as to promote sustainable usage of plant genetic resources for food and agriculture (PGRFA) and, in particular, of CWR. This may best be achieved through national PGRFA programmes and appropriate legislation. Legal issues are central to the conservation and use of CWR as far as they affect ownership of and access to these resources and provide mechanisms for sharing them (Engels *et al.*, 2000).

Armenia's strategy for biodiversity conservation, as identified in the National Environmental Action Plan and Biodiversity Strategy and Action Plan, focuses on sustainable development of landscapes, building human capital and increasing financial investments to achieve improvements in four key areas: (i) institutional and community activities in sustainable development and its enabling legal

Table 4.1. List of principal crop wild relatives of Armenia.

Crop name	Species name	Crop name	Species name
Cereals		*Legumes*	
Wheats	*Triticum araraticum* Jakubz.	Lentil	*Lens orientalis* (Boiss.) Schmalh.
	T. boeoticum Boiss.		*L. ervoides* (Brign.) Grande
	T. urartu Thum. ex Gandil.	Liquorice	*Glycyrrhiza glabra* L.
Aegilops	*Aegilops crassa* Boiss.		*G. echinata* L.
	A. tauschii Cosson	Pea	*Pisum sativum* L. subsp. *humile* (Holmb.) Greut., Matthäs & Risse
	A. umbellulata Zhuk.		*P. elatius* M. Bieb.
	A. cylindrica Host		*Vavilovia formosa* (Steven) Fed.
	A. triuncialis L.	Grass pea	*Lathyrus cicera* L.
	A. biuncialis Vis.	Bitter vetch	*Vicia ervilia* (L.) Willd.
	A. triaristata Willd.	*Oil and/or fibre crops*	
	A. columnaris Zhuk.	Safflower	*Carthamus oxyacanthus* M. Bieb.
	A. mutica (Boiss.) Eig.		*C. gypsicola* Iljin
Rye	*Secale vavilovii* Grossh.	Turnip	*Brassica rapa* L.
	S. montanum Guss.	Rape	*B. napus* L.
Barley	*Hordeum spontaneum* C. Koch	Flax	*Linum bienne* Mill.
	H. glaucum Steud.	Gold of pleasure	*Camelina sativa* L.
	H. murinum L.	Hemp	*Cannabis sativa* L.
	H. geniculatum All.	*Condiments*	
	H. marinum Huds.	Thyme	*Thymus kotschyanus* Boiss. & Hohen
	H. violaceum Boiss. et Huet	Summer savory	*Satureja hortensis* L.
	H. bulbosum L.	Tarragon	*Artemisia dracunculus* L.
	H. hrasdanicum Gandil.	Sumac	*Rhus coriandra* L.
Fruit crops		Wormwood	*Artemisia absinthium* L.
Mountain ash	*Sorbus aucuparia* L.	Lemon balm	*Melissa officinalis* L.
	S. haiastana Gabr.	Caraway	*Carum carvi* L.
	S. takhtadjanii Gabr.	Oregano	*Origanum vulgare* L.
	S. subfusca (Ledeb.) Boiss.	Brown mustard	*Brassica juncea* (L.) Czern.
Crataegus	*Crataegus orientalis* Pallas ex M. Bieb.	Hop	*Humulus lupulus* L.
	C. pontica C. Koch	Coriander	*Coriandrum sativum* L.
Apple	*Malus orientalis* Uglitzk.	Mints	*Mentha longifolia* (L.) L.
Grapevine	*Vitis sylvestris* C.C. Gmelin		*M. pulegium* L.
Currants	*Ribes biebersteinii* Berland. ex DC.		*M. arvensis* L.
	R. armenum Pojark.	*Vegetables*	
Diospyros	*Diospyros lotus* L.	Spinach	*Spinacia tetrandra* Steven ex M. Bieb.
Plum	*Prunus domestica* L.	Beet	*Beta vulgaris* subsp. *maritima* (L.) Arcang
	P. cerasifera Ehrh.		*B. lomatogona* Fisch. et C.A.Mey
	P. spinosa L.		

Continued

Table 4.1. *Continued*

Crop name	Species name	Crop name	Species name
Pear	*Pyrus caucasica* Fed.		*B. macrorrhiza* Steven
	P. syriaca Boiss.		*B. corolliflora* Zoss. ex Battler
	P. takhtadzhianii Fed.	Carrot	*Daucus carota* L.
	P. medvedevii Rubtzov	Asparagus	*Asparagus officinalis* L.
Medlar	*Mespilus germanica* L.		*A. verticillatus* L.
Cornelian cherry	*Cornus mas* L.		*A. persicus* Baker
		Garden cress	*Lepidium sativum* L.
Pomegranate	*Punica granatum* L.	Chicory	*Cichorium intybus* L.
Silver berries	*Elaeagnus angustifolia* L.	Leek	*Allium ampeloprasum* L.
	E. orientalis L.	Purslane	*Portulaca oleracea* L.
Fig	*Ficus carica* L.	Sorrel	*Rumex acetosa* L.
Wood strawberry	*Fragaria vesca* L.		*R. crispus* L.
		Watermelon	*Citrullus colocynthis* (L.) Schrad.
Raspberry	*Rubus idaeus* L.		
Quince	*Cydonia oblonga* Mill.	Melon	*Cucumis melo* L. subsp. *agrestis* (Naud.) Pangalo
Apricot	*Armeniaca vulgaris* Lam.		
Sea buckthorn	*Hippophae rhamnoides* L.	Radish	*Raphanus raphanistrum* L.
Jujube	*Ziziphus jujuba* Mill.	*Nut crops*	
Rosa	*Rosa hemispherica* J. Herrm.	Almond	*Amygdalus nairica* Fed. & Takht.
Sweet cherry	*Cerasus avium* (L.) Moench		
Sour cherry	*C. vulgaris* Mill.		*A. fenzliana* (Fritsch) Lipsky.
Bird cherry	*Padus racemosa* (Lam.) Gilib.	Hazel	*Corylus avellana* L.
		Walnut	*Juglans regia* L.
Gooseberry	*Grossularia reclinata* (L.) Mill.		

framework; (ii) public awareness and participation; (iii) protected area network planning and management; and (iv) mainstream biodiversity conservation into agriculture, forestry and other economic sectors.

Although the existing legislation needs to be improved, it can serve as a reliable legal basis for the implementation of activities envisaged by National Action Plan. A number of laws related to CWR to a greater or lesser extent have been developed and adopted by the National Assembly over recent years: Law on Expertize to Access the Impact on Environment (1995); Law on Payments for Bio-resources Use (1998); Law on Flora (1999); Law on Seeds (2005); The Forest Code (2005); Law on Protected Areas (1991); Law on Lake Sevan (2001); Law on Rehabilitation of Lake Sevan Ecosystem, its Maintenance, Reproduction and Utilization (2001), as well as other laws of the Republic of Armenia and statutes of protected areas. The legislative basis for biodiversity conservation has also been improved due to several developed documents of strategic nature: 'Strategy on Developing Specially Protected Areas and National Action Plan' (adopted in 2003); 'Strategy on Access to Genetic Resources and Benefit-Sharing'; and 'Strategy on Taxonomic Investigations and Development of Biodiversity Monitoring' ('Assessment of Biodiversity Priority

Capacity Building Needs and Establishment of Clearing House Mechanism in Armenia' project).

There is no special national programme or strategy on CWR conservation and use in the republic. This is seen as part of the responsibilities of the national PGRFA programme, and in Armenia it involves several institutions in different ministries and departments. Lack of national coordination hampers the rational preservation and effective use of genetic resources in a country and undoubtedly threatens irreplaceable natural resource and natural ecosystems (Spillane *et al.*, 1999). However, plant genetic resource (PGR) conservation national programmes in the republic are thought to be stable and provide a long-term basis for prevention of the ongoing degradation of natural resources, loss of biodiversity and desertification, providing reliable conservation and sustainable use of PGRs, as well as a means of implementing international agreements, such as the Global Plan of Action on Conservation and Sustainable Utilization of PGRFA, Convention on Biological Diversity, Framework Convention on Climate Change and Convention to Combat Desertification.

Although a coordinated biodiversity monitoring network is absent and regular ecogeographic surveys of biodiversity are not performed, different research institutions implement CWR monitoring in nature in compliance with their research themes and objectives of the research projects. Lack of available information concerning recent population changes in natural habitats together with the limited numbers of botanical expeditions that are possible make CWR gap analyses almost impossible to implement. Specifically in respect of CWR, great expectations are placed upon the interregional (Armenia, Bolivia, Madagascar, Sri Lanka and Uzbekistan) UNDP/GEF project '*In situ* Conservation of Crop Wild Relatives Through Enhanced Information Management and Field Application'. It is hoped that through the information management system, which will be created as part of the project, dispersed information on CWR held by different institutions in the republic will be brought together, the status of CWR will be determined, priority conservation actions will be developed and tested based on decision-making procedures. Means of identifying CWR priority species, as well as important plant areas for CWR will be developed and applied as another project component.

The largest CWR seed collection is located in the laboratory of PGR within Armenian Agrarian University, but the storage conditions do not correspond to contemporary standards and currently do not provide sustainable long-term conservation (Table 4.2). Seed collections of other research institutions are not rich in accessions of CWR (Sarikyan, 2003). Currently, a national gene bank is being established with the support of the International Center for Agricultural Research in the Dry Areas (ICARDA) and undoubtedly CWR will represent an important part of the seed collection, as it is the national wealth of the country. The wild cereals accessions currently existing in the Armenian Agrarian University could then be transferred to the gene bank for reliable long-term conservation.

Armenia's national system of protected areas encompasses approximately 311,000 ha, representing approximately 10% of the country's territory or 6% of the actual land base. In Armenia, there are 28 specially protected nature areas, 3 reserves (1.33% of the territory), 2 national parks (5.84%) and 23

Table 4.2. Number of accessions of crop wild relative seed collections in Armenia.

	Number of accessions stored	
Crop genus	Wild species	Total
Armenian State Agrarian University, Plant Genetic Resources Laboratory		
Triticum L.	243	893
Secale L.	500	504
Hordeum L.	420	696
Aegilops L.	1550	1550
Beta L.	31	31
Daucus L.	10	10
Spinacia L.	12	12
Coriandrum L.	12	12
Lathyrus L.	2	2
Lens Mill.	2	2
Vicia L.	20	20
Trifolium L.	2	2
Medicago L.	3	3
Amygdalus L.	27	27
Total	2834	3764
Scientific Centre of Vegetable-Melons and Industrial Crops		
Lycopersicon Mill.	4	190
Solanum melongena L.	3	151
Melo Adans.	3	67
Citrullus	1	69
Total	11	477
Scientific Centre of Agronomy and Plant Protection		
Triticum L.	15	1547
Hordeum L.	2	2102
Total	17	3649
Grand Total	2862	7890

reservations (3.46%). Two state reserves (Khosrov Forest and Shikahogh) are of particular forest protection significance (Khanjyan and Sharbatyan, 1999), containing rare and endemic species of *Juniperus*, *Quercus*, *Sorbus*, *Fagus*, *Taxus*, *Rhododendron*, *Malus*, *Jasminum*, *Platanus* and *Pyrus*. The Erebuni Reserve was established in the vicinity of Yerevan in 1981 especially to protect wild-growing cereals. Being the smallest reserve (89 ha) of Armenia, it is exceptionally significant for humankind. It is located 8–10 km from Yerevan at 1300–1400 m in the transition between semi-desert and mountain-steppe zones on tertiary red lay soil. It contains 278 vascular plant species belonging to 176 genera and 42 families (Voskanyan and Arevshatyan, 1983). The families *Asteraceae* (54 species), *Poaceae* (27) and *Fabaceae* (27) are represented by numerous species and seven plant species are registered in the Red Data Book of Armenia. But the wild-growing wheats are the gem of the reserve, with good populations of *T. boeoticum*, *T. urartu* and *T. araraticum*. The latter two were originally discovered and described in Armenia and each contains high intraspecies diversity (Ghandilyan and Avagyan, 1998). The wild cereals rye (*S. vavilovii*), several species of *Aegilops* and barley, as well as the extremely rare *A. muticum* (Boiss.) Eig and *Rhizocephalus orientalis* Boiss, are also present.

According to the 'National Strategy and Action Plan on the Development of Specially Protected Areas in the Republic of Armenia' a system of protected areas designed to ensure international standards in the protection, reproduction and sustainable development of unique landscapes and biodiversity should be established. The national action plan drafted for 2003–2010 envisages the establishment of new protected areas (biosphere reserves and natural parks), specifically for the conservation of most vulnerable plant species, including CWR. Also, within the framework of 'Natural Resources Management and Poverty Reduction Project' supported by World Bank, it is planned to develop management plans for the Sevan and Dilijan National Parks and introduce a monitoring system, as well as improve the legislation regulating the management of specially protected areas. The planned activities will be designed not only to promote the effective conservation of Armenian CWR, but also to ensure their sustainable use.

References

Avagyan, I. (1992) Genus *Hordeum* L. in Armenia. PhD thesis, University of Yerevan, Yerevan, Republic of Armenia.

Engels, J.M.M., Withers, L., Raymond, R. and Fassil, H. (2000) Towards sustainable national plant genetic resources programmes – policy, planning and coordination issues. In: Engels, J.M.M. (ed.) *Proceedings of an International Workshop. The Importance of PGR and Strong National Programmes*. International Plant Genetic Resources Institute, Rome, Italy, pp. 12–18.

Gabrielian, E.T. (1991) Wild relatives of cultivated plants in Armenia. *Botanika Chronika* 10, 475–479.

Gabrielian, E.T. and Zohary, D. (2004) Wild relatives of food crops native to Armenia and Nakhichevan. *Flora Mediterranea* 14, 5–80.

Ghandilyan, P.A. (1972) On wild-growing species of *Triticum* in Armenian SSR. *Botanical Magazine* 57(2), 1–4.

Ghandilyan, P. and Avagyan, A. (1998) Conservation of wild relatives of wheat in Armenia. In: Gass, T., Frese, L., Begemann, F. and Lipman E. (eds) *Implementation of the Global Plan of Action in Europe – Conservation and Sustainable Utilization of Plant Genetic Resources for Food and Agriculture*. International Plant Genetic Resources Institute, Rome, Italy, pp. 53–55.

Harutyunyan, M. (1991) Botanical Ecological Characterization of the Genus *Aegilops* L. in Armenia. PhD thesis, University of Yerevan, Yerevan, Republic of Armenia.

Khanjyan, N.S. and Sharbatyan, M.I. (1999) *Flora of Dilijan Reserve*. Ministry of Nature Protection of the Republic of Armenia, Yerevan, Republic of Armenia.

Matevosyan, H. (1987) Genus *Secale* L. in Armenia. PhD thesis, University of Yerevan, Yerevan, Republic of Armenia.

Melikyan, A. (2001) *Biological Peculiarities and Possibilities of Use of a Number of Wild Vegetable Plants Growing in Armenia*. University of Yerevan, Yerevan, Republic of Armenia.

Sarikyan, K. (2003) Solanaceae genetic resources in Armenia. In: Daunay, M., Maggioni, L. and Lipman, E. (eds) *Solanaceae Genetic Resources in Europe*. Report of two meetings – 21 September 2001, Nijmegen, The Netherlands, 22 May 2003, Skierniewice, Poland. International Plant Genetic Resources Institute, Rome, Italy, pp. 10–12.
Spillane, C., Engels, J.M.M., Fassil, H. Withers, L. and Cooper, D. (1999) *Strengthening National Programmes for Plant Genetic Resources for Food and Agriculture: Planning and Coordination*. Issues in Genetic Resources, No. 8. IPGRI Rome, Italy, p. 51.
Takhtajan, A.L. (1992) Botanical observations in littoral Albania. *Flora, Vegetation and Plant Resources of Armenia* 12.
Voskanyan, V.E. and Arevshatyan, I.G. (1983) Flora of vascular plants of Erebuni reserve. *Armenian Biological Magazine* 36, 6.

II Establishing Inventories and Conservation Priorities

5 Crops and Wild Relatives of the Euro-Mediterranean Region: Making and Using a Conservation Catalogue

S.P. Kell, H. Knüpffer, S.L. Jury, B.V. Ford-Lloyd and N. Maxted

5.1 Why Catalogue the Crop Resources of Europe and the Mediterranean?

The combined European and Mediterranean region (the Euro-Mediterranean region) is an important centre for the diversity of crops and their wild relatives – a major socio-economic resource and the cornerstone of agrobiodiversity for the region. Major food crops, such as wheat (*Triticum aestivum* L.), barley (*Hordeum vulgare* L.), cabbage (*Brassica oleracea* L.) and olive (*Olea europaea* L.), originated in the Euro-Mediterranean and the wild relatives of these crops, along with several other major crops that have wild relatives in the region, are an important genetic resource for crop improvement and food security. Many minor crops have also been domesticated and developed in the region, such as chickpea (*Cicer arietinum* L.), lentil (*Lens culinaris* Medik.), sugarbeet (*Beta vulgaris* L.), almond (*Prunus dulcis* (Mill.) D.A. Webb) and apple (*Malus domestica* Borkh.). Other crops of socio-economic importance with wild relatives in the region are forestry species such as *Abies alba* Mill., *Populus nigra* L. and *Quercus ilex* L., ornamentals such as species of *Dianthus* L., *Euphorbia* L., *Geranium* L. and *Primula* L. and medicinal and aromatic plants such as species of *Anemone* L., *Campanula* L., *Helianthemum* Mill., *Orchis* L. and *Verbascum* L. Although it is acknowledged that populations of crop wild relatives (CWR) are under threat in the Euro-Mediterranean region, their conservation has historically received relatively little systematic attention. Creating a CWR inventory is the first step in the conservation and effective use of these vital resources – to tackle CWR conservation, we need to know how many taxa there are, what they are and where they are.

Taxon inventories provide the baseline data critical for biodiversity assessment and monitoring, as required by the Convention on Biological Diversity (CBD) (CBD, 1992), the Global Strategy for Plant Conservation (GSPC) (CBD, 2002), the European Plant Conservation Strategy (EPCS) (Council of Europe and Planta Europa, 2002) and the International Treaty on Plant Genetic Resources for Food and Agriculture (ITPGRFA) (FAO, 2001). They provide the essential foundations for the formulation of strategies for *in situ* and *ex situ* conservation and on the species' current and potential uses as novel crops or gene donors. Some species may already be included in areas managed for conservation purposes, but their status as CWR may be unknown and they may not be actively monitored and managed. We already know that relative to the number of crops conserved *ex situ* in European gene banks, the number of CWR conserved are few (see Maxted *et al.*, Chapter 1, this volume). Inventories are needed to establish which species are already conserved, where the gaps are in their conservation and to provide the data needed for integrating CWR into existing conservation initiatives.

At regional level, a CWR inventory provides policy makers, conservation practitioners, plant breeders and other user groups with an international view of CWR species' distributions and a means of prioritizing conservation activities (see Ford-Lloyd *et al.*, Chapter 6, this volume). A regional inventory provides the basis for monitoring biodiversity change internationally, by linking CWR information with information on habitats, policy and legislation and climate change. It also serves to highlight the breadth of CWR diversity available in the region, which may include important resources for CWR conservation and use in other parts of the world. Furthermore, a regional inventory provides the backbone for the creation of national CWR inventories (e.g. see Scholten *et al.*, Chapter 7, this volume; Maxted *et al.*, in press).

The creation of CWR inventories within Europe has been tackled in some cases at country level – for example, Schlosser *et al.* (1991) for the former German Democratic Republic, and Mitteau and Soupizet (2000) for France – and at regional level, for Europe – especially those proposed by Zeven and Zhukovsky (1975), Heywood and Zohary (1995) and Hammer and Spahillari (1999). However, a comprehensive and systematic approach has not yet been proposed and applied, and previously there has not been a coordinated effort focusing on the production of a comprehensive online Euro-Mediterranean Catalogue.

This chapter summarizes a methodology for establishing a regional catalogue of crops and their wild relatives for the Euro-Mediterranean region (see Kell *et al.*, 2007, unpublished data, for a full explanation of the methodology). The Catalogue (Kell *et al.*, 2005a) is made available through the web-enabled Crop Wild Relative Information System (CWRIS) (PGR Forum, 2005), which provides access to CWR information to a broad user community, including plant breeders, protected area managers, policy makers, conservationists, taxonomists and the wider public (see Kell *et al.*, Chapter 33, this volume) – information that is vital for the sustainable utilization and conservation of CWR. The Catalogue has been created using a systematic approach that can accommodate changes in nomenclature and status, and can be applied at both regional and national levels in any part of the world.

In addition to providing an online information resource, the actual Catalogue data can be analysed to provide statistics on the crop and CWR taxa of the region. This chapter provides information on the number of crop and CWR taxa in the region and how many are native and endemic; the number of crop and CWR species present in individual nations and intranational regions; the number of species within and shared by the different crop groups; the number of worldwide crop genera that are found in the region; the major and minor food crops of the world that are native to the Euro-Mediterranean region and those that have wild relatives in the region. The Catalogue data can also be compared with taxon lists from existing conservation initiatives to establish which species are currently conserved and/or have undergone conservation assessment as a step towards the recognition and inclusion of CWR in current conservation programmes – some examples of this are given here.

5.2 Creating the Catalogue

5.2.1 Scope and basic methodology

The scope of the Catalogue is all species of direct socio-economic importance and their wild relatives – including food, fodder and forage crops, medicinal plants, condiments, ornamental and forestry species, as well as plants used for industrial purposes, such as oils and fibres. Applying the broad definition proposed by Maxted *et al.* (2006), a CWR includes any taxon belonging to the same genus as a crop species – it is upon this premise that the methodology for the creation of the CWR Catalogue is based.

In its simplest terms, the process of creating the Catalogue involves creating a list of genera containing crops, matching these with the genera contained in the flora of the country or region and selecting the taxa within the matching genera from the flora to create the Catalogue (see Kell *et al.*, 2007, unpublished data, for a detailed explanation of the methodology). For example, taking the crop species, *B. oleracea* L. (cabbage) as an example, because taxa within the genus *Brassica* L. occur in the Euro-Mediterranean region, we include all the accepted *Brassica* taxa that occur in the region in the CWR Catalogue – in this case, 34 species and 54 subspecies. All taxa, whether cultivated, wild, native or introduced, are included. For example, the introduced, cultivated taxon, *B. napus* L. subsp. *napus*, is included in the Catalogue, along with native or introduced wild-occurring taxa – for example, *B. tournefortii* Gouan (native) and *B. elongata* Ehrh. subsp. *elongata* (mainly introduced but possibly native in some countries) – and native, cultivated taxa – for example, *B. macrocarpa* Guss.

The reason for including both cultivated and wild taxa in the Catalogue is that we are providing an information resource as a tool for the conservation of plant genetic resources (PGR) of socio-economic importance (i.e. both the crops and their wild relatives). It is not only the wild relatives that may harbour useful genes for crop improvement, but also the crops themselves, particularly in the case of locally adapted forms or landraces. There is also a strong argument for including native and introduced taxa in the Catalogue – populations of crops or wild relatives that are not native may still be an important genetic resource and worthy of conservation efforts, particularly in cases where native populations of taxa have suffered from genetic erosion. While countries may choose to conserve their native flora above the introduced flora, at regional level, in terms of conservation of crop genetic resources, the need to actively conserve introduced populations in some areas may be justified. Ultimately, the CWR Catalogue is a comprehensive information resource, which policy makers, conservation practitioners and crop germplasm user groups can use as an aid to conservation planning and sustainable use. Therefore, the more comprehensive the Catalogue is, the greater its uses will be.

5.2.2 Data sources

The Catalogue is primarily derived from two major databases: Euro+Med PlantBase (Euro+Med PlantBase, 2005), which provides the taxonomic core, and Mansfeld's World Database of Agricultural and Horticultural Crops (Hanelt and IPK Gatersleben, 2001; IPK Gatersleben, 2003), which provides lists of genera containing agricultural and horticultural crops and the crop species themselves.

Euro+Med PlantBase is an online database and information system for the vascular plants of the Euro-Mediterranean region. The database comprises names and associated data from *Flora Europaea*, the MedChecklist database, the Flora of Macaronesia data set and published Floras from the Euro-Mediterranean region. Euro+Med PlantBase includes native species, naturalized aliens, frequently occurring casuals, frequent and well-characterized hybrids, crop weeds and plants that are conspicuously cultivated outdoors. The geographical area covered includes all of Europe,[1] the Caucasus, Asiatic Turkey and the East Aegean Islands, Syria, Lebanon, Israel, Jordan, Cyprus, Egypt, Libya, Tunisia, Algeria, Morocco and Macaronesia.

Mansfeld's World Database of Agricultural and Horticultural Crops (Hanelt and IPK Gatersleben, 2001; IPK Gatersleben, 2003) contains more than 6100 cultivated species of agricultural and horticultural plants worldwide, including medicinal and aromatic plants, but with the exception of ornamental and forestry plants. The database also includes cultivated algae and fungi, pteridophyta and gymnosperms.

Genus lists for forestry and ornamental species and additional medicinal and aromatic plant taxa were drawn from other sources. For forestry taxa, a list of genera was extracted from the 'enumeration of cultivated forest plant species' (Schultze-Motel, 1966). For ornamentals, a list of taxa was provided by the Community Plant Variety Office (CPVO, 2001), which is the organization responsible for implementing the 'system for the protection of plant variety rights' established by European Community legislation, allowing intellectual property rights to be granted for plant varieties within the European Union

[1] The eastern boundary of Europe in Russia and Kazakhstan follows the definition of *Flora Europaea* (Tutin *et al.*, 1968–1980, 1993): from the Arctic Ocean along the Kara River to 68°N, along the crest of the Ural Mountains (following administrative boundaries) to 58°30'N, then by an arbitrary straight line to a point 50 km east of Sverdlovsk, and by another arbitrary straight line to the headwaters of the Ural River (south of Zlatoust); and finally along the Ural River to the Caspian Sea.

(EU). This list contains taxa for which the title had been granted and all active applications as of July 2003 (T. Kwakkenbos, France, 2003, personal communication). For medicinal and aromatic plants, a genus list was extracted from the database, Medicinal and Aromatic Plant Resources of the World (MAPROW) (U. Schippmann, Bonn, 2004, personal communication), which includes wild-harvested as well as cultivated medicinal and aromatic plant species (the cultivated ones are also included in Mansfeld's Database), thus broadening the scope of the CWR Catalogue.

Accepted and synonymous genus names were selected from Mansfeld's Database in order to capture as wide a range of agricultural and horticultural crop and CWR taxa in the Catalogue as possible; thus, when a genus name is considered a synonym in Mansfeld's Database but is accepted by Euro+Med PlantBase, it is included in the CWR Catalogue in addition to accepted genus names that match. Only accepted genus names were selected from Schultze-Motel (1966); since the data was not previously digitized, extraction of synonyms in addition to accepted names was not possible with the available resources. However, it is unlikely that this would have a significant effect on the number of species included in the Catalogue overall, since analysis shows that 95% of forestry species are common to the species in the list of agricultural and horticultural crops. The CPVO and MAPROW do not adopt specific accepted taxonomies; therefore, no distinction was made in these data sets between accepted and synonymous genus names – the genus names were thus used as provided by these data sources. However, again, the list of agricultural and horticultural crop and CWR species shares 90% of its taxa with the ornamental list and 92% with the medicinal and aromatic plants list, thus, taking into account the synonymy in Mansfeld's Database captures the majority of species in all groups. For a detailed discussion on dealing with synonymy in the creation of the CWR Catalogue, readers are referred to Kell *et al.* (2007, unpublished data).

The crop genus list contains 7363 genera in total. Table 5.1 summarizes the number of genera attributable to each data source. Note that some genera are common to two or more sources; for example, Mansfeld's Database contains 68% of the CWR genera sourced from the other crop data sources (forestry, ornamental, medicinal and aromatic genera combined). When the crop genera are matched with Euro+Med PlantBase to select those taxa that occur in Europe and the Mediterranean, Mansfeld's Database is found to contain 82% of the CWR genera sourced from the other crop data sources.

5.2.3 Euro+Med PlantBase data filtering

Euro+Med PlantBase (version September 2005) provides the taxonomic backbone to the CWR Catalogue. The database contains more than 45,000 accepted species and infraspecific taxa (of which more than 33,000 are species and nearly 12,000 are infraspecific taxa) and more than 39,000 specific and infraspecific synonyms (Table 5.1). Only accepted names in Euro+Med PlantBase were used to create the CWR Catalogue. However, the online Catalogue can be searched on any taxon name to find its associated data.

Table 5.1. Summary statistics: CWR Catalogue data sources.

Data source	No. of records	Data source/notes
Euro+Med PlantBase		a
Euro+Med PlantBase: accepted species	33,471	
Euro+Med PlantBase: accepted infraspecific taxa	11,989	
Euro+Med PlantBase: synonyms (species and infraspecific taxa)	39,924	
Crop genera		
Agricultural and horticultural crop genera	1,983	b
Forestry genera	338	c
Ornamental genera	366	d
Medicinal and aromatic genera	1,057	e
Total crop genera	2,539	f
Crop species		
Euro+Med PlantBase species coded 'cultivated'	1,299	
Agricultural and horticultural crop species	6,076	g
Forestry crop species	1,038	h
Ornamental crop species	300	

[a]Euro+Med PlantBase (www.euromed.org.uk) version September 2005.
[b]Mansfeld's World Database of Agricultural and Horticultural Crops (Hanelt and IPK, 2001; http://mansfeld.ipk-gatersleben.de) – accepted genus names. This list includes, amongst others, genera containing cultivated medicinal and aromatic plants. Note that accepted and synonymous genus names from Mansfeld's Database (6914 taxa) were matched with accepted names in Euro+Med PlantBase to create the Catalogue (see Kell *et al.*, 2007, unpublished data).
[c]'Enumeration of cultivated forest plant species' (Schultze-Motel, 1966) – accepted names only.
[d]Community Plant Variety Office (www.cpvo.eu.int) (T. Kwakkenbos, France, 2003, personal communication) – no accepted taxonomy.
[e]Medicinal and Aromatic Plant Resources of the World (MAPROW) (Schippmann, Bonn, 2004, personal communication) – no distinction between accepted names and synonyms. These genera cover all species known to be utilized for medicinal purposes, whether wild-harvested or cultivated.
[f]The four groups listed form the crop genus list, containing 2539 genera (7363, including the synonymous genus names from Mansfeld's Database (see note 2). Note that some genera are common to two or more sources.
[g]Mansfeld's World Database of Agricultural and Horticultural Crops (Hanelt and IPK, 2001; http://mansfeld.ipk-gatersleben.de) – accepted species only. Note that accepted and synonymous species names from Mansfeld's Database (24,578 taxa) were matched with the Catalogue to tag the cultivated species (see Kell *et al.*, 2007, unpublished data).
[h]Figure from the preface of Schultze-Motel's (1966) preliminary worldwide account of cultivated forestry species.

Therefore, if a user searches for a synonym of an accepted taxon name in the Catalogue, CWRIS takes the user to the accepted name and the data associated with it.

Euro+Med PlantBase uses the 'Plant Occurrence and Status Scheme' (WCMC, 1995) – a Standard of the International Working Group on Taxonomic Databases (TDWG) – to record the status of taxa within each geographical unit (Table 5.2). Some taxa are recorded as 'extinct', 'recorded as present in error' or 'absent' – taxon records with these codes were therefore excluded from the Catalogue. Where there is any doubt about the presence

Table 5.2. Codes used in the fields 'native', 'introduced', 'cultivated' and 'status unknown' in Euro+Med PlantBase. (Adapted from Euro+Med PlantBase Secretariat, 2002.) Original data standard: WCMC (1995).

Code	Value	Explanation
Native status		
N	Native	The taxon is native (autochthonous) within the area concerned (as contrasted with 'introduced' and 'cultivated' defined below).
S	Assumed to be native	Assumed to be native to the area concerned.
D	Doubtfully native	There is doubt as to whether the status of the plant in the area concerned is native or not.
E	Formerly native (extinct)	The plant is native, doubtfully native or assumed to be native in the area concerned and has become extinct as such.
A	Not native	The plant is definitely not native.
F	Recorded as native in error	The plant has been recorded as native in the area concerned, but all such records have been disproved or discounted.
Introduced status		
I	Introduced	The plant has been recorded growing in an area that is outside of its assumed true and normal distribution. This implies evidence that the plant did not formerly occur in the area and also that the plant is either established and successfully reproducing (either sexually or asexually) or a frequently occurring casual. The plant must not be in cultivation: it does not mean (or include) 'introduced to cultivation'. The means of introduction, whether by man or any natural means, is irrelevant and may be unknown.
S	Assumed to be introduced	Assumed to be introduced to the area concerned.
D	Doubtfully introduced	There is doubt as to whether the status of the plant in the area concerned is introduced, as defined above, or not. All records about the introduced status of the plant in the area are in doubt.
E	Formerly introduced (extinct)	The plant is introduced, doubtfully introduced or assumed to be introduced in the area concerned and has become extinct as such. The criterion of extinction is that the plant was not found (as an introduction) after repeated searches of known and likely areas (i.e. sites within the area covered by the record), even though the plant may be extant elsewhere.
A	Not introduced	The plant is definitely not introduced (as defined above) in the area concerned.
F	Recorded as introduced in error	The plant has been recorded as introduced in the area concerned, but all of those records have been disproved or discounted. A known fallacious introduced record must have been made, and it must be known that the plant does not occur as an introduction in the area.

Continued

Table 5.2. *Continued*

Code	Value	Explanation
Cultivated status		
C	Cultivated	The plant is established in outdoor cultivation in the area concerned. Only plants that are conspicuously cultivated outdoors should be included (includes crops planted on a field-scale and street and roadside trees).
S	Assumed to be cultivated	Assumed to be cultivated in the area concerned.
D	Doubtfully cultivated	There is doubt as to whether the status of the plant is cultivated or not in the area concerned. All records about the cultivated status of the plant in the area are in doubt.
E	Formerly cultivated (extinct)	The plant was at one time cultivated, doubtfully cultivated or assumed to be cultivated in the area concerned and has become extinct in cultivation in this area, even though it may be extant elsewhere.
A	Not cultivated	The plant is definitely not cultivated (as defined above) in the area concerned.
F	Recorded as cultivated in error	The plant has been recorded as cultivated in the area concerned, but all of those records have been disproved or discounted. A known fallacious record of cultivation must have been made, and it must be known that the plant is not cultivated in the area.
Status unknown		
P	Present	The plant is present in the area and meets the criteria for inclusion in Euro+Med PlantBase; i.e. it is a native species, naturalized alien, frequently occurring casual, frequent and well-characterized hybrid, crop weed or a plant that is conspicuously cultivated outdoors (either a crop planted on a field-scale or street tree, but not a commonly grown park or garden plant). Adventives, casuals, etc. are not included although noxious weeds (other than those that have become naturalized which will be included for that reason) may be recorded.
S	Assumed present	It is highly probable that the plant does occur in the area.
D	Doubt about presence	There is doubt about whether the plant presently occurs in the area. This might be because all records are very old, locality details are uncertain, etc.
E	Extinct	The plant was once in the area (P or S) or may once have been in the area (D), but is now extinct in the area.
F	Recorded as present in error	The plant has been recorded as present in the area concerned, but the record has been discounted or disproved.
A	Absent	There are no records to suggest that a plant has ever occurred in the area concerned.

of a taxon, the record is maintained in the Catalogue until such time as the Euro+Med PlantBase records for that taxon are updated and the status is confirmed (note that the Catalogue is updated automatically by linking directly to the Euro+Med PlantBase data set). Inclusion of these records in the Catalogue makes very little difference to the overall number of species. After filtering, the number of accepted species names in Euro+Med PlantBase is reduced from 33,471 to 30,983; these species are contained within 218 families and 2437 genera (Table 5.3). These taxa form the base taxonomy for the CWR Catalogue.

Table 5.3. Creation of the CWR Catalogue: summary statistics. The total number of families, genera and species are shown for the filtered version of Euro+Med PlantBase (E+Mf), Mansfeld's World Database of Agricultural and Horticultural Crops and for each crop group after matching the crop genus list with Euro+Med PlantBase. The total number of crop taxa in the Euro-Mediterranean region and the number of crop and CWR native and endemic to Europe and the Euro-Mediterranean region are given.

	No. of taxa		
Plant taxa present in the Euro-Mediterranean region	Families	Genera	Species
Total no. of plant taxa (E+Mf)	218	2,437	30,983
Agricultural and horticultural taxa	166	1,109	23,513
Forestry taxa	57	143	2,843
Ornamental taxa	90	230	7,499
Medicinal + aromatic taxa	146	618	19,784
CWR Catalogue for Europe and the Mediterranean (total no. of crop and CWR taxa)	183	1,239	25,687
Crop taxa[a]			
Agricultural and horticultural crops	147	754	1,994
Forestry crops	41	102	282
Ornamental crops	62	104	131
Other crops[b]	66	166	486
Total crop taxa	155	817	2,204
Native and endemic species			
Crop and CWR species native to Europe and the Mediterranean	–	–	23,216
Crop and CWR species endemic to Europe and the Mediterranean	–	–	14,994
Crop and CWR species native to Europe	–	–	15,656
Crop and CWR species endemic to Europe	–	–	8,624

[a]Taxa known to be cultivated worldwide and not necessarily cultivated in the Euro-Mediterranean region. It is not possible to create a list of medicinal and aromatic crops using this data because MAPROW includes wild-harvested taxa and Mansfeld's Database does not contain a single data field that categorizes crop species according to their use.
[b]Other crops are species recorded by Euro+Med PlantBase as cultivated in the region that are not already included in the lists of agricultural and horticultural, forestry and ornamental crops.
– Not applicable.

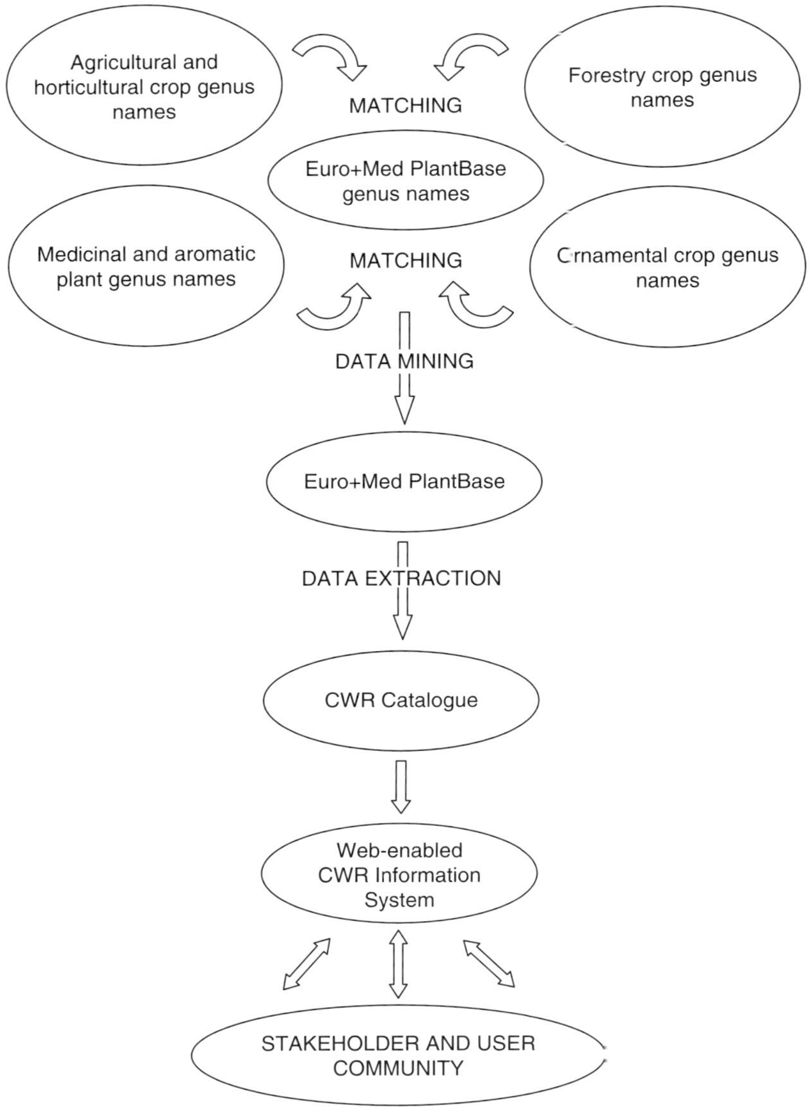

Fig. 5.1. Flow chart showing the basic methodology for the creation and utilization of the CWR Catalogue for Europe and the Mediterranean.

5.2.4 Mining and extraction of crop and CWR taxa from Euro+Med PlantBase

The genera in the filtered version of Euro+Med PlantBase corresponding with the crop genus list described earlier were selected. Following the genus name matching, the accepted taxa within the harmonized genera were selected, forming the CWR Catalogue. Figure 5.1 is a simplified flow chart illustrating the basic methodology, which could be utilized in any region or country. The chart shows the four crop name sources forming the crop genus list, which is matched with the genera contained in the flora of the country or region – in this case, the flora of Europe and the Mediterranean. The flora is then mined for the accepted taxa contained in the matching genera and these are extracted to form the CWR Catalogue.

5.2.5 Coding crop species in the Catalogue

We generally refer to the Catalogue as the 'CWR Catalogue'; however, the Catalogue also contains the crop taxa themselves. To distinguish the crop taxa in the Catalogue, all taxa coded 'C' (cultivated) in Euro+Med PlantBase were selected and tagged. These include plants that are conspicuously cultivated outdoors, such as crops planted on a field-scale and street and roadside trees (Euro+Med PlantBase Secretariat, 2002). In addition, species names from Mansfeld's World Database of Agricultural and Horticultural Crops (Hanelt and IPK Gatersleben, 2001; IPK Gatersleben, 2003), the 'enumeration of cultivated forest plant species' (Schultze-Motel, 1966) and the CPVO ornamental list (T. Kwakkenbos, France, 2003, personal communication) matching species listed in the Catalogue were tagged as crops. To capture as wide a range of crop species as possible, matching between synonymous species in Mansfeld's Database and species in the Catalogue was carried out.

Mansfeld's Database is inclusive of a very wide range of cultivated species, so the agricultural and horticultural species tagged as crops in the Catalogue are wide-ranging. For example, in addition to food, fodder, forage, medicinal, aromatic and industrial crops, plants cultivated for soil improvement, sand dune fixation, hedging, grafting stock, shade and support are included; thus, a broad definition of a 'crop' is adopted. On the other hand, the list of species used to tag the cultivated ornamental species in the Catalogue cannot be considered representative of the extensive number of species utilized in the ornamental plant industry. The reasons for this are that the ornamental genera from the CPVO varieties list were deliberately chosen to keep the ornamental component of the Catalogue to a reasonable minimum, since the use of plant species in the ornamental industry is extremely wide-ranging, and the CPVO does not use a standard nomenclatural system, therefore, many cultivars are listed without inclusion of the specific epithet. A better coverage of cultivated ornamental species could be provided by matching the species in the Catalogue with a more comprehensive database such as the RHS Horticultural Database (Royal Horticultural Society, 2006), which was not completed and thus not available during the time that the CWR Catalogue was created.

It is important to point out that not all the species tagged as crops are necessarily cultivated in the Euro-Mediterranean region – some crop species may occur in the region, but only in their wild form. For example, 1313 species of agricultural and horticultural crops that occur in the region are not actually recorded by Euro+Med as being cultivated. However, knowledge that a cultivated taxon occurs as a wild relative in a country where it is not cultivated may be important for crop security, because the wild material may be utilized in breeding for crop improvement. Table 5.1 summarizes the number of crop species from each data source used to code species in the Catalogue as cultivated.

5.3 What Does the Catalogue Tell Us about Crops and CWR in the Region?

5.3.1 Analysing the Catalogue data

The Catalogue data can be analysed in numerous ways to provide both broad brush-stroke statistics about the crop and CWR species present in the region and more detailed analysis about the species present at national level and about individual crops or crop groups. Results of the following data analyses are presented here:

- The number of crop and CWR species within the Euro-Mediterranean region and within Europe alone, including the number of species native and endemic to the regions;
- The number of crops and their wild relatives within the different crop groups;
- The number of species shared by the different crop groups;
- The number of worldwide crop genera that are found in the region;
- National species richness;
- Which major and minor food crops of the world are native and endemic to the Euro-Mediterranean region and which have wild relatives in the region.

However, the role of the Catalogue goes far beyond provision of interesting statistics on the crop and CWR species of the region – one of its most important functions is to provide a basis for creating comprehensive national inventories (e.g. see Scholten *et al.*, Chapter 7, this volume; Maxted *et al.*, in press) and to aid CWR conservation gap analysis. For example, a regional or national inventory can be compared with protected area inventories (where the data is available), to establish which CWR species are already included within existing protected areas. Detailed gap analysis is beyond the scope of this chapter; however, we have undertaken some preliminary analysis to investigate which CWR taxa are included in: (i) the IUCN Red List of Threatened Species; (ii) the EC Habitats Directive; (iii) Important Plant Areas (IPAs); and (iv) the Plant Search Database of world botanic garden collections, to begin to build up a picture of to what extent CWR have been assessed and included in existing conservation initiatives.

5.3.2 Numbers of crop species and their wild relatives in Europe and the Mediterranean

The CWR Catalogue contains 25,687 of the 30,983 plant species recorded by Euro+Med PlantBase as present in the region. This indicates that approximately 83% of the Euro-Mediterranean flora consists of crops and their wild relatives; in other words, more than three-quarters of plant species in the region have a current or potential direct use to humankind. Ninety percent (23,216 species) are native to the Euro-Mediterranean region and 58% (14,994) are endemic (Table 5.3). However, taking into account synonymy and issues of taxonomic uncertainty, this is probably a slightly artificially large number of species (Kell *et al.*, 2007, unpublished data). Therefore, for the purposes of argument, we may conclude that around 80% of the flora of the region is of current or potential direct use.

Forty-nine percent of genera containing agricultural, horticultural, forestry and ornamental crops and medicinal and aromatic plants worldwide are found in the Euro-Mediterranean region and at least 2204 species in the CWR Catalogue (9%) are known to be cultivated worldwide (Table 5.3). As noted earlier, not all these species are necessarily cultivated within the Euro-Mediterranean region. At least 8% of the species listed in the CWR Catalogue are agricultural and horticultural crops in the Mansfeld sense (see Hanelt and IPK Gatersleben, 2001; IPK Gatersleben, 2003), while at least 1% are forestry crops as recorded by Schultze-Motel (1966). At least 8% of agricultural and horticultural and 10% of forestry crop and CWR species are cultivated worldwide. Although a taxon can be both cultivated and a wild relative (i.e. in some places it might be cultivated, while in others it may occur in its wild form), we can say that approximately 90% of the species in the agricultural, horticultural and forestry groups are wild relatives. In the CPVO (ornamental) list, 131 species match the names in the CWR Catalogue; however, this is not representative of the number of cultivated ornamental species. As explained earlier, if another source of data were consulted, such as the RHS Horticultural Database (Royal Horticultural Society, 2006), the figures for ornamental crop species would undoubtedly increase significantly.

Table 5.4 shows the total number of crop and CWR species in each of the four socio-economic groups: agricultural and horticultural crops, forestry species, ornamentals and medicinal and aromatic plants (note that the medicinal and aromatic species list includes wild-harvested plants and their wild relatives, as well as cultivated species). The percentage of the total number of Euro-Mediterranean crop and CWR species (25,687) attributable to each group is given. Table 5.5 is a matrix showing the percentage of species common to all four groups. Note that very high percentages of crop and CWR species extracted from the genus list derived from Mansfeld's World Database of Agricultural and

Table 5.4. Total number of crop and CWR species in the Euro-Mediterranean region and the numbers and percentages of species in each group.

	Crops	CWR	Total crop and CWR species	Total species per group as percentage of Catalogue
Agricultural and horticultural species[a]	1,994	21,519	23,513	92%
Forestry species	282	2,561	2,843	11%
Ornamental species	131	7,368	7,499	29%
Medicinal and aromatic species[b]	–	–	19,784	77%
Total Euro-Mediterranean species	2,204[c]	23,483	25,687	–

[a]The agricultural and horticultural species list includes cultivated medicinal and aromatic plants.
[b]The medicinal and aromatic species list includes wild-harvested plants and their relatives, as well as cultivated species.
[c]Includes 486 'other' crop species recorded as cultivated in Euro+Med PlantBase (see Table 5.3).
– Not applicable or data not available.

Table 5.5. Matrix showing the percentage of crop and CWR species shared by each of the four groups. The bottom left side of the matrix shows the percentage of species shared by each group in the left-hand column as a percentage of the species in each group given across the top row. The top right side of the matrix expresses the percentages in reverse. For example, 11% of species in the agricultural and horticultural list are also found in the forestry list; and conversely, 95% of forestry species are found in the agricultural and horticultural list. Note that the medicinal and aromatic species list includes wild-harvested plants and their wild relatives, as well as cultivated species.

	Agricultural and horticultural (%)	Forestry (%)	Ornamental (%)	Medicinal and aromatic (%)
Agricultural and horticultural (%)	–	95	90	92
Forestry (%)	11	–	17	14
Ornamental (%)	29	45	–	33
Medicinal and aromatic (%)	77	95	88	–

– Not applicable.

Horticultural Crops are common to the other three socio-economic groups – i.e. 95% of the species in the forestry list, 90% in the ornamental list and 92% in the medicinal and aromatic plant list. This can be explained by the fact that many crop species have several uses, as do ornamental plants (e.g., medicinal and vegetable), and that cultivated medicinal and aromatic plants are also included in the Mansfeld's Database. Moreover, there are many species within the same genera as the agricultural and horticultural crop genera that have been classified within one of the other three socio-economic groups; thus, these groups will share many of the same CWR. The high percentages of medicinal and aromatic plant species common to the other three groups are also notable (i.e. 77% of agricultural and horticultural crops – though as observed earlier, Mansfeld's Database also includes cultivated medicinal and aromatic plants – 95% of forestry species and 88% of ornamental species). This illustrates the extremely broad use of plants for medicinal and aromatic purposes, many of which are species harvested from the wild. Perhaps not surprisingly, the forestry group has the lowest percentages of species common to the other three groups, with 11% of species common to the agricultural and horticultural crops, 17% to the ornamental species and 14% to the medicinal and aromatic plants.

Looking at Europe alone (as defined by Hollis and Brummitt, 2001), there are 17,495 crop and CWR species; therefore, 68% of crop and CWR species found across the Euro-Mediterranean region are found in Europe alone. Of these, 15,656 species (89%) are native to Europe and 8624 (49%) are endemic. As many as 1078 (42%) worldwide crop genera are found in Europe.

5.3.3 National species richness

Data in Euro+Med PlantBase are recorded within 130 geographical units, representing 58 nations. The number of crop and CWR species of each nation is shown in Table 5.6. Four nations contain more than 20% of the species in the region: Turkey, Spain, Italy and France. The nation with the highest CWR species richness is Turkey,

Table 5.6. List of Euro-Mediterranean nations, showing the total number of crop and CWR species per nation in descending order. The right column shows the number of species as a percentage of the total number of crop and CWR species in the region.

Nation	No. of crop and CWR species	Percentage of Euro-Mediterranean crop and CWR species
Turkey	7235	28
Spain	6669	26
Italy	5712	22
France	5528	22
Greece	4818	19
Ukraine	4265	17
Russia	4259	17
Germany	4211	16
Slovakia	3873	15
Bulgaria	3619	14
Austria	3563	14
Czech Republic	3526	14
Romania	3484	14
Croatia	3436	13
Switzerland	3413	13
Morocco	3409	13
Portugal	3296	13
Albania	3030	12
Algeria	2911	11
Poland	2751	11
Hungary	2639	10
Lebanon	2577	10
Slovenia	2533	10
Syria	2421	9
Sweden	2362	9
Serbia	2359	9
Norway	2276	9
Armenia	2235	9
United Kingdom	2169	8
Israel	2084	8
Denmark	2056	8
Tunisia	1882	7
Georgia	1882	7
Moldova	1795	7
Finland	1771	7
Egypt	1745	7
Belgium	1730	7
The Netherlands	1723	7
Libya	1547	6
Estonia	1501	6
Lithuania	1477	6
Cyprus	1448	6
Latvia	1323	5
Ireland	1299	5
Azerbaijan	882	3

Continued

Table 5.6. *Continued*

Nation	No. of crop and CWR species	Percentage of Euro-Mediterranean crop and CWR species
Belarus	754	3
Malta	738	3
Kazakhstan	592	2
Iceland	540	2
Andorra	504	2
Jordan	474	2
Bosnia-Herzegovina	241	1
Montenegro	185	1
Serbia and Montenegro	148	1
Luxembourg	118	<1
Liechtenstein	43	<1
San Marino	8	<1

with 7235 species – 28% of the crop and CWR species of the Euro-Mediterranean region. As might be expected, the proportion of the flora of these four countries that comprises crops and their wild relatives is fairly consistent with the overall proportion of the flora of the region: Turkey – 83%, Spain – 81%, Italy – 84% and France – 86%. Nineteen nations contain between 10% and 20% of the crop and CWR flora of the region, 31 between 1% and 10% and three less than 1%.

We can also look at which crop groups are most prevalent in individual countries and the number of crop species present. For example, of the 2276 crop and CWR species recorded in Norway, 2084 species (92%) are included in the agricultural and horticultural crop group, 345 (15%) in the forestry group, 782 (34%) in the ornamental group and 1855 (82%) in the medicinal and aromatic plant group. Also, 633 of these species (28%) are known to be cultivated worldwide and these comprise: agricultural and horticultural crops – 550 species (87%); forestry crops – 113 species (18%); ornamental crops – 46 species (7%). Euro+Med PlantBase indicates that at least 95 of these species (15%) are cultivated in Norway – of these species, 56 (59%) are agricultural and horticultural crops, 45 (47%) are forestry crops and 10 (11%) are ornamental crops. By comparison, taking a southern European example, of the 6669 crop and CWR species found in the Spanish territories, 5947 species (89%) are included in the agricultural and horticultural crop group, 659 (10%) in the forestry group, 2073 (31%) in the ornamental group and 4829 (72%) in the medicinal and aromatic plant group. Of these, 1279 (19%) are known to be cultivated worldwide (agricultural and horticultural crops – 1172 species (92%); forestry crops – 173 species (14%); ornamental crops – 92 species (7%)) and of the 215 species recorded by Euro+Med PlantBase as cultivated in Spain, 194 (90%) are agricultural and horticultural crops, 54 (25%) are forestry crops and 20 (9%) are ornamental crops. Notable are the significantly different percentages of agricultural and horticultural crops and forestry species cultivated in Norway and Spain.

Because Euro+Med PlantBase is organized into geographical units, it is also possible to look at the proportion of crop and CWR species within different intranational regions, where they exist. This is particularly interesting for those nations that include islands – especially, the oceanic islands such as the Canary Islands (Spain) and the Azores (Portugal) – and also other islands such as Sicily and Malta (Italy) and Corsica (France). Islands exhibit high levels of endemism due to their isolation from continental areas, so they are natural reservoirs of unique genetic diversity. However, it is widely recognized that island populations are also extremely vulnerable to genetic erosion because of the disruption caused by human colonization and associated biological invasions; for example, see Loope and Mueller-Dombois (1989), Schofield (1989), Bramwell (1990), Vitousek (1992) and Simberloff (1995). Taking Spain as an example, around 10% of the crop and CWR taxa of the Spanish territories occur in the Canary Islands – taxa that are not found in mainland Spain – and, of these, an estimated 249 species and 162 subspecies are endemic.[2] The islands of Sicily and Malta also contain a large proportion of the crop and CWR species of Italy – 2404 out of a total of 5712 species. Of the species found in Sicily and Malta, 277 are not found in mainland Italy and of these, 24 are recorded as endemic.[3] Of these endemic species, 23 fall into the agricultural and horticultural group, 3 in the forestry group, 13 in the ornamental group and 21 in the medicinal and aromatic group. As these taxa are endemic to small islands, their conservation may be considered of high priority due to their potential use for crop improvement in the future, combined with their innate vulnerability as island populations.

It is therefore possible to extract a list of crop and CWR taxa for each nation in the Euro-Mediterranean region and to provide a breakdown of the taxa for each geographical unit per nation, for those nations where this occurs. National crop and CWR lists have already been sent to each National PGR Coordinator in the region. Individual nations can then use these lists as a basis for conservation planning, once the list has been checked and verified to account for any potential errors. In turn, nations can feed back any errors they have found and their proposed corrections to the Euro+Med PlantBase Secretariat. Any changes that are made to Euro+Med PlantBase will automatically be made in the CWR Catalogue, which will remain available through the Internet. The Catalogue can be utilized not only to aid national conservation planning, but also to estimate the distribution of crops and their wild relatives within the region – for example, to aid regional conservation planning within the EU. Furthermore, the data can be used to target those taxa that have limited distributions (i.e. they occur in one to a few nations or intranational regions) (see Ford-Lloyd *et al.*, Chapter 6, this volume). For example, of the 25,687 crop and CWR species in the Euro-Mediterranean region, at least 2873 (11%) are endemic to one nation.[4] One of

[2] Estimates are based on taxa only recorded as occurring in the Canary Islands and endemic to the geographical unit 'Macaronesia'.

[3] Analysis of the Catalogue data indicates that there are probably significantly more endemic CWR species in Sicily (possibly as many as 86). However, the data are not complete; therefore, using the current data set, we cannot be certain of the exact number.

[4] This is a conservative estimate because there are more species recorded in Euro+Med PlantBase in only one country (6867) than are recorded as endemic to the Euro-Mediterranean region, but the data have not yet been verified and we cannot be certain that these taxa do not occur in other countries.

the major reasons for providing an information resource on where crop and CWR taxa can be found and for conserving these taxa is for their utilization as gene donors for crop improvement. The CWR Catalogue provides the information needed for plant breeders to source new material and for conservationists to collect material from as wide a range of a taxon's distribution as possible.

5.3.4 Major and minor food crops

So far, we have looked at the number of species within four socio-economically important plant groups: agricultural and horticultural crops, forestry crops, ornamentals and medicinal and aromatic plants. This is useful information, but many people might ask, how many species are found in the region in the major crop groups or within the world's food crops? This is a very good question and one which we have at least partially addressed by looking at the major and minor food crops of the world. Using the food crops of major significance (major food crops) and secondary or local importance (minor food crops) listed by Groombridge and Jenkins (2002), an analysis was undertaken to ascertain how many taxa (cultivated and wild, native and introduced) are found in the Euro-Mediterranean region within the major and minor food crop groups.

Of the 28 major food crop genera of the world, 22 occur in the Euro-Mediterranean region – 15 (54%) of these encompassing wild relatives (Table 5.7). There are 219 species and 100 subspecific taxa (subspecies and varieties) within these major food crop genera which can be found growing in the region. Of these, 106 species are known to be cultivated worldwide and at least 44 species and 24 subspecies are recorded by Euro+Med PlantBase as being cultivated in the region. National-level analysis is required to ascertain the exact number of cultivated and wild-occurring taxa within this list; however, even those taxa that are cultivated, whether also found in their wild form or not, may be a useful, if not vital source of germplasm for crop improvement, especially locally adapted forms or landraces. Four (11%) of the 38 major food crops of the world are native to the Euro-Mediterranean region: cereals – *H. vulgare* L. (barley) and *T. aestivum* L. (wheat); leaf vegetables – *B. oleracea* L. (cabbage); and oil crops – *O. europaea* L. (olive).[5] Three of these crops are native to Europe (as defined by Hollis and Brummitt, 2001): wheat, cabbage and olive.

Within the 28 major food crop genera of the world, 57 species are endemic to the Euro-Mediterranean region. Of these, at least 11 species are endemic to only one nation[6] and many of these are limited to islands

[5] *Vigna unguiculata* (L.) Walp. is also recorded by Euro+Med PlantBase as native to Egypt, but its native distribution is probably limited to sub-Saharan Africa; therefore, it is probably naturalized in Egypt.

[6] Estimate based on Euro+Med PlantBase (version September 2005) data only. There are likely to be further species within the major and minor food crop genera recorded in Euro+Med PlantBase in only one country, but the data have not yet been verified and we cannot be certain that these taxa do not occur in other countries.

Table 5.7. Major food crops of the world with wild relatives in the Euro-Mediterranean region (including both native and introduced taxa), the number of species and subspecific taxa within each genus (including crops) and the major food crop species native to the region.

	Crop[a]	Genus	No. of species	No. of subspecific taxa[b]	Native crop species
	Barley	*Hordeum* L.	13	8	*H. vulgare* L.
	Beans	*Vigna* Savi	4	1	–[c]
	Cabbage	*Brassica* L.	34	54	*B. oleracea* L.
	Millet	*Echinochloa* P. Beauv.	11	2	–
	Millet	*Eleusine* Gaertn.	5	2	–
	Millet	*Panicum* L.	21	3	–
	Millet	*Pennisetum* Rich.	11	5	–
	Millet	*Setaria* P. Beauv.	16	7	–
	Olive	*Olea* L.	4	5	*O. europaea* L.
	Potato	*Solanum* L.	60	6	–
	Rye	*Secale* L.	6	3	–
	Sorghum	*Sorghum* Moench	8	0	–
	Sunflower seed	*Helianthus* L.	12	0	–
	Wheat	*Triticum* L.	13	4	*T. aestivum* L.
	Yam	*Dioscorea* L.	1	0	–
Total	11	15	219	100	4

[a]Major food crops based on food crops of major significance listed by Groombridge and Jenkins (2002).
[b]Subspecies and varieties.
[c]*Vigna unguiculata* (L.) Walp. is recorded by Euro+Med PlantBase as native to Egypt but its native distribution is probably limited to sub-Saharan Africa; therefore, it is probably naturalized in Egypt.
– Not applicable.

(Table 5.8). For example, *Brassica balearica* Pers. is endemic to the Balearic Islands (Spain), *B. rupestris* Raf., *B. macrocarpa* Guss. and *B. villosa* Biv. are endemic to the islands of Sicily and Malta (Italy), *B. hilarionis* Post is endemic to Cyprus and *Solanum patens* Lowe and *S. trisectum* Dunal are endemic to Macaronesia (possibly endemic to the island of Madeira). In addition, 46 subspecies within the 28 major food crop genera of the world are endemic to the Euro-Mediterranean region and at least 22 of these are endemic to only one nation (Table 5.8). Again, some of these taxa are limited to islands; for example, *B. oleracea* subsp. *bourgeaui* (Webb) Gladis & K. Hammer and *O. europaea* subsp. *guanchica* P. Vargas, J. Hess, Muñoz Garm. & Kadereit are only found in the Canary Islands (Spain).

Of the 51 minor food crop genera of the world (listed by Groombridge and Jenkins, 2002), 39 (76%) occur in the Euro-Mediterranean region – 35 (69%) of these encompassing wild relatives (Table 5.9). Within these minor food crop genera, 938 species and 372 subspecific taxa (subspecies and varieties) can be found growing in the region. Of these, 382 species and 46 subspecies are endemic and at least 99 species and 41 subspecies are endemic to only one nation (Table 5.8). Of the 69 minor food crops of the world, 23 (33%) are native to the Euro-Mediterranean region and 22 are native to Europe.

Table 5.8. Taxa within the 28 major and 51 minor food crop genera of the world endemic to one nation in the Euro-Mediterranean region.[a]

Crop	Taxon	Endemic to
Wild relatives of major food crops		
Cabbage	*Brassica balearica* Pers.	Baleares (Spain)
	B. cadmea O.E. Schulz	Greece
	B. cretica subsp. *laconica* M.A. Gust. & Snogerup	Greece
	B. desertii Danin & Hedge	Sinai (Egypt)
	B. desnottesii Emb. & Maire	Morocco
	B. elongata subsp. *imdrhasiana* Quézel	Morocco
	B. elongata subsp. *subscaposa* (Maire & Weiller) Maire	Morocco
	B. fruticulosa subsp. *numidica* (Coss.) Maire	Algeria
	B. fruticulosa subsp. *pomeliana* Maire	Algeria
	B. fruticulosa subsp. *radicata* (Desf.) Batt.	Algeria
	B. hilarionis Post	Cyprus
	B. macrocarpa Guss.	Sicily with Malta (Italy)
	B. nivalis subsp. *jordanoffii* (O.E. Schulz) Akeroyd & Leadlay	Bulgaria
	B. nivalis Boiss. & Heldr. subsp. *nivalis*	Greece
	B. oleracea subsp. *bourgeaui* (Webb) Gladis & K. Hammer	Canary Islands
	B. repanda subsp. *almeriensis* Gómez-Campo	Spain
	B. repanda subsp. *blancoana* (Boiss.) Heywood	Spain
	B. repanda subsp. *cadevallii* (Font Quer) Heywood	Spain
	B. repanda subsp. *cantabrica* (Font Quer) Heywood	Spain
	B. repanda subsp. *dertosensis* Molero & Rovira	Spain
	B. repanda subsp. *diplotaxiformis* (Maire) Gómez-Campo	Morocco
	B. repanda subsp. *galissieri* (Giraudias) Heywood	France
	B. repanda subsp. *glabrescens* (Poldini) Gómez-Campo	Italy
	B. repanda subsp. *gypsicola* Gómez-Campo	Spain
	B. repanda subsp. *humilis* (DC.) O. Bolòs & Vigo	France
	B. repanda subsp. *silenifolia* (Emb.) Greuter & Burdet	Morocco
	B. rupestris Raf.	Sicily with Malta (Italy)
	B. spinescens Pomel	Algeria
	B. villosa Biv.	Sicily with Malta (Italy)

Olive	*Olea europaea* subsp. *cerasiformis* G. Kunkel & Sunding	Madeira, Porto Santo, Desertas (Portugal)
	O. europaea subsp. *guanchica* P. Vargas, J. Hess, Muñoz Garm. & Kadereit	Canary Islands (Fuerteventura with Lobos, Tenerife, Lanzarote with Graciosa, La Palma, Hierro, Gomera, Gran Canaria) (Spain)
Potato	*Solanum patens* Lowe	Macaronesia (possibly endemic to Madeira (Portugal))
	S. trisectum Dunal	Macaronesia (possibly endemic to Madeira (Portugal))
Wild relatives of minor food crops		
Onion, garlic	*Allium autumnale* P. H. Davis	Cyprus
	A. bourgeaui subsp. *creticum* Bothmer	Crete (with Karpathos, Kasos and Gavdhos) (Greece)
	A. callimischon Link subsp. *callimischon*	Greece
	A. chrysantherum Boiss. & Reut.	Turkey
	A. chrysonemum Stearn	Spain
	A. circinnatum Sieber	Crete (with Karpathos, Kasos and Gavdhos) (Greece)
	A. corsicum Jauzein, J.-M. Tison, Deschâtres & H. Couderc	Corsica (Italy)
	A. cupani subsp. *cyprium* Meikle	Cyprus
	A. czelghauricum Bordz.	Turkey
	A. deciduum Özhatay & Kollmann subsp. *deciduum*	Turkey
	A. deciduum subsp. *retrorsum* Özhatay & Kollmann	Turkey
	A. djimilense Regel	Turkey
	A. eldivanense Özhatay	Turkey
	A. favosum Zahar.	Greece
	A. fuscum Waldst. & Kit. subsp. *fussii*	Romania
	A. guttatum subsp. *dilatatum* (Zahar.) B. Mathew	Crete (with Karpathos, Kasos and Gavdhos) (Greece)
	A. gorumsense Boiss.	Turkey
	A. grosii Font Quer	Ibiza with Formentera (Spain)
	A. heldreichii Boiss.	Greece
	A. hierochuntinum Boiss.	Israel
	A. hierosolynmorum Regel	Israel
	A. humbertii Maire	Algeria
	A. hymettium Boiss. & Heldr.	Greece
	A. ilgazense Özhatay	Turkey
	A. incensiodorum Radić	Croatia

Continued

Table 5.8. *Continued*

Crop	Taxon	Endemic to
	A. insubricum Boiss. & Reut.	Italy
	A. integerrimum Zahar.	Greece
	A. junceum subsp. *tridentatum* Kollmann, Özhatay & M. Koyuncu	Turkey
	A. karamanoglui Koyuncu & Kollmann	Turkey
	A. kastambulense Kollmann	Turkey
	A. kurtzianum Kollmann	Turkey
	A. lenkoranicum Miscz.	Azerbaijan
	A. leonidi Grossh.	Azerbaijan
	A. luteolum Halácsy	Greece
	A. macedonicum Zahar.	Greece
	A. mareoticum Bornm. & Gauba	Egypt
	A. mariae Bordz.	Azerbaijan
	A. materculae Bordz.	Azerbaijan
	A. negrianum Maire & Weiller	Libya
	A. nemrutdaghense Kit Tan & Sorger	Turkey
	A. olympicum Boiss.	Turkey
	A. palentinum Losa & P.Monts.	Spain
	A. paniculatum subsp. *antiatlanticum* (Emb. & Maire) Maire & Weiller	Morocco
	A. paniculatum subsp. *breviscapum* Litard. & Maire	Morocco
	A. paniculatum subsp. *exaltatum* Meikle	Cyprus
	A. parnassicum (Boiss.) Halácsy	Greece
	A. phthioticum Boiss.	Greece
	A. pruinatum Spreng.	Portugal
	A. pruinatum var. *bulbiferum* Cout.	Portugal
	A. regnieri Maire	Morocco
	A. robertianum Kollmann	Turkey
	A. rouyi Gaut.	Spain
	A. ruhmerianum Asch.	Libya
	A. seirotrichum Ducellier	Algeria
	A. sintenisii Freyn	Turkey

	A. talyschense Miscz.	Azerbaijan
	A. tardans Greuter & Zahar.	Crete (with Karpathos, Kasos and Gavdhos) (Greece)
	A. trichocnemis J. Gay	Algeria
	A. valdecallosum Maire & Weiller	Morocco
	A. vuralii Kit Tan	Turkey
	A. willeanum Holmboe	Cyprus
Sugarbeet	*Beta nana* Boiss. & Heldr.	Greece
	B. patula Aiton	Madeira, Desertas (Portugal)
Mustard seed, rape seed	See major crops, cabbage (*Brassica* spp.)	
Chickpea	*Cicer graecum* Boiss.	Greece
Hazel, filbert	*Corylus cervorum* Petrov	Azerbaijan
Artichoke	*Cynara alba* DC.	Spain
	C. baetica (Spreng.) Pau	Spain
	C. cyrenaica Maire & Weiller	Libya
	C. hystrix Ball	Morocco
Carrot	*Daucus carota* subsp. *gadecaei* (Rouy & E.G. Camus) Heywood	France
	D. tenuisectus Coss.	Morocco
Fig	*Ficus hyrcana* Grossh.	Azerbaijan
Mate	*Ilex perado* Aiton subsp. *perado*	Madeira (Portugal)
Lettuce	*Lactuca* × *impure* Maire	Morocco
	L. livida Boiss. & Reut.	Spain
	L. longidentata DC.	Sardinia (Italy)
	L. reviersii Litard. & Maire	Morocco
	L. seticuspis Boiss.	Syria
	L. triquetra (Labill.) Benth. & Hook. f.	Lebanon
	L. viminea subsp. *alpestris* (Gand.) Feráková	Crete (with Karpathos, Kasos and Gavdhos) (Greece)
	L. virosa subsp. *cornigera* (Pau & Font Quer) Emb. & Maire	Morocco
	L. watsoniana Trel.	Azores (Faial, Pico, São Jorge, São Miguel, Terceira) (Portugal)
Almond, apricot, plum, cherry	*Prunus lusitanica* subsp. *azorica* (Mouill.) Franco	Azores (Terceira, São Miguel, São Jorge, Pico) (Portugal)
	P. ramburii Boiss.	Spain
	P. spinosa subsp. *insititioides* (Ficalho & Cout.) Franco	Portugal

Continued

Table 5.8. *Continued*

Crop	Taxon	Endemic to
Pear	*Pyrus communis* subsp. *mamorensis* (Trab.) Maire	Morocco
	P. complexa Rubtzov	Armenia
	P. elata Rubtzov	Slovenia
	P. hakkiarica Browicz	Turkey
	P. magyarica Terpó	Hungary
	P. mamorensis Trab.	Morocco
	P. medvedevii Rubtzov	Armenia
	P. nutans Rubtzov	Armenia
	P. raddeana Woronow	Armenia
	P. rossica A.D. Danilov	Russia
	P. sosnovskyi Fed.	Armenia
	P. tamamschianae Fed.	Armenia
	P. voronovii Rubtzov	Armenia
	P. zangezura Maleev	Armenia
Blackcurrant, redcurrant	*Ribes multiflorum* subsp. *sandalioticum* Arrigoni	Sardinia (Italy)
	Ribes sardoum Martelli	Sardinia (Italy
Aubergine	See major crops, potato (*Solanum* spp.)	
Broad bean	*Vicia bifoliolata* Rodr.	Baleares (Spain)
	V. capreolata Lowe	Madeira, Desertas (Portugal)
	V. costae A. Hansen	Madeira, Porto Santo (Portugal)
	V. ferreirensis Goyder	Madeira, Porto Santo (Portugal)
	V. glauca subsp. *giennensis* (Cuatrec.) Blanca & F. Valle	Spain
	V. pectinata Lowe	Madeira (Portugal)
	V. sativa subsp. *devia* J.G. Costa	Madeira, Porto Santo (Portugal)
	V. sinaica Boulos	Sinai (Egypt)

[a]Estimate based on Euro+Med PlantBase (version September 2005) data only. There are likely to be more single country endemic taxa within the major and minor food crop genera, but these are not verified as single country endemic taxa in this data set.

Table 5.9. Minor food crops of the world with wild relatives in the Euro-Mediterranean region (including both native and introduced taxa), the number of species and subspecific taxa within each genus (including crops), and the minor food crop species native to the region.

Crop[a]	Genus	No. of species	No. of sub-specific taxa[b]	Native crop species
Almond	*Prunus*	41	24	*P. dulcis* (Mill.) D.A. Webb
Apple	*Malus*	12	4	*M. domestica* Borkh.
Apricot	*Prunus*	41	24	*P. armeniaca* L.
Artichoke	*Cynara*	10	3	*C. scolymus* L.
Aubergine	*Solanum*	60	6	–
Avocado	*Persea*[c]	1	0	–
Blackcurrant	*Ribes*	18	5	*R. nigrum* L.
Broad bean	*Vicia*	141	73	–
Carrot	*Daucus*	26	18	*D. carota* L.
Cherry	*Prunus*	41	24	*P. avium* L.
Chickpea	*Cicer*	17	0	*C. arietinum* L.
Cucumber	*Cucumis*	7	2	–
Date	*Phoenix*	3	0	*P. dactylifera* L.[d]
Fig	*Ficus*	10	4	*F. carica* L.
Filbert	*Corylus*	11	0	*C. maxima* Mill.
Fonio	*Digitaria*	11	2	–
Garlic	*Allium*	276	76	–
Grape	*Vitis*	10	2	*V. vinifera* L.
Hazel	*Corylus*	11	0	*C. avellana* L.
Lentil	*Lens*	8	0	*L. culinaris* Medik.
Lettuce	*Lactuca*	31	11	–
Lupin	*Lupinus*	15	8	–
Mate	*Ilex*	4	8	–
Melon	*Cucumis*	7	2	*C. melo* L.
Melon seed/ watermelon	*Citrullus*	2	0	*C. lanatus* (L.) Schrad.
Mustard seed	*Brassica*	34	54	–
Oats	*Avena*	29	17	–
Onion	*Allium*	276	76	–
Pea	*Pisum*	2	5	*P. sativum* L.
Pear	*Pyrus*	49	16	*P. communis* L.
Pistachio	*Pistacia*	7	5	–
Plum	*Prunus*	41	24	*P. domestica* L.
Quinoa	*Chenopodium*	51	16	–
Rapeseed	*Brassica*	34	54	*B. napus* L.
Redcurrant	*Ribes*	18	5	*R. rubrum* L.
Sesame seed	*Sesamum*	2	0	–
Spinach	*Spinacia*	2	0	–
Strawberry	*Fragaria*	12	3	–
Sugar beet	*Beta*	14	6	*B. vulgaris* L.
Sweet potato	*Ipomoea*	13	1	–

Continued

Table 5.9. *Continued*

	Crop[a]	Genus	No. of species	No. of sub-specific taxa[b]	Native crop species
	Taro	*Colocasia*	1	0	–
	Tomato	*Lycopersicon*	2	2	–
	Walnut	*Juglans*	6	0	*J. regia* L.
Total	43	35	938[e]	372[e]	23

[a]Minor food crops based on food crops of secondary or local importance listed by Groombridge and Jenkins (2002).
[b]Subspecies and varieties.
[c]*Persea indica* (L.) Spreng. – occurs in the Azores only.
[d]All crop species native to the Euro-Mediterranean region are native to Europe, except *Phoenix dactylifera*.
[e]The total number of species and subspecific taxa within genera containing minor food crops of the world (i.e. not the column totals).
– Not applicable.

The major and minor food crop groups that can be found in the Euro-Mediterranean region, along with other crops of high socio-economic value that are not included in this analysis, for example, forage and fodder crops, are an important genetic resource which may contribute to crop improvement in the future. Taxa that have limited distributions, particularly those that are endemic to one country should be a high priority for conservation and steps need to be taken to assess their conservation status, both *in situ* and *ex situ* (see Ford-Lloyd *et al.*, Chapter 6, this volume, for further discussion about prioritization).

5.3.5 How many CWR are included in the IUCN Red List of Threatened Species?

The answer to this question is simple – currently, very few. The CWR Catalogue data were cross-checked with the 2004 IUCN Red List of Threatened Species to reveal only 161 species and 23 subspecific CWR taxa that occur in the Euro-Mediterranean region are included in the global Red List[7] (Table 5.10). The majority of these taxa are trees and the explanation for this is that much work has been undertaken in the past decade to assess the conservation status of the world's trees; for example, see Oldfield *et al.* (1998) and Farjon (2001). Of the CWR taxa included, 130 are native to the region and 76 are endemic. At least 13 of these are endemic to only one country and of these, one is extinct in the wild (*Betula szaferi* Jentys-Szaferowa & Staszk.) and two are critically endangered (*Abies nebrodensis* (Lojac.) Mattei, endemic to Sicily (Italy), and *Salix tarraconensis* Pau, endemic to Spain). Of the CWR species included in the Red List, 120 fall into the agricultural and horticultural crop group,[8] 152 in the forestry group, 124 in the ornamental group and 148 in the medicinal

[7] Matching carried out with accepted names in the Catalogue only.
[8] Although most of the CWR species in the Red List are trees, they are included in the agricultural and horticultural crop group because Mansfeld's Database includes a very wide range of cultivated plants.

Table 5.10. The number of CWR taxa (species, subspecies and varieties) that occur in the Euro-Mediterranean region that are included in the 2004 IUCN Red List of Threatened Species.[a]

Red List category[b]	No. of taxa	No. of native taxa[c]	No. of endemic taxa[c]	Taxa endemic to one nation[d]
Extinct in the wild	1	1	1	1
Critically endangered	14	10	6	2
Endangered	9	9	3	0
Vulnerable	33	28	14	6
Least concern	2	2	2	0
Lower risk/near threatened	31	27	17	0
Lower risk/conservation dependent	11	8	5	2
Lower risk/least concern	77	40	25	0
Data deficient	6	5	3	2
Total	184	130	76	13

[a]Analysis based on taxa matching accepted names in the CWR Catalogue only.
[b]The taxa listed have been assessed using the 1994 Categories and Criteria (IUCN, 1994).
[c]Taxa native and endemic to the Euro-Mediterranean region.
[d]Taxa verified as endemic according to Euro+Med PlantBase (version September 2005).

and aromatic group, so at least we know that the small number of CWR included have a wide range of uses. Only one taxon, *O. europaea* subsp. *cerasiformis* is a wild relative of a major food crop (olive) – 16 taxa are wild relatives of the minor food crops: almond, apricot, avocado, cherry, date, mate, pear and plum.

While it is interesting to look at which CWR taxa are included in the global Red List, we cannot draw any firm conclusions from this analysis, except to state obviously that there are currently very few taxa included. We must not assume that only few CWR are under threat, because although it is the Red List of Threatened Species, not all species listed are under threat – they have simply been assessed using the IUCN criteria. A Red List assessment may show that a taxon is not threatened, but the taxon will still appear in the Red List. It is only those taxa assigned the categories 'critically endangered', 'endangered' and 'vulnerable' that are considered threatened – the other categories present the conservation status of the taxon and provide a reference point for future monitoring. In fact, of this small number of assessed CWR taxa, 30% have been categorized as threatened and 42% as lower risk or least concern (Table 5.10). We cannot take this small sample of global Red List assessments as representative of CWR in general, but it would be interesting to review the percentage of threatened CWR over time, as more taxa are assessed and added to the List.

One reason for the lack of CWR taxa included is likely to be that the vast majority of plant taxa listed in the 1997 IUCN Red List of Threatened Plants (Walter and Gillett, 1998) have not yet been evaluated against the revised Red List Criteria and are therefore not included in the 2004 Red List. Analysis of the 1997 Red List would probably provide a more realistic picture of progress with Red Listing of CWR, but to ascertain how many CWR are included in the 1997 Red List, we would need access to the electronic data set, which was not available for this analysis (except through an online search facility – see WCMC and RBG Edinburgh, no date). However, analysis of IPA data indicates that at least 488 European CWR species were categorized as globally threatened in the 1997 Red List.

Another reason for the lack of CWR species in the Red List may be that, historically, there has not been a group of specialists taking CWR Red Listing in hand. The establishment of the CWR Specialist Group (CWRSG) of the IUCN Species Survival Commission should rectify this (see Dulloo and Maxted, Chapter 48, this volume). Ultimately, while it is useful to have global Red List assessments available for CWR taxa (or any plant taxa), it may be more useful to investigate which taxa have been assessed at national level. Again, national Red Listing, or investigating which CWR taxa are already included on national Red Lists, could be an important role for the CWRSG.

5.3.6 Does the EU Habitats Directive aid CWR conservation?

In 1992, the European Community adopted Council Directive 92/43/EEC on the conservation of natural habitats and of wild fauna and flora (the EU Habitats Directive). The provisions of the Directive require EU member states to introduce a range of measures, including the protection of species listed in the Annexes, to undertake surveillance of habitats and species and produce a report every 6 years on the implementation of the Directive. Annexes I and II list natural habitat types and plant (and animal) species of community interest, 'whose conservation requires the designation of special areas of conservation', Annex IV lists plant (and animal) species of community interest 'in need of strict protection' (most species listed in Annex II are also listed in Annex IV) and Annex V lists plant (and animal) species of community interest 'whose taking in the wild and exploitation may be subject to management measures' (European Communities, 1995–2007). Species of community interest are those that are: (i) endangered, except those species whose natural range is marginal in that territory and which are not endangered or vulnerable in the western Palaearctic region; or (ii) vulnerable (i.e. believed likely to move into the endangered category in the near future if the causal factors continue operating); or (iii) rare (i.e. with small populations that are not at present endangered or vulnerable, but are at risk); the species are located within restricted geographical areas or are thinly scattered over a more extensive range; or (iv) endemic and requiring particular attention by reason of the specific nature of their habitat and/or the potential impact of their exploitation on their habitat and/or the potential impact of their exploitation on their conservation status (European Communities, 1995–2007).

Each member state is required to prepare and propose a national list of sites for evaluation in order to form a European network of sites of community importance (SCIs). Once adopted, these are designated by member states as special areas of conservation (SACs) and, along with special protection areas (SPAs) classified under the EC Birds Directive, form a network of protected areas known as Natura 2000.

Species listed in Annexes II, IV and V (as of March 2007, including data from all 27 member states) were cross-checked against the Catalogue to see how many CWR are included (Table 5.11).[9] There are 641 plant species listed

[9] Matching carried out with accepted names in the Catalogue only.

Table 5.11. CWR of the European Union member states included in Annexes II, IV and V of the EU Habitats Directive.

Species list	No. of species in the four crop groups: Agricultural and horticultural	Forestry	Ornamental	Medicinal and aromatic	Total no. of species	Percentage of EU CWR species	Percentage of vascular plant species in Annexes II, IV and V of the Habitats Directive
EU CWR species[a]	14,515	2,126	4,785	12,448	16,052	–	–
Vascular plant species listed in Annexes II, IV and V of the EU Habitats Directive	–	–	–	–	641	–	–
EU CWR in HD Annex II[b]	331	18	120	275	380	2	59
EU CWR in HD Annex IV[c]	370	21	137	312	422	3	66
EU CWR in HD Annex V[d]	15	2	4	18	18	<1	3
EU CWR HD priority species[e]	117	9	42	105	141	1	22
Total no. of EU CWR included in Annexes II, IV and V of the EU Habitats Directive	385	23	141	330	440	3	69

[a]Includes all crop and CWR species that occur within the territories of the 27 EU member states.
[b]Annex II includes plant (and animal) species of community interest whose conservation requires the designation of special areas of conservation. Most species listed in this Annex are also listed in Annex IV.
[c]Annex IV lists plant (and animal) species of community interest in need of strict protection.
[d]Annex V lists plant (and animal) species of community interest whose taking in the wild and exploitation may be subject to management measures.
[e]Priority species are endangered species for which the Community has particular responsibility in view of the proportion of their natural range which falls within the territory.
– Not applicable.

in Annexes II, IV and V – 440 (69%) of these are included in the CWR Catalogue. Of these, 385 species (60%) fall into the agricultural and horticultural crop group, 23 species (4%) in the forestry group, 141 species (22%) in the ornamental group and 330 species (51%) in the medicinal and aromatic plant group. A high percentage of priority species (endangered species for which the Community has particular responsibility in view of the proportion of their natural range which falls within the territory) are in the agricultural and horticultural, and medicinal and aromatic plant groups (83% and 74%, respectively). It is notable that only four species included in the Habitats Directive Annexes II, IV and V are wild relatives of major food crops: three *Brassica* species and one *Solanum* sp. This is out of a total of 153 wild relative species of major food crops that occur in the EU territories. A further 13 species are included in the minor food crop group, out of a total of 542.

It is not surprising that quite a high percentage of species listed in Annexes II, IV and V of the Habitats Directive are CWR because more than three-quarters of the flora of the region is of current or potential socio-economic use. What is striking is the relatively small percentage of CWR species listed overall as a proportion of the CWR flora of the region (3%); however, this equates almost exactly to the proportion of vascular plant species that occur in the EU territories included in the Habitats Directive Annexes (641 species out of an estimated total of 19,020). Perhaps this raises a question about the overall effectiveness of the Habitats Directive for plant conservation, let alone the conservation of CWR. Certainly, a small number of CWR in the major and minor food crop groups that are listed in the Habitats Directive Annexes is a strong indication that *in situ* CWR conservation of the most important groups is not being adequately addressed within the EU territories.

It is important to stress that the above analysis only takes into account the species listed in the Habitats Directive Annexes II, IV and V – there are, of course, many more species included within the habitats that are designated for conservation within the Natura 2000 network. As for any *in situ* conservation area, site inventories are required to find out which species are included. At EU level, these data are not available; however, it is possible to look at which CWR species are mentioned as characteristic of the habitats listed in the European Nature Information System (EUNIS) Database (EEA, 2007), some of which are included in the Habitats Directive Annex I (natural habitat types of community interest whose conservation requires the designation of SACs). Here, 1665 CWR species that occur in the EU territories are included (10% of the CWR flora of the EU) – 54 of these species are included in Annex II, 55 in Annex IV and five in Annex V. Of these, 91% are in the agricultural and horticultural crop group, 17% in the forestry group, 36% in the ornamental group and 78% in the medicinal and aromatic plant group. Nine wild relatives in the major food crop genera and 57 in the minor food crop genera, are included. Although not all these habitats are necessarily included in the Natura 2000 network, it is useful to discover that around 10% of the CWR flora of the EU is mentioned as characteristic of the habitats, because many of these habitats are included in the network – however, we cannot assume that these species are actively conserved.

5.3.7 Are CWR important in Important Plant Areas?

IPAs are natural or semi-natural sites exhibiting exceptional botanical richness and/or supporting an outstanding assemblage of rare, threatened and/or endemic plant species and/or vegetation of high botanical value (PlantLife International, no date). IPAs are not legal site designations, but a framework for identifying and highlighting the best sites for plants, and by implication, their conservation. Site selection is based on three criteria: threatened species, botanical richness and threatened habitats – a site qualifies as an IPA if it fulfils one or more criteria.

The CWR Catalogue data for Europe (as defined by Hollis and Brummitt, 2001) were compared with the list of species included in IPAs (designated under Criterion A) as of May 2005 (Table 5.12).[10] Criterion A sites hold significant populations of one or more species that are of global or European conservation concern. Criterion A is further divided into four categories: A(i) – the site contains globally threatened species; A(ii) – the site contains regionally threatened species; A(iii) – the site contains national endemic species with demonstrable threat not covered by A(i) or A(ii); A(iv) – the site contains near endemic or limited range species with demonstrable threat not covered by A(i) or A(ii) (Anderson, 2002). Species included under Criteria A(iii) and A(iv) are nationally threatened species from Belarus, Czech Republic, Slovakia, Estonia, Slovenia, Poland and Romania only, which were the first seven countries in Europe to identify IPAs (see Anderson *et al.*, 2005).

Nine hundred and twelve CWR species of Europe are included in the IPAs – 51% of the vascular plant species included in the IPAs and 5% of the CWR flora of Europe. Of these, 488 (54%) are globally threatened species[11] and 426 (47%) are regionally threatened. The endemic species included under Criteria A(iii) and A(iv) (Belarus, Czech Republic, Slovakia, Estonia, Slovenia, Poland and Romania only) represent around 10% of the CWR species included in the IPAs. Three percent of the agricultural and horticultural crops and CWR of Europe are included under the globally threatened Criterion A (i). Likewise, 2% of species in the forestry group, 4% in the ornamental group and 2% in the medicinal and aromatic group are included under this criterion. Looking at the overall number of European CWR species included in the IPAs, 5% of species in the agricultural and horticultural crop group are included, 3% in the forestry group, 7% in the ornamental group and 5% in the medicinal and aromatic plant group.

As for the CWR species included in the EU Habitats Directive, a relatively small percentage of the CWR species of Europe are included in IPAs (5%); however, this is in the context of the proportion of vascular plant species of Europe included in IPAs – 912 species out of an estimated total of 20,590 – around 4%. Again, the number of CWR in the major and minor food crop groups included in the IPAs may be an indication of how much attention is being paid to CWR in the context of this conservation initiative. With only three out of the 152 species in the major food crop genera that occur in Europe included and none of the 559 species in the minor food crop genera, we might conclude that more needs to be done to ensure that CWR are represented in IPAs.

[10] Matching carried out with accepted names in the Catalogue only.

[11] Based on the 1997 IUCN Red List of Threatened Plants (Walter and Gillett, 1998).

Table 5.12. CWR of Europe included in Important Plant Areas (IPAs).

Species list	No. of species in the four crop groups				Total no. of species	Percentage of total European CWR	Percentage of total vascular plant species in IPAs	Percentage of total CWR species in IPAs
	Agricultural and horticultural	Forestry	Ornamental	Medicinal and aromatic				
European CWR species[a]	15,828	2,267	5,123	13,727	17,495	–	–	–
Vascular plant species included in IPAs	–	–	–	–	1,803	–	–	–
Criterion A(i) European CWR species (globally threatened)	400	52	214	338	488	3	27	54
Criterion A(ii) European CWR species (regionally threatened)	379	16	138	328	426	2	24	47
Criterion A(iii) European CWR species (national endemic species not covered by A(i) or A(ii))[b]	86	16	41	69	95	<1	5	10
Criterion A(iv) European CWR species (near endemic or restricted range species not covered by A(i) or A(ii))[b]	83	2	22	73	86	<1	5	9
Total European CWR species in IPAs[c]	791	75	349	668	912	5	51	–

[a]Includes all crop and CWR species that occur in Europe (Europe as defined by Hollis and Brummitt (2001)).
[b]Species included under Criteria A(iii) and A(iv) are nationally threatened species from Belarus, Czech Republic, Slovakia, Estonia, Slovenia, Poland and Romania only.
[c]Included under Criterion A only (A(iii) and A(iv) species from Belarus, Czech Republic, Slovakia, Estonia, Slovenia, Poland and Romania only).
– Not applicable.

5.3.8 Are botanic gardens' living collections helping to conserve crop resources?

Using data extracted from the Plant Search database managed by Botanic Gardens Conservation International (BGCI, 2007), which is a database compiled from lists of living collections submitted to BGCI by the world's botanic gardens, an analysis of the number of crop and CWR taxa in cultivation in botanic gardens around the world was undertaken (Table 5.13).

Table 5.13. Crop and CWR species in botanic gardens' living collections.[a]

	No. of species in the four crop groups				
Species in Plant Search (BGCI, 2007)	Agricultural and horticultural	Forestry	Ornamental	Medicinal and aromatic	Total no. of species
Total no. of species	–	–	–	–	89,803
Crop and CWR species	54,828	12,199	22,522	38,375	62,746
Species cultivated worldwide[b]	6,388	–	–	–	–
Total species in the major food crop genera[c]	–	–	–	–	791
Crop species in the major food crop genera	–	–	–	–	323
Total species in the minor food crop genera[d]	–	–	–	–	2,668
Crop species in the minor food crop genera	–	–	–	–	633
Crop and CWR species in Europe and the Mediterranean[e]	9,107	1,312	3,631	7,553	9,948
Euro-Mediterranean species in the major food crop genera	–	–	–	–	152
Euro-Mediterranean species in the minor food crop genera	–	–	–	–	521

[a]Based on analysis of data contained in Plant Search (BGCI, 2007).
[b]Species in Plant Search matching species in Mansfeld's Database (accepted names and synonyms). Mansfeld's Database includes cultivated medicinal and aromatic plants.
[c]Based on food crops of major significance, listed by Groombridge and Jenkins (2002).
[d]Based on food crops of secondary or local importance, listed by Groombridge and Jenkins (2002).
[e]Matching accepted species in the CWR Catalogue for Europe and the Mediterranean. Total no. of species in the Catalogue – 25,687.
– Not applicable, or data not available.

Initial results indicate that botanic gardens may be the storehouses of important crop resources and other species of socio-economic importance. Of the 25,687 accepted species in the Euro-Mediterranean Catalogue, 9948 (39%) are recorded in Plant Search as being cultivated in botanic gardens around the world. Of these, 92% are included in the agricultural and horticultural crop group, 13% in the forestry group, 36% in the ornamental group and 76% in the medicinal and aromatic group.

The above analysis only takes into account the socio-economically important species in the Euro-Mediterranean region. Taking a global view, of the 89,803 species included in Plant Search, 62,746 (70%) are species within the combined list of genera containing crops and wild-harvested medicinal and aromatic plants of the world (including synonymous genera in Mansfeld's Database) – at least 10% of these species are known to be agricultural and horticultural species cultivated worldwide. Breaking this list of 62,746 species down into the four crop groups, 87% are in the agricultural and horticultural group, 19% in the forestry group, 36% in the ornamental group and 61% in the medicinal and aromatic group – fairly consistent with the ratios of Euro-Mediterranean crop and CWR species in the database.

Although the total number of species housed in the botanic gardens' living collections that are included in the Plant Search database is not wholly representative of the world flora, if we assume that they are a representative sample, the figure of 70% is not far off what might be expected, since the results of the Euro-Mediterranean analysis indicate that at least three-quarters of the flora of the region are of current or potential socio-economic use. Of course, we cannot confirm this conclusion without further detailed analysis. Other possible explanations for the large proportion of species of socio-economic importance in cultivation in botanic gardens' living collections are that: (i) historically, some botanic gardens were physic gardens and therefore almost exclusively housed medicinal plants; (ii) some gardens were used as repositories and/or quarantine centres for the early movement of crops around the world; and (iii) many gardens have educational displays of crop plants to show visitors what they look like and how they grow; for example, coffee, tea, banana and coconut.

If we look at the major and minor food crop groups (as defined earlier in the chapter) we find that 791 species in the 28 major food crop genera of the world and 2668 in the 51 minor food crop genera can be found in cultivation in the botanic gardens whose collections are recorded in Plant Search – not a vast number, but significant none the less. It is notable that 41% of the species in the major food crop genera and 24% in the minor food crop genera are cultivated species listed in Mansfeld's Database. Perhaps the high proportion of cultivated species in the major food crop groups may be attributable to the fact that botanic gardens often maintain educational displays of important food crops and other cultivated plants.

So, what does this tell us about the potential role of botanic gardens' living collections in crop genetic resources conservation? Taxonomically (i.e. looking at the number of species included), this preliminary analysis indicates that botanic gardens may harbour important resources that could have a role to play in providing germplasm for crop improvement.

However, the analysis does not inform us of the quantity or quality of the plant material in cultivation.[12] Botanic gardens' living collections are sometimes accused of effectively being plant 'museums' because they frequently maintain only one or a few accessions of a taxon in cultivation. None the less, although they may not always conserve genetically representative samples of a taxon or population, the germplasm that is maintained may still be of some value, especially in cases where a taxon is severely threatened in the wild. Another common criticism of botanic gardens' living collections is that once plants have been kept in cultivation for several years, they may no longer resemble the genetic make-up of the wild form that was originally collected. This may be so, but only genetic analysis could reveal the true picture (i.e. if there is still wild material available to compare the cultivated material with). Furthermore, many botanic gardens are focusing their efforts on the conservation of threatened populations and these days are more aware of the need to collect and maintain representative samples.

Even if the germplasm itself is of limited use to plant breeders, perhaps the associated information contained in botanic gardens' collections databases, such as details on locations and habitats, may be a useful resource to the conservation and user community in itself. This, of course, is dependent on the quality and efficiency of botanic gardens' information management systems. Finally, we should acknowledge the important role that botanic gardens' living collections play in educating the public. Many botanic gardens already provide educational information about the importance of directly utilized plants to society – perhaps this role could be extended to include educational information about the wild relatives of crop plants, their role in future food security and what needs to be done to conserve them.

5.4 Conclusions

The Catalogue of CWR for Europe and the Mediterranean (Kell *et al.*, 2005a) is the first comprehensive CWR Catalogue at a continental scale and, through extraction, for the countries included. It provides an informative regional overview of crop and CWR diversity and acts to raise awareness about the importance of crop genetic resources in the region, both within the professional PGR community and other interest groups. Furthermore, it provides the baseline data needed to monitor biodiversity change and to improve access to germplasm for the CWR user community. The Catalogue can be used as the basis for creating national crop and CWR inventories, as a vehicle for conservation gap analysis and for integrating CWR conservation into existing conservation initiatives. It is a core data set providing an opportunity for linking to and building on existing taxon data, such as information on uses, population biology, threats and *in situ* and *ex situ* conservation activities. The Catalogue is available online through CWRIS (PGR Forum, 2005), where users can search by

[12] This information could be obtained by contacting individual botanic gardens.

taxon names and geographical units to obtain this information. To read more about CWRIS and for examples of use cases, see Kell *et al.* (Chapter 33, this volume).

The methodology used for creating the Euro-Mediterranean Catalogue can be applied in any part of the world, either at regional or national level. Although digitized floras are not immediately available in all parts of the world, increasingly, countries are working to create biodiversity databases, particularly in response to the requirements of the provisions of the CBD. Even without a digitized flora, it is possible to undertake the analysis, although this would obviously take more time.

An important and fundamental application of the CWR Catalogue is to aid gap analysis for CWR conservation – for example, by analysing which taxa are already included within existing protected areas and *ex situ* collections and to ascertain how many taxa are included in other conservation databases, such as the IUCN Red List of Threatened Species (IUCN, 2006). Some examples of how the data can be used in this way have been provided in this chapter. Although these are preliminary and largely broad brush-stroke investigations, results do indicate that we may not be paying sufficient attention to CWR in current conservation endeavours. We strongly urge policy makers and conservationists to give greater credence to the inclusion of crops and wild relatives within existing or new conservation initiatives (including legislation), both at regional and national level. For example, by creating a priority list of CWR for the Euro-Mediterranean region (see Ford-Lloyd *et al.*, Chapter 6, this volume), combined with the formulation of national priority lists, the conservation status of these taxa could initially be assessed and a more detailed gap analysis undertaken. Building on the data that are now available, networks of national genetic reserves can be established, following the guidelines provided by the draft Global Strategy for CWR Conservation and Use (see Heywood *et al.*, Chapter 49, this volume).

A more systematic approach to complementary CWR conservation is certainly needed. Looking, for example, at the number of species included in botanic gardens' living collections, we find that there are a significant number of CWR in cultivation around the world. However, it is likely that these were collected for diverse reasons, rather than specifically because of their value as gene donors for crop improvement. National PGR Coordinators and regional and international conservation organizations could do more to put in place a coordinated approach to CWR conservation. A combined approach targeting existing protected areas and establishing new *in situ* conservation sites where necessary, and encouraging managers of *ex situ* collections (gene banks and botanic gardens' living collections) to take a more systematic approach to CWR conservation is needed.

There is undoubtedly an urgent need to undertake Red List assessments for Euro-Mediterranean CWR and most likely for CWR worldwide. Red Listing could initially be undertaken in three phases: (i) the CWR taxa listed in the 1997 IUCN Red List of Threatened Plants could be reassessed using the 2001 Criteria (IUCN, 2001) and assessments submitted for inclusion in the IUCN Red List of Threatened Species; (ii) single country endemic taxa could be assessed and submitted for inclusion in the IUCN Red List; and (iii) national PGR Coordinators could establish which CWR are included in national Red Lists and make these data available for regional and global assessments.

Further investigation can be carried out to provide an indication of to what extent CWR are already conserved, both within the Euro-Mediterranean region and elsewhere in the world. Many taxon data sets are available electronically – it is simply a matter of working together and making the data accessible. For example, global protected area data are available and, using the CWR Catalogue for Europe and the Mediterranean (or other regional CWR inventories as they become available), analysis can be undertaken to assess how many species are afforded some level of protection *in situ*. At national level, the data can also be compared with protected area inventories and *ex situ* collections, which would provide a more detailed picture of CWR conservation within any given region. It would also be interesting to compare CWR inventories with the data contained in EURISCO (European Internet Search Catalogue of *Ex Situ* PGR Accessions) (ECPGR, no date), though this is not straightforward because the data within EURISCO do not currently follow a standard taxonomy.

Sharing and cross-checking conservation data sets is one way of assisting CWR conservation gap analysis. Another way is to bring CWR information together through the Internet, which provides a unique opportunity to link any number of information sources together. CWRIS (PGR Forum, 2005) (see Kell *et al.*, Chapter 33, this volume), which was created under the auspices of the EC-funded project, PGR Forum (see Maxted *et al.*, Chapter 1, this volume; PGR Forum, 2003–2005), goes some way towards achieving this goal. The Catalogue data housed in CWRIS is linked to a number of selected online information resources, such as the Germplasm Resources Information Network (GRIN) (USDA, ARS, National Genetic Resources Programme, 2006), IUCN Red List of Threatened Species (IUCN, 2006), Survey of Economic Plants for Arid and Semi-Arid Lands (SEPASAL) (Royal Botanic Gardens, Kew, 1999), International Legume Database and Information Service (ILDIS, 2007) and FAO Worldwide Information System on Forest Genetic Resources (REFORGEN) (FAO, no date). With the appropriate financial resources, the opportunity exists to develop CWRIS further as a sophisticated online tool to provide access to CWR information at both taxon and geographic level to cater for a wide range of user groups (Kell *et al.*, Chapter 33, this volume).

The results presented in this chapter are based on data extracted from Euro+Med PlantBase (version September 2005). Euro+Med PlantBase is undergoing a process of critical review and updating by taxon experts on a family by family basis. Although it is not anticipated that the overall number of species included in the Catalogue will change significantly once the updates to Euro+Med PlantBase have been incorporated, there are likely to be some changes, particularly with regard to the number of single country endemic species. Currently, the coding system used in the database to record endemic species makes it difficult to gain a reliable estimate. However, crop and CWR lists extracted from the Catalogue have already been sent to National PGR Coordinators throughout the region. These lists can be used as a basis for the development of national CWR Catalogues and this may provide an opportunity to ascertain more accurately how many single country endemic species exist. Data from National PGR Coordinators could be fed back to the Euro+Med PlantBase Secretariat to be considered for inclusion in the database, and in turn, the data in the CWR Catalogue for Europe and the Mediterranean will be automatically updated.

The Catalogue shows that a large proportion of the Euro-Mediterranean flora is of current or potential socio-economic use, both within the region and elsewhere in the world. These resources need to be conserved to benefit the environment and humankind in the future. Knowing what occurs in nature in the region is a first step in CWR conservation. The next steps are to use the Catalogue data to establish conservation priorities, both regionally and nationally, then to ascertain which species are conserved and to what extent they are protected. This should be part of a coordinated systematic approach to the complementary conservation of CWR. This is likely to involve the establishment of new *in situ* sites or at least the adaptation of existing site management plans to accommodate monitoring and management of CWR populations, and systematic collection and *ex situ* conservation of genetically representative CWR population samples.

Results of this analysis confirm the direct and indirect use values of a high proportion of the vascular flora of the Euro-Mediterranean region. We may confidently assume that a similar proportion of the world's flora has the same current or potential use. The method used to create the Euro-Mediterranean Catalogue can be repeated in other regions of the world and/or nationally as a first step in putting in place a systematic complementary global approach to CWR conservation to ensure that these vital resources are maintained for the benefit of society worldwide. The Global Strategy for CWR Conservation and Use, which was a significant outcome of the First International Conference on CWR Conservation and Use (see Kell *et al.*, 2005b; Heywood *et al.*, Chapter 49, this volume) is already being taken forward as an adjunct to the ITPGRFA. This will provide the much-needed guidance and framework for a coordinated approach to the conservation and sustainable utilization of CWR.

Acknowledgements

We are indebted to the following people who provided access to data, without which the creation of the Catalogue of CWR for Europe and the Mediterranean and associated data analysis would not have been possible: Tarik El Atechi, Euro+Med PlantBase Secretariat, University of Reading, United Kingdom; Werner Greuter, Anton Güntsch and Eckhard von Raab-Straube, Euro+Med PlantBase Secretariat, Botanic Garden and Botanical Museum, Berlin-Dahlem, Germany; Norbert Biermann, Institute of Plant Genetics and Crop Plant Research, Gatersleben, Germany; Ton Kwakkenbos, Community Plant Variety Office, Angers, France; Uwe Schippmann, Fachgebiet Botanik und Naturschutz, Bundesamt für Naturschutz, Bonn, Germany; Dominique Richard, Grégoire Loïs and Doug Evans, European Topic Centre on Biological Diversity, Paris, France; Craig Hilton-Taylor and Caroline Pollock, IUCN Red List Programme,

Cambridge, United Kingdom; Liz Radford and Seona Anderson, PlantLife International, Salisbury, United Kingdom; Diane Wyse Jackson and Suzanne Sharrock, Botanic Gardens Conservation International, Kew, United Kingdom.

The concepts discussed in this chapter were stimulated by PGR Forum (the European crop wild relative diversity assessment and conservation forum – EVK2-2001-00192) (see PGR Forum, 2003–2005), funded by the EC Fifth Framework Programme for Energy, Environment and Sustainable Development.

References

Anderson, S. (2002) *Identifying Important Plant Areas*. PlantLife International, London.

Anderson, S., Kušík, T. and Radford, E. (eds) (2005) *Important Plant Areas in Central and Eastern Europe*. PlantLife International, London.

BGCI (2007) *Plant Search*. Botanic Gardens Conservation International, London. Available at: http://www.bgci.org/plant_search.php/

Bramwell, D. (1990) Conserving biodiversity in the Canary Islands. *Annals of Missouri Botanical Garden* 77, 28–37.

CBD (1992) *Convention on Biological Diversity: Text and Annexes*. Secretariat of the Convention on Biological Diversity, Montreal, Canada. Available at: http://www.biodiv.org/convention/convention.shtml (accessed 12 April 2007)

CBD (2002) *Global Strategy for Plant Conservation*. Secretariat of the Convention on Biological Diversity, Montreal, Canada. Available at: http://www.biodiv.org/decisions/?lg=0&dec=VI/9 (accessed 3 April 2007)

Council of Europe and Planta Europa (2002) *European Plant Conservation Strategy*. The Hague, The Netherlands. Available at: http://www.plantaeuropa.org/pe-EPCS-what_it_is.htm (accessed 5 April 2007)

CPVO (2001) *Community Plant Variety Office website*. Available at: http://www.cpvo.eu.int/index.php (accessed 13 April 2007)

ECPGR (no date) *European Internet Search Catalogue of* Ex Situ *PGR Accessions*. Bioversity International. Available at: http://eurisco.ecpgr.org/ (accessed 5 April 2007)

EEA (2007) *European Nature Information System (EUNIS)*. European Environment Agency, Copenhagen K, Denmark. Available at: http://eunis.eea.europa.eu/index.jsp (accessed 13 April 2007)

European Communities (1995–2007) *Council Directive 92/43/EEC of 21 May 1992 on the Conservation of Natural Habitats and of Wild Fauna and Flora*. Available at: http://eur-lex.europa.eu/LexUriServ/LexUriServ.do?uri=CELEX:31992L0043:EN:HTML (accessed 2 April 2007)

Euro+Med PlantBase (2005) *Euro+Med PlantBase: The Information Resource for Euro-Mediterranean Plant Diversity*. Dipartimento di Scienze botaniche ed Orto botanico, Università degli Studi di Palermo. Available at: http://www.emplantbase.org/home.html (accessed 13 April 2007)

Euro+Med PlantBase Secretariat (2002) *Preparation of the initial checklist: data standards version 2.8, 5 July 2002.*

FAO (no date) *REFORGEN – the FAO Forestry Database on Forest Genetic Resources*. Food and Agriculture Organization of the United Nations, Rome, Italy. Available at: http://www.fao.org/forestry/site/39116/en/ (accessed 3 April 2007)

FAO (2001) *International Treaty on Plant Genetic Resources for Food and Agriculture*. Food and Agriculture Organization of the United Nations, Rome, Italy. Available at: http://www.fao.org/ag/cgrfa/itpgr.htm (accessed 4 April 2007)

Farjon, A. (2001) *World Checklist and Bibliography of Conifers*, 2nd edn. World Checklists and Bibliographies, 3. Royal Botanic Gardens, Kew, London.

Groombridge, B. and Jenkins, M.D. (2002) *World Atlas of Biodiversity*. Prepared by the UNEP World Conservation Monitoring Centre. University of California Press, Berkeley, California.

Hammer, K. and Spahillari, M. (1999) *Alternative Crops for Sustainable Agriculture*. Research progress COST 814. Workshop held at BioCity, Turku, Finland. Office of Official Publications of the European Communities, EUR-OP, Luxembourg.

Hanelt, P. and IPK Gatersleben (eds) (2001) *Mansfeld's Encyclopedia of Agricultural and Horticultural Crops*. 6 vols. 1st English edition. Springer, Berlin/Heidelberg/New York, 3645 pp.

Heywood, V.H. and Zohary, D. (1995) A catalogue of the wild relatives of cultivated plants native to Europe. *Flora Mediterranea* 5, 375–415.

Hollis, S. and Brummitt, R.K. (2001) *World Geographical Scheme for Recording Plant Distributions*. Plant Taxonomic Database Standards No. 2. 2nd edn. Published for the International Working Group on Taxonomic Databases for Plant Sciences (TDWG) by the Hunt Institute for Botanical Documentation, Carnegie Mellon University, Pittsburgh. Available at: http://www.tdwg.org/standards/109/ (accessed 20 March 2007)

ILDIS (2007) *International Legume Database and Information Service*, University of Southampton, UK. Available at: http://www.ildis.org/ (accessed 4 April 2007)

IPK Gatersleben (2003) *Mansfeld's World Database of Agricultural and Horticultural Crops*. Leibniz Institute of Plant Genetics and Crop Plant Research, Gatersleben, Germany. Available at: http://mansfeld.ipk-gatersleben.de/mansfeld/mf-inf_e.htm (accessed 13 April 2007)

IUCN (2001) *The IUCN Red List of Threatened Species: 2001 Categories and Criteria (v. 3.1)*, Gland, Switzerland. Available at: http://www.iucnredlist.org/info/categories_criteria2001 (accessed 4 April 2007)

IUCN (2006) *The IUCN Species Survival Commission 2006 Red List of Threatened Species*, Gland, Switzerland. Available at: http://www.redlist.org (accessed 27 March 2007)

Kell, S.P., Knüpffer, H., Jury, S.L., Maxted, N. and Ford-Lloyd, B.V. (2005a) *Catalogue of Crop Wild Relatives for Europe and the Mediterranean*. University of Birmingham, Birmingham, UK. Available online via the Crop Wild Relative Information System (CWRIS – http://cwris.ecpgr.org/) and on CD-ROM.

Kell, S.P., Heywood, V. and Maxted, N. (2005b) Towards a global strategy for crop wild relative conservation and use. *Crop Wild Relative* 5, 11.

Loope, L.L. and Mueller-Dombois, D. (1989) Characteristics of invaded islands, with special reference to Hawaii. In: Drake, J.A., Mooney, H.A., di Castri, F., Groves, R.H., Kruger, F.J., Rejmánek, M. and Williamson, M. (eds) *Biological Invasions a Global Perspective*. Wiley, Chichester, UK, pp. 257–280.

Maxted, N., Ford-Lloyd, B.V., Jury, S.L., Kell, S.P. and Scholten, M.A. (2006) Towards a definition of a crop wild relative. *Biodiversity and Conservation* 15(8), 2673–2685.

Maxted, N., Scholten, M.A., Codd, R. and Ford-Lloyd, B.V. (in press) Creation and use of a national inventory of crop wild relatives. *Biological Conservation*.

Mitteau, M. and Soupizet, F. (2000) Preparation of a preliminary list of priority target species for *in situ* conservation in Europe. In: Laliberté, B., Maggioni, L., Maxted, N. and Negri, V. (compilers). *ECP/GR* In situ *and On-farm Conservation Network Report of a Task Force on Wild Species Conservation in Genetic Reserves and a Task Force on On-farm Conservation and Management: Joint meeting*, 18–20 May 2000, Isola Polvese, Italy, pp. 32–42.

Oldfield, S., Lusty, C. and MacKinven, A. (1998) *The World List of Threatened Trees*. World Conservation Press, Cambridge.

PGR Forum (2003–2005) *European Crop Wild Relative Diversity Assessment and Conservation Forum*. University of Birmingham, Birmingham, UK. Available at: http://www.pgrforum.org/ (accessed 3 April 2007)

PGR Forum (2005) *Crop Wild Relative Information System (CWRIS)*. University of Birmingham, Birmingham, UK. Available at: http://cwris.ecpgr.org/ (accessed 3 April 2007)

PlantLife International (no date) *Introduction to Important Plant Areas (IPAs)*, Salisbury, UK. Available at: http://www.plantlife.org.uk/international/plantlife-ipas-about.htm (accessed 23 March 2007)

Royal Botanic Gardens (1999) *Survey of Economic Plants for Arid and Semi-Arid Lands (SEPASAL) database*, Kew, London. Available at: http://www.rbgkew.org.uk/ceb/sepasal/internet/ (accessed 3 April 2007)

Royal Horticultural Society (2006) *The RHS Horticultural Database*, London. Available at: http://www.rhs.org.uk/databases/summary.asp (accessed 5 April 2007)

Schlosser, S., Reichhoff, L. and Hanelt, P. (1991) *Wildpflanzen Mitteleuropas. Nutzung und Schutz*. Deutscher Landwirtschaftsverlag Berlin GmbH, Berlin.

Schofield, E.K. (1989) Effects of introduced plants and animals on island vegetation: examples from the Galápagos Archipelago. *Conservation Biology* 3(3), 227–238.

Schultze-Motel, J. (1966) *Verzeichnis forstlich kultivierter Pflanzenarten* [*Enumeration of cultivated forest plant species*]. Kulturpflanze Beiheft 4.

Simberloff, D. (1995) Why do introduced species appear to devastate islands more than mainland areas? *Pacific Science* 49(1), 87–97.

USDA, ARS, National Genetic Resources Programme (2006) *Germplasm Resources Information Network – (GRIN)* [Online Database]. National Germplasm Resources Laboratory, Beltsville, Maryland. Available at: http://www.ars-grin.gov/cgi-bin/npgs/html/index.pl (accessed 3 April 2007)

Vitousek, P.M. (1992) Effects of alien plants on native ecosystems. In: Stone, C.P., Smith, C.W. and Tunison, J.T. (eds) *Alien Plant Invasions in Native Ecosystems of Hawaii: Management and Research*. Cooperative National Park Resources Studies Unit, University of Hawaii, Honolulu, Hawaii, pp. 29–41.

Walter, K.S. and Gillett, H.J. (eds) (1998) *1997 IUCN Red List of Threatened Plants*. Compiled by the World Conservation Monitoring Centre. IUCN – The World Conservation Union, Gland, Switzerland and Cambridge, UK, pp. 862

WCMC (1995) *Plant Occurrence and Status Scheme: a Standard for Recording the Relationship between a Plant and a Place*. A Taxonomic Databases Working Group (TDWG) Standard. World Conservation Monitoring Centre, Cambridge. Available at: http://www.tdwg.org/poss_standard.html (accessed 5 April 2007)

WCMC and RBG Edinburgh (no date) *Plants of Global Conservation Concern: 1997 IUCN Red List of Threatened Plants*. World Conservation Monitoring Centre and Royal Botanic Gardens, Edinburgh. Available at: http://www.unep-wcmc.org/species/plants/plants-by-taxon.htm (accessed 13 April 2007)

Zeven, A. and Zhukovsky, P. (1975) *Dictionary of Cultivated Plants and Their Centres of Diversity. Excluding Ornamentals, Forest Trees and Lower Plants*. PUDOC, Wageningen, The Netherlands.

6 Establishing Conservation Priorities for Crop Wild Relatives

B. Ford-Lloyd, S.P. Kell and N. Maxted

6.1 Introduction

Conservation planning has often been criticized for lacking a systematic approach (Maxted *et al.*, 1995; Margules and Pressy, 2000). It has as often been based upon subjective personal experiences (Partel *et al.*, 2005), focused upon national requirements without taking into account regional or global conservation needs (Maxted *et al.*, 1995) and rarely aimed at conserving the full range of plant diversity (taxonomic and genetic) that can directly support human life – diversity that is largely represented by crop wild relatives (CWR).

Most conservation planning activity has involved the designation of particular sites, reserves, habitats or even landscapes for conservation followed by sometimes inadequate conservation management. Such conservation planning, aimed at selecting geographical locations, has resulted in BirdLife's Important Bird Areas (Heath and Evans, 2000), Endemic Bird Areas (ICBP, 1992; Stattersfield *et al.*, 1998), Centres of Plant Diversity (WWF and IUCN, 1994), PlantLife's European Important Plant Areas (Palmer and Smart, 2000), Ramsar Sites (Ramsar Convention, 1971), WWF–US Global 200 (Olsen and Dinnerstein, 1998) and Terrestrial Biodiversity Hotspots (Myers *et al.*, 2000). Sites for conservation are, by and large, subjectively selected by experts on the basis of where important 'assemblages' of taxa are thought to coexist: for instance, 'exceptional concentrations of endemic species undergoing exceptional loss of habitat'. Only infrequently are sites selected for 'gene pools of plants of current or potential value to humans' or 'site adapted species' (WWF and IUCN, 1994).

Margules and Pressy (2000) proposed that priority sites should fulfil two main roles; they should sample or represent the biodiversity of each region or locality and they should aim to separate this biodiversity from the processes that threaten its persistence. On this basis, prioritization could be undertaken using the following steps: setting conservation targets, identifying the selection unit, identifying the site selection criteria, habitat classification, setting the selection thresholds, determining what scoring mechanisms or algorithms should be used and investigating the potential options in situations where data are lacking. Of considerable significance is the step involving the identification of the site selection criteria. Sites themselves could be prioritized according to many different selection criteria, but need to be inclusive of priority taxa, however defined.

With finite funding available, the conservationist always has to prioritize actions: deciding which sites and taxa to focus on scarce conservation resources. Establishing the relative 'economic value of the taxa' is an important criterion for dealing with CWR because they are defined by their potential ability to transfer desired traits to socio-economically important species. The more economically important the crop, the greater the priority could be to conserve its wild relatives. Much prioritization is also based upon some estimate of rarity (e.g. IUCN Red List assessment). However, we now know that the number of CWR species in the Euro-Mediterranean region (Euro+Med) is very substantial (more than 25,000 – Kell *et al.*, Chapter 5, this volume). This means that while prioritization based on relative socio-economic value of CWR may be realistic in the short term, regional threat assessment for all 25,000 CWR taxa has not been undertaken, would take a considerable amount of time and is unlikely in anything but the long term. Maxted *et al.* (1997) drew attention to 11 factors considered important when selecting plant genetic resource (PGR) targets. Among the more important of these were current conservation status, potential economic use, threat of genetic erosion, genetic distinctiveness, ecogeographic distribution, biological importance and cultural importance. They envisaged that these prioritizing factors could effectively be used to compare a limited number of species competing for limited conservation resources, but certainly not systematically prioritizing 25,000 Euro+Med CWR species. The problem with trying to prioritize such a large number of species is the lack of universal data sets for the prioritizing factors that could be applied to all species. Without such data sets, effective prioritization of all CWR cannot take place; however, prioritization is required now. This CWR prioritization is discussed in this chapter.

6.2 The CWR Catalogue and the Need for Prioritization

Kell *et al.* (Chapter 5, this volume) describe the methodology for creating the CWR Catalogue for Europe and the Mediterranean (Kell *et al.*, 2005). It had been agreed early in the process that the intention should be to be as inclusive as possible and to include all categories of plants of socio-economic importance: food, fodder or forage, industrial, forestry, ornamental, herbs and spices, medicinal and aromatic. Although even with hindsight this remains the most logical approach at one level, it could be argued that this has created a problem by way of the sheer numbers of CWR that are listed in the Catalogue. However, to have made the decision at the outset to be exclusive, for instance, not to have included ornamental CWR, would have resulted in a somewhat arbitrary *a priori* prioritization, which would not have been to everyone's satisfaction. On a much more positive note, we should be encouraged that there are more than 25,000 CWR species in the Euro+Med region that are important in some way or other for sustaining human life.

6.2.1 Criteria for immediate prioritization

It is obvious that systematic conservation planning for more than 25,000 species is impossible, at least in the short term; therefore, species must be prioritized so that those which require the most urgent attention and conservation planning can be dealt with as soon as possible within a European framework. There is considerable agreement that threat assessment by way of Red Listing is an important means of prioritizing for conservation and this happens regularly at national level. Many European countries have national Red Lists indicating that threat assessment or rarity is widely regarded as an important criterion for prioritization of conservation action. However, in the context of the activities of the PGR Forum project (see Maxted *et al.*, Chapter 1, this volume and available at: http://www.pgrforum.org/) and this book, we are concerned with the conservation specifically of the wild relatives of crops because of their actual or potential value to human well-being and the provision of food security.

Some form of threat assessment on the one hand and an assessment of socio-economic value on the other might therefore be a good starting point for the prioritization of CWR species for active conservation. But how realistic is this for the 25,000 CWR taxa in the Euro+Med region? If those governments, organizations and individuals involved in conservation planning for PGR in the region wish to plan effectively for conservation, a prioritized list of taxa needs to be available as soon as possible; this in turn means that in order to prioritize the CWR list, complete data are needed now, rather than sometime in the future. Are such data readily available to undertake a complete and immediate threat assessment and socio-economic assessment? For all 25,000 CWR species, the answer could be negative unless we are able to utilize some simple proxy measures. As part of the work of PGR Forum we have been able, in part, to achieve this by taking the numbers of geographical units in which a taxon has been recorded as a proxy of threat assessment, and we have also taken steps towards a second level of prioritization using simple socio-economic criteria. The first has already been achieved, and the second can be accomplished very quickly.

6.2.2 Prioritization using geographical unit occurrence as a proxy for threat

By analysing the data contained in Euro+Med PlantBase (available at: http://www.euromed.org.uk/), which forms the taxonomic core of the CWR Catalogue, it has been possible to determine in how many geographical units each taxon has been recorded. The argument for using this as a proxy for threat is that the more geographic units in which a taxon occurs, the less likely it is to be under threat in all of the units simultaneously and in need of urgent conservation action; generally speaking, if a taxon has been recorded in several countries in the Euro+Med region, it is unlikely to be in danger overall. Taxa which are endemic to a particular geographic unit or country may or may not actually be threatened at national level, but should receive conservation attention to ensure that their future is secured within the region.

What are the advantages and disadvantages of using this proxy criterion? The advantage is that this is a simple, but robust method for prioritization. It has already proved to be simple (see Fig. 6.1), and a prioritized list of CWR which represents around 25% of the total European and the Mediterranean CWR list can be identified if three geographical units are used as a cut-off point. This would then allow us to rather more easily examine in more detail what conservation action is required for these 6274 species in the region. Within this still rather large list of species, the greatest and most urgent attention could be given to those species that only occur in a single geographic unit (endemics), of which there are at least 2746 or 10.7% of the species in the Catalogue (see Kell *et al*., Chapter 5, this volume, for comment on the conservative nature of this estimate).

While such a simple proxy indicator might be intuitively attractive, is there a scientific justification for using it? A key survey by Hamrick and Godt (1990) found that endemic plant species in general possess less than half the genetic diversity of widespread species. Even considering this at a superficial level, it means that in general, endemic species are more vulnerable to loss of genetic diversity below a threshold of population sustainability than widespread species. It could be argued that if they have not already passed through a bottleneck, they are much more likely to do so in the immediate future than widespread

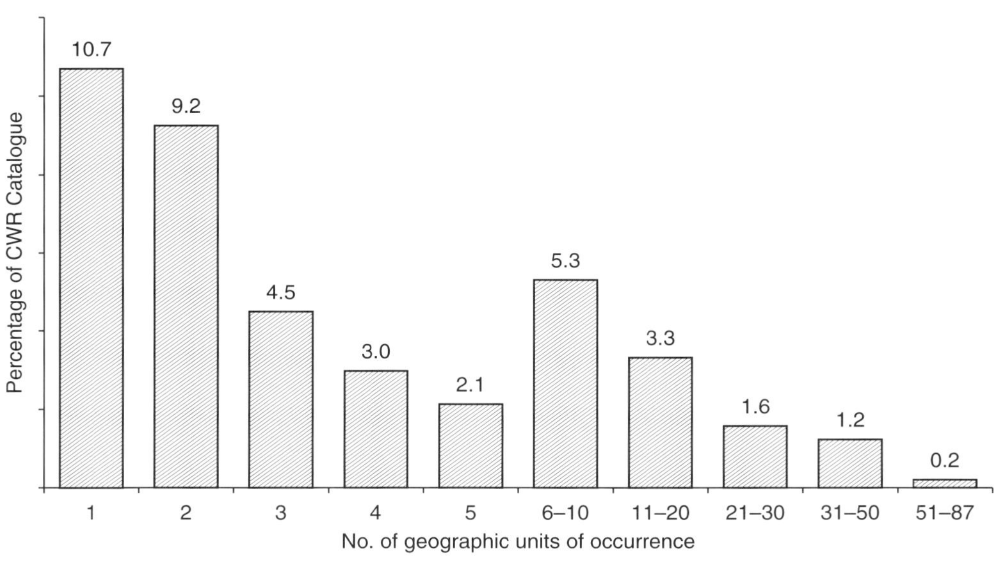

Fig. 6.1. Numbers of European crop wild relatives found in restricted numbers of geographic units.

species. Since they are genetically more vulnerable we should prioritize them for conservation. The same argument applies to narrowly distributed species, which only possess 67% of the genetic diversity of widely distributed species (Hamrick and Godt, 1990). The pattern is similar if genetic diversity is analysed at the population level – populations of endemic species contain only 40% of the diversity of widespread species, and narrowly distributed species contain 66% (Hamrick and Godt, 1990). Recent predictions of the effects of climate change on plant diversity in Europe (Thuiller *et al.*, 2005) also indicate the need to focus on endemic or narrowly distributed species. Almost by definition, these are the species least likely to be able to counter global warming by migration. Species that are currently not widely distributed will be those that are restricted to the environmental zones seen to be at greatest threat from global warming. None of this is to argue that widespread species or populations are unimportant genetically, but that without data supporting the contrary, there is good reason not to prioritize them as highly in terms of conservation action.

There are assumptions that have to be made, which may incur disadvantages in this approach. The unequal sizes of the geographical units will undoubtedly affect the reliability of this proxy indicator (Maxted *et al.*, 1997). Taxa that are recorded as present in two small island units will be under greater potential threat than taxa that occur in two large countries. In contrast, it is possible that some taxa, which are widely distributed among several or many geographical units – and hence would be deemed not to require conservation action – would actually be under threat from global warming, the same specific environmental change taking place across a wide geographical area. Similarly, the same socio-economic pressure could exist because of Europe-wide policies and legislations being implemented across several or many countries bringing a similar threat to certain taxa. For example, a littoral species of the Mediterranean region is likely to be threatened throughout by tourist development. The likelihood of these scenarios occurring cannot easily be assessed, but would represent a problem for this proxy indicator of threat. The reverse situation might also be true: a single country endemic which would be targeted for conservation action using our proxy indicator might not actually require such action, as being very widely distributed throughout that country. However, this situation would be relatively easily detected and conservation priorities would be adjusted accordingly, as would be the case for single country endemics within the Euro-Mediterranean region that occurred widely in countries outside the region.

6.2.3 Prioritization with simple use criteria for socio-economic value

Our proposal for a simple prioritization procedure which can be applied to the CWR Catalogue immediately also includes some form of assessment of potential or actual use. We are focused here on a particular sector of biodiversity, identified specifically within the CBD 2010 Biodiversity Target (CBD, 2002) because of their value to humankind. It is therefore unrealistic not to attempt to prioritize the CWR taxa in some way according to this value. Various economic value criteria could be developed, but this information is not universally available for all crops. However, based on economic value, there would be an implied link between the crop and its wild relatives; thus, the wild relatives of crops with the highest economic value would have higher priority than those of crops of lower economic value. In this case, proxy criteria based upon various FAO crop production statistics (FAO, 2006) might serve. However, this is impractical in the short term, not least, because these statistics do not suggest one single measurement of production that could be used. They cannot be easily applied to the 25,000 CWR species without direct access to the data. The 'value' of CWR taxa could also be assessed by the numbers of citations occurring for each in the CAB Abstracts database (available at: http://www.cabi.org/data page.asp?iDocID = 165). Again, this is not easily achieved and the result might not be acceptable to everyone as there would be clear anomalies generated by taxa which are the focus of the 'omics' technologies. Gaining broad acceptability and achieving a consensus are likely to be extremely difficult whichever way this particular prioritization is performed. During the lifetime of PGR Forum we undertook a very small, unrepresentative survey (n = 30) of PGR experts based upon the use categories of world crops. PGR Forum participants were asked to rank the eight categories of socio-economic plants in order of priority; Table 6.1 shows not only the ranking, but also the range of rankings, which clearly indicates the diversity of opinion that exists as to the so-called value of each crop group (with the exception of the food crop category, voted top priority by everyone). Such a ranking is relatively straightforward and could be undertaken with the CWR Catalogue by applying the rank scores for each category to the genera within the list. All taxa falling within a crop genus would receive the ranking in accordance with that particular crop. When the uses of a particular crop fall into more than one use category, then it would be necessary to apply the highest use ranking. So the issue is not whether such a prioritization can be achieved quickly and easily, but whether any ranking system and set of criteria would be acceptable to a majority. Although this question will be difficult to answer, it is certainly possible to prioritize the CWR taxa in terms of use in the first instance as a cock-shy and to very easily modify it subsequently by changing the rank values.

Table 6.1. Result of expert 'straw pole' on importance of use categories in prioritizing crop wild relative taxa.

Use category	Score	Range
Food crop	8	8
Fodder or forage	7	7–2
Industrial (e.g. fibre, oil, sugar, tobacco)	6	7–3
Forestry	5	7–2
Medicinal	4	7–1
Ornamental	3	7–1
Minor or underutilized crops	2	6–1
Spice or condiment	1	5–1

6.3 Further Considerations

One issue is related to the order in which these simple prioritizations are applied. It is much easier to apply the proxy threat indicator (number of geographic unit occurrences) first, simply because it is easier to determine where the cut-off should lie and then to prioritize the taxa above the cut-off according to use. This results in a substantially reduced list of taxa ordered by use value. There is no need to apply a cut-off for use value – the list can simply be assessed and used as and when necessary.

Another benefit of prioritizing on geographic unit occurrence first is that this could very easily be followed by sorting according to different uses (food, forage, medicinal, etc.), providing separate prioritizations for different 'users'. This possibility, together with that of being able to provide national lists prioritized according to a European regional criterion, is very attractive. However, subdividing CWR into different components according to use does deflect from the need to be able to refer within the Euro+Med region to a prioritized list of all CWR taxa, independent of national interests or interests of specific specialist groups.

Any prioritized list of CWR taxa will need to be scrutinized for serious discrepancies and adjustments made accordingly. For instance, it would be unacceptable if the list did not include all of the CWR occurring in the region that fall within the genera included in the FAO International Treaty on Plant Genetic Resources for Food and Agriculture (FAO, 2001). The same argument would apply to the CWR taxa listed in the EC Habitats Directive (European Council, 1992), the CITES Appendices (CITES Secretariat, 1993) and the IUCN Red List of Threatened Species (IUCN, 2006).

Where does a simple prioritization such as this take us? The next important step would be to ensure that IUCN Red Listing was undertaken for all the prioritized taxa, starting with the most highly prioritized. This should be combined with more detailed use of climate change scenarios for Europe, where, even at the present time, it is possible to predict which species, environmental and geographic zones will be most threatened in the coming decades (Thuiller *et al.*, 2005). Information on the ease with which taxa on the list can be utilized could be added. This would take the form of acquiring information based upon the Gene Pool Concept (Harlan and de Wet, 1971) or the Taxon Group Concept (Maxted *et al.*, 2006). Information on whether these prioritized taxa occur within existing protected areas could be sought. All this information would be important in guiding conservation planning both *in situ* and *ex situ*. Identification of new sites for genetic reserves or gaps in existing *ex situ* collections could be substantially rationalized to ensure that the most important genetic resources in the Euro+Med region were being effectively conserved.

However, perhaps it is appropriate to stress that although there is an overriding need to establish conservation priorities for taxa, there is no one ideal way of prioritizing any taxa, let alone CWR. Maxted *et al.* (1997) conclude that the conservation priority ascribed to a particular taxon will be influenced by the mandate and priorities of the agency commissioning the conservation, because they will give different weights to the kinds of criteria discussed earlier. For example, if the CWR are being prioritized by staff of the ministries of agriculture, forestry and the environment, the species priorities are likely to be quite specific, as will those of ecologists, park managers, plant breeders, population geneticists, agriculturalists and taxonomists. It should also be noted that the discussion of prioritization given earlier was generated by a pan-European project, PGR Forum, so the discussion has focused specifically on how to prioritize at the continental level, given the starting point of a comprehensive CWR catalogue as is available for the Euro+Med region, but the criteria appropriate for establishing priorities at the subregional or national levels may differ from those at the continental level. In some senses, it is easier to prioritize at national level because many of the data sets that might be useful for prioritizing are generated nationally and the person doing the prioritizing will know which data sets are available and whether they can be accessed. The problem with continental or even world data sets is in obtaining globally comparable data or even knowing the full range of data sets available.

In addition, as suggested by Maxted *et al.* (1997), there is a need for a multiple-tier approach to prioritizing. They provide the example of occasionally cultivated *Vicia bithynica* (L.) L. (*Leguminosae*), which is native throughout central and southern Europe and the Mediterranean Basin. The extreme north-western edge of its distribution is the south coast of Britain, where the species is found at a few locations, all subject to increased levels of tourism and natural coastal erosion. These populations are severely threatened and, in the near future, the species may even become extinct in the United Kingdom. However, it is thriving in its centre of diversity and is clearly not threatened at an international level. So, at European level, prioritizing this species would be unlikely to justify active conservation, but in a floristically depauperate country like the United Kingdom, active conservation of this species is important because it is a nationally threatened species.

6.4 Conclusions

1. There is a need to prioritize European CWR taxa for conservation at the continental level as soon as possible.
2. Immediate prioritization can be achieved using two simple criteria for which complete data sets are available.
3. Prioritization according to the number of geographic units in which a taxon is recorded (using the Euro+Med PlantBase data) should be the first criterion, and this has already been applied to the CWR Catalogue at species level. There is good information on the distribution of genetic diversity as well as information on the effects of climate change to back this up.
4. A cut-off of fewer than four geographical units leads to an initial prioritized list, which is approximately 25% of the full CWR Catalogue taxa, but within this, most emphasis could be given to endemics that make up just more than 10% of the CWR Catalogue.
5. A second simple 'use' prioritization, where CWR related to food crops are highest priority, will lead to a second list ordered in terms of priority.

6. A check needs to be made to determine whether the list is inclusive of taxa associated with the genera listed in the FAO International Treaty, European Habitats Directive and the CITES Appendices.
7. The reduced, prioritized list will enable more effective Europe-wide conservation planning to be based upon the most deserving or important CWR species in the Euro+Med region. Subsequent activities should include:

- IUCN Red Listing;
- Assessment of ease of use of applying the Gene Pool Concept or Taxon Group Concept for further prioritization;
- Genetic conservation gap analysis of the occurrence of CWR in existing protected areas and gene banks;
- Genetic Reserve planning for *in situ* conservation and germplasm collection for *ex situ* conservation.

8. Prioritization needs to be applied at both regional and national levels.

Acknowledgements

The concepts discussed in this chapter were stimulated by PGR Forum (the European crop wild relative diversity assessment and conservation forum – EVK2-2001-00192 – available at: http://www.pgrforum.org/), funded by the EC Fifth Framework Programme for Energy, Environment and Sustainable Development. We are grateful to Marianne Mitchell for assistance with the data analysis.

References

CBD (2002) *2010 Biodiversity Target*. Secretariat of the Convention on Biological Diversity, Montreal. Available at: http://www.biodiv.org/2010-target/default.aspx (accessed 3 April 2007)

CITES Secretariat (1993) *Convention on International Trade in Endangered Species of Wild Fauna and Flora*, Washington, DC. Available at: http://www.cites.org/ (accessed 27 March 2007)

European Council (1992) *Directive 92/43/EEC of 21 May 1992 on the Conservation of Natural Habitats and of Wild Fauna and Flora*. Habitats Directive. Available at: http://europa.eu.int/comm/environment/nature/nature_conservation/eu_nature_legislation/habitats_directive/index_en.htm) (accessed 27 March 2007)

FAO (2001) *International Treaty on Plant Genetic Resources for Food and Agriculture*. Food and Agriculture Organization of the United Nations, Rome, Italy. Available at: http://www.fao.org/ag/cgrfa/itpgr.htm (accessed 03 April 2007).

FAO (2006) *FAOSTAT*. Available at: http://faostat.fao.org/ (accessed 27 March 2007)

Hamrick, J.L. and Godt, M.J.W. (1990) Allozyme diversity in plant species. In: Brown, A.H.D., Clegg, M.T., Kahler, A.L. and Weir, B.S. (eds) *Plant Population Genetics, Breeding, and Genetic Resources*. Sinauer Associates, Massachusetts, pp. 43–63.

Harlan, J. and de Wet, J. (1971) Towards a rational classification of cultivated plants. *Taxon* 20, 509–517.

Heath, M.F. and Evans, M.I. (2000) *Important Bird Areas in Europe: Priority Sites for Conservation*. 2 Vols. Birdlife International, Cambridge, UK.

ICBP (1992) *Putting Biodiversity on the Map: Priority Areas for Global Conservation*. International Council for Bird Preservation, Cambridge, UK.

IUCN (2006) *IUCN Red List of Threatened Species*. Available at: http://www.iucnredlist.org/ (accessed 27 March 2007)

Kell, S.P., Knüpffer, H., Jury, S.L., Maxted, N. and Ford-Lloyd, B.V. (2005) *Catalogue of Crop Wild Relatives for Europe and the Mediterranean*. Available online via the Crop Wild Relative Information System (CWRIS – http://cwris.ecpgr.org/) and on CD-ROM. University of Birmingham, Birmingham, UK.

Margules, C.R. and Pressy, R.L. (2000) Systematic conservation planning. *Nature* 405, 243–253.

Maxted, N., van Slageren, M.W. and Rihan, J. (1995) Ecogeographic surveys. In: Guarino, L., Ramanatha Rao, V. and Reid, R. (eds) *Collecting Plant Genetic Diversity: Technical Guidelines*. CAB International, Wallingford, UK, pp. 255–286.

Maxted, N., Hawkes, J.G., Guarino, L. and Sawkins, M. (1997) The selection of taxa for plant genetic conservation. *Genetic Resources and Crop Evolution* 44, 337–348.

Maxted, N., Ford-Lloyd, B.V., Jury, S.L., Kell, S.P. and Scholten, M.A. (2006) Towards a definition of a crop wild relative. *Biodiversity and Conservation* 15(8), 2673–2685.

Myers, N., Mittermier, R.A., Mittermier, C.D., da Fonseca, G.A.B. and Kent, J. (2000) Biodiversity hotspots for conservation priorities. *Nature* 403, 853–858.

Olson, D. and Dinnerstein, E. (1998) The Global 200: a representation approach to conserving the earth's most biologically valuable ecoregions. *Conservation Biology* 12, 502–515.

Palmer, M. and Smart, J. (2000) *Guidelines to the Selection of Important Plant Areas in Europe*. PlantLife, London.

Partel, M., Kalamees, R., Reier, U., Tuvi, E.-L., Roosaluste, E., Vellak, A. and Zobel, M. (2005) Grouping and prioritization of vascular plant species for conservation: combining natural rarity and management need. *Biological Conservation* 123, 271–278.

Ramsar Convention (1971) *The Ramsar Convention on Wetlands*. Ramsar Convention Secretariat, Gland, Switzerland.

Stattersfield, A.J., Crosby, M.J., Long, A.J. and Wedge, D.C. (1998) *Endemic Birds Areas of the World. Priorities for Biodiversity Conservation*. BirdLife Conservation Series No. 7. BirdLife International, Cambridge, UK.

Thuiller, W., Lavorel, S., Araujo, M.B., Sykes, M.T. and Prentice, I.C. (2005) Climate change threats to plant diversity in Europe. *Proceedings of the National Academy of Sciences of the United States of America* 102(23), 8245–8250.

WWF and IUCN (1994) *Centres of Plant Diversity. A guide and strategy for their conservation*. 3 Vols. IUCN Publications Unit, Cambridge, UK.

7 Creation of a National Crop Wild Relative Strategy: a Case Study for the United Kingdom

M. Scholten, N. Maxted, S.P. Kell and B.V. Ford-Lloyd

7.1 Introduction

Crop wild relatives (CWR) are identified as a critical component of plant biodiversity required for future food security and environmental sustainability in the 21st century (Maxted *et al.*, 1997a; Prescott-Allen and Prescott Allen, 1983; Hoyt, 1988; Meilleur and Hodgkin, 2004; Stolton *et al.*, 2006). However, like other wild species, CWR are increasingly threatened by deleterious human actions that negatively impact on both the taxonomic (species level) and genetic diversity of these important biological resources (Maxted *et al.*, 1997a; Hawkes *et al.*, 2000). More concerted conservation action is required if future exploitation options are not to be seriously restricted (CBD, 1992; FAO, 2001). CWR are thus identified as a critical and seriously threatened global resource.

Within the Global Strategy of Plant Conservation (CBD, 2002), the CBD established a series of targets to be achieved by 2010 as a means of invigorating global biodiversity conservation action. The United Kingdom progress towards achieving these targets for plants has recently been reviewed by Plantlife International (2004a) and is further reviewed in relation to United Kingdom CWR diversity in Table 7.1.

Considerable progress has been achieved in United Kingdom plant conservation, but it is recognized that CWR conservation needs to be integrated into national conservation programmes as a priority. Table 7.1 shows considerable progress for CWR with regard to CBD 2010 Targets 1, 2 and 8; there is a checklist of United Kingdom CWR, International Union for the Conservation of Nature (IUCN) threat assessment has been enacted for the complete national flora (including CWR species) and 95% of the national flora has now been conserved *ex situ* by the Millennium Seed Bank Project at the Royal Botanical Gardens, Kew, United Kingdom. However, aspirations to meet the 2010 targets for CWR conservation are most explicitly addressed under Target 9. United Kingdom plant genetic resource policy has been reviewed (DEFRA, 2001), leading to the creation of an

Table 7.1. Goals and progress towards the CBD 2010 Targets for United Kingdom CWR species.

Global strategy for plant conservation target	Achievement	Status for CWR
Target 1: working list of species	Checklist of all 2300 vascular plant species	CWR checklist complete
Target 2: threat assessment	100% of British flora to assess	Priority CWR species assessed
Target 3: develop conservation methods	77 United Kingdom BAP species	47 conservation action plans for priority CWR species
Target 5: 50% of important plant areas protected	10.5% of total land area will fall within 7191 reserves	Six specific CWR IPAs identified
Target 6: 30% of plants within productive land conserved	10% productive land under agroenvironmental schemes	No active inclusion of CWR species
Target 7: 60% of threatened species conserved *in situ*	77 out of 220 with action plans, many more in protected areas	62% of 47 priority CWR species included in the six CWR IPA sites identified
Target 8: 80% of threatened species *ex situ*	94% of United Kingdom flora	All priority CWR species with at least one *ex situ* sample
Target 9: 70% of genetic diversity of crops (landraces and CWR) conserved	Started national inventory	Not attempted yet

inventory of United Kingdom CWR species (Maxted *et al.*, in press). However, although we may have at least one population of each CWR species conserved *ex situ*, it could not be claimed that this is a genetically representative sample of within-species diversity, let alone 70% of that diversity. Therefore, initial steps have been taken to address this target for United Kingdom CWR species by developing and applying a national CWR conservation strategy for the United Kingdom.

The steps involved in the generation of a national CWR strategy have recently been explored in a methodology for creation of a national CWR conservation strategy proposed by Maxted *et al.* (in preparation b) (Fig. 7.1). The objective of this chapter is to illustrate how the proposed methodology might be applied as a means of developing a national CWR conservation strategy using the United Kingdom as an exemplar.

7.2 Development of a United Kingdom CWR Conservation Strategy

7.2.1 Constructing the National Inventory of CWR

Both the CBD and IT place national obligations on sovereign states to conserve their plant genetic resources and as such the development of the National Inventory for the United Kingdom was initiated by the Department of Food, Environment and

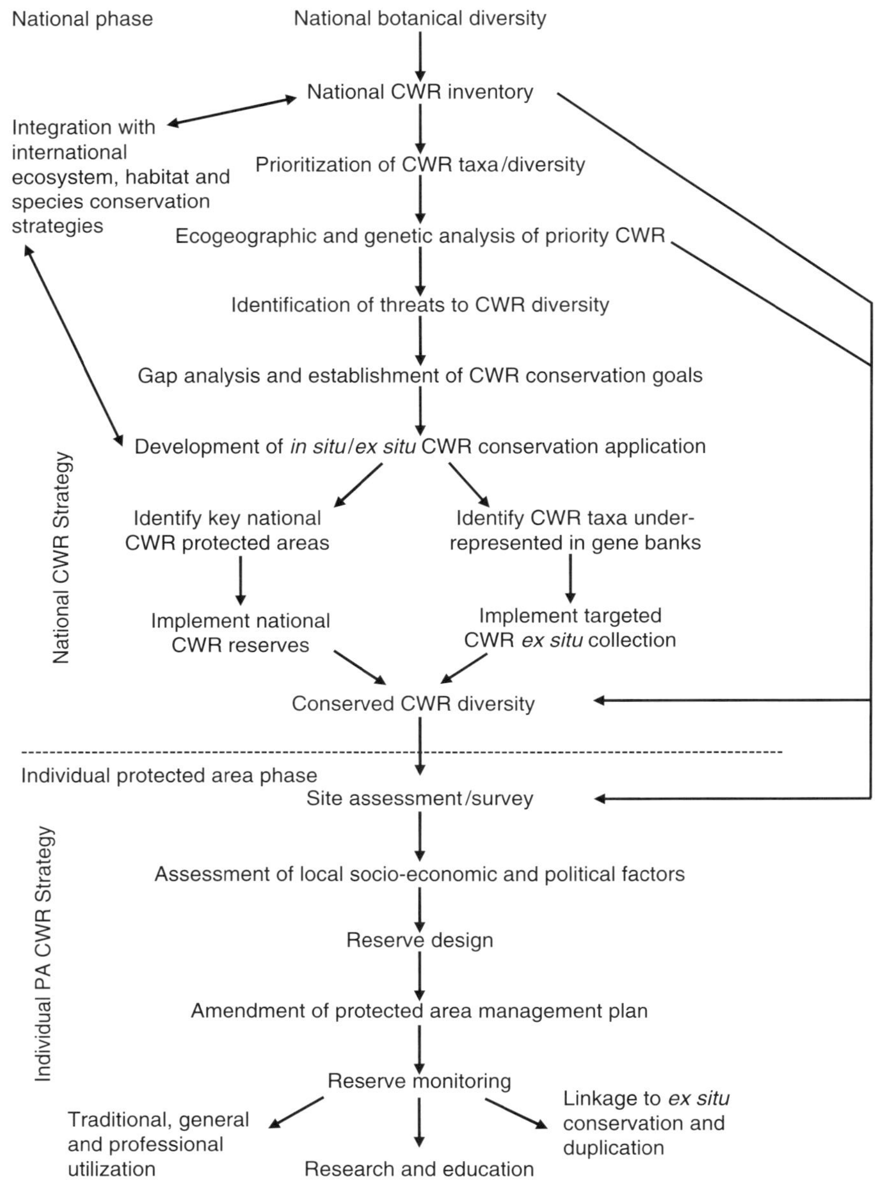

Fig. 7.1. Model for the development of National CWR conservation strategies. (From Maxted *et al.*, in preparation b.)

Rural Affairs (DEFRA) which has the responsibility for food and agriculture. A necessary first step is the establishment of national CWR inventories (Maxted *et al.*, 1997a). The United Kingdom Country Report for the FAO Technical Conference in Leipzig listed 60 native wild progenitors of current or potential important commercial crops for food and agricultural as being present in the United Kingdom

(MAFF, 1995). However, it was recognized that a further more systematic assessment of national genetic resources for food and agriculture was required and this was undertaken by the DEFRA (2002). Forestry resources were assessed separately as part of the Global Forestry Resources Assessment (Forestry Commission, 2005). One of the key recommendations of the DEFRA assessment (DEFRA, 2001) was the urgent need to create a national inventory of United Kingdom plant genetic resources for food and agriculture, as a step towards developing a coherent conservation strategy as part of the United Kingdom commitment to the CBD. DEFRA commissioned this research in 2003/04. As there was already a baseline available for *ex situ* PGRFA, through the United Kingdom Focal Point for the European PGRFA Inventory (EURISCO), it was agreed to focus on preparing inventories of United Kingdom landraces and CWR species. For the United Kingdom CWR inventory, three sources of data were available: (i) national ethnobotanical studies on wild genetic resources (Mabey, 1977, 1996, 2003); (ii) lists of commercial crop species cultivated in the United Kingdom (DEFRA, 2003a); and (iii) European regional lists of CWR (Zeven and Zhukovsky, 1975; Heywood and Zohary, 1995) from which the United Kingdom CWR taxa could be extracted. The latter focused mainly on major food crops and listed about 600 United Kingdom CWR species. The most comprehensive catalogue to date is the 'PGR Forum CWR Catalogue for Europe and the Mediterranean' (Kell *et al.*, 2005), which includes agricultural, horticultural, ornamental, forestry, medicinal and aromatic plants.

Broadly speaking, CWR are defined here as the non-crop taxa within the same genus as a crop (Maxted *et al.*, 2006). Thus, the CWR Catalogue for Europe and the Mediterranean (Kell *et al.*, 2005) was generated by a process of data harmonization and cross-checking between existing databases. A list of genera containing agricultural and horticultural crops (Mansfeld's World Database of Agricultural and Horticultural Crops – Hanelt and IPK Gatersleben, 2001; IPK Gatersleben, 2003), forest trees (*Enumeration of Cultivated Forest Plant Species* – Schultze-Motel, 1966), ornamental species (Community Plant Variety Office) and medicinal and aromatic plants (Medicinal and Aromatic Plant Resources of the World – MAPROW) was matched against the taxa found in those genera within Euro+Med PlantBase (2005) to produce the Catalogue (see Kell *et al.*, Chapter 5, this volume). Using this Euro-Mediterranean catalogue, it is possible to download the CWR species for each phytogeographical unit. In this way, the United Kingdom national inventory of CWR was created. This resulted in a catalogue of 2644 CWR species, microspecies and subspecies in Britain, the majority of which fell into more than one use category: 85% of which are wild relatives of medicinal and aromatic plants, 82% of agricultural and horticultural crops, 15% of forestry plants and 30% of ornamentals out of flora of over 4100 plant taxa listed in the *New Flora of the British Isles* (Stace, 1997). In terms of species there are 2300 plant species of the United Kingdom[1] and 78% of them can be considered CWR.

[1] The number of native species in the United Kingdom botanical literature varies between 1400 and 2300, reflecting different treatments of critical species and/or native status. This 'maze of species numbers' (Good, 1964) reflects different taxonomic treatments of the critical species as well as a high level of botanical activity.

Once the initial United Kingdom CWR inventory was generated, two modifications were required in terms of geographical unit and nomenclature. The *CWR Catalogue for Europe and the Mediterranean* like the Euro+Med PlantBase flora from which it is derived divides the United Kingdom into two phytogeographical units, Great Britain and Ireland. To transform these into one political unit, the United Kingdom, was straightforward as Northern Ireland does not contain any endemics or plants that are not present in Great Britain. So the checklist of CWR for Great Britain would also be valid for the United Kingdom. The second adaptation required harmonization of the nomenclature used in Euro+Med PlantBase, and therefore the CWR Catalogue for Europe and the Mediterranean, with that used by Stace in the *New Flora of the British Isles* (Stace, 1997). Rather than giving one nomenclatural system precedence over the other, both (European and the United Kingdom) systems were included in the final database, with the exception of a few critical genera. The latter includes 11 genera with 830 species, more than 300 of which are *Rubus*, 280 are *Hieracium* and 229 are *Taraxacum* spp. (Rich, 1999). The majority of these species are native, with the exception of some *Taraxacum*. Harmonizing these critical taxa between the European and the United Kingdom catalogues proved problematic as, for example, with *Rubus*, where only 24% of microspecies names showed congruence between the European and the United Kingdom nomenclature (Edees and Newton, 1988).

The United Kingdom CWR Inventory is available at: http://grfa.org.uk/search/plants/index.html. The generation of the Inventory was not an end in itself, but was to act as the taxonomic backbone of the United Kingdom CWR database containing broader baseline conservation data; usage, occurrences and trends, legal status, IUCN threat assessment status and conservation action plans were collated (Table 7.2). The information on usage was derived from the world economic botany references (Wiersema and Leon, 1999; Hanelt and IPK Gatersleben, 2001) and British studies (Mabey, 1996, 2003). The TDWG economic botany standard for usage (Cook, 1995) was applied throughout. Species status, occur-

Table 7.2. United Kingdom CWR inventory structure. (Available at: http://grfa.org.uk/search/plants/index.html.)

Field name	Description	Data source
EURGENUS/EURSPECIES/EURSUBTAXA	European taxonomic names	Kell *et al.* (2005)
FAMILY	Family name	Stace (1997)
GENUS/SPECIES/SUBTAXA	United Kingdom taxonomic names	Stace (1997)
SPAUTHOR	Full citation	Stace (1997)
STATUS	Status	Preston *et al.* (2002)
USECODE	TDWG usage code	Cook (1995)
OCCUR	Number of 10×10 km^2 grid squares	Preston *et al.* (2002)
TREND	Change index	Preston *et al.* (2002)
COMMONNAME	English common name	Stace (1997)
IUCNSTATUS	Threat assessment	Cheffings (2004)
LEGAL	WCA-schedule 8 listed	www.jncc.org.uk
CONSER	Existence of biodiversity action plan	www.ukbap.org.uk

Table 7.3. Overview of data sources and numbers on plant genetic resources of the United Kingdom.

Category of PGR	Taxa	References
National studies		
United Kingdom medicinal plants	440	Allan and Hatfield (2004)
Scottish useful plants (*Flora celtica*)	398	Milliken and Bridgewater (2004)
Wild and traditionally harvested plants in England and Wales	140	Sanderson and Prendergast (2002)
United Kingdom forestry resources	66	Forestry Commission (2005)
Regional lists		
Species occurring in the United Kingdom and listed in the European Catalogue	126	Heywood and Zohary (1994)
Species occurring in the United Kingdom and listed as Crops and CWR of Euro-Siberian and Mediterranean centres of diversity	152	Zeven and Zhukovsky (1973)
United Kingdom CWR (as part of the CWR Catalogue for Europe and the Mediterranean)	2644	Kell *et al.* (2005)
United Kingdom flora		
Native and introduced taxa of the British Isles	>4100	Stace (1997)

rences and trends were taken from *The New Atlas of the British and Irish Flora* (Preston *et al.*, 2002). Common names were taken from the *New Flora of the British Isles* (Stace, 1997). Threat assessment or IUCN threat categories was obtained from Cheffings (2004) and Cheffings *et al.* (2005), while legal status and conservation assessment were based on the Joint Nature Conservation Committee, the official statutory organ for nature conservation in the United Kingdom.

Allied to the United Kingdom CWR Catalogue in the context of GSPC Target 13 (decline of plant genetic resources and associated indigenous knowledge), it is worth drawing attention to: (i) the list of commercially used wild plants and traditionally managed plants of contemporary direct use (Sanderson and Prendergast, 2002; Prendergast and Sanderson, 2004), which includes medicinal plants; (ii) current commercial and ethnobotanical plants of Scotland (Milliken and Bridgewater, 2001, 2004); and (iii) the 440 ethnobotanical medicinal plants listed by Allan and Hatfield (2004). In addition, at least 82 further medicinal plant species are traded in the United Kingdom although not necessarily of United Kingdom origin (Dennis, 1997), because as noted by Plantlife International (2002a) few firms cultivate the plants they market. An overview of these surveys is given in Table 7.3.

7.2.2 Assigning priorities for conservation

Whatever method is used to create a national checklist of CWR, not all listed taxa are likely to have the same actual or potential value; nor are all taxa equally in need of active conservation. With the inclusion of 2644 CWR taxa, the United Kingdom CWR Catalogue represents a comprehensive and complete list of gene sources for all potential plant genetic resources of United Kingdom importance. However, the generation of such an extensive list does mean that the development of the national conservation strategy will include assigning conservation priorities to taxa. Prioritizing plant taxa commonly involves the estimation of endemism, relative rarity or threat (Department of the Environment, 1996; Maxted *et al.*, 1997b; Rich *et al.*, 2005), applied in either a strictly national or wider European or world context (Ford-Lloyd *et al.*, Chapter 6, this volume). However, in the national context, Walters (1984) compared British with European conservation priorities and found that the rare British species *Cephalanthera rubra* (L.) Rich. warranted conservation priority in the United Kingdom, but as the species occurs throughout Europe it would not be a priority species from a European perspective. On the other hand, *Hyacinthoides nonscripta* (L.) Chouard ex. Rothm. is widespread in the British Isles, but a regional endemic on the continent and is therefore a European but not United Kingdom conservation priority (Walters, 1984). Relative national abundance is also regarded as a prioritizing factor and demands special status (international responsibility) if 25–40% of the world population of a taxon occurs within the United Kingdom. Among the 113 taxa for which the United Kingdom has international responsibility are included the following CWR: *Brassica oleracea* L. (wild cabbage), the endemic flax relative *Linum perenne* subsp. *anglicum* (Mill.) Ockendon, 14 *Sorbus* taxa and no less than 29 *Hieracium* spp. (Cheffings *et al.*, 2005).

Prioritizing on economic or use value

When prioritizing on what is perhaps the most important criterion specifically for CWR conservation, namely economic value for food and agriculture, a comprehensive list of crops in current cultivation is required; no such list was available for the United Kingdom. Therefore, the first step in prioritization for economic value was to create such a checklist of United Kingdom crop species or taxa used in the United Kingdom food and agriculture. DEFRA's Basic Horticultural Statistics (DEFRA, 2003a)[2] and *United Kingdom 2004, The Official Yearbook of the United Kingdom of Great Britain and Northern Ireland* (Anonymous, 2004) were used to produce a list of 30 crop species, with a further 25 forage or fodder crop species included in the Seed Traders Annual Returns (DEFRA, 2003b). Potential value to plant breeding as evidenced by inclusion on the National List of Agricultural and Horticultural varieties (DEFRA and PVO, 2005) and the Common Catalogue of Agricultural and Horticultural Crops (EU, 2005a,b) was the final defining element of perceived economic value. Ornamentals were considered by definition not to be associated with food and agriculture and so were excluded. It is worth noting that a drawback of basing prioritization on official economic statistics is that smaller herbs and minor crops are often collated into a catch-all 'other' category and so in terms of official statistics do not appear as distinct crop species themselves. This explains the complete absence of the *Labiatae* species from the list, but where it is known that *Mentha*, *Thymus* and *Rosmarinus* are all cultivated herbs, economically important to some extent.

Using the broad definition of a CWR given above, all wild species in the same genus as the crop were included. However, for species in genera with known inter-

[2] statistics.defra.gov.uk/esg/publications/bhs

generic hybridization (e.g. those within the *Brassicaceae* and *Poaceae*) additional genera with which hybridization occurs were included. For *Brassica* species, the genera *Diplotaxis*, *Sinapis*, *Hirschfeldia*, *Cakile*, *Coincya*, *Rapistrum* and *Crambe* were also included and for *Lolium* and *Festuca* species, *Vulpia* was added.

This procedure resulted in a list of 57 major crop genera in the United Kingdom containing 227 CWR species (Table 7.4). Of the top ten economic crops in the United Kingdom (wheat, barley, potato, oilseed rape, sugarbeet, cabbage, peas, carrot, onion and lettuce), only wheat and peas have no native or naturalized wild relatives in the United Kingdom. The highest CWR generic diversity is found in the grasses and crucifers, while the highest CWR species diversity is found within *Trifolium* and *Festuca*. The crop species known to be fully interfertile with their native CWR are found in *Beta*, *Apium*, *Lactuca*, *Humulus*, *Nasturtium* or *Rorippa*, *Brassica*, *Trifolium* and *Agrostis* (Smartt and Simmonds, 1995). The native crop species are largely vegetables, forages and small fruits, e.g. cabbages, sugarbeet, carrot, raspberry, celery, hops, beetroot, blackcurrants, watercress, asparagus, parsnip, clover, fescues and bent grasses.

The vast majority of the 227 United Kingdom CWR is native. However, some families show significant numbers of introduced taxa, which reflects their history of cultivation in the United Kingdom. Most of the listed fruits, for example, apples, pears and plums, are long-established naturalized species. Legumes

Table 7.4. Major United Kingdom food families and genera (hybrids excluded).

Family	Crop genera	Crop genera names with associated species numbers
Poaceae	14	*Agrostis* (6), *Alopecurus* (6), *Arrhenaterum* (1), *Avena* (3), *Bromus* (8), *Cynodon* (1), *Dactylis* (1), *Festuca* (13), *Hordeum* (3), *Lolium* (2), *Phalaris* (1), *Phleum* (5), *Poa* (15), *Triticum* (1), *Vulpia* (5)
Brassicaceae	10	*Brassica* (3), *Sinapis* (2), *Nasturtium/Rorippa* (8), *Raphanus* (1), *Diplotaxis* (2), *Coincya* (2), *Hirschfeldia* (1), *Cakile* (1), *Crambe* (1), *Rapistrum* (1)
Apiaceae	7	*Apium* (4), *Anthriscus* (3), *Petroselinum* (2), *Carum* (2), *Foeniculum* (1), *Daucus* (1), *Pastinaca* (1)
Fabaceae	7	*Trifolium* (23), *Vicia* (13), *Lotus* (5), *Onobrychis* (1), *Medicago* (5), *Lotus* (5), *Lupinus* (2)
Rosaceae	5	*Fragaria* (2), *Malus* (2), *Prunus* (7), *Pyrus* (2), *Rubus* (7)
Asteraceae	3	*Cichorium* (1), *Lactuca* (3), *Scorzonera* (1)
Liliaceae	2	*Allium* (9), *Asparagus* (1)
Papaveraceae	1	*Papaver* (6)
Solanaceae	1	*Solanum* (5)
Ericaceae	1	*Vaccinium* (5)
Grossulariaceae	1	*Ribes* (6)
Valerianaceae	1	*Valerianella* (4)
Linaceae	1	*Linum* (3)
Chenopodiaceae	1	*Beta* (1)
Cannabaceae	1	*Humulus* (1)

have the highest number of taxa with both native and introduced status, e.g. subterranean clover, crimson clover, sainfoin and broad bean. The greatest United Kingdom CWR species richness is found in forages with *Trifolium*, *Poa*, *Festuca* and *Vicia* each with more than 10 native CWR, except for *Rubus* with 200 endemic microspecies (Rich, 1999).

Occurrence and rarity prioritization

The United Kingdom recording system is one of the most complete and oldest in the world (Harding, 1991, 1995). Biodiversity population data are commonly available online through the National Biodiversity Network (NBN) Gateway[3] and over nine million plant records can be checked at regional, country and county level for different time periods. Overlays of distribution data with protected area boundaries can be created online with interactive maps. Species summary ecological data are also available for 1407 native species through the ecological database of the University of York[4] (Fitter and Peat, 1994), as well as the PLANTATT data (Hill *et al.*, 2004) related to status, size, life history, geography and habitats of British and Irish plants. Further, for 240 taxa detailed biological monographs are published in the series *Biological Flora of the British Isles*. A chromosome database of the British flora at the University of Leicester contains 8151 records with localized chromosome counts of native material of 77% of the British flora.

There have also been national botanical surveys in 1962 and 2002 (Perring and Walters, 1962; Perring and Sell, 1968; Preston *et al.*, 2002) covering 2951 species in 3880 in $10 \times 10\,km^2$ grid squares (hectads) over the British Isles, which now form the main data sources for plant occurrence and change data. Since the publication of the first atlas (Perring and Walters, 1962) 120 county or regional floras have been published, many of which have distribution data of a higher resolution (Preston *et al.*, 2002). There have been over 40 regional surveys since 1962 with a resolution of $5 \times 5\,km^2$ or higher (Pearman, 1996; Preston *et al.*, 2002). There has also been progress with IUCN Red List assessment for rare and scarce taxa, which were undertaken at a scale of $1 \times 1\,km^2$ by Wiggington (1999). Currently, there are four threat assessment surveys at the national level: BSBI Local Change (tetrad level), the Common Plant Survey, the Annual Single Plant Survey and the Rare Plant Recording. Each of these surveys is being undertaken jointly by BSBI and Plantlife International to complete the botanical assessment and to widen public involvement. In addition, a national hybrid survey was launched by the BSBI in September 2005. Distribution data for protected areas are present although not centralized, based on species monitoring by the area managers in the so-called Common Standards Monitoring. Much of these data are partially publicly accessible, through the NBN gateway. Although the quantity of floristic data is large, the data are heterogeneous, both in time and space. Their scale and format, $10 \times 10\,km^2$ in presence or absence format, has restricted their use for conservation planning purposes (Griffiths *et al.*, 1999). The discrepancy in

[3] http://www.nbngateway.co.uk
[4] http://www.york.ac.uk/res/ecoflora/cfm/ecofl/index.cfm

Table 7.5. Occurrence frequencies of priority CWR based on *New Atlas* data (Preston *et al.*, 2002).

Occurrence category	Number of grids	Number of taxa
Very common	50% or more	54
Common	25–50%	33
Scarce	17–100	51
Rare	4–16	12
Very rare	3 or less	6

scale of floristic data and protected area data, and the problems it poses for conservation planning, is discussed under Genetic Conservation Gap Analysis.

Occurrence data as collated in the National Inventory Baseline Conservation Database provide a first indication of conservation needs for CWR. Occurrence data from the *New Atlas of the British and Irish Flora* (Preston *et al.*, 2002) were used to categorize CWR according to frequency of occurrence (Table 7.5). This simple analysis of distribution patterns can be interpreted as a first indication of active conservation needs, which for >40% (common) taxa is a low priority.

An additional source of information for setting priorities and assessing conservation needs is data on species trends, primarily decline. The *New Atlas* (Preston *et al.*, 2002) provides a qualitative description of species trends using data from the two national surveys. However, interpreting changes is complicated by several factors: (i) differences in taxonomic treatment; (ii) differences in taxonomic emphasis (such as a greater interest in alien species); (iii) recording intensity between the two surveys; and (iv) recording bias in general (Rich and Woodruff, 1996). The *New Atlas* lists a change index for each species, which is calculated as a deviation from a regression line calculated over total number of species (Telfer *et al.*, 2002). The change index gives another indication of conservation needs. For example, the significantly negative change index for *Hordeum marinum* Huds. represents a sharp shrinkage in the species distribution over the last 30 years. Of the priority taxa with known change index, 12 had a significantly negative change index.

Two indicators have been used in the United Kingdom to assess threat to CWR, namely rarity and the IUCN Red Listing status, which is in turn based on relative population size, decline or fluctuation, and geographical range size and fragmentation (IUCN, 2001). Increasingly, the IUCN threat assessment is used as a conservation planning tool (Rodrigues *et al.*, 2006) and the most recent United Kingdom Red Listing (Cheffings *et al.*, 2005) was the first to use strict IUCN Red List Criteria. The *New Atlas* (Preston *et al.*, 2002) was used for the most recent Red Listing combined with monitoring data from the United Kingdom protected area network (Common Standards Monitoring) and expert knowledge of individual species (Cheffings *et al.*, 2005). The result was that almost 20% of the United Kingdom flora were considered threatened, out of a total of 1756 assessed taxa. Interestingly, the rarer species were shown to be better covered by current conservation practice than more widespread ones (Cheffings *et al.*, 2005). The Red List included archaeophytes (defined as species that were naturalized in the United Kingdom before AD 1500) for the first time. The inclusion of archaeophytes identified by Preston *et al.* (2004) is highly relevant in terms of CWR conservation as many are obsolete crops; 146 out of the 149 archaeophytes listed are included in the United Kingdom CWR Inventory. Many archaeophytes are also naturalized relatives of current crops with species present in *Brassica*, *Foeniculum*, *Petroselinum*, *Armoracia*, *Carum*, *Allium*, *Sinapis* and *Asparagus officinalis* subsp. *officinalis* or important historic crops that currently are of minor use (*Chenopodium bonus-henricus*, *Isatis tinctoria* L., *Camelina sativa* (L.) Crant.).

Of the 227 CWR species assigned the highest conservation value or priority status on the basis of their comparative economic value, 23 taxa are also threatened doubly justifying their active conservation. These include: *Lolium temulentum* (critically endangered – CR), and *A. officinalis* subsp. *officinalis*, *Carum carvi* L., *H. marinum* Huds., *Valerianella rimosa* Bastard and *V. dentata* (L.) Pollich (endangered – EN) and 16 other species (vulnerable – VU) each of which requires active conservation (Table 7.6).

The background of decline in the United Kingdom flora has been analysed and widely reported (Haines-Young *et al.*, 2000; Preston *et al.*, 2003), and it is expected that there will be a negative impact on CWR. There is a high proportion of arable weeds among the 23 threatened priority taxa and loss of traditional arable and horticultural habitats (Pilgrim *et al.*, 2004) will have significance, so their conservation is an urgent priority (Anonymous, 2004). The relatively high number of threatened archaeophytes compared with native plants had also been observed in the *New Atlas* (Preston *et al.*, 2002). The causes of the decline of archaeophytes may in part be associated with the cessation of the cultivation of traditional crops, such as *C. carvi* (Preston *et al.*, 2003).

7.2.3 Genetic gap analysis and CWR conservation goal setting

Thus, the presence of CWR has not thus far been a site selection criterion for protected areas in the United Kingdom (Franks, 1999). However, CWR may occur in protected areas in relation to their habitat or species association preferences. The problem for CWR is that populations of CWR in protected areas are unlikely to be monitored and so their survival as passively conserved taxa is uncertain (Maxted *et al.*, 1997c). To ensure that the CWR taxa are actively and effectively conserved there is a need for some form of gap analysis. Gap analysis has traditionally been used to identify habitats under-represented by protected areas (Margules *et al.*, 1988; Margules and Pressey, 2000), but Maxted *et al.* (in preparation c) suggested in the genetic conservation context that it is important to adopt a complementary approach where both *in situ* and *ex situ* sampled conserved populations are reviewed for their effectiveness in conserving the maximum range of genetic diversity of the target taxa.

In situ CWR conservation review

There has been no systematic review of United Kingdom CWR distribution and overlap with the United Kingdom protected area network. Although, it has

Table 7.6. Current United Kingdom conservation legislation and policy for 23 target CWR.

Taxonomy		Phytosociology	Threat	Legal status	Conservation plan		Species monitoring	Conservation planning
Taxon scientific name	Status	National vegetation community categories	IUCN category	Legal protection[a] WCA[b] or CROW[c]	Species Action Plan Biodiversity Action Plan (BAP)	Habitat Action Plan Criteria[d] Species listed	Common Standards Monitoring (in PA)	IPA Planning Agricultural Habitats (Plantlife International, 2004b)
Allium oleraceum L.	Native		VU	–				
Allium sphaerocephalon L.	Native	CG	VU	WCA8			Suite 12	
Apium repens (Jacq.) Lag.	Native	MG	VU	WCA8 CROW	BAP + SAC	Annex I, IV, II	As species	
Asparagus officinalis subsp. *prostratus* (Dumort.) Corb.	Native	MC	EN	CROW	BAP	Maritime cliffs and slopes		
Bromus secalinus L.	Archaeophyte		VU	–				Targeted
Carum carvi L.	Archaeophyte		EN	–		(Annex 1 Habitat)		
Coincya wrightii (O.E. Schulz) Stace	Native		VU	WCA8	BAP		As species	
Hordeum marinum Huds.	Native	MG O SM W	VU	–		Ramsar wetlands	Suite 14	
Lactuca saligna L.	Native	SM O	EN	WCA8				
Lolium temulentum L.	Archaeophyte		CR	–				
Medicago minima (L.) Bartal.	Native	OV U SD MC	VU	–			Suite 5, 7, 14	
Papaver argemone L.	Archaeophyte		VU	–				Targeted
Poa flexuosa Sm.	Native		VU	–		Annex 1 Habitat	Suite 13	
Poa glauca Vahl.	Native	OG U	VU	–		Annex 1 Habitat		
Pyrus cordata Desv.	Native	W MG O	VU	WCA8	Managed		Suite 2	

Continued

Table 7.6. *Continued*

Taxonomy		Phytosociology	Threat	Legal status	Conservation plan		Species monitoring	Conservation planning
Taxon scientific name	Status	National vegetation community categories	IUCN category	Legal protection[a] WCA[b] or CROW[c]	Species Action Plan Biodiversity Action Plan (BAP)	Habitat Action Plan Criteria[d] Species listed	Common Standards Monitoring (in PA)	IPA Planning Agricultural Habitats (Plantlife International, 2004b)
Scorzonera humilis L.	Native	MG	VU	WCA8	Monitored at one site			
Trifolium bocconei Savi	Native	MC	VU	–				
Trifolium incarnatum subsp. *molinerii* (Balb. ex Hornem.) Ces.	Native	MC	VU	–				
Trifolium strictum L.	Native	MC	VU	–				
Valerianella dentata (L.) Pollich	Archaeophyte		EN	–	Conservation concern for BAP	Cereal field margins		Targeted
Valerianella rimosa Bastard	Archaeophyte	OV	EN	CROW	BAP		Suite 8	Targeted
Vicia bithynica (L.) L.	Native	MC W O	VU	–			Suite 2, 4	
Vicia parviflora Cav.	Native	MC OV G O	VU	–			Suite 2, 4	Targeted

[a]For Northern Ireland: The Wildlife Order 1985.
[b]Wildlife and Countryside Act, Schedule 8.
[c]Countryside and Right of Way Act 2000, Schedule 74.
[d]Bern Convention 1979.

been suggested that current United Kingdom protected areas probably sufficiently represent United Kingdom diversity (Plantlife International, 2004a), with 95% of major plant taxa represented (Hopkinson *et al.*, 2000), which implies that United Kingdom CWR species are also likely to be adequately represented. However, a recent study has identified the ten priority special areas for conservation (SAC) and sites of special scientific interest (SSSI) which maximize United Kingdom CWR conservation (Maxted *et al.*, in press). These ten sites contain 128 (57%) of the 226 priority United Kingdom CWR species.

Some caution is necessary regarding this high percentage of conservation as the analysis was based on extrapolated not actual presence in protected areas and the viability of populations was not considered. As in most countries the terrestrial protected areas in England and Wales are disproportionately occurring in hills and mountains and tend to under-represent lowland areas where the establishment of protected areas would impinge on profitable land use and land purchase would be expensive (Oldfield *et al.*, 2003). The very small size of most United Kingdom protected areas, on average 0.2 km^2 for 4000 English SSSI and 1.1 km^2 for a national nature reserve (NNR) (Oldfield *et al.*, 2004), also presents problems. For instance, at tetrad level (4 km^2) there is only a 0.05 probability of a floristic record being recorded from a 0.2 km^2 SSSI. Therefore, GIS overlay analyses using species data in hectads or tetrads are likely to produce inflated species occurrences in protected areas.

To illustrate the point, when the National Biodiversity Network Gateway is queried for the very rare, endangered and legally protected *Lactuca saligna* ... the resultant distribution map, based on floristic data, can then be overlaid with protected area boundaries (the SSSI of Rye Harbour in East Sussex[5]). Although the floristic record indicates presence of *L. saligna* within the 100 km^2 grid square much of the grid is sea and therefore the species is likely to be present within the SSSI boundary because the largest part of the terrestrial surface within this grid falls within the SSSI. This was verified by consulting the Rye Harbour SSSI designation report (English Nature, 1992),[6] which lists the presence of *L. saligna* as a reason for notification of the site, among other uncommon plants *Crambe maritime* L., *Vulpia ciliate* subsp. *ambigua* (Le Gall) Stace & Auquier and *Lathyrus japonicus* Willd. Site target species monitoring is undertaken regularly and the reports are available online[7] and show among the 23 target taxa that *L. saligna* has a healthy population at the site. This site is nationally important because it is one of the three United Kingdom locations where *L. saligna* occurs, the species having declined dramatically since the 1930s (Wiggington, 1999). The fact that *L. saligna* occurs at the site, along with other rare CWR, would indicate that it is a possible candidate for a key national CWR reserve. Thus, in this case the overlay analysis would generate an accurate picture of CWR species and protected area occurrence coincidence, but this need not be the case.

[5] www.nbngateway.org.uk
[6] http://www.english-nature.org.uk/citation/citation_photo/1001954.pdf
[7] http://www.english-nature.org.uk/citation/citation_photo/1001954.pdf

Performing *in situ* gap analysis of United Kingdom protected areas and species occurrence for United Kingdom CWR was not feasible within the framework of this study, due to the scale incompatibility problem of botanical recording, mentioned above, and the lack of centralized protected areas species occurrence data. However, a considerable coincidence between CWR occurrence and protected areas is assumed on the basis of regional data where the necessary data are available, e.g. for the United Kingdom county Dorset where the majority of records of our target species were within protected areas (Edwards and Pearman, 2004).

Given the difficulty in undertaking *in situ* gap analysis and the prevalence of a wide range of conservation measures in the United Kingdom including 319 single Species Action Plans, 45 Habitat Action Plans and 162 Local Biodiversity Action Plans, the *in situ* gap analysis was focused on assessing the degree of integration of the target CWR taxa into existing conservation measures. An overview of existing conservation legislature, national and international (European Union – EU) conservation policy as it applies to the 23 prioritized taxa is given in Table 7.7. The United Kingdom legal instrument for plant protection is the

Table 7.7. Summary of current United Kingdom *ex situ* accessions of the 23 priority CWR taxa.

Taxon	Millennium seed bank United Kingdom collections	EURISCO United Kingdom collections
Allium oleraceum L.	Uncertain	0
Allium sphaerocephalon L.	2	0
Apium repens (Jacq.) Lag.	1	0
Asparagus officinalis subsp. *prostratus* (Dumort.) Corb.	1	3 (species level)
Bromus secalinus L.	2	0
Carum carvi L.	1	10 (unknown source)
Coincya wrightii (O.E. Schulz) Stace	1	0
Hordeum marinum Huds.	3	0
Lactuca saligna L.	1	2
Lolium temulentum L.	3	0
Medicago minima (L.) Bartal.	2	0
Papaver argemone L.	1	0
Poa flexuosa Sm.	1	0
Poa glauca Vahl	1	0
Pyrus cordata Desv.	1	0
Scorzonera humilis L.	1	0
Trifolium bocconei Savi	2	0
Trifolium incarnatum subsp. *molinerii* (Balb. ex Hornem.) Ces.	1	0
Trifolium strictum L.	2	0
Valerianella dentata (L.) Pollich	1	0
Valerianella rimosa Bastard	1	0
Vicia bithynica (L.) L.	2	0
Vicia parviflora Cav.	2	0

Wildlife and Countryside Act (WCA) Schedule 8 and the Countryside and Right of Way Act (CROW) 2000 Schedule 74. Of the 23 target CWR taxa, eight are listed under the WCA or CROW and only *Apium repens* (Jacq.) Lag. is covered by international conservation regulations, under the EC Habitat and Species Directive IIb and IVb, under the Bern Convention Appendix II and under Cites Appendix II, besides also being legally protected in the United Kingdom by the WCA and CROW. There are also prioritized CWR taxa without specific United Kingdom national legal protection that are listed under international criteria, e.g. *Poa glauca* Vahl and *P. flexuosa* Sm. are listed under the EC Habitat and Species Directive Annex I Habitat species; and *H. marinum* is listed as a Ramsar wetlands species. It should also be noted that regional and national priorities may not concur, e.g. at European level *C. carvi* is listed under the EC Habitat and Species Directive Annex I as a priority mountain hay meadows species, but this species is currently not prioritized in the United Kingdom.

Another source of information on the degree of a taxon's integration in current conservation activities is its phytosociological status. United Kingdom conservation planning and site designation criteria are based on national phytosociological classifications to a significant extent. The majority of the 23 target CWR taxa are members of one or more national vegetation communities. However, seven are not included as members of a specific national vegetation community and six of these are archaeophytes. Only one of the latter, *V. rimosa*, has United Kingdom legal protection and has a Biodiversity Action Plan. Furthermore ten of the 23 target CWR taxa are statutorily monitored within protected areas under Common Standard Monitoring criteria; the majority of these are indicator species for a certain habitat or vegetation. *A. repens* (Jacq.) Lag. and *Coincya wrightii* (O.E. Schulz) Stace are also monitored as species in their own right.

CWR species are often associated with arable habitats and disturbed places, and their wider distribution and genetic diversity may be under-represented in current conservation action focusing on climax community conservation. Recently, arable habitats have been included in the United Kingdom Biodiversity Action Plan and 29 species were targeted; cereal field margins are regarded as a priority habitat with *Bromus secalinus* L., *V. rimosa* and *Vicia parviflora* Cav. identified as priority CWR taxa within these habitats. Nine protected areas (SSSIs) have been specifically designated for conservation of arable plants in England and 50 SSSIs support some arable species (Wilson and King, 2003). Examples of the conservation carried out on farms with or without statutory designation or financial aid are given by Wilson and King[8] and CWR species conserved in this way include: *Valerianella dentata* and *V. rimosa*, *B. secalinus*, *Papaver argemone* L., *V. parviflora* and *L. temulentum*. Important plant areas for arable plants have been designated by Plantlife International (Plantlife International, 2004b; Byfield and Wilson, 2005) for 150 sites and five pilot sites are being experimentally managed for arable species conservation. Conservation management guidelines for arable weeds have been developed and are currently being tested on trial farms (Plantlife International, 2002b; Wilson and King, 2003).

[8] http://www.arableplants.fieldguide.co.uk/?P=home&SHC=16&PSD=1

Of the 23 target CWR taxa, four have specific Biodiversity Action Plans and ten are currently being monitored within protected areas. The recent updating of the United Kingdom IUCN Red List assessment resulted in a sharp increase of threatened taxa, which has in turn made new conservation plans necessary. An updated list of prioritized threatened taxa is due in March 2006[9] (English Nature, 2006) and it is to be expected that a larger proportion of the 23 target taxa will be targeted for conservation action.

Ex situ CWR conservation review

Comparing population samples conserved *ex situ* with *in situ* occurrence should form an essential element of gap analysis (Maxted *et al.*, in preparation c). The United Kingdom is the most advanced country in the world in terms of *ex situ* conservation of the native flora, and greater than 95% of the native higher plant flora have been collected and one or two accessions per species have been stored *ex situ* by the Millennium Seed Bank, Wakehurst Place.[10] Specifically, United Kingdom CWR forage genera such as *Lolium*, *Festuca*, *Dactylis* and *Trifolium* as well as amenity grasses such as *Agrostis* and *Poa* have been collected by the Institute of Grassland and Environmental Research (Humphreys, 2003) and United Kingdom native CWR of vegetables have been collected at Warwick Horticulture Research International (HRI). The current United Kingdom *ex situ* conservation status of 23 target CWR taxa is summarized in Table 7.7.

7.3 Discussion

This chapter illustrates how a national CWR conservation strategy can be developed using a model proposed by Maxted *et al.* (in preparation b) for the United Kingdom. It should be emphasized that the model presents one scenario but there are undoubtedly other routes whereby national CWR conservation strategies may be developed, and the model itself should be seen as a way of thinking about the issues involved rather a procedure to be followed unthinkingly. It is surely a truism that the generation of each national CWR conservation strategy is likely to be unique because each country will have different types of background data, resources available for the exercise and end goals. However, whatever procedure is used it will involve answering a series of questions to define the conservation targets, sites and units, for example:

- Which taxa in the national flora are CWR?
- Which of the CWR have the highest conservation value?
- Which CWR have the highest conservation urgency?
- What are the genetic and geographic units of conservation for the target taxa?
- What is the genetic–geographic distribution of the target taxon?

[9] http://www.ukbap.org.uk/library/brig/BRIGReviewSummary_07_05.pdf

[10] Flynn, S., Turner, R.M., and Dickie, J.B. 2004. Seed Information Database (release 6.0, October 2004). http://www.rbgkew.org.uk/data/sid

- How many prioritized target taxa occur in existing protected areas and are covered by (adequate) conservation management?
- Where are the country's CWR diversity hot spots?
- Which existing protected areas are associated with CWR diversity hot spots?
- Which, where and how can CWR diversity be conserved outside protected areas?

The essential first step is to distinguish the national CWR taxa from the more general flora assuming that the national flora itself is well known. Ennos *et al.* (2005) makes the point that prior to answering the question of where the eco-geographic–genetic diversity of a target taxon is found, the issue of the genetic *unit* of conservation has to be addressed; in many cases, even in a well-studied flora such as the United Kingdom, apomictic, hybrid or other critical species occur as 'taxonomically complex groups' and lack of taxonomic clarity can undoubtedly hinder their conservation.

However, assuming that the flora is known, distinguishing CWR taxa from other flora will involve the use of a CWR definition and for the United Kingdom national CWR inventory the definition used was the broad definition proposed by Maxted *et al.* (2006), the non-crop taxa within the same genus as a crop. Some may feel this definition to be too broad as it resulted in a CWR list of over 2600 taxa for a country not known for its botanical diversity. However, application of the more precise definition also provided by Maxted *et al.* (2006) 'CWR belonging to gene pools 1 or 2, or taxon groups 1 to 4 of the crop' would have been impossible to apply for the full United Kingdom flora because of the general lack of knowledge of the precise taxonomic relationships between the taxa, and sheer quantity of time and resources required to obtain the necessary information. *A posteriori* the application of the broad definition for CWR when developing a national CWR conservation strategy is the only practical and objective option and is by definition inclusive of all CWR taxa.

However, it necessitates a process of prioritization of the national CWR inventory to balance the number of taxa targeted against the resources available for conservation. Whatever the number of CWR taxa and the resources available for their conservation, it would also be wasteful of resources to attempt to conserve all CWR taxa because not all taxa are equally in need of conservation due to their degree of abundance or threat. Prioritization of the national CWR inventory will always be required and it is important to ensure that the means of prioritization is objective and relevant to the needs of the conserving agency.

IUCN Red Listing provides a well-founded tool to separate taxa requiring conservation from non-threatened taxa. Thus, national application of the IUCN Red List Criteria is likely to be a key means of prioritization, but most national Red Lists are far from complete and rarely employ the most recent set of Red List Criteria (IUCN, 2001). The production of the United Kingdom national CWR priority list of 227 taxa was greatly aided by the recent United Kingdom Red Data List (Cheffings *et al.*, 2005), but the absence of such a list for Ireland impeded production of a comparable list for Ireland and hindered their CWR conservation planning. Some may advocate a wider, European-based, geographic context for prioritization (Walters, 1984; Hodgson, 1986; Ford-Lloyd *et al.*, Chapter 6, this volume).

The other important means of prioritizing CWR taxa is the agronomic value of the associated crop. It can prove surprisingly difficult to obtain estimates of agronomic value from national official statistics or even in the EU context. Prioritization based on inclusion in the EU Common Catalogues for Agricultural and Horticultural varieties resulted in the identification of 57 United Kingdom crop genera with 227 associated United Kingdom CWR taxa. However, this process of prioritization does not take into account relative values of, for instance, the wild relatives of barley compared to the wild relatives of corn-salad. The information required to make this distinction is currently not readily available in the United Kingdom.

Perhaps another factor that should also be taken into account at some point during prioritization is knowledge of the national and international amplitude and pattern of genetic diversity, highlighted by De Hond *et al.* (2005) as well as by English Nature (2001) for the United Kingdom taxa. The supranational context of national conservation is acknowledged in the increasing practice of assigning international responsibility for species with a significant proportion of their world population based within a country. For the United Kingdom, this would include taxa such as *B. oleracea* and the endemic *L. perenne* subsp. *anglicum* (Mill.) Ockendon, where the United Kingdom hosts a significant proportion of the taxon range and gene pool (Cheffings *et al.*, 2005). So, following prioritization and the generation of a more restricted list of CWR taxa (227 for the United Kingdom) using threat or rarity and economic value, there may be a need to collate ecogeographic and genetic diversity as a basis for further detailed conservation planning (Ford-Lloyd *et al.*, Chapter 6, this volume). Although these data may be critical for effective conservation, it is generally available only for a minority of, usually rare, taxa. Therefore, there is a general need for a better understanding of CWR ecogeographic data and particularly their genetic diversity in critical cases and in certain situations.

Finally in the context of the Global Strategy for Plant Conservation and in response to the CBD 2010 Biodiversity Target of a significant reduction in the current rate of loss of diversity, the creation and application of a national CWR inventory is undoubtedly a major contribution to achieving this goal. Whether or not the United Kingdom as an exemplar case is typical in terms of CWR conservation planning is difficult to assess; the United Kingdom has arguably the most well-studied and conserved flora in the world, but the illustration provided will enable others to consider the issues involved in developing a national CWR conservation strategy.

Acknowledgements

The United Kingdom National Inventory of plant genetic resources for Food and Agriculture was part of the project *The establishment of an inventory for Genetic resources for Food and Agriculture and scope a GRFA information system* funded by the United Kingdom DEFRA. Subsequent work on United Kingdom CWR was funded by *PGR Forum European CWR Diversity and Conservation Assessment* (no. EVK2-2001-00192). The BSBI provided conservation baseline data; specifically Drs Chris Preston, Henry Arnold (BSBI), Alex Lockton (BSBI), Bob Ellis (BSBI) and Tim Rich (NHMW) provided further information; the NBN Gateway help desk provided assistance; and the United Kingdom Plant Genetic Resources Group gave advice and suggestions.

References

Allan, D.E. and Hatfield, G. (2004) *Medicinal Plants in Folk Tradition: an Ethnobotany of Britain and Ireland*. Timber Press, Portland, Oregon.

Anonymous (2004) *UK 2004, the Official Yearbook of the United Kingdom of Great Britain and Northern Ireland*. UK National Statistics, London.

Byfield, A.J. and Wilson, P.J. (2005) *Important Arable Plant Areas: Identifying Priority Sites for Arable Plant Conservation in the United Kingdom*. Plantlife International, Salisbury, UK.

CBD (1992) *Convention on Biological Diversity: Text and Annexes*. Secretariat of the Convention on Biological Diversity, Montreal, Canada. Available at: http://www.biodiv.org/convention/convention.shtml (accessed 12 April 2007)

CBD (2002) *Global Strategy for Plant Conservation*. Secretariat of the Convention on Biological Diversity, Montreal, Canada. Available at: http://www.biodiv.org/decisions/?lg=0&dec=VI/9 (accessed 3 April 2007)

Cheffings, C.M. (2004) New Plant Status Lists for Great Britain. *BSBI News* 95, 36–43.

Cheffings, C.M., Farrell, L., Dines, T.D., Jones, R.A., Leach, S.J., McKean, D.R., Pearman, D. A., Preston, C.D., Rumsey, F.J. and Taylor, I. (2005) The Vascular Plant Red Data List for Great Britain. *Species Status* 7, 1–116.

Cook, F.E.M. (1995) *Economic Botany Data Collection Standards*. Royal Botanic Gardens, Kew, UK.

De Hond, L., Iriondo, J.M. and Kell, S.P. (2005) *Report of Workshop 4: Population Management Methodologies*. PGR Forum, University of Birmingham, Birmingham, UK. Available at: http://www.pgrforum.org/Documents/WP_4_Documents/WS4_Report.pdf (accessed 12 April 2007)

DEFRA (2001) *Review of Policy on Genetic Resources for Food and Agriculture*. DEFRA, London.

DEFRA (2002) *Working with the Grain of Nature*. DEFRA, London.

DEFRA (2003a) *Basic Horticultural Statistics*. DEFRA, London.

DEFRA (2003b) *Plant Variety Rights Office and Seeds Division Seed Trader's Annual Return for Year Ended 30 June 2003*. DEFRA, London.

DEFRA and PVO (2005) National listing of varieties of agricultural and vegetable crops in the UK. *Plant Varieties and Seeds Gazette*, 461.

Dennis, F.R. (1997) *The Trade in Medicinal Plants in the United Kingdom*. UK World Wide Fund for Nature, Godalming, UK.

Department of the Environment (1996) *Towards a Methodology for Costing Biodiversity Targets in the UK*. Department of the Environment, London.

Edees, E.S. and Newton, A. (1988) *Brambles of the British Isles*. Ray Society, London.

Edwards, B. and Pearman, D.A. (2004) *Dorset Rare Plant Register*. Dorset Environmental Records Centre and the Botanical Society of the British Isles, Dorchester, UK.

English Nature (1992) *The Management of Lowland Grassland for Selected Rare and Scarce Vascular Plants: a Review*. Research Report No 216. English Nature, Peterborough, UK

English Nature (2001) *Proceedings of a Conservation Genetics Workshop* (held at the Royal Botanic Gardens, Kew, 27 November 2001). Research Report No 607. English Nature, Peterborough, UK.

English Nature (2006) *Priorities for Vascular Plant Conservation in England. Part 1: A Threat/Responsibility Analysis of Taxa Recognised as Threatened, Near Threatened or Nationally Scarce in Great Britain*. English Nature, Peterborough, UK.

Ennos, R.A., French, G.C. and Hollingsworth, P.M. (2005) Conserving taxonomic complexity. *Trends in Ecology and Evolution* 20(4), 164–168.

EU (2005a) *Common Catalogue of agricultural plant species and supplements. Official Journal of the European Community*. Available at: http://eur-lex.europa.eu/en/index.htm (accessed 12 April 2007)

EU (2005b) *Common Catalogue of vegetable plant species and supplements. Official Journal of the European Community*. Available at: http://eur-lex.europa.eu/en/index.htm (accessed 12 April 2007)

Euro+Med PlantBase (2005) *Euro+Med PlantBase: the information resource for Euro-Mediterranean plant diversity*. Dipartimento di Scienze botaniche ed Orto botanico, Università degli Studi di Palermo. Available at: http://www.emplantbase.org/home.html (accessed 13 April 2007)

FAO (2001) *International Treaty on Plant Genetic Resources for Food and Agriculture*. FAO, Rome, Italy.

Fitter, A.H. and Peat, H. (1994) The Ecological Flora Database. *Journal of Ecology* 82, 415–425.

Forestry Commission (2005) *Global forestry resources assessment 2005 – UK Country Report*. Forestry Commission, Edinburgh, UK.

Franks, J. (1999) *In situ* conservation of plant genetic resources for food and agriculture: a UK perspective. *Land Use Policy* 16, 81–91.

Good, R. (1964) *The Geography of Flowering Plants*, 3rd edn. Longman Scientific and Technical, Harlow, UK.

Griffiths, G.H., Eversham, B.C. and Roy, D.B. (1999) Integrating species and habitat data for nature conservation in Great Britain: data sources and methods. *Global Ecology and Biogeography* 8, 229–345.

Haines-Young, R.H., Barr, C.J., Black, H.I.J., Briggs, D.J., Bunce, R.G.H., Clarke, R.T., Cooper, A., Dawson, F.H., Firbank, L.G., Fuller, R.M., Furse, M.T., Gillespie, M.K., Hill, R., Hornung, M., Howard, D.C., McCann, T., Morecroft, M.D., Petit, S., Sier, A.R J., Smart, S.M., Smith, G.M., Stott, A.P., Stuart, J.W. and Watkins, J.W. (2000) *Accounting for Nature: Assessing Habitats in the UK Countryside*. DETR, London.

Hanelt, P. and IPK Gatersleben (eds) (2001) *Mansfeld's Encyclopedia of Agricultural and Horticultural Crops*. 6 vols. 1st English edition. Springer, Berlin/Heidelberg/New York.

Harding, P.T. (1991) Famous laboratories: the Biological Records Centre. *Biologist* 37, 162–164.

Harding, P.T. (1995) Data for nature conservation in statutory and voluntary heritage agencies. In: Berry, A.Q. and Brown, I.W. (eds) *Managing Ancient Monuments: an Integrated Approach*. Clwyd County Council, Wrexham, UK, pp. 110–123.

Hawkes, J.G., Maxted, N. and Ford-Lloyd, B.V. (2000) *The* Ex Situ *Conservation of Plant Genetic Resources*. Kluwer Academic, Dordrecht, The Netherlands.

Heywood, V.H. and Zohary, D. (1995) A catalogue of the wild relatives of cultivated plants native to Europe. *Flora Mediterranea* 5, 375–415.

Hill, M.O., Preston, C.D., and Roy, D.B. (2004) *PLANTATT Attributes of British and Irish Plants: Status, Size, Life History, Geography and Habitats*. Centre for Ecology and Hydrology, Huntingdon, UK.

Hodgson, J.G. (1986) Commonness and rarity in plants with special reference to the sheffield flora. Part IV: a European context with particular reference to endemism. *Biological Conservation* 36, 297–314.

Hopkinson, P., Travis, J.M.J., Prendergast, J.R., Evans, J., Gregory, R.D., Telfer, M.G. and Williams, P.H. (2000) A preliminary assessment of the contribution of nature reserves to biodiversity conservation in Great Britain. *Animal Conservation* 3(4), 311–320.

Hoyt, E. (1988) *Conserving the Wild Relatives of Crops*. IBPGR/IUCN/WWF, Rome, Italy.

Humphreys, M.O. (2003) Utilization of plant genetic resources in breeding for sustainability. *Plant Genetic Resources Characterization and Evaluation* 1, 11–18.
IPK Gatersleben (2003) *Mansfeld's World Database of Agricultural and Horticultural Crops*. Leibniz Institute of Plant Genetics and Crop Plant Research, Gatersleben, Germany. Available at: http://mansfeld.ipk-gatersleben.de/mansfeld/mf-inf_e.htm (accessed 13 April 2007)
IUCN (2001) *IUCN Red List Categories and Criteria, Version 3.1*. IUCN Species Survival Commission. IUCN Gland, Switzerland/Cambridge, UK.
Kell, S.P., Knüpffer, H., Jury, S.L., Maxted, N. and Ford-Lloyd, B.V. (2005) *Catalogue of Crop Wild Relatives for Europe and the Mediterranean*. Available online via the Crop Wild Relative Information System (CWRIS – http://cwris.ecpgr.org/) and on CD-ROM. University of Birmingham, Birmingham, UK.
Mabey, R. (1977) *Plants with a Purpose: a Guide to the Everyday Uses of Wild Plants*. Collins, London.
Mabey, R. (1996) *Flora Britannica*. Chatto & Windus, London.
Mabey, R. (2003) *Food for Free*. HarperCollins Publishers, London.
MAFF (1995) *Country Report for the UK*. Ministry of Agriculture, Fisheries and Food, London.
Margules, C.R. and Pressey, R.L. (2000) Systematic conservation planning. *Nature* 405, 243–253.
Margules, C.R., Nicholls, A.O. and Pressey, R.L. (1988) Selecting networks of reserves to maximize biological diversity. *Biological Conservation* 43, 63–76.
Maxted, N., Ford-Lloyd, B.V. and Hawkes, J.G. (1997a) *Plant Genetic Conservation: the* In Situ *Approach*. Chapman & Hall, London.
Maxted, N., Hawkes, J.G., Guarino, L. and Sawkins, M. (1997b) The selection of taxa for plant genetic conservation. *Genetic Resources and Crop Evolution* 44, 337–348.
Maxted, N., Ford-Lloyd, B.V. and Hawkes, J.G. (1997c) Complementary conservation strategies. In: Maxted, N., Ford-Lloyd, B.V. and Hawkes, J.G. (eds) *Plant Genetic Conservation: the* In Situ *Approach*. Chapman & Hall, London, pp. 20–55.
Maxted, N., Ford-Lloyd, B.V., Jury, S.L., Kell, S.P., Scholten, M.A. (2006) Towards a definition of a crop wild relative. *Biodiversity and Conservation* 15(8), 2673–2685.
Maxted, N., Scholten, M., Codd, R. and Ford-Lloyd, B.V. (in press) Creation and use of a national inventory of crop wild relatives. *Biological Conservation*.
Maxted, N., Scholten, M., Kell, S.P. and Ford-Lloyd, B.V. *Developing a National Plant Genetic Resource Strategy: Crop Wild Relatives* (in preparation b).
Maxted, N., Dulloo, E., Ford-Lloyd, B.V., Iriondo, J. and Jarvis, A. *Gap Analysis: a Tool for Effective Genetic Conservation Assessment of Agrobiodiversity* (in preparation c).
Meilleur, B. and Hodgkin, T. (2004) *In situ* conservation of CWRs: status and trends. *Biodiversity and Conservation* 13, 663–684.
Milliken, W. and Bridgewater, S. (2001) *Flora Celtica: Sustainable Development of Scottish Plants*. Scottish Executive Central Research Unit, Edinburgh.
Milliken, W. and Bridgewater, S. (2004) *Flora Celtica: Plants and People in Scotland*. Birlinn Limited, Edinburgh, UK.
Oldfield, T.E.E., Smith, R., Harrop, S.R. and Leader-Williams, N. (2003) Field sports and conservation in the United Kingdom, *Nature* 423, 531–533.
Oldfield, T.E.E., Smith, R., Harrop, S.R. and Leader-Williams, N. (2004) A gap-analysis of terrestrial protected areas in England and its implications for conservation policy. *Biological Conservation* 120, 307–313.
Pearman, D. (1996) Towards a new definition of rare and scarce plants. *Watsonia* 21, 231–251.
Perring, F.H. and Walters, S.M. (1962) *Atlas of the British flora*. Thomas Nelson & Sons, London.
Perring, F.H. and Sell, P.D. (1968) *Critical Supplement to the Atlas of the British Flora*. Thomas Nelson & Sons, London.
Pilgrim, E.S., Crawley, M.J. and Dolphin, K. (2004) Patterns of rarity in the native British flora. *Biological Conservation* 120, 161–170.

Plantlife International (2002a) *Herbal Harvest with a Future. Towards Sustainable Sources for Medicinal Plant*. Plantlife International, Salisbury, UK.
Plantlife International (2002b) *England's Green Unpleasant Land*. Plantlife International, Salisbury, UK.
Plantlife International (2004a) *Plant Diversity Challenge – the UK's Response to the Global Strategy for Plant Conservation*. Plantlife International, Salisbury, UK.
Plantlife International (2004b) *Farming for Arable Flowers*. Plantlife International, Salisbury, UK.
Prendergast, H.D.V. and Sanderson, H. (2004) *Britain's Wild Harvest: the Commercial Uses of Wild Plants and Fungi*. Royal Botanic Gardens, Kew, UK.
Prescott-Allen, R. and Prescott Allen, C. (1983) *Genes from the Wild: Using Wild Genetic Resources for Food and Raw Materials*. Earthscan Publications, London.
Preston, C.D., Pearman, D.A. and Dines, T.D. (2002) *New Atlas of the British and Irish Flora*. Oxford University Press, Oxford, UK.
Preston, C.D., Telfer, M.G., Roy, D.B., Carey, P.D., Hill, M.O., Meek, W.R., Rothery, P., Smart, S.M., Smith, K. and Walker, J. (2003) *The Changing Distribution of the Flora of the United Kingdom: Technical Report CEH PROJECT No: C01093*. Centre for Ecology and Hydrology, Monks Wood, Abbots Ripton, UK.
Preston, C.D., Pearman, D.A. and Hall, A.R. (2004) Archaeophytes in Britain. *Botanical Journal of the Linnean Society* 145, 257–294.
Rich, T.C.G. (1999) List of plants endemic to the British Isles. *BSBI Newsletter* 80, 23–27.
Rich, T.C.G. and Woodruff, E.R. (1996) Changes in the vascular plant floras of England and Scotland between 1930–1960 and 1987–1988: the BSBI Monitoring Scheme. *Biological Conservation* 75, 217–229.
Rich, T.C.G., Hack, V. and McMeechan, F. (2005) Plant conservation evaluation criteria. In: Hill, D., Fasham, M., Tucker, G., Shewry, M. and Shaw, P. (eds) *Handbook of Biodiversity Methods*. Cambridge University Press, Cambridge, pp. 303–321.
Rodrigues, A., Rodrigues, S.L., Pilgrim, J.D., Lamoreux, J.F., Hoffmann, M. and Brooks, T.M. (2006) The Value of the IUCN Red List for conservation. *Trends in Ecology and Evolution* 21(2), 71–76.
Sanderson, H. and Prendergast, H.D.V. (2002) *Commercial Uses of Wild and Traditionally Managed Plants in England and Scotland*. Royal Botanic Gardens, Kew, UK.
Schultze-Motel, J. (1966) Verzeichnis Forstlich Kultivierter Pflanzenarten [Enumeration of Cultivated Forest Plant Species]. *Kulturpflanze Beiheft* 4, 486.
Smartt, J. and Simmonds, N.W. (1995) *Evolution of Crop Plants*. Longman Scientific & Technical, Harlow, UK.
Stace, C.A. (1997) *New Flora of the British Isles*. Cambridge University Press, Cambridge.
Stolton, S., Maxted, N., Ford-Lloyd, B.V., Kell, S.P. and Dudley, N. (2006) *Food Stores: Using Protected Areas to Secure Crop Genetic Diversity*. WWF Arguments for protection series. Gland, Switzerland.
Telfer, M.G., Preston, C.D. and Rothery, P. (2002) A general method for measuring relative change in range size from biological atlas data. *Biological Conservation* 107, 99–109.
Walters, S.M. (1984) The relationship between the British and the European floras. *New Phytologist* 98, 3–13.
Wiersema, J.H. and Leon D. (1999) *World Economic Plants – A Standard Reference*. CRC Press, Boca Raton, Florida.
Wiggington, M.J. (1999) *British Red Data Books: 1. Vascular Plants*. Joint Nature Conservation Committee, Monks Wood, UK.
Wilson, P. and King, M. (2003) *Arable Plants – a Field Guide*. English Nature and WILD Guide, Peterborough, UK.
Zeven, A.C. and Zhukovsky, P.M. (1975) *Dictionary of Cultivated Plants and Their Centres of Diversity. Excluding Ornamentals, Forest Trees and Lower Plants* PUDOC, Wageningen, The Netherlands.

8 National Crop Wild Relative *In Situ* Conservation Strategy for Russia

T.N. Smekalova

8.1 Introduction

The increasing threat of genetic erosion for plant biodiversity caused by negative anthropogenic effects and subsequent global climatic changes calls for a vital need to develop a universal strategy of agrobiodiversity conservation – both in genebanks (*ex situ*) and in natural plant communities (*in situ*) including maintenance of local populations representing cultivated (on farm) and wild species (in nature). In most of the states, biodiversity conservation is implemented by many different organizations:

- Botanical gardens, nurseries, *in vivo* collections and genebanks preserve separate components of plant biodiversity.
- Scientific institutions on the basis of their research efforts give recommendations on the protection of individual plant species (by publishing 'Red Books', issuing special lists, etc).
- Ministries (of nature, ecology and resources), agencies and committees work out legal acts, adopt resolutions concerning the protection of separate plant species and populations, establish protected natural areas of various categories, and launch special programmes and projects to conserve separate biodiversity components.

At the national level, the strategy of *in situ* conservation and sustainable utilization of biodiversity components is regulated by a *national programme*, developed with regard to national specific features and adopted by the government.

8.2 Development of a National Plant Genetic Resource *In Situ* Conservation Strategy for Russia

For Russia, as for all other countries, the problems of plant genetic resource (PGR) *in situ* conservation are of primary importance. On the one hand, the country experiences constant and ceaseless depletion of agrobiodiversity components. Specifically:

- In the past decades local crop populations have been constantly and often unreasonably substituted by modern breeding cultivars.
- The present situation, after the disintegration of the Soviet Union, is marked by drastic changes in the structure of agricultural production, land management and territorial distribution of crops between plant cultivation areas.
- Replacement of the traditional assortment of cultivated varieties is accompanied by disappearance of associated weedy plants, most of which are close relatives of cultivated crops.
- Increasing anthropogenic effect reduces the genetic potential of the species representing wild relatives of cultivated plants (WRCP), which serve as an indispensable source of continuous enrichment of crop gene pools with valuable genes.

At the same time, it is true that:

- Local administrative authorities are often uninterested in the establishment of new protected areas within their territories, being much more concerned with further intensification of commercial land management (construction, profit-yielding, agriculture, etc).

All these processes require urgent and thorough examination of the current status of agrobiodiversity and, if necessary, development of specific recommendations and taking measures to preserve its separate components.

Nowadays, however, especially after the transformation of the USSR, the country lacks clear coordination between different agencies involved in plant diversity conservation and needs a unified national conservation strategy. None of these agencies and organizations is specifically responsible for protecting separate components of agrobiodiversity, except for the Vavilov Institute of Plant Industry (VIR), which has recently been working out a scientifically justified integral strategy of PGRFA conservation.

Since the time of its foundation (1894), the VIR has been dealing with the problems of collecting, studying and preserving cultivated plants and their wild relatives both in genetic collections and within natural plant communities. All the materials collected by Vavilov and his colleagues were thoroughly studied at VIR. The results of such research allowed his followers to work out conservation guidelines.

The works by Regel (1922), Vavilov (1962a,b, 1964), Brezhnev and Kovorina (1981), Zhukovsky (1964), Wulf and Maleeva (1969) and other Russian scientists laid the foundation of the modern strategy for *in situ* conservation of PGR.

During the last 5–10 years, diverse countries offered varied models for PGR conservation. We used the model proposed by Maxted *et al.* (1997); in our opinion, it is the most universal and scientifically justified one. This model, if amended and corrected in accordance with national specificity, is quite suitable for a majority of countries.

8.3 Development of the National *In Situ* Crop Wild Relative Conservation Strategy for Russia

In order to implement the *in situ* conservation strategy for crop wild relatives (CWR) in Russia, a number of tasks have been specified:

- Selection of optimal methodologies;
- Preliminary complex study of conservation objects (CWR species, their morphological, ecological, taxonomic, geographic and other features), territories and existing conservation strategies;
- Development of databases (DB) and an information retrieval system (IRS) for storage and analysis of the collected information;
- Definition of conservation priorities (objects, territories);
- Analysis of local populations representing priority species for conservation, with identification of conservation priorities among the populations;
- Monitoring of individual species and individual populations;
- Development of specific recommendations and measures for conservation (management).

8.3.1 Selection of optimal methodologies

The first task and the first phase of work refer to the definition of the concept of CWR as the main research object, i.e. specifying the meaning of the term CWR. There are different concepts of CWR. More common is a narrower understanding of CWR, when they are considered to include the species of natural vegetation that directly participated in the formation of cultivated species. We prefer a broader meaning of CWR, assuming that they should include:

> the species of natural vegetation, evolutionarily and genetically related to cultivated ones, being within the same genus as cultivated plant species, domesticated or potentially capable of domestication, participating or potentially capable of participation in crosses or other applications (as seedling stock) with the purpose of obtaining new cultivars or improving the existing ones
>
> (Smekalova and Chukhina, 2003).

8.3.2 Generation of national CWR inventory for Russia

The next phase is the development of a national CWR inventory for Russia. It is impossible to start conservation work without accurate inventory of CWR species growing in the studied territory. This list continues to be adjusted in the course of further revision of the composition of Russia's flora and domesticated plant species. The development of such a list should be preceded by thorough analysis of the species' nomenclature and taxonomy, when the scope, structure and name of each species are verified. Today, the list comprises 1629 species, representing 43 families and 195 genera. The list is not just an enumeration of updated nomenclature combinations; it provides information on the distribution of the species, their taxonomic composition, growing habits, ways of utilization, etc. Such information on every species is loaded into the CWR database 'Wild Relatives of Cultivated Plants in Russia'. The latest version of IRS contains the data of nomenclature and general characteristics of CWR species occurring in Russia. General characterization includes: the rank according to the degree of relationship with cultivated plants; distribution pattern (general and more specific within Russia); most typical habitats; life form category; ways of utilization; and conservation criteria. The system makes it possible to search and select data concerning nomenclature, geography (regions, administrative provinces and districts of Russia and natural reserves), utilization characteristics, ranking groups according to the degree of their relationship to cultivated plants and conservation criteria.

The analysis of the national CWR inventory found that the largest group of species based on their utilization was the forage plants, followed by food (fruits, berries and vegetables) and industrial plants. Geographic analysis demonstrated non-uniform distribution of CWR species through different regions of Russia. From a floristic point of view, the European part of Russia is very diverse, blending together floras of various origins; that is why the greatest number of CWR species (838) occurs within this territory. Most of these species are widespread throughout the Holarctic floristic realm. The Caucasus, on the whole, cannot be numbered among the largest by its size, but this region is one of the richest as far as WRCP species are concerned (738). The flora of the Russian Far East is influenced by the East Asiatic centre of species' formation; therefore, the local range of WRCP is quite unique and contains numerous species which occur only within this territory (223 out of 598). The least in number are the WRCP species of Western Siberia (529).

8.3.3 Priority species for conservation

In order to select a conservation object we identified criteria of conservation priority for CWR species of Russia. There are two criteria:

1. Criterion of relationship and economic value, consisting of the following parameters:

- Participation in the breeding process (direct utilization, participation in hybridization, utilization as a donor of useful properties or as seedling stock, etc.);
- Extent of applicability for economic purposes;
- Systematic closeness to a cultivated species.

2. Criterion of rarity and vulnerability.

Each species from the CWR list needs to be analysed according to both criteria. The analysis based on criterion (1) has divided (classified) the list into five groups (Table 8.1).

It appears that among all CWR species, only 200 fall into the first ranking group, which means that only one-tenth of the national useful plant genetic diversity is intensively utilized in agricultural production. The plants of the first two groups (261 species) are most actively used in breeding practice. The species of the third

Table 8.1. Ranking groups of CWR in Russia (total = 1629 species).

Ranking group	What species are included?	Number of species
1	The species is cultivated, contains varieties and is economically important	202
2	The species participates in crosses and is used as seedling stock or a source of genes	60
3	Promising for utilization; closely related to a cultivated species (within one section or one subgenus)	160
4	Other useful species of the same genus, objects of plant hunting and folk breeding (no varieties)	297
5	All other species of the genus	910

and fourth groups represent potential reserve for economy; hence, they are also potentially important, interesting for researchers and promising for utilization.

In terms of significance, criterion (2) of rarity and vulnerability is the most important one. CWR species have unequal degrees of rarity, vulnerability, threat of extinction, etc. Some of them are listed in the international and regional 'Red Books' and attributed to different categories of rarity according to the International Union for the Conservation of Nature (IUCN) classification. Such species are subject to top-priority *in situ* conservation. They should also include local endemics and subendemics of various regions as well as, in some cases, relicts of different epochs and CWR species having only a small part of their areas of distribution within Russia.

After analysing the ranking list of CWR according to both criteria, a group of CWR species were identified, which require urgent conservation measures (priority objects for conservation) within their natural coenoses (*in situ*). This group comprises the species of CWR referred to the above-mentioned categories of rarity and local endemics and subendemics of different regions of Russia; the most important from an economic viewpoint are those included in the first and second ranking groups. At present, the list of priority conservation objects contains about 340 CWR species.

8.3.4 Selection of conservation priority sites

In order to determine which territory should be assigned for conservation of priority species, it is necessary to analyse their areas of distribution. Geographic information system (GIS) mapping of the areas of priority species for conservation is based mainly on the geographic data on the labels of herbarium specimens, and other kinds of information. Superposition of the area maps of priority taxa for conservation makes it possible to identify the places of their highest concentration – the 'concentration zones'.

It is impossible to conserve all priority species not only throughout their natural habitats in Russia and adjacent countries, but also within the limits of the identified areas in Russia. There is also no possibility to arrange special

Table 8.2. Ranking scale of CWR in strict natural reserves of Russia.

Rank	Number of species
I	158
II	45
III	112
IV	245
V	587

reserves for their conservation even in the site of their maximum concentration. Meanwhile, a group of CWR species are now under the actual threat of extinction and require urgent protection. The most realistic possibility is, therefore, to preserve these species within the existing network of protected natural territories (PNTs), which has been functioning in Russia. For this, we have made a conjugate analysis of the databases 'Wild Relatives of Cultivated Plants in Russia' and 'Vascular Plants in Russian Reserves', which allowed us to investigate CWR within the national network of natural reserves. It appeared that the territories of 91 reserves (out of 100 functioning in October 2004) harbour 1147 species of wild crop relatives from 39 plant families, i.e. 71.2% of their total number. About 29% of all CWR (463 species) do not occur in any of Russia's natural reserves (at least by now they have not yet been spotted). The ranking scale of the species is presented in Table 8.2.

The CWR most actively used by breeders (those of the first two ranking groups) are represented in the reserves by 203 species (approximately 18% of the total number of CWR). Very high numbers of CWR species (408, or 36% of the total number) were found in the territory of only one of Russia's natural reserves. The greatest numbers of CWR species occur in the reserves of the Caucasus, southern mountains of Siberia, maritime areas of the Far East and the southern areas of the European part of Russia (the total numbers of species in the protected flora are indicated in Table 8.3 in parentheses).

Table 8.3. CWR species in the natural reserves of Russia.

Natural reserves	No. of CWR	Natural reserves	No. of CWR
Kavkazsky	246 (1439)	Bolshekhekhtsyrsky	175 (944)
Altaisky	237 (1357)	Ilmensky	173 (936)
Khopersky	231 (1159)	Sikhote-Alinsky	172 (1064)
Lazovsky	230 (1272)	Privolzhskaya Lesostep	171 (824)
Tsentralno-Chernozemny	228 (1036)	Sokhondinsky	169 (988)
Zhigulevsky	202 (1030)	Khingansky	169 (972)
Belogorye	202 (958)	Ussuriysky	162 (815)
Teberdinsky	196 (1120)	Kedrovaya Pad	160 (909)
Voronezhsky	195 (978)	Sayano-Shushensky	159 (967)
Galichya Gora	190 (885)	Prioksko-Terrasnyi	157 (888)
Kabardino-Balkarsky	175 (1043)		

Table 8.4. CWR species included in the Red Book of the RSFSR (1988).

Allium altaicum Pall.	*Juglans ailanthifolia* Carr.
A. pumilum Vved.	*Lathyrus litvinovii* Iljin
Armeniaca mandshurica (Maxim.) Skvortsov	*Lespedeza tomentosa* (Thunb.) Maxim.
Asparagus brachyphyllus Turcz.	*Medicago cancellata* Bieb.
Corylus colurna L.	*Poa radula* Franch. et Savat.
Elytrigia stipifolia (Czern. ex Nevski) Nevski	*Prinsepia sinensis* (Oliv.) Bean
Festuca sommieri Litard.	*Rheum altaicum* Losinsk.
Ficus carica L	*Secale kuprijanovii* Grossh.
	Staphylea colchica Stev.
	Viburnum wrightii Miq.

Eighteen species of CWR were included in the Red Data Book of Russia (1988), see Table 8.4. Six species were included in the International Red List of Threatened Species: *Allium altaicum* Pall., *Allium pumilum* Vved , *Elytrigia stipifolia* (Czern. ex Nevski) Nevski, *Secale kuprijanovii* Grossh , *Medicago cancellata* Bieb. and *Staphylea colchica* Stev.

Thus, the future of the species within the boundaries of protected areas seems more favourable than the future of the species that occur beyond these boundaries. The species that never occur in any of the reserves have already been identified. Among them are the wild persimmon *Dyospirus lotus* and wild pomegranate *Punica granatum*. The sites outside the protected areas where CWR are concentrated could be transformed into microreserves and used either to establish new protected areas or expand the existing ones. CWR species will be the major objects of protection within these sites.

We consider these components of the strategy more or less developed. Already underway is the next stage of work: development of a monitoring programme in order to formulate specific conservations measures.

8.3.5 Further study with the purpose of formulating conservation measures

Development of specific conservation measures should be preceded by a thorough and comprehensive study of the conservation objects, including their morphological, taxonomic, biological, geographic, ecological and other features. Therefore, in order to work out a monitoring programme it is necessary to conduct complex (geobotanical, phytocoenotic, populational, etc.) research on each of the priority species. Most important are complex populational studies, which should include not only the analysis of quantity, structure and productivity, but also the assessment of the value of a population and prognosis of its viability. The results of these research efforts should be used to develop a system of monitoring for all priority species and specific measures of their conservation. Analysis of populations is to be performed on a yearly basis for annual species, once in 3–5 years for perennial herbaceous plants, and once in 3–7 years for perennial trees and shrubs, depending on the specificity of the object and the status of its population.

8.3.6 Development of specific recommendations and conservation measures

In order to make a list of specific *in situ* conservation measures the following are recommended:

- Talk to the staff of natural reserves to raise the status of CWR conservation within their reserve; on the basis of floristic inventories of the reserves to make lists of CWR and priority species to be given conservation priority within the natural reserves.
- Supply the reserves with information on biological geographical and ecological features of the CWR species to assist their conservation.
- Prepare documentation and other materials for the nature-protecting authorities in order to justify the need to include CWR in the lists of top-priority objects for *in situ* conservation within the protected natural areas, expand the territories of the existing protected areas and, in individual extraordinary cases, establish new protected areas of various ranks where CWR species will be the major conservation objects.
- Assess, on the basis of monitoring results, the viability of local populations of the priority CWR species within the protected natural areas; in the case when a population is reduced in number or threatened, provide local authorities, environmental agencies and staff members of the protected areas with recommendations concerning its conservation, maintenance in a balanced state and, in some cases, restoration of the population's components.
- Issue recommendations for each CWR species concerning the most appropriate choice of conservation strategies – *in situ* or *ex situ*.

8.4 Conclusion

The accumulated experience of the world community in the face of the current extinction crisis is that the application of a *complementary strategy* of *ex situ* together with *in situ* (including on-farm) conservation offers the most appropriate solution, especially for the species whose populations are in a disturbed state or under actual threat of extinction. The strategy outlined above follows this approach and can be used as a model for other countries and territories worldwide.

References

Brezhnev, D.D. and Kovorina, O.N. (1981) *Wild Relatives of Cultivated Plants in the USSR Vegetation*. Kolos, Leningrad.

Maxted, N., Ford-Lloyd, B.V. and Hawks, J.G. (eds) (1997) *Plant Genetic Conservation: the* in situ *Approach*. Chapman & Hall, London.

Red Data Book of Russia (1988) Rosagropromizdat Publisher, Moscow.

Regel, R.E. (1922) *Corns in Russia*. Nauka, Leningrad (*Hleba v Rossii*, in Russian).

Smekalova, T.N. and Chukhina, I.G. (2003) Crop wild relatives conservation strategy in Russia. In: *Botanical Researches in the Asian Russia: Proceeding of the XI Congress of the*

Russian Botanical Society (18–22 August 2003, Novosibirsk-Barnaul). Azbuka, Barnaul, pp.118–120.

Vavilov, N.I. (1962a) *World Plant Resources and its Use in Breeding* (*Miroviye Rastitelniye resursy i ih ispolzovanie v selekcii, in Russian*), Selected publications, vol. III. USSR Academy of Sciences Publishers, Moscow-Leningrad.

Vavilov, N.I. (1962b) *Five Continents* (*Pyat' Kontinentov,* translated from Russian, 1987). IPGRI, Rome.

Vavilov, N.I. (1964) *World Resources of Cereals. Wheat (Miroviye Resursy Hlebnyh Zlakov. Pshenitsa*, in Russian). USSR Academy of Sciences Publishers, Moscow-Leningrad.

Wulf, E.V. and Maleeva, O.F. (1969) *World Resources of Useful Plants. Food, Forage, Medicinal, etc.* Nauka, Leningrad.

Zhukovsky, P.M. (1950) *Cultivated Plants and their Wild Relatives* (translated by P.S. Hudson). CAB International, Wallingford, UK.

Zhukovsky, P.M. (1964) Kulturnye rasteniya i ikh sorodichi [Cultivated plants and their relatives]. Kolos, Leningrad.

9 Diversity and Conservation Needs of Crop Wild Relatives in Finland

H. KORPELAINEN, S. TAKALUOMA, M. POHJAMO AND J. HELENIUS

9.1 Introduction

The wild ancestors as well as the wild relatives of crop plants provide potentially important genetic resources for use in plant breeding and consequent plant production. In addition, some wild plants are suitable for direct exploitation through harvesting from nature. The importance of wild ancestors as raw material for plant breeding is well known and acknowledged. However, the significance of crop wild relatives (CWR – generally defined as taxa belonging to the same genus as a crop plant) has generally been overlooked. Gradually, for the most part of the last 10 years, theoretical, methodological and experimental investigations aimed specifically at CWR have been initiated to discover the genetic resources present among them (Maxted, 2003; Meilleur and Hodgkin, 2004). The uses of CWR are not limited to food production, but may also include uses as fodder, ornamentals, fibre, timber trees, herbs and condiments, and as medicinal plants. At present, cultivated plants form only a small portion of all known species, and it is apparent that there are possibilities for wider exploitation of plant resources.

It is noteworthy that among vascular plants regarded as being threatened to some degree, many are CWR. These plants are not only a component of biological diversity and material for evolution by natural selection, but they also constitute invaluable genetic resources for crop improvement (Tanksley and McCouch, 1997). This is an additional important aspect to consider when prioritizing the conservation of threatened plants. For instance, weeds that have been subjected to coevolution or codomestication with cultivated plants have also developed a wide range of diversity and, occasionally, they possess value for cultivation (Hammer *et al.*, 1997).

This chapter describes the diversity and composition of CWR in Finland, including established species, subspecies, variants and hybrids. A wide definition of CWR is being used, which includes wild plants belonging to the same genus as present crops, wild plants rarely used as crops and their relatives and wild plants with known exploitation (e.g. berries). A special emphasis is placed on such plants which are presently threatened and provide information on their habitats, threats and conservation status and on their use or potential use.

9.2 Diversity of Finnish Crop Wild Relatives

A list of CWR and their conservation situation has recently been compiled to fill the gap in the knowledge of Finnish CWR (Takaluoma, 2005). The compilation was conducted using the existing lists of wild vascular plants (Hämet-Ahti *et al.*, 1998) and crop plants (Räty and Alanko, 2004) as the main sources. Table 9.1 summarizes the results, which revealed that up to 60% of the wild vascular flora of Finland can be classified as CWR, of which more than one-third already have some known use. The list of Finnish CWR includes plants from 63 families. The following seven families comprised 54% of all CWR: *Cyperaceae* 12.8%, *Asteraceae* 11.7%, *Poaceae* 7.7%, *Rosaceae* 6.2%, *Fabaceae* 5.9%, *Brassicaceae* 4.9% and *Caryophyllaceae* 4.8%. Of all CWR with known uses, the great majority are ornamentals, 4.7% have some use as food, 1.1% as aromatic plants, 1% as fodder, 0.6% as medicinal plants and 0.6% as grass plant crops. Just a few taxa are used as a source for fibre or dyes.

Many wild taxa, especially berries, are commonly collected for human consumption, primarily those from the genera *Vaccinium* (*Ericaceae*) and *Rubus* (*Rosaceae*). The most important berry species used as food include *Vaccinium vitis-idaea* L., *V. myrtillus* L., *V. oxycoccus* L., *V. uliginosum* L., *Rubus chamaemorus* L., *R. idaeus* L., *R. arcticus* L. and *Fragaria vesca* L. (*Rosaceae*). The genera *Vaccinium*, *Rubus* and *Fragaria* include a total of six, 15 and five wild taxa, respectively. Other potentially interesting CWR for berry or fruit production are *Malus sylvestris* (L.) Mill. (*Rosaceae*, in addition its hybrids with *M. domestica* Borkh.), *Hippophaë rhamnoides* L. (*Elaeagnaceae*), 14 *Ribes*

Table 9.1. Statistics of the Finnish CWR taxa including species, subspecies, variants and hybrids.

	Number	Percentage of vascular flora	Percentage of CWR
The whole vascular flora	~3200	100	–
CWR taxa	1905	60	100
CWR with no known use	1177	37	62
CWR with some known use	728	23	38
IUCN category of CWR			
Near threatened (NT)	70	2.2	3.7
Vulnerable (VU)	69	2.2	3.6
Endangered (EN)	37	1.2	1.9
Critically endangered (CR)	28	0.9	1.5
Regionally extinct (RE)	5	0.2	0.3

taxa (*Grossulariaceae*), *Prunus spinosa* L. (*Rosaceae*, in all eight wild *Prunus* taxa) and *Empetrum nigrum* L. (*Empetraceae*, in all three wild *Empetrum* taxa).

Exploitable aromatic herbs include *Allium schoenoprasum* L. (*Alliaceae*, in all 11 wild *Allium* taxa), *Origanum vulgare* L. (*Lamiaceae*, the only wild *Origanum* taxon), *Thymus serpyllum* L. (*Lamiaceae*, in all seven wild *Thymus* taxa) and *Angelica archangelica* L. (*Apiaceae*, in all four wild *Angelica* taxa). Prospective legumes (*Fabaceae*) include 19 wild taxa of the genus *Trifolium*, 11 of *Medicago*, eight of *Lotus* and six of *Melilotus*. Potential grass plant crops (*Poaceae*) include five wild taxa of the genus *Phleum*, 16 of *Festuca*, five of *Lolium*, 17 of *Poa*, two of *Phalaris*, nine of *Alopecurus* and the species *Dactylis glomerata* L. Annual cereal crop CWR (*Poaceae*) include taxa from the genera *Elymus* (nine taxa), *Leymus* (one taxon), *Avena* (three taxa), and other annual crop CWR include a number of *Brassicaceae*, such as ten *Brassica*, one *Alliaria*, four *Barbarea*, five *Erysimum*, two *Raphanus* and four *Sinapis* taxa, and there are 12 *Solanum* taxa (*Solanaceae*). Among ornamental plants, the genus *Rosa* (*Rosaceae*) with 14 wild taxa possesses the highest potential for exploitation, primarily concerning the species *R. dumalis* Bechst., *R. majalis* Herrm. and *R. acicularis* Lindl.

9.3 Conservation Needs of Finnish Crop Wild Relatives

9.3.1 Conservation status

Information on the conservation status, habitats and risk factors of the Finnish CWR (Tables 9.1 and 9.2) was obtained from the publications of the Finnish Environment Institute (Rassi *et al.*, 2001; Internet publications, 2004–2005). It was discovered that 11% of the CWR taxa (6.5% of the whole vascular flora) belonged to some International Union for the Conservation of Nature (IUCN) category of threat (near threatened, vulnerable, endangered, critically endangered or regionally extinct). The knowledge of the taxonomic diversity and the degree of rarity among Finnish CWR is quite good. However, what is lacking among the majority of any kind of wild plants is detailed information on their demography, and even more so, on the genetics of their populations.

Threatened CWR having important wild or cultivated crops within the genus include the berries *Fragaria viridis* Duchesne (VU), *Rubus aureolus* Allander (VU), *R. humulifolius* C.A. Mey (RE) and *R. pruinosus* Weihe ex Gűnther *et al.* (VU). Among fruit tree CWR, the threatened taxa include *M. sylvestris* (VU), *P. spinosa* (NT), *Sorbus intermedia* (Ehrh.) Pers. (NT) and *S. meinichii* (Lindeb.) Hedl. (CR), all from the family *Rosaceae*. Other threatened CWR with potentially important exploitation possibilities are *A. schoenoprasum* subsp. *sibiricum* (L.) Čelak (NT), *A. ursinum* L. (NT), *A. vineale* L. (NT), *Trifolium aureum* Pollich (NT), *T. fragiferum* L. (NT), *T. montanum* L. (NT), *T. spadiceum* L. (NT), *Phleum phleoides* (L.) H. Karst. (NT), *P. pratense* subsp. *nodosum* (L.) Arcang. (NT), and *Festuca gigantea* (L.) Vill. (EN). Most of the endangered or critically endangered CWR (Table 9.2) possess potential as ornamental plants, such as *Rosa canina* L. (CR) and *R. sherardii* Davies (EN).

Table 9.2. Habitats, risk factors and uses (food, fodder, ornamental, medicinal, aromatic, dye or grass plant crop) of endangered and critically endangered Finnish CWR taxa (all numbers and uses refer to the situation in Finland).

Species	Origin	IUCN category	Main habitat	Main risk factors	Uses	Wild crops/all wild taxa within genus	Percentage threatened or near-threatened within genus	Farmed crops within genus	Uses within genus
Agrimonia pilosa Ledeb.	Archaeophyte	EN	Cultural habitats	Habitat destruction	Med.	1/3	33	0	Ornam., med.
Anagallis minima (L.) E.H.L. Krause	Native	EN	Shores of the Baltic Sea	Small population numbers, habitat destruction	–	1/2	50	1	Ornam.
Androsace septentrionalis L.	Native Archaeophyte	EN	Cultural habitats	Small population numbers, habitat destruction	–	0/3	33	6	Ornam.
Anemone patens L.	Native	EN	Esker forests	Habitat destruction	Ornam.	8/11	36	15	Ornam.
Anthyllis vulneraria ssp. *polyphylla* D.C. Nyman	Native	CR	Esker forests	Small population sizes	–	1/6	16	1	Ornam.
Arctium nemorusum Lej.	Native	EN	Herb-rich forests, cultural habitats	Small population numbers, habitat destruction	–	1/6	16	0	Food, ornam.
Arenaria norvegica Gunnerus	Native	EN	Alpine habitats	Small population sizes	–	0/3	66	4	Ornam.
Armeria maritima ssp. *elongata* (Hoffm.) Bonnier	Archaeophyte	EN	Cultural habitats, shores of the Baltic Sea	Small population numbers	Ornam.	2/5	60	2	Ornam.

Continued

Table 9.2. *Continued*

Species	Origin	IUCN category	Main habitat	Main risk factors	Uses	Wild crops/all wild taxa within genus	Percentage threatened or near-threatened within genus	Farmed crops within genus	Uses within genus
Armeria maritima ssp. *intermedia* (T. Marsson) Nordh.	Archaeophyte	CR	Cultural habitats	Small population sizes, habitat destruction	–	2/5	60	2	Ornam.
Armeria maritima ssp. *sibirica* (Turcz. ex Boiss.) Nyman	Native	EN	Alpine habitats	Small population sizes	–	2/5	60	2	Ornam.
Artemisia campestris ssp. *bottnica* Lundstr. ex Kindh.	Native	CR	Shores of the Baltic Sea	Small population sizes	–	3/11	9	8	Arom., med., ornam.
Asperula tinctoria L.	Native	EN	Cultural habitats	Small population numbers, habitat destruction	Dye	1/3	33	2	Ornam., dye
Asplenium adulterinum Milde	Native	EN	Serpentine rocks	Small population numbers, habitat destruction	–	1/8	38	1	Ornam.
Astragalus glycyphyllos L.	Established alien	CR	Heath forests	Small population sizes	–	1/6	16	0	Ornam., med., fodder
Bromus benekenii (Lange) Trimen	Native	CR	Herb-rich forests	Small population numbers, habitat destruction	–	1/12	8	0	Ornam., fodder
Cardamine flexuosa With.	Native	EN	Springs, spring fens	Small population numbers, habitat destruction	–	1/10	30	0	Ornam.
Cardamine impatiens L.	Native	EN	Herb-rich forests	Small population sizes	–	1/10	30	0	Ornam.

Cardamine parviflora L.	Archaeophyte	EN	Shores, ditches	Decline in the area of occupancy, habitat destruction	–	1/10	30	0	Ornam.
	Native								
Carex hartmanii A. Cajander	Native	EN	Cultural habitats	Habitat destruction	–	0/148	15	4	Ornam.
Carex lepidocarpa ssp. *lepidocarpa* (Tausch) J. Lange	Native	EN	Rich fens	Habitat destruction	–	0/143	15	4	Ornam.
Carex microglochin Wahlenb.	Native	EN	Open shores, rich fens, alpine wetlands	Habitat destruction	–	0/143	15	4	Ornam.
Carex ornithopoda Willd.	Native	CR	Cultural habitats	Small population numbers, habitat destruction	–	0/143	15	4	Ornam.
Carex remota L.	Native	EN	Herb-rich forests	Small population sizes	–	0/143	15	4	Ornam.
Carex vulpina L.	Archaeophyte	EN	Cultural habitats	Small population numbers, habitat destruction	–	0/143	15	4	Ornam.
Carlina biebersteinii Bernh. ex Hornem.	Archaeophyte	EN	Cultural habitats	Small population numbers, habitat destruction	–	1/2	100	2	Ornam.
Cerastium alpinum L. (Kaavi race)	Native	CR	Serpentine rocks	Small population sizes	Ornam.	3/18	28	3	Ornam.
Cerastium alpinum (Keski-Lappi race)	Native	EN	Serpentine rocks	Small population numbers	Ornam.	3/18	28	3	Ornam.
Cerastium fontanum ssp. *vulgare* var. *kajanense* (Kotilainen & Salmi) Jalas	Archaeophyte	EN	Serpentine rocks	Small population numbers	–	3/18	28	3	Ornam.

Continued

Table 9.2. *Continued*

Species	Origin	IUCN category	Main habitat	Main risk factors	Uses	Wild crops/all wild taxa within genus	Percentage threatened or near-threatened within genus	Farmed crops within genus	Uses within genus
Crepis praemorsa (L.) Tausch	Archaeophyte	EN	Esker forests, cultural habitats	Habitat destruction	–	1/7	29	2	Ornam.
Crepis tectorum ssp. *nigrescens* (Pohle) A. Löve & D. Löve	Native	CR	Rocky habitats	Small population numbers	–	1/7	29	2	Ornam.
Dianthus superbus L.	Native	CR	Serpentine rocks	Small population sizes	Ornam.	4/5	60	17	Ornam.
Draba alpina L.	Native	CR	Alpine habitats	Small population sizes	–	0/12	58	2	Ornam.
Epilobium laestadii Kytövuori	Native	EN	Rich fens, springs	Small population numbers, habitat destruction	–	2/27	11	1	Ornam.
Erica tetralix L.	Native	CR	Fens	Small population sizes, habitat destruction	Ornam.	1/1	100	2	Ornam.
Festuca gigantea (L.) Vill	Native	EN	Herb-rich forests, springs	Small population numbers, habitat destruction	–	6/17	6	5	Ornam., grass, fodder
Galium saxatile L.	Archaeophyte	CR	Cultural habitats	Small population numbers, habitat destruction	–	3/19	21	0	Ornam.
Galium schultesii Vest.	Archaeophyte	CR	Heath forests	Small population sizes	–	3/19	21	0	Ornam.

Gentianella campestris (L.) Börnen	Archaeophyte	EN	Cultural habitats, herb-rich forest	Habitat destruction	–	1/4	100	13	Med.
Hippuris tetraphylla L. f.	Native	EN	Baltic Sea, small ponds	Habitat destruction	–	1/3	33	0	Ornam.
Hypericum montanum L.	Native	CR	Herb-rich forests	Small population sizes	–	2/4	25	2	Ornam.
Lonicera caerulea L.	Native	EN	Lake and river shores, herb-rich forest	Small population sizes	Ornam.	5/5	20	21	Ornam., food
Melica ciliata L.	Native	CR	Rocky habitats	Small population sizes	Ornam.	3/4	75	1	Ornam.
Petasites spurious (Retz.) Rchb.	Established alien	CR	Shores of the Baltic Sea	Small population sizes, habitat destruction	–	2/4	25	0	Ornam.
Pimpinella major (L.) Huds.	Archaeophyte	CR	Cultural habitats	Small population sizes, habitat destruction	Med.	1/3	33	0	Arom., med., ornam.
Polygala comosa Schkuhr	Native	EN	Cultural habitats, shores of the Baltic Sea	Small population numbers, habitat destruction	–	0/3	100	1	Ornam., med.
Polygonum oxyspermum C.A. Mey. & Bunge	Native	CR	Shores of the Baltic Sea	Small population sizes	–	1/8	13	4	Ornam.
Primula farinosa L.	Native	EN	Cultural habitats, shores of the Baltic Sea	Habitat destruction	Ornam.	4/5	60	24	Ornam.

Continued

Table 9.2. *Continued*

Species	Origin	IUCN category	Main habitat	Main risk factors	Uses	Wild crops/all wild taxa within genus	Percentage threatened or near-threatened within genus	Farmed crops within genus	Uses within genus
Primula nutans var. *jokelae* L. Mäkinen & Y. Mäkinen	Native	EN	Shores of the Baltic Sea	Habitat destruction	–	3/5	60	24	Ornam.
Rosa canina L.	Native	CR	Cultural habitats, herb-rich forests	Small population sizes	Food, ornam.	9/14	14	35	Food, ornam.
Rosa sherardii Davies	Native	EN	Cultural habitats, herb-rich forests	Small population sizes	Ornam.	9/14	14	35	Food, ornam.
Rumex maritimus L.	Native	EN	Shores and cultural habitats	Small population sizes	–	4/41	5	2	Food, ornam.
Salix arbuscula L.	Native	EN	Alpine heaths	Small population sizes	–	27/61	7	19	Ornam.
Salix pyrolifolia Ledeb.	Native	CR	Rich fens	Small population sizes	Ornam.	27/56	7	19	Ornam.
Sedum villosum L.	Native	EN	Alpine wetlands	Small population sizes	–	13/15	7	28	Ornam.
Silene furgata ssp. *angustiflora* (Rupt.) Walters	Native	CR	Limestone rocks	Small population sizes	–	5/20	15	10	Ornam.
Sium latifolium L.	Native	CR	Aquatic habitats	Small population sizes	–	0/1	100	1	Food

Sorbus meinichii (Lindeb.) Hedl.	Native	CR	Cultural habitats	Small population sizes	Ornam.	5/10	20	16	Ornam., food
Stellaria crassifolia var. *minor* Wahlenb.	Native	CR	Shores of the Baltic Sea	Small population numbers, habitat destruction	–	3/16	19	0	Ornam.
Thalictrum aquilegiifolium L.	Native	CR	Herb-rich forests	Small population sizes	Ornam.	4/11	36	7	Ornam.
Thalictrum lucidum L.	Archaeophyte	CR	Cultural habitats	Small population sizes	Ornam.	4/11	36	7	Ornam.
Veratrum album L.	Native	CR	Lake and river shores	Small population sizes	Ornam.	1/1	100	1	Ornam.
Veronica alpina ssp. *pumila* (All.) Pennell	Native	CR	Alpine habitats	Small population sizes	–	7/29	10	6	Ornam.
Viola collina Besser	Native	EN	Herb-rich forests	Small population numbers, habitat destruction	–	8/28	18	5	Ornam.
Viola reichenbachiana Jord. ex Boreau	Native	EN	Herb-rich forests	Small population sizes	–	8/28	18	5	Ornam.
Viola uliginosa Besser	Native	EN	Shores, spruce mires, herb-rich forests	Habitat destruction, small population numbers	–	8/28	18	5	Ornam.

9.3.2 Causes of the rarity and decline of highly endangered CWR taxa

The primary habitats among endangered, critically endangered and regionally extinct CWR in Finland include: traditional agricultural habitats; semi-natural and natural grasslands, such as grazing grounds, meadows and forest pastures; herb-rich forests; heath forests; esker forests; serpentine rocks; alpine habitats; springs and spring fens; and the sandy shores of the Baltic Sea (Table 9.2). It is typical for highly threatened taxa that they are habitat specialists and that their populations are very small and occur in restricted areas. Consequently, they are sensitive to environmental changes. About 23% of highly threatened taxa have been classified as threatened because of their small population sizes. Although a lowered level of genetic variation is a major risk and concern in small-sized populations, knowledge of genetic variability is non-existent in most cases.

Habitat destruction is the main risk factor for species occurring in cultural habitats, herb-rich forests and spring areas. Overgrowing of meadows following the cessation of grazing and hay cutting was the main threat in 31% of the cases, forest management practices in 14%, peatland drainage in 9%, increased construction activities (sand and gravel quarrying, mining, road and house building and waterway construction) in 16.5% of the cases and water pollution in 3% of cases.

9.3.3 Conservation needs and actions

Most of Finland lies within the boreal coniferous forest natural vegetation zone with large forested areas where the dominant tree species are Norway spruce (*Picea abies* (L.) H. Karst.) and Scots pine (*Pinus sylvestris* L.). The forests have been exploited for commercial forestry for long periods. At present, about 9% of the total area of the country is protected under the Nature Conservation Act or the Act on the Protection of Wilderness Reserves (Finnish Environment Institute, Internet publications, 2004–2005). Most of these protected areas also belong to the Natura 2000 network. The Finnish government has approved seven specific nature conservation programmes covering the following areas: national parks and strict nature reserves, mires, bird wetlands, eskers, herb-rich woodland, shores and old-growth forests. As many as 137 taxa of vascular plants are also included among the organisms protected by the Nature Conservation Act. The protection of CWR is not specified in any programme, but it is covered by the protection acts aimed at wild plants. Examples of practical monitoring work on threatened vascular plants to assess the need for conservation and management measures are described by Ryttäri *et al.* (2003), who discuss the importance of understanding the biology of the target species, the significance of the timing of population monitoring, the usefulness of permanently marking plots and individuals, the application of size or stage classifications, the usefulness of measuring various morphological characteristics, the interpretation of short-term and long-term results and the recording of environmental parameters.

A relatively high number (18 species, 28%) of the endangered or critically endangered taxa are dependent on cultural habitats (Table 9.2). The majority of the habitats occupied by those plants are abandoned semi-natural dry or mesic grasslands, which are a rapidly declining ecosystem type in Finnish cultural landscapes. The remaining few thousands of hectares of these habitats are small, fragmented and isolated patches within the otherwise intensively managed agricultural land. Besides, these patches are not well represented in the Natura 2000 network in Finland, which focuses more on 'natural' habitat types (Pykälä, 2003). The cultural habitats that are included in the Natura scheme are mainly managed with the agrienvironmental support scheme of the European Union (EU) (Salminen and Kekäläinen, 2000). Pykälä (2003) has shown that efficient restoration of these critical habitats, either with the agrienvironmental scheme or by operationalizing of the EU's article 16 (EC reg. 1257/1999) support, would be possible by reintroduction of grazing by cattle, sheep or horses. It is evident that the quality of management needs to be improved in order to successfully manage grassland types of the EU Habitats Directive (Pykälä, 2003). This seems to be critically important for many of the CWR taxa, and not just in Finland.

9.4 Conclusions

CWR are important components of both natural and cultural habitats, and they contribute significantly to the diversity of ecosystems. It follows that the conservation and sustainable use of CWR taxa is of prime importance for both the maintenance of biodiversity and for the improvement of agricultural production. Based on the present review, up to 60% of the established wild plant taxa of Finland can be considered CWR. Among the CWR taxa, a substantial proportion, 11%, is presently threatened to some degree, mainly due to environmental changes, such as a decrease in traditional agricultural habitats, forest management practices and increased construction activities, and because populations are small and restricted to small areas. Despite quite prominent conservation actions aimed at the Finnish flora, especially plants dependent on cultural landscapes are rather poorly considered in the present programmes.

The main points to consider in the conservation of CWR are: first, the selection of target taxa (i.e. prioritization); and secondly, the understanding of their demography, ecology and population genetics, including the estimation of genetic diversity with the help of available molecular marker analyses. Targeted research should be conducted to identify and manage the populations of CWR that need to be conserved. Effective conservation must involve population monitoring and the maintenance of adequate levels of genetic diversity and sufficient population numbers and sizes within each taxon, as well as the termination of detrimental habitat disturbance. It is vital to investigate and compare populations and their demographic, ecological and genetic characteristics across the taxon's distribution covering its entire ecogeographical range, and, if needed, to take action to promote the viability of the populations.

Plant taxa may be prioritized for conservation according to different criteria, e.g. socio-economic use, current conservation status, ecogeographic distribution, threat of genetic erosion and biological and cultural importance (Maxted *et al.*, 1997). In the case of the Finnish CWR, we suggest a prioritization which combines the conservation status (i.e. threatened, vulnerable, endangered or critically endangered) and the socio-economic importance (i.e. use for food, fodder or medicinal purposes). Based on such criteria, the family *Rosaceae* and its genera *Fragaria*, *Rubus*, *Malus* and *Sorbus*, with threatened but potentially important CWR for berry or fruit production, and the genus *Rosa*, with a high exploitation potential among ornamental plants, emerge as CWR which can be considered an important genera for *in situ* conservation of CWR in Finland.

References

Finnish Environment Institute (2004–2005) Available at: http://www.ymparisto.fi/default.asp?node=6052&lan=en (accessed 12 April 2007)

Hämet-Ahti, L., Suominen, J., Ulvinen, T. and Uotila, P. (eds) (1998) *Retkeilykasvio (Field flora of Finland*, 4th edn). Finnish Museum of Natural History, Botanical Museum, Helsinki, Finland.

Hammer, K., Gladis, T. and Diederichsen, A. (1997) Weeds as genetic resources. *Plant Genetic Resources Newsletter* 111, 33–39.

Maxted, N. (2003) Conserving the genetic resources of crop wild relatives in European protected areas. *Biological Conservation* 113, 411–417.

Maxted, N., Hawkes, J.G., Guarino, L. and Sawkins, M. (1997) The selection of taxa for plant genetic conservation. *Genetic Resources and Crop Evolution* 44, 337–348.

Meilleur, B.A. and Hodgkin, T. (2004) *In situ* conservation of crop wild relatives: status and trends. *Biodiversity and Conservation* 13, 663–684.

Pykälä, J. (2003) Effects of restoration with cattle grazing on plant species composition and richness of semi-natural grasslands. *Biodiversity and Conservation* 12, 2211–2226.

Rassi, P., Alanen, A., Kanerva, T. and Mannerkoski, I. (eds) (2001) *The 2000 Red List of Finnish Species*. Ministry of the Environment, Finnish Environment Institute, Helsinki, Finland.

Räty, E. and Alanko, P. (2004) *Viljelykasvien nimistö (Nomenclature of crop plants)*. Puutarhaliitto, Helsinki, Finland.

Ryttäri, T., Kukk, Ü., Kull, T., Jäkäläniemi, A. and Reitalu, M. (eds) (2003) *Monitoring of Threatened Vascular Plants in Estonia and Finland – Methods and Experiences*. Ministry of the Environment, Finnish Environment Institute, Helsinki, Finland.

Salminen P. and Kekäläinen H. (2000) *The Management of Agricultural Heritage Habitats in Finland*. Report by the Heritage Landscapes Working Group. *The Finnish Environment 443* (in Finnish with an English summary), Finland, pp. 1–161.

Takaluoma, S. (2005) Viljelykasvien luonnonvaraisten sukulaisten uhanalaisuus Suomessa (Conservation status of crop wild relatives in Finland). MSc thesis, University of Helsinki, Finland.

Tanksley, S.D. and McCouch, S.R. (1997) Seed banks and molecular maps: unlocking genetic potential from the wild. *Science* 277, 1063–1066.

10 Crop Wild Relatives in the Netherlands: Actors and Protection Measures

R. HOEKSTRA, M.G.P. VAN VELLER AND B. ODÉ

10.1 Introduction

The importance of wild plant species that are related to crops has already been stressed by Vavilov in the early 20th century. From these wild species, various crops have been domesticated since the beginning of agriculture. Moreover, these wild species are still important in present agricultural systems because of their ability to exchange genes with their related crops. A wild species that is related to a species of direct socio-economic importance (i.e. a crop) can be defined as a crop wild relative (CWR). Depending on the criteria for the matter of relatedness as well as for what one considers to be a crop, several selections of CWR species can be made. In Europe, several CWR species are known that are related to various crops like oats, sugarbeet, apple, annual meadow grass and white clover. Unfortunately, the wild plant species of Europe are often under threat by habitat fragmentation, loss and extinction. The same holds for the flora of the Netherlands.

For an inventory of the state of conservation of wild plant species in the Netherlands that are important candidates for the selection of a list with CWR species, it is important to know the relevant actors that play a role in the conservation of the Dutch flora. This chapter aims at giving an inventory of these actors as well as of protection measures that are taken in the Netherlands for the conservation of wild plant species. Besides organizations that manage the nature conservation areas and *ex situ* collections, organizations that are involved in the policy development as well as the monitoring of the distribution of plant species in the Netherlands are presented. They are important for the protection management of the Dutch flora and fauna by Red Lists and several legal measurements.

By the combination of several data sources in this chapter, a rather broad list of CWR species has been produced. Because most of the Dutch flora appears to be CWR, for the Netherlands it is proposed to use a simple and straightforward approach by concentrating on the total Dutch flora and using the Dutch Red List as such as a guideline for setting priorities in the protection of biodiversity in vascular plants.

10.2 Relevant Actors for the Conservation of Species in the Netherlands

The actors in the Netherlands that are important for the conservation of species can be organizations that own or manage areas important for the conservation of species (e.g. nature reserves) or organizations that are involved in (the support) of policy development on nature conservation. Some of these organizations are involved in the monitoring of the distribution of vascular plants in the Netherlands, thereby supporting the development of conservation policy on the Dutch wild flora.

10.2.1 Organizations that own or manage nature conservation areas

The main organizations that manage nature conservation areas in the Netherlands are (Floron, 2005):

- Twelve provincial nature conservation organizations, which receive financial support from the Dutch government as well as from many private donors, owning about 90,000 ha of nature areas.
- The State Forest Service of the Netherlands which manages the nature areas (246,000 ha) that are owned by the Dutch state.
- The Society for the Preservation of Nature Reserves in the Netherlands, a non-governmental foundation with more than 910,000 members, owning 90,000 ha of nature areas.

Other important public organizations that own nature areas in the Netherlands are the Ministry of Defence (17,500 ha) and municipalities (more than 43,000 ha). Further, several smaller local non-governmental organizations (NGOs) and private landowners own more than 123 000 ha of nature areas in the Netherlands (MNP and CBS, 2005).

More information on the distribution of nature conservation areas in the Netherlands can be found in Fig. 10.1 (adopted from the Environmental Data Compendium; MNP and CBS, 2005).

10.2.2 Organizations maintaining *ex situ* collections of plants

The main actors that conserve vascular plants *ex situ* are the Dutch botanical gardens. Together with several private collections, most of the botanical gardens are collaborating in the Dutch Botanic Garden Collections Foundation (DBGCF). This foundation is the caretaker of the Dutch National Plant Collection, which consists of taxonomically revised plant collections that are either significant from a scientific or cultural perspective or significant to the Netherlands. The goals of

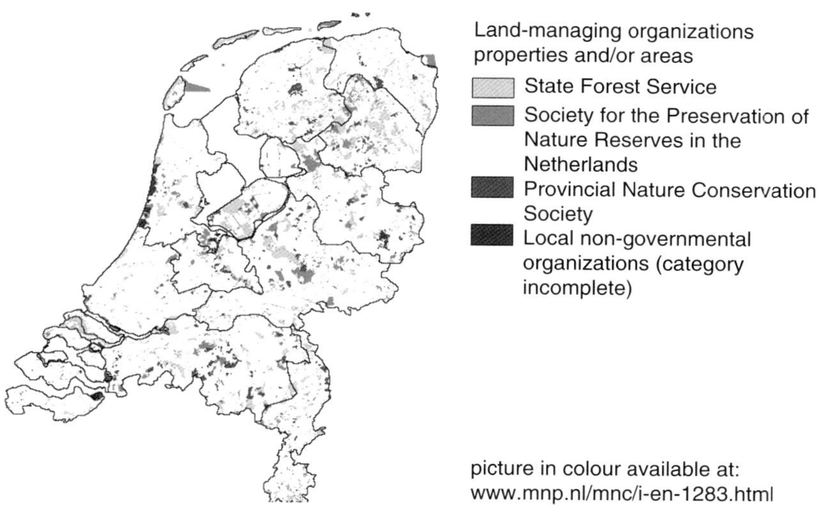

Fig. 10.1. Areas owned by nature conservation organizations including some privately owned areas. (Picture courtesy of RIVM.)

the collaboration in this foundation are: (i) conservation and improvement of the plant collections that are of national importance; (ii) conservation of biodiversity in the light of the Convention on Biological Diversity (CBD); (iii) gene bank functions; and (iv) prevention of unnecessary duplication of collections through assignment of tasks and specialization (DBGCF, 2005). A common administration of the collections and presentation on the web is being planned. Furthermore, the Centre for Genetic Resources, the Netherlands (CGN), maintaining plant genetic resources for food and agriculture, has 214 wild relatives from the Dutch origins of clover, *Lolium* and lettuce in its collection.

10.2.3 Organizations involved in the policy development of nature conservation

Besides the Dutch Ministry of Agriculture, Nature and Food Quality, several governmental and NGOs are also involved in the development of a policy on the conservation of nature in the Netherlands.

The Netherlands Society for Research on Fauna and Flora (VOFF) is the joint venture of 12 data collecting and maintaining NGOs that are specialized in molluscs and marine fauna (Anemoon), mosses and lichens (BLWG), invertebrates (EIS), vascular plants (Floron), insects, millipedes and spiders (NEV), mushrooms (NMV), reptiles, amphibians and fishes (Ravon), birds (Sovon), butterflies and dragonflies (Tinea/Vlinderstichting), mammals (VZZ) and field biology in general (KNNV). For these different species groups approximately 15,000 active volunteers monitor distribution and abundance in the Netherlands.

Together with Statistics Netherlands and the Netherlands Environmental Assessment Agency, these 12 NGOs analyse the monitoring results and calculate trends for different species groups. The Ministry of Agriculture, Nature and Food Quality and other governmental organizations use these trends to develop further policy and legislation on the conservation of species in the Netherlands.

Locality-specific data for different species are also collected from the monitoring activities and are made available online through a web site (www.natuurloket.nl). Through this web site detailed information can be obtained on the occurrence of endangered and protected species at specific localities, which is important for decisions concerning town and country planning (e.g. construction of new roads).

10.2.4 Monitoring the distribution of vascular plants in the Netherlands

In the Netherlands, monitoring of vascular plants was initiated in around 1900 when the National Herbarium in Leiden started a project to produce distribution maps of plants growing there. The first distribution maps were published in 1902 (Mennema *et al.*, 1980). To centralize the results of the vegetation recordings, the Institute for Vegetation Research in the Netherlands (IVON) was founded in 1930, with the task of producing cartographic descriptions of the Dutch flora. In 1950, the Dutch Topographic Service started to distribute topographical maps based on a 1×1 km^2 grid. Before 1950, the florists had to draw a grid by themselves. As a result, the size of the grids used in vegetation research changed from 20.8 to 25 km^2, making it difficult to compare old and newly collected data. The first standard list of the Dutch flora was approved by the florist assembly in 1975 (Arnolds and van der Meijden, 1976). For all the taxa on this list the distribution in the Netherlands had to be determined. As a result of cooperation between IVON and Statistics Netherlands (CBS), three volumes on the distribution of the Dutch flora have been published. The first volume contains information on 332 extinct or very rare taxa and was published in 1980 (Mennema *et al.*, 1980). In addition, the second and third volumes contain the distribution maps of the rare and general taxa (Mennema *et al.*, 1985; Meijden *et al.*, 1989).

Floron was founded in 1989 as the successor of IVON. It has a few staff, coordinating the training and work of hundreds of volunteers who monitor the higher plant species occurrences per 1×1 km^2 grid cell in the Netherlands. Field data are collected in the framework of several projects: (i) a total project, aiming at the coverage of all of the Netherlands; (ii) a Red List project, scoring a limited number of taxa; and (iii) an endangered species project, scoring selected populations on a frequent base. Further, Floron performs monitoring studies of flora for private owners or organizations with nature conservation areas (Floron, 2005). In cooperation with the National Herbarium in Leiden, Floron maintains two databases: FlorBase, containing observations from 1975 onwards in a 1×1 km^2 grid; and FLORIVON, containing historical data from 1900 to 1950. Figure 10.2 has been produced with FlorBase and shows the distribution of the number of vascular plant species in the Netherlands. From this figure it follows that a large number of species can be found in the coastal dunes, along the rivers and in the south of the Netherlands.

Figure 10.3 presents the distribution of endangered vascular plants in the Netherlands. The map is obtained from observations for the period 1975–2000

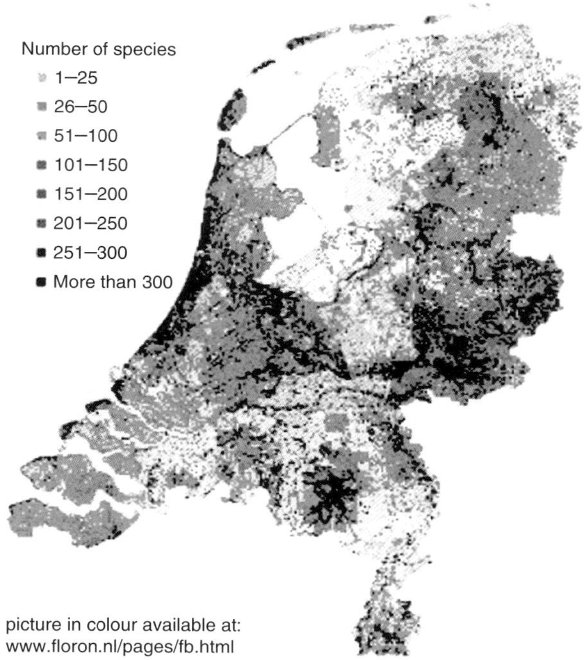

Fig. 10.2. Distribution of the number of vascular plant species in the Netherlands according to Florbase (1975–2001). (Picture courtesy of Floron.)

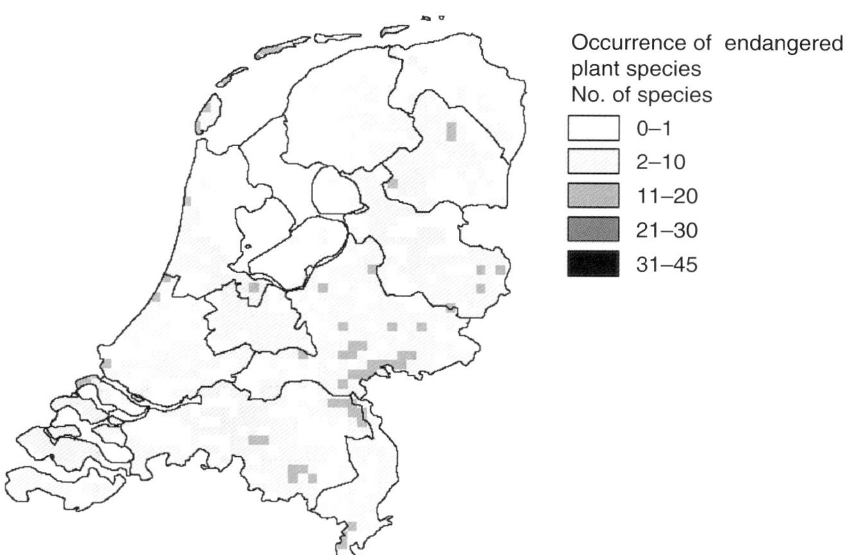

Fig. 10.3. Occurrence of endangered plant species in the Netherlands. (Picture courtesy of RIVM, originating from Floron.)

that are collected in FlorBase. The number of endangered plant species (these are the species in the Red List of vascular plants) has been totalled for each $1 \times 1\,km^2$ grid cell. FlorBase has insufficient data for some important areas, most of which are in the north-west of Overijssel province.

A comparison of Fig. 10.3 with Fig. 10.1 demonstrates that nature conservation areas cover only a limited part of the areas where endangered vascular plant species occur.

10.3 Dutch Policy on Nature Conservation in a National and an International Context

The legislation and policy development on the conservation of species in both a national as well as an international context concentrates on two aspects. Red Lists are developed on an international and a national level in order to promote conservation strategies on endangered species. Besides these Red Lists, legislation exists on an international as well as a national level to protect species and habitats from further fragmentation, loss and extinction.

10.3.1 International and national Red Lists for the protection of endangered species

International Red Lists

The Species Survival Commission (SSC) of the World Conservation Union (International Union for the Conservation of Nature, IUCN) assesses the conservation of taxa on a global scale and highlights those that are threatened and need conservation (IUCN, 2005a,b). In the IUCN Red List of Threatened Species, nine categories are distinguished for the status of the taxa (extinct, extinct in the wild, critically endangered, endangered, vulnerable, near threatened, least concern, data deficient and not evaluated). Before 2003, the 'least concern' taxa did not appear in the Red List, because the main focus was on threatened species. It will take a few years to complete the inclusion of the 'least concern' taxa on this list (IUCN, 2005b).

National Red Lists of European countries result from the 'Convention on the Conservation of European Wildlife and Natural Habitats', which was held in Bern in 1979 and came into force in 1982. The countries that ratified it will take appropriate measures to ensure the conservation of the habitats of wild flora and fauna species, including planning and development policies and pollution control. They will promote education and disseminate general information on the need to conserve species of wild flora and fauna and their habitats. The aim is to ensure the conservation of the species living in these habitats, giving special attention to endangered and vulnerable species, including migratory species. In three appendices, flora (Appendix I) and fauna (Appendices II and III) species are listed for special attention in conservation (Counsel of Europe, 2005). However, these appendices give no legal protection for the listed species.

Red Lists of the Netherlands

The Netherlands ratified the Convention of Bern and has published the following official Red Lists in the Dutch Government Gazette: birds (1994), mammals (1995), butterflies (1995), reptiles and amphibians (1996), mushrooms (1996) dragonflies (1998), crickets and locusts (1998), lichens (1998) and freshwater fishes (1998). In a recent publication of the Dutch Government Gazette (LNV 2004), these Red Lists have been withdrawn and replaced by modified or new Red Lists for: mammals (22 species), birds (78 species), reptiles (6 species), amphibians (nine species), caddis flies (86 species), locusts and crickets (18 species), stoneflies (19 species), dragonflies (27 species), mayflies (39 species), fishes (35 species), bees (188 species), butterflies (48 species), land and freshwater molluscs (68 species), flatworms (four species), vascular plants (501 species), mosses (245 species), lichens (241 species) and mushrooms (1648 species).

Red Lists provide no legal protection status. However, by law it is regulated that the Dutch government will support the protection of Red List species. It is expected from regional authorities (i.e. provinces and municipalities) as well as from NGOs that they take account of Red Lists when managing their terrains (MNP and CBS, 2004).

The Dutch Red List for vascular plants

The Dutch Red List for vascular plants is based upon the list of Meijden *et al.* (2000). The species were evaluated for a combination of rareness and decrease. Based on the total number of 1 × 1 km^2 grid cells in which the species currently occurs, each species was classified into one of five different categories of rareness. The presence of species (measured by the total number of grid cells in which the species occurs) in *c.*1935 has been compared with the situation in *c.*1999 to classify the species into five different categories of trend. The combination of rareness and trend has been used to classify the species in the following IUCN categories: not evaluated, data deficient, not threatened, least concern or low risk, near threatened, vulnerable, endangered, critically endangered and regionally extinct. Species that fall under one of the latter five categories are included in the Red List.

The Red List for vascular plants of the Netherlands totals 498 species with the following distribution in five categories: (i) near threatened: 114 species (23%); (ii) vulnerable: 136 species (27%); (iii) endangered: 103 species (20%); (iv) critically endangered: 97 species (19%); and (v) regionally extinct: 48 species (10%).

In comparison with a former proposed list (Weeda *et al.*, 1990), after correction for changes in criteria for including species in the evaluation for the combination of rareness and decrease, the present Red List contains 10% more species (Meijden *et al.*, 2000).

10.3.2 National and international legislation on the protection of species and habitats

In the European Birds and Habitats Directives, several wild flora and fauna species are protected by the protection of certain habitats as well as by general protection. Under the Habitats Directive, special areas of conservation must be designated for 51 species of invertebrates, fishes, amphibians, mammals, vascular plants and mosses. Under the Birds Directive, such areas have been selected for 42 bird species (MNP and CBS, 2005). The member states of the European Union (EU) are obliged to incorporate the species protection provisions of the directives in national legislation.

For the Netherlands, species protection is obtained by the EU Directives and the Dutch Flora and Fauna Act of 2002, which replaces subsections of the Nature Conservancy Act of 1998 and incorporates the old Hunting and Birds Act. The act provides protection for: birds (all species except exotics), mammals (64 species), reptiles (seven species), amphibians (16 species), fishes (12 species), crustaceans (one species), ants (four species), butterflies (26 species), beetles (five species), dragonflies (eight species), molluscs (two species) and a number of vascular plants (105 species). The act prohibits killing of the listed animals and disturbing their roosts or habitats. The act also imposes a duty on citizens to care for flora and fauna (MNP and CBS, 2005). In the nature management policy plan of 1990, a supplementary species-directed policy was developed in addition to the existing habitat-directed policy.

10.4 Distribution of Crop Wild Relatives in the Netherlands

In the framework of the EU project, PGR Forum (PGR Forum, 2003–2005) CGN has studied the presence and distribution of CWR in the Netherlands.

10.4.1 Development of a database with information on the Dutch flora

Four different databases, available in electronic form, were incorporated into one MS-Access database:

1. The NatuurCompendium 2003 (MNP and CBS, 2004) contains a CD-ROM with Biobase, a database with a table on the higher plants of the Netherlands, including information on the status of legal protection.
2. In November 2004, the latest version of the standard list of the flora of the Netherlands 2003 was published by the National Herbarium and Floron (Tamis *et al.*, 2004). A revised list is available at the web site (Floron, 2005). It contains 1536 vascular plant species in 65 genera that comprise the Dutch wild flora. According to the Dutch Red List for vascular plants, 48 of these taxa are probably extinct in the Netherlands.
3. The Institute of Plant Genetics and Crop Plant Research (IPK, Gatersleben) kindly supplied the Mansfeld's World Database of Agricultural and Horticultural Crops, comprising 25,406 species (6090 accepted) and 6246 genera (1974 accepted). This database does not contain ornamental and forestry plants and is online searchable (IPK, 2004). The synonyms from the Mansfeld's database were used to improve the linking of the tables from different sources.
4. In April 2005, the University of Birmingham provided a list of CWR of the Netherlands extracted from the PGR Forum Catalogue of CWR for Europe and

the Mediterranean (Kell *et al.*, 2005). It contains 1853 taxa. An updated version of this catalogue is searchable through the PGR Forum Crop Wild Relative Information System (CWRIS) (PGR Forum, 2005). The catalogue has been produced through a process of data harmonization between a number of inventories and databases; primarily Euro+Med PlantBase and Mansfeld's World Database of Agricultural and Horticultural Crops.
5. Furthermore, 267 species from 134 genera listed in the Catalogue of the Wild Relatives of Cultivated Plants Native to Europe (Heywood and Zohary, 1995) were put in the database. These authors also included species from Cyprus and the Canary Islands, but did not include Anatolian Turkey, because its flora is as diverse as that of Europe as a whole and would deserve separate treatment.

Within PGR Forum (PGR Forum, 2005) a CWR is being defined as a taxon related to a species of direct socio-economic importance, including food, fodder and forage crops, medicinal plants, condiments, ornamental and forestry species, as well as plants used for industrial purposes, such as oils and fibres. A formal definition of a CWR has been proposed by Maxted *et al.* (2006): 'A crop wild relative is a wild plant taxon that has an indirect use derived from its relatively close genetic relationship to a crop; this relationship is defined in terms of the CWR belonging to gene pools 1 or 2, or taxon groups 1 to 4 of the crop'. (Taxon Group 1 [TG1] = taxa within the same species; TG2 = taxa within the same section or series; TG3 = taxa within the same subgenus; TG4 = taxa within the same genus; TG5 = taxa in different but related genera).

10.4.2 Selection of lists with crop wild relatives that are distributed in the Netherlands

Many of the species on the standard list of the flora of the Netherlands are, to a varying extent, related to cultivated crop species. Using different combinations of data sources three different lists with CWR species that occur in the Netherlands can be determined: (i) a list obtained by the combination of the standard list of the Dutch flora with the Mansfeld's World Database of Agricultural and Horticultural Crops; (ii) one obtained by the combination of the standard list with the PGR Forum Catalogue of CWR for Europe and the Mediterranean; and (iii) one obtained by the combination of the standard list with the list of Heywood and Zohary (1995). These three lists can be produced in a narrow and extensive way by linking the databases on the species or on the genus level. The results are summarized in Table 10.1.

When linked on the genus level, the first list, based on the Mansfeld's database, revealed that about 74% of the species on the standard list are selected as CWR (Table 10.1). Thereby it should be taken into account that the Mansfeld's database does not include ornamental and forestry plants, but, on the other hand, includes non-European crops in addition to the European crops in which PGR Forum is interested.

The second list, based on the PGR Forum Catalogue, indicates that 82% of the Dutch flora is CWR. On the genus level, it selects 151 taxa (from 58

Table 10.1. Percentage of CWR on standard list of the flora of the Netherlands according to Mansfeld, PGR Forum (Kell *et al.*, 2005) and Heywood and Zohary (1995), when linked on the species or genus level.

	Mansfeld (%)	PGR Forum (%)	Heywood and Zohary (%)
Species	31	71	6
Genus	74	82	22

genera) that are not linked by the Mansfeld's database, mainly because the PGR Forum Catalogue does include ornamentals and forestry plants. The Mansfeld's database selects 23 taxa (from 18 genera) that are not selected by the PGR Forum Catalogue.

The third list is a much more restricted CWR list. The Catalogue of the Wild Relatives of Cultivated Plants Native to Europe (Heywood and Zohary, 1995) includes all European species belonging to the primary gene pool of cultivated crops. This results in a list of 322 species (22% of the Dutch flora). A further limited CWR list with only 94 species (6% of the Dutch flora) is obtained by linking the two tables on the species level.

To get the broadest CWR list, we combined lists one and two. As a result, 83% of the Dutch flora can be considered CWR. This combined list was used for Fig. 10.4.

Figure 10.4 shows the frequency of Dutch flora species in ten different classes besides a class 'unknown'. Each species is allocated to one class based on its present abundance in cells of a $1 \times 1\,km^2$ grid of the topological map of the Netherlands. For the 1287 species that have been allocated to a class, almost 4% were not observed in one of the cells. Further, 40% of the species were observed in less than 300 cells. The more general distributed species take up 56% of the Dutch flora. The differences in frequencies between CWR and non-CWR are not significant.

10.5 Discussion

Following the definition of CWR used in PGR Forum, 83% of the taxa in the standard list of the flora of the Netherlands can be regarded as CWR, because they could be linked on the genus level to the Mansfeld's database or the PGR Forum Catalogue. This result is in line with the results of data analysis carried out showing that approximately 80% of the Euro-Mediterranean flora consists of CWR and other utilized species (Kell *et al.*, Chapter 5, this volume).

One of the primary outputs of PGR Forum project will be a conservation gap analysis and recommendations for *in situ* and *ex situ* conservation of European CWR. However, when more than 80% of the flora in the Netherlands can be considered to be CWR, then one can question the usefulness of asking for special attention for the preservation of CWR. Moreover, the restricted selection based on the publication of Heywood and Zohary (1995) is unfortunately missing several important genera and when compared to the Dutch Red

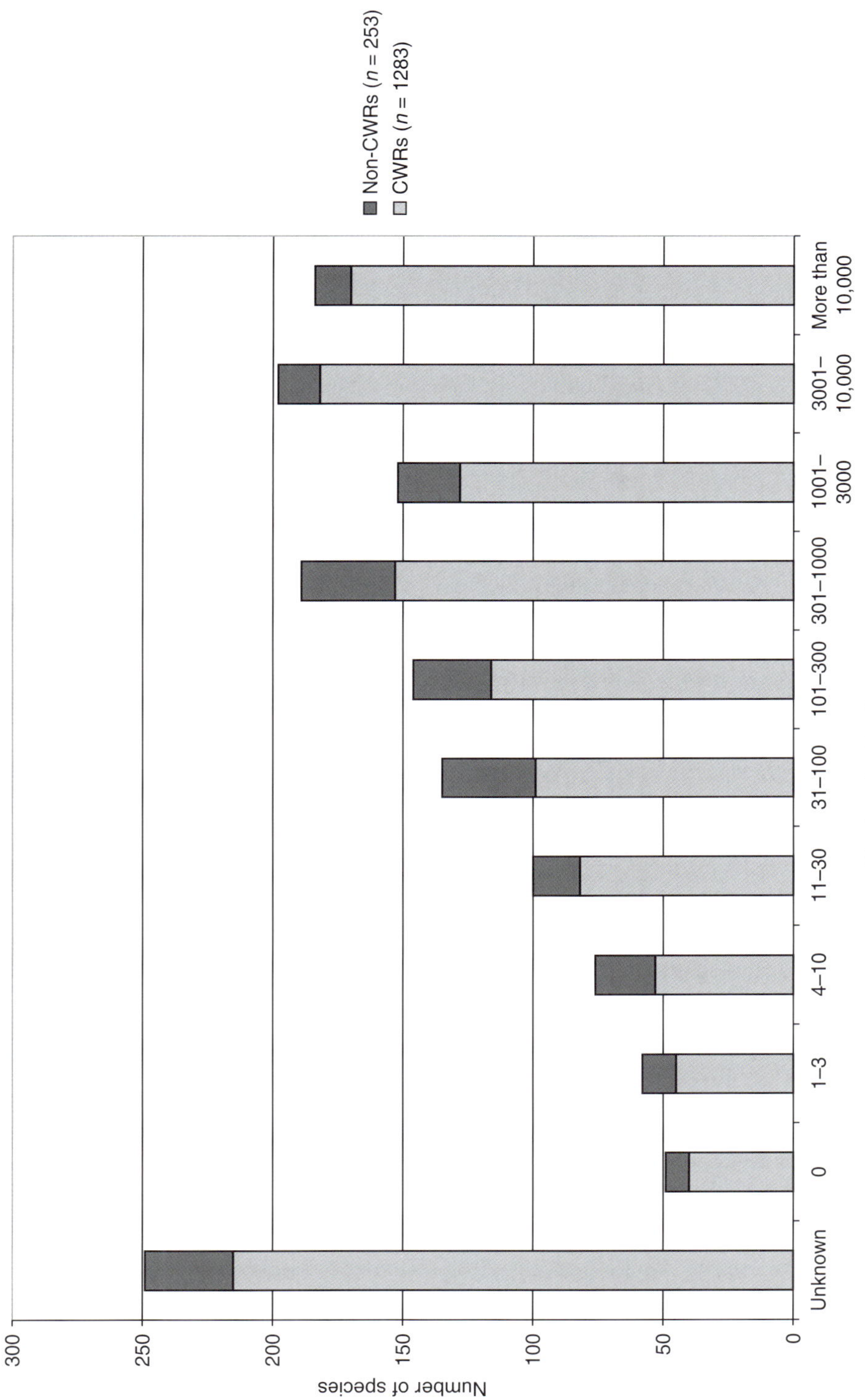

Fig. 10.4. Frequency of the Dutch flora in ten classes for the abundance of each species in cells of a 1 × 1 km² grid projected on to the topological map of the Netherlands, CWR and non-CWR separated.

List, cannot produce an alternative satisfactory priority list. At least in the Netherlands, a more simple and straightforward approach would be to use the Red List as such as a guideline for setting priorities in the protection of the biodiversity in vascular plants. Moreover, since genomics has weakened the barriers for transferring genes between distinct species, also unrelated taxa may contribute valuable genes to our crops.

From the 498 species on the Dutch Red List for vascular plants, only 72 are legally protected by means of the Flora and Fauna Act, which means that 86% of the Red List species are protected without obligations. From its Threatened Species project, Floron (Odé, 2004) reports that many populations, also within nature preserve areas, are only small, often less than 50 plants, and may easily be lost because of habitat destruction, soil acidification, fertilizer excess, drought, etc. In particular, climate change due to global warming will threaten those species adapted to cool and wet habitats. So, even if species are listed in the Flora and Fauna Act or growing in nature reserve areas, these species may still become extinct in the Netherlands in the near future. However, a species endangered in the Netherlands may be growing in other European countries and the need for *ex situ* conservation may be questionable. It is unclear whether the current measures of the Flora and Fauna Act and the nature management policy plan of the Dutch government will be sufficient to ensure the conservation of species in the Red List, whether covered by the Flora and Fauna Act or not. Therefore, *ex situ* conservation might be considered for specific cases. For all remaining species no approach is available yet. Hence, the dilemma of coming to a meaningful and sufficiently limited selection remains. Currently, no institution has taken such responsibility. Whether botanical gardens already maintain endangered species from the Netherlands is unclear, but a web site presenting the accessions of the Dutch National Plant Collection is being planned.

References

Arnolds, E.J.M. and van der Meijden, R. (1976) *Standaardlijst van de Nederlandse Flora 1975*. Rijksherbarium Leiden, Netherlands.

Counsel of Europe (2005) *Convention on the Conservation of European Wildlife and Natural Habitats*. CETS No. 104. Available at: http://conventions.coe.int/

DBGCF (2005) *Web site of the Dutch Botanical Gardens Collection Foundation*. Available at: http://www.nationale-plantencollectie.nl

Floron (2005) *Web site of Floron*. Available at: http://www.floron.nl

Heywood, V.H. and Zohary, D. (1995) A catalogue of the wild relatives of cultivated plants native to Europe. *Flora Mediterranea* 5, 375–415.

IPK (2004) *Web site of the Mansfeld World Database of Agricultural and Horticultural Crops*. Available at: http://mansfeld.ipk-gatersleben.de

IUCN (2005a) *Web site of The World Conservation Union*. Available at: http://www.iucn.org

IUCN (2005b) *IUCN Red List of Threatened Species*. Available at: http://www.redlist.org

Kell, S.P., Knüpffer, H., Jury, S.L., Maxted, N. and Ford-Lloyd, B.V. (2005) *Catalogue of Crop Wild Relatives for Europe and the Mediterranean*. Available online via the Crop Wild Relative Information System (CWRIS – http://cwris.ecpgr.org/) and on CD-ROM. University of Birmingham, Birmingham, UK.

LNV (2004). Besluit Rode lijsten flora en fauna. *Staatscourant* 11, 218.

Maxted, N., Ford-Lloyd, B.V., Jury, S.L., Kell, S.P. and Scholten, M.A. (2006) Towards a definition of a crop wild relative. *Biodiversity and Conservation* 15(8), 2673–2685.

van der Meijden, R., Plate, C.L. and Weeda, E.J. (eds) (1989) *Atlas van de Nederlandse Flora 3. Minder zeldzame en algemene soorten*. Rijksherbarium/Hortus Botanicus, Leiden and Centraal Bureau voor de Statistiek, Voorburg/Heerlen.

van der Meijden, R., Odé, B., Groen, C.L.G., Witte, J.P.M. and Bal, D. (2000) Bedreigde en kwetsbare vaatplanten in Nederland. Basisrapport met voorstel voor de Rode Lijst. *Gorteria* 26(4), 85–208.

Mennema, J., Quené-Boterenbrood, A.J. and Plate, C.L. (eds) (1980) *Atlas van de Nederlandse Flora 1. Uitgestorven en zeer zeldzame planten*. Kosmos, Amsterdam, The Netherlands.

Mennema, J., Quené-Boterenbrood, A.J. and Plate, C.L. (eds) (1985) *Atlas van de Nederlandse Flora 2. Zeldzame en vrij zeldzame planten*. Bohn, Scheltema and Holkema, Utrecht, The Netherlands.

MNP and CBS (2004) *NatuurCompendium 2003. Natuur in cijfers*, p. 494. Available at: http://www.natuurcompendium.nl

MNP and CBS (2005) *Environmental Data Compendium*. Available at: http://www.mnp.nl/mnc/index-en.html

Odé, B. (2004) *Floron dringt aan op extra inspanningen voor bedreigde plantensoorten*. Press release 23 December 2004. Available at: http://www.floron.nl/downloads/PBdec04.pdf

PGR Forum (2003–2005) *European Crop Wild Relative Diversity Assessment and Conservation Forum*. University of Birmingham, Birmingham, UK. Available at: http://www.pgrforum.org/

PGR Forum (2005) *Crop Wild Relative Information System (CWRIS)*. University of Birmingham, Birmingham, UK. Available at: http://cwris.ecpgr.org/

Tamis, W.L.M., van der Meijden, R., Runhaar, J., Bekker, R.M., Ozinga, W.A., Odé, B. and Hoste, I. (2004) Standaardlijst van de Nederlandse flora 2003. *Gorteria* 30, 101–195.

Weeda, E.J., van der Meijden, R. and Bakker, P.A. (1990) Rode Lijst van de in Nederland verdwenen en bedreigde planten (Pteridophyta en Spermatophyta) over de periode 1. I.1980–1.I.1990. *Gorteria* 16, 2–26.

11 European Forest Genetic Resources: Status of Current Knowledge and Conservation Priorities

F. Lefèvre, E. Collin, B. De Cuyper, B. Fady, J. Koskela,
J. Turok and G. von Wühlisch

11.1 Genetic Characteristics of Forest Trees

11.1.1 Forest trees: a 'crop and wild' complex gene pool

Genetic resources of most European forest tree species can be recognized as semi-wild, while gene pools of only a few species can be considered truly wild or domesticated. Both natural and anthropogenic evolutionary forces have shaped forest genetic resources in Europe. Thus, sustainable management of forests must consider genetic resources that have been influenced by a broad range of different factors and conditions in a global perspective. The most advanced domestication situation is observed in poplars (*Populus* L.), where intensive breeding programmes are producing interspecific hybrids between distant species classically used in monoclonal plantations, and in fruit trees (*Prunus* L., *Malus* Mill, etc.). The wildest gene pools are those of rare and scattered tree species, which are neither bred nor even planted (e.g. *Taxus baccata* L.).

The history of breeding programmes in forest trees starts as early as 1759 with the selection of *Pinus sylvestris* L. seed stands in northern Sweden and seed transfer to the southern part of the country (Bouvarel, 1986). Such programmes have generally been carried out for less than three generations of trees under selection, except for fast-growing species (*Populus*, *Salix* L.) that may have reached the fifth generation in the most advanced breeding populations (Bisoffi and Gullberg, 1996). Therefore, the genetic divergence between 'domestic' (i.e. managed) production forests and 'wild' forests is recent and generally low. The status of the forests is also susceptible to change through time. However, for some species, domestication may have occurred much earlier and already impacted their genetic resources; the very low genetic diversity found in European stone pine (*Pinus pinea* L.) is probably related to long-term domestic use comparable with the domestication of agricultural crops (Fallour *et al.*, 1997). Human influence has modified forests and their biodiversity for millennia in all parts of the world, including the seemingly intact and pristine natural tropical forests (e.g. McNeely, 1994). In Europe, human activity has progressively utilized most of the territory since prehistoric times and only very few forests can be assumed to have evolved in a purely natural way. In addition, European tree species are often prone to interactions between 'domesticated' and 'wild' populations, including gene flow and exchange of pathogens.

Due to their biological characteristics and, in particular, the long generation time, tree species that recolonized Europe after the last glaciation cannot be considered to have reached a stable genetic equilibrium. Moreover, the current and next generations of trees will experience major climate change that will probably alter the trajectory of their evolution. Therefore, in forest trees, we must think in evolutionary terms rather than focusing on conserving the present genetic resources. Priority should be given to dynamic gene conservation strategies, both through *in situ* conservation networks specifically dedicated to maintaining the capacity of future adaptation, and through the control of the genetic impact of silvicultural practices in production forests. Static *ex situ* approach is generally applied for the most threatened and endangered species, in addition to efforts to restore some dynamics in the gene pools through *in situ* conservation whenever possible.

Human impacts on forest genetic resources are diverse and must be considered globally (Lefèvre, 2004). The direct control of genetic diversity and demography within managed forests (seed selection and transfer, selection of reproductive individuals, control of mortality and spatial distribution of trees) not only affects the target species, but also accompanying species (El-Kassaby and Benowicz, 2000). In Europe, several scattered broadleaved tree species, occurring as single individuals or small groups in mixed forests, have particularly suffered from silvicultural practices targeted to more dominant species (Oddou-Muratorio *et al.*, 2004). At ecosystem level, disturbances such as pollution, changes in populations of symbiotic or pathogenic species and fragmentation have an impact on both domestic and wild genetic resources. The low divergence between domestic and wild forest gene pools has favoured the development of gene conservation programmes for at least two reasons. First, the value of 'wild' populations as a potential reservoir of genes of interest for human use is largely recognized because genetic improvement and breeding programmes still rely on the use of a very broad genetic base. Secondly, the management of production forests and *in situ* conservation basically rely on the same silvicultural practices. Gene resource conservation programmes have been developed at the national level since the 1980s (e.g. Behm *et al.*, 1997; Teissier du Cros, 2001) and launched at the European scale in the 1990s.

11.1.2 Drivers of genetic diversity in forest trees

Natural dynamics and anthropogenic disturbances interact and shape the genetic diversity of European forest tree populations. Based on recent experimental and theoretical studies, information has become available on the evolutionary processes and the impact of the drivers of genetic diversity in trees (e.g. mutation, genetic drift, gene flow and selection). Little is known about the mutation rate and its impact on the diversity in long-living organisms like trees, but we can assume as a first approximation that it has a lower impact than other forces in the short-term (i.e. a few centuries), at a timescale from the last glaciation to next climate change.

By combining palaeopalynology and molecular biology, postglacial recolonization routes have been clarified for several European tree species (e.g. Petit *et al.*, 2002). The clear geographical patterns observed for some particular markers (chloroplast DNA, maternally inherited in angiosperms) show that current genetic diversity is still impacted by the long-term history before the Holocene and recent migrations (Comps *et al.*, 2001; Petit *et al.*, 2003). These studies showed that migration across the continent occurred rapidly, in a few generations, which implies long-distance gene dispersal events (Le Corre *et al.*, 1997). However, the genetic diversity in forest tree species is very high within populations and differentiation for neutral markers is low (Hamrick *et al.*, 1992). This means that pioneer founder events were not associated with drift effects as expected. By simulations, Austerlitz *et al.* (2000) showed that the long juvenile phase is a life history trait that increases effective gene flow among populations during a colonization process. Thus, gene flow is a major driver of diversity in trees and it contributes to the high level of within-population diversity.

Abundant literature has been published since the beginning of the 19th century about adaptive diversity in forest trees and its geographic structure (e.g. provenance trials established in France by Duhamel du Monceau *c.*1760, by de Vilmorin *c.*1840 (Bouvarel, 1986), and in Germany by Kienitz in 1877 (Kalela, 1937/1938)). Within their natural range, tree populations show a high level of genetic diversity for dendrometric or adaptive traits and marked differences among populations; latitudinal or altitudinal clines are frequently observed (Modrzynski and Eriksson, 2002). Such clinal patterns are supposed to result from natural selection rather than random genetic drift. This means that adaptation developed in only few generations, in spite of important gene flow. Besides, important genetic variation for adaptive traits is still observed within populations (Cornelius, 1993; von Wuehlisch *et al.*, 1995; Savolainen *et al.*, 2004). The maintenance of genetic variation within populations under selection can be explained with the model of quantitative genetics (McKay and Latta, 2002). Given the level of genetic variation observed for adaptive traits within populations, a rapid response to selection is predicted by theoretical models (Rehfeldt *et al.*, 2001). Rapid response to selection was also experimentally evidenced on populations that had been transplanted for seed production (Skrøppa and Kohmann, 1997). Subsequently, current adaptive diversity is not at a stable or optimum equilibrium and this can be seen as an instant situation resulting from a dynamic process combining selection, gene flow and drift. This non-equilibrium would explain the slight adaptation lag observed between theoretically optimal and actual environmental conditions where tree populations are found (Rehfeldt *et al.*, 2002).

Long-lived organisms like trees may experience significant environmental changes within the life cycle of an individual. Therefore, phenotypic plasticity is a major component of the capacity of tree populations to adapt to environmental changes. Plasticity could result either from direct selection, and therefore be considered an adaptive trait, or from the simple effect of different selection pressures in various environments (Via *et al.*, 1995). For trees, plasticity is expected to be adaptive due to temporal environment stochasticity. Genetic variation for plasticity has been observed among forest tree populations (Rehfeldt *et al.*, 2001; Modrzynski and Eriksson, 2002). Further research on a lower scale, i.e. within-population level, is still needed.

11.2 Genetic Aspects of European Forestry

11.2.1 Forest management and forest genetic diversity

There is a particular need to consider the impact of silvicultural practices on genetic diversity in natural and semi-natural forests which have been, or will be, regenerated naturally over several generations. Because the biological situations (pioneer versus climax species, scattered versus stand-forming species, etc.) and the forest management strategies (plantation versus natural regeneration, even-aged versus irregular forest, etc.) are too diverse, we can only briefly summarize the paradigm of current research and conservation programmes. The intensity of genetic erosion within a forest tree population is determined by demographic processes (demographic fluctuations, mating system, variance of reproductive success, generation overlap, etc.) and environmental conditions (selection). For sustainable management, the fundamental principle is that human impact on demographic processes and environmental conditions should not affect the processes that maintain a high level of diversity within tree populations. Controlling human impact requires the use of adequate criteria and indicators (C&I) that still need further development.

Silvicultural practices, especially thinning and regeneration treatments, have significant influence on genetic processes within a stand and ultimately on the genetic basis of the next tree generation. Rotation periods applied for many tree species can be extremely long (hundreds of years), so it is often assumed that present stands of native tree species are 'natural' unless detailed records of their history exist. However, silvicultural systems maintaining land areas with continuous forest cover are becoming increasingly popular for a number of reasons. Subsequently, it is not rare that artificially established stands also end up being regenerated naturally. In this case, the forest reproductive material used to establish such stands in the first place can have greater impact on the genetic diversity of the next tree generation than silvicultural practices. Recently, four comprehensive reviews on the impact of silvicultural practices on forest genetic diversity have been published (see Savolainen and Kärkkäinen, 1992; Finkeldey and Ziehe, 2004; Lefèvre, 2004; Geburek and Müller, 2005).

A number of silvicultural systems are applied in different parts of Europe according to tree species, climatic conditions and socio-economic and environmental considerations. Thinning is the main silvicultural practice applied after the regeneration phase to guide the future development of stands. Most silvicultural systems involve thinning of young, dense stands by removing slow-growing or inferior phenotypes. Finkeldey and Ziehe (2004) concluded that this kind of thinning from below has rather limited impact on genetic diversity of stand-forming tree species because there are still a large number of trees left after the thinning treatment. Loss of rare alleles depends obviously on the intensity of thinning, but often thinning treatments are carried out gradually and thinning from below is likely to raise concerns only if a large number of trees are removed in small populations. However, if inferior phenotypes are associated with particular genotypes, the thinning treatment is likely to have a more profound impact on the genetic structure of a stand.

Thinning from above is usually targeted to mature trees that have reached a certain diameter or are considered to be at the end of their life cycle. This thinning treatment is associated with silvicultural systems emphasizing the maintenance of continuous forest cover and applied to a relatively small area at a time. Geburek and Müller (2005) noted that selective thinning does not cause major differences in genetic diversity between mature trees and their offspring. The short-term effects of selective thinning on genetic diversity may be minor, but its long-term effects can be more significant. Finkeldey and Ziehe (2004) pointed out that selective removal of large trees is more likely to reduce the frequency of rare alleles than random thinning. They concluded that some of these rare alleles in large trees may be associated with superior fitness and expressed concerns about reduced potential of future generations to reach larger diameters.

Silvicultural treatments aiming at natural regeneration do not differ much from selective thinning treatments in uneven-aged forests, which are managed according to the continuous forest cover system. In other systems, however, large-scale removal of mature reproductive trees and subsequent changes in the spatial distribution of the remaining trees has a major impact on the genetic diversity of the seeds establishing the next tree generation (Lefèvre, 2004). Various shelterwood systems typically involve removal of 30–50% of mature trees leaving the stand rather well stocked with reproducing trees. Under shelterwood systems, regeneration takes place during a relatively longer period of time and the genetic composition of mature trees and their offspring is unlikely to be altered significantly (Geburek and Müller, 2005). A regeneration treatment using a seed tree method involves removal of 80–90% of mature trees and thus this treatment has significant genetic consequences. The relatively short regeneration periods and the low number of reproducing trees are likely to reduce genetic diversity in the next generation (Geburek and Müller, 2005). However, the area regenerated is likely to receive pollen and seeds from the neighbouring stands (Streiff *et al.*, 1999) and this reduces the overall loss of genetic diversity. Small clear cuttings can be considered a natural regeneration treatment, but usual clear-cut areas are regenerated artificially by planting seedlings or sowing seeds of target tree species. In this case, the genetic basis of the new stand depends on the quality of forest reproductive material, which commonly originates from outside the planting site. If clear-cut areas are reforested with native tree species, it is likely that the neighbouring stands also contribute some naturally established seedlings to the new generation.

11.2.2 Use of genetic diversity in afforestations and reforestation

Artificial regeneration often involves transfer of reproductive material outside of the area where it originates and is adapted to. Forest reproductive material is also imported from other countries. The genetic composition of artificial stands may, therefore, be considerably different from the one originated from natural regeneration. Artificial forests may also have reduced levels of genetic variation as seeds are often collected from a limited number of trees or collection guidelines may not be followed carefully. Unless the material is clearly documented and the user is aware of its appropriate use, the risk of establishing forests with unknown or poorly adapted material is high. This can create problems for both long-term sustainability of the production forests and conservation of genetic resources within or nearby these forests.

The use of proper material for afforestation and reforestation is a prerequisite of sustainable forest management (SFM) including conservation of genetic resources. Concerns about the importance of origin, quality and diversity of seed source for adaptation and growth of progeny have been raised at various occasions. Examples of measures that can ensure the use of proper material in forestry include approval of a sufficient number of seed sources, guidelines for collection from a minimum population size, rules for a minimum number of clones in seed orchards, rules for clonal mixtures and restrictions on the use of certain exotic species.

To ensure proper use of forest genetic resources in afforestation and reforestation, several countries have developed national guidelines for the movement of forest reproductive material based on environmental and particularly climatic similarities (called regions of provenance or seed zones). As pointed out by Hattemer (1987), there are fundamental biological reasons making legal regulations on trading forest reproductive material among countries necessary. In fact, the European Council Directive 1999/105/EC (Anonymous, 1999) regulates the marketing of forest reproductive material in the territory of the European Union (EU). It more specifically concerns labelling of forest reproductive material, and not the forest practices in forest units. However, for plants that are sold or traded, a market directive strongly influences seed collection, seed or plant production and their use in the forest. The regulation requires that the material traded within the EU is documented according to certain standards and encourages appropriate use of the material (see Ackzell and Turok, 2005).

The use of material for afforestation in the natural range of the species, which is the most common situation, creates new interactions between local and transplanted gene pools. The global impact on forest genetic diversity is highly dependent on the relative importance of local and introduced resources and on the 'maladaptedness' of introduced material, which should be avoided or at least minimized through the regulations (Lefèvre, 2004). Several complex processes are involved in balancing positive and negative effects (Lenormand, 2002). The positive effects can be a demographic rescue (when local population is declining), an increase of genetic diversity or a reduction of inbreeding. The negative effects can be the introduction of maladapted genes, a general reduction in fitness when all offspring result from hybridization, an increase of inbreeding or decrease of diversity when all offspring result from hybridization and the migrant pool has a limited genetic base (in particular for widely spread cultivars). The negative effects are reduced by several factors: when the local population is not rare, reproductive barriers are strong, generation time is long, selfing or vegetative propagation occurs and differential selection is enhanced (Wolf *et al.*, 2001). If migration is planned or cannot be avoided, through seed transfer or pollen contamination, gene resource managers can play on these factors to reduce the risk of negative genetic impact.

11.3 European Collaboration for Conservation of Forest Genetic Resources

11.3.1 Ministerial conferences on the protection of forests in Europe

In 1990, a forest policy process was initiated at the first Ministerial Conference on the Protection of Forests in Europe (MCPFE) held in Strasbourg, France. Ministers responsible for forests adopted six resolutions to initiate immediate action to improve the state of the forests as a response to concerns on the impacts of environmental pollution and genetic erosion on forest ecosystems in Europe. Strasbourg Resolution 2 specifically addressed conservation of forest genetic resources and urged participating countries to implement appropriate policies in this area. This resolution also called for establishment of a functional but voluntary instrument of international cooperation on forest genetic resources in Europe to promote and coordinate development of *in situ* and *ex situ* conservation methods, exchanges of reproductive materials and monitoring of progress in these fields.

After the Strasbourg Conference, a group of experts from four countries developed a proposal for the establishment of the collaborative mechanism in collaboration with the FAO Forestry Department, the International Plant Genetic Resources Institute (IPGRI) and the European Commission (EC). The proposal was then endorsed by the second Ministerial Conference in Helsinki, Finland in 1993 and the European Forest Genetic Resources Programme (EUFORGEN) was launched in October 1994. The major focus of the Helsinki Conference was SFM and biodiversity conservation in Europe following the adoption of the global Convention on Biological Diversity (CBD) in 1992. The Helsinki Conference accepted the CBD definition on biological diversity and noted that biodiversity conservation should be an essential operational component in SFM. The 'Forest Principles', endorsed by the United Nations Conference on the Environment and Development (UNCED) in 1992, initiated the development of a broader concept for SFM.

In Europe, the implementation of the global commitments on the conservation of forest biodiversity as part of SFM made significant progress during the third Ministerial Conference held in Lisbon, Portugal in 1998. This conference adopted pan-European criteria, indicators and operational level guidelines for SFM. Ever since, conservation of forest biodiversity and promotion of SFM have been receiving increasing attention also at practical level among European countries. In 2003, the fourth Ministerial Conference in Vienna, Austria made further resolutions in these two areas and countries also committed themselves specifically to promote the conservation of forest genetic resources as an integral part of SFM and to continue the pan-European collaboration in this area.

11.3.2 European forest genetic resources programme

EUFORGEN is a collaborative programme among European countries to promote conservation and sustainable use of forest genetic resources. It is also an implementation mechanism of the MCPFE Resolutions addressing forest genetic resources. The Programme is fully financed by the participating countries and coordinated by the IPGRI in technical collaboration with the Food and Agriculture Organization (FAO) of the United Nations. Most of the EUFORGEN work is carried out through Networks which bring together scientists, policy makers and managers from the participating countries to exchange information, discuss needs and develop strategies and methods for better management of forest genetic resources in Europe.

At the beginning of phase I (1995–1999), EUFORGEN started its activities with four pilot networks for black poplar, cork oak, noble hardwoods and Norway spruce. A fifth network for social broadleaves (temperate oaks and beech) was also launched in 1997. These species and the pilot networks were chosen to reflect the different geographic and ecological conditions in Europe. At the end of phase I, the EUFORGEN networks had established themselves as dynamic platforms for pan-European collaboration on forest genetic resources.

During phase II (2000–2004), the EUFORGEN work continued through the five species-oriented networks, but the scope of the networks were broadened and some of them were renamed. The networks focusing on Norway spruce and cork oak evolved into the Conifers and the Mediterranean Oaks networks, respectively. In 2002, the Social Broadleaves network was renamed as the Temperate Oaks and Beech network. During phase II, the number of the participating countries grew to 32.

Following Vienna Resolution 4 (conserving and enhancing forest biological diversity in Europe), phase III (2005–2009) of the Programme was designed to promote appropriate use of forest genetic resources as an integral part of SFM. Subsequently, the Steering Committee adopted new objectives and a new network structure of the Programme. During phase III, EUFORGEN has focused on:

1. Promoting practical implementation of gene conservation and appropriate use of genetic resources as an integral part of SFM;
2. Facilitating further development of methods to conserve genetic diversity of European forests;
3. Collating and disseminating reliable information on forest genetic resources in Europe.

The Steering Committee also established a new thematic Forest Management Network to address genetic issues relevant to practical forest management and policy development within national forest programmes (NFPs). The species-oriented work on noble hardwood and black poplar will be carried out through the Scattered Broadleaves network, and the Stand-forming Broadleaves network continues the efforts of the earlier Mediterranean Oaks and Temperate Oaks and Beech networks.

The EUFORGEN networks have produced many concrete outputs such as technical guidelines for genetic conservation and use of forest trees and long-term forest genetic resources conservation strategies. They have also developed descriptors and databases, as well as facilitated establishment of collections of genetic material. In addition, the EUFORGEN work has contributed to the strengthening of national efforts on forest genetic resources and the development of new programmes and policies of the EU. Furthermore, EUFORGEN has provided a useful platform for developing bilateral and multilateral cooperation and large research projects on forest genetic resources in Europe.

11.3.3 EUFORGEN work on conservation priorities and inventories

The species-oriented networks have developed long-term conservation strategies for different target species or groups of species. These target species include 12 so-called noble hardwood species (Norway maple (*Acer platanoides* L.), sycamore (*A. pseudoplatanus* L.), black alder (*Alnus glutinosa* (L.) Gaertn.), chestnut (*Castanea sativa* Mill.), ash (*Fraxinus* L. spp.), walnut (*Juglans regia* L.), wild fruit trees (*Prunus avium* (L.) L., *Malus sylvestris* Mill., *Pyrus pyraster* Burgsd.), mountain ash (*Sorbus* L. spp.), lime (*Tilia cordata* Mill.) and elms (*Ulmus* L. spp.)), which commonly grow in small stands or are found scattered in mixed forests. European black poplar (*Populus nigra* L.), a pioneer tree species forming metapopulations, is an example of another kind of scattered broadleaved species for which EUFORGEN has developed the conservation strategies. The same has been developed for several stand-forming broadleaved species (European white oaks (*Quercus petraea* (Matt.) Liebl. and *Q. robur* L.) and cork oak (*Q. suber* (L.))) and widely occurring conifers (Norway spruce (*Picea abies* (L.) H. Karst.)).

The gene conservation strategies are based on the concept of dynamic conservation, which emphasizes the maintenance of evolutionary processes to ensure continuous adaptation of forest trees under changing environmental conditions (Eriksson *et al.*, 1993). This means either managing tree populations at their natural sites within the environment to which they are adapted (*in situ*), or artificial but dynamically evolving populations elsewhere (*ex situ*). Climate change, and the environmental changes it is predicted to bring along, makes the dynamic approach even more important.

Countries have followed the gene conservation strategies while implementing gene conservation efforts in practice and establishing networks of gene conservation units for different tree species in natural and presumably locally adapted forests. The units are typically *in situ* conservation stands or gene reserve forests. The selection of the units often relies on, in the absence of detailed data on genetic variation at molecular level, information about patterns of adaptive traits and climatic conditions within the geographical distribution range of a species. Natural regeneration is a preferred method in regenerating the units, but they can be also regenerated artificially if the reproductive material is collected from the same units.

During the past years, the EUFORGEN Networks have initiated the development of 'common action plans' to facilitate national gene conservation efforts of several target tree species from the pan-European perspective. The common action plans aim at sharing of responsibilities for conservation of forest genetic resources among European countries. They would also promote better use of limited human and financial resources available for gene conservation. The development of these plans has created an urgent need to strengthen the inventories and documentation work on forest genetic resources in Europe. Earlier EUFORGEN has developed descriptors for inventories of *in situ* conservation stands for several tree species to help countries in their national inventories on forest genetic resources. The next step is to clarify the definition of a 'dynamic gene conservation unit' and establish minimum requirements for the units at pan-European level. This would help enormously the collection of georeferenced data on the existing dynamic gene conservation units of forest trees throughout their entire distribution ranges in Europe for further analyses and formulation of the common action plans.

Recognizing the particular importance of forest biodiversity in south-eastern Europe, EUFORGEN has also provided a platform for launching, in 1997, a multilateral cooperation project aimed at inventory of *in situ* forest genetic resources in Bulgaria, Moldova and Romania (Blada *et al.*, 2002). Ukraine joined in 2001 with a comprehensive inventory project of the existing system of genetic reserves for all broadleaved tree species carried out throughout the country, with particular emphasis on rare species. This part of Europe is characterized by autochthonous broadleaved forest resources valued for their quality and natural diversity. A large number of stands were described according to common standards, identified and selected in 11 priority species of the genera *Acer* L., *Fagus* L., *Fraxinus* and *Quercus* L. Maps of the distribution areas were also digitized and printed. Recommendations for conservation and sustainable use of genetic resources, including silvicultural practices promoting natural regeneration, were formulated and disseminated in the national languages. The data from the inventories have been made available as part of the information and documentation resources maintained by EUFORGEN. The 5-year project, financially supported by the government of Luxembourg, resulted in an improved knowledge and capacity, and helped to set priorities for *in situ* conservation of forest genetic resources in the four partner countries.

11.3.4. Criteria and indicators for sustainable forest management and conservation of forest biodiversity in Europe

C&I emerged during the 1990s as a means to conceptualize SFM and evaluate its implementation. C&I are assessment tools with four hierarchical levels (principle, criterion, indicator and verifier) (Prabhu *et al.*, 1996). In Europe,

Box 11.1. Pan-European criteria for sustainable forest management. (Available at: www.mcpfe.org)

Criterion 1: Maintenance and appropriate enhancement of forest resources and their contribution to global carbon cycles
Criterion 2: Maintenance of forest ecosystem health and vitality
Criterion 3: Maintenance and encouragement of productive functions of forests (wood and non-wood)
Criterion 4: Maintenance, conservation and appropriate enhancement of biological diversity in forest ecosystems
Criterion 5: Maintenance and appropriate enhancement of protective functions in forest management (notably soil and water)
Criterion 6: Maintenance of other socio-economic functions and conditions

after the Helsinki Conference in 1993 had first made a resolution to promote SFM, the MCPFE process made a series of efforts to develop pan-European C&I for SFM. A first set of the criteria and quantitative indicators were developed in 1994 and the improved pan-European indicators were adopted in 2002 (Box 11.1).

Criterion 4 emphasizes conservation of forest biological diversity at the ecosystem, species and genetic levels. One indicator (indicator 4.6) refers specifically to genetic resources of forest trees: 'Area managed for conservation and utilization of forest tree genetic resources (*in situ* and *ex situ* gene conservation) and area managed for seed production.' However, there is no requirement for species-specific data or direct measurement of genetic diversity. Thus, it is difficult to assess how effective this indicator is in promoting genetic sustainability of forest management. The MCPFE process is not the only regional process or C&I initiative that has found development of genetic C&I a complex and difficult task. A recent review concluded that the current genetic indicators developed by various processes are generally not effective for measuring the status and trends in genetic diversity in forest ecosystems (FAO, 2002a, unpublished data).

In general, there are several operational and conceptual problems in using C&I to assess genetic diversity as part of SFM. To be operational and practical, genetic C&I assessment needs to be practical and field measurements easy to carry out and cheap. This does not obviously favour direct quantitative measurements of genetic diversity using biochemical techniques which are time consuming and rather expensive. Any C&I system needs to be based on an appropriate conceptual framework, obtain information from multiple sources and define which threshold values should be applied when assessing indicators (Boyle, 2000). In case of genetic C&I, additional problems include the choice of species to assess, how to measure genetic variation and how to detect temporal variation in genetic parameters (Boyle, 2000). Subsequently, it is understandable that various C&I initiatives are using indirect or surrogate measures to assess the genetic sustainability of forest management.

As an approach to develop more comprehensive genetic C&I, Namkoong *et al.* (1996) proposed conservation of the processes that maintain genetic variation as a criterion together with four indicators (i.e. levels of genetic variation, directional changes in gene or genotypic frequencies, gene migration between populations and mating system processes). The proposed verifiers for these indicators included both demographic and genetic ones for keystone species at population or forest management unit level. However, this proposal does not meet the operational requirements from a practical point of view and it has been recognized that further research is needed to develop efficient direct and surrogate measures of genetic diversity (FAO, 2002b, unpublished data).

It is also important to consider the scale at which genetic C&I will be applied (Boyle, 2000). Typically, C&I are applied at forest management unit level, which is a forest area managed according to a long-term management plan to meet an objective or a set of objectives (Prabhu *et al.*, 1996). In European conditions, this means that C&I are applied at a local scale as a forest management unit usually covers a few tens to several hundred hectares. However, the distribution ranges of European forest trees often include different ecosystems or habitats across neighbouring countries or even throughout the continent. Therefore, a holistic approach at larger geographical scale (landscape or continental scale) is of particular importance for developing genetic C&I in addition to assessing genetic diversity in individual forest management units. It has also been questioned whether direct quantitative assessments of genetic diversity can ever be feasibly used at a practical level (FAO, 2002a, unpublished data). It is likely that indirect assessments of demographic verifiers or ecological processes will continue to be the most practical surrogates of genetic diversity. However, this does not remove the need for the development and implementation of specific gene conservation strategies at a larger geographical scale.

Between 1998 and 2000, the BEAR project (indicators for forest biodiversity in Europe, EU FAIR) carried out further work on the development of forest biodiversity indicators at pan-European level. However, the BEAR project acknowledged the importance of genetic diversity, but did not include any genetic indicator into its proposal to improve forest biodiversity indicators and methodologies for the data collection (Larsson *et al.*, 2001). Subsequently, in 2003, the European Forest Institute (EFI) assisted the European Topic Centre for Nature Protection and Biodiversity (ETC/NPB) to develop reporting methods for genetic aspects of forest conservation and proposed two indicators on forest genetic resources (Schuck and Rois, 2003, unpublished data). The first proposed genetic indicator referred to the participation of European countries in EUFORGEN and the integration of national programmes on FGR into the overall NFPs. The second one focused on forest reproductive material and the implementation of Council Directive 1999/105/EC on the marketing of forest reproductive material as part of national legislation. The European Environmental Agency (EEA) is currently using these indicators while it is preparing a subreport on biodiversity loss for its report on the state of environment in Europe (SOER 2005).

In spring 2005, EEA also initiated a streamlining European biodiversity indicators (SEBI2010) process in collaboration with the European Centre for Nature Conservation and the UNEP-World Conservation Monitoring Centre.

The tenth meeting of the Subsidiary Body on Scientific, Technical and Technological Advice (SBSTTA 10) of the CBD recommended that the headline indicators in six relevant areas should be developed for assessing progress towards the 2010 target at the global level.

In 2004, the EU decided to establish the SEBI2010 process as a response to this CBD decision. This decision is also in line with the requirements of the UNECE Environment for Europe process and the Pan-European Biological and Landscape Diversity Strategy (PEBLDS). Six expert groups have been established in Europe to develop these indicators and one of them focuses on trends in genetic diversity of domesticated animals, cultivated plants (including forest trees) and fish species of major socio-economic importance. The expert group has started its work by focusing on domesticated animals and crops because there are more data readily available on their genetic resources. It remains to be seen what kind of indicators the expert group can produce for forest genetic resources.

11.4 Perspectives

Gene conservation of forest trees, not dissimilar from wild crop relatives as well, should be based on maintaining the evolutionary process in natural populations. This makes *in situ* conservation a preferable method, but both *in situ* and *ex situ* conservation approaches should be applied in such a manner that they complement each other, depending on the conservation needs and biological characteristics of a tree species. Practising SFM should support conservation of forest genetic resources (e.g. Koski *et al.*, 1997). With minor changes, silvicultural practices can also greatly promote gene conservation of rare or scattered tree species (e.g. Rotach, 1999). But genetic C&I still need further development before the adequate assessment of the genetic sustainability of forest management.

Besides direct management of tree populations through silviculture, ecosystem-based approaches aiming at gene resource conservation are progressively developed (e.g. Lefèvre *et al.*, 2001). These approaches are particularly needed for pioneer species whose regeneration relies mainly on the dynamics of the ecosystem, like in the riparian forest, and for scattered tree species that cannot be monitored on a small geographical scale. Habitat or ecosystem management also offers the possibility to consider several species at a time, a challenge for CWR conservation. Biodiversity conservation and related C&I development needs to extent its focus from habitat and species conservation to gene conservation.

Global change is a major factor that will affect the evolution of genetic resources, in particular for long-lived organisms. Climate change has already impacted the actual distribution range of species (Root *et al.*, 2003). The response of populations will be acclimatization, genetic adaptation or migration. Predictions of changes in potential distribution range are made for several species, but we still have no clear idea of their capacity of migration towards these new margins in the context of global change (Higgins *et al.*, 2003).

Moreover, the potential of adaptive evolution is not yet included in the models. This context was not considered initially when gene conservation strategies were developed. Fundamental research in these fields is urgently needed to develop good predictive models of evolution. Some of these research questions will be addressed by EVOLTREE, a new EU-funded forest genetics research network. Then, silvicultural guidelines should be refined so that more attention is paid to maintaining genetic diversity in production forests to ensure that forestry can be practised in view of the future climate change.

European countries have made good progress in their efforts to implement SFM and conserve forest genetic resources. However, these efforts need to be strengthened by closer integration of gene conservation into NFPs, which is the major tool to promote implementation of SFM at national level (Koskela *et al.*, 2004).

Better information systems are also needed to monitor genetic sustainability of forest management and develop pan-European gene conservation strategies for forest trees. In this regard, a new EU project (TREEBREEDEX) will collect information from tree breeders on their collections of trees and experiments. Within the framework of EUFORGEN, similar effort is under development to collect information on dynamic gene conservation units of European forest trees and to define minimum requirements for these units at pan-European level.

References

Ackzell, L. and Turok, J. (2005) International legislation with implications on the exchange of forest genetic resources. In: Geburek, T. and Turok, J. (eds) *Conservation and Management of Forest Genetic Resources in Europe*. Arbora Publishers, Zvolen, Slovakia, pp. 75–88.

Anonymous (1999) *Council Directive 1999/105/EC of 22 December 1999 on the marketing of forest reproductive material*. Official Journal L 011 (15/01/2000): 17–40. Available at: http://europa.eu.int/eur-lex/en/lif/dat/1999/en_399L0105.html

Austerlitz, F., Mariette, S., Machon, N., Gouyon, P.H. and Godelle, B. (2000) Effects of colonization processes on genetic diversity: differences between annual plants and tree species. *Genetics* 154, 1309–1321.

Behm, A., Becker, A., Doerflinger, H., Franke, A., Kleinschmit, J., Melchior, G.H., Muhs, H.J., Schmitt, H.P., Stephan, B.R., Tabel, U., Weisgerber, H. and Widmaier, T. (1997) Konzept für die Erhaltung Forstlicher Genressourcen in der Bundesrepublik Deutschland [Concept for the Conservation of Forest Genetic Resources in the Federal Republic of Germany]. *Silvae Genetica* 46, 24–34.

Bisoffi, S. and Gullberg, U. (1996) Poplar breeding and selection strategies. In: Stettler, R.F., Bradshaw, H.D., Heilman, P.E., and Hinckley, T.M. (eds) *Biology of* Populus *and its Implications for Management and Conservation*. NRC Research Press, Ottawa, Canada, pp. 139–158.

Blada, I., Alexandrov, A., Postolache, G., Turok, J. and Donita, N. (2002) Inventories for *in situ* conservation of broadleaved forest genetic resources in south-eastern Europe. In: Engels, J.M.M., Ramanatha Rao, V., Brown, A.H.D. and Jackson, M. (eds) *Managing Plant Genetic Diversity*. CAB International, Wallingford, UK, pp. 217–227.

Bouvarel, P. (1986) L'amélioration Génétique des Arbres Forestiers – Essai d'une Histoire. *Revue Forestière Française* 38(Suppl.), 7–11.

Boyle, T.J. (2000) Criteria and indicators for the conservation of genetic diversity. In: Young, A., Boshier, D. and Boyle, T. (eds) *Forest Conservation Genetics: Principles and Practice.* CSIRO Publishing, Collingwood, Australia, pp. 239–251.

Comps, B., Gömöry, D., Letouzey, J., Thiébaut, B. and Petit, R.J. (2001) Diverging trends between heterozygosity and allelic richness during postglacial colonization in the European beech. *Genetics* 157, 389–397.

Cornelius, J. (1993) Heritabilities and additive genetic coefficients of variation in forest trees. *Canadian Journal of Forest Research* 24, 372–379.

El-Kassaby, Y.A. and Benowicz, A. (2000) Effects of commercial thinning on genetic, plant species and structural diversity in second growth Douglas fir (*Pseudotsuga menziesii* (Mirb.) Franco) stands. *Forest Genetics* 7, 193–203.

Eriksson, G., Namkoong, G. and Roberts, J.H. (1993) Dynamic gene conservation for uncertain futures. *Forest Ecology and Management* 62, 15–37.

Fallour, D., Fady, B. and Lefèvre, F. (1997) Study on isozyme variation in *Pinus pinea* L.: evidence for low polymorphism. *Silvae Genetica* 46, 201–207.

Finkeldey, R. and Ziehe, M. (2004) Genetic implications of silvicultural regimes. *Forest Ecology and Management* 197, 231–244.

Geburek, T. and Müller, F. (2005) How can silvicultural management contribute to genetic conservation? In: Geburek, T. and Turok, J. (eds) *Conservation and Management of Forest Genetic Resources in Europe*. Arbora Publishers, Zvolen, Slovakia, pp. 651–666.

Hamrick, J.L., Godt, M.J.W. and Sherman-Broyles, S.L. (1992) Factors influencing levels of genetic diversity in woody plant species. *New Forests* 6, 95–124.

Hattemer, H.H. (1987) Are the EEC Directives on forest reproductive material genetically adequate? *Silvae Genetica* 36, 94–102.

Higgins, S.I., Clark, J.S., Nathan, R., Hovestadt T., Schurr, F., Fragoso, J.M.V., Aguiar, M.R., Ribbens, E. and Lavorel, S. (2003) Forecasting plant migration rates: managing uncertainty for risk assessment. *Journal of Ecology* 91, 341–347.

Kalela, A. (1937/1938) Zur Synthese der Experimentellen Untersuchungen über Klimarassen der Holzarten. *Communicationes Instituti Forestalis Fenniae* 26, 1–445. Berichtigungen und Nachtraege.

Koskela, J., de Vries, S.M.G., Gil, L., Mátyás, C., Rusanen, M. and Paule, L. (2004) Conservation of forest genetic resources and sustainable forest management in Europe. In: Beaulieu, J. (ed.) *Silviculture and the Conservation of Genetic Resources for Sustainable Forest Management*. Proceedings of the Symposium of the North American Forest Commission, Forest Genetic Resources and Silviculture Working Groups and the International Union of Forest Research Organizations (IUFRO), 21 September 2003, Quebec City, Canada, Information Report LAU-X-128, pp. 9–19.

Koski, V., Skrøppa, T., Paule, L., Wolf, H. and Turok, J. (1997) *Technical Guidelines for Genetic Conservation of Norway Spruce* (Picea abies *(L.) Karst.*). International Plant Genetic Resources Institute, Rome, Italy.

Larsson, T.-B., Dias, S., Frank, G., Puumalainen, J., Richard, D., Tømmerås, B.A., Watt, A. and Wolfslehner, B. (2001) *Assessing Forest Biodiversity on a Pan-European scale*. BEAR Technical Report 7, EU FAIR Project.

Le Corre, V., Machon, N., Petit, R.J. and Kremer, A. (1997) Colonization with long-distance seed dispersal and genetic structure of maternally inherited genes in forest trees: a simulation study. *Genetical Research* 69, 117–125.

Lefèvre, F. (2004) Human impacts on forest genetic resources in the temperate zone: an updated review. *Forest Ecology and Management* 197, 257–271.

Lefèvre, F., Barsoum, N., Heinze, B., Kajba, D., Rotach, P., de Vries, S.M.G. and Turok, J. (2001) *Technical Bulletin: In Situ Conservation of* Populus nigra. International Plant Genetic Resources Institute, Rome, Italy.

Lenormand, T. (2002) Gene flow and the limits to natural selection. *Trends in Ecology and Evolution* 17, 183–189.

McKay, J.K. and Latta, R.G. (2002) Adaptive population divergence: markers, QTL and traits. *Trends in Ecology and Evolution* 17, 285–291.

McNeely, J.A. (1994) Lessons from the past: forests and biodiversity. *Biodiversity and Conservation* 3, 3–20.

Modrzynski, J. and Eriksson, G. (2002) Response of *Picea abies* populations from elevational transects in the Polish Sudety and Carpathian mountains to simulated drought stress. *Forest Ecology and Management* 165, 105–116.

Namkoong, G., Boyle, T., Gregorius, H.-R., Joly, H., Savolainen, O., Wickneswari, R. and Young, A. (1996) *Testing Criteria and Indicators for Assessing the Sustainability of Forest Management: Genetic Criteria and Indicators*. Working Paper No. 10, Centre for International Forestry Research, Bogor, Indonesia.

Oddou-Muratorio, S., Demesure-Musch, B., Pélissier, R. and Gouyon, P.H. (2004) Impacts of gene flow and logging history on the local genetic structure of a scattered tree species *Sorbus torminalis* L. Crantz. *Molecular Ecology* 13, 3689–3702.

Petit, R.J., Brewer, S., Bordacs, S., Burg, K., Cheddadi, R., Coart, E., Cottrell, J., Csaikl, M., van Dam, B., Deans, J.D., Espinel, S., Fineschi, S., Finkeldey, R., Glaz, I., Goicoechea, P.G., Jensen, J.S., König, A.O., Lowe, A.J., Madsen, S.F., Matyas, G., Munro, R.C., Popescu, F., Slade, D., Tabbener, H. and Kremer, A. (2002) Identification of refugia and post-glacial colonisation routes of European white oaks based on chloroplast DNA and fossil pollen evidence. *Forest Ecology and Management* 156, 49–74.

Petit, R.J., Aguinagalde, I., de Beaulieu, J.L., Bittkau, C., Brewer, S., Cheddadi, R., Ennos, R., Fineschi, S., Grivet, D., Lascoux, M., Mohanty, A., Müller-Starck, G., Demesure-Musch, B., Palmé, A., Martin, J.P., Rendell, S. and Vendramin, G.G. (2003) Glacial refugia: hotspots but not melting pots of genetics. *Science* 300, 1563–1565.

Prabhu, R., Colfer, C.J.P., Venkateswarlu, P., Tan, L.C., Soekmadi, R. and Wollenberg, E. (1996) *Testing Criteria and Indicators for the Sustainable Management of Forests: Phase 1 Final Report*. CIFOR Special Publication, Bogor, Indonesia.

Rehfeldt, G.E., Wykoff, W.R. and Ying, C.C. (2001) Physiological plasticity, evolution, and impacts of a changing climate on *Pinus contorta*. *Climatic Change* 50, 355–376.

Rehfeldt, G.E., Tchebakova, N.M., Parfenova, Y.I., Wykoff, W.R., Kuzmina, N.A. and Milyutin, L.I. (2002) Intraspecific responses to climate in *Pinus sylvestris*. *Global Change Biology* 8, 912–929.

Root, T.L., Price, J.T., Hall, K.R., Schneider, S.H., Rosenzweig, C. and Pounds, J.A. (2003) Fingerprints of global warming on wild animals and plants. *Nature* 421, 57–60.

Rotach, P. (1999) *In situ* conservation and promotion of noble hardwoods: silvicultural management strategies. In: Turok, J., Jensen, J., Palmberg-Lerche, C., Rusanen, M., Russel, K., de Vries, S. and Lipman, E. (compilers). *EUFORGEN Noble Hardwoods Network, Report of the third meeting*, 13–16 June 1998, Sagadi, Estonia. International Plant Genetic Resources Institute, Rome, Italy, pp. 39–50.

Savolainen, O. and Kärkkäinen, K. (1992) Effect of forest management on gene pools. *New Forests* 6, 329–345.

Savolainen, O., Bokma, F., Garcia-Gil, R., Komulainen, P. and Repo, T. (2004) Genetic variation in cessation of growth and frost hardiness and consequences for adaptation of *Pinus sylvestris* to climatic change. *Forest Ecology and Management* 197, 79–89.

Skrøppa, T. and Kohmann, K. (1997) Adaptation to local conditions after one generation in Norway spruce. *Forest Genetics* 4, 171–177.

Streiff, R., Ducousso, A., Lexer, C., Steinkellner, H., Glöessl, J. and Kremer, A. (1999) Pollen dispersal inferred from paternity analysis in a mixed oak stand of *Quercus robur* L. and *Q. petraea* (Matt.) Liebl. *Molecular Ecology* 8, 831–841.

Teissier du Cros, E. (2001) *Forest Genetic Resources Management and Conservation. France as a Case Study*. Ministry of Agriculture and Fisheries, Bureau of Genetic Resources, Commission of Forest Genetic Resources. INRA-DIC, Paris, France.
Via, S., Gomulkiewicz, R., De Jong, G., Scheiner, S.M., Schlichting, C.D. and van Tienderen, P.H. (1995) Adaptive phenotypic plasticity: consensus and controversy. *Trends in Ecology and Evolution* 10, 212–217.
von Wuehlisch, G, Krusche, D. and Muhs, H.J. (1995) Variation in temperature sum requirement for flushing of beech Provenances. *Silvae Genetica* 44, 343–346.
Wolf, D.E., Takebayashi, N. and Rieseberg, L.H. (2001) Predicting the risk of extinction through hybridization. *Conservation Biology* 15, 1039–1053.

12 Using GIS Models to Locate Potential Sites for Wheat Wild Relative Conservation in the Palestinian Authority Areas

S. Allahham and H. Hasasneh

12.1 Introduction

Wheat is the major food crop in the world in terms of production. Of the total food production by the world's top 30 crops (based on dry matter), about 23.4% comes from wheat, followed by maize (21.5%) and rice (16.5%) (Harlan, 1995). Wheat is the most important cereal traded on international markets. The total world trade in wheat and wheat flour (in grain equivalent) is close to 95 million tonnes, with the developing countries accounting for about 80% of imports. Wheat is the major commodity provided as food aid. In 1992/1993, for example, wheat accounted for more than half of the 15,200 t of cereals shipped as food aid to developing countries.

At the local level, about 19,000 ha planted with wheat crops are concentrated in the Hebron and Jenin districts of the West Bank. Jenin district constitutes 23% of the total area, where the ecological space in terms of soil, topography and climate is preferable for wheat production. In Jenin district, the productivity of wheat, in good years, reaches 3500 kg/ha while in Hebron district where the climate is arid and the ecosystem is fragile and vulnerable to drought, the productivity does not exceed 1500 kg/ha. The majority of wheat crops in Jenin district are improved varieties while in Hebron district local varieties are predominant.

The first cultivation of cereals, when hunter-gatherers became farmers, began about 10,000–12,000 years ago. In the transition, people gained a more abundant and dependable source of food, including their daily bread, and changed the world forever. Archaeologists and historians agree that the rise of agriculture, along with the domestication of animals for food and labour, produced the most important transformation in human culture since the last ice age, and perhaps since the control of fire, farming and herding has led to the growth of large, settled human populations and increasing competition for productive lands, possibly leading to organized warfare. Excavations at more than 50 sites over the last half century have established the Fertile Crescent of the Middle East as the homeland of the first farmers. This arc of land, broadly defined, extends from Palestine through Lebanon and Syria, then through the plains and hills of Iraq and southern Turkey and all the way to the head of the Persian Gulf. Among its 'founder crops' were wheat, barley, various legumes, grapes, melons, dates, pistachios and almonds. The region also produced the first domesticated sheep, goats, pigs and cattle. New genetic studies suggest the Karacadag Mountains, in southeast Turkey at the upper fringes of the Fertile Crescent, as the site where einkorn wheat was first domesticated from a wild species around 11,000 years ago.

Genetic resources are fundamental to sustain global wheat production at present and in the future. They embody a wide range of genetic diversity that is critical to enhance and to maintain the yield potential of wheat, for they provide new sources of resistance and tolerance to biotic and abiotic stresses. Modern high-yielding wheat cultivars are an assembly of genes or gene combinations pyramided by breeders using, in most cases, well-adapted cultivars from their regions. International agricultural research has enormously expanded the availability of widely adapted germplasm that is genetically diverse (i.e. descended from more sources). However, introgression of additional variation found in genetic resources is necessary to increase yield stability and further improve wheat. For example, the genus *Aegilops* is closely related to *Triticum* (Kerby and Kuspira, 1988). Interest has developed in recent years in exploiting *Aegilops* spp. as important genetic resources for wheat improvement (Comeau *et al.*, 1992; William *et al.*, 1993; Farooq *et al.*, 1996). *Ae. geniculata* Roth (= *Ae. ovata* L.) is an annual, selfing (Hammer, 1980) allotetraploid species ($2n = 4x = 28$) with MU genome (Van Slageren, 1994). Among the 22 species of the genus *Aegilops*, it is particularly interesting as a source of disease and pest resistance (Valkoun *et al.*, 1985; Dimov *et al.*, 1993). Some information is also available concerning its response to drought (Rekika *et al.*, 1998) and salinity (Farooq *et al.*, 1996), and suggests that this species could represent a valuable reservoir of genes for resistance to these stresses.

In Palestine, there are 13 species of wild relatives of wheat, while in Palestinian Authority (PA) areas there are eight species namely: *Ae. biuncialis* Vis., *Ae. geniculata*, *Ae. kotschyi* Boiss., *Ae. longissima* Schweinf. & Muschl, *Ae. triuncialis* L., *Ae. searsii* Feldman & Kislev., *Ae. peregrina* (Hack.) Maire & Weill. and *Triticum dicoccoides* (Körn. ex Asch &Graebn.) Schweinf.

Wheat wild relatives are essential components of natural and semi-natural habitats, as well as agricultural systems and are critical for maintaining ecosystem health. Their conservation and sustainable use can lead to improving agricultural production, increasing food security and maintaining the environment. Genetic resources are fundamental to the world's food security and central to efforts to alleviate poverty, contribute to the development of sustainable production systems and supplement the natural resource base. The germplasm conserved is especially rich in wild crop relatives, traditional farmer cultivars and old cultivars, which represent an immense reserve of genetic diversity.

Genetic resources are fundamental to sustaining global wheat production today and in the future.

Although considerable wild and landrace wheat germplasm has been collected since 1989, there are still large gaps in world collections. Wild wheat and landraces, especially material adapted to microhabitats, are rapidly disappearing because of the introduction of agronomically superior new cultivars. Severe overgrazing by huge flocks of sheep and goats in the Near East and Central Asia can, in a very few years, wipe out late-flowering forms. About 640,000 accessions of *Triticum* spp., *Aegilops* spp. and *Triticosecale* can be found in collections around the world.

The West Bank of Palestine harbours a tremendous diversity of climates and biomes, providing suitable habitats for over 2500 species of wild plants, 550 species of birds and 117 species of mammals. This unique rich diversity of Palestinian biota has long captured the interest of ecologists and scientists. The climate is unique in that the wet season coincides with the low sun or winter period. Summers are dry. Total annual precipitation ranges between 350 and 720 mm per year. Temperatures are those of the subtropics moderated by maritime influence and fog associated with the cold ocean currents. The result is a very limited, but predictable, growing season.

The severe overgrazing, urbanization, land clearing, overexploitation of species and lack of land use planning have led to land degradation in a lot of areas of the West Bank and thus led to genetic erosion of important wild relatives. In addition, replacement of the traditional farming systems by modern agricultural practices are endangering wheat wild relatives. Food demands and market forces have encouraged the replacement of the locally adapted varieties (land races and local varieties) of both fruit trees and field crops with highest yielding cultivars, hence reducing the gene pools of crops.

For the high rate of genetic erosion and the high risks of anthropogenic impacts (urbanization, overgrazing and desertification), it is critical to map the distribution of the wheat wild relatives and to devise proper conservation plans to avoid the further loss of this natural resource.

Our objectives were to use 1349 geo-referenced observations of wild relatives of wheat (eight species) to predict the potential distribution areas identified as primary suitable conservation sites with respect to particular climatic variables, rather than the actual range of distribution, and in addition to map the distribution of species and species richness for wheat relatives in West Bank areas using grid cells. This type of study can provide baseline data for further geographic information system (GIS) analysis for exploration, conservation and use of germplasm of wild crop relatives, as well as for studies of factors that explain geographic distribution of these species.

12.2 Materials and Methods

This section will cover the spatial and non-spatial data, software and methodologies used to generate the species richness maps to predict suitable sites for the studied species.

12.2.1 Data sets used

Spatial and non-spatial data and previous technical reports were used in this study.

1. Climatic data:[1] the data were extracted from a global interpolated climate database. The global climate layers (grids) were at 1 km resolution. The data layers were generated through interpolation of average monthly climate data from weather stations on a 30 arc-second resolution grid (often referred to as '1 km' resolution). Variables included were monthly total precipitation, and monthly mean, minimum and maximum temperatures.
2. Altitude: 1 km resolution.
3. Geographical distribution of eight species of wheat crop wild relatives.
4. West Bank and district boundary.
5. Five bioclimatic parameters:

- Annual mean temperature;
- Maximum temperature of the warmest month;
- Minimum temperature of the coldest month;
- Annual precipitation;
- Precipitation of the wettest month.

12.2.2 Software used

ARC VIEW 3.2a was used for GIS (vector spatial analysis), while DIVA-GIS (GARP and DIVA-GIS, 2003) was used as the GIS for the analysis of the biodiversity data using the techniques developed by Hijmans and Spooner (2001), Hijmans *et al.* (2001, 2004) and Jarvis *et al.* (2003).

12.2.3 Description of data set

The data presented were derived from the Agrobiodiversity Project botanical database and the Hebrew University of Jerusalem BioGIS Project available at: http://www.biogis.huji.ac.il. The data are a compilation of 1349 unique observations of wheat wild relative germplasm accessions and herbarium specimens. Basic statistics of the distribution of these point observations were calculated to determine the number of *Aegilops* species and *T. dicoccoides* and the geographical distribution of these species in West Bank and historical Palestine to assess surveyed areas and those areas that were not properly surveyed. For each species, we mapped the number of observations and species richness, using grid cells. Species richness was used because it is a simple, widely used, well-understood and useful measure of taxonomic diversity (Gaston, 1996). We used DIVA-GIS to predict the potential suitable areas of wheat wild relative species in the study area based on bioclimatic variables. DIVA-GIS software is available at www.diva-gis.org for predicting the distribution of organisms in the wild when little or

[1] The data sources of climate data from www.diva-gis.org site.

nothing is known of the physiology of the species involved. It is assumed that the climate at the point of observation of a species is representative of the environmental range of the organism.

12.3 Modelling Capabilities

DIVA-GIS uses the BIOCLIM model (bioclimatic analysis and prediction system) to generate a predictive map showing the potential spatial distribution range of wheat wild relative species based on five bioclimatic variables of the sampled data. The main idea of BIOCLIM modelling is to find a single rule that identifies all areas with similar climate to the various geographic regions for the species of interest. Such analysis is done by finding the climatic range of the points on each climatic variable. The DIVA-GIS program was run with the Bioclim classic option (six groups) for identifying the excellent, very high, high, medium and low suitability and not suitable areas.

To generate the predictive model, two stages are involved. The first stage in modelling was extracting the 'climatic envelope'[2] of the species from the data. Second stage was projecting the climatic envelope from the multidimensional climatic space into two-dimensional geographic space. The bioclimatic variables used in modelling the predictive map are:

1. Set of climatic variables:
   - Annual mean temperature;
   - Maximum temperature of the warmest month;
   - Minimum temperature of the coldest month;
   - Annual precipitation;
   - Precipitation of the wettest month.
2. Digital Elevation Model (DEM) (cell size = 1 km^2).
3. Geographical points of wheat wild relative distribution data set.

The modelling process is as follows:

1. The points (longitude and latitude) of each selected species in the West Bank were taken as points of occurrence of the species.
2. For each climatic variable, value of the sites where the species was recorded is ranked in numerical order.
3. In addition, the minimum and the maximum percentile values of each climatic variable are determined in which these values define the climatic envelope of the species.
4. Cells falling within the climatic envelope of the species for all variables are marked by comparing the climatic characteristics of each grid cell on the grid map of the West Bank and the climatic envelope determined for the species.

[2] A climatic envelope is the range of the species distribution within a multidimensional space determined by a set of climatic variables.

12.4 Results and Discussion

This section will cover the results of spatial pattern of species distribution of wheat wild relatives in the study area, species diversity of wheat wild relatives in the West Bank, and the results of the prediction model for primary potential suitability sites of wheat wild relatives.

12.4.1 Species distribution

In this study, 1349 geo-referenced observations were used for eight species of wheat wild relatives distributed in Palestine. Basic statistics on the distribution of observations in the study area indicate a strong bias in collection along the roads (70–80%) (Fig. 12.1). The second important point about the distribution

Fig. 12.1. The relationship between wheat wild relative species distribution, species richness per district and distance to main roads.

Table 12.1. Species of wheat wild relatives and number of species observations in the study area.

Species	No. of observations	Percentage
Aegilops biuncialis	24	1.8
Aegilops geniculata	310	23
Aegilops kotschyi	226	16.8
Aegilops longissima	154	11.4
Aegilops triuncialis	18	1.3
Aegilops searsii	7	0.5
Aegilops peregrina	540	40
Triticum dicoccoides	70	5.2
Total	1349	100

of species is that the northern part is not so thoroughly surveyed compared to the southern part of the West Bank. On the other hand, the species richness and diversity in northern parts have the highest biodiversity index. Table 12.1 reveals that the highest number of observations is for *Ae. peregrina* (40.0%), while the lowest number is for *Ae. searsii* (0.5%).

Simple analysis was used to calculate the number of species observed by district and the Shannon biodiversity index; the results show that the Jerusalem district has the highest species richness because: (i) this district is heavily studied and surveyed; and (ii) the management and conservation status is relatively good compared to other areas. While Jericho and Tubas districts contain the lowest number of species, this may be attributed to the climate conditions in this area which are arid to hyper-arid and to the huge numbers of livestock (Table 12.2).

Table 12.2. Biodiversity indices and number of species per district.

District	Area (km²)	No. of species	NOB/NOS	Percentage from total	MENHINICK	SHANNON	SIMPSON
Jenin	573	4	1	1	2.00	1.39	1.00
Tulkarm	245	3	2	2	1.23	1.10	0.80
Tubas	366	2	8	5	0.50	0.56	0.40
Nablus	614	5	3	5	1.29	1.23	0.68
Qalqiliya	174	3	1	1	1.50	1.04	0.83
Jericho	609	3	34	34	0.30	0.62	0.36
Salfit	202	3	4	4	0.87	0.96	0.62
Ramallah	849	5	3	5	1.25	1.04	0.53
Jerusalem	354	8	9	24	0.96	1.27	0.63
Bethlehem	608	4	4	5	1.07	1.25	0.75
Hebron	1068	5	8	14	0.77	1.20	0.65
Total	5661						

NOB/NOS: number of observations/number of species.

12.4.2 Species richness grid-based distribution

By using the grid analysis functions in DIVA-GIS the species richness in the study area was calculated by using the circular neighbourhood method. In this method, the calculations are made for a circle with its centre in the middle of a grid cell. A grid cell of 1 km size was used. The species richness of wild wheat ranged from 1 to 7 per grid, the majority of grid values were 1–3 and one grid cell has the highest value of 7. The area of the highest grid is located in Jerusalem district at the border area. It is worth mentioning that this area is protected from the anthropogenic activities due to its sensitivity location.

12.4.3 Bioclimatic analysis and prediction system

Based on the number of observations and genetic importance, four species were selected to use in the bioclimatic analysis and prediction system model, namely *Ae. geniculata*, *Ae. kotschyi*, *Ae. peregrina* and *T. dicoccoides*. These species have more than 200 observations except for *T. dicoccoides* which was included for its importance. The other species have fewer observations making it difficult to run the model and predict the suitable sites for them.

The results of the simulation model using DIVA-GIS suggested excellent suitable sites for *T. dicoccoides* in the northern part of the West Bank in Jenin district and eastern part of Nablus (Fig. 12.2). These areas represent the potential distribution, and do not take into account potential anthropogenic effects that may have destroyed wild wheat habitats. It is worth mentioning that this area is considered the main area of wheat production in Palestine. More than 60% of the West Bank area is not at all suitable for *T. dicoccoides*.

GIS also suggests that Hebron and Bethlehem, Jericho and Tubus districts are not likely to be suitable for this species. The geographical distribution areas are proved by the climatic characterization of the collecting sites. For all observed species sites, the annual rainfall range is 413–784 mm.

The bioclimatic model was run for three species of *Aegilops*: *Ae. geniculata*, *Ae. kotschyi* and *Ae. peregrina*. The prediction map suggested that the majority of the West Bank areas except the Jordan valley (Jericho district) area are suitable for *Ae. geniculata*. The excellent suitable area is located in the northern region of the West Bank in the western part of Jenin, Nablus and Ramallah district (Fig. 12.3). It is worth mentioning that wheat improvement programmes focus on the use of *Ae. geniculata* to expand genetic variability, develop alternative plant types and physiological processes and increase drought stress and high temperature tolerance in durum wheat and bread wheat. *Ae. geniculata* grows in the Mediterranean regions (Van Slageren, 1994) characterized by a dry summer season with high temperature and irradiance.

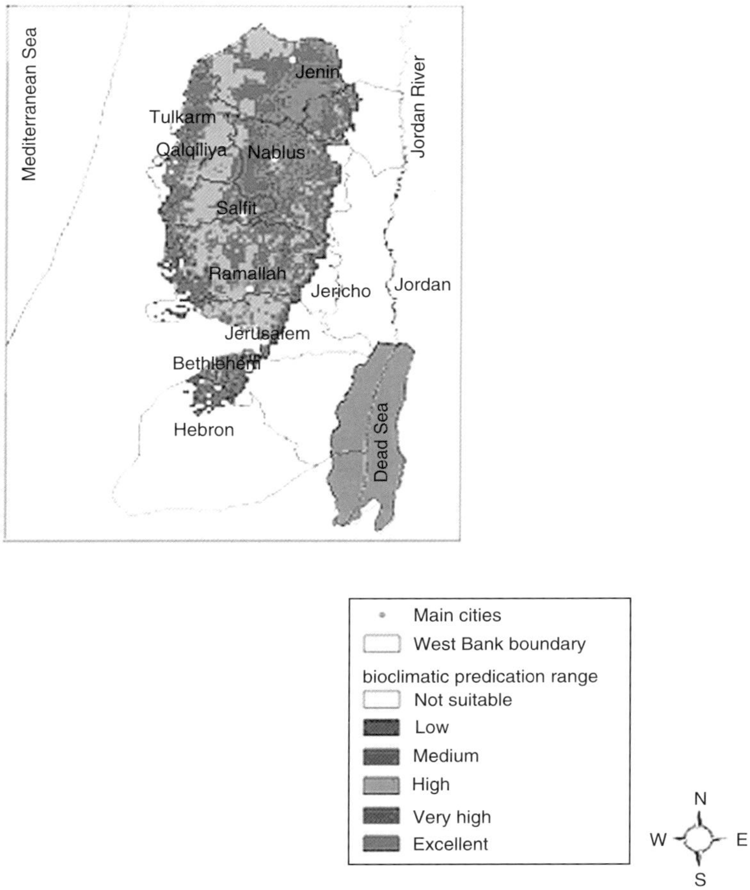

Fig. 12.2. Predicted suitable sites for *Triticum dicoccoides* in the West Bank.

The prediction map suggested that the southern part of Hebron district is considered suitable for *Ae. kotschyi* (Fig. 12.4). *Ae. kotschyi* is a drought-resistant species which can exist in annual rainfall range between 141 and 550 mm (Fig. 12.4). *Ae. kotschyi* grows in arid regions characterized by a dry summer season with high temperature and high irradiance. As with other wild species, it can acclimate to these constraints by escape, avoidance and tolerance (Blum, 1988).

The prediction map shows that the majority of the West Bank areas are suitable for *Ae. peregrina*, while the most suitable areas are located in the middle (eastern part of Ramallah and in the northern part of the West Bank – in Jenin district and eastern part of Salfit and Nablus) (Fig. 12.5). The prediction area for *Ae. peregrina* is similar or overlapping with *Ae. geniculata*. *Ae. peregrina*

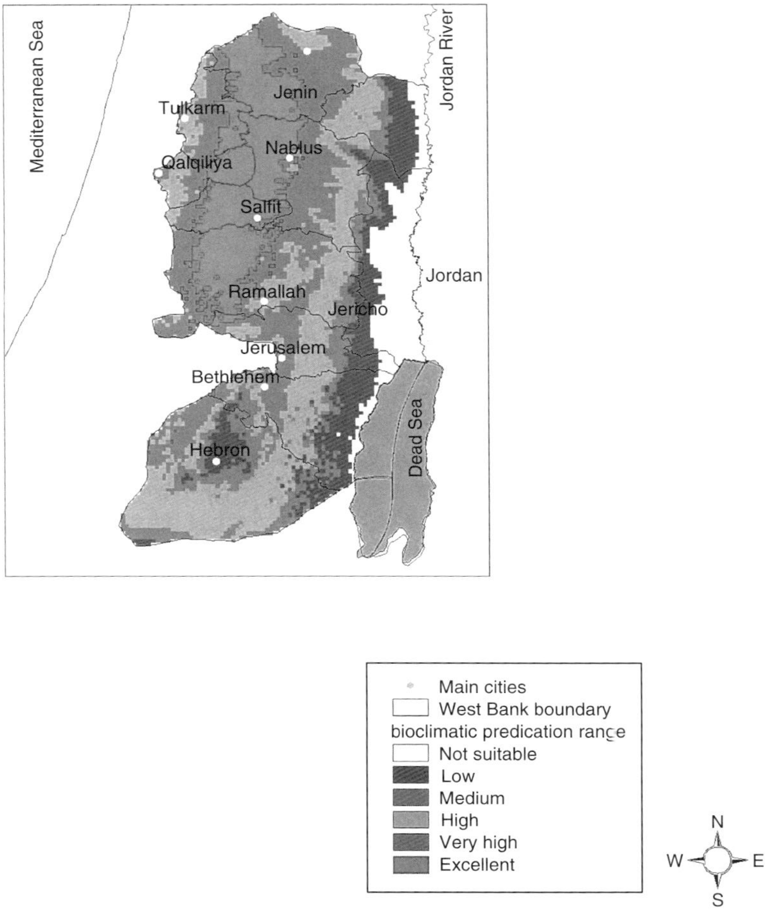

Fig. 12.3. Predicted suitable sites for *Aegilops geniculata* spec es in the West Bank.

is the most dominant species of wheat wild relative, 40% of observations are *Ae. peregrina*. It exists in 207–784 mm range of rainfall.

12.5 Conclusions

In Palestine, some *Aegilops* species still exist in a wide area in the central highlands. The predicted primary suitable sites for wheat wild relatives are not necessarily potential sites because the analysis did not take into consideration the interspecies competition or some physical environmental factors, including soil type, soil depth, soil texture and erosion risk. This necessitates field surveys and

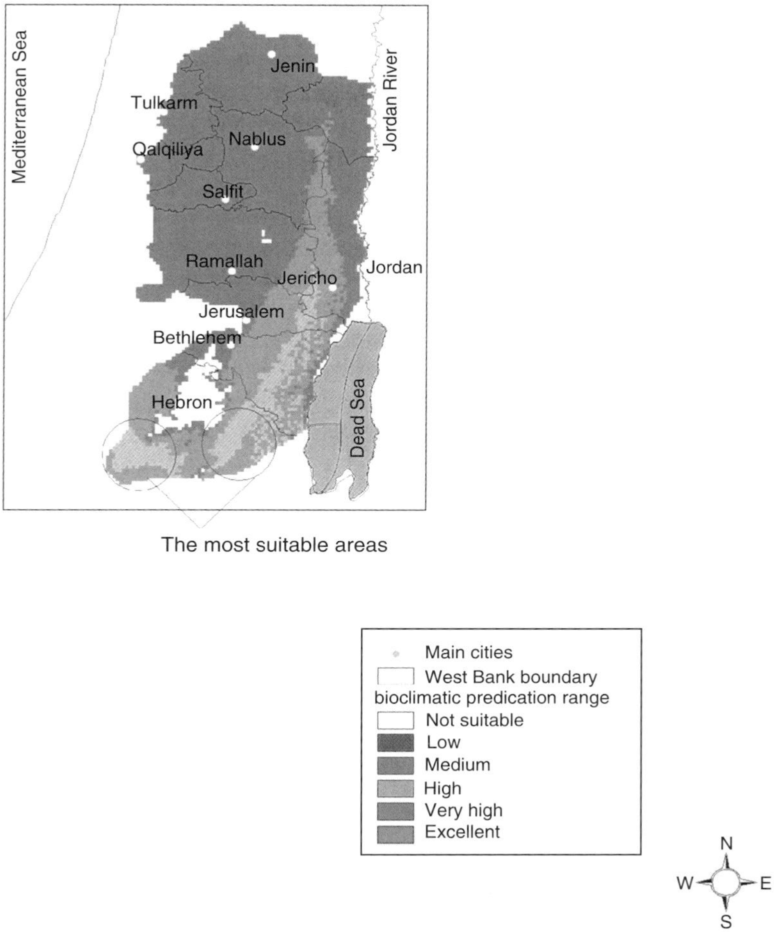

Fig. 12.4. Predicted suitable sites for *Aegilops kotschy* in the West Bank.

studies to assess the suitability of a particular area. However, the data presented in this work provide a good idea about how to use GIS platforms in unison for more concrete spatial analysis and can also provide a low-scale map to pinpoint large areas with high similarity of the environmental factors. Ecogeographic survey is important for pinpointing the suitability of areas for *T. dicoccoides* and the reintroduction of this species is an urgent priority, since in the last 10 years or more the species has not been observed. Future botanical survey missions should focus in the areas which are classified as excellent areas, and the survey should be more representative and not biased. *In situ* conservation management plans for the area, which has the highest biodiversity richness, should be initiated and proposed, along with conservation management plans.

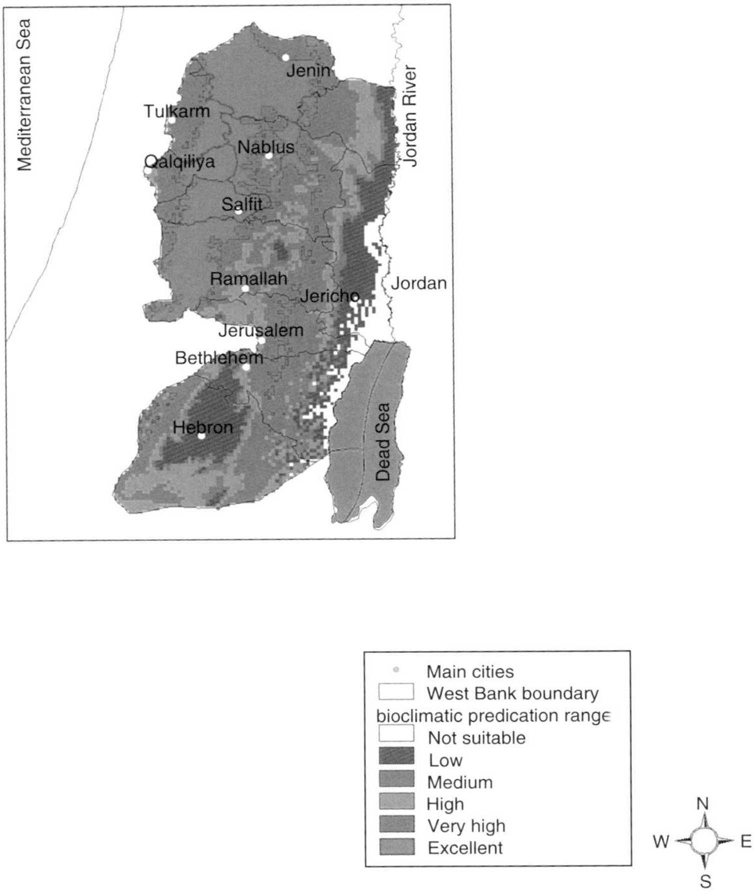

Fig. 12.5. Predicted suitable sites for *Aegilops peregrina* in the West Bank.

References

Blum, A. (1988) *Plant Breeding for Stress Environments*. CRC Press, Boca Raton, Florida.

Comeau, A., Nadeau, P., Plourde, A., Simard, R., Maes, O., Kelly, S., Harper, L., Lettre, J., Landry, B. and St-Pierre, C.-A. (1992) Media for the *in ovulo* culture of proembryos of wheat and wheat-derived interspecific hybrids or haploids. *Plant Science* 81, 117–125.

Dimov, A., Zaharieva, M. and Mihova S. (1993) Rust and powdery mildew resistance in *Aegilops* accessions from Bulgaria. In: Damania, A.B. (ed.) *Biodiversity and Wheat Improvement*. Wiley, Chichester, UK, pp. 165–169.

Farooq, S., Shah, T.M. and Asghar, M. (1996) Intergeneric hybridization for wheat improvement: V. production of and metaphase 1 chromosome analysis in F1 hybrids of wheat (*Triticum aestivum*) with *Aegilops ovata* L. *Cereal Research Community* 24, 155–161.

GARP and DIVA-GIS (2003) Predicting the potential geographical distribution of sugarcane woolly aphid. *Current Science* 85, 1526–1528.

Gaston, K.J. (1996) Species richness: measure and measurement. In: Gaston, K.J. (ed.) *Biodiversity: A Biology of Numbers and Difference*. Blackwell Science, London, pp. 77–113.

Hammer, K. (1980) Vorarbeiten zur monographischen darstellung von wildpflanzen-sortimenten: *Aegilops* L. *Kulturpflanze* 28, 33–180.

Harlan, J.R. (1995) *The Living Fields – Our Agricultural Heritage*. Cambridge University Press, Cambridge.

Hijmans, J. and Spooner, D.M. (2001) Geographical distribution of wild potato species. *American Journal of Botany* 88(11), 2101–2112.

Hijmans, R.J., Guarino, L., Cruz, M. and Rojas, E. (2001) Computer tools for spatial analysis of plant genetic resources data. *Plant Genetic Resources Newsletter* 127, 15–19.

Hijmans, R.J., Guarino, L., Bussink, C., Mathur, P., Cruz, M., Barrentes, I. and Rojas, E. (2004) *DIVA Manual, a Geographic Information System for the Analysis of Biodiversity Data*. CIP, Lima, Peru.

Jarvis, A., Ferguson, M.E., Williams, D.E., Guarino, L., Jones, P.G., Stalker, H.T., Valls, J.F.M., Pittman, R.N., Simpson, C.E. and Bramel, P. (2003) Biogeography of wild arachis: assessing conservation status and setting future priorities. *Crop Science* 43(3), 1100–1108.

Kerby, K and Kuspira, J. (1988) Cytological evidence bearing on the origin of the B genome in polyploid wheat. *Genome* 30, 36–43.

Rekika, D., Zaharieva, M., Stankova, P., Xu, X., Souyris, I. and Monneveux, P. (1998) Abiotic stress tolerance in *Aegilops* species. In: Nachit, M.M., Baum, M., Porceddu, E., Monneveux, P. and Picard, E. (eds) *Durum Research Network*, Proceeding of the SEWANA, South Europe, West Asia and North Africa, ICARDA, Aleppo, Syria, pp. 113–118,

Valkoun, J., Hammer, K., Kucerova, D. and Bartos, P. (1985) Disease resistance in the genus *Aegilops* L. – stem rust, leaf rust, and powdery mildew. *Kulturpflanze* 33, 133–153.

Van Slageren, M.W. (1994) *Wild Wheats: a Monograph of* Aegilops *L. and* Amblyopyrum *(Jaub and Spach) Eig (Poaceae)*. Agricultural University, Wageningen – International Center for Agricultural Research in Dry Areas, Aleppo, Syria.

William, M.D.H.M., Pena, R.J. and Mujeeb-Kazi, A. (1993) Seed protein and isozyme variation in *Triticum tauschii* (*Aegilops squarrosa*). *Theoretical and Applied Genetics* 87, 257–263.

III Threat and Conservation Assessment

13 IUCN Red Listing of Crop Wild Relatives: is a National Approach as Difficult as Some Think?

J. MAGOS BREHM, M. MITCHELL, N. MAXTED,
B.V. FORD-LLOYD AND M.A. MARTINS-LOUÇÃO

13.1 Introduction

13.1.1 Crop wild relatives

Crop wild relatives (CWR) are taxa related to species of direct socio-economic importance including food, fodder and forage crops, medicinal plants, condiments, ornamental and forestry species, as well as plants used for industrial purposes such as oils and fibres. CWR also include plants harvested directly from the wild and minor crops or underutilized species (Kell *et al.*, 2004). Maxted *et al.* (2006) provided a workable definition using the gene pool concept of Harlan and de Wet (1971) together with a new taxon group concept based on the taxonomic hierarchy to be used where genetic information is unavailable. Thus, according to Maxted *et al.* (2006), CWR are wild plant species that have an indirect use derived from their relatively close genetic relationship to a crop. The importance of CWR mainly lies in the fact that they are an extremely important source of genetic variation for the development of new varieties (Jain, 1975) and the improvement of domesticated species (Schoen and Brown, 1993). Furthermore, CWR are essential components of natural habitats and agricultural systems and they naturally cross-breed with the crops keeping them genetically diverse and healthy (Prescott-Allen and Prescott Allen, 1983; Hoyt, 1988; Maxted *et al.*, 1997a; IPGRI, 2000a).

CWR have been identified as an essential target for conservation for more than 80 years (Vavilov, 1926; Prescott-Allen and Prescott Allen, 1983; Maxted *et al.*, 1995; Meilleur and Hodgkin, 2004). Yet, these species are often disregarded when formulating conservation priorities because the responsibility for their conservation has historically fallen between the ecological and agricultural conservation communities. Only recently the conservation focus on these taxa has increased, mainly because of an increasing concern over global food security which has focused attention on the means of seeking genetic material that might be used to enhance productivity, disease resistance and tolerance to various environmental conditions (Heywood, 1997). On the other hand, the habitats that support many CWR are under threat (IPGRI, 2003). In particular, these species are traditionally found in agroecosystems and their occurrence is decreasing due to agricultural industrialization (Jain, 1975; IPGRI, 2000b).

In global terms the Convention on Biological Diversity (CBD) (CBD, 1992), the Global Strategy for Plant Conservation (GSPC) (CBD, 2002), the European Plant Conservation Strategy (EPCS) (Planta Europa and the Council of Europe, 2002), the European Community Biodiversity Strategy (ECBS) (European Commission, 2000), the Global Plan of Action for the Conservation and Sustainable Utilization of Plant Genetic Resources for Food and Agriculture (FAO, 1996) and the International Treaty on Plant Genetic Resources for Food and Agriculture (ITPGRFA) (FAO, 2001), each stress the need for plant genetic resources (PGRs) conservation including the wild relatives of crops. In particular, the CBD refers to 'wild relatives of domesticated or cultivated species' in the list given in Annex I as one of the categories of organisms to be identified and monitored (CBD, 1992). The GSPC in its Targets for 2010 (CBD, 2002) also addresses the need for conserving the 'genetic diversity of crops, livestock, and of harvested species of trees (...)' (Target 3.1), and the 'biological resources that support sustainable livelihoods, local food security and health care' (Target 8.2) (CBD, 2002). The EPCS recognizes the economic importance of wild gene pools of domesticated European species (UNEP/CBD, 2002) and the ITPGRFA refers to the PGRs for food and agriculture as 'any genetic material of plant origin of actual or potential value for food and agriculture' (Article 2) (FAO, 2001), which would necessarily include CWR.

13.1.2 Species threat assessment

CWR, although a vital resource to humankind, are increasingly threatened by the direct and indirect careless actions of humankind; therefore, CWR species require active conservation. Species conservation measures aim to avoid or reduce biodiversity losses and are usually focused on those species most at risk (Keith and Marion, 2002). This is mainly because these threatened species are good indicators of the habitat quality (Keith and Marion, 2002). Consequently, the assessment of species threat is a critical and essential step in order to allocate resources for conservation. It is not possible to conserve every component of diversity; therefore, priorities need to be established.

There have been considerable efforts in studying the actual processes of extinction in individual species as well as attempting to find consistent methods of assessing the threats species are facing. The latter are usually measured as the risk of extinction, the most commonly used indicators being rarity, decline and fragmentation of populations (Hartley and Kunin, 2003). Rarity is a term that has been widely used for a variety of applications and has numerous definitions (see Rabinovitz *et al.*, 1986). However, this concept is not always associated with threat as in some cases rare populations may persist for long periods

(Mace and Kershaw, 1997). On the other hand, the concept of decline is simpler and might be measured at several different levels, e.g. change in number of mature individuals, number of locations and overall decrease in geographic range over time (Dobson, 1998). Fragmentation is considered an indicator of threat in two ways: too much fragmentation (edge effect) and too little fragmentation (too few locations) (Hartley and Kunin, 2003).

Burgman *et al.* (1999) suggest the protocols to assign conservation status and to set priorities for conservation come under four headings: (i) qualitative assessment – species are assigned subjectively to qualitatively defined categories of risk; (ii) point scoring procedures – points are allocated for a number of attributes and a total score for the species is obtained by summing points over all attributes which determines the rank of the species relative to the scores for other species (e.g. Millsap *et al.*, 1990), in turn, a relative weighting of variables might be applied (Keith *et al.*, 1997); (iii) rule sets – using explicitly defined attributes, the species are assessed by determining whether they meet a set of logical conditions defined by thresholds for each of the attributes (e.g. IUCN, 2001); and (iv) population viability analysis (PVA) – by building models of interaction of a species with the environment (e.g. Boyce, 1992; Iricondo, 1996; Dobson, 1998; Menges, 2000). According to Mace and Lande (1991) the principal weakness of most of these methods is the lack of explicit guidelines for assigning taxa to categories of risk and as a result there is no guarantee of consistency (Lunney *et al.*, 1996) and reliability (Burgman *et al.*, 1999) often leading to conflicting opinions. Nevertheless, each of these approaches has particular strengths and weaknesses and it is unlikely that any single protocol will be the best in all cases (Burgman *et al.*, 1999).

With several ways of carrying out threat assessment the main requirement is that these should be objective and comparable. Therefore, a standardized system of classification is needed to help categorize species according to the risk of extinction they are facing, and thereby, to help prioritize efforts and allocation of resources for the most endangered species.

13.1.3 IUCN Red Listing

One objective and comparable approach to threat assessment is to undertake IUCN Red Listing. The Species Survival Commission (SSC) of IUCN – The World Conservation Union (IUCN) has taken the lead in evaluating the conservation status of species. Initially, Red List assessments were subjective and relied heavily on the experience of experts (Burton, 2003). In recent years, the nature of the assessments changed considerably and now follows a strict protocol with objective criteria for estimating the extinction risk (Rodrigues *et al.*, 2006). The first relatively subjective *IUCN Red List Categories and Criteria* were published in 1994 following 6 years of research and broad consultation (IUCN, 1994). They were applied to a large number of species in particular animals, but were criticized for their inherent subjectivity, which resulted in more discussion and the new *2001 IUCN Red List Categories and Criteria* (version 3.1) (IUCN, 2001).

Currently, the *2001 IUCN Red List Categories and Criteria* represent the most accepted and widely used method of producing Red Lists, at national (Spain by Bañares *et al.*, 2004; Great Britain by Cheffings and Farrell, 2005; Sweden by Gärdenfors, 2005; Luxembourg by Colling, 2005) and international level (IUCN, 2004). Red Listing-associated 'myths', such as that the threat assessment is exclusively based on experts' opinions, that the Red List is solely a classification of species into threat categories, or that few species have been assessed so it cannot be used as a conservation tool, are no longer valid (Lamoreux *et al.*, 2003).

This system is commonly used as a conservation tool that complements the setting of conservation priorities. It is an objective, consistent, transparent, repeatable, quantifiable and standardized system that provides clear guidance on how to evaluate different factors which affect the risk of extinction. It uses explicitly defined attributes which are assessed by determining whether a species meets a set of logical conditions defined by thresholds for each of the attributes (IUCN, 2005a). It also uses the precautionary principle, which was established as a global standard at the Earth Summit in 1992. This principle states that scientists are encouraged to provide their own opinions even in the absence of scientific certainty. Matsuda (2003) discusses the precautionary principles in the context of the IUCN Red List Criteria and emphasizes 'that the IUCN should encourage the collection of better data and should not discourage the development of models that make use of these data' (IUCN/SSC, 1999).

In recent years, the Red Listing assessment process itself has developed considerably, extending the value of the Red List far beyond the assignment of threat status (Rodrigues *et al.*, 2006). Recently, Rodrigues *et al* (2006) showed how Red List assessments are vehicles for the compilation, synthesis and dissemination of data related to a wide range of species. Red Lists can assist in the prioritization of species for conservation, help to identify priority sites for conservation, guide management of natural resources and help to evaluate the state of biodiversity as well as to monitor change (Butchart *et al.*, 2004). On the other hand, Domínguez Lozano *et al.* (2004) emphasize the importance that the process of developing a list of threatened taxa has in reducing the gap between conservation and plant taxonomy because it gives the opportunity to the taxonomic experts to participate in the conservation process. These authors have concluded that even though species are usually taxonomically stable, taxonomy changes or new species have an effect on relative threat assessment. Further, Callmander *et al.* (2005) emphasize the role taxonomists can play prior to assessment by identifying potentially threatened taxa from data available to them and by including preliminary Red List assessments in all their taxonomic work. Therefore, conservation assessments will boost the development of taxonomic work (Moreno Saiz *et al.*, 2003; Siebert and Smith, 2004) and taxonomists may increase the number of assessments available for this conservation initiative (Callmander *et al.*, 2005).

Global assessment

The *2001 IUCN Red List Categories and Criteria* can be applied to any well-defined taxonomic unit at or below species level, down to wild populations inside their natural range' (IUCN, 2001) and to any geographical or political area (OECD, 1996). The global categories defined by the IUCN Species Survival Commission (2001) are extinct (EX), extinct in the wild (EW), critically endangered (CR), endangered (EN), vulnerable (VU), near threatened (NT), least concern (LC), data deficient (DD) and not evaluated (NE) (see IUCN, 2001, 2005a). The three categories, CR, E and VU, are counted as 'threatened categories (IUCN, 2001, 2005a). The criteria by which these categories are defined are clear, comprehensive (Rodrigues *et al*., 2006), much more objective than before (OECD, 1996), but flexible enough to handle uncertainty (Akçakaya *et al*., 2000). These are: (i) declining population (past, present and/or projected) (criterion A); (ii) geographic range size, and fragmentation, decline or fluctuations (criterion B); (iii) small population size and fragmentation, decline, or fluctuations (criterion C); (iv) very small or restricted population (criterion D); and (v) quantitative analysis of extinction risk (e.g. PVA) (criterion E) (see IUCN, 2001, 2005a). They use quantitative measurements as much as possible (OECD, 1996) and include (sub)population size, mature individuals, generation reduction, continuing decline, extreme fluctuation, severe fragmentation, extent of occurrence (EOO), area of occupancy (AOO), location and quantitative analysis studies (IUCN, 2001). These measurements can be observed, estimated, projected, inferred or suspected by rates of habitat loss (IUCN, 2005a).

A species is assigned to a particular category of risk if its attributes fall within the relevant thresholds (Keith *et al*., 2000). Although a species should be tested against all criteria, only one criterion needs to be met for a species to qualify as threatened. However, even for these criteria the application is open to some degree of subjectivity and misinterpretation might occur. To facilitate the assessments and to improve standardization, IUCN has developed the *Guidelines for Using the IUCN Red List Categories and Criteria* (IUCN 2005a) and these are regularly updated enabling the improvement of consistency and quality of the assessments (Rodrigues *et al*., 2006). Also, the assessments should always be backed up with data, justifications, sources and estimates of uncertainty and data quality (Rodrigues *et al*., 2006).

Regional assessment

National boundaries are irrelevant to populations, so if a particular species whose distribution goes beyond the limits of a geopolitical border defining the region which is being studied, there might be genetic flow to or from other conspecific populations beyond the border; this will obviously affect the extinction risk of the regional population (Keller and Bollmann, 2004). The thresholds under each criterion of the 2001 IUCN Categories and Criteria will be incorrect because the unit being assessed is not the same as the whole population or subpopulation (Gärdenfors *et al*., 2001; IUCN, 2003). Therefore, after using the 2001 IUCN Categories and Criteria to assess the species' risk of extinction as if the populations were isolated, their risk of extinction is reassessed in that particular region within the light of its overall distribution. The *Guidelines for Application of IUCN Red List Criteria at Regional Levels* (IUCN, 2003) were developed for this purpose. However, when the regional population is isolated from conspecific populations, global criteria can be used

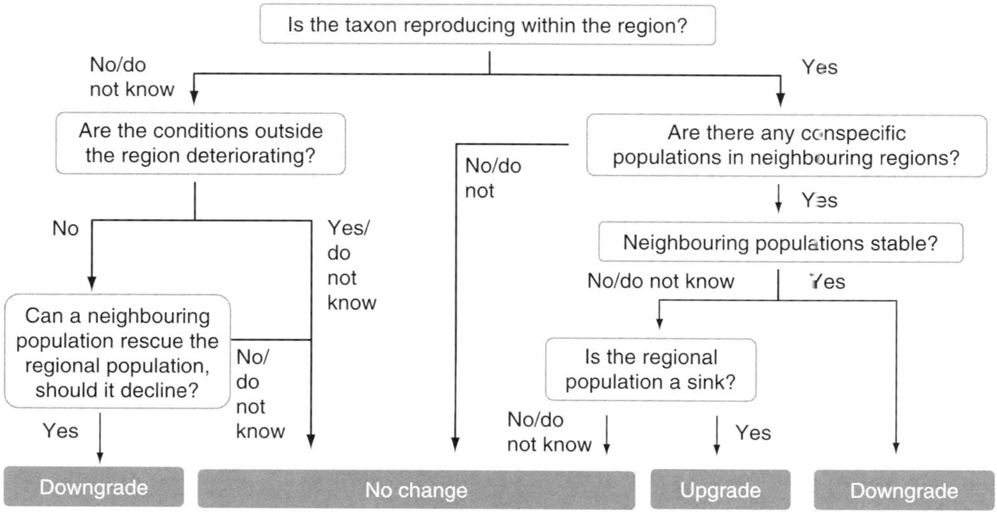

Fig. 13.1. Basic scheme adapted from IUCN (2003) of how to undertake a regional assessment. (Adapted from IUCN, 2003.)

without modification (Gärdenfors *et al*., 2001). It should be added that 'region' is defined by IUCN (2003) as any subglobal geographically defined area (e.g. continent, country or province).

The regional categories are the same as the global except for the fact that there are two new categories: regionally extinct (RE) and not applicable (NA) (IUCN, 2003). The regional assessments are the result of downgrades and upgrades from global assessments and they are based on a series of questions essentially concerning conspecific populations outside the region and the status of regional populations as sinks (IUCN, 2003) (see Fig. 13.1).

13.1.4 National Red listing: achieving the global strategy for plant conservation 2010 targets

Target 2 of the GSPC refers to the need to produce 'a preliminary assessment of the conservation status of all known plant species, at national, regional and international levels' (CBD, 2002). This would then constitute a biodiversity baseline against which assessments of the rate of change can be judged. This in turn could help to monitor whether the 2010 Biodiversity target of 'a significant reduction of the current rate of biodiversity loss at the global, regional and national level' has been achieved (CBD, 2002). Thus, national Red Lists have their importance in the fact that they highlight the threatened species in their countries which should be the first targets for future conservation assessments and actions.

More recently, the resolution RESWCC3.013 adopted during the World Conservation Congress (WCC), in Bangkok in November 2004 (IUCN, 2005b), emphasizes: (i) that the inclusion of species in national legislature should require information on the level of threat, on the types of threatening processes and on the conservation measures needed; (ii) 'that conservation action is not automatically linked to the inclusion of a species in any particular threatened category of the *IUCN Red List*, but must rather follow a careful analysis of threats and the measures needed to counteract these'; (iii) 'that the data in the *IUCN Red List of Threatened Species* can be used to develop indices on trends in the status of biodiversity at the species level'; (iv) that IUCN members and others should be encouraged 'to make use of data included in the *IUCN Red List of Threatened Species* to assist in conservation planning such as site-based approaches implemented at the national level, combining *Red List* data with other data sets'; and (v) the importance of governments and research institutions to encourage research on species listed as threatened by IUCN in order to enhance our understanding of the biology and conservation needs of these species.

In 1996, the First WCC was held in Montreal, Canada, and the WCC resolution D. 1.25 adopted requesting 'the SSC, within available resources, to complete the development of guidelines for using the IUCN Red List Categories at the regional level as soon as practicable'. Several draft versions of the guidelines were published (Gärdenfors *et al.*, 1999, 2001). The main proposal was to determine the national Red List category by a two-step procedure: first, the population is assessed using the 2001 IUCN Categories and Criteria as though the population was endemic to the country or completely isolated from other conspecific populations, and second, it should be considered whether the target population is part of a more widely distributed population (Gärdenfors 2001). If the latter is true, the Red List category should be adjusted (downgraded or upgraded) to one that appropriately reflects the long-term extinction risk of the subpopulation.

In 2003, the current version of the Regional Guidelines (3.0) (IUCN, 2003) was finally made available. These were developed to assist in the application of the 2001 IUCN Categories and Criteria at regional levels (Gärdenfors *et al.*, 2001; IUCN, 2003). However, there are still few countries that actually have applied the regional criteria to produce national Red Lists.

It should be remembered that assessment of extinction risk and setting conservation priorities are two related but different processes. Assessment of extinction risk, such as the assignment of IUCN Red List Categories, generally precedes the setting of priorities (Gärdenfors *et al.*, 2001). Possingham *et al.* (2002) pointed out that it is inappropriate to use threatened species lists for resource allocation of species conservation and that other additional criteria should also be taken into consideration. These often include consideration of extinction risk, but may also comprise consideration of costs, logistics, chances of success (IUCN, 1994; Maxted *et al.*, 1997b), taxonomic, genetic or ecological distinctiveness (Williams *et al.*, 1993; Maxted *et al.*, 1997b), legal frameworks for conservation of threatened taxa, ecological, phylogenetic, historical and cultural preferences for some taxa over others (Maxted *et al.*, 1997b; Gärdenfors *et al.*, 2001), what proportion of the global population occurs within the country and what is the risk of extinction in other parts of the world (Avery *et al.*, 1995). Such conservation priorities will be established in different ways in different countries and, in contrast to the objectively scientific process of assigning species to Red Lists, will often involve political and logistical considerations (Maxted *et al.*, 1997b; Gärdenfors, 2001).

13.2. Aims of this Exercise

The main aim of this chapter is to provide an overview of the Red Listing process as well as to answer to our initial question of whether CWR national Red Listing is as difficult as some of the PGR community may think it is, as well as giving examples of its application in the PGR context and interpretation of its use. The IUCN Red List Criteria are quite useful and fairly easy to apply when data are available. However, in most cases, there is a considerable lack of data and a formal IUCN Red List assessment is generally thought to be impossible. In resume, the chapter reviews the process of Red Listing as well as methodological issues that are raised in order to help discussion on the application of the *2001 IUCN Red List Categories and Criteria* and to contribute to the refinement and improvement of this system.

13.3. Case Study: Portuguese Crop Wild Relatives

13.3.1 Methodology

The exemplar IUCN Red Listing was undertaken for the CWR species of mainland Portugal (Magos Brehm *et al.*, submitted) for which the required data were available.

Data sources and types of available data

The available information for Red Listing was of two kinds: (i) information on species distribution; and (ii) other biological information, including population numbers and number of mature individuals. Regarding the current taxa distributions, data from several sources were collated: from a herbaria and gene bank survey undertaken in 2004 for 66 *Leguminosae* species (Magos Brehm, 2004), personal contacts (Pedro Ivo Arriegas, Instituto da Conservação da Natureza, Portugal), reports (e.g. Ministério do Ambiente, 1999, 2000; Magos Brehm, 2003) and papers (Rosselló-Graell *et al.*, 2003) for certain species. The information was completed with the data available from the Proyecto Anthos (ANTHOS, 2005). The distribution data were loaded in ARCVIEW 3.3 (ESRI, 2002) and distribution maps were produced. IUCN advises the use of $2 \times 2\,km^2$ grid cells for Red List assessments (IUCN, 2005a); however, in this study such precise information was not available and a much coarser scale ($10 \times 10\,km^2$ hectads) was used. Only data recorded in the last 20 years were taken into account for the Red Listing in order to decrease the inclusion of species localities that no longer exist.

Data on the species biology and ecology, abundance, threats and conservation status were obtained from the web site of the Instituto da Conservação da Natureza (Portugal) (ICN, 2006a) as well as from personal contacts (Pedro Ivo Arriegas, Instituto da Conservação da Natureza, Portugal). Information on the occurrence of fires between 1994 and 2004 was also considered (DGRF and ISA–DEF, 1990–2004). Table 13.1 shows some examples of the data types collated to undertake the Red List assessments.

Table 13.1. Examples of data types collated in order to undertake the Red List assessment.

Species name	No. of 10 × 10 km^2 grids	Fragmentation	No. of locations	No. of mature individuals	Threats	Other facts	References
Lathyrus tingitanus L.	1	–	1	–	Urbanization	The species occurs in the city of Lisbon	Magos Brehm (2004)
Iris lusitanica Ker-Gawler	9	Yes	–	–	• Urbanization • Overcollection of flowers for commercial purposes • Fires	• Threatened, with reduced occupancy area • Wide distribution perimeter with centres of abundance from where less frequent and smaller subpopulations derive • Abundance unknown	Barreto Caldas *et al.* (1996); Pinto *et al.* (1996a,b); De Koe *et al.* (1998); Albuquerque *et al.* (2003); DGRF and ISA–DEF (1990–2004)
Narcissus asturiensis (Jordan) Pugsley	11	Yes	4	Thousands of plants in Serra da Estrela Natural Park	• Commercial exploitation (ornamental species) • Fires	–	DGRF and ISA–DEF (1990–2004); ICN (2006c)
Narcissus cyclamineus DC.	4	No	2	• Low number of plants • Isolated and sparse or small subpopulations	• Overcollection for commercial exploitation • Habitat degradation (agricultural exploitation of [illegible])	Very rare, in danger of extinction	ICN (2006d)

Continued

Table 13.1. *Continued*

Species name	No. of 10 × 10 km^2 grids	Fragmentation	No. of locations	No. of mature individuals	Threats	Other facts	References
				• Fires	• Human disturbance (music festival in Côrno do Bico Natura 2000 site where the species also occurs)		
Plantago algarbiensis Samp.	1	–	1	• One of the subpopulations (Barrocal Algarvio Natura 2000 site) <10,000 individuals • Biggest subpopulation = several thousands of plants by botanists	• Mining • Disorganization of the drainage network • Urban expansion • Forest exploitation • Trampling by humans • Grazing • Overcollection	• Rare • One of the subpopulations (Barrocal Algarvio Natura 2000 site) = 0.05 ha • Biggest subpopulation = 29 ha	DGRF and ISA–DEF (1990–2004); ICN (2006e)

Application of the global 2001 IUCN Red List Categories and Criteria

Once the data were collated, the process of applying the criteria was started. The *2001 IUCN Categories and Criteria* (IUCN, 2001) were adopted and the *Guidelines for Using the IUCN Red List Categories and Criteria* (IUCN, 2005a) were used. The taxa were assessed under each criterion (A–E), but for a taxon to be assigned a category only one criterion had to be met. The most highly threatened assessment was then taken to be the overall category (see Fig. 13.2) (IUCN, 2005a).

The category LC was not applied in this study. A taxon is LC when it has been evaluated against the criteria but does not qualify for any of the threatened categories (IUCN, 2005a). As most of the species assessed in this study were already thought to be rare and because the available information was insufficient to assess the species against all the criteria, any taxa that fell into the LC category were assigned as DD.

Regarding the NT category, a taxon qualifies for this category if it is close to qualifying for a V category (IUCN, 2005a). However, IUCN does not stipulate any quantitative thresholds. Few authors have suggested quantitative thresholds to facilitate the assignment of the NT category (e.g. Cheffings and Farrell, 2005; Eaton *et al.*, 2005). In our study we have adopted a combination of the thresholds suggested by Cheffings and Farrell (2005) for *The Vascular Plant Red Data List for Great Britain* and by Eaton *et al.* (2005). These are: for the criterion A ≥20% decline between the two date classes (Cheffings and Farrell, 2005); for the criterion B ≤30 locations and continuing decline (Cheffings and Farrell, 2005) or meeting the threshold for restricted range (20,000 km^2) but not fulfilling two subcriteria or conversely fulfilling the

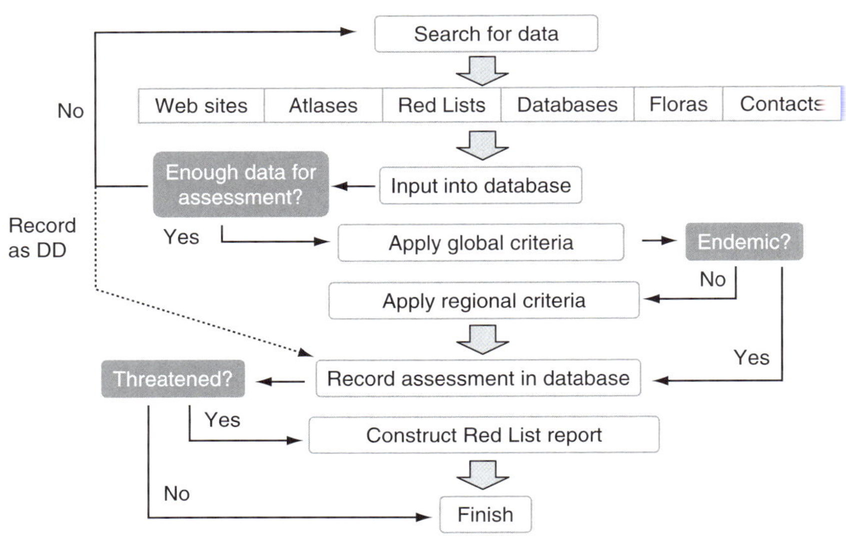

Fig. 13.2. Scheme of global and regional assessment. (From Mitchell, 2004.)

subcriteria but narrowly exceeding the restricted range thresholds (Eaton *et al.*, 2005); and for the criterion C ≤10,000 individuals.

The DD category was used where there was inadequate information about a species' current status or possible threats in order to assign a threat category although the guidelines state that it is 'important to make positive use of whatever data are available' (IUCN, 2005a).

Application of the 2003 IUCN List Criteria at regional levels
After the global criteria have been applied, those taxa that are not endemic to Portugal were assigned a threatened category using the *2003 IUCN Red List Criteria at Regional Levels*. This involved collating information on the species' distribution and status in Spain. The sources of information include the web site of the Proyecto Anthos (ANTHOS, 2005), the *Atlas y Libro Rojo de la Flora Vascular Amenazada de España* (Bañares *et al.*, 2004) which assesses species according to the *2001 IUCN Categories and Criteria* and *Lista Rioja de la Flora Vascular Española* (VV.AA., 2000) which uses the version of 1994. Subsequently, the species were subjected to a series of questions which aim to determine whether this taxon's category should be upgraded, downgraded or remain the same (see Fig. 13.1).

13.4 Results and Discussion

13.4.1 Assessments

In this study, 216 Portuguese CWR species were assessed using the *2001 IUCN Red List Categories and Criteria*. One species was considered to be EX (*Astragalus algarbiensis* Bunge) because it has not been detected since the beginning of the 20th century (see Magos Brehm, 2004). As many as 48 species were assessed as EN, 47 as DD, 43 as VU, 45 as NT and 22 as CR (Fig. 13.3). Figure 13.4 shows the number of species assessed under each of the criteria. Criterion B (geographic range size and fragmentation, decline or fluctuations) was primarily used to ascribe categories. In the case of southern African plant species, Golding (2004) concluded that the VU category was the most frequently assigned category on the basis of criterion B. The author ascribed this fact to the broad numerical parameters associated with this criterion that allowed for uncertainty, to the subjective determination of the extent of fragmentation (B1a or B2a) and the lack of a specific level for continuing decline (B1b or B2b).

Criterion D was also used to a great extent when assessing the VU species. This may be due in part to the assignment of VU-D2 which can be given based on a single threshold being met, namely that of occurring in less than 20 km^2 or in fewer than five locations (IUCN, 2001). Keller and Bollmann (2004) observed that a fair number of species were assessed under this criterion for one of the threat categories. However, not all the species showed decline and they explained that these species might have always been rare and are unlikely to increase their populations because suitable habitat is lacking. The species to which they are referring are associated with populations at or close to the edge of their geographical ranges (Samways, 2003). These populations are typically small, sparse

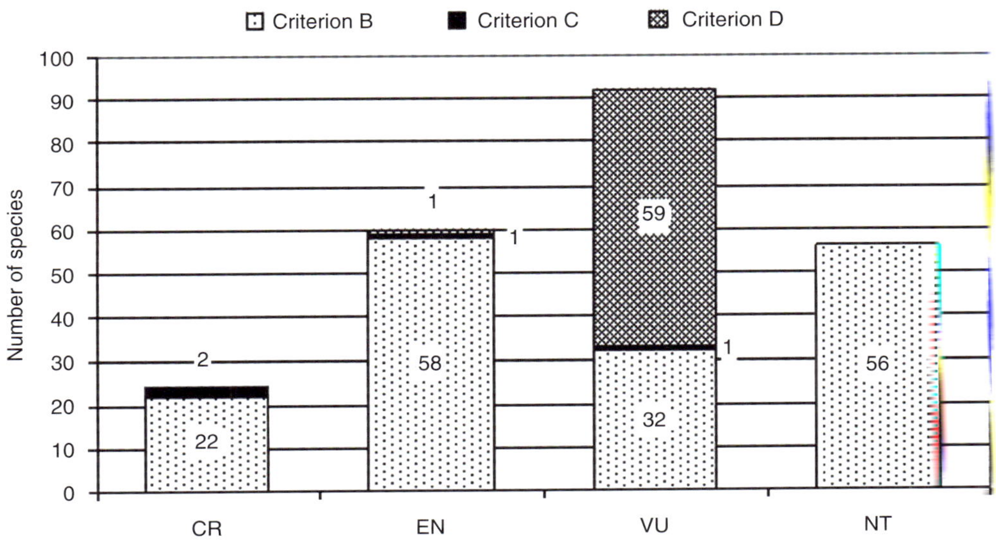

Fig. 13.3. Number of Portuguese CWR taxa assessed using criteria B–D.

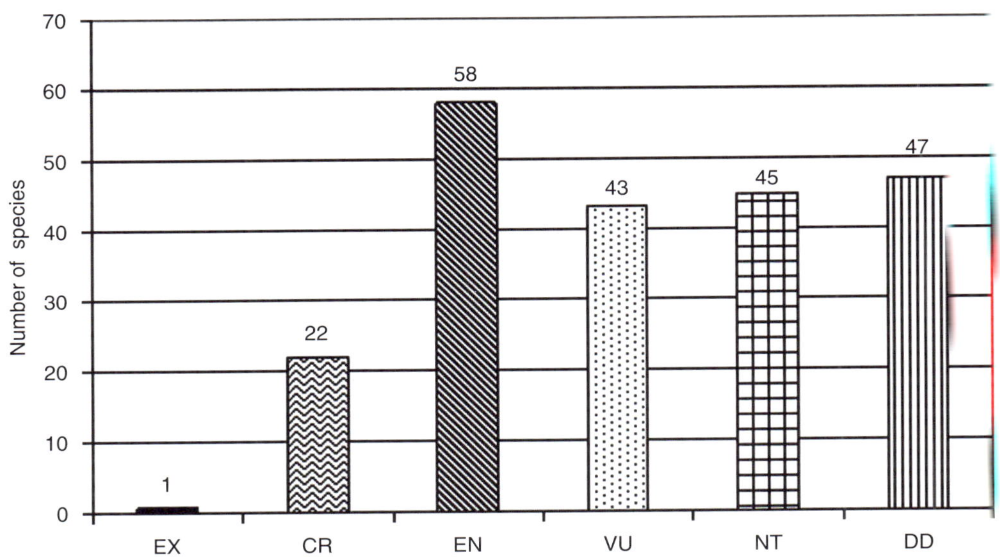

Fig. 13.4. Number of Portuguese CWR taxa assessed under each category on the basis of criteria B–D.

and isolated from one another (Lawton, 1993) and are highly susceptible to habitat change (Samways, 2003). We should be aware that this might also be the case for some of the species assessed during this study because for many species Portugal is the extreme western edge of their distribution. Conversely, Keller and Bollmann (2004) pointed out that these species should not receive the same conservation attention as other species that became rare after serious decline. To some extent IUCN has already tried to reduce the risk of these naturally rare species being included by reducing the threshold of VU-D2 from 100 km^2 (IUCN, 1994) to 20 km^2 (IUCN, 2001).

The Regional IUCN Red Listing was undertaken on only 18 species out of 216 because only these were appropriate for regional assessment and had available data. For all of them there was no change of the global assessment (Table 13.2). Of the species assessed 62 are endemic and therefore need not be regionally assessed.

13.4.2 IUCN Red List Criteria not applied

Using the available data it was not possible to apply criteria A and E. Criterion A (past, present and/or future reduction in population size) requires measurements of decline calculated for more than 10 years or three generations, or projected to experience a significant decline in the near future (IUCN, 2005a), regardless of the current range or abundance (Cheffings and Farrell, 2005). Systematic information about the features of populations in the past is lacking, and therefore the percentage of decline was not possible to estimate from the available data. In addition, as observed by Magos Brehm (2004) for wild legume species, extrapolation of these results to the rest of the Portuguese taxa is not possible due to inconsistent collecting effort throughout the years. Fresh field collecting is often a reflection of the presence, capability and enthusiasm of local botanists working within the study area rather than the abundance of the taxon itself (MacDougall *et al.*, 1998). Therefore, the available data do not allow estimation of population decline. Furthermore, as Cheffings and Farrell (2005) also emphasized there is no confidence that detected changes are reversible, understood and that the decline has ceased.

To make an assessment under criterion E a quantitative analysis must be conducted in order to determine the probability of extinction over a given time period (IUCN, 2005a). There are several techniques that can be used and the choice of the appropriate method depends not only on the availability of data, but also on the ecology of the taxon (IUCN, 2005a). The techniques include: PVA (see Boyce, 1992; Akçakaya and Sjögren-Gulve, 2000; Menges, 2000; Kindvall and Gärdenfors, 2003) including metapopulation modelling or other spatially explicit models when data on the spatial distribution are available (see Menges, 2000); models of occupancy, when presence–absence information from a number of locations is available (see Sjögren-Gulve and Hanski, 2000); unstructured dynamic models if census information from a number of years is available (see Dennis *et al.*, 1991); structured models if data for different age classes or stages (e.g. juvenile and adult) are available (see Akçakaya, 2000); and individual-based models if detailed data at the individual level are available (see Lacy, 2000). The types of data that can be used in order to undertake a criterion E assessment include spatial distributions of suitable habitat, local populations or individuals, patterns of occupancy and extinction in habitat patches, presence–absence data, habitat relationships, abundance estimates from surveys and censuses, vital rate (fecundity and survival) estimates, as well as temporal variation and spatial covariation in these parameters

Table 13.2. The 18 species that were regionally Red Listed using the 2003 Guidelines for Application of IUCN Red List Criteria at Regional Levels.

Species name	Global assessment, Portugal (IUCN, 2001)	Global assessment, Spain	Reassessment
Antirhinum molle L. subsp. *lopesianum* (Rothm.) P. Silva	EN – B1ab(ii,iii)	EN – B2ab(ii,iii,iv); C2a(ii)[a]	No change (keep the global assessment)
Asplenium hemionitis L.	CR – B1ab(iii)	VU – D2[b]	No change (keep the global assessment)
Digitalis purpurea L. subsp. *heywoodii* P. & M. Silva	NT	VU – D2[b]	No need
Eryngium viviparum Gay	CR – B1ac(ii,iii)	EN – A1c; B2ab(ii,iv,v)[a]	No change (keep the global assessment)
Genista ancistrocarpa Spach	EN – B1ac(ii,iii)	CR – B1ab(i,ii,iii,iv,v) + 2ab(i,ii,iii,iv,v)[a]	No change (keep the global assessment)
Holcus annuus C. A. Meyer subsp. *duriensis* (P. Silva) Franco & Rocha Afonso	EN – B1b(ii,iii,iv,v) c(ii,iii,iv)	DD[b]	No change (keep the global assessment)
Iris boissieri Henriq.	EN – B1ab(iii,v) c(ii,iii); C2a(i)	CR – B1 + 2b(i,ii,iv,v) c(iv)[a]	No change (keep the global assessment)
Iris xiphium L. subsp. *lusitanica* (Ker-Gawler) Franco	VU – B1ab(ii,iv,v) c(ii,iii)	DD[b]	No change (keep the global assessment)
Isatis platyloba Link ex Steud	VU – D2	VU – B2ac(iv); C2b[a]	No change (keep the global assessment)
Linaria coutinhoi Valdés	EN – B1ac(i,ii,iii,iv)	EX (RE)[a]	No change (keep the global assessment)
Linaria lamarckii Rouy	DD	CR – A3c; B1ab(i,ii,iii,v) + 2ab(i,ii,iii,v); C1 + 2a(i,ii,iii,v)[a]	No need
Marsilea batardae Launert	VU – B1b(ii,iii,iv,v)c(ii)	EN – A2ace + 3ace + 4ace[a]	No change (keep the global assessment)
Marsilea quadrifolia L.	CR – B1ab(i,ii,iii,iv,v)	EW(RE)[a]	No change (keep the global assessment)
Myrica gale L.	VU – B1ab(ii,iii)	DD[b]	No change (keep the global assessment)

Continued.

Table 13.2. Continued.

Species name	Global assessment, Portugal (IUCN, 2001)	Global assessment, Spain	Reassessment
Narcissus cyclamineus DC.	EN – B1ab(iii)c(ii,iii)	DD[b]	No change (keep the global assessment)
Rhododendron baeticum L. *ponticum* (Boiss. & Reuter) Hand.-Mazz.	EN – B1ac(ii,iii)	VU – B2c + 3d[b]	No change (keep the global assessment)
Thymus carnosus Boiss.	VU – B1ab(iii); D2	VU – D2[b]	No change (keep the global assessment)
Veronica micrantha Hoffm. & Link	VU – B1ab(iii); D2	VU – B2d + 3d;C2a[b]	No change (keep the global assessment)

[a]Bañares *et al.* (2004).
[b]VV.AA (2000).

(IUCN, 2005a). However, IUCN (2005a) emphasizes that when there are not sufficient data or when the available information is too uncertain, it is risky to use this criterion. Therefore, as with the Red Listing undertaken for Great Britain very recently (Cheffings and Farrell, 2005), the criterion E was not used in this study.

13.4.3 IUCN Red List Criteria applied

The IUCN Red List Criteria B, C and D were the three used in the study and their application will be demonstrated using examples from the study. Criterion B (geographic range) was designed to identify populations with restricted distributions that are also severely fragmented, undergoing a form of continuing decline, and/or exhibiting extreme fluctuations (either in the present or near future) (IUCN, 2005a). Under this criterion either EOO (B1) or AOO (B2) were estimated. EOO is defined as 'the area contained within the shortest continuous imaginary boundary which can be drawn to encompass all the known, inferred or projected sites of present occurrence of a taxon, excluding cases of vagrancy' (IUCN, 2005a). 'It can be measured by a minimum convex polygon which is the smallest polygon in which no internal angle exceeds 180° and which contains all the sites of occurrence' (IUCN 2001). On the other hand AOO is defined as 'the area within its "extent of occurrence" which is occupied by a taxon, excluding cases of vagrancy' (IUCN, 2001). This measurement takes into consideration the fact that usually the area of a taxon's EOO may contain unsuitable or unoccupied habitats. The size of the AOO should be at a scale appropriate to relevant biological aspects of the taxon, the nature of threats and the available data (IUCN 2001).

However, some complications arose during the course of this study in implementing this criterion, which are discussed below.

Problems of scale

There are several ways of estimating AOO, but the basic one is obtained by counting the number of occupied cells in a uniform grid that covers the entire range of a taxon and totalling them giving the total area of all occupied cells (IUCN, 2005a). According to IUCN (2001) 'the choice of scale at which AOO is estimated may thus, itself, influence the outcome of Red List assessments and could be a source of inconsistency and bias'. It should be noted that scales of 3.2 km grid size or coarser (larger) are inappropriate as they do not allow any taxon to be listed as CR (where the threshold AOO under criterion B is 10 km^2 (IUCN, 2005a). Scales of 1 km grid size or smaller tend to list more taxa at higher threat categories than these categories imply (IUCN, 2005a). In this study, however, the available distribution data are at $10 \times 10\,km^2$ grid cells. Usually, this is considered to be too coarse to calculate AOO (B2) because the result will be that some parts of whole cells will be unoccupied by the taxon. As a result only species occurring in one grid cell (100 km^2) will be CR and all the others will fall into other threatened categories. To solve this problem IUCN (2005a) suggests the use of equations to scale down where necessary. However, Mitchell (2004) showed that neither scaling up nor scaling down can be relied upon for plant populations, while Cheffings and Farrell (2005) found that the equations only worked for those taxa present in more than 30 hectads and therefore this procedure was not undertaken. To resolve this problem, as an alternative, the value of AOO (B2) was used as an estimate of EOO (B1) (C. Hilton-Taylor, Cambridge, 2005, personal communication).

Identification of fragmentation

'Severe fragmentation' was not always easy to identify. Fragmentation refers to the way that habitats or populations within a landscape divide into small or large patches (Willis *et al.*, 2003). Severe fragmentation was included as an alternative risk factor to the 'number of locations' in the subcriterion (a). Cheffings and Farrell (2005) investigated the possibility of using the average distance between populations as a potential measure of fragmentation, but decided that this parameter should include population measures as well as distances between populations. Also, the minimum viable population size was not taken into consideration since this information is not available for the Portuguese flora. For these reasons, this subcriterion was only used when populations showed an obvious disjunctive distribution, as in Fig. 13.5.

Number of locations

The actual number of locations using subcriterion (a) was difficult to define. According to the IUCN (2001) the term 'location' is defined as a 'geographically or ecologically distinct area in which a single threatening event can rapidly affect all individuals of the taxon present'. It should be noted that 'number of locations' is not the same as 'number of populations' (or subpopulations). A 'location' may include part of one or many subpopulations, and conversely if a single subpopulation covers an area affected by several single events, it must be counted as more than one location (IUCN, 2005a).

Despite the difficulties explained above, this criterion was the most used so that 86.6% of the species were assessed with criterion B. Figures 13.5 and 13.6 show two examples of how this criterion was applied.

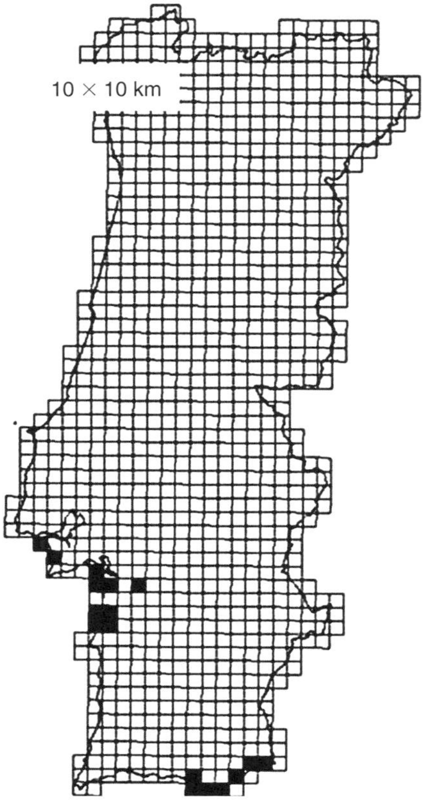

Thymus carnosus Boiss.

- 17 grids: 1700 km^2 = E00 < 5000 km^2
- Fragmented
- Coastal dunes; very sensitive to habitat changes
- Trampling by tourists and off-road vehicles, trash deposition, exotic and invasive species (declining in habitat quality)

(Source: ICN, 2006b)

Endangered (EN): Blab(iii)

Fig. 13.5. Global assessment for *Thymus carnosus* Boiss. as an example of the use of criterion B solely.

Criterion C (small population and size) has been designed to identify taxa with small populations that are currently declining or may decline in the near future (IUCN, 2005a). Fulfilment of this criterion requires population counts (enumeration of all individuals) and/or estimation of decline, which is usually difficult to obtain if specific studies have not been undertaken. This criterion was generally difficult to apply because most of the time information concerning the percentage of population decline (C1, C2) was not available. Only four species were assessed with this criterion and this was due to the use of number of mature individuals in each subpopulation (C2a(i)) or the percentage of individuals in at least one subpopulation (C2a(ii)). Figure 13.7 shows an example of how this criterion was applied.

Criterion D identifies very small or restricted populations (IUCN, 2005a). This was applied to 28.7% of the species because in some cases the number of mature individuals was known (D1) and also because the number of locations (D2), although difficult, was possible to obtain. Figure 13.8 shows how this criterion can be used with limited information. The main problems in applying this criterion to the Portuguese CWR species were related to the subcriterion D2 for the category of VU. This subcriterion defines a VU species as occurring

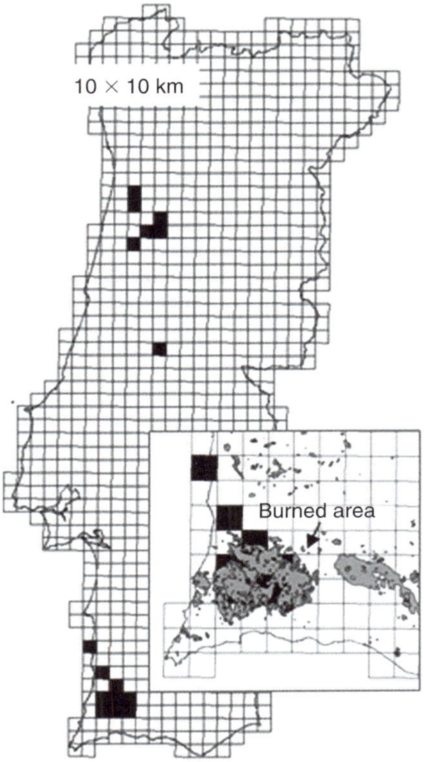

Fig. 13.6. Global Red List assessment for *Rhododendron ponticum* L. subsp. *baeticum* (Boiss. & Reuter) Hand.-Mazz. as an example of the use of criterion B solely.

in five or fewer locations or having AOO < 20 km^2. The calculation of AOO in order to qualify for this threshold was not possible to use because the scale that was used was 10×10 km^2 grid cells. Again we faced the problem of defining the 'number of locations' (see criterion B).

13.4.4 Regional assessments

In order to assess the species at regional level, information on species status in adjacent areas from the same regions is required. In this particular case, information of the status of the species in Spain was obtained. However, in order to undertake the regional assessments some assumptions had to be made. These were: (i) if there was any Spanish conspecific population near a Portuguese one, the migration was likely (either by dispersal or pollination); (ii) if a species was assessed as being threatened in Spain, we either had the information that the regional population was a sink or assumed that the conditions outside the region were deteriorating; and (iii) if a species was extinct in Spain but not in Portugal, there was no immigration to Portugal since Portugal is isolated from

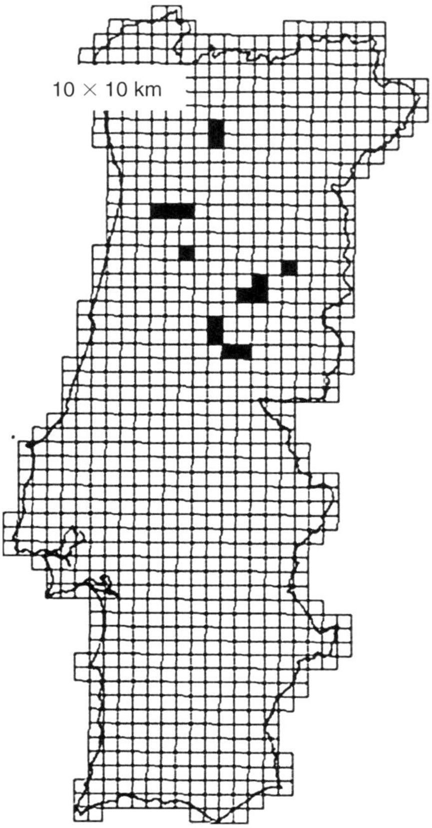

Teucrium salviastrum Schreber

- 100,000 plants in just one population; therefore the total number of plants of the species >10,000 (Source: Fidalgo, 1996; Pinto-Gomes, 1996).

Least concern (LC)

Fig. 13.7. Red List assessment of *Teucrium salviastrum* Schreber as an example of the use of criterion C only.

any other land mass and the extent of migration due to rare cases of drift seed is unknown. Hence, if the species was considered to be threatened in Spain, the risk of extinction of the Portuguese populations increased. Figure 13.9 shows an example of how the regional assessment can be carried out.

Concerning the process of regional Red Listing itself (Fig. 13.2), Eaton *et al.* (2005) have identified a number of stages at which different assessors may diverge in application which can result in considerable differences in the Red Lists produced. This is mainly due to the flexibility in the criteria and guidelines, variation in interpretation and the availability of different data sources. The main point is related to the fact that visiting taxa should be assessed but not vagrant taxa. They suggested that the *2003 IUCN Red List Criteria at Regional Levels* should be more prescriptive on the filtering of species before assessment. Furthermore, Mitchell (2004) suggested a schema to assist the decision of whether a species should be assessed or excluded. In this study, we have not faced this problem because all the assessed species were native. It should also be mentioned that the inclusion of uncertainty in most of the steps of the regional Red Listing process helps the assessments immensely.

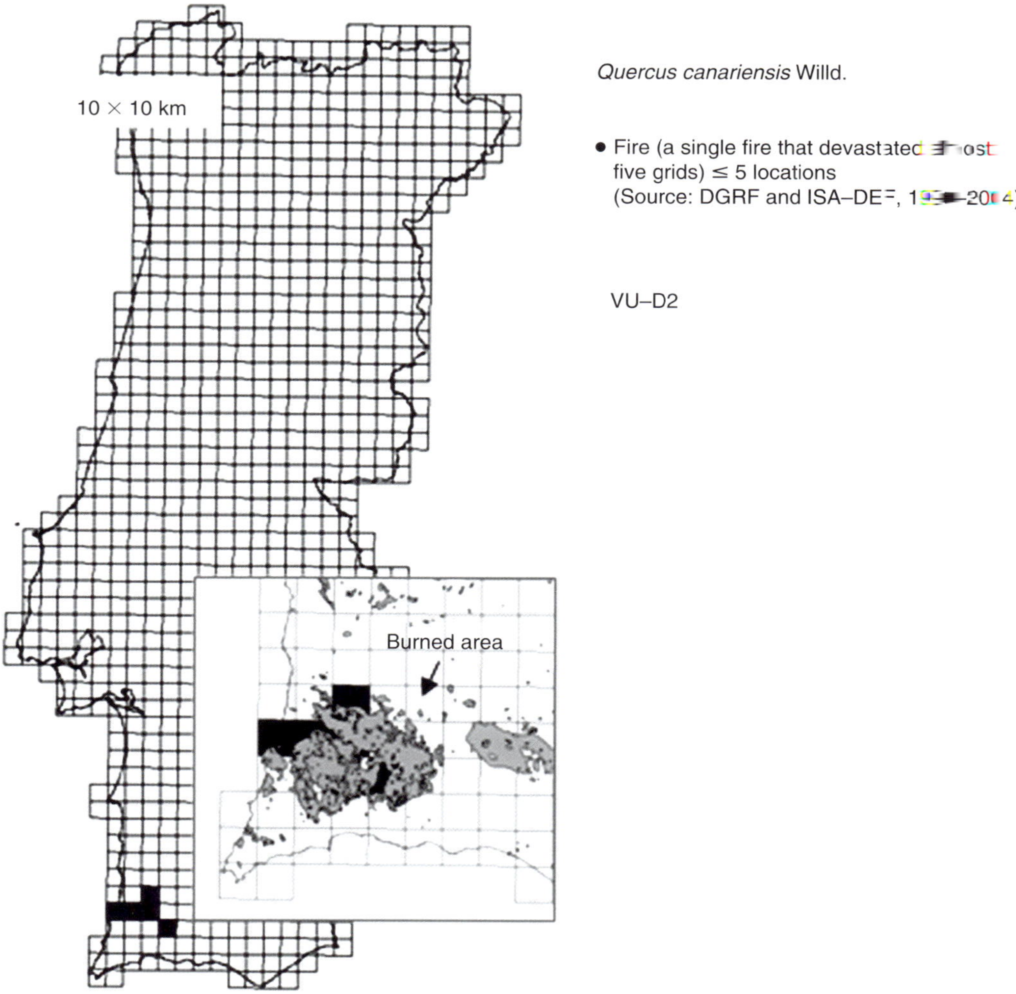

Fig. 13.8. Global Red List assessment for *Quercus canariensis* Willd. as an example of the use of criterion D only.

The need for data from other regions or countries in order to create regional assessments is a limiting factor. That means it is impossible to apply such assessments for regions whose surrounding areas have no equivalent data; this is a fact that highlights the need for cooperation between neighbouring areas that are considering compiling regional red lists. In this case, despite the huge amount of work developed by the Spanish assessors, both with the *Atlas y Libro Rojo de la Flora Vascular Amenazada de España* (Bañares *et al.*, 2004) and the *Lista Rioja de la Flora Vascular Española* (VV.AA., 2008), unfortunately most of the species with accessible information did not match those species assessed in this exercise.

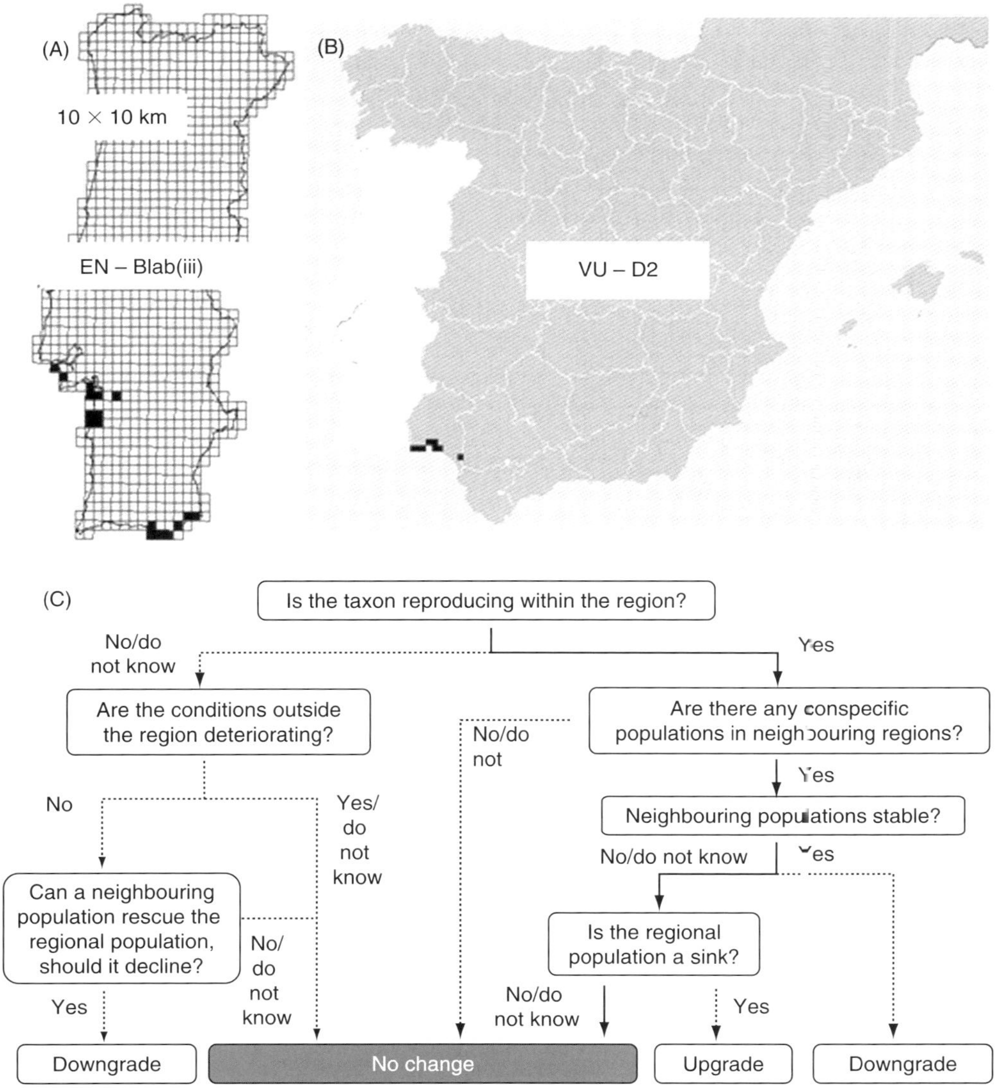

Fig. 13.9. An example of how a regional assessment was undertaken (*Thymus carnosus* Boiss.). (A) Assessment using the 2001 IUCN Red List Criteria for Portugal, (B) assessment of the same species in Spain; (C) regional Red Listing (2003), where filled arrows indicate the pathway taken to assess the species. (From ICN, 2006b; VV.AA., 2000.)

13.4.5 Some limitations of Red Listing

The IUCN Red List system (IUCN, 2001) requires the application of 'the best available evidence' to perform an assessment. Often, herbaria and gene bank collections provide the only source of information for threat assessment and must therefore qualify as 'best available evidence' (Willis *et al.*, 2003). Ter Steege *et al.* (2000) consider that these data are sufficiently reliable to enable conservation decisions and MacDougall *et al.* (1998) emphasize that these can also help in defining priorities for conservation efforts. Conversely, some authors consider that this type of data is unreliable and therefore the use of specimen information should be regarded as provisional because it can result in an inaccurate assignment of Red List status of poorly known species (Golding 2004). While the latter may well be strictly true where other data are unavailable, it does permit the production of at least tentative assessments.

It is obviously wise to be aware of the limitations in using herbarium data (gene bank data are even more limited because of the relatively recent commencement of systematic germplasm collection). First, while collections made over the last 50 years may provide data about scientific name, vernacular name, morphology, locality, habitat, ecology, date of collection, uses, collector name and collector number, more often historical specimens (before or early 20th century) contain only a few handwritten details of the plant name, collector and locality and therefore are of limited value for conservation assessments (Maxted *et al.*, 1995). Second, even today herbarium specimen labels rarely include detailed latitude and longitude records and locality coordinate data inferred from collecting localities recorded on herbarium specimen data are often only a rough approximation of species distribution and are inadequate to truly represent the species' distribution, although some authors consider this type of data a good starting point to assess species' threat of extinction (Maxted *et al.*, 1995; Willis *et al.*, 2003). Third, the level of the precision of the specimen geo-referencing may also be problematic if not consistent (Willis *et al.* 2003). Also, herbarium specimen information cannot be used to estimate infer or predict population demographic change such as the number of mature individuals, birth and mortality rates or fluctuations in a population (Golding 2004) and most of the time ecological information is often difficult to obtain from herbarium specimens (Maxted *et al.*, 1995; Willis *et al.*, 2003). Herbarium data alone can be used to determine categories of threat using criterion B and in some cases the number of locations as criterion D2, but only if the threats are exactly specified (Willis *et al.*, 2003).

To overcome the difficulties associated with the use of herbarium specimens in the current study only distribution data were obtained from them. In addition, we have only used the specimens collected during the last 20 years in order to avoid considering historical locations where the taxa no longer exist.

Keith *et al.* (2004) compared several different protocols to forecast species extinction, including the *2001 IUCN Red List Categories and Criteria*. Among other things they have concluded that the use of the IUCN Red Listing was prone to high levels of operator error. However, this can be greatly overcome by paying careful attention to all the definitions provided by IUCN (2001, 2005a). On the other hand, the value of the IUCN Red Listing also relies on the fact that the assessments are peer reviewed by at least two 'evaluators' assigned by the competent 'Red List Authority' (Rodrigues *et al.*, 2006) so the results can be checked for consistency. Keith *et al.* (2004) also concluded that using the 2001 IUCN Categories and Criteria is less exposed to risk underestimation than other protocols because whenever enough data were not available the category of DD was assigned.

When applying regional assessments, Eaton *et al.* (2005) noted that after the strict application of quantitative measurements with the 2001 IUCN Criteria, the regional Red Listing is far more subjective and introduces greater variability into the outcome. They believe that the regional assessment requires significant knowledge of the taxa across a broad area and subjective and arbitrary decisions are required, which is contrary to what IUCN assessment procedure ought to be, that is, based on quantitative data (Keller and Bollmann, 2004).

However, the application of the IUCN Red Listing system improves with experience when its structure and thresholds are fully understood (Keith *et al.*, 2004). The protocol should be re-evaluated in order to better accommodate missing data and uncertainty in parameter estimates (Keith *et al.*, 2004). Willis *et al.* (2003) suggested the creation of an index of fragmentation to ensure that its recognition is easier and that comparison can be undertaken effectively. Provisional assessments should be backed up by field verification wherever possible (Willis *et al.*, 2003). There is a need for coherent and objective quantitative thresholds for the NT category.

Eaton *et al.* (2005) comment that with the increasing application of regional Red Listing, along with the associated discussion and debate, the Red Listing process will be refined, so as to improve uniformity and ease of use. Nevertheless, as Akçakaya *et al.* (2000) point out, 'to adequately assess a species' extinction risk requires a detailed review of its population, trend, range, and ecology, and the data used for this should be assessed critically, with explicit consideration of sources of uncertainty'. Therefore Red Listing will always be a good estimate rather a precise record.

13.4.6 Basic schema for National Red Listing of CWR

At the national level, most countries still need to undertake threat assessment for their PGR so that they can meet the Biodiversity 2010 Targets. The IUCN Red Listing has proven to be a good method to undertake this task in order to elucidate the current state of PGR to help with the prioritization of these taxa for conservation actions. Table 13.3 gives an indication of the kind of data that is required for Red List assessments and Fig. 13.10 outlines the steps needed in order to undertake the IUCN Red Listing for CWR at a national level starting from the creation of a CWR inventory through assessment and on to prioritization. It is hoped that this schema might be of aid to those initiating a CWR Red Listing project.

13.5 Conclusions

In conclusion, we found although the *2001 IUCN Red List Categories and Criteria* based on quantitative thresholds are relatively flexible when applied even for relatively poorly known species. For these species the assessment relies on data inference and projection but the results remain biologically valid. Despite the

Table 13.3. Basic data necessary to undertake the IUCN Red Listing assessment.

Population reduction (% reduction over time)	Geographic range	Locations	Extreme fluctuations	Population size	Decline	Biology
Past[a]	EOO	Number	EOO[a]	Number of mature individuals	AOO[a]	Seed dormancy/ viability
Present[a]	AOO	Fragmentation	AOO[a]	Numbers at subpopulation level	EOO[a]	Generation time/lifespan
Projected			Mature individuals[a]		Mature individuals[a]	Habit
			Number of locations[a]		Habitat quality[a]	Migration (how and where to and from)

[a]Requires data from at least two time points.
EOO = extent of occurrence, AOO = area of occupancy

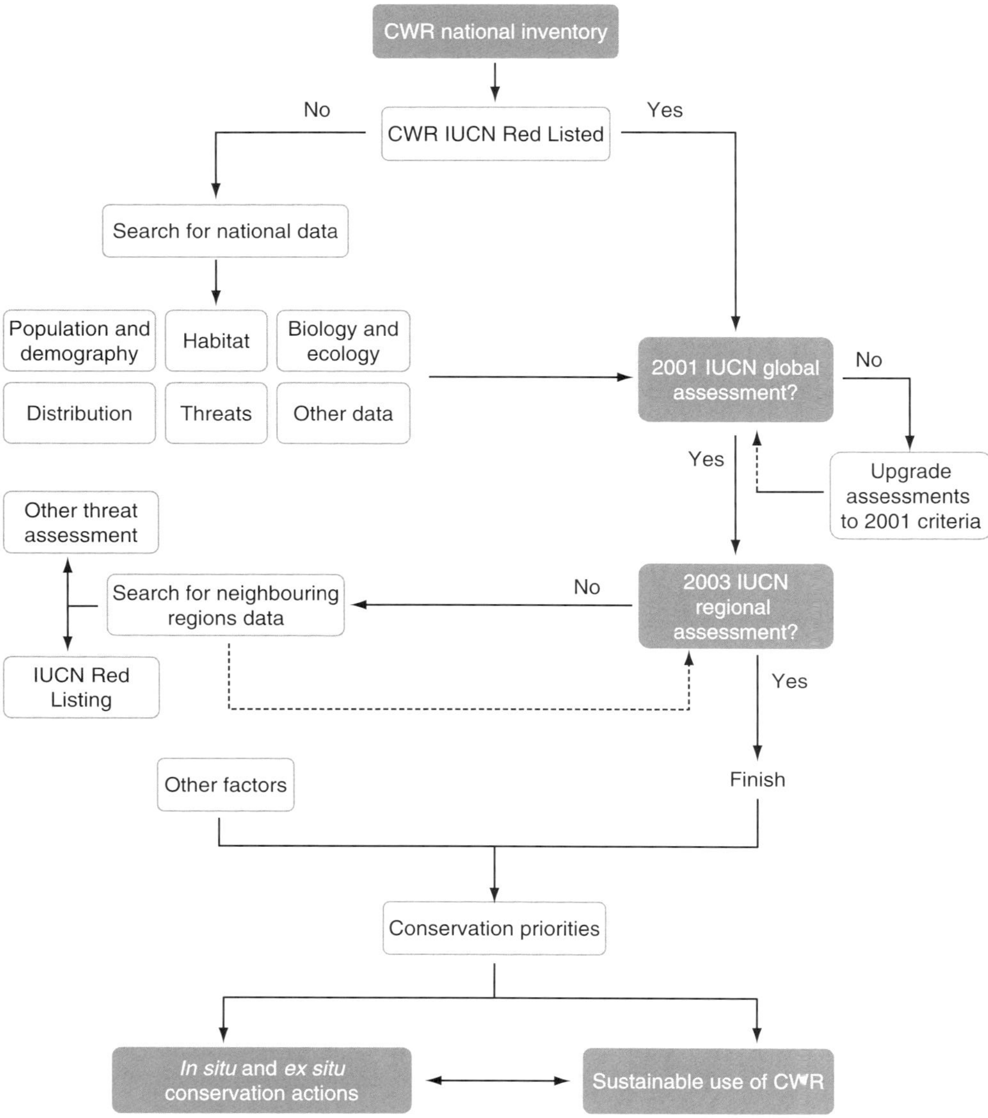

Fig. 13.10. Schema explaining the basic steps in order to undertake a CWR IUCN Red List national assessment.

reservations discussed in this chapter, the 2001 IUCN Categories and Criteria still provide the 'best' most useful objective framework for the adequate assessment of a species' extinction risk at a global and regional scale for a very broad range of taxa, including CWR. It is unique in its efforts to be both quantitative and comprehensive, to reduce the potential for misuse (Lamoreux *et al.*, 2003) and still remains one of the most effective tools for conservation planners.

In answer to our initial question 'is a national approach to IUCN Red Listing of CWR as difficult as some think?' we conclude it is not. Using the example of applying the criteria for Portuguese CWR species we conclude that the global assessment is possible, fairly easy and quick to be undertaken once the data are collated, even in data-poor situations. However, regional assessment can be seriously limited by lack of comparable data sets from adjacent regions. Regional assessment is more difficult to undertake than the global assessment, especially because more knowledge about species' populations is needed. Nevertheless, a certain level of uncertainty is possible and this is probably one of the strongest points in data-poor situations.

Acknowledgements

The authors would like to thank P. Arriegas (ICN, Portugal) for providing the majority of the data, P. Cardoso (Bio3, Portugal) for his precious help with GIS, Shelagh Kell (University of Birmingham, United Kingdom) for help provided with the list of CWR for Portugal, Dr C. Hilton-Taylor (IUCN, United Kingdom) for training and advice, and Maria Scholten (University of Birmingham, United Kingdom) for her useful suggestions and ideas.

References

Akçakaya, H.R. (2000) Population viability analyses with demographically and spatially structured models. *Ecological Bulletins* 48, 23–38.

Akçakaya, H.R. and Sjögren-Gulve, P. (2000) Population viability analysis in conservation planning: an overview. *Ecological Bulletins* 48, 9–21.

Akçakaya, H.R., Ferson, S., Burgman, M.A., Keith, D.A., Mace, G.M. and Todd, C.A. (2000) Making consistent classifications under uncertainty. *Conservation Biology* 14, 1001–1013.

Albuquerque, C., Fonseca, J.P. and Bernardes, C. (2003) *Relatório Técnico do Plano de Ordenamento do Parque Natural de Sintra/Cascais*. PNSC, Instituto da Conservação da Natureza.

ANTHOS (2005). Anthos. Sistema de información de las plantas de España. Real Jardín Botánico, CSIC – Fundación Biodiversidad. Available at: http://programanthos.org (accessed 16 September 2005)

Avery, M., Gibbons, D.W., Porter, R., Tew, T., Tucker, G. and Williams, G. (1995) Revising the British Red Data List for birds: the biological basis of UK conservation priorities. *Ibis* 137, 232–239.

Bañares, Á., Blanca, G., Güemes, J., Moreno Saiz, J.C. and Ortiz, S. (eds) (2004) *Atlas y Libro Rojo de la Flora Vascular Amenazada de España*. Dirección para la Biodiversidad, Publicaciones del O.A.P.N., Madrid.

Barreto Caldas, F., Séneca, A.M., Almeida da Silva, R., Carvalho, A.C., Branco, H.A., Alves, H. and Honrado, J.J. (1996) *Distribuição Geográfica e Estatuto de Ameaça das Espécies da Flora a Proteger. Relatório final*. Núcleo de Ecologia Vegetal, ICETA. Universidade do Porto, Porto, Portugal.

Boyce, M.S. (1992) Population viability analysis. *Annual Review of Ecology and Systematics* 23, 481–506.

Burgman, M.A., Keith, D.A. and Walshe, T.V. (1999) Uncertainty in comparative risk analysis for threatened Australian plant species. *Risk Analysis* 19 (4), 585–598.

Burton, J. (2003) On Red Lists and IUCN. *PlantTalk* 32, 4–5.
Butchart, S.H.M., Stattersfield, A.J., Bennun, L., Shutes, A.S.M., Akçakaya, H.R., Baillie, J.E.M., Stuart, S.N., Hilton-Taylor, C. and Mace, G.M. (2004) Measuring global trends in the status of biodiversity: Red List indices for birds. *PLoS Biology* 2(12), e383. Available at: http://biology.plosjournals.org/perlserv/?request=get-document&doi=10.1371/journal.pbio.0020383#AFF3 (accessed 25 September 2005)
Callmander, M.W., Schatz, G.E and Lowrey II, P.P (2005) IUCN Red List assessment and the Global Strategy for Plant Conservation: taxonomists must act now. *Taxon* 54(4), 1047–1050.
CBD (1992) *Convention on Biological Diversity: Text and Annexes*. Secretariat of the Convention on Biological Diversity, Montreal, Canada.
CBD (2002) *Global Strategy for Plant Conservation*. Secretariat of the Convention on Biological Diversity, Montreal, Canada.
Cheffings, C.M. and Farrell, L. (eds) (2005) *The Vascular Plant Red Data List for Great Britain. Species Status* 7, 1–116. Joint Nature Conservation Committee, Peterborough, UK.
Colling, G. (2005) Red List of the Vascular Plants of Luxembourg. *Ferrantia* 42. Available at: http://www.mnhn.lu/recherche/ferrantia/publications/Ferrantia42.pdf (accessed 18 August 2005)
De Koe, T., Sequeira, M.M., Bento, J., Aranha, J.T. and Varejão, E. (1998) *Douro Internacional (incl. Bacia da Ribeira de Mós). Habitats Naturais e de Espécies da Flora de Portugal (Continente). Relatório final (adenda e fichas)*. Universidade de Trás-os-Montes e Alto Douro, Vila Real, Portugal.
Dennis, B., Munholland, P.L. and Scott, J.M. (1991) Estimation of growth and extinction parameters for endangered species. *Ecological Monographs* 61, 115–143.
DGRF (Direcção-Geral dos Recursos Florestais) and ISA–DEF (Instituto Superior de Agronomia – Departamento de Engenharia Florestal) (1990–2004) *Cartografia das áreas queimadas*, Ministério da Agricultura, do Desenvolvimento Rural e das Pescas.
Dobson, A.P. (1998) *Conservation and Biodiversity*. Scientific American Library, New York.
Domínguez Lozano, F., Moreno Saiz, J.C. and Sainz Ollero, H. (2004) Conservation listing and taxonomy in the Iberian vascular flora. Are we making progress? *Planta Europa – 4th European Conference on the Conservation of Wild Plants Valencia (Spain) 17–20th September*. Available at: http://www.nerium.net/plantaeuropa/Download/Procedings/Dominguez.pdf (accessed 10 January 2006)
Eaton, M.A., Gregory, R.D., Noble, D.G., Robinson, J.A., Hughes, J., Procter, D., Brown, A.F. and Gibbons, D.W. (2005) Regional IUCN Red Listing: the process as applied to birds in the United Kingdom. *Conservation Biology* 19(5), 1557–1570.
European Commission (2000) *Communication of the European Commission to the Council and to the Parliament on a European Community Biodiversity Strategy*. Available at: http://biodiversitychm.eea.eu.int/convention/ECBS/fulltext.html (accessed 22 April 2004)
ESRI (2002) ArcView 3.3. ESRI, Redlands, California.
FAO (Food and Agriculture Organization of the United Nations) (1996) *Global Plan of Action for the Conservation and Sustainable Utilization of Plant Genetic Resources for Food and Agriculture and the Leipzig Declaration. Adopted by the International Technical Conference on Plant Genetic Resources*. FAO, Leipzig.
FAO (Food and Agriculture Organization of the United Nations) (2001) *International Treaty on Plant Genetic Resources for Food and Agriculture*. FAO, Rome, Italy.
Fidalgo, J.P.P. (1996) *Plantas Prioritárias na Serra da Estrela*. Relatório interno. Instituto da Conservação da Natureza.
Gärdenfors, U. (2001) Classifying threatened species at national versus global levels. *Trends in Ecology and Evolution* 16(9), 511–516.
Gärdenfors, U. (ed.) (2005) *The 2005 Red List of Swedish Species*. ArtDatabanken, SLU, Uppsala, Sweden.

Gärdenfors, U., Rodríguez, J.P., Hilton-Taylor, C., Hyslop, C., Mace, G., Molur, S. and Poss, S. (1999) Draft guidelines for the application of the IUCN Red List criteria at national and regional levels. *Species* 31–32, 58–70.

Gärdenfors, U., Hilton-Taylor, C., Mace, G.M. and Rodríguez, J.P. (2001) The application of the IUCN Red List Criteria at regional levels. *Conservation Biology* 15(5), 1206–1212.

Golding, J.S. (2004) The use of specimen information influences the outcomes of Red List assessments: the case of southern African plant specimens. *Biodiversity and Conservation* 13, 773–780.

Harlan, J. and de Wet, J. (1971) Towards a rational classification of cultivated plants. *Taxon* 20, 509–517.

Hartley, S. and Kunin, W.E. (2003) Scale dependency of rarity, extinction risk and conservation priority. *Conservation Biology* 17(6), 1559–1570.

Heywood, V.H. (1997) Introduction. In: Valdés, B., Heywood, V.H., Raimondo, F.M. and Zohary, D. (eds) Proceedings of the Workshop on 'Conservation of the Wild Relatives of European Cultivated Plants'. *Bocconea* 7, 15–18.

Hoyt, E. (1988) *Conserving the Wild Relatives of Crops*. IBPGR–IUCN–WWF, Rome, Italy.

ICN (Instituto da Conservação da Natureza) (2006a) Plano Sectorial da rede Natura 2000. Fichas de caracterização ecológica e de gestão. Version January 2006. Available at: http://www.icn.pt/psrn2000/caract_flora.htm (accessed 26 January 2006)

ICN (Instituto da Conservação da Natureza) (2006b) Plano Sectorial da rede Natura 2000. Fichas de caracterização ecológica e de gestão. *Thymus carnosus* Boiss. Version January 2006. Available at: http://www.icn.pt/psrn2000/caracterizacao_valores_naturais/flora/Thymus%20carnosus.pdf (accessed 26 January 2006)

ICN (Instituto da Conservação da Natureza) (2006c) Plano Sectorial da rede Natura 2000. Fichas de caracterização ecológica e de gestão. *Narcissus asturiensis* (Jordan) Pugsley. Version January 2006. Available at: http://www.icn.pt/psrn2000/caracterizacao_valores_naturais/flora/Narcissus%20asturiensis.pdf (accessed 26 January 2006)

ICN (Instituto da Conservação da Natureza) (2006d) Plano Sectorial da rede Natura 2000. Fichas de caracterização ecológica e de gestão. *Narcissus cyclamineus* DC. Version January 2006. Available at: http://www.icn.pt/psrn2000/caracterizacao_valores_naturais/flora/Narcissus% 20cyclamineus.pdf (accessed 26 January 2006)

ICN (Instituto da Conservação da Natureza) (2006e) Plano Sectorial da rede Natura 2000. Fichas de caracterização ecológica e de gestão. *Plantago algarbiensis* Samp. Version January 2006. Available at: http://www.icn.pt/psrn2000/caracterizacao_valores_naturais/flora/Plantago%20algarbiensis.pdf (accessed 26 January 2006)

IPGRI (International Plant Genetic Resources Institute) (2000a) *In situ* conservation in natural ecosystems (Crop Wild Relatives). Available at: http://www.ipgri.cgiar.org/themes/in_situ_project/home/insituhome.htm (accessed 7 April 2004)

IPGRI (International Plant Genetic Resources Institute) (2000b) *In situ* conservation of crop wild relatives through enhanced information management and field application. A UNEP/GEF PDF B Project linking countries and international conservation agencies. Available at: http://www.ipgri.cgiar.org/themes/in_situ_project/wild_relatives/gef.htm (accessed 7 April 2004)

IPGRI (International Plant Genetic Resources Institute) (2003) High priority for crop wild relatives. Available at: http://ipgri-pa.grinfo.net/nucleus/plugins/print/print.php?itemid=8 (accessed 3 March 2004)

Iriondo, J.M. (1996) The survey and modelling of small plant populations as a basis for developing conservation strategies. *Bocconea* 5, 151–157.

IUCN (International Union for Conservation of Nature and Natural Resources) (1994) *IUCN Red List Categories*. IUCN Species Survival Commission, Gland, Switzerland/Cambridge. Available at: http://www.redlist.org/info/categories_criteria1994 (accessed 8 May 2005)

IUCN (International Union for Conservation of Nature and Natural Resources) (2001) *IUCN Red List Categories and Criteria: Version 3.1*. IUCN Species Survival Commission, Gland, Switzerland/Cambridge, UK. Available at: http://www.iucn.org/themes/ssc/redlists/redlist-cat senglish.pdf (accessed 5 January 2005)
IUCN (International Union for Conservation of Nature and Natural Resources) (2003) *Guidelines for Application of IUCN Red List Criteria at Regional Levels: Version 3.0*. IUCN Species Survival Commission, Gland, Switzerland/Cambridge, UK. Available at: http //app.iucn.org/webfiles/doc/SSC/SSCwebsite/Red_List/regionalguidelinesEn.pdf (accessed 5 January 2005)
IUCN (International Union for Conservation of Nature and Natural Resources) (ed.) (2004) IUCN Red List of Threatened Species. IUCN. Species Survival Commission, Gland, Switzerland/Cambridge, UK. Available at: http://www.redlist.org/ (accessed 5 January 2005)
IUCN (International Union for Conservation of Nature and Natural Resources) (2005a) *Guidelines for Using the IUCN Red List Categories and Criteria*. IUCN. Species Survival Commission, Gland, Switzerland and Cambridge, UK. Available at: http://app.iucn.org/webfiles/doc/SSC/RedList/RedListGuidelines.pdf (accessed 15 May 2005)
IUCN (International Union for Conservation of Nature and Natural Resources) (ed.) (2005b) Resolution 3.013: The uses of the IUCN Red List of Threatened Species. In: *Resolutions and Recommendations: World Conservation Congress*, Bangkok, Thailand, 17–25 November 2004, pp. 14–16. Available at: http://app.iucn.org/congress/members/WCC_Res_Recs_ENGLISH.pdf (accessed 14 May 2005)
IUCN/SSC (International Union for Conservation of Nature and Natural Resources, Species Survival Committee) (1999) *The IUCN Criteria Review: Report of the Marine Workshop*. Report of a workshop held at the Asia centre of Japan, Tokyo on 16–17 January 1999, part of the review of the IUCN Criteria for listing threatened species. Available at: http://app.iucn.org/webfiles/doc/SSC/RedList/CriteriaReview_Marine_Workshop.pdf (accessed 14 May 2005)
Jain, S.K. (1975) Genetic reserves. In: Frankel, O.H. and Hawkes, J.G. (eds) *Crop Genetic Resources for Today and Tomorrow*. Cambridge University Press, Cambridge, UK, pp. 379–396.
Keith, P. and Marion, L. (2002) Methodology for drawing up a red list of threatened freshwater fish in France. *Aquatic Conservation: Marine and Freshwater Ecosystems* 12, 169–179.
Keith, D.A., Chalson, J.M. and Auld, T.D. (1997) Assessing the status of threatened plants: a new methodology and an application to the vascular flora of New South Wales, unpublished report, NSW NPWS. In: Burgman, M.A., Keith, D.A. and Walshe, T.V. (1999) Uncertainty in comparative risk analysis for threatened Australian plant species. *Risk Analysis* 19(4), 585–598.
Keith, D.A., Auld, T.D., Ooi, M.K.J. and Mackenzie, B.D.E. (2000) Sensitivity analyses of decision rules in World Conservation Union (IUCN) Red List criteria using Australian plants. *Biological Conservation* 94, 311–319.
Keith, D.A., McCarthy, M.A., Regan, H., Regan, T., Bowles, C., Drill, C., Craig, C., Pellow, B., Burgman, M.A., Master, L.L., Ruckelshaus, M., Mackenzie, B., Andelman, S.J. and Wade, P.R. (2004) Protocols for listing threatened species can forecast extinction. *Ecology Letters* 7, 1101–1108.
Kell, S.P., Maxted, N., Ford-Lloyd, B.V., Hilton-Taylor, C., Pollock, C. and Strahm, W. (2004) *Crop Wild Relatives: A Vital Resource for a Sustainable Future*. Poster presentation given at the 3rd IUCN World Conservation Congress, Bangkok, Thailand, November. Available at: http://www.pgrforum.org/Documents/Poster_presentations/WCC_2004.pdf (accessed 9 March 2006)
Keller, V. and Bollmann, K. (2004) From Red Lists to species of conservation concern. *Conservation Biology* 18(6), 1636–1644.
Kindvall, O. and Gärdenfors, U. (2003) Temporal extrapolation of PVA results in relation to the IUCN Red List Criterion E. *Conservation Biology* 17(1), 316–321.
Lacy, R.C. (2000) Considering threats to the viability of small populations with individual-based models. *Ecological Bulletins* 48, 39–51.

Lamoreux, J., Akçakaya, H.R., Bennun, L., Collar, N.J., Boitani, L., Brackett, D., Bräutigam, A., Brooks, T.M., da Fonseca, G.A.B., Mittermeier, R.A., Rylands, A.B., Gärdenfors, U., Hilton-Taylor, C., Mace, G., Stein, B.A. and Stuart, S. (2003) Value of the IUCN Red List. *Trends in Ecology and Evolution* 18(5), 214–215.

Lawton, J.H. (1993) Range, population abundance and conservation. *Trends in Ecology and Evolution* 8, 409–413.

Lunney, D., Curtin, A., Ayers, D., Cogger, H.G. and Dickman, C.R. (1996) An ecological approach to identifying the endangered fauna of New South Wales. *Pacific Conservation Biology* 2, 212–231.

MacDougall, A.S., Loo, J.A., Clayden, S.R., Goltz, J.G. and Hinds, H.R. (1998) Defining conservation priorities for plant taxa in southeastern New Brunswick, Canada using herbarium records. *Biological Conservation* 86, 325–338.

Mace, G.M. and Kershaw, M. (1997) Extinction risk and rarity on an ecological timescale. In: Kunin, W.E. and Gaston, K.J. (eds) *The Biology of Rarity: The Causes and Consequences of Rare–Common Differences*. Chapman & Hall, London, pp. 131–149.

Mace, G.M. and Lande, R. (1991) Assessing extinction threats: towards a re-evaluation of IUCN threatened species categories. *Conservation Biology* 5, 148–157.

Magos Brehm, J. (2003) *Criação de um Banco de Sementes Representativo da Flora Afectada pela Construção da Barragem do Alqueva*. Relatório Final de Bolsa de Iniciação Científica. Banco de Sementes António Belo Correia, Museu, Laborátorio e Jardim Botânico da Universidade de Lisboa, Lisboa, Portugal.

Magos Brehm, J. (2004) Conservation of Wild Legumes in Portugal. MSc Dissertation, School of Biosciences, University of Birmingham, UK.

Matsuda, H. (2003) Challenges posed by the precautionary principle and accountability in ecological risk assessment. *Environmetrics* 14, 245–254.

Maxted, N., van Slageren, M.W. and Rihan, J. (1995) Ecogeographic surveys. In: Guarino, L., Ramanatha Rao, V. and Reid, R. (eds) *Collecting Plant Genetic Diversity: Technical Guidelines*. CAB International, Wallingford, UK, pp. 255–286.

Maxted, N., Hawkes, J.G., Ford-Lloyd, B.V. and Williams, J.T. (1997a) A practical model for *in situ* genetic conservation. In: Maxted, N., Ford-Lloyd, B.V. and Hawkes, J.G. (eds) *Plant Genetic Conservation: The* In Situ *Approach*. Chapman & Hall, London, pp. 545–592.

Maxted, N., Hawkes, J.G., Guarino, L. and Sawkins, M. (1997b) The selection of taxa for plant genetic conservation. *Genetic Resources and Crop Evolution* 44, 337–348.

Maxted, N., Ford-Lloyd, B.V., Jury, S.L., Kell, S.P. and Scholten, M. (2006) Towards a definition of a crop wild relative. *Biodiversity and Conservation* 15(8), 2673–2685.

Meilleur, B.A. and Hodgkin, T. (2004) *In situ* conservation of crop wild relatives: status and trends. *Biodiversity and Conservation* 13, 663–684.

Menges, E.S. (2000) Population viability analyses in plants: challenges and opportunities. *TREE* 15(2), 51–56.

Millsap, B.A., Gore, J.A., Runde, D.E. and Cerulean, S.I. (1990) Setting the priorities for the conservation of fish and wildlife species in Florida. *Wildlife Monographs Supplement, Journal of Wildlife Management* 54, 5–57.

Ministério do Ambiente (1999) *Plano de Bacia Hidrográfica do Rio Sado. 1a fase, Analise e Diagnostico da Situacao Actual. Anexo Temático 1. Análise Biofísica. Parte 5. Análise da Fauna e da Flora e Vegetação*. Ministério do Ambiente, Lisboa, Portugal.

Ministério do Ambiente (2000) *Plano de Bacia Hidrográfica do Rio Sado. Volume III. Análise – Parte 3. Subsistema Ambiental* (Rev. [n°00] – data [31–05–00]). Ministério do Ambiente, Lisboa, Portugal.

Mitchell, M. (2004) Regional Red List Assessment of Crop Wild Relatives in Europe. MSc Dissertation, School of Biosciences, University of Birmingham, UK.

Moreno Saiz, J.C., Domínguez Lozano, F. and Sainz Ollero, H. (2003) Recent progress in conservation of threatened Spanish vascular flora: a critical review. *Biological Conservation* 113, 419–431.

OECD (Organisation for Economic Co-operation and Development) (1996) *Saving Biological Diversity: Economic Incentives*. OECD, Paris.

Pinto, M.J.G., Cotrim, H. and Draper, D.M. (1996a) *Distribuição Geográfica e Estatuto de Ameaça das Espécies da Flora a Proteger*. 5° Relatório de Progresso e 3° Relatório Adicional ao Protocolo de Colaboração. Museu, Laboratório e Jardim Botânico da Universidade de Lisboa, Lisboa, Portugal.

Pinto, M.J.G., Cotrim, H. and Draper, D.M. (1996b) *Distribuição Geográfica e Estatuto de Ameaça das Espécies da Flora a Proteger*. 6° Relatório de Progresso e 4° Relatório Adicional ao Protocolo de Colaboração (relatório final do sub-projecto 'Lagoas'). Museu, Laboratório e Jardim Botânico Universidade de Lisboa, Lisboa, Portugal.

Pinto-Gomes, C. (1996) *Distribuição Geográfica e Estatuto de Ameaça das espécies da Flora a Proteger*. Relatório final. Universidade de Évora, Évora.

Planta Europa and the Council of Europe (2002) *European Plant Conservation Strategy*. Council of Europe and Planta Europa, The Hague, The Netherlands.

Possingham, H.P., Andelman, S.J., Burgman, M.A., Medellín R.A., Master, L.L. and Keith, D.A. (2002) Limits to the use of threatened species lists. *Trends in Ecology and Evolution* 17 (11), 503–507.

Prescott-Allen, R. and Prescott Allen, C. (1983) *Genes from the Wild: Using Wild Genetic Resources for Food and Raw Materials*. Earthscan Publications, London, p. 112.

Rabinovitz, D., Cairns, S. and Dillon, T. (1986) Seven forms of rarity and their frequency in the flora of British Isles. In: Soule, M.E. (ed.) *Conservation Biology: Science of Scarcity and Diversity*. Sinauer Associates, Sunderland, Massachusetts, pp. 182–204.

Rodrigues, A.S.L., Pilgrim, J.D., Lamoreux, J.F., Hoffmann, M. and Brooks, T.M. (2006) The value of the IUCN Red List for conservation. *Trends in Ecology and Evolution* 12(2), 71–76.

Rosselló-Graell, A., Marques, I. and Draper, D. (2003) Segunda localidad de *Narcissus cavanillesii* A. Barra & G. López (Amaryllidaceae) para Portugal. *Acta Botanica Malacitana* 28, 196–197.

Samways, M.J. (2003) Marginality and national Red Listing of species. *Biodiversity and Conservation* 12, 2523–2525.

Schoen, D.J. and Brown, A.H.D. (1993) Conservation of allelic richness in wild crop relatives is aided by assessment of genetic markers. *Proceedings of the National Academy of Sciences of the United States of America* 90(22), 10623–10627.

Siebert, S.J. and Smith, G.F. (2004) Lessons learned from the SABONET Project while building capacity to document the botanical diversity of southern Africa. *Taxon* 53(1), 119–126.

Sjögren-Gulve, P. and Hanski, I. (2000) Metapopulation viability analysis using occupancy model. *Ecological Bulletins* 48, 53–71.

Ter Steege, H., Jansen-Jacobs, M.J. and Datadin, V.K. (2000) Can botanical collections assist in a National Protected Area Strategy in Guyana? *Biodiversity and Conservation* 9, 215–240.

Vavilov, N.I. (1926) Studies on the origin of cultivated plants. *Bulletin of Applied Botany, Genetics and Plant Breeding* 16, 1–248.

VV.AA. (2000) Lista Rioja de la Flora Vascular Española. *Conservación Vegetal* 6, 11–38.

Williams, P.H., Humphries, C.J. and Vane-Wright, R.I. (1993) Measuring biodiversity: taxonomic relatedness for conservation priorities. *Australian Systematic Botany* 4, 665–679.

Willis, F., Moat, J. and Paton, A. (2003) Defining a role for herbarium data in Red List assessments: a case study of *Plectranthus* from eastern and southern tropical Africa. *Biodiversity and Conservation* 12, 1537–1552.

14 Traditional Farming Systems in South-eastern Turkey: the Imperative of *In Situ* Conservation of Endangered Wild Annual *Cicer* Species

S. Abbo, C. Can, S. Lev-Yadun and M. Ozaslan

14.1 Introduction

Domesticated chickpea (*Cicer arietinum* L.) is one of the 'founder crops' of the Neolithic Near Eastern farming (Zohary and Hopf, 2000). Based on meiotic chromosome pairing and seed storage protein profiles, the wild progenitor of domesticated chickpea was identified as *C. reticulatum* Ladiz. (Ladizinsky and Adler, 1975; 1976a). Unlike the wild progenitors of domesticated wheat, barley, pea or lentil, most wild annual *Cicer* species are characterized by relatively narrow distribution and confinement to specific habitat types within their respective range (van der Maesen, 1972; Zohary and Hopf, 2000; Abbo *et al.*, 2003; Berger *et al.*, 2003). Recently, it was hypothesized that the narrow geographic range of *C. reticulatum* represents (or stems from) its narrow ecological adaptation (Abbo *et al.*, 2003). Interestingly, domesticated chickpea is also characterized by narrow adaptation compared with other cool season legumes of Near Eastern origin like pea and lentil (Miller *et al.*, 2002; Abbo *et al.*, 2003).

The fact that chickpea physiological adaptations and agronomic performance are both limited in its wild form as well as a crop plant under domestication compared with other crop plants of similar Near Eastern origin suggests a strict genetic control of the respective adaptation traits (Abbo *et al.*, 2003). These authors used the wider ecophysiological adaptation of pea (or barley), and the parallel wider genetic diversity (in terms of DNA markers and other traits) available in domesticated pea (and barley) compared with chickpea to explain the evolutionary reasons underlying the genetic basis of chickpea adaptation (therein).

A certain degree of reduction in the allelic repertoire (or genetic erosion), usually referred to as the 'domestication founder effect', is an almost universal feature of plant domestication (Ladizinsky, 1985, 1998). In essence, this means that from the very early days of agriculture, the domesticated stocks encompassed only a limited subset of the genetic variation present in the wild progenitors that gave rise to crop plants (Ladizinsky, 1998). A further reduction in the genetic variation of many crop plants occurred during the last century with the introduction of modern plant breeding methodologies and the replacement of traditional (and genetically diverse) local landraces with modern (and genetically uniform) high yielding crop varieties (Harlan, 1992; Tanksley and McCouch, 1997).

The ability of plant breeders to further improve adaptation and thereby promote higher and more stable yield of crop plants is entirely dependent upon the allelic variation within the respective crop's germplasm. As long as breeders are able to find the required allelic variation for yield increase and for combating biotic and abiotic stresses within adapted genetic backgrounds they will mostly employ such germplasm in their crossing schemes. This is because in most cases the progeny from domesticated × wild crosses lack the basic agronomic features required from adapted cultivars. If, however, no adequate allelic variants can be identified within modern varieties, breeders will seek the solution by first screening and crossing with exotic landraces, and if needed, with wild accessions from the crop's wild progenitor or other closely related taxa.

14.2 Wild *Cicer* Species and Their Relevance to Improvement of the Chickpea Crop

The idea of using wild relatives for crop improvement is some 100 years old and was discussed in numerous publications (e.g. Aaronsohn, 1910; Zamir, 2001). The long history of the idea explains the everlasting interest of crop physiologists, geneticists, phytopathologists and breeders in wild relatives (e.g. Evans, 1993). The genus *Cicer* holds more than 40 species, and the majority of them are perennials native to alpine ecosystems of the Balkans, Asia minor and Central Asia. No information is available on the genetic affinity and crossability of the perennial *Cicer* species. However, crosses were attempted between the cultigen and most wild annual *Cicer* species, and they were assigned into crossability groups based on the success rate of the crosses, the viability of the F_1 zygotes and the fertility of the F_1 hybrid plants (Ladizinsky and Adler, 1976b). The first crossing group contains *C. arietinum*, its wild progenitor *C. reticulatum* and *C. echinospermum* P.H. Davis. The second group contains *C. judaicum* Boiss., *C. pinnatifidum* Jaub. & Spach and *C. bijugum* K.H. Rech. The third group contains the rest of the annual taxa, which do not cross with any of the above species. Meagre information is available on the genetic relations among the species of the latter group (*C. chorassanicum* (Bunge) Popov, *C. cuneatum* Hochst. ex A. Rich., *C. yamashitae* Kitam.) (for a detailed review of the subject see Croser *et al.*, 2003).

The rarity of resistances to biotic and abiotic stresses in domesticated chickpea has inspired screening of wild *Cicer* species regardless of their immediate relevance to breeding (see two excellent reviews by Singh (1997) and Croser *et al.* (2003)). Many attempts were also made to overcome crossability barriers to facilitate transfer of desirable alleles from wild species to domesticated chickpea (e.g. Mallikarjuna, 2003).

However, it should be borne in mind that successful utilization of wild germplasm depends on the availability of a wide array of genotypes sampled across wide ecogeographic amplitude. Unfortunately, only very limited numbers of independent accessions are available from most wild annual *Cicer*. For example, 34 accessions of *C. judaicum* are available, but only three, two and one are held in the world collections from *C. yamashitae*, *C. chorassanicum* and *C. cuneatum*, respectively (see Berger *et al.*, 2003). Accordingly, intensive attempts are currently being made to increase the number of wild *Cicer* accessions for both comparative physiological studies and for genetic analyses of adaptive traits (Ben-David, 2005; Ben-David and Abbo, 2005).

14.3 Niche Preferences of Wild *Cicer*

The literature, as well as our own personal observations, indicates that most wild *Cicer* species grow in stony or rocky niches, and on rubble slopes (van Der Maesen, 1972; Ben-David, 2005). Three species, however, grow as tolerated weeds in traditional farming systems. These are *C. cuneatum* in Eritrea and Ethiopia, and *C. bijugum* and *C. echinospermum* in eastern Turkey (and probably neighbouring territories). Unlike most other *Cicer* species, these three taxa were reported from deep fine-textured (heavy) soils, mostly of basaltic origin.

In eastern Turkey, both *C. echinospermum* and *C. bijugum* occur on the plains west of Diyarbakir, around Siverek and Hilvan, and also elsewhere in similar habitats. During the last decades, traditional agriculture in large parts of this area was transformed into highly mechanized modern farming. In addition to the introduction of modern machinery for dryland grain cropping, vast tracts of land in eastern Turkey were brought under irrigated (e.g. cotton) farming with the availability of huge water cubature from Ataturk and Karakaya dams established on the Euphrates River. Companion weeds are a common feature of traditional grain cropping, especially if such weeds are not aggressive and pose no threat to the regular crop rotation. Indeed, both *C. echinospermum* and *C. bijugum* are such plants. Farmers in eastern Turkey recognize both as 'Yabani Nohut' (wild chickpea in Turkish) and eat them when green or partly mature. When encountered at (manual) harvest time, these plants are left in the field to shed their seeds and are not weeded out. In this way, the traditional farm operations allow the existence of the crop–(tolerated) weed complex.

Contrary to the above, introduction of modern farm operations that include deep ploughing, herbicide application (as evident from the relatively weed-free cereal fields west of Diyarbakir) or careful weeding to ensure high cotton quality is likely to change the weed flora of the respective areas. We are unaware of any report of *Cicer* species as weeds in modern agrosystems (Abbo *et al.*, 2005); therefore, we forecast that the extensive modernization of agriculture in eastern Turkey will most likely eliminate the typical weeds of traditional farming including *C. echinospermum* and *C. bijugum*.

At least two other wild annual *Cicer* species are native to the region, namely *C. pinnatifidum* and *C. reticulatum*. These two species have strong affinity to stony and rocky habitats. Declaration of nature reserves or natural parks, often associated with some relaxation of grazing pressure; is most likely to help in preserving the two latter taxa. However, nature reserves rarely include large tracts of arable land which have been under farming for many years. Therefore, conventional nature protection operations will not be useful in protecting either *C. echinospermum* or *C. bijugum*.

14.4 Why and How to Protect *C. echinospermum* and *C. bijugum*

Both taxa were screened for a number of biotic and abiotic stress agents. Useful traits (like disease and pest resistance) were reported from both species (Singh, 1997; Croser *et al*., 2003). In addition, *C. echinospermum*, which is cross-compatible with domesticated chickpea, was already utilized in Australia to introduce nematode resistance by R. Knights, Tamworth, NSW, 2001 (personal communication). Attempts are also being made to optimize introgression protocols involving crosses between domesticated chickpea and *C. bijugum* (e.g. Mallikarjuna, 2003). Therefore, there is no doubt that these two wild *Cicer* species are of great importance both from pure floristic as well as agronomic perspectives.

Several *Cicer* species are listed in the Red Book of Turkish plants. *Cicer bijugum* is not mentioned among them, but *C. echinospermum* is listed as vulnerable, i.e. they may face danger in the near future (Ekim *et al*., 2000). The Turkish government policy concerning the preservation of the plants listed in the Red Book requires that the Environment and Forestry Ministry takes great care of such taxa. Indeed, the Turkish General Directorate of Nature Preservation and National Parks runs a project concerning capacity building aiming to formulate and apply rules for nature protection management. Public support for such action also comes from the Turkish Association for the Conservation of Nature. Indeed, Turkey is a partner in an on-farm conservation research project (see IPGRI, no date); however, in the framework of this project there is no reported activity concerning wild *Cicer* species.

A considerable number of on-farm conservation activities were reported in recent years (e.g. Piegiovanni and Laghetti, 1999; Maxted *et al*., 2002; Negri and Tosti, 2002). In this context, it is important to note that there is a growing understanding that weeds should be seen as an integral and essential component of agrobiodiversity (e.g. Spahillari *et al*., 1999). It should be borne in mind that the species composition and diversity of the companion weeds of crop plants are directly related to the agrotechnical practice in any farming system. Consequently, any agrotechnical change is likely to result in a change in the weed flora. Therefore, if certain companion weeds of traditional farming systems are of interest, special measures should be taken to conserve them. Specifically, great care should be taken prior to the introduction of any changes or modernization of farming operations, like the introduction of deep ploughs, irrigation or herbicides.

We are also aware that any attempt to maintain traditional farming in an era of globalization, and volatile commodity prices must take into account the welfare of the local farmers and their scope for improved income and higher standard of living. Inference on ecotourism, and agroeconomic planning aiming at both wild plant protection and sustainable development is beyond our expertise and is outside the scope of this chapter. It should be borne in mind, however, that raising awareness is not enough to protect these two important *Cicer* species.

References

Aaronsohn, A. (1910) *Agricultural and Botanical Explorations in Palestine*. USDA, Bureau of Plant Industry, Bulletin no. 180, Washington.

Abbo, S., Berger, J. and Turner, N.C. (2003) Evolution of cultivated chickpea: four genetic bottlenecks limit diversity and constrain crop adaptation. *Functional Plant Biology* 30, 1081–1087.

Abbo, S., Lev-Yadun, S., Rubin, B. and Gopher, A. (2005) On the origin of near eastern founder crops and the 'dump-heap hypothesis'. *Genetic Resources and Crop Evolution* 52, 491–495.

Ben-David, R. (2005) Eco-geography and demography of wild *Cicer judaicum* in Israel. MSc thesis, Faculty of Agricultural, Food and Environmental Quality Sciences, The Hebrew University of Jerusalem, Rehovot, Israel. Hebrew (English summary).

Ben-David, R. and Abbo, S. (2005) Phenological variation among Israeli populations of *Cicer judaicum* Boiss. *Australian Journal of Agricultural Research* 56, 1219–1225.

Berger, J., Abbo, S. and Turner, N. (2003) Ecogeography of annual wild *Cicer* species: the poor state of the world collection. *Crop Science* 43, 1076–1090.

Croser, J.S., Ahmad, F., Clarke, H.J. and Siddique, K.H.M. (2003) Utilization of wild *Cicer* in chickpea improvement – progress, constraints, and prospects. *Australian Journal of Agricultural Research* 54, 429–444.

Ekim, T., Koyuncu, M., Vural, M., Duman, H., Aytac, Z. and Adlguzel, N. (2000) *Red Data Book of Turkish Plants*. Ankara, Turkey, 246 pp.

Evans, L. (1993) *Crop Evolution, Adaptation and Yield*. Cambridge University Press, Cambridge, UK.

Harlan, J.R. (1992) *Crops and Man*, 2nd edn. American Society of Agronomy, Crop Science Society of America, Madison, Wisconsin.

IPGRI (no date) Agrobiodiversity management in production systems: Turkey. Available at: http://www.ipgri.cgiar.org/themes/in_situ_project/countries/insitutur.htm (accessed 28 April 2006)

Ladizinsky, G. (1985) Founder effect in crop plant evolution. *Economic Botany* 39, 191–199.

Ladizinsky, G. (1998) *Plant Evolution Under Domestication*. Kluwer Academic, Dordrecht, The Netherlands.

Ladizinsky, G. and Adler, A. (1975) The origin of chickpea as indicated by seed protein electrophoresis. *Israel Journal of Botany* 24, 183–189.

Ladizinsky, G. and Adler, A. (1976a) The origin of chickpea *Cicer arietinum* L. *Euphytica* 25, 211–217.

Ladizinsky, G. and Adler, A. (1976b) Genetic relationships among the annual species of *Cicer* L. *Theoretical and Applied Genetics* 48, 197–208.

Mallikarjuna, N. (2003) Wide hybridization in important food legumes. In: Jaiwal, P.K. and Singh, R.P. (eds) *Improvement Strategies of* Leguminosae *Biotechnology*. Kluwer Academic, Dordrecht, The Netherlands, pp. 155–170.

Maxted, N., Guarino, L., Mayer, L. and Chiwona, E.A. (2002) Towards a methodology for on-farm conservation of plant genetic resources. *Genetic Resources and Crop Evolution* 49, 31–46.

Miller, P.R., McConkey, B.G., Clayton, G.W., Brandt, S.A., Staricka, J.A., Johnston, A.M., Lafond, G., Schatz, B.G., Baltensperger, D.D. and Neill, K. (2002) Pulse crop adaptation in the northern great plains. *Agronomy Journal* 94, 261–272.

Negri, V. and Tosti, N. (2002) *Phaseolus* genetic diversity maintained on-farm in central Italy. *Genetic Resources and Crop Evolution* 49, 511–520.

Piergiovanni, A.R. and Laghetti, G. (1999) The common bean landraces from Basilicata (Southern Italy): an example of intergrated approach applied to genetic resources management. *Genetic Resources and Crop Evolution* 46, 47–52.

Singh, K.B. (1997) Chickpea (*Cicer arietinum* L.). *Field Crops Research* 53, 161–170.

Spahillari, M., Hammer, K., Gladis, T. and Diederichsen, A. (1999) Weeds as part of agrobiodiversity. *Outlook on Agriculture* 28, 227–282.

Tanksley, S.D. and McCouch, S.R. (1997) Seed banks and molecular maps: unlocking genetic potential from the wild. *Science* 277, 1063–1066.

van Der Maesen, L.J.G. (1972). *Cicer* L., a monograph of the genus, with special reference to the chickpea (*Cicer arietinum* L.), its ecology and cultivation. *Mededelingen Landbouwhogeschool Wageningen* 72(10), 1–342.

Zamir, D. (2001) Improving plant breeding with exotic genetic libraries. *Nature Reviews Genetics* 2, 983–989.

Zohary, D. and Hopf, M. (2000) *Domestication of Plants in the Old World*, 3rd edn. Clarendon Press, Oxford, UK.

15 Ecogeographical Representativeness in Crop Wild Relative *Ex Situ* Collections

M. Parra-Quijano, D. Draper, E. Torres and J.M. Iriondo

15.1 Introduction

As crop wild relatives (CWR) are presently a key component of plant genetic resources (PGRs) the scientific community has been calling attention to the need to improve their representation in both *ex situ* and *in situ* conservation approaches (Noy-Meir *et al.*, 1989; Tanksley and McCouch, 1997; Tewksbury *et al.*, 1999). The FAO's first report on the state of the world's PGRs (FAO, 1996) alerted us to the low number of CWR accessions within gene banks, reporting that CWR and other wild species represent only 15% of total gene bank accessions compared to 74% for cultivated species, although these numbers may vary depending on the species and on the type of *ex situ* collection (governmental, CGIAR or private). In order to balance this wild/cultivated conservation ratio, efforts need to be focused on increasing the number of CWR accessions. However, additional aspects need to be considered.

Successful CWR *ex situ* conservation at the gene bank level is threatened by intrinsic factors related to conservation itself. These problems can be grouped into two categories: (i) at the time of collection; and (ii) during conservation. At the time of collection, the chosen samples may not adequately represent the genetic and/or ecogeographic diversity in the natural populations. Nevertheless, genetic representativeness (GR) and ecogeographical representativeness (ER) of the samples are very important features for the collection. The relationship between GR and ER has been explained by Greene and Hart (1999) based on the ecogeographical factors that contribute to genetic differentiation. Additionally, several research works have found that geographical and genetic distances appear to be correlated (Del Rio *et al.*, 2001; Gupta *et al.*, 2002; Baek *et al.*, 2003). Here, representativeness refers to the relative amount of genetic or ecogeographic species variation conserved with respect to that existing in nature. During conservation, the most important risks are genetic erosion, genetic drift and bottleneck effects. These problems are not solved by merely increasing the number of accessions. Here, the approach must be qualitative rather than quantitative. Hence, collections should be evaluated to assess representativeness (GR and ER) and to check the genetic integrity of their accessions. Although cultivated species conserved in gene banks may also experience a similar situation, the magnitude of these problems is normally greater in CWR species (Brown *et al.*, 1997; Greene *et al.*, 1999a,b).

In the last century, efforts in morphological and molecular gene bank characterizations have made it possible to accurately assess GR, while the assessment of ER still remains an incomplete task. The importance of assessing ER in germplasm collections has been underlined since the 1990s with contributions like that of Steiner and Greene (1996) who introduced the term 'ecological descriptors' to refer to the environmental conditions at the collection site. Factors that may alter ER (biases) have also been assessed (Hijmans *et al.*, 2000). However, few additional works have been carried out in this sense due to a lack of specifically designed computer software, heterogeneous or incomplete passport data and inaccessible environmental information. Nevertheless, the development of methodologies such as geographic information systems (GIS), multivariate analysis or predictive distribution models and their recent integration in specialized software offers PGR researchers new possibilities for carrying out ecogeographical studies of germplasm using georeferenced collection sites from passport data (Guarino *et al.*, 2002). Nowadays, several aspects of germplasm collection have been improved, resulting in better quality passport data and optimized collecting efforts. Some examples may be the detection of biases in *ex situ* collections (Hijmans *et al.*, 2000), the optimization of germplasm collection for *ex situ* conservation (Greene *et al.*, 1999a) and definition of uniform ecogeographical patches to optimize collecting missions oriented to gathering representative samples of the genetic diversity of target species occurring in a particular area (Draper *et al.*, 2003).

This chapter describes three methodological approaches for assessing the ER of a gene bank with emphasis on collection of CWR species. Before that, we explain aspects which should be considered in ER studies like homogeneity, spatial or taxonomical resolution together with the data inputs necessary for carrying out any ER study.

15.2 Environmental Representativeness Studies Applied to CWR Species

15.2.1 Why assess environmental representativeness in CWR species?

Data from both CWR and cultivated species conserved in gene banks can be used to assess representativeness from an ecogeographical point of view. However, CWR have certain characteristics which make ER studies especially appropriate. These characteristics are:

- CWR have a natural distribution which does not depend on man, although it may be influenced by man.
- The fact that a CWR accession was collected at a specific site indicates an adaptation of the CWR population to the environment at the site, which is not

always true in the case of an accession of a cultivated species as agricultural practices usually compensate for a lack of adaptation (Brown *et al.*, 1997).

- The localities of CWR species can be represented in herbaria and botanic bibliography. Thus, their distribution is often better documented than that of cultivated species. Nevertheless, one must note that botanic and bibliographic data may have their own bias.

Wild forms of some cultivated species still survive in their natural environment, especially in centres of origin or diversity. In these cases, the study of ER has the same characteristics as in the case of a wild species with the additional advantage of being able to compare the ecogeographical diversity between wild and cultivated forms. On the other hand, the biogeographic and modelling methodologies developed to determine real and potential distribution of wild species (Guisan and Zimmermann, 2000) can be directly applied to the study of CWR species.

15.2.2 Homogeneity

An important consideration when undertaking an ER study is the homogeneity of the data, whether the source be gene bank collection sites or other sites where the presence of CWR populations has been detected (herbaria, scientific literature, databases, etc.). Data-collecting efforts should be homogeneous enough to provide a reasonable certainty that no regions have been deliberately under- or oversampled (Hijmans *et al.*, 2000).

15.2.3 Spatial range and resolution

With regard to the size of the study area, ER can be calculated for extremely large areas (continents or subcontinents), countries, provinces, special interest areas or simply for the area needed to work with all the gene bank collection sites for a specific species or group of species.

In any case, the 'ecological descriptors' (Steiner and Greene, 1996) and their sources, environmental variables, should cover the entire study area. The spatial resolution of the maps must be similar for the different variables. Resolution usually becomes higher as the size of the study area decreases. It is important to define an optimum balance between the cover of the analysis, resolution and uniformity of information. These questions were further detailed by Hart (1999).

15.2.4 Taxonomical resolution

Another consideration is the taxonomical level of work. ER can be assessed at the interspecific level, normally between species of the same genus, or at the intraspecific level (accessions, populations, etc.). In ecogeographical studies of gene banks, the unit of study is commonly the species. These studies consider groups of CWR species that do not necessarily encompass an entire genus (see Hijmans *et al.*, 2000; Hijmans and Spooner, 2001; Berger *et al.*, 2003; Jarvis *et al.*, 2003; Ferguson *et al.*, 2005). Studies at the intraspecific level are scarcer (see Lobo Burle *et al.*, 2003).

15.3 Elements of an Ecogeographic Representativeness Analysis

15.3.1 Data inputs

Georeferenced gene bank collection sites and presence data

This is the essential information for any ER study. The passport data that accompany each accession conserved in a gene bank provide information on the location of the collection site. However, the quality of this information often varies between accessions and gene banks. Some accessions may have no information on the collection site, others may only have name references (country, region or municipality), while others may have a description of the collection site (i.e. the code of a highway or the point in kilometres). In some cases, especially in the most recent accessions, the passport data may provide detailed geographic coordinates (latitude–longitude or regional coordinates). This heterogeneity in passport data is the first obstacle when analysing ER in a gene bank. For cases where geographic coordinates are not available, there are ways to georeference collection sites with the aid of gazetteers (databases with names of geographic locations and their respective coordinates) or manual tools such as MaNIS (Wiecorzek *et al.*, 2004), MaPSTeDI (Murphey *et al.*, 2004) or INRAM (available at: http://www.inram.org/). The obtained geographic coordinates can then be validated through GIS (Hijmans *et al.*, 1999; Guarino *et al.*, 2002).

Information parallel to that in gene banks on populations of species of interest is available in herbaria, bibliography and floristic databases (Dulloo *et al.*, 1999). The georeferencing of a population's location as well as additional ecogeographic information that can be extracted from these sources is especially important. However, the information on the ecogeographic distribution of the species obtained from different sources may be different, although complementary, as the design of germplasm-collecting missions and botanical expeditions is substantially different due to the differences in their objectives. It is important to keep in mind that the information compiled may also have the same problems as the records from gene banks including the old age of some references or limitations in georeferencing.

Ecological descriptors

Ecological descriptors are ecogeographical attributes that describe the environment of a collection site. This type of descriptors correspond to environmental variables that are classified based on our own criteria and previous classifications (Auricht *et al.*, 1995) in:

- Geophysic – altitude, aspect, slope, longitude, latitude.
- Climatic – temperature and rainfall (monthly, annual, mean, maximum and minimum), solar irradiation, thermal amplitude, relative humidity, potential evapotranspiration (ETP), etc.
- Climatic or temporal – warm period, cold period, dry period, etc. (number of days or months above or below a temperature or rainfall threshold).
- Bioclimatic indices – Dantin–Revenga index (thermopluviometric) (Dantin and Revenga, 1940), Emberger index (Mediterranean influence) (Emberger,

1932), Gorczynsky index (continentality) (Gorczynsky, 1920), etc. These indices cannot be applied worldwide, but should be selected according to the characteristics of the studied area.
- Edaphic – soil type, soil pH, texture, geologic substrate, etc.
- Vegetation – forest maps, type of predominant vegetation.
- Anthropogenic – urban areas, roads, land use, political and administrative boundaries, etc.
- Agroclimatic or bioclimatic zones – maps with land classifications made from the fusion of climatic, bioclimatic and anthropogenic variables and vegetation maps.

There are simple descriptors represented by only one environmental variable (e.g. temperature, rainfall and pH) and complex descriptors which involve different environmental variables obtained by means of mathematical functions (e.g. ETP or bioclimatic indices) or more interpretative features (e.g. soil type, agroclimatic or bioclimatic zones).

The environmental variables come from different sources such as the interpretation of satellite images (urban areas, land use and vegetation maps), digital elevation models (DEMs) (altitude, slope and aspect) or the geostatistic interpolation of information from weather stations (climatic descriptors).

15.3.2 Methodological approaches for assessing ecogeographical representativeness

ER studies can be based on the comparison of gene bank passport data with external sources, the ecogeographic characterization of a gene bank or the use of ecogeographical land characterization maps.

Comparing gene bank data to data from external sources

Comparing the collecting sites from gene bank passport data and sites obtained from external sources (other gene banks, herbaria, literature reports or botanical databases) through gap analysis is one alternative for assessing ER. Using this approach, sites of population occurrence determined by external sources are superimposed on the collecting sites from the target gene bank to search for mismatches. The sites obtained from external sources that do not match the target gene bank's collecting sites can be considered gaps in the gene bank collection (Guarino, 1995) and, thus, in its representativeness. In fact, comparing different gene bank collections may help us to detect gaps and identify collection biases.

LUPINUS SPECIES IN SPAIN: A CASE STUDY. We carried out a gap analysis to detect gaps in the representativeness of the *Lupinus* species collection at the Spanish Central Plant Genetic Resources gene bank (CRF). In order to do this, we compiled gene bank, herbarium and bibliographic data from different sources for our spatial range of Peninsular Spain. Georeferenced collecting sites were extracted from passport data from CRF and Australian Plant Genetic Resources Information System (AusPGRIS) (AUS). These two gene banks hold the most important collections of *Lupinus* worldwide. Georeferenced herbarium data were obtained from the Real Jardín Botánico de Madrid (MA herbarium) and bibliographic data from the ANTHOS project botanic database (available at: http://www.anthos.es/). We compiled a total of 1887 records (usually latitude or longitude points) which we transformed into 1851 1 × 1 km squares in a Universal Transverse Mercator (UTM, European Datum 1950 zone 30 north) coordinate system. The total number of records for each *Lupinus* species was as follows: *L. albus* L. (291), *L. angustifolius* L. (1073), *L. consentinii* Guss. (5), *L. hispanicus* Bois. & Reut. (298), *L. luteus* L. (154) and *L. micranthus* Guss. (30). A grid of 10 × 10 km UTM squares that cover all the spatial range was superimposed on *Lupinus* 1 × 1 km UTM squares. Then, for each species, 10 × 10 km UTM squares were classified according to the matches with each *Lupinus* species 1 × 1 km UTM square as follows: (i) none; (ii) only records from CRF; (iii) only records from gene banks (CRF + AUS); (iv) only records from external sources (MA herbarium + ANTHOS database); (v) records from both CRF and external sources; and (vi) records from both gene bank and external sources (Fig. 15.1).

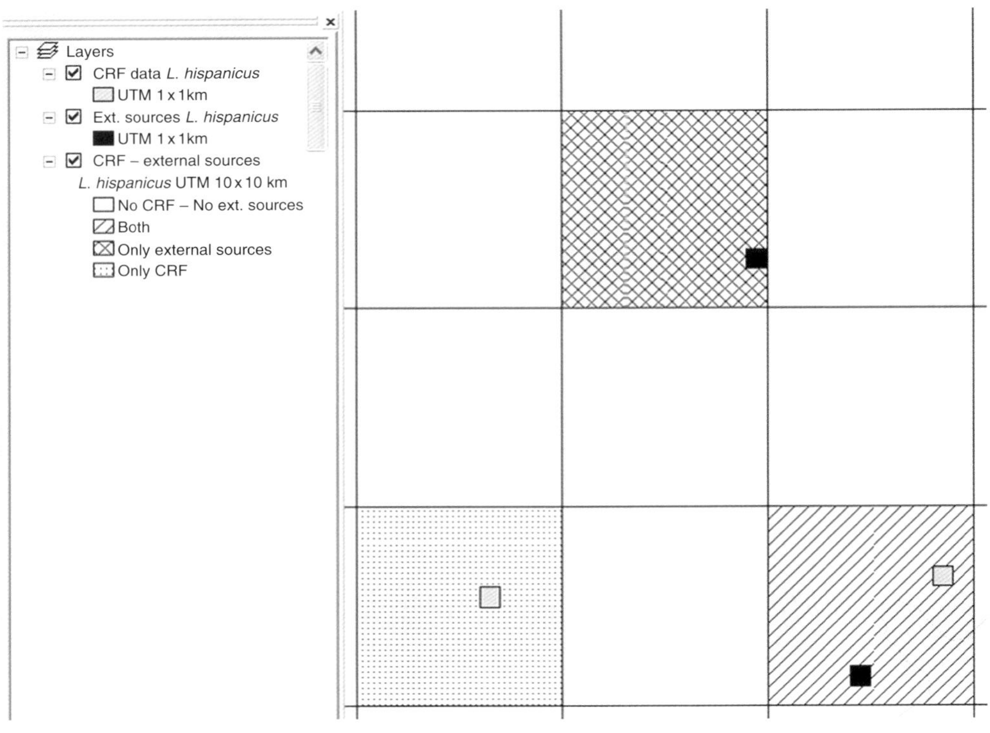

Fig. 15.1. Methodology to find gaps in CWR gene bank collections. Classification of 10 × 10 km UTM squares of a target territory according to the presence or absence of 1 × 1 km UTM squares within, representing populations collected by the gene bank or visited by external sources. UTM squares of 10 × 10 km that contain only 1 × 1 km UTM squares from external sources can be considered gaps in CRF or gene bank representativeness.

We will illustrate the comparison between gene bank and external source data using *L. hispanicus*. The 10 × 10 km UTM squares were classified into the classes (i)–(vi) (described above) and we compared:

- CRF collection sites to external sources (MA herbarium + ANTHOS database);
- Gene bank (CRF + AUS) collection sites to external sources (MA herbarium + ANTHOS database).

In both comparisons, 10 × 10 km UTM squares that contain only 1 × 1 km UTM squares from external sources were considered gaps in the CRF collection (Fig. 15.2) or gene bank representativeness (Fig. 15.3). However, some considerations should be kept in mind when carrying out a gap analysis. In the CRF case, collecting germplasm only in areas determined as gaps in CRF representativeness may give rise to duplication between CRF and AUS, at least from an ecogeographical point of view. To avoid duplication in *ex situ* conservation, it is important to include the greatest number of collection sites from the most important gene banks in the study. The results of the gap analysis for *L. hispanicus*, measured in percentage of characterized 10 × 10 km UTM squares, can be seen in Fig. 15.4. As high

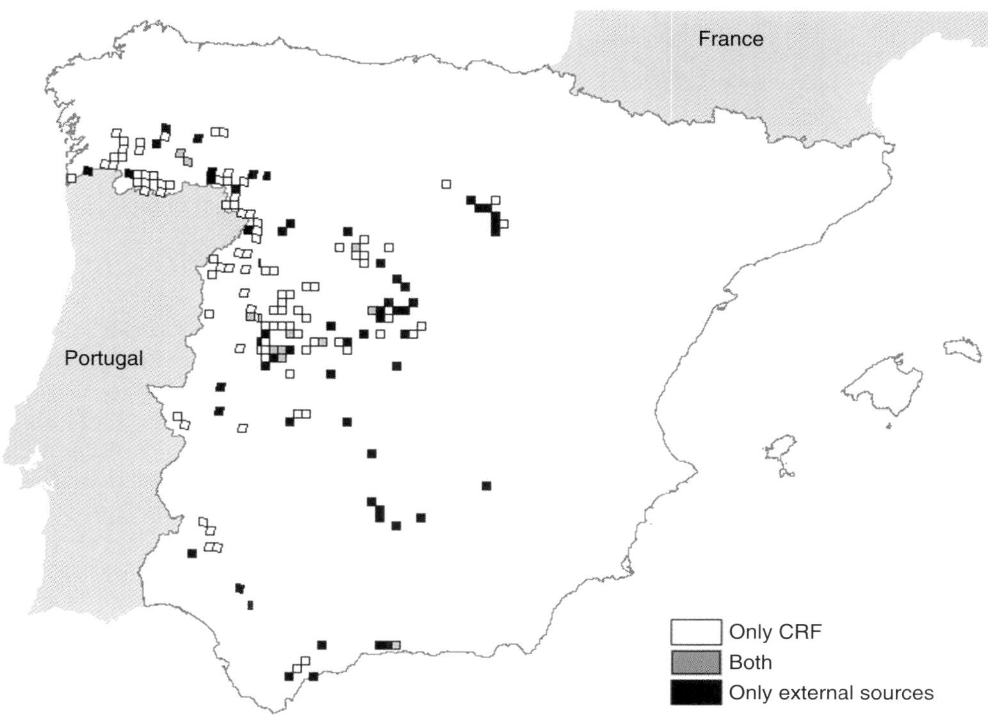

Fig. 15.2. 10 × 10 km UTM squares classified according to the presence of 1 × 1 km UTM squares within, representing *Lupinus hispanicus* accessions from CRF, presence data from external sources (herbarium + bibliographic databases) or both. Gaps in CRF gene bank representation are squares in black.

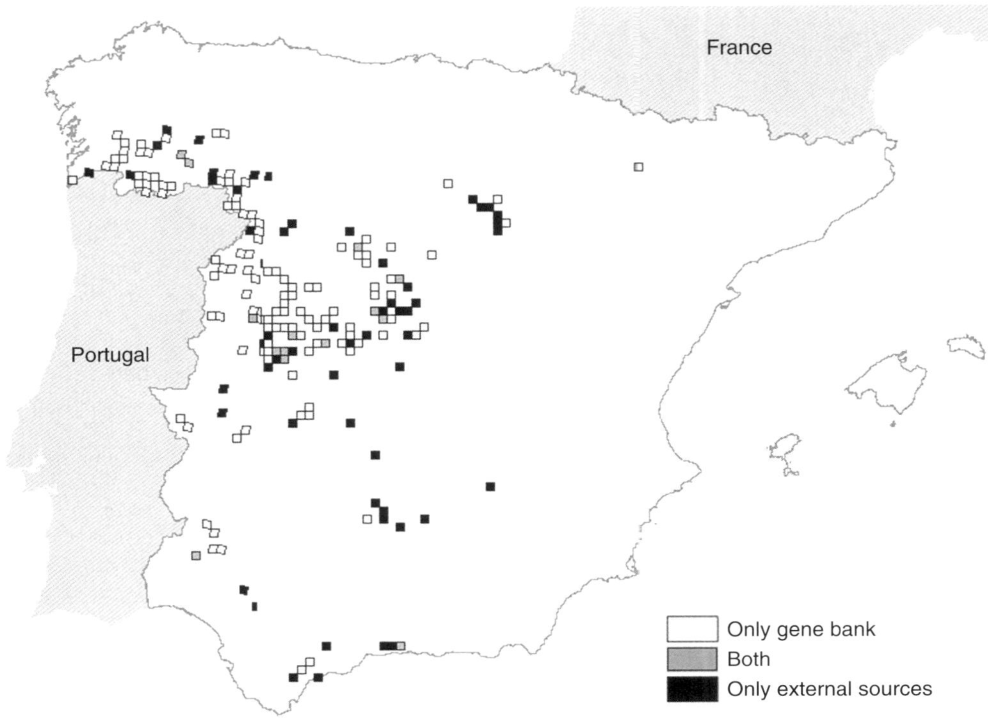

Fig. 15.3. 10 × 10 km UTM squares classified according to the presence of 1 × 1 km UTM squares within, representing *Lupinus hispanicus* accessions from gene banks (CRF + AUS), presence data from external sources (herbarium + bibliographic databases) or both. Gaps in gene bank representation are squares in black.

as 36% of the 10 × 10 UTM squares are gaps in *L. hispanicus* representation in the CRF collection, but this percentage decreases to 30% when AUS data are included. In conclusion, the comparison of herbaria and bibliographic records and gene bank data suggests that the CRF collection has great gaps in the representation of *L. hispanicus* known distribution. However, as the information on *Lupinus* populations occurrence compiled from herbarium and floristic databases is not 100% reliable, additional analyses including predictive species distribution models can help to improve the gap analysis and set more accurate priorities.

Ecogeographical characterization of a gene bank

Just as collections can be characterized at the morpho-agronomic, biochemical or DNA levels, among others, can they be characterized from an ecogeographical point of view. In this approach, environmental variables are compiled and some of them are selected as 'ecological descriptors' (Steiner and Greene, 1996). In a multivariate analysis, ecological descriptors work in the same way as morphological descriptors or data from molecular markers, considering their quantitative or qualitative nature. Different approaches differ in the methodology used to obtain ecological descriptors. These can be obtained directly from passport data (Hazekamp,

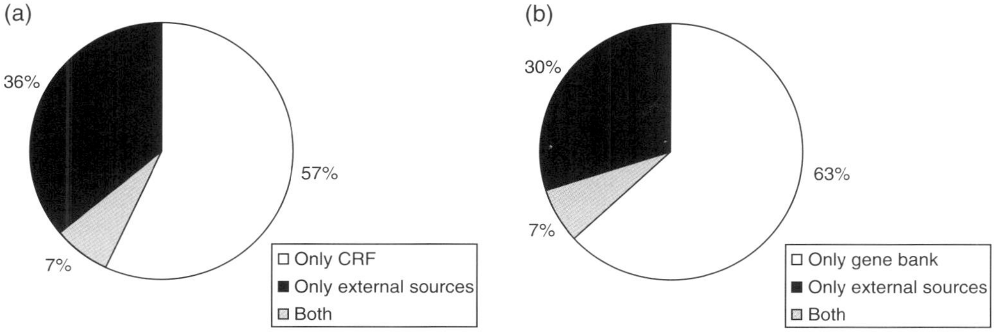

Fig. 15.4. Percentage of gaps (black) for *Lupinus hispanicus* in Spain: (a) gaps between CRF and external sources (herbarium + bibliographic databases) and (b) gaps between gene bank (CRF + AUS) and external sources.

2002) or indirectly by georeferencing collection sites (Guarino *et al.*, 2002). In the first case, ecogeographic information needs to be complete, homogeneous and objective, with regard to collection conditions. Given that these conditions are difficult to meet in the passport data of any gene bank, the best alternative is to extract ecogeographical information by superimposing collection sites on georeferenced environmental information, visualized as GIS layers. GIS allows overlap and the correct extraction of ecogeographic information. In consequence, the environmental information extracted is homogeneous for all accessions.

Ecogeographic characterization usually involves more than one variable. This leads to the use of multivariate analysis techniques. These techniques allow us to assess ecogeographic similarities or differences between the different operative taxonomic units (OTUs) (Sneath and Sokal, 1973), as well as to form similar ecogeographic groups and to determine which ecological descriptors represent the greatest proportion of ecogeographic variability. This analysis is usually visualized by means of dendrograms or biplots.

In the ecogeographic characterization of a gene bank, the spatial range of the analysis is the area that the accessions to be analysed occupy given that only collection site information is used. With regard to taxonomic resolution and multivariate analysis in interspecific studies, OTU corresponds to a species, and its value for each ecological descriptor is a trend statistic which summarizes the information of the corresponding populations of that species (see Ferguson *et al.*, 2005). In the case of intraspecific studies, OTU is the accession (population), and its value is directly extracted from the ecological descriptors. Assuming that in an interspecific study a species is represented by more than one population, both interspecific and intraspecific studies can be carried out simultaneously to obtain a more complete, integrated perspective.

With ecogeographic characterization at the interspecific level, a representative environment can be described for each species and ecogeographic affinities can be defined between species or groups of species (Bennet and Bullita, 2003). At the intraspecific level, ecogeographic characterization can determine the environments represented within the gene bank, the number of accessions which represent each environment and the most relevant ecological descriptors for the species.

Examples

A first approach to the ecogeographical characterization of a gene bank quite frequently found in literature is assigning a value to the OTU according to its collection site and previously determined environmental zones visualized in maps. The environmental zones are the product of combining environmental variables and anthropogenic categorizations (a more objective method of generating these maps is later described in 'using ecogeographical land characterization'). In this way, the accession or the species have one or very few ecogeographic information values and multivariate analysis cannot be applied. The objective of this type of study is the constitution of core collections with high representativeness. This ecogeographic information is used to classify the collection into groups of accessions collected in the same environmental zones. Additional groups can then be determined according to data from morpho-agronomic, biochemical or molecular characterizations. Examples of this type of work are the constitution of nuclear collections in sorghum (Grenier *et al.*, 2001), sesame (Xiurong *et al.*, 2000), barley (Igartua *et al.*, 1998), rice (Yawen *et al.*, 2003) or sweet potato (Huamán *et al.*, 1999).

More detailed studies extract ecological descriptors from collecting sites through GIS previously fed with environmental variables. Using a distance coefficient, they obtain matrices of ecogeographic distances. The use of this ecogeographic distance matrix is varied. In some cases, it has been used for ER studies or simply to calculate its correlation with other types of characterizations. An example of the ecogeographic characterization of a gene bank using this methodology and GIS is the study of wild *Arachis* at the interspecific level by Ferguson *et al.* (2005). Bennett and Bullitta (2003) carried out the ecogeographic characterization of six *Trifolium* species, in which they classified the accessions into ecogeographic groups and determined the most important ecological descriptors for each group.

LUPINUS SPECIES IN SPAIN: A CASE STUDY. We carried out the ecogeographic characterization of the collection of *Lupinus* accessions collected in Spain and conserved in the CRF. Here, we considered the interspecific and intraspecific level. Data from five *Lupinus* species were used: three CWR species in Spain (*L. angustifolius*, *L. hispanicus* and *L. micranthus*), one cultivated species with wild forms (*L. luteus*) and one cultivated species with wild and naturalized forms (*L. albus*). *L. angustifolius* is an already domesticated species, but only wild forms grow in Spain. *L. hispanicus* is endemic to the Iberian Peninsula and *L. luteus* has its centre of origin and diversity in the same area. Collection sites (longitude and latitude) were extracted from the passport data and used to create a point distribution map for each species. The coordinate points were then revised in search of errors as described by Hijmans *et al.* (1999). As many of the CRF collection sites have also been reported by other sources (herbaria, botanists and other gene bank collections), all available collection site information was contrasted and cases of divergence were registered for potential use in the interpretation of the ecogeographic characterization. The number of CRF records for each species was *L. albus* (248), *L. angustifolius* (489), *L. hispanicus* (175), *L. luteus* (104) and *L. micranthus* (5).

The point maps and the environmental variables were then transformed into the same coordinate system (UTM, European Datum 1950 zone 30 north) and converted into vectorial ARCGIS 9.0 compatible format. The point maps were overlapped on 43 environmental variables and 27 ecological descriptors were extracted. The selection of ecological descriptors was based on *a priori* knowledge about their biological meaning for *Lupinus* species. We also tried to ensure equilibrium among the different types of environmental information. The selected ecological descriptors (Table 15.1) were standardized for subsequent analysis. Tables with accessions and the corresponding ecological descriptors were created. Ecogeographical distances between all pairs of accessions were obtained using the Gower similarity coefficient (Sneath and Sokal, 1973). A cluster analysis was then carried out using distance matrix and unweighted pair grouping method of agglomeration (UPGMA) method. The results were dendrograms that represent ecogeographical similarities between accessions. The ecogeographical groups represented in the gene bank were then visualized and each group could be described based on ranges of original ecological descriptors. As the 27 ecological descriptors provided a great amount of information, which might be difficult to analyse, we carried out a principal component analysis (PCA) to reduce the original descriptors to two or three synthetic variables and simplify the analysis. The new synthetic variables were linear combinations of ecological descriptors where some descriptors have more weight than others.

Table 15.1. Ecological descriptors used in the ecogeographical characterization of *Lupinus luteus* collection from CRF gene bank.

Type	Ecological descriptors	Units	Source
Climatic	Temperature: October, December and mean minimal annual	°C	Sánchez Palomares *et al.* (1999)
	Rainfall: April, July and mean annual	mm	Sánchez Palomares *et al.* (1999)
	Evapotranspiration	mm	Tragsatec – Spanish Ministry of Agriculture
	Thermic amplitude	°C	Based on Sánchez Palomares *et al.* (1999)
Climatic/ temporal	Dry, cold and warm period	months	Tragsatec – Spanish Ministry of Agriculture
Bioclimatic indices	Emberger		Based on Sánchez Palomares *et al.* (1999)
	Gorczynski		Based on Sánchez Palomares *et al.* (1999)
	Dantin–Revenga		Based on Sánchez Palomares *et al.* (1999)
Geophysical	Altitude	m	Draper *et al.* (2003)
	Aspect		Draper *et al.* (2003)
	Slope		Draper *et al.* (2003)
	Longitude		Draper *et al.* (2003)
	Latitude		Draper *et al.* (2003)
Edaphic	Soil type (USDA classification)	8 classes	Instituto Geográfico Nacional (1992)

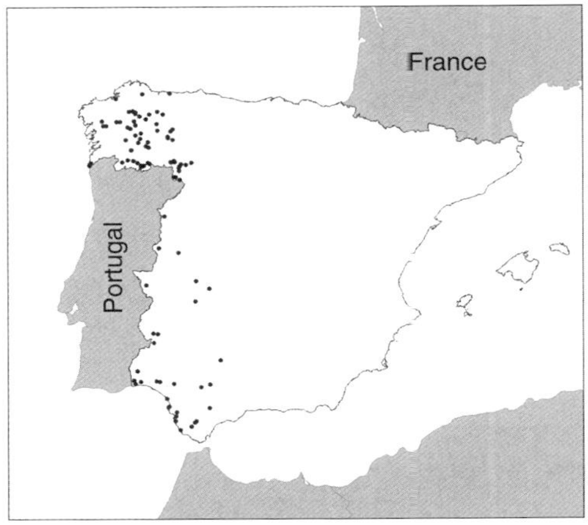

Fig. 15.5. Spatial distribution of *Lupinus luteus* accessions from CRF.

To make an interpretation we can describe them in terms of the ecological descriptors with the greatest weight in each of the principal components.

Figure 15.5 shows the case of *L. luteus*, which has an intermediate number of accessions (104). For this species we considered the status of the accessions, in other words, if the accessions were collected from cultivated, weed or wild populations. Dendrograms from a cluster analysis illustrate the ecogeographical relationship of *L. luteus* (Figs 15.6 and 15.7). This analysis identified four ecogeographical groups (EG).

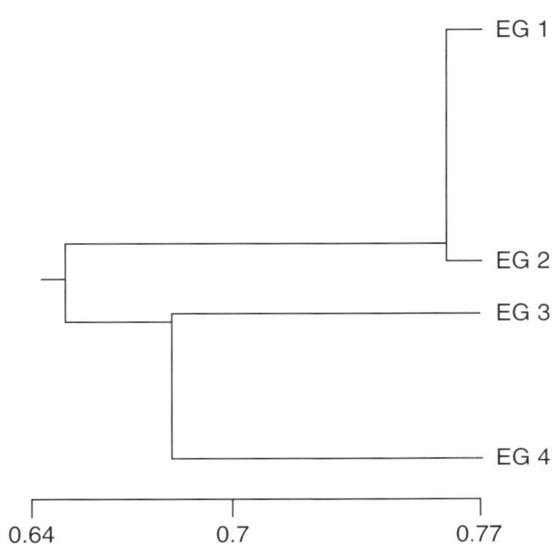

Fig. 15.6. General ecogeographical grouping dendrogram of *L. luteus* accessions. Four ecogeographical groups (EG) were detected within the CRF collection (EG1–EG4).

The PCA extracted three new synthetic variables that represent 64% of the total variance and correspond to the three first principal components. When we analysed the eigenvalues, we detected that the first new variable is related to thermopluviometric factors, the second to temperature factors and the third to edaphic factors. Then, we characterized each EG from the cluster analysis using the new synthetic variables (Fig. 15.8). For this characterization, we also used the ecological descriptors with the greatest weight for the three first principal components (Fig. 15.9). Finally, we assigned each accession its value from the corresponding EG and visualized them in a map (Fig. 15.10).

This ecogeographical characterization provided us with valuable knowledge on the conserved germplasm of *L. luteus* and its environmental variability. It is very interesting to note how ecological descriptors can distinguish between wild and cultivated accessions, classifying them into different EG. This study allowed

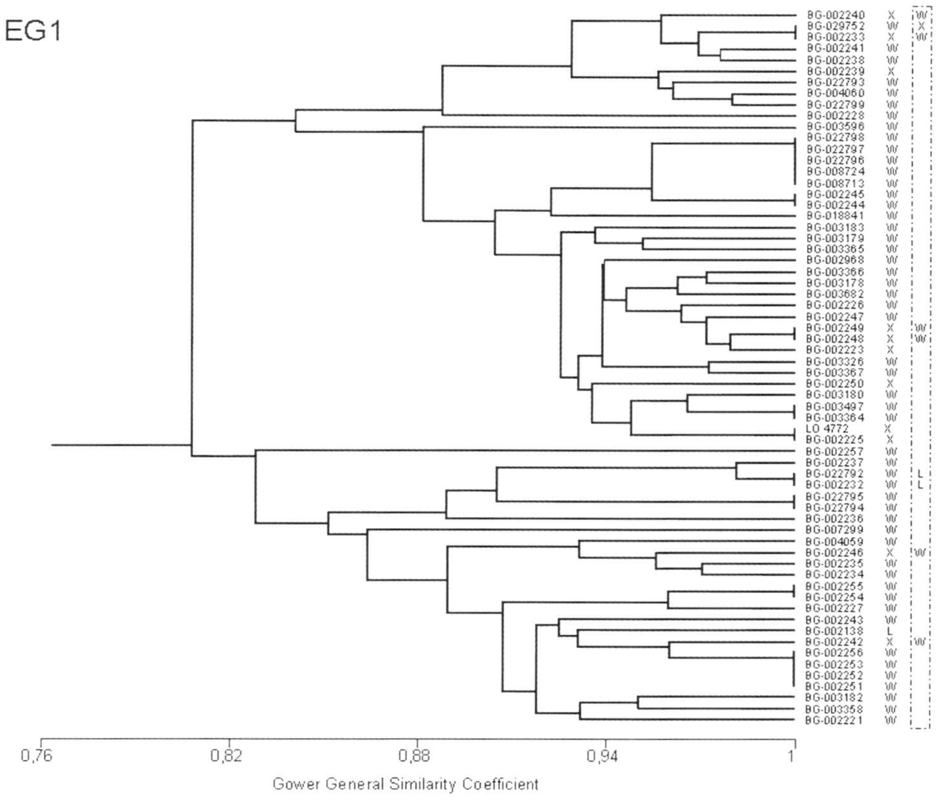

Fig. 15.7. Detailed dendrograms for each ecogeographical group (EG) derived from the general dendrogram (Fig. 15.6). On the right side of each CRF accession code, there are two columns with information about the population status (X = weed, W = wild and L = landrace). The first column corresponds to CRF passport status and the second (dotted line square) to other external sources collected at the same site but have recorded a different status.

(*continued*)

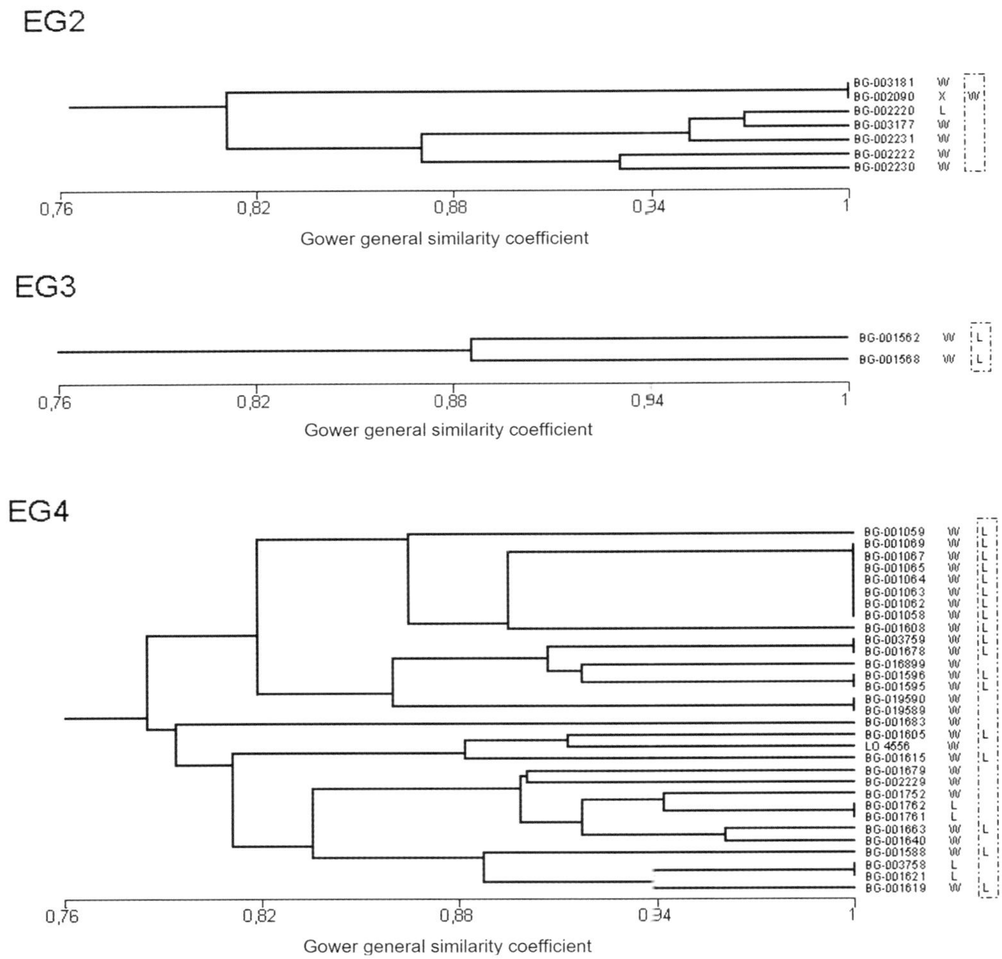

Fig. 15.7. *Continued*

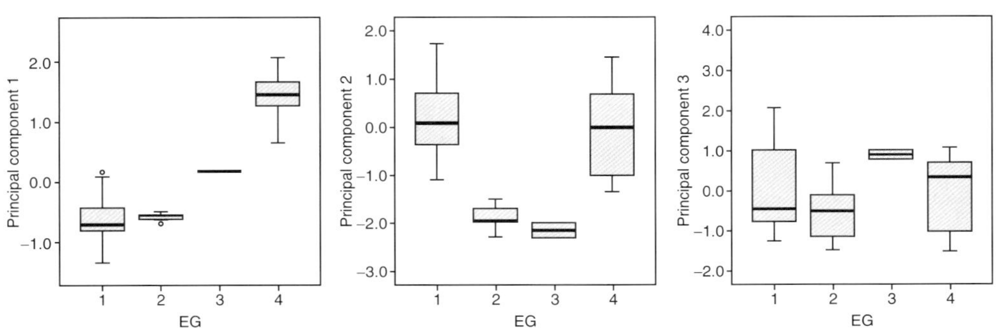

Fig. 15.8. Differences between ecogeographical groups (EG considering ecological descriptors with the greatest weight in the first three principal components.

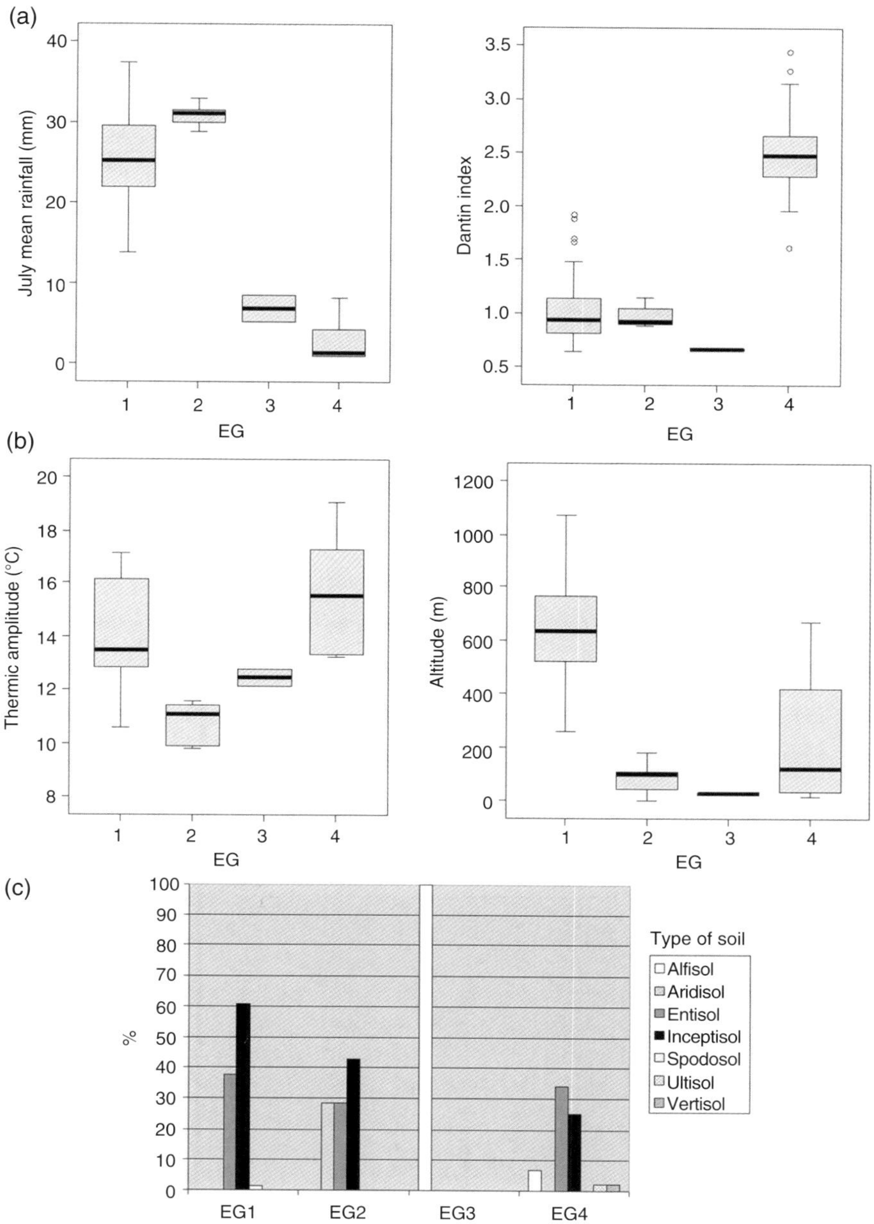

Fig. 15.9. Differences between ecological groups (EG) considering the ecological descriptors that are more correlated to the first three principal components: (a) first principal component: July mean rainfall (eigenvalue = –0.956) and Dantin index (eigenvalue = 0.951); (b) second principal component: thermic amplitude (eigenvalue = –0.835) and altitude (eigenvalue = 0.718); (c) third principal component: inceptisol end entisol (from USDA soil type classification) (eigenvalue = –0.869 and 0.840, respectively) (also showing soil types for each EG).

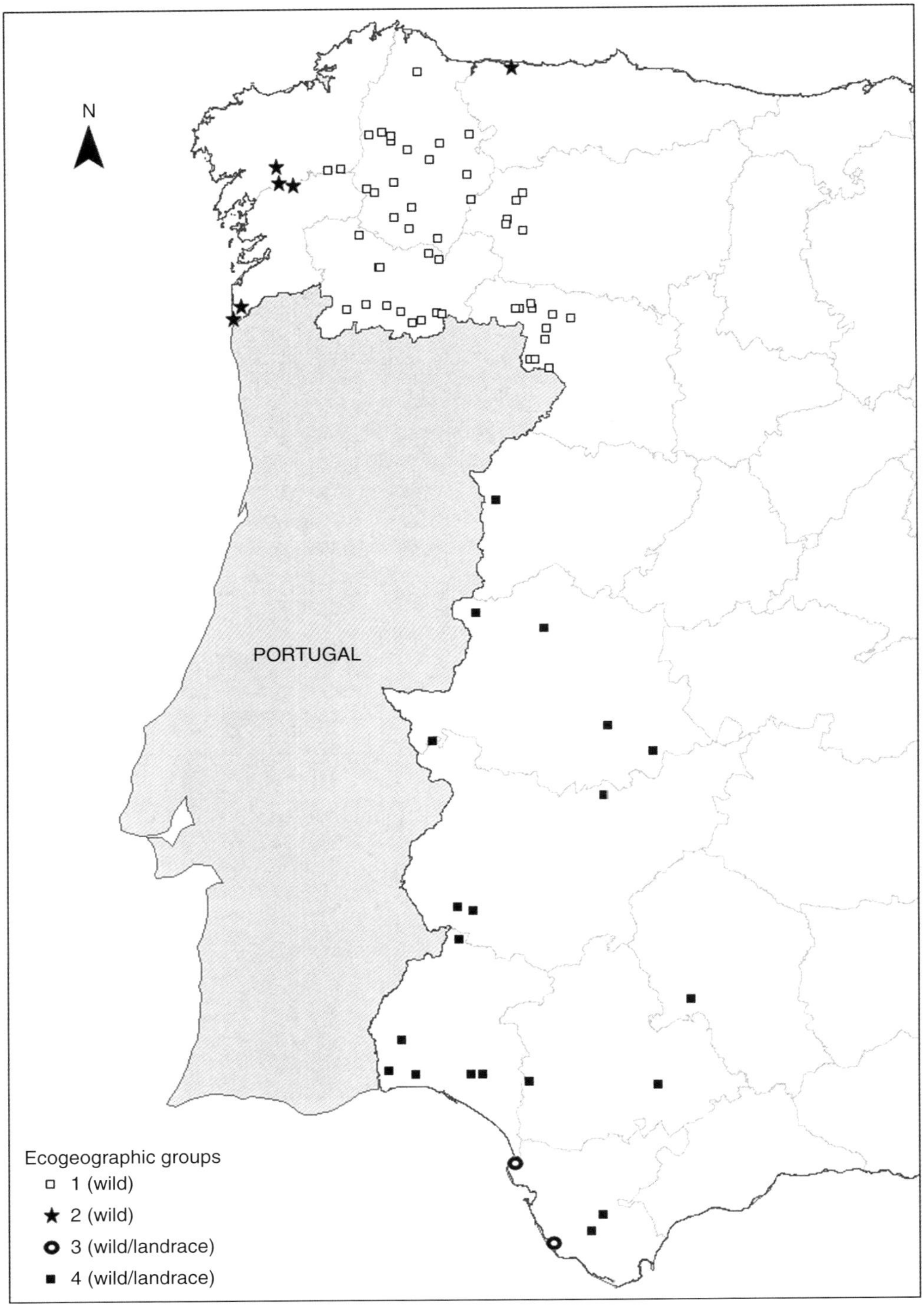

Fig. 15.10. Spatial distribution of *Lupinus luteus* accessions from CRF, categorizing them into the four ecogeographical groups (EG).

us to explore the potential of ecogeographical characterization and its degree of definition. On the other hand, EGs from cluster analyses are useful in identifying the different environments represented in a gene bank and detecting deficiencies. According to our results, it is advisable to improve the representation of EG 3 by collecting *L. luteus* for CRF in the south-east of Andalucía (Spain). Before this analysis, knowledge on the ecogeographical distribution and environmental adaptation of *L. luteus* was scarce and/or subjective. As a further step, studies including the whole Iberian Peninsula (Spain and Portugal) would be desirable.

Using ecogeographical land characterization

Another alternative for assessing ER in a gene bank is through ecogeographical land characterization. In this method, the land areas where the gene bank has collected germplasm are characterized based on ecological or environmental variables. The characterization product is a map of ecogeographical zones. This map must be as objective and accurate as possible. Point maps of collecting sites are then superimposed on the map of ecogeographical zones and each accession is assigned a value according to the ecogeographical zone where it was collected. In the section 'ecogeographical characterization of a gene bank' we cited some examples of studies that used maps of environmental zones which had a poor resolution and/or were highly subjective. Most ecogeographical studies for establishing core collections use this kind of map, although there are some exceptions like the establishment of cassava core collections in Brazil (Lobo Burle *et al.*, 2003). In this work, they used a previously developed map of ecogeographical regions made especially for cassava and other climatic, vegetation and agroecological maps. These maps were combined through GIS software to obtain smaller, more detailed ecogeographical units. The combination of biotic (i.e. type of vegetation) and abiotic (i.e. climatic) factors may make results more difficult to interpret. While species occurrence may or may not be explained by our knowledge about the occurrence of other species (represented by available biotic data), species occurrence can be related to abiotic conditions more easily. Thus, it is advisable to carry out studies including only abiotic variables as a first step in assessing ER through ecogeographical land characterization.

LUPINUS CASE STUDY. Ecogeographical land characterization was carried out to obtain a map of ecogeographical zones for Peninsular Spain and the Balearic Islands to assess the ER of the CRF *Lupinus* collection. We superimposed the map of *Lupinus* collection sites on the map of ecogeographical zones and then assigned an ecogeographical value to each accession.

To generate the map of ecogeographical zones we used methodologies based on GIS and clustering procedures. First, we divided the spatial range (Peninsular Spain and the Balearic Islands) into 500,949 1×1 km UTM squares. GIS software (ARCGIS 9.0) extracted 33 ecological descriptors for each square using previously gathered environmental data layers. The descriptors were then classified into three main groups: geophysic, climatic and edaphic descriptors (Table 15.2). We carried out a two-step cluster analysis for each group of descriptors (SPSS, 2003). Two-step cluster analysis allowed us to:

Table 15.2. Groups of ecological descriptors used to characterize Peninsular Spain and Balearic Islands.

Groups of descriptors	Number of descriptors	Descriptors involved
Climatic	28	Mean monthly rainfall and temperatures, mean annual rainfall and temperature, mean annual minimum and maximum temperatures
Soil/edaphic	2	USDA soil classification (nine orders) and geology (46 types of mineral predominant materials)
Geophysical	3	Altitude, slope and aspect

- Carry out the clustering of very large data sets.
- Work with continuous and categorical variables at the same time.
- Define the resulting clusters using Bayesian information criterion (BIC). This implies that the researcher does not influence the clustering or the final number of ecogeographical clusters.

This clustering procedure, as indicated by its name, has two steps: (i) to pre-cluster the data into many small subclusters; and (ii) to cluster the subclusters resulting from step 1 into a BIC-determined number of clusters. In our case, the two-step clustering analysis automatically produced three ecogeographical clusters for each group of descriptors (geophysic, climatic and edaphic). Unique combinations of each ecogeographical cluster (33) produced 27 ecogeographical categories (Fig. 15.11). Additionally, descriptive statistics for each ecogeographical cluster were obtained allowing us to quantitatively describe the environments represented by each ecogeographical category. Finally, each of the 500,949 1 × 1 km UTM squares was assigned its corresponding ecogeographical category. In the map of the categorized 1 × 1 km UTM squares (Fig. 15.12), the ecogeographical zones, defined as adjacent squares with the same ecogeographical category, can be visualized. The resulting map of ecogeographical zones for Peninsular Spain and the Balearic Islands showed homogeneous, small and discontinuous environmental areas.

We then superimposed the map of points corresponding to collection sites (gene banks) or presence data (external sources) of *Lupinus* on the map of ecogeographical zones from the two-step cluster analysis. We will illustrate this methodology using the *L. angustifolius* collection from CRF as the target of the ER study. In addition to the CRF passport data for *L. angustifolius* collection sites (489 georeferenced records), we also used collection sites from AUS (447 records for Peninsular Spain) and other external sources (137 records from MA herbarium and ANTHOS database). The GIS software assigned each collecting site or presence datum an ecogeographical zone depending on its location on the map. Finally, we obtained the frequency and percentage of each of the 27 ecogeographical zones represented in the gene banks (CRF and AUS) and external sources (MA and ANTHOS) (Fig. 15.13). Individual and comparative analyses of these frequencies and percentages allow us to assess the ER of the CRF gene bank.

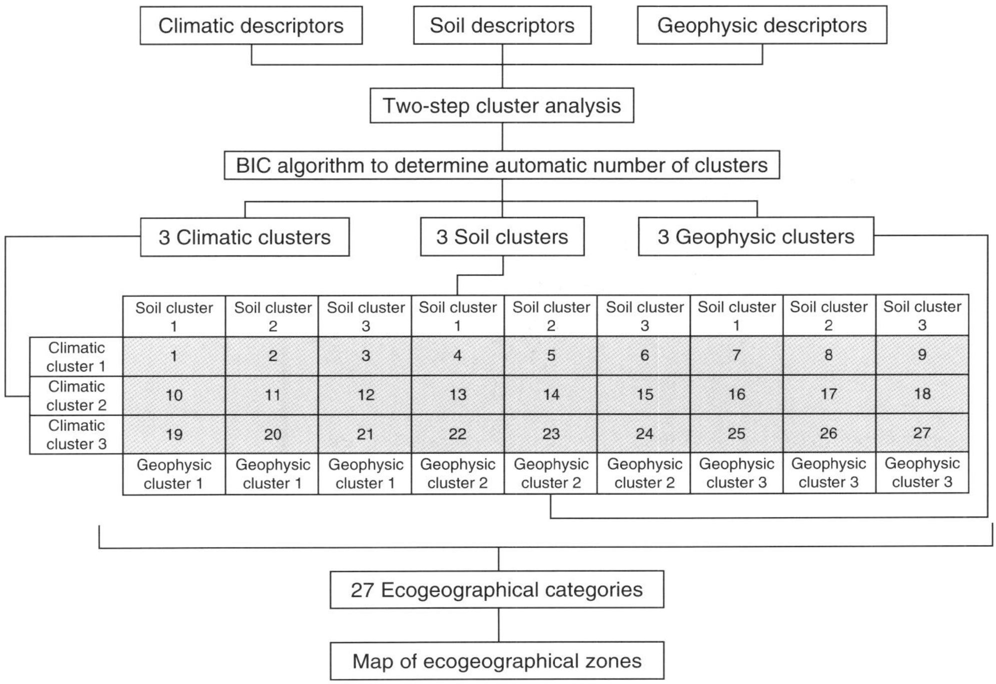

Fig. 15.11. Methodology to obtain the map of ecogeographical zones for Spain based on 27 ecogeographical categories and two-step cluster analysis.

Our analysis showed that 24 of the 27 ecogeographical categories are represented in the CRF collection. Within the 24 categories represented, five might be considered 'rare' (1, 2, 3, 11 and 19) due to their low representativeness (less than 1%). In contrast, three categories (4, 5 and 8) are highly represented, with each category representing more than 10% of the total number of *L. angustifolius* CRF accessions.

The occurrence of different percentages in the representation of ecogeographical categories in a gene bank can be explained from several perspectives. In the case of a species that had no preference for any particular environment, a gene bank with a good representativeness should have similar representation percentages for all ecogeographical categories. However, as this situation does not occur in nature, the most likely result is that one category or group of categories will be better represented than others, as in the case of *L. angustifolius*. Another reason that supports heterogeneity in the representation of ecogeographical categories is the fact that each category has different frequencies of occurrence in the framework. So, categories with a low frequency of occurrence in nature have a lower probability of being visited or having germplasm collected there, but may enclose environmentally distinct characteristics extremely interesting in gene bank collecting (Draper *et al.*, 2003). The lower probability of being visited can be minimized by weighting each category according to the total area that it covers. Biases in

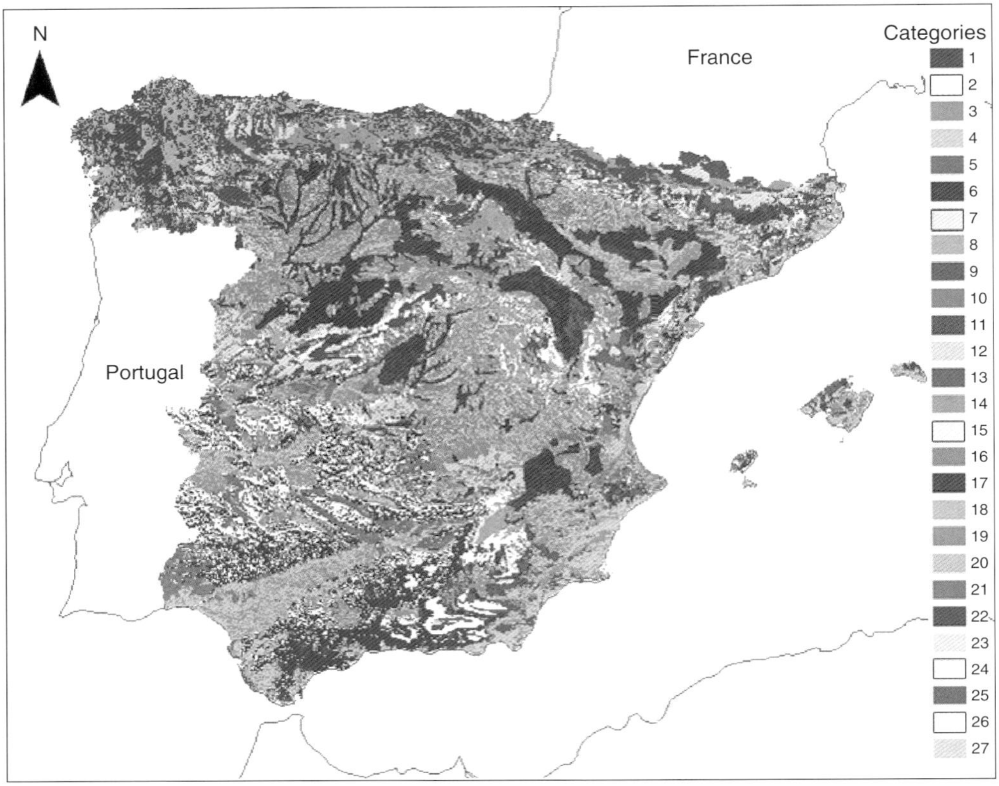

Fig. 15.12. Map of ecogeographical zones based on 27 ecogeographical categories for Peninsular and Balearic Spain. Categories are represented in different grey tones. The zones are 1 × 1 km UTM squares adjacent to the same category value.

collecting germplasm (Hijmans *et al.*, 2000) can also be an important factor in increasing the representation of certain categories while reducing that of others. We can try to identify the effect of biases by comparing the representation of ecogeographical categories in our gene bank to that of other gene banks (which may have similar biases) and external sources such as herbaria and botanic databases (which may have different biases due to different objectives (see 'comparing external sources and gene bank data'). For this reason, we compared the representation of ecogeographical categories in the CRF collection to AUS and external sources (Fig. 15.13). This comparison detected some categories which are not represented in the CRF collection, but which are represented in AUS (10, 12 and 18) or external sources (10 and 18). There are also categories (2, 11, 15 and 19) with a very low representation in CRF, but with a high representation in AUS and external sources. These results show that the ER of the CRF collection for *L. angustifolius* can be improved if new collecting trips are carried out in ecogeographical categories 2, 10, 11, 12, 15 and 18. It would also be advisable to collect *L. angustifo-*

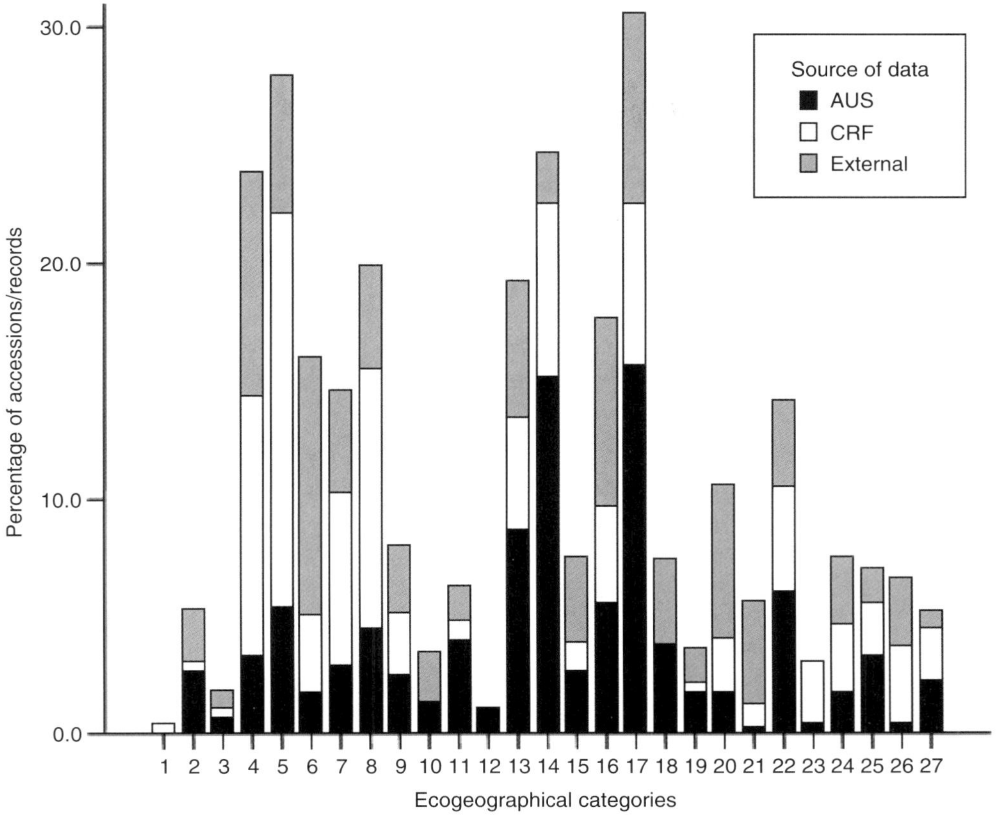

Fig. 15.13. Frequency of the 27 ecogeographical categories represented on *Lupinus angustifolius* collections from CRF and AUS gene banks and external sources (ANTHOS and MA).

lius for CRF in the ecogeographical zones (with the values of the mentioned categories) where AUS has not collected to avoid duplication between gene banks.

15.4 Conclusion

The use of an ecogeographical or environmental approach in the characterization of PGR is more than just a cheap alternative to genetic characterization. Here, our aim is to show how ER studies can be as detailed, consistent and accurate as we wish, thanks to the development of spatial data, geostatistics, multivariate analysis and GIS-based methodologies.

ER studies in gene banks are still scarce, while several studies on GR have been carried out since the molecular marker boom in the 1990s. A bibliographic search for works that involve ecogeographical aspects in characterization of PGRs in gene bank collections reveals that, save a few exceptions, curators and scientists use this alternative only to explain their genetic results and many times only geographical distances are used.

According to our results, ecogeographical land characterization offers a different, complementary perspective with respect to the ecogeographical characterization of a gene bank. The main differences between these two approaches, in addition to the target of the characterization itself, are:

- Computational capacity – land characterization often requires special skills needed to create the map of ecogeographical zones.
- Interpretation of results – gene bank characterization requires more knowledge about the species.
- Multispecies analyses – once the map of ecogeographical zones has been obtained, working with many species is very fast and inexpensive. Gene bank characterization takes considerable time for each species.

On the other hand, comparing external sources and gene bank data through gap analysis is another interesting alternative for assessing ER. This method is based on the differences between the sources in terms of objectives, *modus operandi* and biases. Its results can be directly used to establish collecting objectives, as this method detects spatial gaps in the gene bank and indicates the exact location of sites with a high probability of species occurrence. In this case, we used the comparison to evaluate gene bank representation and point out priorities for improving ER. We think this analysis can be used as a first step before a subsequent ecogeographical characterization of a gene bank collection or ecogeographical land characterization because it is simpler and more intuitive. Only a basic knowledge of GIS and low computational capacity is necessary to carry out the comparison.

In summary, the three alternatives combined can offer us a highly accurate ER for any gene bank and any species. However, if the target species is a CWR, the potential of ER studies is even greater.

We believe that the importance of environmental features of the collecting sites is currently underestimated. This information is very useful in the selection of accessions for plant breeding and should be considered just as important as other types of information such as morphological or molecular data. In consequence, ecogeographical studies should be a routine procedure in every gene bank. The economic cost of ER studies will decrease with time, thanks to the greater availability of software and georeferenced environmental information. In developing countries, the greatest obstacle to carrying out ER studies is often the lack of qualified personnel. Further efforts in this direction should be a priority for NGOs, international organizations and governments in order to promote the conservation of PGRs.

Acknowledgements

We would like to thank Lori De Hond for her linguistic assistance and MA herbarium and CRF-INIA personnel for facilitating our work. Mauricio Parra-Quijano was supported by BSCH-UPM fellowship programme.

References

Auricht, G.C., Reid, R. and Guarino, L. (1995) Published information on the natural and human environment. In: Guarino, L., Ramanatha Rao, V. and Reid, R. (eds) *Collecting Plant Genetic Diversity. Technical Guidelines*. CAB international, Wallingford, UK, pp. 131–151.

Baek, H.J., Beharav, A. and Nevo, E. (2003) Ecological-genomic diversity of microsatellites in wild barley, *Hordeum spontaneum*, populations in Jordan. *Theoretical and Applied Genetics* 106, 397–410.

Bennett, S.J. and Bullitta, S. (2003) Ecogeographical analysis of the distribution of six *Trifolium* species in Sardinia. *Biodiversity and Conservation* 12, 1455–1466.

Berger, J., Abbo, S. and Turner, N.C. (2003) Ecogeography of annual wild *Cicer* species: the poor state of the world collection. *Crop Science* 43, 1076–1090.

Brown, A.H.D., Brubaker, C.L. and Grace, J.P. (1997) Regeneration of germplasm samples: wild versus cultivated plant species. *Crop Science* 37, 7–13.

Dantin, J. and Revenga, A. (1940) Una nueva relación climatológica: El índice termopluviométrico. Avance al estudio de la aridez en España. Asociación Española para el Progreso de las Ciencias. Congreso de Zaragoza.

Del Rio, A.H., Bamberg, J.B., Huamán, Z., Salas, A. and Vega, S.E. (2001) Association of ecogeographical variables and RAPD marker variation in wild potato populations of the USA. *Crop Science* 41, 870–878.

Draper, D., Roselló-Graell, A., Garcia, C., Tauleigne, C. and Sérgio, C. (2003) Application of GIS in plant conservation programmes in Portugal. *Biological Conservation* 113, 337–349.

Dulloo, M.E., Maxted, N., Guarino, L., Florens, D., Newbury, H.J. and Ford-Lloyd, B.V. (1999) Ecogeographic survey of the genus *Coffea* in the Mascarene islands. *Botanical Journal of the Linnean Society* 131, 263–284.

Emberger, L. (1932) Sur une formule climatique et ses applications en botanique. *La Météorologie* 423–432.

FAO (1996) *Global Plan of Action for the Conservation and Sustainable Utilization of Plant Genetic Resources for Food and Agriculture*. FAO, Rome, Italy.

Ferguson, M.E., Jarvis, A., Stalker, H.T., Williams, D.E., Guarino, L., Valls, J.F.M., Pittman, R.N., Simpson, C.E. and Bramel, P.J. (2005) Biogeography of wild *Arachis* (*Leguminosae*): distribution and environmental characterisation. *Biodiversity and Conservation* 14, 1777–1798.

Gorczynsky, L. (1920) Sur le calcul du degré du continentalisme et son application dans la climatologie. *Geografiska Annaler* 2, 324–331.

Greene, S.L. and Hart, T.C. (1999) Implementing a geographic analysis in germplasm conservation. In: Greene, S.L. and Guarino, L. (eds) *Linking Genetic Resources and Geography: Emerging Strategies for Conserving Crop Biodiversity*. ASA and CSSA, Madison, Wisconsin, pp. 25–38.

Greene, S.L., Hart, T.C. and Afonin, A. (1999a) Using geographical information to acquire wild crop germplasm for *ex situ* collections: I. Map development and field use. *Crop Science* 39, 836–842.

Greene, S.L., Hart, T.C. and Afonin, A. (1999b) Using geographical information to acquire wild crop germplasm for *ex situ* collections: II. Post-collection analysis. *Crop Science* 39, 843–849.

Grenier, C., Bramel-Cox, P.J. and Hamon, P. (2001) Core collection of *Sorghum*: I. Stratification based on eco-geographical data. *Crop Science* 41, 234–240.

Guarino, L. (1995) Mapping the ecogeographic distribution of biodiversity. In: Guarino, L., Ramanatha Rao, V. and Reid, R. (eds) *Collecting Plant Genetic Diversity. Technical Guidelines*. CAB international, Wallingford, UK, pp. 287–327.

Guarino, L., Jarvis, A., Hijmans, R. and Maxted, N. (2002) Geographic Information Systems (GIS) and the conservation and use of plant genetic resources. In: Engels, J.M.M., Ramanatha Rao, V., Brown, A.H.D. and Jackson, M.T. (eds) *Managing Plant Genetic Resources*. CAB International, Wallingford, UK, pp. 387–404.

Guisan, A. and Zimmermann, N.E. (2000) Predictive habitat distribution models in ecology. *Ecological Modelling* 135, 147–186.

Gupta, P.K., Sharma, P.K., Balyan, H.S., Roy, J.K., Sharma, S., Beharav, A. and Nevo, E. (2002) Polymorphism at rDNA loci in barley and its relation with climatic variables. *Theoretical and Applied Genetics* 104, 473–481.

Hart, T.C. (1999) Scale considerations in mapping for germplasm acquisition and the assessment of *ex situ* collections. In: Greene, S.L. and Guarino, L. (eds) *Linking Genetic Resources and Geography: Emerging Strategies for Conserving Crop Biodiversity*. ASA and CSSA, Madison, Wisconsin, pp. 51–61.

Hazekamp, Th. (2002) The potential role of passport data in the conservation and use of plant genetic resources. In: Engels, J.M.M., Ramanatha Rao, V., Brown, A.H.D. and Jackson, M.T. (eds) *Managing Plant Genetic Diversity*. CAB international, Wallingford, UK, pp. 185–194.

Hijmans, R.J. and Spooner, D.M. (2001) Geographic distribution of wild potato species. *American Journal of Botany* 88(11), 2101–2112.

Hijmans, R.J., Schreuder, M., De la Cruz, J. and Guarino, L. (1999) Using GIS to check co-ordinates of gene bank accessions. *Genetic Resources and Crop Evolution* 46, 291–296.

Hijmans, R.J., Garret, K.A., Huamán, Z., Zhang, D., Schreuder, M. and Bonierbale, M.W. (2000) Assessing the geographic representativeness of gene bank collections: the case of Bolivian wild potatoes. *Conservation Biology* 14, 1755–1765.

Huamán, Z., Aguilar, C. and Ortiz, R. (1999) Selecting a Peruvian sweetpotato core collection on the basis of morphological, eco-geographical, and disease and pest reaction data. *Theoretical and Applied Genetics* 98, 840–845.

Igartua, E., Gracia, M.P., Lasa, J.M., Medina, B., Molina-Cano, J.L., Montoya, J.L. and Romagosa, I. (1998) The Spanish barley core collection *Genetic Resources and Crop Evolution* 45, 475–481.

Instituto Geográfico Nacional (1992) Atlas Nacional de España, sección II, grupo7, edafología, Madrid, Spain.

Jarvis, A., Ferguson, M.E., Williams, D.E., Guarino, L., Jones, P.G., Stalker, H.T., Valls, J.F.M., Pittman, R.N., Simpson, C.E. and Bramel, P. (2003) Biogeography of wild *Arachis*: assessing conservation status and setting future priorities. *Crop Science* 43, 1100–1108.

Lobo Burle, M., Torres Cordeiro, C.M., Fonseca, J.R., Palhares de Melo, M., Neves Alves, R. and Abadie,T. (2003) Characterization of germplasm according to environmental conditions at the collecting site using GIS – two case studies from Brazil. *Plant Genetic Resources Newsletter* 135, 1–11.

Murphey, P.C., Guralnick, R.P., Glaubitz, R., Neufeld, D. and Ryan, J. A. (2004) Georeferencing of museum collections: a review of problems and automated tools, and the methodology developed by the Mountain and Plains Spatio-Temporal Database-Informatics Initiative (Mapstedi). *PhyloInformatics* 3, 1–29.

Noy-Meir, I., Anikster, Y., Waldman, M. and Ashri, A. (1989) Population dynamics research for *in situ* conservation: wild wheat in Israel. *Plant Genetic Resources Newsletter* 75/76, 9–11.

Sánchez Palomares, O., Sánchez Serrano, F. and Carretero Carretero, M.P. (1999) *Modelos y cartografía de estimaciones climáticas para la España peninsular*. Instituto Nacional de Investigación y Tecnología Agraria y Alimentaria. Ministerio de Agricultura, Pesca y Alimentación, Madrid, Spain.

Sneath, P.H.A. and Sokal, R.R. (1973) *Numerical Taxonomy. The Principles and Practice of Numerical Classification*. W.H. Freeman, San Francisco, California.

SPSS (2003) *SPSS Base 12.0 User's Guide*. SPSS, Chicago, Illinois.

Steiner, J.J. and Greene, S.L. (1996) Proposed use of ecological descriptors and their utility for plant germplasm collections. *Crop Science* 36, 439–451.

Tanksley, S.D. and McCouch, S.R. (1997) Seed banks and molecular maps: unlocking genetic potential from the wild. *Science* 277, 1063–1066.

Tewksbury, J.J., Nabhan, G.P., Norman, D., Suzán, H., Tuxill, J. and Donovan, J. (1999) *In situ* conservation of wild Chiles and their biotic associates. *Conservation Biology* 13, 98–107.

Wiecorzek, J.W., Guo, Q. and Hijmans, R.J. (2004) The point-radius method for georeferencing locality and calculating associated uncertainty. *International Journal of Geographical Information Science* 18, 745–767.

Xiurong, Z., Yingzhong, Z., Yong, C., Xiangyun, F., Qingyuan, G., Mingde, Z. and Hodgkin, T. (2000) Establishment of sesame germplasm core collection in China. *Genetic Resources and Crop Evolution* 47, 273–279.

Yawen, Z., Shiquan, S., Zichao, L., Zhongyi, Y., Xiangkun, W., Honglian, Z. and Guosong, W. (2003) Ecogeographic and genetic diversity based on morphological characters of indigenous rice (*Oryza sativa* L.) in Yunnan, China. *Genetic Resources and Crop Evolution* 50, 567–577.

IV Genetic Erosion and Genetic Pollution

16 Genetic Erosion and Genetic Pollution of Crop Wild Relatives: the PGR Forum Perspective and Achievements

E. BETTENCOURT, B.V. FORD-LLOYD AND S. DIAS

16.1 Introduction

Genetic erosion and genetic pollution affect not only crop wild relatives (CWR), but crops as well. Although many of the causes and effects of genetic erosion have been known to scientists for sometime, the perception of genetic pollution is a more recent concept and cause of concern.

The project 'European Crop Wild Relative Diversity Assessment and Conservation Forum' (PGR Forum) elected, as one of the thematic backbones of the project, to discuss how to assess genetic erosion and genetic pollution from conventionally and biotechnologically bred crops for European CWR populations. This was the subject of a thematic workshop held with the objective to develop methodologies for the assessment and prediction of genetic erosion and genetic pollution, as they become an increasing risk to the *in situ* genetic conservation of European CWR.

The workshop, held under the theme 'Genetic erosion and pollution assessment methodologies', debated how genetic erosion and genetic pollution might be predicted and assessed, analysing existing methodologies and making recommendations on the more suitable ways forward.

Genetic erosion and genetic pollution present many different forms and their causes are very diverse. Genetic erosion, which is the 'permanent reduction in the number, evenness and distinctness of alleles, or combinations of alleles, of actual or potential agricultural importance, in a defined geographical area' (FAO, 1999), can be caused by anthropogenic and/or natural changes. The country reports collated for the preparatory process of the State of the World's Plant Genetic Resources for Food and Agriculture (PGRFA) (FAO, 1998) identify the following as the main causes of genetic erosion (in descending order of importance):

- Replacement of local varieties;
- Land clearing (including deforestation and bush fires);

- Overexploitation of species;
- Population pressure (including urbanization);
- Environmental degradation;
- Overgrazing;
- Legislation or policy;
- Changing agricultural systems;
- Pests, weeds or diseases;
- Civil strife;
- Reduced fallow (in the case of shifting cultivation).

It is interesting to note that such important causes of plant genetic diversity loss, such as the introduction of exotic and alien species, human-caused disasters (water, air and soil pollution), natural calamities (soil erosion, floods, drought and landslides) and destruction and fragmentation of natural ecosystems, are not mentioned in the country reports in spite of the fact that their importance and effects are abundantly referred to and documented in the scientific literature (Hawkes *et al.*, 2000).

Genetic pollution, meaning 'the gene flow from conventionally and biotechnologically bred crops and introduced exotic and alien species to natural populations', although a more recent concept, is a growing risk and concern as it presents a threat to the *in situ* conservation of genetic diversity and eventual effects in the environment and health (see Wilkinson and Ford, Chapter 17, this volume).

The precise number of flowering plants in existence is not yet known. However, it is estimated as being between 300,000 and 500,000, of which only 250,000 have been identified or described (FAO, 1998). On the other hand, Walters and Gillett (1998), based on a 20-year study, which involved 16 research institutions, estimated that 34,000 species, representing about 12.5% of the world's flora, are currently under the threat of extinction, but admitting that this estimate, due to the lack of and impreciseness of data, could be even higher.

Humankind faces the possibility of plant species or even entire plant families disappearing before it is even known they existed. However, the scientific evidence of extinction can only be assumed much later after plant species disappear, as a species can only be scientifically considered as extinct when there is no reasonable doubt that the last individual has died (IUCN, 2001) or according to CITES (1993) species may be cited as extinct if not recorded for at least 50 years.

We are in the presence of a contradictory and conflicting situation. On the one hand, there is still a wealth of untapped resources (unknown species or unquantified genetic diversity within known species) in which humankind could invest in study to firstly identify and quantify them; and secondly, to assess the usefulness and value of those resources for exploitation thus contributing to the world's food security. On the other hand, we assist in processes leading to the diminution and disappearance of these invaluable and vital resources, even before they are made known to humanity. Prance (1997), estimates that only about 10% of plants have been evaluated for their medicinal or agricultural potential and so there are certainly many new drugs and new crops yet to be discovered. For the vast majority of these we have little or no idea of the extent of their individual genetic diversity – what range of useful genes are present within these species.

Humankind has much to be concerned about the problems that affect the maintenance, even the survival, of the plant diversity, as it depends much on it for its own survival. The world's food security is at stake and urgent and vigorous actions are imperiously needed. Genetic erosion and genetic pollution afflicting any plant species in any part of the world (at local, national, regional or global level) is bound to have repercussions globally. The reduction of biodiversity (erosion) or its contamination (pollution), by whatever causes and means, have a 'domino effect' as, in the present times, agriculture in practically every country in the world is, to a great extent, dependent on species originating from other regions.

Interdependency, which is 'the production of crops based on species originating from other regions', is a bind that ties up the world's food production and thus the world's food security. Regions' interdependence on PGRFA is estimated from a maximum of 100% (Australia and North America) to a minimum of 30.8% (West Central Asia). In Europe, percentage of food production of major crops based on species originating from other regions is estimated as 90.8% (FAO, 1998).

Very few crops feed the world. As high as 90% of the world's calorie consumption is based on 30 crops of which, only three (rice, maize and wheat) contribute to 60% of the total (FAO, 1998). In spite of the heavy dependency on very few crops, agrobiodiversity, described as 'that part of biodiversity which nurtures people and which are nurtured by people' (FAO, 1998), has been very badly affected by genetic erosion and, more recently, by genetic pollution.

Farmers are the main contributors and the guardians of the diversity of PGRFA or, in other words, agrobiodiversity. However, the changes in the agricultural systems dictate changes in land use thereby heavily influencing the future level of PGRFA diversity in the farmers' fields. These changes in the agricultural systems, the main one being caused by its intensification and the utilization of new technologies leading to landrace replacement, overexploitation of genetic resources and habitat destruction, are not determined by a farmer's decisions alone but largely by socio-economic factors, development of relevant market opportunities and policies (Virchow, 1999). It is estimated that 75% of the genetic diversity of agricultural crops has been lost during the 20th century (FAO, 1998).

How serious and how big is the problem concerning genetic erosion and genetic pollution? As a rough indication, during the preparation of this chapter (August 2005), an Internet search using the search string ('genetic erosion' + plant) gave 20,000 hits, while a search using the search string ('genetic pollution' + plant) gave 17,800 hits. This gives an idea of the number of posted documents with reference to these problems, and thus the magnitude of the concern.

Public awareness, around the world in general and in Europe in particular, has been raised for these matters and a number of legal instruments form a legal framework concerning genetic erosion, namely:

- Bern Convention;
- Bonn Convention;
- International Treaty on Plant Genetic Resources for Food and Agriculture (ITPGRFA);
- Convention on International Trade in Endangered Species of Wild Fauna and Flora (CITES);
- Council Directive no. 92/43/CEE, 21 May (known as the Habitat Directive);
- Natura 2000;
- Important plant areas;
- Sites of special scientific interest;
- Convention on Biological Diversity (Art. 8, h).

Although the legal instruments mentioned above are tailored to the protection of species through measures for their conservation and the prevention of introduction of alien species, in an indirect way they expedite the reduction of the risk of genetic erosion by the promotion of its conservation.

As for genetic pollution, the legal instruments that form the legal framework in Europe are:

- EC Directive 2001/18/EC and Appendix VII;
- EC Regulation 1829/2003;
- Convention on Biological Diversity (Art. 8, g).

Calling for mandatory monitoring of genetically modified organisms (GMOs) placed on the market in order to:

- Trace and identify eventual effects of the placing on the market of GMO on environment and health;
- Give feedback to the risk assessment procedure.

In 1998, the European Commission (EC) launched a programme directed to supporting activities in the field of energy, environment and sustainable development in which activities aimed at the conservation and promotion of the sustainable utilization of European socio-economically important species such as CWR were also addressed. The Fifth Framework Programme for Energy, Environment and Sustainable Development (1998–2002) aimed to: 'contribute to sustainable development by focusing on key activities for social well-being and economic competitiveness in Europe'.

The programme was structured on several 'key actions' relevant to the conservation and to promoting the sustainable utilization of European socio-economically important species. Key Action 2 'Global change, climate and biodiversity' states that: 'the assessment of biodiversity loss should be made in the context of sustainable development and international treaties', while aiming to: 'develop scientific, technological and socio-economic basis and tools necessary for the study and understanding of changes in the global or regional environment'. Within this context, Key Action 2.2, while having the objective: 'to understand and assess the interaction and effects of anthropogenic pressure and global changes on key ecosystems, carbon and nutrient cycles and biodiversity', also related to: 'assessment of ecosystem change and vulnerability' and . . . 'rational methods for conservation of biodiversity'. Key Action 2.2.3 'Assessing and conserving biodiversity' aimed to: 'develop methods to assess biodiversity, quantify and understand its status and maintain biodiversity'. Thus, it was within the Fifth Framework Programme for Energy, Environment and Sustainable Development, that the 'European Crop Wild Relative Diversity Assessment and Conservation Forum' (PGR Forum) was placed.

The PGR Forum elected, as one of the thematic backbones of the project, to discuss how to assess genetic erosion and genetic pollution from conventionally and biotechnologically bred crops for European CWR populations. A number of recommendations, which included the identification of further steps, needed for genetic erosion and pollution assessment thus arose.

16.2 Legal and Socio-political Issues and Recommendations

The advent of increased use of GMOs, threat of genetic pollution and likelihood of climate change should give momentum to studies on genetic erosion or pollution and the need for conservation activities in terms of the public and policy makers. Another issue that needs to be addressed is that traditionally, protected areas are not established with CWR in mind at all, and there is a need to establish legislation which promotes CWR or landrace conservation in the same way as for rare breeds of domestic animals.

There is also a need to prioritize (see below) key CWR species and focus surveying and monitoring on them. As part of this process, criteria (economic importance, level of threat, etc.) need to be identified that would enable a list of CWR for Europe in general and each country in particular to be protected and managed in protected areas (see Ford-Lloyd *et al.*, Chapter 6, this volume).

Other legislation, although perhaps unlikely, could be aimed at ensuring that GMOs are not grown near protected areas containing CWR. This would require, however, the determination whether the crop and wild relatives share the same close gene pool, making them more or less likely to exchange genes giving rise to genetic pollution.

There is an urgent need to prioritize CWR at the national level, the European level and the global level. This prioritization should take the form of an assessment of the ease with which a crop and its wild relative can cross and therefore exchange genes. Those CWR that are genetically isolated from their closest crop relative, or where the crop relative does not occur in the country or region, would receive zero priority (e.g. potatoes in the United Kingdom). Those, which easily formed hybrids, would receive greatest attention (e.g. sugarbeet in mainland Europe).

16.3 Monitoring of Genetic Erosion or Genetic Pollution

At the taxonomic level, monitoring of genetic pollution can certainly be best undertaken by way of IUCN Red List Assessment, providing that infraspecific categories (subspecies) can be accommodated. The ability to accurately identify the threat to, or loss of subspecies or species, is extremely important as this will serve as a clear proxy indicator of genetic erosion, without the need for extensive population biology or population genetic assessments (see Magos-Brehm *et al.*, Chapter 13, this volume).

Justifiable only after CWR have been thoroughly prioritized, more detailed and specific monitoring, at and around population level may involve assessment of relative taxon rarity, and taxon-specific characters such as the breeding system, mode of dispersal, life form, longevity of soil seed bank, population characteristics (minimum viable population – MVP), number and isolation and extent of occurrence or area of occupancy. These will also provide criteria for assessing genetic erosion by the way of proxy assessments of intermediate complexity or difficulty, but, it is important to prioritize the CWR taxa on which such studies are undertaken. Other much more specific but genetically imprecise indicators of erosion can be identified. These may be socio-economic such as whether CWR are object of complementary conservation strategies (e.g. *ex situ*), the extent and form of socio-economic use, vulnerability to landscape management changes, and susceptibility to genetic pollution (see earlier). Threat-specific criteria such as absence of CWR in protected areas, vulnerability to natural disaster (flood and desertification), as well as habitat-specific criteria, namely threatened habitats, geographic location relative to anthropogenic developments and whether CWR occupy rare or restricted habitats, all may play a part in our ability to monitor genetic erosion and/or pollution.

16.4 Monitoring at the Gene Level

While various parameters may well act as convenient and relatively easy tools to estimate genetic erosion in proxy, in the strictest sense, estimation of genetic erosion may demand a more accurate assessment of the actual loss of genes or alleles from populations or taxa. For this, there is a need to identify the available techniques, which may enable us to do the job. It is also important to identify which molecular genetic markers to use and which population genetic parameters are informative for our purposes. As stressed earlier, it will also be extremely important to identify under which circumstances it will be acceptable and appropriate to use such complex tools, and once again this will require careful prioritization of CWR taxa (see Ford-Lloyd *et al.*, Chapter 6, this volume)

16.5 Some Key Population Genetic Issues and Parameters

Reproductive fitness is the measure of an individual's ability to contribute offspring or progeny to the subsequent generation. Reduction in reproductive fitness could either be an indicator of ensuing genetic erosion, or be a consequence of it. Maintenance of heterozygosity in a population, which is not necessarily of adequate minimum size, and also of its fitness, can be achieved by gene flow from other populations, provided that the 'genetic diversity' that is introduced in this way does not contain maladaptive genes. This may represent a situation where so-called genetic pollution is actually beneficial.

The other factor to take into account is that the expression of fitness traits will be dependent upon the environment (and fitness traits generally have low heritability), so changes in environmental conditions will obviously affect fitness, and may therefore give rise to genetic erosion. Environmental change may gradually eliminate a population, or the population may be able to maintain its fitness by adapting to the change, but this adaptative capability will depend upon adequate levels of genetic variation existing in the population.

Effective population size can broadly be defined as 'the size of an idealized (hypothetical) population that would lose genetic diversity (genetic erosion) at the same rate as the actual population (the one under study)' (after Frankham *et al.*, 2002). Effective population size in relation to actual population size will always be smaller for inbreeding species than outbreeding, smaller if just a few plants contribute most to the next generation, and smaller if the actual population size drops substantially even for just one generation, and this is, therefore, probably the most important criterion which is useful for indicating both genetic erosion and deleterious genetic pollution.

Nei's index of diversity (Nei, 1973) is one estimator based upon the expected heterozygosity, which can be used to assess 'diversity' occurring within populations, reserves or protected areas, or within different geographical regions. Its estimation will enable comparisons of genetic diversity to be made for monitoring what happens to diversity in a reserve over a particular time period – and therefore for estimating genetic erosion.

The estimation of gene flow between populations or subpopulations, inbreeding within populations or subpopulations and, differentiation between subpopulations (F statistics) can be important and informative for measuring genetic erosion and pollution (Frankham *et al.*, 2004). Particularly, important questions to be answered might be: is the level of inbreeding within a conserved population increasing, giving rise to genetic erosion, or is there gene flow from a crop to a wild population giving rise to advantageous or disadvantageous genetic pollution?

Finally, what is the minimum size of a population needed to remain genetically viable and maintain genetic variation and heterozygosity? This minimum size, allowing for substantial variation in definition, is commonly referred to as the MVP size. It could tell us whether a population has suffered so much genetic erosion as to be no longer viable.

16.6 Choice of Molecular Markers for Monitoring Erosion and Pollution

The range of molecular markers that can now be relatively easily used to indicate genetic levels of diversity is quite extensive. Reviews of these techniques are plentiful (Newbury and Ford-Lloyd, 1997; Westman and Kresovich, 1997; Henry, 2001). For genetic monitoring of populations, a co-dominant marker system will often be most appropriate, which means microsatellites (simple sequence repeats – SSRs). However, out of 160 random CWR taxa (genera) surveyed, only 29% had SSR primers currently available for easy use. Therefore, sacrifice of a co-dominant marker system might be necessary in favour of one where prior knowledge of primer sequences is not required, such as amplified fragment length polymorphism (AFLP).

16.7 Using the CWR Catalogue to Monitor Genetic Erosion or Genetic Pollution

The CWR catalogue collated by PGR Forum has over 25,000 species (Kell *et al.*, 2005). In reality, it will be neither possible nor desirable to undertake detailed assessment of genetic diversity on such a large number of species. Key issues are therefore:

- How can we assess the majority of our CWR species using more easily obtainable data, with the aim of minimizing genetic erosion or pollution and maximizing genetic diversity in *in situ* conservation?
- How do we prioritize a much smaller number of taxa where there may actually be a need for molecular population genetic intensive study utilizing the tools described above? Ford-Lloyd *et al.* (Chapter 6, this volume) address this question in detail, but some other issues can be addressed as well (see below).

Information on breeding systems will be important – the greatest amount of diversity is found *within* populations of outbreeders, whereas most diversity is found *among* populations of inbreeders. While effective and actual population sizes are rarely the same, actual population size can be a useful rough guide, and will give us an idea about erosion if actual population size is decreasing. This however demands resampling – obtaining data at more than one time point, but if populations are stable concerning size, molecular population genetic analysis may be needed only once. If population size is decreasing, there may be a need for resampling. Other simple guides might be:

- Taxonomic diversity – assuming diversity is spread across taxa – ensuring that subspecific taxa that are conserved should ensure that diversity is conserved.
- Ecogeographic diversity – populations that have different adaptive norms will be genetically diverse.
- Red data listing – can reveal important genetic information.

Only after using such proxy indicators or tools should we start to ask questions requiring more detailed studies:

- Are any subspecific taxa seriously threatened?
- Are any major habitats or regions threatened?
- Are most populations' sizes declining (outbreeding species)?
- Are some populations' sizes declining (inbreeding species)?
- Do sampled populations contain significant genetic diversity?

16.8 Monitoring Genetic Pollution

The Gene Pool Concept provides an indicator of the CWR species that are vulnerable to genetic pollution. Twenty-two out of 25 of the world's most important crops have evidence of natural hybridization with one or more wild relatives. This could extrapolate to over 22,000 (90%) of our CWR taxa. So, can genetic pollution affect genetic diversity?

- Gene flow can cause change in genetic diversity – in 12 different studies, diversity in introgressed populations was greater (Ellstrand, 2003).
- Can gene flow cause extinction? – more data are needed to answer this question, but it is 'speculated' that hybridization might have caused extinction among the CWR of capsicum, date palm, hemp, maize and sweet pea (Small, 1984).

Assessment may require the measure of gene flow, involving population genetics and molecular markers. It is possible to assess:

- Occurrence of hybrids and hybrid derivatives (morphological);
- Fitness of hybrids and hybrid derivatives;
- Spread of hybrids and hybrid derivatives.

This needs to be measured over a large timescale, large geographical area and large sample size.

16.9 Conclusions

One of the most important issues is, given finite resources and time, when and why should we undertake molecular population genetic studies instead of, or in addition to, the use of other proxy indicators which may be relatively more simple, easy and quicker to use?

It may be important to undertake a study using molecular markers and population genetic analyses if *most* outbreeding populations' sizes are declining, or if *some* key inbreeding populations' sizes are declining. This will be important if any one major habitat or region is known to be threatened or if any subspecific taxon within the target species is seriously threatened. Under these circumstances, it might be important to sample and do molecular population genetics to establish whether populations in protected areas are adequately maintained, or which populations should be subject to the greatest degree of protection.

Although no proxy indicators can be guaranteed to provide totally reliable estimates of genetic erosion, in many instances the estimates will be sufficient to plan conservation of CWR. Given that time is running fast towards 2010, indicators of loss of genetic diversity that provide quick and easy baseline data which can be accumulated at more than one time point may be much more useful than restricted data derived from detailed experimental studies which are incomplete and available for only one time point.

References

CITES (1993) *Convention on International Trade in Endangered Species of Wild Fauna and Flora*. CITES Secretariat, Geneva, Switzerland.

Ellstrand, N.C. (2003) *Dangerous Liaisons? When Cultivated Plants Mate With Their Wild Relatives*. Johns Hopkins University Press, Baltimore, Maryland.

FAO (1998) *The State of the World's Plant Genetic Resources for Food and Agriculture*. FAO, Rome, Italy.

FAO (1999) *Report of the Technical Meeting on the Methodology of the World Information and Early Warning System on Plant Genetic Resources*. Research Institute of Crop Production, Prague, Czech Republic. 21–23 June 1999. FAO, Rome, Italy.

Frankham, R., Ballou, D. and Briscoe, D.A. (2002) *Introduction to Conservation Genetics*. Cambridge University Press, Cambridge.

Frankham, R., Ballou, D. and Briscoe, D.A. (2004) *A Primer of Conservation Genetics*. Cambridge University Press, Cambridge.

Hawkes, J.G., Maxted, N. and Ford-Lloyd, B.V. (2000) *The* Ex Situ *Conservation of Plant Genetic Resources*. Kluwer Academic, Dordrecht, The Netherlands.

Henry, R.J. (ed.) (2001) *Plant Genotyping, the DNA Fingerprinting of Plants*. CAB International, Wallingford, UK.

IUCN (2001) *IUCN Red List Categories and Criteria Version 3.1*. International Union for Conservation of Nature and Natural Resources, Gland, Switzerland/Cambridge.

Kell, S.P., Knüpffer, H., Jury, S.L., Maxted, N. and Ford-Lloyd, B.V. (2005) *Catalogue of Crop Wild Relatives for Europe and the Mediterranean*. Available online via the Crop Wild Relative Information System (CWRIS – available at: http://cwris.ecpgr.org/) and on CD-ROM. University of Birmingham, Birmingham, UK.

Nei, M. (1973) Analysis of gene diversity in subdivided populations. *Proceedings of the National Academy of Sciences of the United States of America* 70, 3321–3323.

Newbury, H.J. and Ford-Lloyd, B.V. (1997) Estimation of genetic diversity. In: Maxted, N., Ford-Lloyd, B.V. and Hawkes, J.G. (eds) *Plant Genetic Conservation: The* In Situ *Approach*. Chapman & Hall, London. pp. 192–206.

Prance, G.T. (1997) The conservation of botanical diversity. In: Maxted, N., Ford-Lloyd, B.V. and Hawkes, J.G. (eds) *Plant Genetic Conservation: The* In Situ *Approach*. Chapman & Hall, London. pp. 3–14.

Small, E. (1984) Hybridization in the domesticated-weed-wild complex. In: Grant, W.F. (ed.) *Plant Biosystematics*. Academic Press, Toronto, Canada, pp. 195–210.

Virchow, D. (1999) *Conservation of Genetic Resources – Costs and Implications for a Sustainable Utilization of Plant Genetic Resources for Food and Agriculture*. Springer, Heidelberg/New York.

Walters, K.S. and Gillett, H.J. (1998) *1997 IUCN Red List of Threatened Plants*. Compiled by the World Conservation Monitoring Centre. IUCN – The World Conservation Union, Gland, Switzerland/Cambridge, pp. 1–862.

Westman, A.L. and Kresovich, S. (1997) Use of molecular marker techniques for description of plant genetic variation. In: Callow, J.A., Ford-Lloyd, B.V. and Newbury, H.J. (eds) *Biotechnology and Plant Genetic Resource*. CAB International, Wallingford, UK, pp. 9–48.

17 Assessing the Potential for Ecological Harm from Gene Flow to Crop Wild Relatives

M.J. WILKINSON AND C.S. FORD

17.1 Introduction

Genetic exchange between crops and their wild relatives is nothing new and has undoubtedly occurred in both directions since the dawn of agriculture. Indeed, the introduction of genes into crops from crop wild relatives (CWR) has been a feature of plant breeding for around a century, with modern plant breeders using a range of increasingly sophisticated techniques in order to transfer desirable features from CWR. These efforts have naturally focused on the targeted introduction of genes providing traits of agronomic interest, particularly those relating to disease resistance and tolerance of abiotic stress. Examples are manifold and include (amongst others) the introduction of potato virus X (PVX), potato virus A (PVA) and potato virus Y (PVY) resistance from *Solanum brevidens* into potato by somatic hybridization (e.g. Valkonen and Rokka, 1998) and the recruitment of drought tolerance into wheat from its relatives by bridging crosses (e.g. Valkoun, 2001).

However, it was not until the advent of genetically modified (GM) crops that serious concerns were raised over the possible environmental consequences of gene flow from crops to their relatives. The key justification for the sudden interest in gene flow between crops and their relatives appears to lie in the unknown ecological consequences of the introduction into natural habitats of genes that are new to the organism and probably new to the family. It could have been argued previously that recurrent rounds of selection meant that the cultivated crop almost inevitably contained less genetic diversity than its CWR and so is unlikely to act as a source of a new gene or allele that confers enhanced ecological fitness to the CWR. This is obviously not the case for GM crops since the transgenes may originate from any group of organisms or could even be synthetic. However, the first generation of GM crops are significant for the lack of diversity in the transgenes that they carry and in the number of crops carrying them. Indeed, four crops (maize, soybean, cotton and rapeseed) and just two transgene types (insect resistance and herbicide tolerance) (James, 2004) currently dominate the commercial GM varieties grown worldwide.

This simple situation has allowed for a considerable literature to amass on the environmental risks posed by a relatively small number of scenarios. This situation seems set to change in the near future: both recent trends towards the generation of GM lines with multiple inserts (Wilkinson *et al.*, 2003) and substantial diversification of the transgenes being introduced (Dunwell, 2002) means that regulation of GM crops will need to become increasingly targeted to accommodate the new set of potential hazards that diversification brings (Raybould and Wilkinson, 2005). At the same time, dramatic improvements have been made in the capacity of molecular technologies other than genetic modification to introduce new genes into crops. These include amongst others asymmetric protoplast fusion (e.g. Zhou and Xia, 2005), bridging crosses (e.g. Ortiz and Ehlenfeldt, 1992), quantitative trait loci (QTL) analysis (e.g. Herrmann *et al.*, 2006) and marker-assisted breeding (e.g. Steele *et al.*, 2006). These strategies have allowed breeders to extend the range of germplasm that is exploitable for breeding purposes and now mean that many modern varieties possess genes from several distantly related CWR that are not ordinarily cross-compatible with the crop or even with each other. At the same time, extensification of land used for agricultural purposes and the gradual replacement of local landraces for high-yielding modern varieties (Hammer and Laghetti, 2005) has increased the scope for contact between modern varieties containing genes from many, rather distant relatives and scarce relatives that have restricted distributions. In this way, endangered CWR are becoming increasingly exposed to modern varieties containing large numbers of exotic genes, with a much smaller number of populations being exposed to transgenic varieties containing a relatively small number of highly divergent exotic genes (i.e. transgenes). Viewed in this context, the ecological and conservational risks posed by transgenes should be viewed as an extension of the risks presented by modern varieties containing genes outside the primary gene pool. We shall therefore examine the conservational and ecological risks posed by gene flow from all modern crops to CWR, and explore the extent to which the principles developed for the risk assessment of GM crops can be applied to this broader set of problems.

17.2 Preparing for Risk Assessment

The first task for any risk assessment process is to define the nature of the concern. This is known as 'hazard specification'. The main aim is to state exactly what we are concerned about and wish to evaluate. In the case of gene flow from crops to their wild relatives, this can include a wide range of possible unwanted outcomes that can be broadly categorized as follows: (i) gene flow will change abundance of the CWR within its community; (ii) allow the CWR to invade new habitats or geographic ranges; (iii) cause change to the genetic diversity of the CWR (by introducing genetic sweeps); (iv) dilute the genetic integrity of the CWR (genetic swamping); (v) change fitness aspects of the CWR such that it impacts on the abundance of associated flora, fauna or disease-causing agents.

A helpful start prior to the risk assessment process is to first consider the list of CWR of the crops grown in a region, since many modern crops are now grown so far from their centres of origin and diversity that no cross-compatible CWR exist over much of their cultivated range. Of the major food crops grown within Europe, for instance, neither maize (*Zea mays* L.) nor potato (*Solanum tuberosum* L.) have CWR with which they can form hybrids. However, this pre-screen can still leave us with a very large number of potential crop–CWR pairings for a particular region and an even greater number of possible environmental consequences.

17.3 Risk Assessment Terminology

At this point, it is useful to define some terms that are central to risk assessment. Risk is a balanced evaluation based on the severity of a particular unwanted outcome and the likelihood that it will occur and can be usefully defined by the formula: Risk = *f* (hazard, exposure).

The term 'hazard' represents the severity of the unwanted environmental change and often relates to a particular species. This element inevitably involves some subjectivity and is usually semi-qualitatively represented (e.g. severe, moderate or low). In essence, the hazard element defines how unwanted the environmental change is.

The term 'exposure', on the other hand, represents the probability that the hazard will occur and so should be quantifiable, provided the hazard is adequately defined. We quantify the risk by combining a well-defined hazard with the probability of occurrence (exposure). Viewed in this manner, the process of 'risk assessment' involves the cumulative assessment of all 'risks' that are related to a specified crop in a particular area, with each risk being an assessment of 'how bad' a particular hazard is and how likely it is to occur.

17.4 Focusing on Important Scenarios

It is clearly possible to assemble a long list of hazards for all CWR given sufficient time and imagination, although many of these may ultimately prove to be trivial or else extremely unlikely. Whilst it is clearly a matter of subjective judgement as to which hazards qualify as 'significant' and which are effectively irrelevant, it is possible to broadly rank or categorize hazards by imposing a series of criteria that are related to a particular purpose. The criteria applied will vary according to application or to the purpose for which the risk assessment is being performed. In relation to the consequences of gene flow from crops to CWR, we can consider four main purposes for which divergent priorities could be set.

1. Conservation;
2. Genome evolution;
3. Threat to exploitable genetic resources;
4. Agriculture.

For conservational purposes, the first and foremost concern centres on the possible direct and indirect impacts of gene flow on the abundance of endangered organisms or communities. The threatened organisms under consideration may be the recipients of gene flow (i.e. the CWR themselves) or the flora and fauna that associate with the CWR. Priorities drawn up with conservation as the primary concern will obviously tend to place emphasis on scenarios that cause change to threatened species over those that affect more abundant species.

A second and so far largely overlooked area that could be used as a basis for prioritizing hazards relates to the long-term evolutionary implications of gene flow. The introduction of new genes into a species could potentially provide a long-term platform that allows the CWR to radiate and create new forms. In this context, the ranking of hazards will be based on the identity and properties of the alien genes carried by the crop compared with those contained within the genome of the CWR.

The third purpose relates to the long-term conservation of CWR genetic resources for future crop improvement. Here, the emphasis may be on the potential agronomic value of the genes contained by the CWR and so may sometimes place greater weight on a common CWR that possesses a high genetic diversity for traits of agronomic importance (e.g. disease resistance) over an endangered CWR that contains little diversity and no traits of value.

The fourth category relates to the more immediate needs of the agricultural industry. Here, greatest concern may gravitate around consequences of gene flow that affect farm practice or productivity. This means, for instance, that higher priority is given to gene flow of herbicide tolerance into an existing CWR weed than to a CWR in native communities that lack the capacity to compete in an agricultural setting. Similarly, the introduction of legislative measures aimed at increasing on-farm biodiversity may increase emphasis on gene flow that influences the abundance of associated species within the agricultural field.

17.5 Oilseed Rape in the United Kingdom: a Case Study

17.5.1 Ranking of hazards

The ranking of hazards is seemingly reliant on the perspectives of the individual or group performing the task. It should nevertheless be possible to provide crude guidance on the relative importance of a particular crop in a specified region for each of these categories by reference to the literature. We can illustrate this point by reference to oilseed rape in the United Kingdom as a case study.

There are 16 species reported to be capable of hybridization with oilseed rape in the United Kingdom including both weeds and wild species occupying natural and semi-natural habitats (Table 17.1). On this basis alone, there is apparently considerable potential for this crop to cause unwanted conservational or ecological change in the United Kingdom by altering the abundance of one of its many CWR or (probably more importantly from a conservational

Table 17.1. Ranking CWR of *Brassica napus* in the United Kingdom. Species are ranked in three ways: according to their ease of hybridization (Scheffler and Dale,[a] 1994; Raybould and Gray, 1993;[b] Warwick *et al.*, 2003[c]); by crude abundance (Stace, 2001; Preston *et al.*, 2003); and by the number of associated species with conservational status (Rodwell, 1991; Cheffings and Farrell, 2005).

Recipient species	Hybridization ranking[a]	CWR scarcity in United Kingdom	Associate ranking
Brassica oleracea[b]	3	1	1
Diplotaxis muralis[a]	7	2	2
Raphanus raphanistrum[a]	6	2	3
Sinapis arvensis[a]	8	3	3
Brassica nigra[b]	5	2	4
Brassica rapa[b]	1	3	5
Brassica juncea[a]	2	1	5
Brassica carinata[a]	4	1	5
Diplotaxis erucoides[a]	7	2	5
Sinapis alba[b]	8	3	5
Brassica tournefortii[a]	9	1	5
Diplotaxis tenuifolia[a]	9	2	5
Eruca vesicaria[a]	9	1	5
Raphanus sativus[a]	9	3	5
Erucastrum gallicum[c]	10	1	5
Hirschfeldia incana[b]	10	1	5

viewpoint) of their associates. There is also scope for the transfer of specific genes into one of these CWR through gene flow providing a platform for evolutionary change. However, for transgenic crops this will depend entirely on the transgene that they contain and must be considered on a case-by-case basis.

There are currently no GM varieties of oilseed rape in the United Kingdom and so exposure to transgenes is limited to field trials. The conventional crops are more problematic since there is a long history of using wild germplasm in the breeding of oilseed rape (e.g. Crouch *et al.*, 1994; Wang *et al.*, 2006) and the identity of genes or even genome regions that have been introduced from wild germplasm is rarely well characterized. In terms of the threat to future CWR genetic resources for the breeding of new oilseed rape varieties, however, it should be noted that the United Kingdom does not represent the centre of diversity or origin for any of the CWR of oilseed rape. Oilseed rape to CWR gene flow in the United Kingdom therefore presents only a minor threat to the future capacity to broaden the genetic base of the crop, largely regardless of any ecological consequences that may occur locally. Likewise, the scope for serious agricultural problems arising from gene flow from the crop to one of these CWR is very limited since only one of the CWR is a significant agricultural weed (*Sinapis arvensis* L.) and gene flow into this species is extremely rare (Bing *et al.*, 1996; Lefol *et al.*, 1996; Moyes *et al.*, 2002; Chevre *et al.*, 2003; Daniels *et al.*, 2005).

Overall, oilseed rape in the United Kingdom appears to have considerable scope for causing conservational harm, only moderate potential to cause significant evolutionary change to the group (and this depends on the introduced gene), and poses only relatively minor hazards relating to the preservation of CWR genetic resources for breeding or to the agricultural industry. For this reason, emphasis of early work should probably concentrate on the ranking of hazards relating to conservational and ecological consequences.

17.5.2 Ranking of crop: CWR pairings

There are numerous approaches that could be adopted to rank the relative importance of hazards relating to gene flow from a crop grown in a particular region. Perhaps the simplest approach is to first rank the potential CWR recipient species. Historically, the most widely adopted strategy for this purpose is to order cross-compatible CWR recipient species according to the ease with which they hybridize with the crop. If this tactic is applied to CWR of oilseed rape in the United Kingdom, the species forming most hybrids (*Brassica rapa* L.) appears highest on the list and distant relatives like *Hirschfeldia incana* (L.) Lagr.-Foss. lowest (Table 17.1). However, the fundamental problem with this as an initial screen lies in the presumption of a direct and causal relationship between the frequency of hybrids and the severity and/or number of ecological consequences. Here, ranking is based on likelihood of hybrid occurrence (i.e. an exposure term) rather than its importance (i.e. a term relating to the hazard). Moreover, it does not necessarily follow that 'more hybrids means more harm'. For instance, it is entirely plausible that even extensive hybrid formation between a crop and an alien weed species may result in little or no long-term ecological damage to natural communities simply because the CWR recipient does not grow outside the agricultural environment.

Arguably, a more attractive criterion may be to rank CWR taxa according to their potential to suffer from 'ecologically damaging' gene flow on the basis of the scarcity of the CWR itself – CWR with a more restricted distribution being more at risk. If this system is used to rank the United Kingdom relatives of oilseed rape, then *B. juncea* is identified as the recipient of greatest concern, with the common and widespread *S. arvensis* and *Raphanus raphanistrum* L. warranting little interest. Now, difficulty is created by the fact that most of the relatives of oilseed rape are alien introductions into the United Kingdom and so have little importance from a conservational or even ecological viewpoint. Removal of non-native, cross-compatible CWR from the list leaves only *B. oleracea* L., followed by the more abundant *B. nigra* (L.) W.D.J. Koch. It is worth noting that neither of these species is listed as having conservational importance on either a national or international scale. Should this criterion be applied strictly, the conservational importance of gene flow from oilseed in the United Kingdom could be deemed to be low since there are no listed native CWR of the crop. However, this argument crucially ignores the possibility that gene flow from the crop results in a changed interaction between the recipient CWR and the plants and animals with which it associates. Given the large number of plant and insect species potentially associated with each CWR recipient, it seems inevitable that this indirect form of ecological consequence is likely to provide the highest number of hazards and therefore, in all likelihood, the most probable and the most serious.

Difficulty lies in the lack of information on bilateral and multilateral interactions between CWR and other species within their communities. One must therefore adopt a simplified approach in which one first screens lists of associates for species that have importance in their own right in terms of conservational status, ecological importance (e.g. keystone species) or even cultural significance (e.g. the bald eagle in the USA). In some cases, it is possible to collate species associations for cross-compatible CWR from the community association literature (e.g. Rodwell, 1991) or else by direct surveys. Interestingly, when lists of associates from the CWR of oilseed rape are screened on the combined basis of their national scarcity (Cheffings and Farrell, 2005), *B. oleracea* L. now ranks as the most important CWR, followed by *Diplotaxis muralis* (L.) DC., with the weeds (e.g. *B. juncea* (L.) Czern.) appearing towards the base of the list. Such ranking clearly identifies which CWR have greatest capacity to cause conservational harm or cultural harm (if associate lists were screened for species of cultural importance) but lack consideration of crude measures of exposure (answering the question: is it at all realistic that hybrids will form in this region?) or of the significance of the CWR species themselves.

In reality, therefore, it is probably advisable to take cognizance of all three sources of information when ranking CWR on the basis of their scope for inflicting unwanted conservational harm. We would advocate that when assessing the scope for conservational harm, greatest weighting should be applied to the abundance of associates of known conservational, ecological or cultural importance, followed by a second-level weighting based on the relative importance assigned to the CWR itself and with abundance of hybrids providing the finest level of discrimination. It is only after the highest ranking the crop–CWR pairings are identified so that specific hazards should be identified and the complex process of risk assessment should be started.

17.6 Conclusions

The processes required for a risk assessment of the possible environmental consequences of all forms of gene flow from a crop to its CWR share many features in common with protocols developed for GM crop risk assessment. The main additional difficulty lies in the lack of information relating to the genes and alleles contained within the crop and their source of origin.

As with GM crops, there can be a bewildering array of possible outcomes for a particular crop in a named geographic region. These hazards can range widely in their severity and in the likelihood of occurrence. This complexity, coupled with the cost and effort required to perform comprehensive risk assessments necessitates that a system is applied to target attention towards the hazards that are both serious and most likely to occur. We suggest that a three-stage preliminary screen should be performed to allow crude ranking of crop–CWR combinations within a given geographic location. First, comprehensive lists should be compiled of all cross-compatible CWR within the specified geographic region. Second, the crop–location combination should be explored to identify which broad categories of hazards (conservation or ecological, evolutionary, genetic resources or agricultural) are most germane and merit greatest attention for the purpose of which risk assessment is being performed.

Finally, CWR should be ranked according to their scope to inflict harm. For conservational purposes, this can be achieved by initially emphasizing CWR that grow in species-rich habitats containing associates, with secondary weighting on the conservational importance of the CWR itself and finally on the crude assessment of the likelihood and probable frequency of spontaneous hybrids and introgression. It is only at this point that a list of specific hazards should be assembled (e.g. decline or local extinction of named associates) and ranked (presumably, the most threatened species first). High ranking hazards can then be the subject of a full risk assessment.

References

Bing, D.J., Downey, R.K. and Rakow, G.F.W. (1996) Hybridizations among *Brassica napus, B. rapa* and *B. juncea* and their two weedy relatives *B. nigra* and *Sinapis arvensis* under open pollination conditions in the field. *Plant Breeding* 115, 470–473.

Cheffings, C. and Farrell, L. (eds) (2005) *The Vascular Plant Red Data List for Great Britain*. Joint Nature Conservation Committee, Peterborough, UK.

Chevre, A.M., Eber, F., Jenczewski, E., Darmency, H. and Renard, M. (2003) Gene flow from oilseed rape to weedy species. *Acta Agriculturae Scandinavica Section B-Soil and Plant Science* 53 (Suppl. 1), 22–25.

Crouch, J.H., Lewis, B.G., and Mithen, R.F. (1994) The effect of A-genome substitution on the resistance of *Brassica napus* to infection by *Leptosphaeria maculans. Plant Breeding* 112, 265–278.

Daniels, R., Boffey, C., Mogg, R., Bond, J., and Clarke, R. (2005) *The Potential for Dispersal of Herbicide Tolerance Genes from Genetically Modified, Herbicide Tolerant Oilseed Rape Crops to Wild Relatives*. Final report to DEFRA, contract reference EPG 1/5/151. Defra, London.

Dunwell, J.M. (2002) Future prospects for transgenic crops. *Phytochemistry Reviews* 1, 1–12.

Hammer, K. and Laghetti, G. (2005) Genetic erosion – examples from Italy. *Genetic Resources and Crop Evolution* 52, 629–634.

Herrmann, D., Boller, B., Studer, B, Widmer, F. and Kolliker, R. (2006) QTL analysis of seed yield components in red clover (*Trifolium pratense* L.). *Theoretical and Applied Genetics* 112, 536–545.

James, C. (2004) Preview: *Global Status of Commercialized Biotech/GM Crops: 2004. ISAAA Briefs* 32. ISAAA, Ithaca, New York.

Lefol, E., Danielou, V., and Darmency, H. (1996) Predicting hybridization between transgenic oilseed rape and wild mustard. *Field Crop Research* 45, 153–161.

Moyes, C.L., Lilley, J.M., Casais, C.A., Cole, S.G., Haeger, P.D. and Dale, P.J. (2002) Barriers to gene flow from oilseed rape (*Brassica napus*) into populations of *Sinapis arvensis. Molecular Ecology* 11, 103–112.

Ortiz, R. and Ehlenfeldt, M.K. (1992) The importance of endosperm balance number in potato breeding and the evolution of tuber-bearing *Solanum. Euphytica* 60, 105–113.

Preston, C.D., Pearman, D.A. and Dines, T.D. (2003) *New Atlas of the British and Irish Flora*. Oxford University Press, Oxford, UK.

Raybould, A.F. and Gray, A.J. (1993) Genetically-modified crops and hybridization with wild relatives – a UK perspective. *Journal of Applied Ecology* 30, 199–219.

Raybould, A.F. and Wilkinson, M.J. (2005) Assessing the environmental risks of gene flow from GM crops to wild relatives. In: Poppy, G.M. and Wilkinson, M.J. (eds) *Gene Flow from GM Plants*. Blackwell, Oxford, UK, pp. 169–185.

Rodwell, J.S. (ed.) (1991) *British Plant Communities: 1–5*. Cambridge University Press, Cambridge.

Scheffler, J.A. and Dale, P.J. (1994) Opportunities for gene-transfer from transgenic oilseed rape (*Brassica napus*) to related species. *Transgenic Research* 3, 263–278.

Stace, C.A. (2001) *New Flora of the British Isles*, 2nd edn. Cambridge University Press, Cambridge.

Steele, K.A., Price, A.H., Shashidhar, H.E. and Witcombe, J.R. (2006) Marker-assisted selection to introgress rice QTLs controlling root traits into an indian upland rice variety. *Theoretical and Applied Genetics* 112, 208–221.

Valkonen, J.P.T. and Rokka, V.M. (1998) Combination and expression of two virus resistance mechanisms in interspecific somatic hybrids of potato. *Plant Science* 131, 85–94.

Valkoun, J.J. (2001) Wheat pre-breeding using wild progenitors. *Euphytica* 119, 17–23.

Wang, Y.P., Sonntag, K., Rudloff, E., Wehling, P. and Snowdon, R.J. (2006) GISH analysis of disomic *Brassica napus-Crambe abyssinica* chromosome addition lines produced by microspore culture from monosomic addition lines. *Plant Cell Reports* 25, 35–40.

Warwick, S.I., Simard, M.J., Legere, A., Beckie, H.J., Braun, L., Zhu, B., Mason, P., Seguin-Swartz, G. and Stewart, C.N. (2003) Hybridization between transgenic *Brassica napus* L. and its wild relatives: *Brassica rapa* L., *Raphanus raphanistrum* L., *Sinapis arvensis* L., and *Erucastrum gallicum* (Willd.) OE Schulz. *Theoretical and Applied Genetics* 107, 528–539.

Wilkinson, M.J., Elliott, L.J., Allainguillaume, J., Shaw, M.W., Norris, C., Welters, R., Alexander, M., Sweet, J. and Mason, D.C. (2003) Hybridization between *Brassica napus* and *B. rapa* on a national scale in the United Kingdom. *Science* 302, 457–459.

Zhou, A.F. and Xia, G.M. (2005) Introgression of the *Haynaldia villosa* genome into Gamma-ray-induced asymmetric somatic hybrids of Wheat. *Plant Cell Reports* 24, 289–296.

18 Reciprocal Introgression between Wild and Cultivated Peach Palm (*Bactris gasipaes* Kunth, *Arecaceae*) in Western Ecuador

J.-C. Pintaud, T.L.P. Couvreur, C. Lara B. Ludeña and J.-L. Pham

18.1 Introduction

Gene flow between crops and their wild relatives has been a major evolutionary factor since the beginning of agriculture. In agroecosystems where both wild and cultivated plants are present, gene flow from the wild to the cultivated type, if permitted by the seed system or the farmers' seed management, will contribute to nurture crop diversity. This has been documented in the literature for several tropical crops, not only for cereals (e.g. for rice, Chu and Oka, 1970), but also for tuber crops like yam (Scarcelli, 2005; Scarcelli *et al.*, Chapter 21, this volume).

The intensification of agriculture and its technological advances have somehow troubled the idyllic image of crop wild relatives (CWR)–crop interaction. In pearl millet in Niger, Robert *et al.* (2004) studied the impact on yield of weedy millet that originates from hybridization between wild and cultivated millet. The frequency of hybridization might result from increasing land pressure, making cultivated fields closer to wild populations. Most concerns, however, are relevant to the crop-to-wild gene flow. This is illustrated by the lively debate on the use of genetically modified organisms and their potential impact on biodiversity. This debate has hidden a more general question of the genetic impact of crop cultivation on the diversity of wild species. To our knowledge, this question has not been widely addressed.

Forest conservationists seem to have given the most attention to the question of the impact of genetic introduction on the diversity of native species (e.g. Cagelli and Lefèvre, 1995; Barbour *et al.*, 2003). Forest conservation has also addressed the issue of habitat erosion and fragmentation, a concern shared by most genetic conservationists. These consequences of human activities affect the viability of wild populations.

This chapter presents a study on *Bactris gasipaes* Kunth, the only domesticated palm in South America. We studied the genetic dynamics of the crop–wild complex of this species in western Ecuador. Although *B. gasipaes* was traditionally used for its fruits, the development of monoculture for palm heart production impacted considerably on the evolutionary dynamics of the complex. We studied the overall genetic structure of this species complex and the genetic consequences of the introduction of non-indigenous *B. gasipaes* seeds in the areas of occurrence of wild populations.

Domestication of the peach palm (*B. gasipaes* var. *gasipaes*) from its wild relative *B. gasipaes* var. *chichagui* (H. Karst.) A.J. Hend. (Henderson, 2000) possibly began in western Amazonia, where many tropical American crops such as cassava, cocoa, pineapple, passion fruit and papaya had originated (Clement, 1989, 1995; Olsen and Schaal, 1999, 2001). Development of agriculture initiated progressively in western Amazonia between 10,000 and 5000 years BP (Lathrap, 1977), and cassava was introduced to coastal Peru 6000 BP (Weber *et al.*, 1986). It is not known when exactly the process of domestication of *B. gasipaes* started. Wild populations are widespread in South America and have also been reported from Central America (Mora Urpí, 1999). Archaeological data suggest that cultivation of the peach palm was established in Costa Rica between 2300 and 1700 BP (Corrales and Mora Urpí, 1990). Early archaeological remains of *B. gasipaes* endocarps suggest that the first domesticated plants produced fruits of about 10 g, which is already a significant increase from wild fruits, weighing 0.5–3 g. Subsequently, the selection process led to macrocarp races producing fruits of up to 200 g (Mora-Urpí *et al.*, 1997).

Traditional cultivation of the peach palm for its edible fruits has decreased since the arrival of the Europeans, with the introduction of new crops like rice and plantain. Since a few decades, modern cultures for the production of palm heart were established in many Latin American countries. In such plantations, palms are grown at high density and maintained in a subherbaceous state by cutting young stems every 12–18 months for palm heart harvest. The products are mostly canned and exported to northern countries.

B. gasipaes var. *chichagui* is distributed along the Andes up to 1800 m elevation from Venezuela to Bolivia, extending to the Pacific lowlands of Ecuador and Colombia, and north to Panama and Costa Rica (Mora Urpí, 1999). On the eastern side of the Andes it is largely distributed in the southern periphery of the Amazon reaching the Xingu River in Pará, Brazil, under a relatively dry climate with up to 6 months of dry season (Da Silva and Clement, 2005). Populations become more scattered in the wet equatorial western Amazon (Fig. 18.1), in Loreto (Peru), Rio Nangaritza valley (Ecuador) and western Amazonas (Brazil). Both traditional and modern-type cultivations of *B. gasipaes* overlap considerably with the natural occurrence of wild populations in South and Central America. Introgression is suspected to exist between these sympatric wild and cultivated populations of *B. gasipaes*, although the extent of this process may vary considerably depending on local situations. Rodrigues *et al.* (2004) considered the impact of introgression to be minor in western Amazonas (Brazil), while Couvreur *et al.* (2006) documented extensive introgression between wild and cultivated plants in western Ecuador.

In western Ecuador, wild and cultivated compartments of *B. gasipaes* are largely sympatric (Fig. 18.1). Natural populations of *B. gasipaes* var. *chichagui*

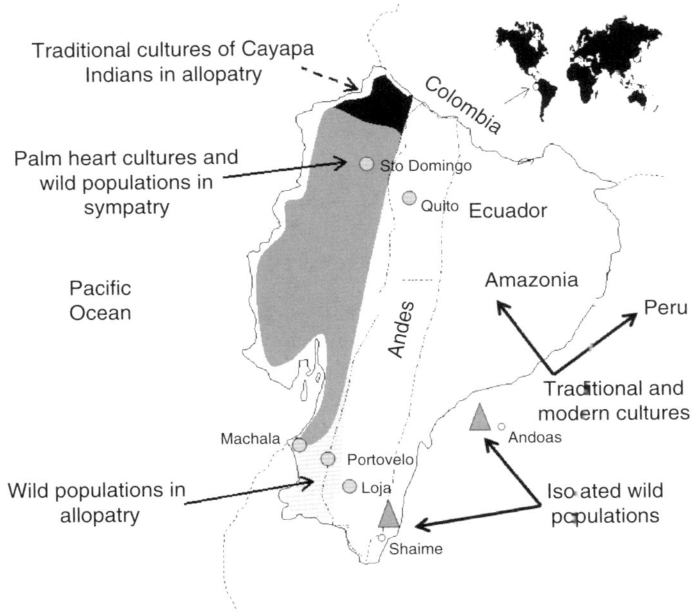

Fig. 18.1. Distribution of wild and cultivated populations of *Bactris gasipaes* in Ecuador and adjacent Peru.

are distributed from sea level to 1000 m over most of the Pacific region, under a seasonal climate. In south-western Ecuador, close to the border with Peru, it thrives in a semi-deciduous forest which undergoes 7 months of dry season. Wild populations are absent in the extreme north of the Pacific region, close to the Colombian border, where the climate is too humid. In this northern region however, exists a traditional landrace of *B. gasipaes*, cultivated by Cayapa Indians for fruit production. Southwards, modern cultivation of *B. gasipaes* for palm heart production developed extensively in the area of occurrence of wild populations. According to local informants, these cultures have been established from seeds and seedlings of Amazonian origin since about 1980, and not from the landrace cultivated by the Cayapa Indians. This modern cultivation does not extend to the extreme south, where only wild populations are found.

18.2 Materials and Methods

We sampled four groups of *B. gasipaes* corresponding to different situations of sympatry or allopatry of wild and cultivated plants (Fig. 18.1): *BgcS* composed of wild plants in allopatry under the dry climate of south-western Ecuador, *BgcN* and *BggN* wild and cultivated plants in sympatry in north-western Ecuador, respectively, and *BggA* composed of cultivated plants from Amazonia and Central America. In order to determine the genetic relationships of these groups, we genotyped individual plants at eight nuclear dinucleotide microsatel-

Table 18.1. Characteristics of microsatellite loci used in this study. Loci from Billotte *et al.* (2004)[a], from Couvreur *et al.* (2006)[b] and from Pintaud *et al.* (2007, unpublished data)[c].

Locus name	Genomic compartment and motif	Primers sequence forward reverse	Ta (°C)	Mean size (bp)	No. of alleles
mBgCIR010[a]	Nuclear (GA)n	GTGGAAATAAAGCGAGTGAG ATCCCCTCGTCCTAAAT	52	217	13
mBgCIR057[a]	Nuclear (GA)n	CCAGCCACAAAAGACATC GTCCTTTTGGTCTCAAGACTA	52	259	17
mBgCIR058[a]	Nuclear (GA)n	TTTGATACCCCAGAGAGA AGCGAGAAACACGAATAC	52	225	23
mBgCIR062[a]	Nuclear (GA)n	CTACAGGGAGTGCATCTAC CCACCATTCAGCAATATTAG	52	180	25
mBgCIR071[a]	Nuclear (GA)n	GGTTGGACGGTTTTCAT CACTGCTTTCTTGGTTACT	52	115	11
mBgCIR087[a]	Nuclear (GA)n	TTGCTTTTCGCAGAGAG GCACCTCAATCAGTTTATC	52	165	17
mBgCIR094[a]	Nuclear (GA)n	ACGACGGTGCTCTTCTTAG TGCGGGAACGGGAATAT	52	161	17
mBgCIR204[b]	Nuclear (GA)n	TGGCAGTTCAAAGTAGTATCAAT TAAGCCACCACCAAGCAGTCC	51	202	21
mEgtrnQrps16[c]	Chloroplastic (GATA)n	CGTCCCGAACAATCT TT TACCATAACATTCCTCTAATT	52	398	5
Q16-T[c]	Chloroplastic (T)n	TTTTTCTTTCATTTGGTTTCTCA AGGCCTTTCTTTCATTCACATTTT	48	252	3

lite loci (Billotte *et al.*, 2004; Couvreur *et al.*, 2006), and two chloroplast microsatellite loci (a mononucleotide and a tetranucleotide) isolated from the trnQ-rps16 intergenic spacer (Hahn, 2002). The characteristics of these ten loci are reported in Table 18.1. In addition, we conducted a morphometric analysis of the fruits from different groups (Table 18.2). The number of specimens from each of the four groups used in each experiment (nuclear and chloroplast microsatellites, morphometrics) is indicated in Table 18.3.

Table 18.2. Coding of four semi-quantitative and qualitative multistate fruit characters for the neighbour-joining analysis

Volume	Shape	Ornamentation	Colour
0 = 1.0–2.9 cm^3 (wild type only)	0 = ovoid	0 = smooth epicarp	0 = red
1 = 3.0–19.9 cm^3 (hybrids only)	1 = conical	1 = striate epicarp	1 = orange
2 = 20–59 cm^3 (hybrids and cultivated)	2 = obusiform		2 = yellow
3 = 60–200 cm^3 (cultivated type only)	3 = ellipsoid		3 = brown
	4 = obovoid		4 = variegated

Table 18.3. Number of individuals used in nuclear microsatellite (SSR) genotyping, chloroplast microsatellite (CpSSR) genotyping, and morphometric analysis of fruits. In the last analysis, ten fruits from each individual were measured according to Table 18.2.

	BgcS	*BgcN*	*BggN*	*BggA*	Total samples
Nuclear SSR	24	32	13	14	83
Cp SSR	28	87	27	56	198
Morphometrics	5	8	13	16	42

BgcS = *Bactris gasipaes* var. *chichagui* from south-western Ecuador; *BgcN* = *B. gasipaes* var. *chichagui* from north-western Ecuador; *BggN* = *B. gasipaes* var. *gasipaes* from north-western Ecuador; *BggA* = *B. gasipaes* var. *gasipaes* from Amazonia and Central America.

Diversity measures at the nuclear microsatellite loci in each of the four groups include the total number of alleles, non-biased expected heterozygosity, observed heterozygosity, allelic richness, private alleles and pair-wise linkage disequilibrium. Methods of calculation of these measures were described in Couvreur *et al*. (2006). In order to determine the degree of genetic differentiation between groups, we performed a hierarchical analysis of molecular variance (AMOVA) using the ARLEQUIN software (Excoffier *et al*., 2005). The fixation index obtained (Φ_{ST}) incorporates variance in allele size and distribution of allele frequencies in each population. Significance levels for the overall values were determined after 1000 permutations. Individual microsatellite genotype scores were also ordinated in a multidimensional space by principal coordinate analysis (PCOA), computed with STATISTICA 6.0 (StatSoft Inc. 2001). For this analysis, we used the allele-sharing distance D_{AS} calculated with MSA 3.10 (Dieringer and Schlötterer, 2003). We also studied more specifically the interactions between wild and cultivated forms in sympatry in north-western Ecuador (*BgcN* and *BggN*) using Bayesian methods. These methods compute for each individual a probability of belonging to a population, taking into account *a priori* information. Hybrid individuals will have a probability of belonging to more than one population. In a first analysis, using the software GENECLASS (Cornuet *et al*., 1999), we introduced the origin of each individual (wild or cultivated form) as prior information. In a second analysis, implemented in the software STRUCTURE (Pritchard *et al*., 2000), we aimed at detecting individuals with admixed ancestry assuming two populations (K = 2), but no prior information on the origin (wild or cultivated) of the plants was used. The results were based on 1,000,000 iterations following a burn-in period of 100,000 iterations. Individual proportions, q_i, were estimated with their 90% probability intervals.

We analysed the chloroplast microsatellite data with the measures H_T (total diversity), H_S (intrapopulational diversity) and G_{ST} (interpopulational differentiation) using the software HAPLODIV (Pons and Petit, 1995).

For the morphometric study, we measured the volume of ten fruits per plant and calculated the mean and standard deviation for each group. A two-sample unequal t-test was performed between mean values of each group using GENSTAT version 7.1.0.205 (VNS International Ltd, 2003). The volumetric data were also transformed in a semi-quantitative character and analysed jointly with three qualitative characters (Table 18.2) using the neighbour-joining (NJ)

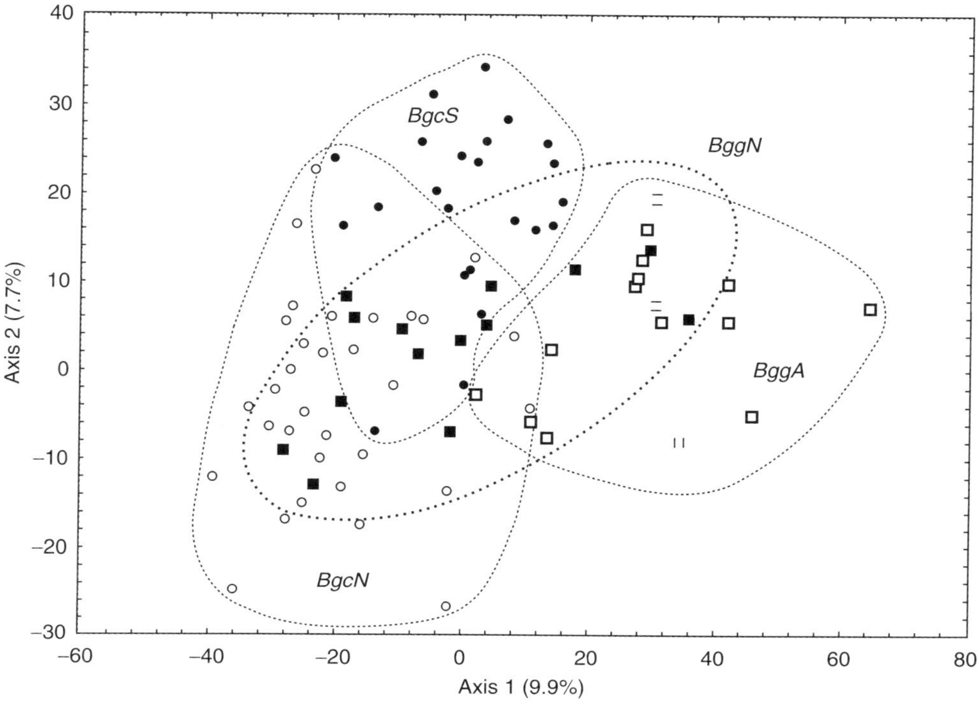

Fig. 18.2. Principal coordinate analysis of nuclear microsatellite data (eight loci) in four groups of samples of *Bactris gasipaes*: wild from south-western Ecuador (*BgcS*), wild from north-western Ecuador (*BgcN*), cultivated from north-western Ecuador (*BggN*) and cultivated from Amazonia and Central America (*BggA*).

method as implemented in the software PAUP* 4.0b10 (Swofford, 2002), with the mean character difference index.

18.3 Results

Analysis of genetic diversity with the nuclear microsatellites revealed important polymorphism at all loci (Table 18.1), with expected heterozygosity values between 0.7 and 0.9. Allelic richness in cultivated groups *BggN* and *BggA* is significantly higher than in wild groups *BgcS* and *BgcN*. Each group has private alleles between 8 and 20 and no linkage disequilibrium was identified. AMOVA indicated that 90.6% of the total variance corresponded to intragroup diversity and 9.6% to intergroup differentiation ($\Phi_{ST} = 0.09$, $P < 0.001$). PCOA (Fig. 18.2) explained 17.6% of total variance on the first two axes. The first axis (9.9%) separates almost completely the wild plants (*BgcS* and *BgcN*) from the cultivated plants in allopatry (*BggA*). The second axis (7.7%) partially separates the wild populations from south-western Ecuador growing under a dry climate from wild populations growing in north-western Ecuador in a humid climate. Cultivated plants of the *BggN* group were originally established from Amazonian seeds and grow in sympatry with wild populations in north-western Ecuador. This group shows a transversal distribution with respect to these two axes, and encompasses both the sympatric wild group (*BgcN*) and the allopatric Amazonian group (*BggA*).

Bayesian analyses provide further information about the situation in the sympatric zone. The analysis with GENECLASS assigns 18 of the 32 plants from the wild compartment (*BgcN*) to the wild type alone, while the 14 other samples are significantly associated with both wild and cultivated types, but with a higher probability of belonging to the wild type. Seven of the 13 individuals from the cultivated group are assigned to the cultivated type alone, while six are associated to both wild and cultivated types. Of these, five have a higher probability of being the cultivated type and one has a higher probability of being the wild type. The analysis with STRUCTURE (Fig. 18.3), assuming two populations, recovered a first group formed mostly of cultivated plants (10/13), and a second group formed almost only of wild plants (14/15). As many as 20 individuals showed ancestry in more than one cluster, and among them, 16 came from the wild form and four from the cultivated form.

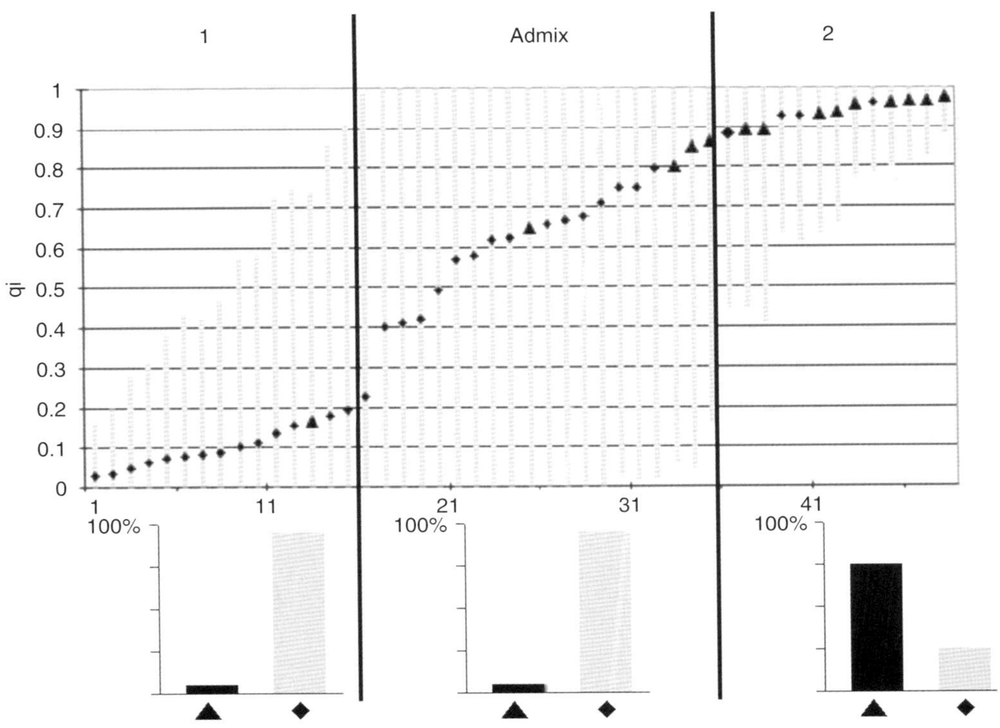

Fig. 18.3. Bayesian analysis of assignment of wild (diamonds) and cultivated (triangles) individuals of *Bactris gasipaes* in sympatry in north-western Ecuador to two populations (1 and 2). Admix individuals have ancestry in both populations.

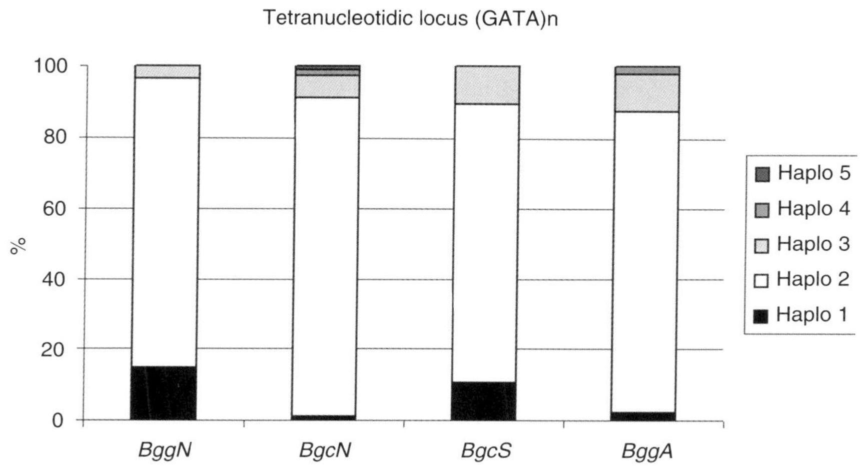

Fig. 18.4. Frequency of haplotypes of a tetranucleotide chloroplastic microsatellite locus in four groups of *Bactris gasipaes* defined as in Fig. 18.2.

Both chloroplast microsatellite loci showed very low intergroup differentiation ($G_{ST} \approx 0.02$). The tetranucleotide locus had five haplotypes ($H_S = 0.3$; $H_T = 0.3$) and the mononucleotide locus had three haplotypes ($H_S = 0.6$; $H_T = 0.6$). The tetranucleotide locus showed one haplotype at very high frequency in all populations (haplotype 2, >80%), but the distribution and frequency of rare haplotypes differed among populations (Fig. 18.4). All populations had the three haplotypes at the mononucleotide locus, with relatively similar frequencies (Fig. 18.5).

The volumetric study of fruits revealed a small mean fruit volume for the wild individuals of south-western Ecuador (*BgcS*, $x = 2.13\,cm^3$), and of north-western

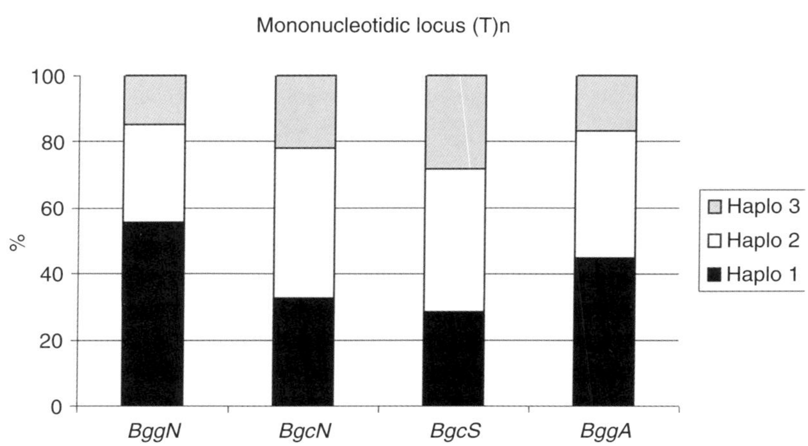

Fig. 18.5. Frequency of haplotypes of a mononucleotide chloroplastic microsatellite locus in four groups of *Bactris gasipaes* defined as in Fig. 18.2.

Ecuador (*BgcN*, $x = 5.451\,\text{cm}^3$). However, *BgcN* showed a much higher variance than *BgcS*. The mean fruit volume of the Amazonian individuals (*BggA*, $x = 70.04\,\text{cm}^3$) was more than ten times the mean fruit volume of the wild individuals, and the largest Amazon fruit ($189\,\text{cm}^3$) was 100 times bigger than the smallest wild fruit ($1.9\,\text{cm}^3$). However, the mean fruit volume of the cultivated individuals of north-western Ecuador (*BggN*, $x = 18.84\,\text{cm}^3$) was intermediate between the autochthonous wild group (*BgcN*) and the Amazon group (*BggA*). Fruit size in *BggN* actually encompasses the upper size range of *BgcN* and the lower size range of *BggA*. All groups had highly significant mean volume differences from one another ($P < 0.001$), except between *BgcS* and *BgcN*, which were only significantly different at $P < 0.01$. The semi-quantitative and qualitative characters were very useful in discriminating the wild, hybrid and cultivated types. Wild-type fruits from populations in allopatry (*BgcS*) always had a volume smaller than $3\,\text{cm}^3$, while all Amazonian cultivated fruits had a fruit volume greater than $20\,\text{cm}^3$. Therefore, individuals of north-western Ecuador with fruits of 5–20 cm^3 are likely to be hybrids. Moreover, wild-type fruits from allopatric populations were all smooth, red and ovoid, and corresponded to type 3 *B. gasipaes* var. *chichagui* known along the Andes from Colombia to Bolivia, and extending to Central America in Panama and Costa Rica (Mora Urpí, 1999; Da Silva and Clement, 2005). Amazonian cultivated plants had fruits of all shapes and colours (Table 18.2) and many were striate. Fruits from north-western Ecuador, too small for the cultivated type (5–20 cm^3), were often striate and with various shapes and colours, therefore indicating their hybrid nature.

The NJ analysis integrating these characters (Fig. 18.6) showed a cluster of monomorphic samples of wild plants from western Ecuador, including all the fruit samples for south-western Ecuador, where the wild populations are grown in allopatry and are therefore presumably pure. The main cluster was divided into two groups: (i) hybrid plants from both wild and cultivated compartments of north-western Ecuador, except for two Amazonian fruit accessions; and (ii) cultivated plants only, including both hybrids and pure cultivated races. The largest fruit races formed a distinct cluster of only Amazonian accessions.

18.4 Discussion

All data suggest an extensive introgression process in the area of sympatry of wild and cultivated plants in north-western Ecuador. The Bayesian analyses of nuclear microsatellite data identified almost half of the individuals sampled from both forms as admixes. The PCOA (Fig. 18.2) shows that the cultivated plants in the area of sympatry (*BggN*) fall into both the sympatric wild group (*BgcN*) and the Amazonian cultivated group (*BggA*), therefore indicating a mixed ancestry. The tetranucleotide chloroplastic locus (Fig. 18.4) revealed three haplotypes in the wild populations in allopatry (*BgcS*), while there were five haplotypes in wild populations in sympatry (*BgcN*), so it is likely that the latter acquired allochthonous haplotypes through introgression. Morphometric data also unambiguously identified hybrid plants producing fruits with a mixture of

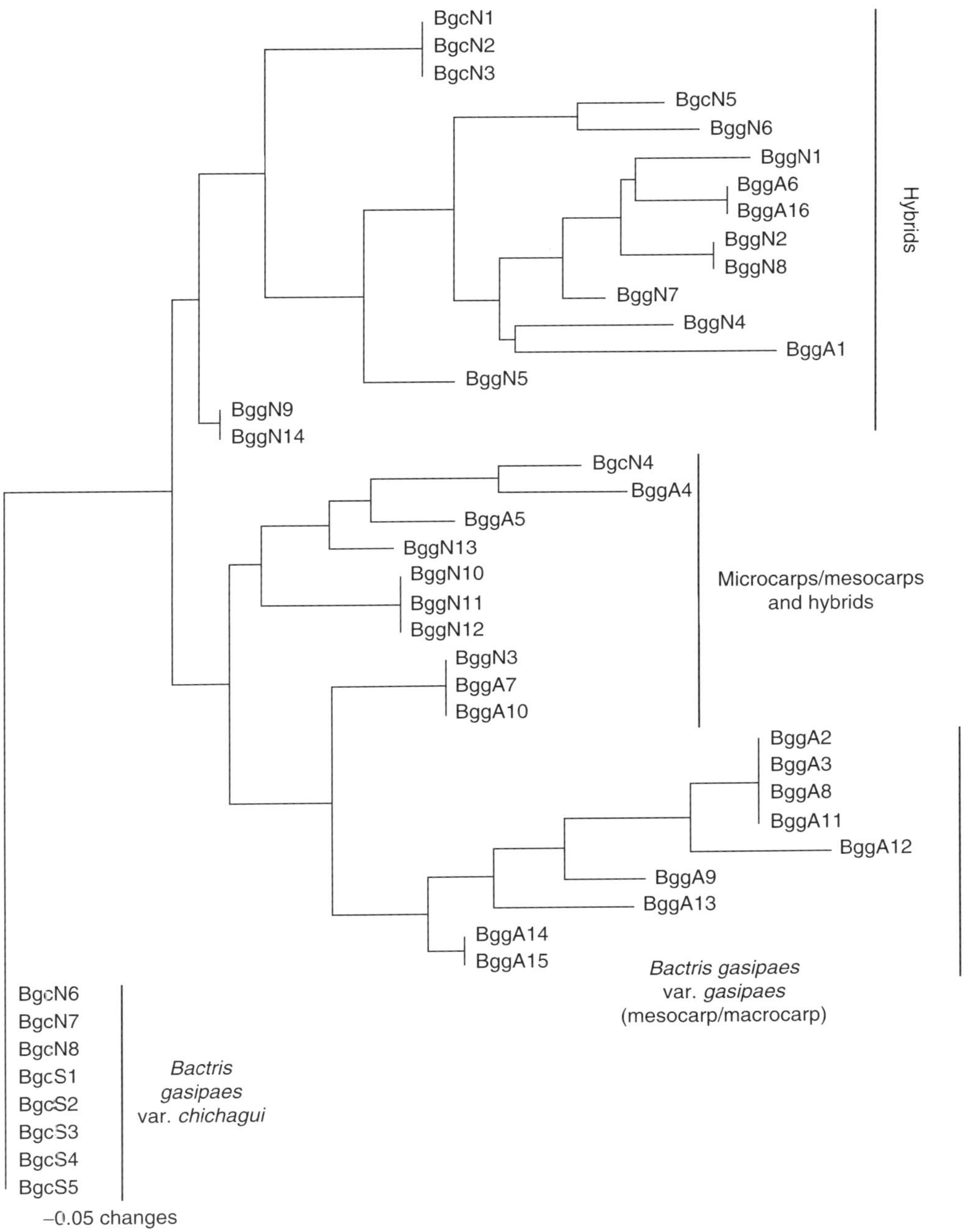

Fig. 18.6. Neighbour-joining analysis of morphological fruit characters in 42 plants of *Bactris gasipaes*. Sample accessions defined as in Fig. 18.2.

wild and cultivated characters (Fig. 18.6). A most interesting result in the Bayesian analysis with STRUCTURE (Fig. 18.3) is that a plant of cultivated origin is assigned to group 1 which corresponds to wild genotypes and four plants of wild origin are assigned to group 2 which corresponds to cultivated genotypes. This means that the status of a plant (cultivated or spontaneous) is relatively independent of the genetic nature of the plant. A 'wild' plant may have a more 'cultivated' genotype than a cultivated plant and *vice versa*. This demonstrates that the wild and the cultivated forms are substantially overlapping in north-western Ecuador.

Couvreur *et al*. (2006) described the cultural practices that generated this situation. Some cultures are established with seedlings coming from Amazonian nurseries and the palms are never grown to maturity. Therefore, these cultures do not interact with the wild forms (type Ia). Many farmers, however, allow a few of these imported seeds to grow to maturity around cultivation near fences, but do not use the seeds for their own plantations. In this case, the cultivated palms, fully grown to maturity, can hybridize with the neighbouring wild palms; the gene flow is unidirectional (type Ib). The palms grown locally can also serve as seed producers for the next plantings. In this case, the mother trees can be pollinated by wild plants and hybrid plants are introduced into the fields (type II). This last practice is convenient and less expensive than buying seedlings, and is therefore common, thus generating extensive introgression. Moreover, surveys made on farms revealed that the farmers were not aware of this process of hybridization and therefore do not control it in any way.

The impact of this phenomenon on the production is not known. It is possible that the introduction of genes from wild plants into the monocultures improves adaptation to local conditions (disease or drought resistance), but it may also alter the quality of the product (palm heart). Mora Urpí *et al*. (1997) described the characteristics of peach palm races most suitable for palm heart production, and it is unlikely that hybrid plants maintain the qualities of elite races.

The impact of introgression on the wild forms is also difficult to evaluate. We sampled the wild plants in and around the fields, and in secondary forests or pastures. Wild plants growing in these conditions are often hybrids, but populations in primary or little disturbed forests are likely to maintain their genetic integrity. However, ongoing extensive deforestation in western Ecuador (Dodson and Gentry, 1991) considerably reduces the extent of such forests, and wild populations of *B. gasipaes* are highly fragmented. Genetic pollution from the cultivated forms is an additional threat to these populations which may represent a useful resource for the improvement of the crop in western Ecuador.

Acknowledgements

This study was supported by the Bureau des Resources Génétiques, France (grant BRG 69–2003) and was realized as part of an agreement between Pontificia Universidad Católica del Ecuador (PUCE) and Institut de Recherche pour le Développement (IRD, France). We thank Dra. Laura Arcos (PUCE) and Dr Francis Kahn (IRD), who established this partnership.

References

Barbour, R.C., Potts, B.M. and Vaillancourt, R.E. (2003) Gene flow between introduced and native eucalyptus species: exotic hybrids are establishing in the wild. *Australian Journal of Botany* 51, 429–439.

Billotte, N., Couvreur, T., Marseillac, N., Brottier, P., Perthuis, B., Vallejo, M., Noyer, J.-L., Jacquemoud-Collet, J.-P., Risterucci, A.-M. and Pintaud, J.-C. (2004) A new set of microsatellite markers for the peach palm (*Bactris gasipaes* Kunth): characterization and across-taxa utility within the tribe *Cocoeae*. *Molecular Ecology Notes* 4, 580–582.

Cagelli, L. and Lefèvre, F. (1995) The conservation of *Populus nigra* L. and gene flow with cultivated poplars in Europe. *Forest Genetics* 2, 135–144.

Chu, Y. and Oka, H. (1970) Introgression across isolating barriers in wild and cultivated *Oryza* species. *Heredity* 23, 1–22.

Clement, C.R. (1989) A center of crop genetic diversity in western Amazonia. *Bioscience* 39, 624–631.

Clement, C.R. (1995) Pejibaye (*Bactris gasipaes*). In: Smartt, J. and Simmonds, N.W. (eds) *Evolution of Crop Plants*, 2nd edn. Longman, London, pp. 383–388.

Cornuet, J.M., Piry, S., Luikart, G., Estoup, A. and Solignac, M. (1999) New methods employing multilocus genotypes to select or exclude populations as origins of individuals. *Genetics* 153, 1989–2000.

Corrales, F. and Mora Urpí, J. (1990) El protopejibaye en Costa Rica. *Boletín Guilielma* 2, 1–11.

Couvreur, T.L.P., Billotte, N., Risterucci, A.-M., Lara, C., Vigouroux, Y., Ludeña, B., Pham, J.-L. and Pintaud, J.-C. (2006) Close genetic proximity between cultivated and wild *Bactris gasipaes* Kunth revealed by microsatellite markers in Western Ecuador. *Genetic Resources and Crop Evolution* (Online first) 1–13.

Da Silva, J.B.F. and Clement, C.R. (2005) Wild pejibaye (*Bactris gasipaes* Kunth var. *chichagui*) in Southeastern Amazonia. *Acta Botanica Brasilica* 19, 281–284.

Dieringer, D. and Schlötterer, C. (2003) Microsatellite analyzer (MSA): a platform independent analysis tool for large microsatellite data sets. *Molecular Ecology Notes* 2, 1–3.

Dodson, C.H. and Gentry, A.H. (1991) Biological extinction in western Ecuador. *Annals of the Missouri Botanical Garden* 78, 273–295.

Excoffier, L., Schneider, S. and Roessli, D. (2005). *Arlequin, version 3.0: A software for population genetic data analysis*. Genetics and Biometry Laboratory, Geneva, Switzerland.

Hahn, W.J. (2002) A phylogenetic analysis of the Arecoid line of palms based on plastid DNA sequence data. *Molecular Phylogenetics and Evolution* 23, 189–204.

Henderson, A. (2000) *Flora Neotropica monograph 79:* Bactris *(PALMAE)*, The New York Botanical Garden, Bronx, New York.

Lathrap, D.W. (1977) Our father the Cayman, our mother the gourd: Spinden revisited, or a unitary model for the emergence of agriculture in the New World. In: Reed, C.A. (ed.) *Origins of Agriculture*. Mouton, The Hague, The Netherlands, pp. 713–752.

Mora Urpí, J. (1999) Origen y domesticación. In: Mora Urpí, J. and Gainza, J. E (eds) *Palmito de pejibaye (*Bactris gasipaes *Kunth): su cultivo e industrialización*. Editorial Universidad de Costa Rica, San José, Costa Rica, pp. 17–24.

Mora Urpí, J., Weber, J.C. and Clement, C.R. (1997) *Peach palm:* Bactris gasipaes *Kunth. Promoting the conservation use of underutilized and neglected crops 20*. Institute of Plant Genetics and Crop Plant Research/International Plant Genetic Resources Institute, Gatersleben/Rome, Italy.

Olsen, K.M. and Schaal, B.A. (1999) Evidence on the origin of cassava: phylogeography of *Manihot esculenta*. *Proceedings of the National Academy of Sciences of the United States of America* 96, 5586–5591.

Olsen, K.M. and Schaal, B.A. (2001) Microsatellite variation in cassava (*Manihot esculenta*, Euphorbiaceae) and its wild relatives: further evidence for a southern Amazonian origin of domestication. *American Journal of Botany* 88, 131–142.

Pons, O. and Petit, R. (1995) Estimation, variance and optimal sampling of gene diversity. I. Haploid locus. *Theoretical and Applied Genetics* 90, 462–470.

Pritchard, J.K., Stephens, M. and Donnelly, P. (2000) Inference of population structure using multilocus genotype data. *Genetics* 155, 945–959.

Robert, T., Luxereau, A., Mariac, C., Ali, K., Allinne, C., Amoukou, A.I., Bani, J., Banoin, M., Beidari, Y., Bezancon, G., Cayeux, S., Couturon, E., Decieu, V., Gamatche, I., Hamidou, A., Moussa, D., Sadou, M., Seydou, M., Seyni, O., Tidjani, M. and Sarr, A. (2004) Gestion de la diversité en milieu paysan: influence de facteurs anthropiques et des flux de genes sur la variabilité génétique des formes cultivées et spontanées du mil (*Pennisetum glaucum*) dans deux localités du Niger. In: Actes du 4ème Colloque National Bureau des Ressources Génétiques – Union des Ressources Génétiques du Berry 'La Diversité et la Ressource', 14–15 Octobre 2002, La Châtre, France. Actes du BRG 4, pp. 223–245.

Rodrigues, D.P., Astolfi Filho, S. and Clement, C.R. (2004) Molecular marker-mediated validation of morphologically defined landraces of Pejibaye (*Bactris gasipaes*) and their phylogenetic relationships. *Genetic Resources and Crop Evolution* 51, 871–882.

Scarcelli, N. (2005) Structure et dynamique d'une plante cultivée à multiplication végétative: le cas des ignames au Bénin. Thèse de doctorat, Université Montpellier II, France.

Swofford, D.L. (2002). PAUP* 4.0b10. *Phylogenetic Analysis Using Parsimony (and other methods)*. Sinauer and Associates, Sunderland, Massachusetts.

Weber, E., Nestel, B. and Campbell, M. (1986) Archeological manioc (*Manihot*) from coastal Peru. *Economic Botany* 40, 78–102.

19 Impoverishment of the Gene Pool of the Genus *Aegilops* L. in Armenia

M. Harutyunyan, A. Avagyan and M. Hovhannisyan

19.1 Introduction

The territory of Armenia belongs to the Western Asia centre of cultivated plants origin; many crop wild relatives (CWR), which are valuable materials for genetic research and breeding, are still growing in the territory of the Republic. However, as a result of changes in land use and development, many species of CWR are now under threat and likely to disappear.

Increased interest in the genus *Aegilops* L. (goat grass) is connected with the cytogenetic evidence concerning its role in the origin of tetraploid and hexaploid wheat (as the source of B and D genomes) (Van Slageren, 1994). Species of goat grass are a rich reservoir of genes for drought resistance, poor soil tolerance and pest and disease resistance. Due to these important features they are considered as good initial breeding material for selection for adaptability to abiotic and biotic factors.

19.2 *Aegilops* Diversity in Armenia

Nine *Aegilops* species with wide intraspecific diversity have been found in Armenia: *A. biuncialis* Vis., *A. columnaris* Zhuk., *A. crassa* Boiss. (= *A. trivialis* Zhuk., ($2n = 42$)), *A. cylindrica* Host, *A. mutica* (Boiss.) Eig. (= *Amblyopyrum muticum* Boiss.), *A. tauschii* Coss. (= *A. squarrosa* L.), *A. triaristata* Willd., *A. triuncialis* L. and *A. umbellulata* Zhuk. (Table 19.1).

A. cylindrica is the prevailing species of *Aegilops* in Armenia. It can be found in nine floristic provinces. *A. triuncialis* is the most polymorphic species. As a rule, *Aegilops* species are found growing in semi-desert and mountainous steppe zones, on dry stony slopes, along roads, along the edges of cereal fields, in cemeteries, borders of forests, rubbish tips and at altitudes from 500 to 2200 m above sea level.

Table 19.1. *Aegilops* spp. growing in Armenia.

Species	Genomic formula	Chromosome number (2*n*)	Floristic provinces where found	Altitude above sea level (m)
Aegilops biuncialis Vis	CM	28	3	790–1400
A. columnaris Zhuk.	CM	28	3	500–1600
A. crassa Boiss.	DDM	42	1	1450
A. cylindrica Host	DC	28	9	500–2100
A. mutica (Boiss.) Eig.	M	14	1	1300–1400
A. tauschii Coss.	D	14	5	500–1800
A. triaristata Willd.	CM	28	1	970–1620
A. triuncialis L.	C	28	5	500–2200
A. umbellulata Zhuk.	C	14	1	1000–1350

19.3 Threats to Armenian *Aegilops* Diversity

As a result of increased urbanization and loss of habitat, impoverishment of the gene pool of *Aegilops* is taking place. Analysis of data received in the course of the most recent expeditions and their comparison with similar indices obtained during the field explorations carried out 10–15 years ago (Figs 19.1 and 19.2) indicates that there has been considerable impoverishment of the *Aegilops* gene pool in the territory of Armenia (Figs 19.3 and 19.4). Land privatization and other anthropogenic factors have a negative influence upon natural populations; heavy impoverishment and partial disappearance of gene pools occurs. As a result of these processes, phytocoenotic (plant community) composition has also been changed.

According to research carried out in the 1980s, populations of *A. mutica* and *A. crassa* were under threat and disappearing, and because of this these two species were included in the Armenian Red Book. *A. biuncialis* and *A. umbellulata* were considered rare species (Ghazaryan, 1989). The results of the last survey have shown drastic changes in the number and composition of these populations.

The only natural habitat of *A. mutica* is the south-east part of Yerevan, among the villages Jrvezh, Vokhchaberd and Mushavan, near the Erebuni State Reserve. Nowadays, only a few plants can be found in this habitat.

The only natural habitat of *A. crassa* in Armenia is the ravine of the Razdan River. At present, this population has completely disappeared as a result of fire. *A. biuncialis* was prevalent in the phytocoenoses in the Ararat Marz (villages Urtzadzor, Shagap, Zovashen and Lusashogh) on stony slopes along roads at altitudes of 1225–1400 m above sea level, but now only a few plants can be found here.

19.4 Discussion of *Aegilops* Conservation

In the light of the above-mentioned points it is very important to take urgent measures for the conservation of endangered species of *Aegilops* both *in situ*

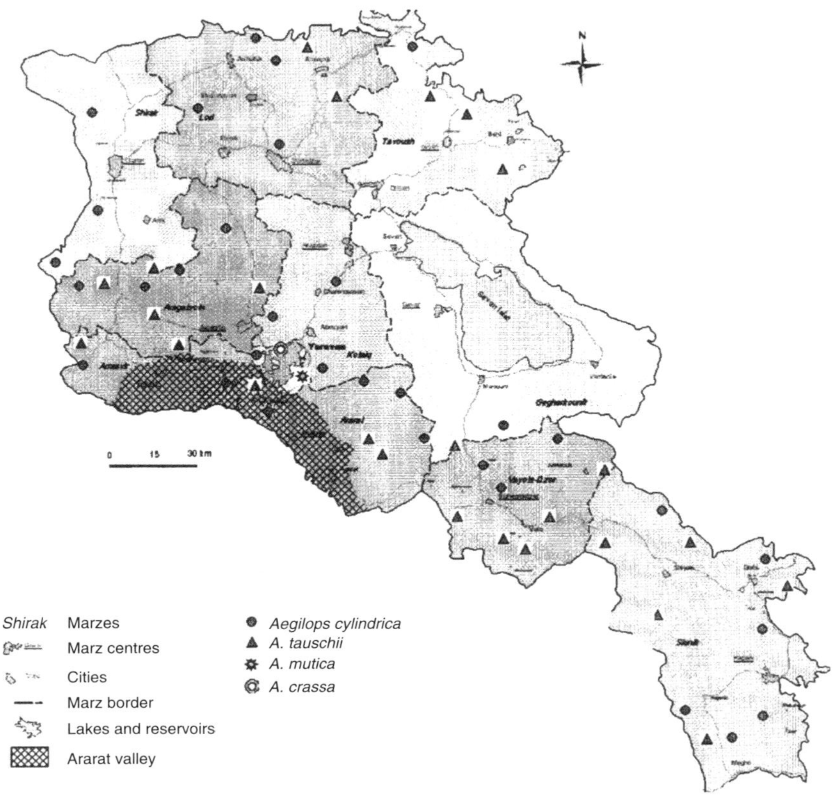

Fig. 19.1. Distribution of *Aegilops* species in Armenia in the 1980s (*A. cylindrica*, *A. tauschii*, *A. mutica* and *A. crassa*).

and *ex situ*. Conservation of genetic diversity of this genus is necessary, because each species and intraspecific form has a large reserve of potentially useful germplasm for stability, quality and productivity of cultivated wheat. More attention is now given to *in situ* conservation of *Aegilops* species.

Four species of *Aegilops* (*A. triuncialis*, *A. cylindrica*, *A. tauschii*, *A. columnari*) grow on the territory of Erebuni Reserve, which was founded in 1981 specifically to protect endemic xerophytes and the unique natural mountainous complex – with hundreds of varieties of wild cereal. This reserve on a territory of 90 ha is located in close proximity to the capital of Armenia on red clay soil. On the territory of the Erebuni Reserve, *Aegilops* species are abundantly found together with other wild wheat. Implementation of appropriate conservation measures in this reserve will contribute to protection of the full range of genetic diversity of these species. *A. mutica*, a species considered to be taxonomically intermediate between *Aegilops* and *Agropyron* Gaertn., was also found near this nature reserve. Unfortunately, overgrazing has degraded several sites of the reserve and actions are needed to save the remaining populations.

In comparatively moist regions of the Republic, such as in dense woods, *Aegilops* species do not occur, as they grow only in dry habitats. In large forest

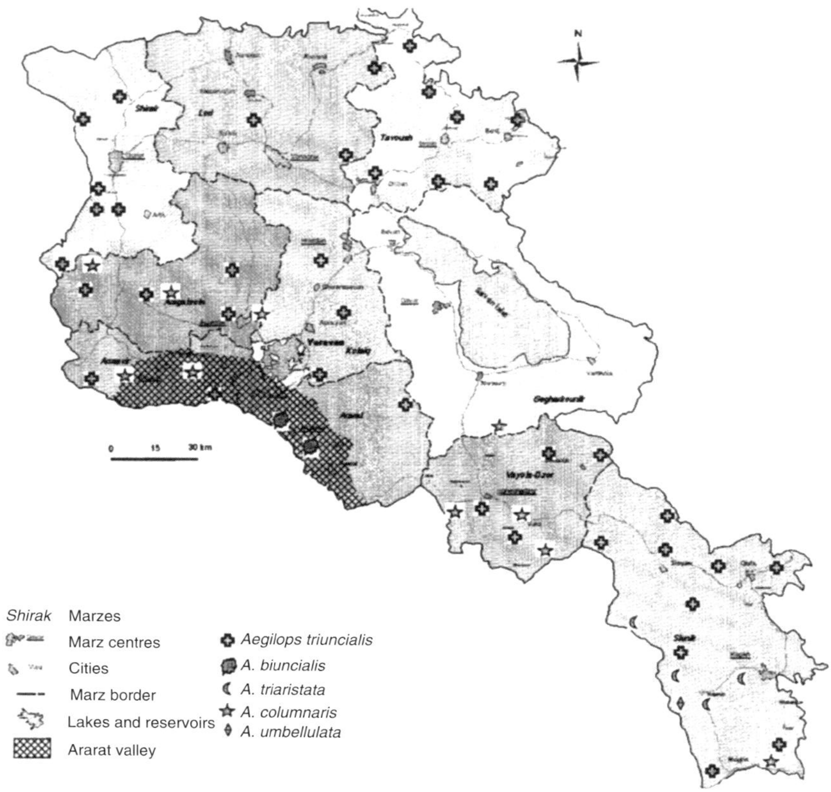

Fig. 19.2. Distribution of *Aegilops* species in Armenia in the 1980s (*A. triuncialis*, *A. biuncialis*, *A. triaristata* and *A. columnaris*, and *A. umbellulata*).

areas, *Aegilops* can be found at the edges. In areas containing *Aegilops* habitats, the species are distributed non-uniformly.

As a rule, several species of *Aegilops* are grown in one coenosis, very rarely phytocoenoses with only one *Aegilops* sp. being found. For example, in the phytocoenosis where *A. triuncialis* is predominant, *A. tauschii*, *A. cylindrica*, *Secale* L., *Helichrysum* Mill., *Teucrium polium* L., *Thymus kotschyanus* Boiss. et Hohen., *Xeranthemum sguarrosum* Boiss. *Artemisia* L., *Hordeum bulbosum* L. and *H. murinum* L. are also growing. *A. columnaris*, *A. biuncialis* and *A. triaristata* are growing in different districts (Yerevan, Zangezur). In the Erebuni Reserve where four species of *Aegilops* are growing together, *Triticum boeoticum* Boiss., *T. araraticum* Jakubz. and *T. urartu* Thum.ex Gandil., *Hordeum glaucum* Steud., *Secale vavilovii* Grossh.s.l., *Gundelia tournefortii* L., *Actinolema macrolema* Boiss. and *Asparagus officinalis* L. can often be found.

Spontaneous hybridizations have great importance in the evolution of the *Aegilops* genus. Interspecific and intraspecific spontaneous hybrids between *T. aestivum* and *A. cylindrica*, *A. triuncialis* and *A. cylindrica*, and *A. columnaris* and *A. cylindrica* are growing in natural habitats. As a result of introgres-

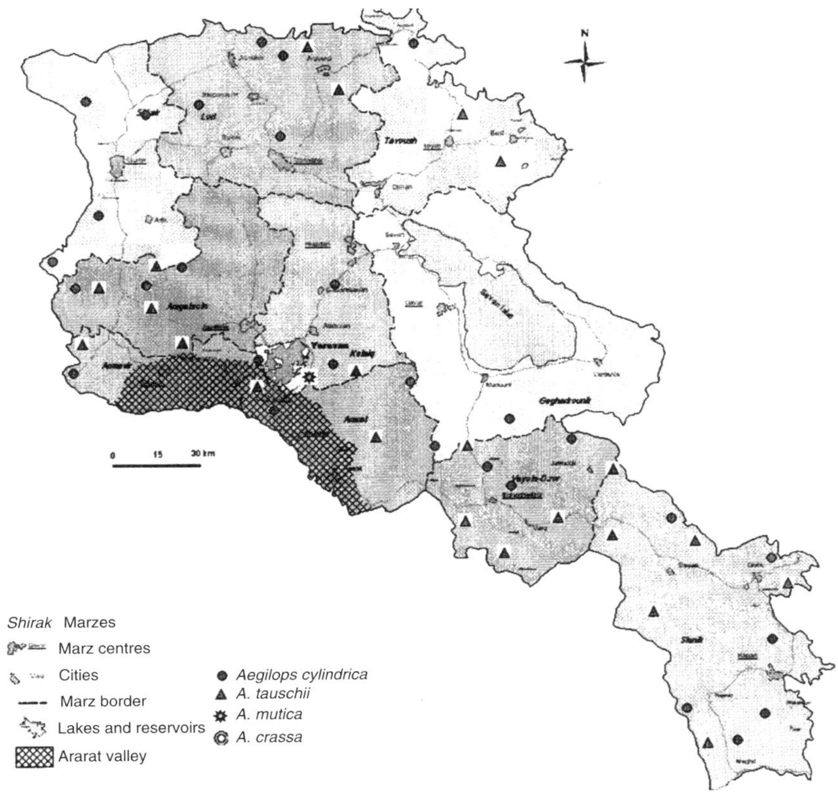

Fig. 19.3. Distribution of *Aegilops* spp. in Armenia in the 2000s (*A. cylindrica*, *A. tauschii*, *A. mutica* and *A. crassa*).

sion new species arise, therefore intergeneric spontaneous hybrids are considered a valuable breeding material.

19.5 Recommendations for *Aegilops* Conservation

Some measures have been taken in order to conserve endangered species of *Aegilops ex situ*. With this purpose, since 1985 thorough investigation of the *Aegilops* gene pool in Armenia has been carried out at the Laboratory of Plant Genetic Resources within the Armenian State Agrarian University. The work is implemented in the following ways:

1. Field explorations:
   - Collection missions – field collection missions are organized every year in different administrative Marzes (administrative divisions) of the Republic, which are located in different floristic provinces (12 floristic provinces in Armenia (Takhtajyan, 1972)).

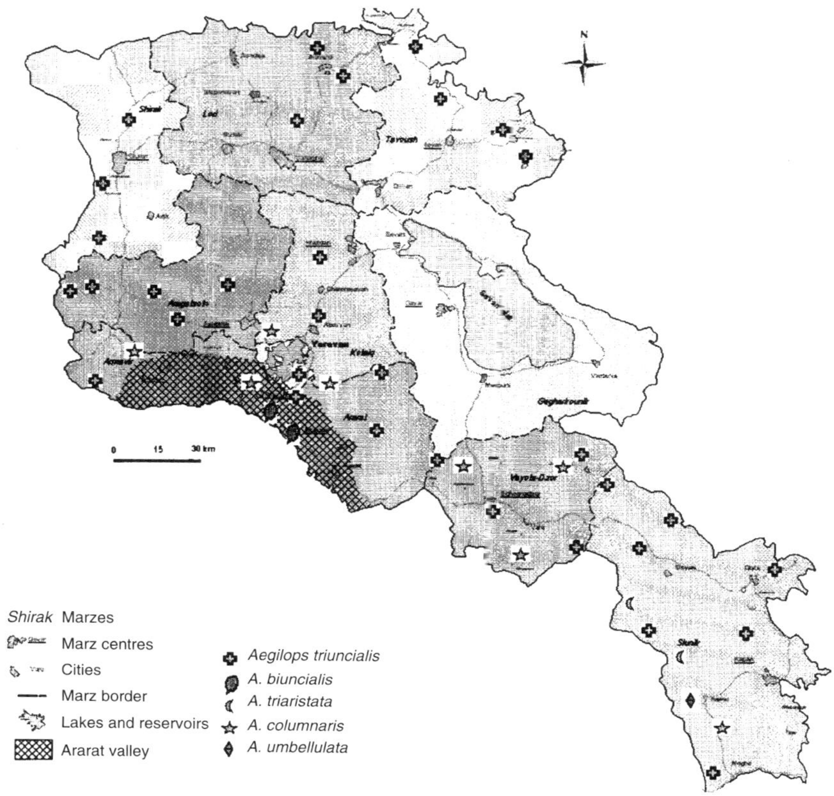

Fig. 19.4. Distribution of *Aegilops* species in Armenia in the 2000s (*A. triuncialis*, *A. biuncialis*, *A. triaristata* and *A. columnaris*, and *A. umbellulata*).

- Discovery of new sites and more precise definition of habitats.
- Observation and control for populations *in situ*.

2. Accurate definition of composition and detailed study of intraspecific diversity.
3. Accession evaluation (morphological, physiological and cytogenetic).
4. Registration, storage: a seed collection of 1610 accessions of *Aegilops* has been established in the Laboratory (Table 19.2). Although seed and spikes are stored under non-controlled conditions this collected material can serve as initial breeding material and after the construction of long-term storage facilities can be moved to the gene bank.
5. Reproduction: the Laboratory staff have established a small nursery as a field gene bank, where sowings, observations, morphological, physiological and cytogenetic investigations are carried out.
6. Providing breeders with genetic material and evaluation data.
7. Breeding (selection of interesting forms, their use in hybridization for breeding new high quality varieties of cereals): a new variety of wheat 'Voskehask'

Table 19.2. Number of *Aegilops* germplasm accessions held by PGR laboratory.

Species	Number of accessions
Aegilops biuncialis Vis	60
A. columnaris Zhuk.	136
A. crassa Boiss.	40
A. cylindrica Host	525
A. mutica (Boiss.) Eig.	31
A. tauschii Coss.	300
A. triaristata Willd.	440
A. triuncialis L.	45
A. umbellulata Zhuk.	33

('Golden Spike') was created in 1985 using *Aegilops tauschii*. This variety is now released for growing in valley and low altitude zones of Armenia.

8. Developing recommendations on conservation of *Aegilops* gene pool: our researchers involved in studying the *Aegilops* gene pool have suggested including *A. umbellulata* and *A. biuncialis* in the Armenian Red Data Book, which should be republished.

References

Ghazaryan, V.H. (ed.) (1989) *Red Book of Armenian SSR*. Sovetakan Grogh Publishers, Yerevan, Armenian SSR.

Takhtajyan, A.L. and Fedorov, A.A. (eds) (1972) *Flora Jerevana: opredelitel' rastuščikn rastenij Araratskoj kotloviny*. Nauka, Leningrad, Moscow.

Van Slageren, M.W. (1994) *Wild Wheats: A Monograph of* Aegilops *L. and* Amblyopyrum *(Jaub and Spach) Eig (Poaceae)*. Agricultural University, Wageningen – International Center for Agricultural Research in Dry Areas, Aleppo, Syria.

V *In Situ* Conservation

20 Crop Wild Relative *In Situ* Management and Monitoring: the Time Has Come

J.M. Iriondo and L. De Hond

20.1 Introduction

As farmers at a global scale have been abandoning their traditional cultivars in favour of more productive, genetically improved cultivars, domesticated biodiversity has experienced a severe loss in terms of genetic erosion (Frankel and Hawkes, 1975). For the last three decades, great efforts have been made both at national and international level to collect and preserve representative holdings of this germplasm. Seedbanks covering all major crops have been established in most countries for this purpose (Chin, 1994). As *ex situ* conservation of crop cultivars and landraces has reached maturity within the community of plant genetic resources (PGRs), in the last decade attention has been shifting to the conservation of wild plants, which are also experiencing a dramatic decline due to human-induced mass destruction of natural habitats (Meilleur and Hodgkin, 2004). The PGR community is especially interested in wild ancestors and relatives of crops as they are apt to have useful traits for breeding (Harlan, 1976, 1984; Hawkes, 1977; Prescott-Allen and Prescott-Allen, 1981). However, as we are dealing with wild plants living in natural conditions, there is a general consensus that these species are best conserved in their natural habitats (Hawkes, 1991).

As a result, the importance of *in situ* conservation of crop wild relatives (CWR) has been included in the most relevant conventions and strategies developed by the international community in the last decade, including the Convention on Biological Diversity (CBD, 1992), the Global Plan of Action for the Conservation and Sustainable Use of Plant Genetic Resources for Food and Agriculture (PGRFA) (FAO, 1996), the European Community Biodiversity Strategy (Anonymous, 2001), the Global Strategy for Plant Conservation (Anonymous, 2002) and the International Treaty PGRFA (FAO, 2002).

Now that the importance of *in situ* conservation is becoming increasingly recognized by the international community and policy makers (Meilleur and

Hodgkin, 2004), the question arises as to how we can go about it. For *ex situ* conservation, seedbanks with cost-efficient, effective protocols have been developed over decades, but what tools and methodologies do we need to implement effective *in situ* conservation?

Several books (Gadgil *et al.*, 1996; Maxted *et al.*, 1997a; Zencirci *et al.*, 1998) and national guidelines (Pavek and Garvey, 1999) have provided theory and methods for *in situ* CWR conservation. However, conservation biology is the scientific discipline that can ultimately provide answers to the questions that arise in each case. This young scientific discipline originated from other basic scientific disciplines as a response to the biodiversity crisis that the Earth is currently experiencing (Soulé, 1985). Conservation biology has fed upon contributions from ecology, genetics, physiology, taxonomy, sociology, economy and philosophy, among others, and has become a meeting point of scientists with many different perspectives but all of whom are concerned with the conservation of biodiversity (Meffe and Carroll, 1997). In response to the pressure of finding effective ways to counteract biodiversity loss, this discipline has experienced great development in recent years and its fundamental principles are in constant evolution. However, conservation biology has developed further in the field of the preservation of animal species. Thus, many of the tools and methodologies applied to plant conservation have largely been derived from the experience in animal conservation and have been assumed without much criticism. Nevertheless, some authors are beginning to question if the management procedures currently accepted in the conservation of plant species are indeed appropriate for the goals that are envisioned (Iriondo *et al.*, 2007a).

To date, most *in situ* plant conservation has focused on the conservation of threatened plant species. The question is whether this experience can largely be applied to the conservation of CWR or fundamental changes need to be made due to the differences in conservation goals and conservation status of CWR.

20.2 Previous Steps in Genetic Reserve Conservation

As we live in a world with limited resources, *in situ* conservation cannot be independently applied to all CWR. Thus, there is a need to prioritize. The first step in genetic reserve conservation in a particular region is to come up with a selection of CWR taxa. This selection can be approached using different criteria although socio-economic value and threat status are two of the most important features that are normally taken into account. As the operative unit of conservation is the population rather than the species, the process of prioritization involves the determination of a network of populations of selected CWR species. These populations should be located, if possible, in protected areas in order to facilitate the implementation of conservation actions. Thus, effective *in situ* conservation of a particular CWR species requires that the maximum genetic diversity of the targeted species be adequately represented in a network of a minimum number of genetic reserves carefully selected within the species' distribution range.

Once the number and location of the genetic reserves have been specified, the next step involves the establishment of a clear set of objectives for the reserves taking into account that we are following a model of taxon-based *in situ* conservation. If we were dealing with the conservation of a threatened species, the main objective of a network of reserves would be to maintain viable populations of the target species as well as populations' natural evolutionary processes, thus allowing new variation to be generated in the gene pool that would allow the species to adapt to gradual changes in environmental conditions (Heywood and Dulloo, 2005). Does the nature of CWR species change the focus of conservation? It certainly does because in this case it is not enough to maintain viable populations. The purpose of CWR conservation is to maintain the potential of using existing genetic diversity in CWR populations for crop breeding to obtain cultivars that better suit the needs of humankind at each moment (Maxted *et al.*, 1997b). Thus, we can conclude that the main objective of a network of genetic reserves for CWR is to maintain the genetic diversity of target populations in order to keep it available for use. The maintenance of viable populations is obviously a prerequisite.

The specific goals that need to be achieved to maintain viable populations in both threatened species and CWR conservations are:

- Minimize the risk of extinction from demographic fluctuations, environmental variation and catastrophes;
- Maintain genetic diversity and potential for evolutionary adaptation;
- Minimize human threats to target populations;
- Support actions that promote a positive balance between births and deaths in target populations.

However, CWR conservation involves additional specific goals to maintain genetic diversity and to make it available for use:

- Minimize the risk of genetic erosion from demographic fluctuations, environmental variation and catastrophes;
- Minimize human threats to genetic diversity;
- Support actions that promote genetic diversity in target populations (provided they do not affect viability in a negative way);
- Ensure access to populations for research and material for breeding;
- Ensure availability of material of target populations that are exploited and/or cultivated by local people.

20.3 *In Situ* Management of Crop Wild Relatives

In general terms, management is a process in which planning, work and resources are effectively combined to obtain a desired goal. Similarly, the *in situ* management of CWR should be considered a process in which these elements are wisely used (Iriondo *et al.*, 2007b). Populations need to be assessed and actively managed in order to achieve clear, measurable conservation goals. This can be obtained through the design and implementation of management plans for the prioritized CWR species of a region.

A management plan is a planning tool that contains a set of prescriptions and interventions to meet the objectives of the genetic reserve (Heywood and Dulloo, 2005). The minimum elements of a management plan have been described in detail by Maxted *et al*. (1997c). Essentially, a management plan is composed of three basic elements:

- Assessment of the taxon, population and site;
- Establishment of management targets;
- Prescription of conservation actions.

Assessment is an essential component for management success. It has the same elements as those found in threatened species recovery programmes, but with special emphasis on genetic diversity. The purpose of assessment is to identify and determine the importance of the factors that condition population viability and the maintenance of genetic diversity in the populations.

The establishment of clear, precise management targets is needed to obtain and evaluate the desired results and to measure success (Elzinga *et al*., 1998). Management and monitoring approaches depend on the definition of specific targets such as what status do we want to achieve in the population? or under what circumstances are we going to take a particular action? Some examples of management targets can be to keep population size over a certain number of individuals, not to allow a decline in population size over a certain percentage in 1 year, or to keep a specific allele that is key for pathogen tolerance in a gene at a frequency above a certain percentage.

Finally, management prescription includes all types of actions that can be executed on the target population, its habitat and the surrounding community to maintain ecological conditions and processes that are compatible with and necessary for the survival and maintenance of the genetic diversity in the target population. Management prescription also includes monitoring, which is a procedure to assess the effects of the management actions that are implemented. Since as a result of monitoring we may be able to conclude that our management actions are not producing the results we wanted, management prescription must also consider what actions to take when things do not go the way they are supposed to.

Thymus herba-barona subsp. *bivalens* Mayol, L. Sáez & Rosselló is a CWR that is severely threatened as it has only one known population located on the island of Mallorca (Spain). The assessment of the taxon has revealed that the habitat of the population has been changing during the last decades due to the abandonment of grazing by goats and sheep, resulting in the transformation of the open herbaceous pasture plant community to a more enclosed formation. The previously open community has partially been taken over by *Ampelodesmos mauritanica* (Poiret) T. Durand et Schinz, a perennial tall grass that forms big tussocks. The critical low size of the population of *T. herba-barona* subsp. *bivalens* with barely over 100 individuals explains why the main management target is not to allow any decline in population size. Management prescription for this taxon includes both *in situ* and *ex situ* actions. The main action in the natural population has been cutting the tussocks of *A. mauritanica* that are inter-

Fig. 20.1. Management prescription in Mallorca (Spain): clearing tussocks of *Ampelodesmos mauritanica* to create an open habitat for the only existing population of target CWR *Thymus herba-barona* subsp. *bivalens* (photograph by Magdalena Vicens).

spersed with the individuals of *T. herba-barona* subsp. *bivalens* at their base in order to reproduce an open habitat that resembles the one kept by grazing activity in the past and is more favourable for the target species (Fig. 20.1).

20.4 Monitoring Target Populations

The main purpose of monitoring is to assess whether the management targets established in the management plan are being properly met (Elzinga *et al.*, 1998). Depending upon the nature of the management targets, three different approaches to monitoring can be considered: demographic monitoring focused on the structure and dynamics of the target population, ecological monitoring to keep track of the changes of the surrounding habitat and community, and genetic monitoring to assess the structure and dynamics of genetic diversity in the target population.

The design of the monitoring scheme is essential for monitoring success. One of the key elements of design is the identification and selection of the variables to be monitored. For the sake of efficacy, variables need to be simple and easy to measure, while being directly associated to the management target. The quality and accuracy of monitoring activities are directly linked to the sampling procedure. Therefore, at the design stage, the monitoring procedure must be clearly specified indicating what, how, when, where, how much and how often to sample.

Since the maintenance of genetic diversity in the populations is one of the main objectives of CWR conservation, genetic monitoring is an especially relevant part of the monitoring scheme. The study of genetic variation in adaptive traits requires specific knowledge and various tests for long periods and at many sites in order to understand the genetic bases. This classical approach is simply impossible to apply on a large scale in the monitoring of CWR populations in genetic reserves (Neale, 1998). The use of molecular markers, such as random amplified polymorphic DNA (RAPD), restriction fragment length polymorphism (RFLP), amplified fragment length polymorphism (AFLP), inter-simple sequence repeat (ISSR) and simple sequence repeat (SSR), provides rapid surveys of genetic variation within and between populations (Nybom, 2004). Good accounts of the ways in which genetic diversity can be measured both at the species and the population levels are included in different texts (Mallet, 1996; Smith and Wayne, 1996; Karp *et al.*, 1997; Newbury and Ford-Lloyd, 1997). However, apart from economic considerations, their value in guiding genetic conservation is limited, as they do not characterize the frequency and distribution of adaptive traits, but simply rely on genetic sequences that are randomly obtained and/or have no phenotypic expression. Thus, genetic markers should be used with care unless combined with observations on quantitative traits such as growth and survival or other complementary approaches (Pefender *et al.*, 2000; Bekessy *et al.*, 2003). Therefore, there is presently an urgent need to find new genetic indicators that may be useful for monitoring genetic diversity in CWR populations. According to Graudal *et al.* (1997), surveys based on the use of ecological data in combination with biochemical markers and data from already established field trials may be a possible approach to the problem.

20.5 Management Scenarios

When dealing with the management of CWR populations in genetic reserves, the different situations that one may find can be classified into two basic scenarios: (i) the target CWR population is in good condition and no specific threats to the population are identified in the short or medium term; and (ii) the target CWR population is threatened by one or several factors and the population and its genetic diversity are likely to experience a significant decline in the future if no action is taken.

Initially, the network of populations for the *in situ* conservation of a CWR species is bound to be selected on the basis of the good condition of its populations in terms of population size, genetic diversity and absence of specific threats. Therefore, most cases will fall into the first scenario where the target CWR population is stable and has no threats. In these cases, the management plan can be greatly simplified. First, it will require an assessment of the taxon, population and habitat to ascertain good status of the population. The management targets will be reduced to maintaining the status quo of the population and the management prescription will only require periodic monitoring of the population and its habitat to confirm the well-being of the population.

However, a significant percentage of CWR species may themselves be globally threatened, and, therefore, the populations selected for conservation may be experiencing serious threats. In this case, the assessment phase of the management plan will be critical to correctly diagnose the status of each population and clearly determine the factors that most greatly condition the viability and genetic diversity of the population. Consequently, the management targets and prescription will focus on the control and/or removal of the factors originating the threats to the population and obtaining genetically diverse viable populations. Monitoring, in this case, will be crucial for assessing and measuring the success of the management actions that have been applied. In general, the more threatened the population is, the more intensive the conservation interventions needed are likely to be.

An illustrative case of severely threatened CWR populations and intensive management actions can be found in *Narcissus cavanillesii* A. Barra & G. López in Portugal. As a wild relative of ornamental *Narcissus* cultivars, one of its most outstanding features is the fact that it flowers in autumn, while most *Narcissus* species and cultivars flower in spring. *N. cavanillesii* is endemic to the southern Iberian Peninsula and North Africa and is included in Annexes II and IV of EU Habitats Directive (92/43/CEE). In Portugal, only two populations of this species are known. These two populations became severely threatened by the construction of the Alqueva dam, the largest dam in Europe. As the population of Montes Juntos was going to be flooded, a translocation action was implemented, which was led by the Natural History Museum and Botanic Garden of Lisbon University, promoted by EDIA, S.A. and co-funded by EDIA, S.A. and FEDER. The operation started with the establishment of a geographic information system in which the geographical coordinates of the subpopulations of Montes Juntos were stored and environmental information on the site and its surroundings was gathered. This information was used to build a predictive distribution model of the species to find adequate alternative sites and finally select the receptor site. At the reception site, major shrubs were removed to obtain a microenvironment suitable for the target population. The subpopulations growing on schists were extracted using drills and other mechanical equipment used in the exploitation of quarries. A 1 m^2 grid was imposed on the subpopulations growing on soil. These subpopulations were then divided into 1 m^2 grid, which were properly identified and numbered, before transporting them to the receptor site. At the receptor site, the spatial structure of the subpopulations was re-created maintaining the same distances and orientation as at the original location. The subpopulations that grew on soil were reconstructed by placing the 1 m^2 grid in the same order and orientation as at the original location (Fig. 20.2). All the translocated subpopulations were intensively cared for during the first few months to restore original conditions as much as possible. After this, a monitoring procedure was established to follow the dynamics of the population and to detect any problems of adaptation to the new site or the appearance of additional threats that were not present in the original population. During the first few years, the flowering percentage of

Fig. 20.2. Intensive management of Portuguese *Narcissus cavanillesii* populations as a result of the construction of the Alqueva dam: extraction of a subpopulation growing on soil before translocation to the receptor site. The subpopulation was divided into a 1 m^2 grid to facilitate transportation. The original arrangement was reproduced at the receptor site (photograph by Isabel Marques).

individuals decreased significantly due to the effects of the translocation. However, in the fifth year, the flowering percentage reached the same values as those that were present before the translocation, a good indicator that the population is recovering from the drastic operation.

20.6 Management Problems

The main problem facing genetic reserve management for the *in situ* conservation of CWR is probably its economic cost. The selection of the populations to be conserved, the assessment of the taxon, its populations and sites, the actions to eliminate operating threats and the establishment of monitoring procedures all require a significant input of economic resources. Apart from showing politicians and stakeholders that this type of investment is profitable in the medium and long term, there is a clear need to design cost-effective networks of genetic reserves. This is the main reason why the *in situ* conservation of a particular CWR species requires that the maximum genetic diversity of the targeted species be adequately represented in a network of a minimum number of genetic reserves carefully selected within the distribution range of the species.

The most practical approach for Europe is to make use of the present network of protected areas. Although this network was established for other purposes, its human and material resources could be readily applied for active conservation of selected CWR. Thus, the establishment of a network of genetic reserves for CWR conservation would be more cost-effective if as many populations as possible were located within the network of protected areas. The participation of society in the process and local involvement, providing local people with opportunities to work in the management and monitoring procedures and to exploit the plant resources conserved in the genetic reserve in a sustainable way, are important elements for successfully overcoming the economic limitations of these operations (Meilleur and Hodgkin, 2004).

Some other problems that can be anticipated in the management of genetic reserves are those that are directly related to differences in scale. This may be the case of CWR species in which metapopulation dynamics have a special relevance and where management operations on a local scale may not be sufficient. In this situation, the scale of management should be widened to include several populations, which may form part of one or several coordinated genetic reserves, and unoccupied sites that may be suitable for the colonization of the species. Similarly, there are also problems originated by climate change or other global changes. Once again, an appropriate management action for these problems exceeds the scale of operation of a single genetic reserve and will require coordinated action by the whole network of genetic reserves established for a particular CWR taxon.

Other problems to take into account in the management of genetic reserves for CWR are those related to biotic interactions. Specific to the case of CWR are problems derived from possible interactions between CWR being conserved in a genetic reserve and related crops grown nearby. These crops can potentially affect the genetic integrity of the conserved CWR populations through cross-pollination and seed rain. The intensity of the problem will depend, to a great extent, on the breeding system, pollination method and seed dispersal mechanism of the target species. In cases where this may pose a severe problem, a buffer zone may need to be established around the genetic reserve where the cultivation of crops related to the target species would not be allowed.

Another management problem linked to species interactions can arise when genetic reserves are established to conserve several CWR species or when genetic reserves are located within protected areas where active measures are being taken to conserve other target plant or animal species. Actions carried out for the conservation of a CWR species can potentially have a negative effect on other target CWR or non-CWR species of the genetic reserve or protected area and vice versa. There is no simple way around this type of problem. However, coordinating the management plans of different CWR taxa in a single genetic reserve or integrating the management plans of the genetic reserve in the management of the protected area where it is located will obviously be very helpful in avoiding or minimizing these effects.

20.7 Conclusions

Despite all the progress experienced in the *ex situ* conservation of PGRs during the last three decades, it is evident that *in situ* genetic conservation has not progressed in the same way and is still in its infancy (Jain, 1975; Hawkes, 1991; Maxted *et al.*, 1997b). Nevertheless, in recent years several local, national and international initiatives have been put forward in CWR *in situ* conservation providing invaluable experience in management and monitoring (Meilleur and Hodgkin, 2004; Heywood and Dulloo, 2006). In many cases, these initiatives are totally independent and unrelated, and somewhat isolated. In order to take synergic advantage of these initiatives, there is a need to organize meetings, workshops and symposia that may facilitate training, communication and coordination in management practices (Meilleur and Hodgkin, 2004). In Europe, the European Crop Wild Relative Diversity Assessment and Conservation Forum (PGR Forum) funded by the European Commission and developed from 2002 to 2005 has been very helpful in focusing needs, developing tools and bringing together local initiatives to a new and greater dimension. In September 2005, the final meeting of PGR Forum served to launch the First International Conference on Crop Wild Relative Conservation and Use, providing the first chance to overview the different projects being developed worldwide. We are, thus, living in a critical and exciting time in the establishment of management and monitoring methodologies for CWR. The time has come to implement wide-scale initiatives for the *in situ* conservation and management of CWR.

References

Anonymous (2001) *European Plant Conservation Strategy*. Council of Europe and Planta Europa, London.

Anonymous (2002) *Global Strategy for Plant Conservation*. Secretariat of the Convention on Biological Diversity, Montreal, Canada.

Bekessy, S.A., Ennos, R.A., Burgman, M.A., Newton, A.C. and Ades, P.K. (2003) Neutral DNA markers fail to detect genetic divergence in an ecologically important trait. *Biological Conservation* 110, 267–275.

CBD (1992) *Convention on Biological Diversity: Text and Annexes*. Secretariat of the Convention on Biological Diversity, Montreal, Canada

Chin, H.F. (1994) Seedbanks: conserving the past for the future. *Seed Science and Technology* 22, 385–400.

Elzinga, C.L., Salzer, D.W. and Willoughby, J.W. (1998) *Measuring and Monitoring Plant Populations*. BLM Technical Reference 1730–1731. Bureau of Land Management, Denver, Colorado.

FAO (1996) *Global Plan of Action for the Conservation and Sustainable Utilization of Plant Genetic Resources for Food and Agriculture*. FAO, Rome, Italy.

FAO (2002) *The International Treaty on Plant Genetic Resources for Food and Agriculture*. FAO, Rome, Italy.

Frankel, O.H. and Hawkes, J.G. (eds) (1975) *Crop Genetic Resources for Today and Tomorrow*. Cambridge University Press, Cambridge.

Gadgil, M., Singh, S., Nagendra, H. and Chandran, M. (1996) *In Situ Conservation of Wild Relatives of Cultivated Plants: Guiding Principles and a Case Study*. FAO, Rome, Italy, and Indian Institute of Science, Bangalore, India.

Graudal, L., Kjaer, E.D., Thomsen, A. and Larsen, A.B. (1997) *Planning National Programmes for Conservation of Forest Genetic Resources*. Danida Forest Seed Centre Series of Technical notes No 48. Danida Forest Seed Centre, Humlebaek, Denmark.

Harlan, J. (1976) Genetic resources in wild relatives of crops. *Crop Science* 16, 329–333.

Harlan, J. (1984) Evaluation of wild relatives of crop plants. In: Holden, J. and Williams, J. (eds) *Crop Genetic Resources: Conservation and Evaluation*. George Allen & Unwin, London, pp. 212–222.

Hawkes, J. (1977) The importance of wild germplasm in plant breeding. *Euphytica* 26, 615–621.

Hawkes, J. (1991) International workshop on dynamic *in situ* conservation of wild relatives of major cultivated plants: summary of final discussion and recommendations. *Israel Journal of Botany* 40, 529–536.

Heywood, V.H. and Dulloo, E. (2005) *In Situ Conservation of Wild Plant Species – a Critical Global Review of Good Practices*. IPGRI-FAO, Rome, Italy.

Iriondo, J.M., Escudero, A. and Albert, M.J. (2007a) Conservation of plant populations: myths and paradigms. In: Valladares, F., Camacho, A., Elosegui, A., Estrada, M., Gracia, C., Senar, J.C. and Gili., J.M. (eds) *Unity in Diversity*. Fundación BBVA, Madrid, Spain.

Iriondo, J.M., Maxted, N. and Dulloo, E. (eds) (2007b) *Conserving Plant Genetic Diversity in Protected Areas*. CAB International, Wallingford, UK.

Jain, S. (1975) Genetic reserves. In: Frankel, O. and Hawkes, J. (eds) *Crop Genetic Resources for Today and Tomorrow*. Cambridge University Press, Cambridge.

Karp, A., Kresovich, S., Bhat, K.V., Ayad, W.G. and Hodgkin, T. (1997) *Molecular Tools in Plant Genetic Resources Conservation: A guide to the Technologies*. IPGRI, Rome, Italy.

Mallet, J. (1996) The Genetics of biological diversity: from varieties to species. In Gaston, K. (ed.) *Biodiversity: Biology of Numbers and Difference*. Blackwell, Oxford, UK, pp. 13–47.

Maxted, N., Ford-Lloyd, B. and Hawkes, J. (eds) (1997a) *Plant Genetic Conservation: The* In Situ *Approach*. Chapman & Hall, London.

Maxted, N., Ford-Lloyd, B. and Hawkes, J. (1997b) Complementary conservation strategies. In: Maxted, N., Ford-Lloyd, B. and Hawkes, J. (eds) *Plant Genetic Conservation: The* In Situ *Approach*. Chapman & Hall, London, pp. 15–39.

Maxted, N., Guarino, L. and Dulloo, M.E. (1997c) Management and monitoring. In: Maxted, N., Ford-Lloyd, B. and Hawkes, J. (eds) *Plant Genetic Conservation: The* In Situ *Approach*. Chapman & Hall, London, pp. 144–159.

Meffe, G.K. and Carroll, C.R. (1997) *Principles of Conservation Biology*. Sinauer Associates, Sunderland, Massachusetts.

Meilleur, B. and Hodgkin, T. (2004) *In situ* Conservation of wild relatives: status and trends. *Biodiversity and Conservation* 13, 663–684.

Neale, D.B. (1998) Molecular genetic approaches to measuring and conserving adaptive genetic diversity. In: Zencirci, N., Kaya, Z., Anikster, Y. and Adams, W. (eds) *Proceedings of International Symposium on In Situ Conservation of Plant Genetic Diversity*. Central Research Institute for Field Crops, Ankara, Turkey, pp. 385–390.

Newbury, H.J. and Ford-Lloyd, B.V. (1997) Estimation of genetic diversity. In: Maxted, N., Ford-Lloyd, B.V. and Hawkes, J.G. (eds) *Plant Genetic Conservation: The* In Situ *Approach*. Chapman & Hall, London, pp. 192–206.

Nybom, H. (2004) Comparison of different nuclear DNA markers for estimating intraspecific genetic diversity in plants. *Molecular Ecology* 13, 1143–1155.

Pavek, D. and Garvey, E. (1999) *The American Wild Relatives of Crops:* In Situ *Conservation Guidelines*. USDA/ARS, Beltsville, Maryland.

Pefender, M.E., Spitze, K., Hicks, J., Morgan, K., Latta, L. and Lynch, M. (2000) Lack of concordance between genetic diversity estimates at the molecular and quantitative-trait levels. *Conservation Genetics* 1, 263–269.

Prescott-Allen, R. and Prescott-Allen, C. (1981) In Situ *Conservation of Crop Genetic Resources. A Report to the International Board for Plant Genetic Resources (IBPGR)*. IUCN, Gland, Switzerland.

Smith, T.B. and Wayne, R.K. (eds) (1996) *Molecular Genetics Approaches in Conservation*. Oxford University Press, New York.

Soulé, M. (1985) What is conservation biology? *BioScience* 35, 727–734.

Zencirci, N., Kaya, Z., Anikster, Y. and Adams, W. (eds) (1998) *Proceedings of International Symposium on In Situ Conservation of Plant Genetic Diversity*. Central Research Institute for Field Crops, Ankara, Turkey.

21 Does Agriculture Conflict with *In Situ* Conservation? a Case Study on the Use of Wild Relatives by Yam Farmers in Benin

N. Scarcelli, S. Tostain, M.N. Baco, C. Agbangla, O. Daïnou, Y. Vigouroux and J.L. Pham

21.1 Introduction

The development of agriculture affected ecosystems as soon as the domestication of plants and animals allowed human populations to colonize and manage habitats for their own use. The threat to wild plant species increased dramatically with the intensification of agricultural practices and population growth.

Crop wild relatives (CWR) in centres of diversity of cultivated plants are also submitted to these threats. Those having an ability to develop weedy habits (e.g. wild rice) can survive in extensive agroecosystems. For those that grow in ecological conditions very different from cultivated fields, such as forest species, the situation is more critical as damage caused by the extension of cultivation areas may not be reversible.

Although protection of particular areas can be set up, agroeconomical pressures make the viability of this approach uncertain. The acceptability of protection plans by both decision makers and local populations can be enhanced by the assessment of expected ecological and genetic benefits.

The role of farmers in preserving crop diversity is now recognized: 'We acknowledge the role played by generations of men and women farmers and plant breeders, and by indigenous and local communities, in conserving and improving plant genetic resources' (FAO, 1996). However, research studies reporting farmers' use of wild relatives to enrich the cultivated pool are scarce. Even studies on teosinte and maize (Wilkes, 1977) in Mexico are not fully conclusive (Wood and Lenné, 1997).

This chapter reports a case study on yam, a tuber crop, and its wild relatives in Benin. It describes the use made by local farmers of yam wild relatives and present results of genetic studies that demonstrate the contribution of farmers' practices to the evolutionary dynamics of this vegetatively propagated crop. It also discusses how such practices should be integrated in *in situ* conservation strategies.

21.2 Yam and its Ennoblement

Yam is mainly cultivated in West Africa (95% of the world production, FAO, 2005). Two different species are cultivated in this region: *Dioscorea alata* L., an Asian species, and *D. rotundata* Poir., an African species. Excepting Côte d'Ivoire, *D. rotundata* is the main cultivated species in West Africa. In Benin, for example, *D. rotundata* accounts for 95% of the total production.

Yam is a vegetatively propagated crop largely cultivated in traditional agrosystems in West Africa; i.e. there is no mechanization of the cultivation processes, use of fertilizers is limited and few improved varieties are available (Baco *et al.*, 2004). Yam farmers use about 30% of the production to sow a new field. A lot of cultivated varieties of *D. rotundata*, a dioecious species, are still able to flower and produce seeds. However, no study has reported the use of seeds by the farmers.

The two wild relatives of *D. rotundata* are grown in Benin. *D. abyssinica* Hochst. ex Kunth grows in savannah in the north and centre of Benin and *D. praehensilis* Benth. in forests of the centre and south of Benin (Hamon *et al.*, 1995). These two species reproduce sexually and are dioecious. In Benin, these two wild species grow in sympatry with cultivated *D. rotundata*, i.e. they grow in areas surrounding the fields of cultivated yams. Controlled crosses have shown that wild and cultivated species are interfertile (Akoroda, 1985), but to our knowledge, no evidence of natural hybridization has ever been reported.

In West Africa, and particularly in Benin, sociological studies have documented a unique farming practice (Dumont and Vernier, 2000; Vernier *et al.*, 2003) that leads farmers to use wild yam species. Following Mignouna and Dansi (2003), this practice is termed as 'ennoblement'. A detailed description of this practice is given by Dumont *et al.* (2005).

The first step of ennoblement is to collect tubers of wild yams. Only a piece of tuber is collected, allowing the survival of the wild plant. Farmers select wild tubers for their likeness to cultivated varieties, e.g. in northern Benin, they look for plants with large green stems, with large tubers and white flesh and lacking spines (Baco, 2000). According to sociological studies, both wild species, *D. abyssinica* and *D. praehensilis*, are collected, depending on which species are grown in the region (Vernier *et al.*, 2003). Tubers are then submitted to different types of stress in order to obtain a change in their morphological characters. One practice consists of harvesting the tubers produced twice: the first harvest is made in June/September and the second harvest in November/December. Another practice consists of putting an obstacle (piece of pottery, stone) in the soil in order to limit the length of the tuber and obtain a tuber with a cultivated-like shape.

In this chapter, the tubers selected by the farmers and undergoing the ennoblement process are termed as 'pre-ennobled' yams. Tubers are stressed during 3–6 years. According to farmers, some of these pre-ennobled yams develop a tuber that is morphologically close to those of cultivated varieties. The biological processes underlying the change in tuber morphology and its maintenance over generations are unknown. Phenotypic plasticity and epigenetic modifications have been proposed as possible explanations, but further analyses are needed to understand this part of the ennoblement process (Scarcelli *et al.*, 2006).

If farmers are satisfied with the morphology of ennobled tubers, they introduce them in a variety with similar tuber morphology or create a new variety. If they are not satisfied, tubers are no longer grown and used.

By the ennoblement practice, farmers attempt to create new varieties or to recover lost varieties (Baco, 2000). Indeed, they claim that each variety exists in a wild form. Some farmers practice ennoblement only for curiosity, to see if it is really possible to change a wild yam in a cultivated yam. Finally, other old farmers just want to explain to young farmers how to practice ennoblement. Ennoblement is practised by both old and young farmers (Baco, 2000).

According to this description, ennoblement is a practice that leads farmers to use wild yam relatives and to introduce diversity in the cultivated pool. This is, from a genetic standpoint, of considerable importance for the *in situ* conservation on farms of yam genetic resources. Theoretically, somatic mutations are the only evolutionary factor in the genetic pool of strictly vegetatively reproduced crops. Genetic studies were therefore needed to assess the genetic impact of ennoblement.

This chapter analyses the use of wild yams by farmers in Benin. First, we investigate *in situ* interspecific hybridization between wild and cultivated species. Then, we study the genetic nature of the pre-ennobled yams, i.e. wild, hybrid or cultivated genotypes, and the existence of wild or hybrid yams in the cultivated pool. These results are presented in detail in Scarcelli *et al.* (2006). Finally, we discuss the need to integrate such practices in the *in situ* conservation of wild yam relatives.

21.3 Materials and Methods

All genetic studies were made in Benin (West Africa). First, the existence of interspecific hybridizations between wild and cultivated species was studied in the north of Benin, in the village of Gorobani (Fig. 21.1). In this village, only the wild species *D. abyssinica* was present. Crop fields were set up in reclaimed wooded savannah. *D. abyssinica* was grown in sympatry with *D. rotundata*, the cultivated species, i.e. grown in the savannah area surrounding the fields of cultivated yams. Some wild and cultivated samples were collected to make two reference populations. The cultivated population consisted of 46 plants collected in the fields of five farmers. The wild population was made up of 105 yams collected in the savannah surrounding these five fields. In addition, 93 seeds from seven cultivated plants and 102 seeds from seven wild plants in the surrounding savannah were collected.

All samples were genotyped with 11 nuclear microsatellites. A paternity analysis was made using a home-made programme, to test whether these seeds resulted from fertilization by a male originating from the wild or the cultivated yam species.

Then, the genetic nature of the pre-ennobled yams and the introduction of wild or hybrid plants in the cultivated pool was analysed. This study was made

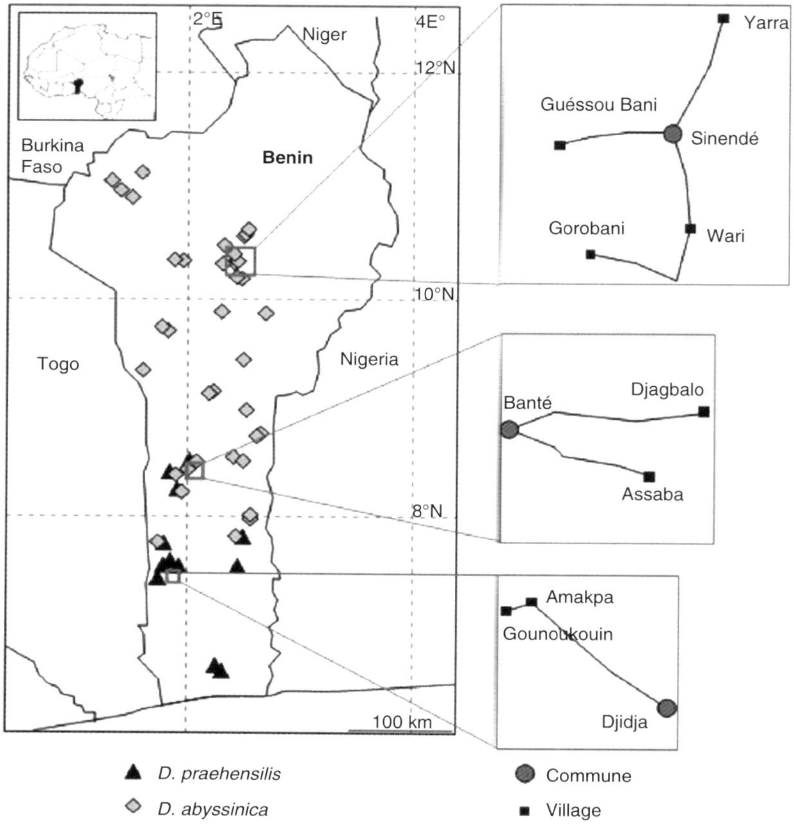

Fig. 21.1. Geographic origin of yam samples. Cultivated and pre-ennobled samples were collected in eight villages (Gorobani, Guéssou Bani, Yarra, Wari, Assaba, Djagballo, Amakpa, Gounoukouin), corresponding to three geographical and four ethno-linguistic regions (Bariba and Gando in the north, Nagot in the centre and Fon in the south). Wild yams were sampled according to the area distribution of the species: *Dioscorea abyssinica* grows in the north and centre of Benin and *D. praehensilis* in the centre and south of Benin (Hamon *et al.*, 1995).

in eight villages (Fig. 21.1). The village sample encompassed a broad range of situations with respect to geography, ethnicity and ecology. In southern Benin, two villages were sampled, Amakpa and Gounoukouin, inhabited by the Fon ethnic group. In this region, the wild species *D. praehensilis* grows in rain forests. In central Benin, two villages of the Nagot ethnic group, Assaba and Djagballo, were sampled. The two wild species *D. praehensilis* and *D. abyssinica* grow in a mosaic of forests and wooded savannah and are parapatric. In northern Benin, the four sampled villages, Gorobani, Guessou Bani, Wari and Yarra, corresponded to the two ethnic groups Bariba and Gando. In this region, only *D. abyssinica* grows in wooded savannah. Thirty tubers of pre-ennobled yams and 56 tubers of *D. rotundata* were collected from these eight villages. The farmers claimed that among them the variety 'Gban' was the result of the ennoblement of *D. praehensilis*. As many as 71 *D. abyssinica* and 33 *D. praehensilis* plants were also sampled (Fig. 21.1) throughout the distribution area of the two wild species in Benin (Hamon *et al*., 1995).

All samples were genotyped with 11 nuclear microsatellites. Then, an assignment analysis was made, using the software STRUCTURE 2.0 (Pritchard *et al*., 2000), to test whether the pre-ennobled yams corresponded to wild, hybrid or cultivated yams and if some plants with a wild or a hybrid signature in the cultivated pool can be found. Results are presented using a principal component analysis (PCA).

21.4 Results

21.4.1 Hybridization study

Seeds collected from wild and cultivated yams were analysed to assess if female parents were fertilized by wild or cultivated male gamete. In the progeny of wild female parents, 93.6% of male gametes were assigned to the wild population and no male gamete was assigned to the cultivated population (Table 21.1). However, in the progeny of cultivated female parents, only 3.2% of male gametes were assigned to the cultivated population, whereas 77.4% were assigned to the wild population (Table 21.1). The latter result showed that spontaneous interspecific hybridizations had occurred between wild and cultivated yams.

21.4.2 Assignment of pre-ennobled yams

The genetic diversity of the three yam species, i.e. *D. abyssinica* and *D. praehensilis*, the wild species, and *D. rotundata*, the cultivated species, was analysed first. Eleven microsatellites were amplified in a total of 160 samples and revealed 196 alleles. Among these alleles, 49% were specific to one of the three species. The PCA (Fig. 21.2a) showed that the three species are well separated.

Pre-ennobled tubers were assigned to one of the three species, *D. abyssinica*, *D. praehensilis* and *D. rotundata*, using STRUCTURE 2.0. Results are presented in PCA (Fig. 21.2b and c). Out of the 30 pre-ennobled yams, four (13%)

Table 21.1. Assessment of the progeny of seven cultivated and seven wild female yams. For each offspring collected on a female parent, the paternity analysis assessed if the male parent came from the wild (*Dioscorea abyssinica*) or the cultivated (*D. rotundata*) species.

	Female parent	
Male parent	Wild	Cultivated
Wild	95	68
Cultivated	0	3
Unknown	7	22
Total	102	93

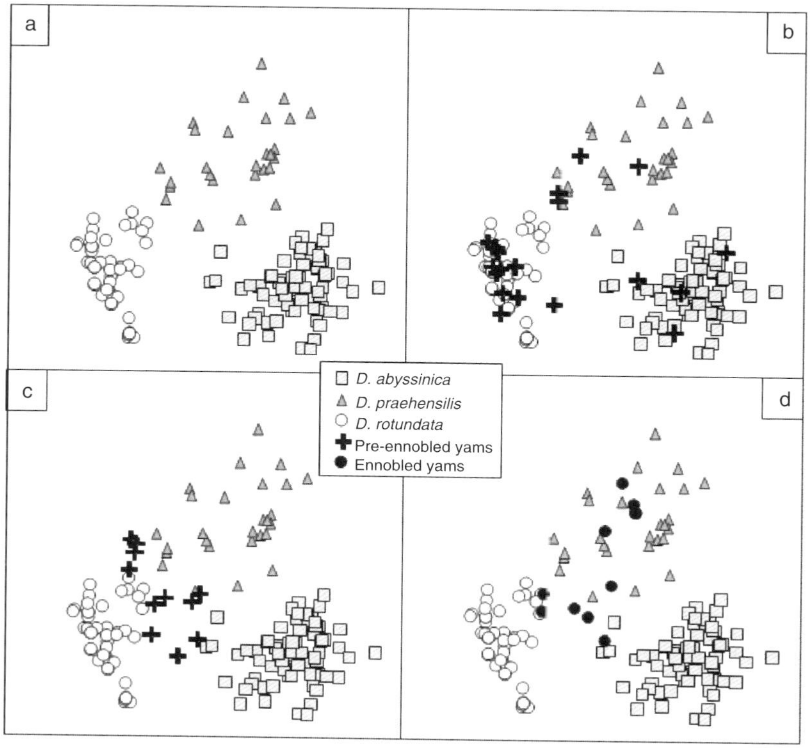

Fig. 21.2. Principal component analysis (PCA) of genetic diversity of wild, cultivated and pre-ennobled yams. (a) Genetic diversity of wild (*Dioscorea abyssinica* and *D. praehensilis*) and cultivated species (*D. rotundata*); (b) Position of pre-ennobled yams that were assigned to the wild or the cultivated species; (c) Position of pre-ennobled samples that had a hybrid origin; (d) Position of cultivated varieties created by ennoblement (ennobled yams). These varieties had a wild or a hybrid genotype.

were assigned to the *D. abyssinica* group and four (13%) to the *D. praehensilis* group (Fig. 21.2b). These samples were interpreted as wild yams. Eleven (37%) pre-ennobled yams had admixed ancestry and were considered to be hybrids (Fig. 21.2c). Out of these 11 samples, four corresponded to *D. abyssinica* × *D. rotundata* hybrids, five to *D. praehensilis* × *D. rotundata* hybrids and two to *D. praehensilis* × *D. abyssinica* hybrids. Finally, 37% of the pre-ennobled yams were assigned to the *D. rotundata* group (Fig. 21.2b). Pre-ennobled yams that presented a wild or a hybrid genotype were found in the four ethnic groups analysed.

21.4.3 Introduction of wild or hybrid yams in the cultivated pool

Nine cultivated samples, representing four different genotypes, were assigned to the wild group *D. praehensilis* (Fig. 21.2d), which means that some

D. praehensilis genotypes were present in the cultivated varieties. This suggested the successful ennoblement of *D. praehensilis* individuals, i.e. the 'Gban' variety. Farmers presented this variety as ennobled from *D. praehensilis*. The genetic analysis confirmed the farmers' statement. Six other cultivated samples corresponded to *D. abyssinica* × *D. rotundata* hybrids and *D. praehensilis* × *D. rotundata* hybrids. This result suggested successful ennoblement of hybrid individuals.

21.5 Discussion

The genetic analysis of the ennoblement process brings data that demonstrate the importance of this traditional practice for the dynamics of yam diversity in Benin. Our results showed that a gene flow takes place from the wild to the cultivated species thanks to ennoblement, and thanks to the natural hybridizations that spontaneously occur between wild and cultivated yams. The first consequence is that farmers maintain the diversity of the cultivated species by using the yam wild relatives and then introducing new alleles and new genetic combinations. The second important consequence is that farmers unconsciously use the sexual reproduction of the wild and cultivated yam species. This use of sexuality in this asexually propagated crop has some evolutionary consequences.

Indeed, in a vegetatively propagated crop such as yam, mutation is the only factor that creates diversity. The adaptive potential of yam is therefore limited by the existing diversity. However, sexuality produces new genetic combinations by recombination during meiosis and by combining genes from both parents (Barton and Charlesworth, 1998). Our results show that yam farmers combine the advantages of both sexuality and asexuality. By testing and selecting new combinations created by sexuality, they maintain the potential for future adaptation, and, at the same time, they preserve their best yam genotypes from recombination by using asexual reproduction.

These results suggest that the use of wild yam diversity through the ennoblement practice should be maintained in order to maintain the genetic diversity of cultivated yams. Thus, the on-farm conservation of cultivated yams should also address both the *in situ* conservation of wild yam diversity and the conservation of traditional knowledge and farmers' practices.

In Benin, wild yam diversity is threatened. The demographic pressure is increasing. Forests and savannahs are destroyed to produce intensive cultures like maize or cash crops like cotton. Moreover, traditional beliefs that protect wild habitat are progressively lost. For example, in southern Benin, a lot of forests where *D. praehensilis* grows are sacred and their exploitation is strictly governed by traditional chiefs (Allomasso, 2001). However, the authority of these traditional chiefs is eroding and these forests are more and more exploited and destroyed.

Only the conservation of the natural environments will allow the conservation of the wild yam diversity. However, it is not reasonable to preserve wild areas by prohibiting their exploitation by local populations. First, life and well-being of local populations depend on the exploitation of these natural areas. Moreover, such programmes would be too expensive to be viable (Louafi and Tubiana, 2005). According to these authors, without the participation of local populations and users and without taking into account their respective interests, the biodiversity has little chance to be preserved on the long term. Moreover, according to Kaimowitz (2005), it is not sufficient to preserve threatened species, it is also important that these species remain available where the populations need them.

In the case of yam wild relatives in Benin, there is a clear need to link conservation and use. The best way to achieve *in situ* conservation of yam wild relatives through a sound management of wild habitat is to develop awareness of local farmers and extension services of the value added to yam germplasm by the ennoblement practice. The fact that yam breeders have not been able to use wild germplasm so far to improve yam varieties shows how ennoblement is a remarkable practice. Preserving and developing this practice will have a positive impact on the *in situ* conservation of yam wild species, as the negative consequences of the erosion of wild habitats will be obvious to those ennobling wild yams.

Therefore, maintaining and promoting the ennoblement practice appears crucial. However, the future of the ennoblement practice in Benin is uncertain. Indeed, only 5% of the farmers practise ennoblement at the present time and this percentage seems to be decreasing (Baco, 2000; Dumont and Vernier, 2000). One problem is that ennoblement is sometimes considered a shaming practice (Baco, 2000): in Benin, the social position of farmers depends, in part, on yam production. A farmer who collects wild yam is considered a bad producer because others think that he needs to use wild yam to complete his production. Thus, action has to be taken to promote ennoblement as a wise practice. However, knowledge must be gained on the social and cultural factors that have been contributing to the survival of this practice.

Developing strategies for the *in situ* conservation of yams in Benin is a challenging task. Our genetic studies strongly support the idea that the *in situ* conservation of wild yams and on-farm conservation of cultivated yams should be linked to each other. This suggests giving more attention to the issue of *in situ* conservation of the wild–cultivated species complex in locations where the wild and the cultivated forms genetically interact.

References

Akoroda, M.O. (1985) Pollination management for controlled hybridization of white yam. *Tropical Agriculture (Trinidad)* 60, 242–248.

Allomasso, T. (2001) *Conservation Des Ressources Génétiques Forestières du Département de l'Atlantique: Stratégies de conservation de l'igname sauvage* Dioscorea praehensilis *(Benth) dans les forêts sacrées et étude de sa domestication*. DESS, Université Nationale du Bénin, Cotonou, Bénin.

Baco, M.N. (2000) La 'Domestication' des Ignames Sauvages dans la Sous-Préfecture De Sinendé: Savoirs Locaux et Pratiques Endogènes d'Amélioration Génétique des *Dioscorea abyssinica* Hochst. Thèse d'ingénieur. Université Nationale du Bénin, Cotonou, Bénin.

Baco, M.N., Tostain, S., Mongbo, R.L., Daïnou, O. and Agbangla, C. (2004) Gestion dynamique de la diversité variétale des ignames cultivées (*Dioscorea cayenensis-D. rotundata*) dans la Commune de Sinendé au Nord Bénin. *Plant Genetic Resources Newsletter* 139, 18–24.

Barton, N.H. and Charlesworth, B. (1998) Why sex and recombination? *Science* 281, 1986–1990.

Dumont, R. and Vernier, P. (2000) Domestication of yams (*Dioscorea cayenensis-rotundata*) within the bariba ethnic group in Benin. *Outlook on Agriculture* 29, 137–142.

Dumont, R., Dansi, A., Vernier, P. and Zoundjihèkpon, J. (2005) *Biodiversité et Domestication des Ignames en Afrique de l'Ouest. Pratiques Traditionnelles Conduisant à* Dioscorea rotundata *Poir*. CIRAD–IPGRI Collection Repère, Rome, Italy.

FAO (1996) *Global Plan of Action for the Conservation and Sustainable Utilization of Plant Genetic Resources for Food and Agriculture*. FAO, Rome, Italy.

FAO (2005) *FAOSTAT. FAO Statistical Databases*. Available at: http://www.faostat.fao.org

Hamon, P., Dumont, R., Zoundjihèkpon, J., Tio-Touré, B. and Hamon, S. (1995) *Les Ignames Sauvages d'Afrique de l'Ouest*. Éditions de l'ORSTOM, Paris, France.

Kaimowitz, D. (2005) Au service des plus pauvres. *Courrier de la Planète* 75, 16–18.

Louafi, S. and Tubiana, L. (2005) Conservation et développement. *Courrier de la Planète* 75, 4–7.

Mignouna, H.D. and Dansi, A. (2003) Yam (*Dioscorea* ssp.) domestication by the Nago and Fon ethnic groups in Benin. *Genetic Resources and Crop Evolution* 50, 519–528.

Pritchard, J.K., Stephens, M. and Donnelly, P. (2000) Inference of population structure using multilocus genotype data. *Genetics* 155, 945–959.

Scarcelli, N., Tostain, S., Vigouroux, Y., Agbangla, C., Daïnou, O. and Pham, J.L. (2006) Farmers' use of wild relatives and sexual reproduction in a vegetatively propagated crop. The case of yam in Benin. *Molecular Ecology* 15, 2421–2431.

Vernier, P., Orkwor, G.C. and Dossou, A.R. (2003) Studies on yam domestication and farmers' practices in Benin and Nigeria. *Outlook on Agriculture* 32, 35–41.

Wilkes, H.G. (1977) Hybridization of maize and teosinte, in Mexico and Guatemala and the improvement of maize. *Economic Botany* 31, 254–293.

Wood, D. and Lenné, J.M. (1997) The conservation of agrobiodiversity on-farm: questioning the emerging paradigm. *Biodiversity and Conservation* 6, 109–129.

22 Management Plans for Promoting *In Situ* Conservation of Local Agrobiodiversity in the West Asia Centre of Plant Diversity

N. Al-Atawneh, A. Amri, R. Assi and N. Maxted

22.1 Introduction

The West Asia region contains one of the three major megacentres of diversity for crop of global significance. The region is equally important in terms of the diversity of the crop wild relatives (CWR) of wheat, barley, lentil, chickpea, almond, pistachio, pear and many species of *Lathyrus* L., *Medicago* L., *Trifolium* L. and *Vicia* L. Almost 10,000 years ago the Near East was the centre of origin for a wide range of temperate crops and tree species that still underpin the present day agriculture. However, individual populations of these species are now highly threatened by overexploitation, mainly by overgrazing and natural habitat destruction due to unchecked agricultural and urban development. These species constitute a socio-economically vital genetic resource for future wealth creation, food security and environmental sustainability both regionally and globally, therefore they deserve more systematic conservation. Due to the sheer number of species present and the lack of resources for extensive *ex situ* collection and conservation, *in situ* conservation through the management of genetic reserves offers the most logical solution.

22.2 Dryland Agrobiodiversity Conservation in the Fertile Crescent

Agrobiodiversity may be defined as the diversity of agroecosystems, agriculture-related plants, animals, birds, insects, microbes and genetically modified organisms. This is also known as agricultural biodiversity, and is the raw material for agricultural industry activities. It is the key to sustainable food security because it enables farmers to adapt crops to their ecological needs and cultural traditions. The Dryland Agrobiodiversity Project (Conservation and Sustainable Use of Dryland Agrobiodiversity in Jordan, Lebanon, Syria and the Palestinian Authority) was a Global Environment Facility Project funded to promote the conservation of CWR and land races of important agricultural species in the Near East region. Its main application was to introduce and test *in situ* and on-farm mechanisms and techniques to conserve and sustain the use of agrobiodiversity of the region. The project was divided into a regional component and four national components, one for each of the four participating countries: Jordan, Lebanon, the Palestinian Authority and Syria. The regional component led by the International Centre for Agricultural Research in the Dry Areas (ICARDA), in cooperation with the International Plant Genetic Resources Institute (IPGRI) and the Arab Centre for the Study of Arid Zones and Dry Lands (ACSAD) provided technical assistance and training to the national programmes and served to integrate the activities regionally through coordination, networking, monitoring of project activities and impact assessment.

The West Asia Dryland Agrobiodiversity Project was implemented in two target areas in Jordan, Lebanon, Palestine and Syria, and in each the CWR element focused on the location and implementation of genetic reserves. The writing and implementation of a genetic reserve management plan is an essential step in efficient and effective *in situ* genetic conservation of CWR diversity. The raw materials of plant genetic conservation are the genes and alleles within gene pools that constitute the total genetic diversity of the particular target taxa being conserved. Therefore, the primary focus of genetic reserve management plan is to ensure the maintenance or enhancement of the genetic diversity of the target taxa within the reserve. Along with this primary goal the management plan should also help to ensure all seven basic management goals are met:

- Maintain maximum genetic diversity of the target taxa along with key associated species within the reserve;
- Promote general conservation of biodiversity;
- Maintain local ecological and evolutionary processes;
- Minimize external threats to biodiversity;
- Implement appropriate but minimally intrusive site management;
- Promote public awareness for the need of genetic and protected area conservation;
- Ensure diversity is available for actual or potential utilization.

Target taxa were selected depending on criteria such as their importance for food and agriculture as potential gene donors, the socio-economic value of the species and the degradation level of the species. The following target taxa were chosen to be under the process of conservation:

- Forage legumes – *Medicago* L. spp., *Vicia* L. spp., *Trifolium* L. spp., *Lathyrus* L. spp. and *Lens* Mill. spp.;
- Cereal – *Triticum* L. spp., *Avena* L. spp., *Hordeum* L. spp. and *Aegilops* L. spp.;
- Fruit trees – *Amygdalus* L. spp., *Prunus* L. spp., *Pyrus* L. spp., *Pistacia* L. spp. and *Olea* L. spp.;
- Vegetables – *Allium* L. spp.

The project targeted two areas in each country depending on the abundance of the target species and the two areas selected were deliberately chosen to represent different ecosystems and permit the maximum diversity of the target species. The areas selected were: Arsal and Balbak in Lebanon, Hefa and Sewaida in Syria, Mowaqer and Ajloon in Jordan, Hebron and Jenin in the Palestinian Territories.

An informal ecogeographic survey was carried out in each country of the project (Lebanon, Syria, Jordan and Palestine) using the standard methodology (Maxted *et al.*, 1995). These surveys included site visits and data collection for the species, as well as meetings with indigenous farmers. Once the broad areas for the reserve establishment were identified, project implementation followed a series of steps:

Step 1: Factors considered when selecting site

- Plant diversity (species richness) present at the site;
- Area of the rain-fed land at the site;
- Area of the irrigated land at the site;
- Area of the common land at the site;
- Area of the unreclaimed land at the site;
- Area of the rangeland at the site;
- Lack of hybrid crop varieties;
- Relative distance of the site from industrial areas;
- Prevalence of the agricultural machines at the site;
- Degree of the women's participation in agricultural activities in the site;
- Annual rainfall rate at the site;
- Annual temperature rate at the site;
- Presence of farmers' organizations at the site;
- Number of the herders at the site;
- Total number of sheep and goats at the site.

Step 2: Selection procedure

Each site was given values from 0 to 2 for each criterion depending upon the applicability of each criteria in the site, 0 means that the criteria is inapplicable in the site or very weak, 1 means that the criterion is moderate or medium in the site and 2 means that the criterion is strongly applicable or present in the site. Each site in the target area was given the suitable mark for each criterion in the weighing matrix.

Step 3: Selection weighting

Each criteria was given a significant weight, which reflected the negative and the positive effects of the presence of each criterion on the agrobiodiversity at the site, and ranked from 1 to 3. Rank 1 means that the presence of the criteria in the site has negative effect on the agrobiodiversity, rank 2 means medium effect and rank 3 means positive effect on the agrobiodiversity in the site.

Step 4: Selection calculation

Finally, the values of the criterion with the significant weight of each criterion for each site were calculated and two sites with the highest accumulative marks were selected.

In addition to this site selection protocol, the project staff also gathered information on villages and residential areas (sites) in the target areas, and conducted several site visits for the same purpose. A thorough analysis of the baseline criterion selection and the field visits lead to the final selection of the two target sites in each country. One of these was designated as a genetic reserve site. Thus, Abu Taha Site in Nabaha village in Northern Lebanon, Sale-Fsheida nature reserve in Swedia south of Syria and Wadi Sair in Hebron in the south of Palestine were designated genetic reserves for the target species of the project. The final selection of these sites was a result of compromise between abundance of the target species, genetic diversity in the target population at the sites and the socio-economic–political–ethnographic environment at the site.

22.3 Why Is a Management Plan Required?

The selected reserves in the partner countries had been chosen because they contain abundant and hopefully genetically diverse populations of the target taxon; these target populations were subject to different kinds of degradation factors, but predominantly the misuse or rather overuse by farmers and the local communities. As a result the sites required immediate intensive, active management interventions to maintain diversity of the target populations and to ensure the genetic reserve long-term sustainability. The only effective means of organizing this intense management is through a carefully constructed management plan.

Maxted *et al*. (1997a) recognized that CWR species are undoubtedly conserved in numerous existing formally designated protected and non-protected areas worldwide, such as areas of wasteland, field margins, primary forest and national parks, but in each of these cases the existence of a particular species is likely to be coincidental because the site is managed for agriculture, recreation or conservation of habitat diversity. In terms of conservation this may be termed 'passive', where healthy CWR species populations occur coincidentally and not through the action of active species management by conservationists. These passively conserved populations are not actively monitored and, as such, are inherently more vulnerable to extinction, i.e. any deleterious environmental trend that would impact on a particular species would be less likely to be noted. To ensure the CWR population health it is necessary to enact positive actions to promote the sustainability of the target taxa and the maintenance of the natural or artificial (e.g. agricultural) ecosystems which contain them. This implies the need for associated target species and habitat monitoring, along with 'active' management and protection.

For CWR species it could be argued that the need for a management plan is more critical than for traditional habitat-based conservation because potentially, with maintenance of genetic diversity as a goal, it would be possible for a CWR population to maintain normal population characteristics (density, frequency and cover) while losing genetic diversity. As such the management is likely to be more 'active' than for a more ecosystem-based protected area conservation management where intrataxon diversity is not the focus of the conservation effort. It is likely to be the case that the more active and complex the management the greater the need for a conservation blueprint, the genetic conservation management plan to direct conservation actions and management interventions.

A management plan is also likely to be required because, as noted earlier, the location of genetic reserves will often result from a compromise between biological best practice and socio-political–ethnographic expediency. This compromise in locating genetic reserves means that the management plan can act as a useful tool to balance competing priorities. For example, the reserve area may be too small or fragmented and isolated to support the ideal minimum viable population or permit natural immigration to balance local extinctions. Therefore, intervention management is necessary to increase or maintain populations at viable levels or to translocate individuals between management areas. All of these can be discussed and recommendations made within the management plan.

The reserve may also be surrounded by hostile anthropogenic environments that result in regular introduction of invasive species (weeds, diseases and generalist predators) and degrading processes (siltation and pollution). Therefore, intervention management outlined in the plan is necessary to minimize or remove such negative influences. A reserve established in or near an urban environment may be under pressure for development, for release of their natural resources for human use or for use as agricultural lands in rural areas to feed rapidly increasing and desperately poor human populations, in which case it is only through the application of the management plan that competing biological and socio-political–ethnographic factors may be objectively evaluated.

CWR are most often associated with disturbed, pre-climax communities (Maxted *et al.*, 1997a). As such, the target taxon may have evolved and developed a 'healthy' population at the reserve location with regular disturbance and may be effectively held at an intermediate stage of succession by the site's management. In this case, removal of the disturbance (grazing, fire, mowing, etc.) or allowing succession to the climax community is likely to be to the detriment of the target taxon population. The ecological dynamics of the target taxon population and the interactions with other taxa within the reserve need to be understood and the appropriate active management interventions should be written into the management plan. Lack of active management may result in the lack of the necessary disturbance or succession that, in turn, is likely to harm population diversity.

Once established the management plan will serve multiple purposes in aiding the conservation of the target taxon population. It will describe the physical, socio-political–ethnographic and biological environment of the reserve, thus providing a reference work for site management. It will articulate the general conservation objectives and specific goals of the individual reserve and how that reserve accommodates within institutional, national and regional conservation strategy, thus ensuring consistency of implementation. It will, through its analysis and statement of socio-political–ethnographic and biological environment, facilitate the anticipation of any natural or anthropogenic conflict or problems associated with managing the reserve. It will describe the management objectives and, in as much detail as possible, the management interventions required for management effectiveness, as well as the monitoring practices to be implemented. It will assist in organizing human and financial resources, and act as a training guide for new reserve staff. It will facilitate communication and collaboration among the individual reserve sites and other genetic reserves, protected areas and *ex situ* conservation facilities. It will also act as a guideline for diversity utilization of the target taxon.

22.4 Constructing the Management Plan

Once various competing locations for the reserve site have been assessed and the final reserve location selected, the structure of the reserve can be designed and the target taxon can be described; the next stage in establishing the genetic reserve is the formulation of a formal management plan (Maxted *et al.*, 1997b). This step involves clarifying the conservation context, studying and possibly researching the abiotic, biotic and anthropogenic characteristics of the site, describing the various target taxa and their populations present in the reserve site and on the basis of all this information generating the management prescription which details the management interventions required at the site. The reserve site has been selected because of its abundant and hopefully genetically diverse populations of the target taxon, and to maintain this diversity there will be a need to observe and describe the anthropogenic, biotic and abiotic qualities and dynamics of the site. Once the ecological dynamics of the reserve are known and understood, a management plan and intervention regime that promotes these elements, at least as they relate to the target taxon, can be proposed.

The general long-term goal of the genetic reserve of the target species is to maintain target taxa diversity and dynamics, and this can only be achieved by having a regime of minimum, effective management interventions that are detailed in the management prescription. Therefore, the first step in formulating the prescription will be to observe various dynamics of the site. It should be surveyed so that the species present in the ecosystem are known and the ecological interactions within the reserve are understood. A clear conservation goal should be decided and a means of implementation agreed. The process of formulating a genetic reserve management plan is summarized in Fig. 22.1.

The West Asia Dryland Agrobiodiversity Project was implemented in two target areas in each of Jordan, Lebanon, Palestine and Syria where the distribution, frequency and density of the target species were surveyed within selected monitoring areas over the period of 2000–2004 and the major factors causing degradation were identified. Within the holistic strategy developed by the project, management plans were developed for each site to promote the community-driven *in situ* conservation. The plans focused on the conservation of target species and their natural habitat, by demonstrating techniques, management, add-value and alternative sources of income, as well as policy and legislation options.

This management plan is the first of its kind dealing with management and the sustainable use of a designated site to conserve distinct species of global importance for food and agriculture. The plan has followed the frame proposed by Maxted *et al.* (1997b, 2006).

Following the previously proposed process of writing a genetic reserve management plans, West Asia Dryland Agrobiodiversity Project prepared the management plan for each selected reserve in Palestine, Lebanon and Syria.

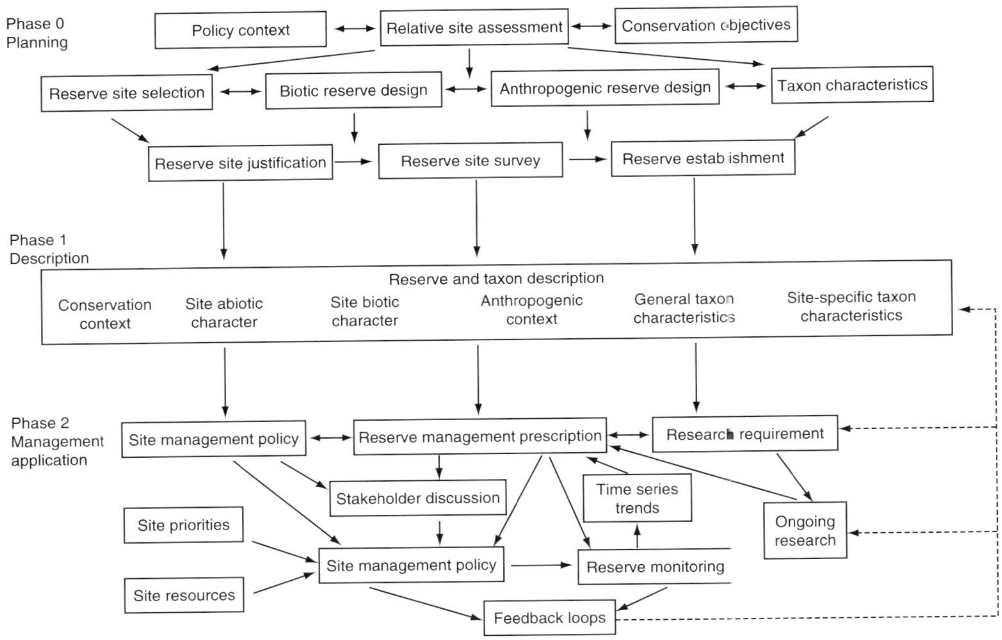

Fig. 22.1. Process of compiling a genetic reserve management plan.

The suggested management plans are part of the exit strategy of the project to ensure the sustainability of its goals and to ensure the long-term conservation of the target taxa in the selected reserves. We can summarize the steps of the management plan writing up as follows.

22.4.1 Phase 0: planning

Conservation objectives

The project stated its object for the conservation and the sustainable use of the target species in four countries and has the following specific objectives:

1. Ensuring long-term conservation and availability of agrobiodiversity especially the target species of the project – forage legumes (*Medicago*, *Lathyrus*, *Vicia*, etc.), cereals (wheat and barley) and some fruit trees (apple, almonds, pear and pistachio);
2. Promoting alternative land uses which promote the conservation of target species while improving the livelihoods of local communities;
3. Increasing national capacity and providing training on *in situ* conservation techniques;
4. Encouraging the local authorities and empowering local communities to contribute to integrated management of the ecosystems including the preservation of local agrobiodiversity;

5. Assessing and monitoring the trends in agrobiodiversity and better understanding of the major causes of its loss;
6. Combining conservation and site management with ecotourism as an alternative source of income for livelihood improvement.

Relative site assessment

As a result of ecogeographic survey, two target areas in each country were designated to apply *in situ* conservation techniques for the target species: Arsal and Balbak in Lebanon, Hefa and Sewaida in Syria, Mowaqer and Ajloon in Jordan, and Hebron and Jenin in the Palestinian Areas. However, the selected areas primarily represented areas rich in landraces and where marginal agriculture favoured maintenance of CWR.

Policy context

The partner countries review all the old policies and legislations that are related to the biodiversity conservation and establish new policies and legislations for the agrobiodiversity conservation with new policy options to encourage the *in situ* conservation of the CWR in the area.

Reserve site selection

As a result of the ecogeographic survey carried out in each country of the project (Lebanon, Syria, Jordan and Palestine), the following target sites were designated as genetic reserves in the selected target areas: Abu Taha site in Nabaha village in Northern Lebanon, Sale-Rsheida nature reserve in Sewaida south of Syria and Wadi Sair in Hebron, south of Palestine. Different kinds of criteria are used in the selection process.

The selection of these sites is a compromise between abundance of the target species and genetic diversity in the target populations and the sophisticated socio-economic–political–ethnographic environment.

Reserve design

Genetic reserves are usually designated in areas which have the minimum size of the target populations. There are two methods of designing the reserve: establishing a single large reserve and several small reserves. The selected reserves in the project are single large reserves and each reserve has a buffer, transition and core area.

Taxon characterization

The target taxa are the same in the four countries, however, the occurrence of these taxa vary from site to site. The characterization of the taxa is mentioned in the plan of each site.

Reserve sites justification

In general, the sites are selected to conserve the genetic diversity of the target species for the long term; however, each site has its own private justification in terms of its selection.

Reserve sites survey

The project used the same survey methodology in surveying the sites. Socio-economic and botanical surveys were carried out at each selected site in each country to determine the socio-economic situation of the communities settled near the target sites; the botanical survey also draws a clear view about the botanical cover in the sites.

Reserve establishment

Based upon the previous information each country designates the suitable reserve to be controlled and managed for conserving the genetic diversity of the target species.

22.4.2 Phase 1: Reserve and taxon description

Conservation context

The selected reserves in each country are some of the most biodiversity-rich spots remaining in West Asia. The loss of local agrobiodiversity and biodiversity, as well as land degradation in the sites would prove a major blow for the region. The establishment of genetic reserves to conserve the significant national, regional and even global genetic resources, through appropriate site management, will ensure sustainable use of this biodiversity for future generations, while facilitating continued natural evolution of diversity. *In situ* conservation in these sites will complement the Ministries of Agriculture *ex situ* efforts in each country to collect representative samples of the target species. The conservation of this globally significant agrobiodiversity in these particular reserves in Palestine, Lebanon and Syria will facilitate the conservation of the target species and will work in a complementary manner to other habitat-targeted natural reserves in the region.

The ownership of the land in most of the selected reserves sites is private. This may affect the sustainability of the reserve and is considered as one of the important risks associated with the project. However, the management plan for the reserve was developed and will be implemented in partnership with land owners, and this associated with the vital public awareness programme and incentives is addressed in the management plan. Overgrazing, species replacement and land cultivation are considered the main threats to genetic diversity in the target sites. The management plan with the applied monitoring system has been designed in each case to play an effective role in mitigating these threat factors.

Site abiotic character

The Sale-Rsheida reserve is located in Sewaida province, Syria, in the plateau area (latitude 32° 39' 116", longitude 36° 47' 538", altitude 1425 m and rainfall 200–250 mm). The whole area is a flat plateau with minor slopes not exceeding 2%. The site is characterized as agricultural marginal land with remnants of degraded evergreen oak forest, now dominated by grassland and dwarf shrubs with few remnants and isolated trees of *Crataegus* L. and *Amygdalus kotschyi* Hohen. ex Spach.

Wadi Sweid reserve in Lebanon is located in the lower part of Aarsal (near the residential area), with altitudes ranging from 1400 to 1600 m. The average annual rainfall is around 300 mm; the mean annual temperature is around 12°C. The site is composed of a series of valleys and hills. The bottoms of the valleys are used for agriculture (especially, fruit trees and vineyards) and the hills are kept as wooded grasslands.

The Sair reserve in Palestine is located in the north-eastern region of the Hebron area. The area is dominated by strongly dissected hills. The main agricultural activities in this area are the cultivation of grapes, stone fruits, field crops and grazing of the natural grasslands. The Wadi Sair genetic reserve is located in the east of Sair town and extends from the middle of Sair village to Al-Baqaa plain and Tequoe in the east. The valley has a fertile soil and the site has an area of about 2 km^2. The area has a steep-sided north slope and a less steep south slope.

Site biotic character

The Sale-Rsheida reserve in Syria contains large populations of herbaceous species, including some rare and endemic species such as *Triticum dicoccoides* (Körnicke) G. Schweinfurth with a few plants of *T. urartu* Thumanjan ex Gandilyan. Most of the species are in early stages of growth due to long winters and snow coverage, and additional species could be found later in the season. Seven species classified as rare in the West Asia region are found in the site including several *Iris* species. Many of the species present are palatable to animals and could be used as resources for genetic improvement, particularly the *Triticum* and *Hordeum* wild species which are widely used for breeding programmes in the world.

The Wadi Sweid reserve in Lebanon contains cultivated lands which are protected from grazing. The rest of the area is only partially protected since there are no local guards (Natours), who regulated grazing at the community level until recently. The site is divided into 'arable land' and 'wooded grassland'. The human activities in the area include: cultivation of perennial crops, grazing (very limited) and collection of some wild and medicinal plants. The following wild trees are found in the site: *Amygdalus communis* L., *A. orientalis* Mill., *Pistacia atlantica* DC, *P. palaestina* Boiss., *Prunus microcarpa* C.A. Mey., *Pyrus syriaca* Boiss., *Rhus coriaria* L., *Quercus calliprinos* Webb, *Crataegus aronia* Decne., *C. azarolus* L., *Juniperus excelsa* M. Bieb., *Cotoneaster* Rupp. sp., *Lonicera* L. sp. and *Acer* L. sp., and the following herbs: *Aegilops biuncialis* Vis., *A. triuncialis* L., *A. ovata* L., *Medicago radiata* L., *Triticum uraratu* Thumanjan ex Gandilyan, *Allium* L. sp., *Vicia hybrida* L., *Vicia* L. sp., *Trifolium* L. sp., *Lens* Mill. sp., *Hordeum bulbosum* L., *H. spontaneum* K. Koch, *Avena* L. sp., *Bromus* L. sp., *Astragalus* L. sp. and *Fibigia* Medik. sp. In addition, apricot, cherry, almond, fig, wheat, barley, chickpea and lentil are cultivated in the arable land that has mostly been reclaimed from natural habitats over the years.

In the Wadi Sair reserve in Palestine, target taxa are distributed throughout the site area, however, the density and frequency of the species vary from place to place. In general, large populations are found near the field borders and in the non-grazing areas. The pests of the target taxa are found predominantly along the valley, which has deep and fertile soil with high moisture content. Fortunately, the largest populations are almost all protected from animal grazing by local customs because they are surrounded by the fruit tree orchards in the site. Smaller populations are found at the hill tops, where there are shallow, non-fertile soils. These plants also suffer from severe overgrazing because most of the hill top areas are not cultivated and thus not protected from grazing.

Anthropogenic context
The complexity of land tenure is one of the main complexities of site management; all land is privately owned and the sites employ traditional practices for grazing, medicinal plants collection, cutting and land reclamation to introduce new hybrid crop species. The management plan attempts to control these activities to benefit the target taxon populations. Thus, discussion with and education of the local communities is essential and involving local people in implementing the management plan and creating income for them from the reserve utilization is vital.

General taxon characteristics
The genetic reserves established in the West Asia countries will focus on conservation of the project target species (detailed earlier), however, the prepared management plan will facilitate the conservation of all ecosystem components.

Site-specific taxon characteristics
Each site has viable populations of the target species, and the frequency and the abundance of these species is well above the minimum viable population levels. However, population levels vary from species to species in each site and from site to site in the countries. For each site, a botanical survey was carried out and a large amount of data for each species recorded and stored in the regional project database. This includes information on the taxonomy, frequency and abundance of each species, as well as all ecogeographical data that are related to each site.

22.4.3 Phase 2: management application

Site management policy
The ultimate aim of the site management and therefore the management plan is to maintain a demographically and genetically healthy target population within the reserve site, within a self-sustaining ecosystem with all other components. This will serve as a model for ecological restoration work in similar ecosystems in Aarsal and other sites. It will also serve as sites for research and studies to be conducted by students and researchers in areas related to natural sciences.

The target sites are very rich in genetic diversity of wild relatives and landraces of important agricultural crops. Farmers are still using landraces of cereals and legumes and local varieties of fruit trees and some vegetables because of their adaptation to the harsh conditions of the area (rain-fed agriculture). However, the genetic diversity of the predominant species within the sites is being reduced. The factors of degradation include overgrazing, wood cutting and land reclamation (destruction and conversion of natural habitats), and it is these threats that the management plan is targeted to address.

The main objectives for conserving the sites are to preserve the species of global importance and genetic resources with value for the livelihood of the local community. Creating a pilot and model of patch conservation can lead to the conservation of the remaining biodiversity patches. Local and national authorities should work with local communities for co-management of these areas.

Reserve management prescription

The management options are selected and discussed with local communities and the owners of the areas to be protected in order to combine conservation aspects with sustainable use and ensure the benefits to the local users. These options include technical packages, institutional arrangements, add-value and alternative sources of income that can enhance the livelihood of local communities and preserve local agrobiodiversity and all the ecosystem functions.

Research requirement

Different kinds of researches have been highlighted as required at the reserve site, including:

- Noting of the socio-economic changes associated with implementation of the management plan;
- Studying the conservation status of the target species within the site;
- Recording the frequency and abundance of the target populations;
- Reviewing the efficiency of the monitoring system applied within the sites;
- Reviewing the efficiency of the implemented ecotourism programme initiated.

Stakeholder discussion

Various meetings, workshops and brainstorming sessions were held with disparate stakeholder groups at the national level within each country and at the regional level between the countries. National and international consultants were involved in the process of writing up the management plan. These discussions led to the preparation of a practical and effective management plan.

Time series trend

Each country has established a monitoring schedule and the local authority in each country (in each case the Ministry of Agriculture) is responsible for collating and analysing the time series data. The management and even the monitoring of the site will be modified according to the results of monitoring. The project is still relatively young and so no examples of changes can be cited at this early stage.

Site priorities
As mentioned above, the site priority is the *in situ* conservation of globally significant agrobiodiversity, which is also a national priority for the Ministries of Agriculture in each country as well as for the regional CG Centre, ICARDA.

Site resources
Each site has its own natural resources that play an important role in the conservation process. The presence of water springs in some of the sites encourages the ecotourism programme, which helps create income for the local population. In addition, the presence of wild species and landraces only metres from cultivated fruit trees plays an important role in educating the local and broader communities. The financial resources for site management are in the hands of the respective Ministries of Agriculture but it is vital that sufficient funds are made available to ensure effective sustainability of the site and guarantee these resources for future generations.

Prescription implementation
Each site management plan has its own unique prescription and implementation of each reserve management plan is likely to be equally unique. It is at least at first, likely to be tentative in the sense that it does not radically alter the previous site management under which conditions the site facilitated the growth of a healthy target taxon population. However, with time the implementation becomes more experimental, altering the previous site management possibly to better suit the target taxa.

Reserve monitoring
There would be no point in establishing and managing a genetic reserve unless it was regularly monitored. The objectives of the monitoring are to periodically record key population characteristics, analyse the time series data collected and thus detect any detrimental changes in the target population.

Feedback loops
The main reason for establishing a monitoring system is the feedback into changes in the management plan and to review whether management options and activities are going well or not. Depending on the result of the analysis of the monitoring data the management plan can be modified, thus providing a feedback loop.

Within the Dryland Agrobiodiversity Project the process of selecting the locations and implementing the genetic reserves was assisted by international, national and local experts, who worked alongside the project staff.

22.4.4 Stakeholder discussion

No genetic reserve is established in an anthropogenic vacuum (Maxted *et al.*, 1997b), it is always a fundamental element for locating and implementing a genetic reserve to consult the different stakeholders and ensure that their wishes and aspirations are incorporated into the management plan.

Local farmers, who are often the landowners, are the main participants in the implementation of the management options. The Dryland Agrobiodiversity Project staff organized several meetings and workshops with the local farmers to discuss the management options for the sites. This was an important point as the land is privately owned by the local farmers, and this would have been impossible without the complete agreement of the farmers. Where there are potential points of conflict there is a need for more discussion, compromise and possible provision of in-kind incentives to encourage farmers' cooperation. Experience showed that the smaller the number of owners, the easier it was to reach an agreement. Local institutions and local farmer's bodies should also be involved in discussing the management plans, so that the municipality as a whole is consulted. Thus, the community is a real partner in the conservation planning and action. Broader national non-governmental organizations (NGOs) may also be consulted to facilitate broader support.

In each country the leading government agency was the Ministry of Agriculture, but economic development, education, environment, finance, forestry, rangeland, rural development, tourism, trade, transport and water ministries were also included in project discussions. The responsibilities were shared through the National Steering Committee (NSC) for the project which was chaired by the Ministry of Agriculture.

At the national level, the leading research institutions in the project and accordingly in preparing the management options were the National Centre for Agriculture Research and Technology Transfer (NCART) in Jordan, the Lebanese Agriculture Research Institute (LARI), the Extension General Department in Palestine and the General Commission for Scientific and Agriculture Research (GCSAR) in Syria.

At the regional level, the lead institutions were ICARDA, IPGRI and ACSAD. These agencies played the pivotal role of providing technical support, network building and dissemination for the project and the implementation of the activities. In addition, universities in each country helped in providing technical support for the project.

The Dryland Agrobiodiversity Project also worked with some members of the private sector, particularly as a means of increasing their role in the sustainable use of agriculture and for the suitable and sustainable management of the targeted reserves. The project also contacted unions and associations of unions, cooperatives and consumer groups.

22.4.5 Drafting the management plan

The four countries applied the same approach in developing the management plans for the selected reserves; all developing activities were coordinated at the regional level by ICARDA as the coordination agency for the project. The drafted management plans were reviewed by the international consultant.

The major difference between countries was the ownership of the land. The land in all four countries was privately owned, but the number of owners varied from one or two farmers in Syria and Lebanon to Palestine where the land was owned by more than 50 farmers. The latter made reaching agreement on site management challenging! To overcome this problem the Ministry of Agriculture in Palestine created a committee of ten persons to represent the farmers and follow up all of the management options with the specialized staff from the Ministry.

The Ministry of Agriculture in each country took the responsibility for sustainability of the selected sites and the management and long-term monitoring of the site. Each country established a new governmental body with a clear terms of reference to follow up all the agrobiodiversity conservation issues.

Local farmers also have a very crucial role in terms of the reserve sustainability; the awareness of those farmers of the importance of the genetic resources in the site for the future generation is an important subject. The continued contact between the government through the Ministry of Agriculture and the local farmers is vital for the continued successful implementation of the management regime.

22.4.6 Monitoring analysis and feedback

As the goal of a genetic reserve is to maintain genetic diversity of the target taxa at the site, the monitoring system should not only monitor population demographics, but also include an estimation of genetic diversity. The management of a genetic reserve will involve an element of experimentation and evolution; it is not likely that the ideal management regime will be known in full *a priori*. Therefore, the populations of the target species in the reserve and possibly the competitors and other associated species will need to be assessed regularly in terms of their size, genotypic composition and/or overall genetic diversity in order to be able to detect changes. If deleterious change is detected, the management prescription will need to be reviewed.

In the Dryland Agrobiodiversity Project the monitoring system was designed at the regional level by the project staff with close cooperation from ICARDA experts and international consultants. The monitoring objectives were clearly defined and understood before the start of the monitoring regime. As the reserve objective is to maintain or enhance the target populations representing the target species in the quality and quantity of their genetic diversity, so it follows that the objectives of the monitoring are to detect any detrimental changes in the target population characterization.

The project used the stratified random strategy, and three transects of 200 m were recorded in each site. Five quadrates (plots) of $1\times1\,m^2$ were taken along each transect with 25 m between each quadrate. The first quadrate was placed on a random distance from the beginning of transects and the other four were placed in a systematic way. The species survey information was collected in the quadrates. Thus, 15 plots were recorded for the herbaceous species inventory at each site. For each quadrate the following features were recorded: species name (scientific and Arabic), family, cover, frequency, abundance, phenology and vitality. For the fruit trees in each of the project sites two transects were placed along the slope gradient. A total of five quadrates of $20\times20\,m^2$

were placed on each transect. The first quadrate was placed at a random distance from the beginning of transect and the other four were selected systematically. For each quadrate the following features were surveyed in relation to species information: species name (scientific and Arabic), family, cover, height, perimeter (circumference of the tree trunk) at 50 cm, age, phenology and vitality. The collected data were entered into a computerized geographic information system (GIS) database and analysed. On the basis of the analysis a report summarizing the results and the changes found in the population in the sites was published, and changes to the management interventions were suggested.

22.5 Elements of a Genetic Reserve Management Plan

The management plan for a genetic reserve is likely to differ from a more ecologically based protected area management plan where the aim is to conserve the entire ecosystems within the protected area. The difference is associated with the level of biodiversity being addressed; for ecosystem-based protected area management the goal is commonly broader in terms of taxa. As such, the taxonomic focus is likely to be either the keystone or the threatened wild species, while for genetic reserve management the focus will be the priority CWR species, which have a socio-economic value associated with their actual or potential use as gene donors, and the full range of their genetic diversity. The need for maintenance of diversity is key to the International Union for the Conservation of Nature (IUCN) definition of a viable population as one which maintains its genetic diversity and potential for evolutionary adaptation, and is at minimal risk of extinction from demographic fluctuations, environmental variations and potential catastrophe, including overuse (IUCN, 1993). Having made these distinctions, with the obvious exception of the genetic diversity elements of the plan, the actual format of the management plan will have obvious similarities.

The actual content or style of a genetic reserve management plan will vary depending on the location, target species, organization, staff, etc. There is no standard format, but issues commonly addressed are: conservation context and objectives, site abiotic, biotic and anthropogeneric description, taxon description, necessary research agenda and management prescription. Possible elements of a genetic reserve management plan were summarized by Maxted *et al.* (1997b), these were discussed and amended during the PGR Forum Population Management Methodologies workshop in Menorca (De Hond *et al.*, 2004) and further developed by Maxted *et al.* (2006). The proposed elements of a genetic reserve management plan are summarized in Box 22.1.

The management plan includes technological, institutional and policy options, as well as alternative sources of income which will allow full participation of local communities in the conservation and management of local agrobiodiversity. The technological aspects may include reseeding and replanting with native species during rangeland improvement and for reforestation, use of water harvesting techniques, introduction of alternative feed resources and management of grazing and livestock. Among the options for improving the

Box 22.1. Genetic reserve management plan content (Maxted *et al.*, 2007).

1. *Preamble*: conservation objectives, site ownership and management responsibility, reasons for location of reserve, evaluation of populations of the target taxon, reserve sustainability, factors influencing management (legal, constraints of tenure and access).
2. *Conservation context*: place reserve within broader national conservation strategy for the responsible conservation agency and target taxon, likely interaction between target taxon and climate change at site, externalities (e.g. political considerations), obligations to local people (e.g. allowing sustainable harvesting), present conservation activities (*ex situ* and *in situ*), general threat of genetic erosion.
3. *Site abiotic description*: location (latitude, longitude, altitude), map coverage, photographs (including aerial), detailed physical description (geology, geomorphology, climate and predicted climate change, hydrology, soils).
4. *Site biotic description*: general biotic description of the vegetation, flora, fauna of the site, focusing on the species that directly interact with the target taxa (keystone species, pollinators, seed dispersers, herbivores, symbionts, predators, diseases, etc.).
5. *Site anthropogenic description*: effects of local human population (both within reserve and around it), land use and land tenure (and history of both), cultural significance, public interest (including educational and recreational potential), bibliography and register of scientific research.
6. *General taxon description*: taxonomy (classification, delimitation, description, iconography, identification aids), wider distribution, habitat preferences, phenology, breeding system, means of reproduction (sexual or vegetative) and regeneration ecology, genotypic and phenotypic variation, local name(s) and uses.
7. *Site-specific taxon description*: taxa included, distribution, abundance, demography, habitat preference, minimum viable population size and genetic structure and diversity of the target taxon within the site, autecology within the reserve with associated fauna and flora (particularly pollinators and dispersal agents), specific threats to population(s) (potential for gene flow between CWR and domesticate).
8. *Site management policy*: site objectives, control of human intervention, allowable sustainable harvesting by local people and general genetic resource exploitation, educational use, application of material transfer agreements.
9. *Taxon and site population research recommendations*: taxon and reserve description, autoecology and synecology, genetic diversity analysis, breeding system, pollination, characterization and evaluation.
10. *Prescription (management interventions)*: details (timing, frequency, duration, etc.) of management interventions, population mapping, impact assessment of target taxon prescriptions on other taxa at the site. Staffing requirements and budget, project register.
11. *Monitoring and feedback (evaluation of interventions)*: demographic, ecological and genetic monitoring plan (including methodology, schedule, etc.), monitoring data analysis and trend recognition. Feedback loops resulting from management and monitoring of the site in the context of site itself and the regional, national and international context.

incomes of local communities, dairy production, cultivation of medicinal plants and ecotourism are investigated with local NGOs and lead farmers. For some highly threatened species such as *Triticum* L. and *Lens* Mill., protected areas might be necessary to conserve the remaining populations.

It is important to stress that both government and institutional support are needed to implement the proposed management plans if species of global importance are to be safeguarded to overcome environment challenges caused by climatic change and biotic stresses. This implementation requires resources, that is a real cost, and it must also be realized that this cost is ongoing if the reserve has a sustainable long-term future. Without the long-term commitment of local government and institutions, the shorter project-based expenditure of resources on establishing the reserve would be a complete waste of resources; a form of conceptual conservation without lasting value.

As the specific focus of establishing the genetic reserve will be to conserve diversity at the taxonomic, demographic and genetic diversity levels for the specific target taxon or taxa, the management plan requires details associated with each of these biodiversity levels. It will be necessary to clarify the recognizable taxonomic elements of the target taxa present in the reserve, i.e. species, subspecies and varieties, and describe their characteristics (e.g. taxonomy, phenology, habitat preference, breeding system and minimum population size). For these taxa it will be necessary to describe their reserve site demography (e.g. mapping of populations and density within the site, autecology within the reserve, synecology with associated fauna and flora). It will also be necessary to describe the genetic diversity within the target taxa. Once the taxonomic, demographic and genetic diversity levels for the specific target taxon or taxa within the reserve are described and details included in the management plan, they form the foundation for future monitoring and assessment of change in diversity.

However, changes in taxon, demographic and genetic diversity are a natural characteristic of community dynamics. The management plan must allow for natural fluctuations due to stochastic (severe weather, floods, fire and epidemics), cyclical (density-dependent interactions, which may be dramatic but their effects do not persist) and successional (directional, which may be halted by management intervention) changes. Stochastic and cyclical changes in the short term may be quite dramatic, but will rarely lead to species extinctions (Hellawell, 1991), although they are likely to result in genetic drift (see Gillman 1997). The management plan should not only describe the taxon, demographic and genetic diversity, but also attempt to establish the normal limits of natural diversity change. Once these are established the limits are set beyond which revision of management intervention is triggered in an attempt to better promote target taxon health. After emphasizing the natural changes seen in plant populations, humans undoubtedly have the most dramatic effect on communities, through incipient urbanization and pollution, or changes in agricultural and forestry practice. Therefore, the management plan must be flexible enough to accommodate superficial anthropogenic factors, and recognize those factors that could seriously threaten the levels of the target population.

Given (1994) stresses that it is important to realize that preserving communities is not necessarily the same as preserving genes. Maxted *et al* (1997a) conclude that it is quite possible to preserve a community type and still lose genetic diversity, if not species. Therefore, it is vital that the reserve is designed and managed in an appropriate manner to maintain genetic diversity of the target taxon or taxa and if this objective is threatened, the corrective action is automatically taken.

As reserve management is unique, experimental and will evolve over time, each management plan itself will have the same characteristics. The management plan is not a document written when the reserve is initially established and set for all time, but it requires regular revision to take account of changes in the policy context and conservation objectives, site biotic and anthropogenic description, research agenda and perhaps most importantly the management prescription that should be seen as being dynamic, changing to meet the target taxa conservation goals as they are better understood.

22.6 Minimum Content of a Genetic Reserve Management Plan

The proposed elements of a genetic reserve management plan summarized in Box 22.1 are comprehensive, but it should be realized that there may be many practical or pragmatic reasons why it is not possible to cover all of these issues in an individual reserve management plan. Much of the taxon and reserve descriptive information may be unavailable; for example, the breeding system, minimum viable population size and genetic structure of the target taxa will often be unknown and intense research may be required to provide this information. It may be necessary in the reserve to balance local community development aspiration with biological expediency, which may therefore compromise pure conservation goals, and finally, the resources required to write such a plan may not be available to the reserve management staff.

Writing a management plan is not purely an academic exercise, but it aims to facilitate site management and target taxa conservation, therefore, the more detailed the management plan, the more useful it will be. However, practically where it is not possible to complete all elements of the genetic reserve management plan as envisaged in Box 22.1, it would still be necessary to record in the management plan the accessible information or that which can be collated with the limited resources available. As such, even when writing a minimum genetic reserve management plan, it would be necessary to address the conservation context, the abiotic, biotic and anthropogeneric characteristics of the site, the target taxa included and their population characteristics, as well as outlining the management prescription.

The proposed regional management plans differed from the ideal management plan as might be expected there is variation between the theory and the practice. Variation is associated with the target species included at each site, the need for the Dryland Agrobiodiversity Project to integrate wild plants with landrace conservation, the challenges associated within the project with locating reserves on privately owned rather than governmental land. All of these made the management option implemented differ from the ideal options. Certainly, the establishment of the reserves would have been more straightforward if the reserve had been sited on governmental lands. However, it could be argued that locating genetic reserve primarily on privately owned land meant that the project was forced to pay more than just lip service to working with the local communities and local authorities, as well as ensuring that they understood the importance of the need to conserve *in situ* agrobiodiversity.

Examples of the management plans actually written and applied as part of the Dryland Agrobiodiversity Project can be obtained from the senior author on request.

22.7 Genetic Management Outside of Protected Areas

CWR are commonly found in disturbed, pre-climax plant communities (Maxted *et al.*, 1997a), and as such many may be excluded from or marginalized in established protected areas, which more often aim to conserve pristine habitats, ecosystems or landscapes, or animal species that are now restricted to these environments. Therefore, the genetic conservation of CWR in non-protected areas should be addressed.

These non-protected areas where CWR thrive may include roadsides, field margins, orchards and even fields managed using traditional agrosilvicultural practices. In each case, these sites are not managed for biodiversity conservation and the CWR that they contain are purely incidental. If these sites are to be considered suitable for sustainable conservation the management they currently receive and that has permitted the existence of a healthy CWR population must be consistent. It might be argued that these sites are more vulnerable to sudden, radical change. The threat of radical management change would be less likely in protected areas because the *raison d'état* is already conservation, so any management change is more likely to be conservative in nature. Examples of the additional threats faced by non-protected area sites would include widening of roads, scrubbing out of hedgerows and introduction of herbicides. Therefore, for non-protected area sites there is a need to establish some level of protection or the conservation action will be unsustainable. It would be essential that a management agreement is reached with the non-protected area site owner and/or manager to ensure that the current site management is not radically changed, adversely affecting the CWR diversity. The management agreement will need to be predicated on an understanding of the conservation context, the abiotic, biotic and anthropogenic characteristics of the non-protected area site, the target taxa included and their population characteristics, and an understanding of the existing management of the site that has resulted in the healthy target taxon population that can be formalized into a site prescription. The prescription will then form the basis of the management agreement between the conservation agency and the land owner.

In other words, it is just as important to have a management plan for a non-protected area site where CWR are to be sustainably conserved, as it is for a more formal genetic reserve. Sustainability is central to CWR conservation and lack of a management plan and management agreement is likely to impede the sustainability of non-protected area conservation.

The project found that many areas neighbouring the target sites contained significant agrobiodiversity that might also be conserved. These populations are often associated with field edges, habitat patches or roadsides, e.g. the base of the Beqaa Valley in Lebanon is industrially cultivated but significant populations of target and non-target species are found along the roadside. In the Hebron area of the southern West Bank healthy, viable populations of the target species are often found along field borders, and in Jabal Al-Doroze in Sweda, Syria, target CWR populations that are common in modern apple orchards where traditional varieties of apple have been lost. In fact in Palestine, *P. syriaca* Boiss. is found only as scattered trees and never as a coherent population; therefore, this species was always conserved outside of the traditional protected area system. This means that special attention was drawn to these isolated trees, a leaflet was prepared to help raise awareness of this resource and individual trees were mapped using a GIS system so that their long-term presence was easier to monitor.

As part of the Dryland Agrobiodiversity Project special training at the regional and the national level was given to farmers and local communities in how best to conserve these important threatened habitat corridors and how these populations might be sampled for complementary *ex situ* conservation.

22.8 Integration of Local Development and Reserve Management

Increasing the productivity of the reserve while conserving its intrinsic nature should also be targeted as an ancillary management option to the focus on the target taxa. In the Dryland Agrobiodiversity Project this was achieved through:

- Application of low-input technological packages combining integrated pest management, use of manure and low rates of fertilizers and herbicides. Use of treated and cleaned seed lots for field crops and of healthy seedlings for fruit trees.
- Application of water harvesting techniques and planting fruit trees without much reclaiming of the surrounding land.
- Development of a community-based seed production unit and fruit tree nursery focusing on multiplying local varieties, for rangeland and forest ecosystems.
- Development of grazing management plans either through rotational or deferred grazing.
- Designation of a core area rich in plant biodiversity, both herbaceous and wild fruit trees species, for protection and to serve for seed multiplication to improve the land cover
- Establishment and promotion of more holistic livestock management practices that benefit both wild and agrobiodiversity.
- Investigating the possibility of using fruits from wild species in the production of feed block. This technology along with plantation of native shrubs and reseeding with native species could improve the degraded land and contribute to the management of grazing.
- Cultivation and reseeding with local, native species collected in a sound scientific way to present different populations was done for rehabilitation of the degraded habitats.

- Collecting seed from rare species and isolated wild fruit trees for *ex situ* conservation and for rehabilitation purposes.
- Regulating land destruction for urbanization and agricultural purposes (this might require continued awareness efforts of local communities along with arrangements with government for better planning of land uses).
- Avoiding the extension of quarries to biodiversity rich areas.

It is appropriate that local communities should see the implementation of *in situ* conservation projects as a means of helping them meet their development aspirations. It is true that if the local community can see direct personal benefit to them in having a genetic reserve sited locally, they are much more likely to enthusiastically support the initiative. Therefore, the integration of local development aspirations and reserve management should not be seen as an optional add-on to conservation activities themselves, but as an equally essential component of the whole conservation process.

Within the Dryland Agrobiodiversity Project much effort was devoted to generation of added-value for local products and the identification of alternative sources of income for local communities. These include the following.

Agrofood processing

The project in collaboration with local institutions in the target sites contributed to the training of women in food processing techniques of local products and helped create opportunities for adding value to local products from cereals (Friekeh, Burghul) to fruit trees (syrup, compots, jams and dried fruits). For wild fruit trees, the pioneer experience of the processing of wild plums, hawthorn and pears proved successful and could be extended and advertised more widely.

Production of honey and cultivation of medicinal plants

The region is rich in medicinal plants and mieliflora species, and these offer opportunities for diversification of income of local communities, particularly for women in the Near East through the development of apiculture and introduction of medicinal plants cultivation into home gardens.

Dairy production

The general mobilization of financial resources experienced by project communities allowed the local community to better pack and market their dairy products for sale throughout the region.

Ecotourism

The Dryland Agrobiodiversity Project encourages tourists to visit the target sites to view the target taxa and also to generate income for the local farmers. They have produced an explanatory leaflet about the target sites including excellent guiding information for the visitors, infrastructure, agriculture, educational, water, green cover, population and site map with some distinct photographs.

Precise site management included in the project varied from country to country and from site to site with a country. However, in each case local communities were encouraged to play a pivotal and effective role in managing the genetic diversity and integrate these activities as far as possible with their own community development strategies.

The management of project sites was combined with a comprehensive awareness-raising campaign that focused on the community at different levels, children in the schools, school teachers, women in the villages, local farmers, technical staff in the Ministry of Agriculture, newly graduated agronomists and other groups. The purpose being to increase awareness of the importance of conserving the genetic diversity of the target species in the site, while at the same time raising consciousness of the locals and the decision makers of the importance of future societal health of genetic reserves establishment.

22.9 Conclusion

The West Asia region contains one of the three major megacentres of diversity for crops of global significance. The region is equally important in terms of the diversity of the CWR of wheat, barley, lentil, chickpea, almond, pistachio, pear and many species of *Lathyrus*, *Medicago*, *Trifolium* and *Vicia*. The diversity of these genera is still widely found under natural habitats in rangelands and forest ecosystems, and in field edges. However, for each of these genera, individual populations are highly threatened by overexploitation, mainly by overgrazing and natural habitat destruction for agricultural and urbanization purposes. These species constituting socio-economically vital gene pools for future wealth creation, food security and environmental sustainability in the 21st century need to be conserved both *ex situ* and *in situ*, but farmers and herders do not always realize this genetic value.

The Global Environment Facility funded West Asia Dryland Agrobiodiversity Project, which implemented *in situ* conservation in two target areas in each of the four countries, Jordan, Lebanon, Palestine and Syria. Practically, this involved surveying the distribution, frequency and density of the target species of forage legumes, cereals, fruit trees and vegetables within the selected monitoring areas over the period of 2000–2004, assessing the major factors of their degradation and establishing genetic reserves for *in situ* CWR conservation. An essential step in establishing and managing the genetic reserve is the writing and implementation of the management plans and they were developed for the selected sites to promote target CWR conservation within their natural habitats. These plans are the first of this kind dealing with management and sustainable use of a designated site to conserve distinct CWR species of global importance for food and agriculture.

Management plans include technological, institutional and policy options, as well as alternative sources of income which will allow full participation of local communities in the conservation and management of local agrobiodiversity. For some highly threatened species such as *Triticum* and *Lens*, the establishment of protected areas is seen as critical to maintaining the diversity essential to meet the requirements of future generations. Associated with the application of the management plan is the monitoring of the site to assess significant changes in the population, as well as assessing the effectiveness of the management options on diversity of the target species.

Government and international support are needed to fully implement the proposed management plans if species of global importance are to be safeguarded to overcome environment challenges caused by climatic change and biotic stresses. It must be stressed that *in situ* CWR conservation is not a short-term option. Having established the reserves, it will be important for various actors to ensure that they are implemented, updated and continue to meet the priorities of the West Asian CWR conservation and user communities.

References

De Hond, L., Iriondo, J.M. and Kell, S.P. (compilers) (2004) European Crop Wild Relative Diversity Assessment and Conservation Forum. Report of Workshop 4: Population Management Methodologies. University of Birmingham, Birmingham, UK. Available at: http://www.pgrforum.org/Documents/WP_4_Documents/WS4_Report.pdf (accessed on 13 April 2007)

Gillman, M. (1997) Plant population ecology. In: Maxted, N., Ford-Lloyd, B.V. and Hawkes, J.G. (eds) *Plant Genetic Conservation: The* In Situ *Approach*. Chapman & Hall, London, pp. 114–131.

Given, D.R. (1994) *Principles and Practice of Plant Conservation*. Chapman & Hall, London.

Hellawell, J.M. (1991) Development of a rationale for monitoring. In: Goldsmith, F.B. (ed.) *Monitoring for Conservation and Ecology*. Chapman & Hall, London, pp. 1–14.

IUCN (1993) *The Convention on Biological Diversity: an Explanatory Guide*. Prepared by the IUCN Environmental Law Centre, Draft Text, Bonn, Germany.

Maxted, N., van Slageren, M.W. and Rihan, J. (1995) Ecogeographic surveys. In: Guarino, L., Ramanatha Rao, V. and Reid, R. (eds) *Collecting Plant Genetic Diversity: Technical Guidelines*. CAB International, Wallingford, UK, pp. 255–286.

Maxted, N., Hawkes, J.G., Ford-Lloyd, B.V. and Williams, J.T. (1997a) A practical model for *in situ* genetic conservation. In: Maxted, N., Ford-Lloyd, B.V. and Hawkes, J.G. (eds) *Plant Genetic Conservation: The* In Situ *Approach*. Chapman & Hall, London, pp. 545–592.

Maxted, N., Guarino, L. and Dulloo, M.E. (1997b) Management and monitoring. In: Maxted, N., Ford-Lloyd, B.V. and Hawkes, J.G. (eds) *Plant Genetic Conservation: The* In Situ *Approach*. Chapman & Hall, London, pp. 231–258.

Maxted, N., Iriondo, J., De Hond, L., Dulloo, E., Lefèvre, F., Asdal, A. Kell, S.P. and Guarino, L. (2006) Genetic reserve management. In: Iriondo, J.M., Maxted, N. and Dulloo, E. (eds) *Plant Genetic Population Management*. CAB International, Wallingford, UK.

23 *In Situ* Conservation Strategy for Wild Lima Bean (*Phaseolus lunatus* L.) Populations in the Central Valley of Costa Rica: a Case Study of Short-lived Perennial Plants with a Mixed Mating System

J.-P. BAUDOIN, O.J. ROCHA, J. DEGREEF, I. ZORO BI, M. OUÉDRAOGO, L. GUARINO AND A. TOUSSAINT

23.1 Introduction

A study was conducted in the Central Valley of Costa Rica to support the *in situ* conservation of wild lima bean, *Phaseolus lunatus* L., prone to extinction as a result of growing urbanization and changes in agricultural and land use practices. These populations represent a very important genetic reservoir for the improvement of the various *Phaseolus* bean cultigens, commonly found in many traditional cropping systems not only in Latin America, but also in other tropical regions. *P. lunatus* is also considered a useful plant model due to its reproductive biology. Lima bean is a self-compatible annual or short-lived perennial species with a mixed mating system, e.g. predominantly self-pollinating but with a fair amount of outcrossing, mediated by insects.

In order to achieve the conservation objective, several investigations were conducted in the following areas: (i) ecogeography and metapopulation dynamics; (ii) population demography and phenology; (iii) floral biology and gene flow; (iv) genetic structure of populations using morphological, biochemical and molecular markers; and (v) *in situ* conservation methodologies. A summary of the most relevant information is provided here; further details can be found in Baudoin *et al.* (2004). These investigations were made under the project 'Studies on breeding systems: the case of a short-living perennial, alternatively outbreeder–inbreeder species – *Phaseolus lunatus* – and its consequences for germplasm conservation', which ran in 2–4-year phases from 1992 to 2000 with funding from Belgium's Directorate General for Development Cooperation. It was a collaboration among three partners: International Plant Genetic

Resources Institute (IPGRI)'s Regional Office for the Americas in Cali, Colombia, the Escuela de Biología of the Universidad de Costa Rica (UCR) and the Unité de Phytotechnie tropicale et d'Horticulture of Gembloux Agricultural University (FUSAGx), Belgium.

23.2 The Study Area and the Target Species

The Central Valley is an intermontane valley located in the geographic centre of Costa Rica and enclosing an area of approximately 1500 km^2, with an altitudinal range between 800 and 1800 m above sea level (masl). The maximum length of the area is about 70 km, running from east to west, and the width is about 30 km, running from north to south. Most soils of the valley can be described as deep, rich in organic material and well drained. Great variation in the microgeographical distribution of rainfall results from the orientation of the mountain ranges and the location of wind passes. Rainfall is seasonal, with a well-defined dry season during December–April. In general, annual rainfall is lower in the eastern valley than in the western valley.

Wild populations of lima bean can be found throughout the Central Valley (Standley, 1937; Rocha *et al.*, 1997, 2002). The populations are usually found in open and disturbed areas with grasses and scattered trees or bushy thickets. They also colonize the coffee plantations from perennial fences (usually *Erythrina* L. and euphorbs) bordering the plots and are found where coffee is grown under shade (traditional coffee plantations), as well as in the waste lands around these plantations. Typically, agricultural activities are less intense in this agroecosystem, and do not rely on heavy use of herbicides for the elimination of weeds (Rocha *et al.*, 1997). However, because of changes in agricultural practices and in land use due to urban development, the populations of lima bean in the Central Valley are fragmented and undergo local extinction and recolonization (Rocha *et al.*, 1997).

23.3 Ecogeography and Metapopulation Dynamics

During the course of several surveys in the valley, the geographic location of a total of 565 populations (defined as groups of lima bean individuals isolated at least 500 m from any other) was determined. Studies in subsequent years revealed the appearance of new populations as well as the disappearance of old ones. In order to analyse the physical and ecological attributes of all locations where *P. lunatus* was found, a detailed classification of each site was conducted considering climate, soil, vegetation and topography using a geographical information system (GIS). Lima beans were found in seven different life zones, based on the Holdridge classification (Holdridge, 1966). The species is most abundant in the humid premontane (38% of populations) and the very humid premontane forests (54%). Similarly, the species was observed in 15 different biotic units, these being based on the structure and floristic composition of the plant communities (Gómez, 1986). However, 72% of the populations were

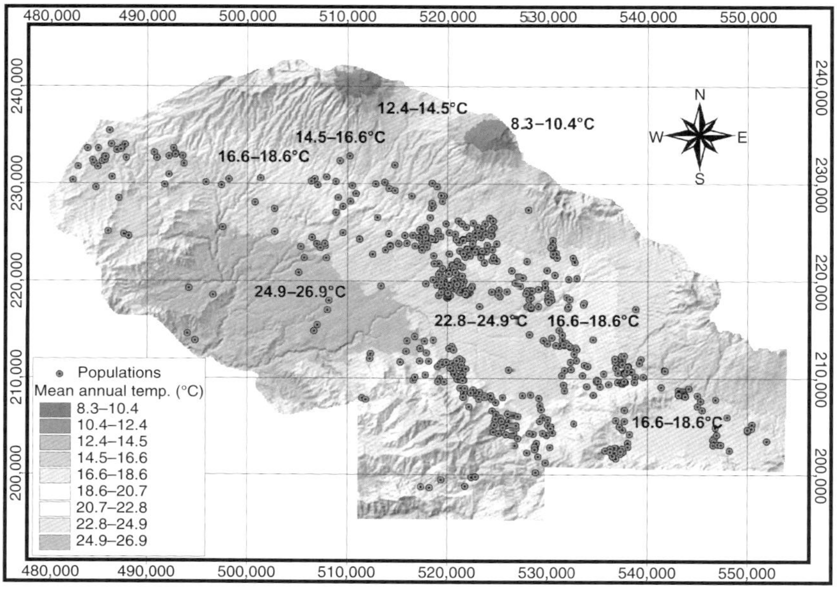

Fig. 23.1. Distribution of wild lima bean populations in the Central Valley of Costa Rica in relation to mean annual temperature. Dots indicate lima bean populations.

found in only three of these, i.e. humid, temperate and subtropical areas with a marked dry season lasting 3–6 months (Baudoin *et al.*, 2004). Figure 23.1 indicates the distribution of wild lima bean populations in the Central Valley in relation to mean annual temperature.

The numerous censuses and surveys conducted in the valley clearly demonstrate that wild lima bean populations undergo episodes of local extinction and recolonization in the study area (Rocha *et al.*, 1997, 2002). Therefore, a metapopulation approach would be the most suitable way to study population genetic structure and dynamics. For this purpose, six sampling transects were established along main roads in the valley. In 1994, 103 populations were found along these transects. All populations were visited every 2 weeks from January 1995 to April 2000. During each visit, the phenological status of each population was recorded by counting the number of individual plants that had foliage, flower buds, flowers, immature fruits and/or mature fruits with seeds. In addition, any disturbance experienced by each population was recorded such as fire, weeding (manually or with herbicides) and habitat destruction due to urban development.

This study confirmed that local extinction of lima bean populations is a common event in the Central Valley. Two types of extinction may be recognized:

1. Transient extinction, when all plants at the location disappeared but recolonization occurred during the same year;
2. Effective extinction, when all plants at the location disappeared and no recolonization occurred by the end of the year.

Most locations where extinctions occurred were recolonized during the same year, indicating that the soil seed bank plays a major role in restoring populations. In addition, the number of plants growing at a location appears to be negatively associated with the risk of local extinction. In general, most of the effective extinction events were observed at locations with fewer than ten plants. On the contrary, local extinction was not observed at locations with more than 50 plants. Moreover, once a group of plants became extinct at a given location, the risk of experiencing another extinction after recolonization was also high. In other words, the risk of extinction is not evenly distributed among the locations which were monitored, indicating different intensities of disturbances in the study area.

Metapopulation dynamics was studied by determining the transition probabilities among five possible population stages: (i) populations that remain vegetative during the year; (ii) populations that flowered during the year; (iii) populations that produced seeds during the year; (iv) populations that became extinct; and (v) populations that recolonized a site. Lefkovitch matrices (Lefkovitch, 1965) were used to describe the dynamics of these populations. The populations differed in their capacity to flower and to produce seeds every year and to maintain a seed reservoir in the soil (Rocha *et al.*, 2002). Populations that produce seeds in a given year have a high probability of producing seeds in future years, while those that become extinct also have a high probability of staying extinct. The UNIFIED LIFE MODEL software program (Legendre and Clobert, 1995) was applied to determine demographic parameters, in particular metapopulation growth rate (λ, the rate at which the number of populations in the metapopulation increases or decreases). Details are given in Rocha *et al.* (2002). In spite of the frequent extinctions recorded in this study, the metapopulation growth rate was rather close to 1 ($\lambda = 0.990$), indicating that the lima bean is not under a clear risk of overall extinction in the Central Valley.

23.4 Demography

Very few data are available about the demographic behaviour of tropical food legumes like *Phaseolus* (Baudoin *et al.*, 2004). However, demography is widely considered to be a key to the formulation of *in situ* conservation strategies (Oostermeijer *et al.*, 1996; Ehrlen and Van Groenendael, 1998). Matrix models have been used to determine which stages of the life cycle of a plant are most vulnerable (Charron and Gagnon, 1991; Damman and Cain, 1998) and which life stage transitions most strongly affect population growth (Caswell, 1989), hence improving our understanding of how plant populations respond to changes in the environment, including the impact of human activities.

23.4.1 Field observations

Demography of five wild Lima bean populations in the Central Valley of Costa Rica was analysed for 3 years. These populations were located in highly disturbed sites along trails, at the border of coffee plantations or in more natural sites such as at the edge of secondary regeneration forests. The following parameters were monitored in each population: (i) seed germination and longevity in the soil seed bank; (ii) survival and growth of seedlings and adult plants; and (iii) reproductive output of individual plants.

Matrix population models were used to compare the responses of the species to different environments and to identify the most critical life stages in each population.

The life cycle of lima bean was described on the basis of monthly field observations at each study site. In all, 49 quadrats (1 m^2 each) were sampled; all adult individuals were present and all seedlings appearing during the course of the study were labelled and identified by a code number. Developmental stage (vegetative or lignified stem), stem diameter (at 2 cm height), fecundity of each individual (number of seeds produced per year) and mortality rate in each quadrate were documented. Germination tests were conducted with painted seeds placed in plastic plates in the vicinity of each quadrate (Degreef *et al.*, 1997; Degreef, 1998).

23.4.2 Seed bank dynamics

Germination rates ranged from 70% to 86% after 1 year in the soil and from 89% to 94% after 2 years, for the five populations. Seed coat dormancy is likely to be the major factor responsible for this delay in germination. This dormancy is induced by drought stress, which may occur on the soil surface just after seed dispersal (Degreef *et al.*, 2002). Hypothesizing that the annual germination rate is similar from year to year, it was estimated that 96–99% of seeds will germinate within 3 years after dispersal. The design of the model was simplified by assuming that all seeds were germinated within this period.

23.4.3 Matrix demographic model

A life cycle graph was prepared for the species, in which each node is associated with a specific stage in the life of each individual. Due to the presence of a soil seed bank, seed classes were identified according to age. Juvenile and adult plants were grouped according to their developmental stage and stem diameter. Results are given in Degreef (1998).

A generalized projection model, using the UNIFIED LIFE MODELS software (Legendre and Clobert, 1995), was developed to describe the demography of lima bean populations. This allowed the determination of the asymptotic growth rate of each population when it reaches its stable structure; this rate can be used as a measure of fitness for the population in its particular environment, indicating either a decrease or an increase in the number of individuals, according to the location in threatened or more isolated areas.

Rocha *et al.* (1997) reported that wild lima bean populations in the Central Valley experience frequent extinctions and population fragmentation. These phenomena were recognized as resulting mainly from human perturbations, but some populations not subject to weeding and other disturbances also showed oscillation in the number of their individuals in the course of time. Therefore, the critical status of some populations, indicated by low values of their asymptotic growth rate, probably results from perturbations occurring at a very early developmental stage and still having an impact on their dynamics. The status of each population, whether declining or recolonizing a site has to be taken into account when designing a conservation plan and making management recommendations.

Elasticity is another parameter given by the model, which is of particular interest to determine which life cycle phase of the individuals is the most critical for the survival of the population, to quantify the contribution of each vital rate to population growth and to evaluate the effects of environmental perturbations on population dynamics (De Kroon *et al.*, 1986; Caswell, 1989; Mesterton-Gibbons, 1993; Degreef, 1998; Menges and Dolan, 1998). Analysis of elasticity values shows the importance of the direct transition from a seed present in the soil to a ligneous, potentially fertile individual, on population growth rate. This fate is only possible if seeds germinate soon after dispersal. Consequently, rapid germination is a key factor in the population dynamics of wild lima beans in the Central Valley of Costa Rica.

The elasticity analysis also reveals the relevance of the soil seed bank in the dynamics of populations. The growth rate of populations that are decreasing in size is particularly influenced by the arrival of new seeds into the soil seed bank. On the contrary, in increasing populations, elasticity for this particular transition is low.

Furthermore, the analysis shows the importance of the growth of ligneous individuals, mainly favoured by adequate air and soil moisture; and, finally, the survival of well-established ligneous and fertile individuals appears to be critical for the growth rate in the two populations located in the most natural sites.

It is generally recognized that the life history component which most strongly affects population growth depends on the habitat where the population grows (Bierzychudek, 1982; Silvertown *et al.*, 1996; Damman and Cain, 1998). Higher elasticities for growth and fecundity are typical of open habitats while higher elasticities for survival are characteristic of closed habitats (Menges and Dolan, 1998). In particular, for wild lima beans, this pattern is best shown by the extent and the frequency of disturbances to which populations are exposed. Higher growth and fecundity elasticities were obtained in populations experiencing perturbations and lower environmental stability. In contrast, populations in more stable habitats had higher survival elasticities. These results are essential for the implementation of an *in situ* conservation programme in the region.

23.5 Genetic Diversity

A major component of this project was to evaluate the genetic diversity of the wild lima bean populations found in the Central Valley. This diversity was studied at both the intra- and inter-population level, with the aim to analyse factors responsible for the genetic structure and the microgeographical patterns of the *P. lunatus* gene pool in the region.

Despite the relatively small size of the study area and the small sample size taken from each population (five seeds from a bulked sample), significant phaseolin variation was found, mainly among the wild populations analysed, all belonging to the meso-american gene pool of *P. lunatus* (Vargas *et al.*, 2000, 2001).

Assessment of genetic variability was made in 330 populations and 1–60 individuals per population using both enzyme and microsatellite markers (Maquet *et al.*, 1996; Zoro Bi, 1999; Ouédraogo, 2003; Zoro Bi *et al.*, 2005). Table 23.1 presents results for 28 populations evenly selected throughout the valley. Microsatellite markers showed more allelic and genetic diversity than enzyme markers. However, the two markers revealed a relatively low percentage of polymorphic loci at the intra-populaton level (probably due to the belonging of the populations to the single Meso-american gene pool), few alleles per locus and a lack of heterozygotes, attributed mainly to self-fertilization. The coefficient of gene differentiation (G_{ST}), reflecting the contribution of among population genetic diversity (D_{ST}) to total genetic diversity (H_T), is relatively high and similar for the two markers. Each wild lima bean population may therefore constitute a valid *in situ* conservation unit. In another study, based on 96 well-scattered wild populations, significant heterogeneity of allele frequencies was found in all enzyme markers (Table 23.2). The Wright consanguinity coefficient (F_{IT}) showed deviation of populations from Hardy–Weinberg equilibrium, due to genetic differentiation among populations (F_{ST}) and non-random mating within populations (F_{IS}).

Table 23.1. Genetic diversity and structure of 28 wild lima bean populations in the Central Valley of Costa Rica with the use of two markers.

	Intrapopulation polymorphism indices (means ± SE)			
Marker	P[a]	A[b]	H_0[c]	H_e[d]
Enzymes	20.57 ± 5.07	1.227 ± 0.051	0.020 ± 0.021	0.080 ± 0.024
Microsatellites	48.89 ± 21.47	1.644 ± 0.384	0.012 ± 0.009	0.143 ± 0.058
	Nei's genotypic diversity indices (means)			
	H_T[e]	H_S[f]	D_{ST}[g]	G_{ST}[h]
Enzymes	0.120	0.083	0.036	0.303
Microsatellites	0.220	0.153	0.067	0.303

[a]Percent of polymorphic loci.
[b]Mean number of alleles per locus.
[c]Observed heterozygosity.
[d]Expected heterozygosity.
[e]Total genic diversity.
[f]Intra-population gene diversity.
[g]Inter-population gene diversity.
[h]Genic differentiation coefficient.

Table 23.2. Heterogeneity of allele frequencies and *F*-statistics estimated for 96 wild lima bean populations of the Central Valley of Costa Rica with the use of enzyme loci.

Locus	G-test (df)	*F*-statistics ± standard deviation		
		F_{IT}[a]	F_{IS}[b]	F_{ST}[c]
Adh-2	529.24 (52)***	0.873 ± 0.026	0.778 ± 0.079	0.433 ± 0.079***
Dia-1	145.00 (16)***	0.921 ± 0.061	0.874 ± 0.111	0.376 ± 0.136*
Gpi-1	26.61 (10)**	1.000 ± 0.000	1.000 ± 0.000	1.000 ± 0.007***
Mdh-2	402.16 (62)***	0.823 ± 0.035	0.747 ± 0.038	0.299 ± 0.077***
Pgm-2	874.11 (48)***	0.917 ± 0.035	0.777 ± 0.065	0.625 ± 0.122***
Mean		0.882 ± 0.026	0.761 ± 0.012	0.504 ± 0.094***

[a]Coefficient of consanguinity.
[b]Coefficient of intra-population consanguinity.
[c]Genetic differentiation between populations.
*$P < 0.05$; **$P < 0.01$; ***$P < 0.001$; the comparisons being based on G-tests for allelic frequencies among populations and Student *t*-tests for F_S.

Despite small population sizes, common alleles at several loci, very few heterozygous individuals and low pollen and seed dispersal (see Section 5), significant polymorphism within populations was observed. This result, apparently contradictory with the low percentage of polymorphic loci at the intra-population level, could be due partly to the existence of gene flow over long distances, and partly to non-random mating within populations. Significant correlation was observed between population size and genetic variability. The loss of genetic diversity in small-sized populations could be attributed to inbreeding and bottleneck effects in some populations.

Through phaseolin characterization, populations at different sites of the valley were arranged in groups that were also congruent in terms of geographical proximity and phenology (Vargas *et al.*, 2001). This could be explained, at least in part, by climatic factors. Using isozyme markers, a similar study of 96 populations showed a high heterogeneous allelic distribution through all the polymorphic loci (Zoro Bi, 1999). Alleles were present in either very few or numerous populations and had an irregular geographic distribution. This non-random spatial distribution of alleles might result from limited gene flow between populations, and/or high localized selection pressure caused by biotic or abiotic stresses.

Genetic variation among populations and its geographical pattern are affected by several factors, including population dynamics and environmental conditions of the valley (Baudoin *et al.*, 2004). For example, wild populations might undergo repeated bottlenecks, as weeding and other agricultural practices only allow a few plants to survive and reproduce. These processes lead to significant reductions in effective population size, and to high levels of inbreeding, favouring the decrease of heterozygotes in the populations. The recurrent reduction in population size will also favour genetic differentiation among populations. The discontinuity of the habitats where wild lima beans are most likely to be found in the valley also promotes genetic differentiation among populations.

Such fragmentation is mainly the result of replacement of traditional coffee plantations by modern, intensive plantations and accelerated urban development. Differences in abiotic (climate and soil) and biotic factors also affect levels and patterns of genetic variation in the valley.

23.6 Gene Flow

Field and laboratory investigations have revealed a considerable amount of information relevant to gene flow in wild lima beans in the valley, such as the high frequency of small population sizes (66% of populations with fewer than 30 individuals), the low allogamy rate ($t \leq 10\%$), the presence of major alleles at several loci, the low frequency of heterozygous individuals and the high intra-population polymorphism (indicated by significant G_{ST} values for both isozyme and microsatellite markers).

The origin of this significant intra-population diversity is in part related to the importance of short- and long-distance gene flow. This can be estimated at the within- and between-populations levels, using direct (field measures) and indirect methods.

23.6.1 Direct methods of gene flow estimation

According to Crawford (1984) and Gliddon *et al.* (1987), estimation of gene flow is based upon the number of individuals in a local random breeding unit, i.e. a 'neighbourhood', defined more precisely as the genetic neighbourhood area (NA) and the effective neighbourhood size (Nb). These two parameters are determined from two equations: (1) $NA = 4\Pi(1/2t \times \sigma_n^2 + \sigma_s^2 + \sigma_v^2)$; and (2) $Nb = NAd(1 + t)/2$, where σ_n^2, σ_s^2 and σ_v^2 are dispersal variances, respectively for pollen, seed and flower (or vegetative growth); t is the outcrossing rate and d is adult plant density.

Gene flow was measured by this method in three selected populations of the valley (Hardy *et al.*, 1997; Baudoin *et al.*, 1998). Pollen grains and seeds were labelled *in vivo* using stains and fluorescent dye. Pollinators were identified by observing each population at different dates during their flowering period and estimating the lima bean pollen load on the insect body.

In the valley, lima bean blooms during the dry season, from about mid-November to mid-February. Mean pollen/ovule ratio is about 863 (Hardy *et al.*, 1997). According to Cruden (1977), this ratio is typical of species with a mating system qualified as facultative allogamy. In the target area, the major pollinator is *Apis mellifera* L. In all studied populations, most pollen transfers occurred across distances of less than 1 m, confirming that common bees disperse pollen mostly over short distances. The frequency of pollen transfers dropped quickly around 1 m, although transfer could reach a maximum value of 5.5 m. The corresponding dispersal variance for pollen was $\sigma_p^2 = 1.7\,m^2$. Measures of flower dispersal showed great variability in the vegetative growth of wild lima bean individuals. Distances separating each inflorescence from its respective plant base ranged between 0.37 and 6.5 m, according to the presence of surrounding vegetation that constituted a support allowing the plant to climb. In the tested populations, the mean flower dispersal variance was $\sigma_p^2 = 2.7\,m^2$. Seeds of wild lima beans are too heavy to be carried by air. The most important contribution to seed dispersal occurs when dehiscent pods open and project their seeds on the ground. In the studied populations, the maximum distance of seed dispersal was 5.5 m from the pod to the site on the ground. The mean seed dispersal variance was $\sigma_s^2 = 1.68\,m^2$.

Using isozymes, Maquet *et al*. (1996) and Hardy *et al*. (1997) calculated a mean outcrossing rate and a mean adult plant density of, respectively, $t = 0.1$ and $d = 0.235$ plants/m². From the two equations (1) and (2), the neighbourhood area (NA) and the effective neighbourhood size (Nb) were, respectively, 56 m² and 7.23 individuals. According to Wright (1946), the low value for Nb (<20 individuals) corresponds to a high probability of random local genetic differentiation within the population. As the populations in the region spread over areas ranging from about 100 m² to more than 1000 m², they could contain several to many neighbourhood areas. Therefore, allelic distribution in a single population is expected to be highly structured and there is a need to sample systematically at many sites of the population for collecting the whole genetic diversity.

23.6.2 Indirect methods of gene flow estimation

A conventional approach to quantify gene flow is to transform measures of population structure into indirect estimates of the average number of migrants exchanged per generation: this can be done by using the island model (Wright, 1951) or the isolation-by-distance model or Slatkin's private alleles model, in which the rate of gene flow is expected to decline monotonically with increasing geographic distance between continuous populations (Slatkin, 1985, 1993). Both methods were applied using enzyme and microsatellite markers and analysing polymorphic loci in populations located at various distances from each other.

Using isozymes, Wright's method produced a G_{ST} value of 0.575, corresponding to a number of migrants per generation (Nm) of 0.18, suggesting restricted gene flow among subpopulations (Nm < 1). By the method of Slatkin using rare alleles, Nm was estimated at 0.08.

Using microsatellites, the fixation index (F_{ST}) was 0.346, and the average inbreeding coefficient within populations was high ($F_{IS} = 0.916$). The number of migrants per population and per generation from Wright's method was 0.47. By the method of Slatkin using private alleles, Nm was estimated at 0.06.

The two models of indirect gene flow estimation used in this study are useful for understanding the evolution of genetic structure in plant species with metapopulation dynamics. Results indicated low to moderate levels of gene flow (0.06–0.47) for the wild populations of the valley, although heterogeneity in the number of individuals per population could cause underestimation (Slatkin, 1985). On the basis of enzyme and microsatellite markers, very high divergence occurs among populations. This is probably due to restricted gene flow, with genetic drift, therefore playing a major role in the genetic structure of lima bean populations in the study area.

Gene flow was not related to geographical distances at larger scales (Baudoin *et al.*, 2004). This could be due to heterogeneity among the gene flow values when all populations are considered or, in the isolation-by-distance model, to the assumption that populations must be in equilibrium between migration and genetic drift.

23.7 *In Situ* Conservation of Wild *P. lunatus* in the Central Valley

On the basis of the data of genetic diversity, gene flow, demography and population dynamics, an *in situ* conservation strategy was developed for the Central Valley. Two complementary options were investigated. One was to identify and protect specific existing populations. The other was to establish new, synthetic populations in protected areas.

23.7.1 *In Situ* conservation areas for existing populations

Life zones and soil type maps were used to identify all possible environmental combinations in which lima beans occur. Two combinations (bh-P and bmh-P life zones on inceptisol) are particularly important, including more than 75% of the total number of wild lima bean populations recorded in the valley. Only sites likely to maintain their integrity were considered. In view of the two main agents of genetic erosion in the valley, i.e. urban expansion and intensifying agricultural practices, and the small size of many lima bean populations, sites prone to human pressure (such as modern coffee plantations, small-scale farms and hedges near roads) were discarded. Some isolated sites, remote from cultivated land or human settlements, mainly located along water streams and deep slopes, and legally protected from cutting and weeding, were identified as potential conservation areas. In addition, with the purpose to preserve the available genetic diversity of wild lima beans, priority was also given to sites where populations are characterized by the presence of localized and private alleles (Zoro Bi, 1999).

In the end, the survey of the Central Valley identified 30 conservation sites, representing 24 combinations of life zone and soil type and distributed at elevations from 340 to 1980 m asl. Most of these sites are located at the margins of secondary forests or near a river or stream. The micro-environment was very often characterized by a diverse overstorey, producing a layered vertical structure and allowing light penetration and support for the climbing individuals.

The populations maintained *in situ* are large, covering an area of 1000 m^2 or more and containing at least 100 individuals. Their field management should follow specific recommendations derived from the results of the demographic studies, and be adapted to the disturbance level of their environment. In perturbed sites, such as in the vicinity of coffee plantations or along trails, conservation must favour the growth of young lignified individuals, which closely depends upon the humidity of the environment. The suggested management is to maintain a mulch on the soil surface at the end of the rainy season or to install a vegetation cover favouring both high air and soil moisture. In addition, recruitment of new individuals from seeds can be promoted by weeding the soil surface just after seed dispersal at the end of the dry season. In contrast, in the more natural, undisturbed sites, mainly located at the border of secondary forests, it is important to favour the survival of large lignified individuals. Selective clearings should be carried out in these sites to maintain potential reserves of lignified adult plants.

23.7.2 Establishment of *in situ* synthetic populations

Interesting populations of smaller size and at risk of anthropogenic disturbance can only be preserved *in situ* in the long term by moving them to protected areas in the previously defined 24 environmental combinations. For this purpose, we developed a process involving synthetic populations. A synthetic population is defined as a group of individuals derived from seeds collected from original sites of wild lima bean populations, then sown in a protected site selected to allow optimal plant development and gene flow within and among populations. The protected sites are located in the same life zones as the original wild lima bean populations, or at least in similar ecological niches. The first synthetic populations were established in June 1998 in protected microsites covering four life zones of the Central Valley.

At each of these sites, synthetic populations were sown with seeds collected from nearby populations found in natural areas. The microconservation plots containing the synthetic populations were designed to fulfil several requirements, with regard to gene flow, plant dispersal, population size and fragmentation, extinction and recolonization processes. Such requirements were identified on the basis of our investigations in the valley and also from the studies made by Given (1993), Maxted *et al.* (1997), Tiebout and Anderson (1997) and Yonezawa (2000). Taking into account the patchy distribution of wild lima beans in the Central Valley, the usually small population size, the twining vegetative growth and the possibility of gene exchange between nearby populations through pollen and seed dispersal, two types of conservation microreserves were designed and established (Meurrens *et al.*, 2001):

- Circular design, consisting of circular cleared patches, linked by corridors. In each patch, seeds from nearby populations were sown at the base of an existing tree or shrub, to provide a support for the vines. The cleared corridors are meant to facilitate gene flow between patches through pollen movement or development of very long twining branches, bearing racemes at leaf nodes.
- Linear conservation design, consisting of several groups of nearby population seeds sown every metre along thickets.

For the two conservation models, the minimum area of the patches varied from 56 to 150 m^2 according to the neighbourhood area calculated in our investigations.

The number of seeds to be introduced in each patch was determined with the aim of reaching a mean density of 0.35 adult plants/m^2, a situation frequently observed in the valley (Degreef, 1998). The seeds sown in each patch should represent the original genetic variability of the populations in its original site, following the sampling recommendations of Zoro Bi *et al.* (1998) regarding number of plants per population and number of seeds per plant.

Considering the results of the demographic studies, the vegetation in each patch was slightly disturbed during the development stage of introduced populations in order to obtain optimal germination and rapid growth of the plantlets and to avoid uncontrolled dispersal.

The effectiveness of this methodology was assessed by comparing the demography of the synthetic populations with that of natural populations located in the same environments. Similar field observations and measurements were taken with the two types of populations. Results of the comparison are presented in details by Meurrens *et al.* (2001), showing a positive impact for at least three important demographic parameters: germination percentage of seeds sown, lignification and death rates.

The increase in germination percentage in synthetic populations can be explained by the management method used to break seed dormancy, i.e. successive weeding after sowing. This practice moves the seeds to the soil surface, where they are exposed to high temperature, a factor favouring the breaking of seed coat dormancy (Degreef *et al.*, 2002). When establishing synthetic populations, the first step is to allow as many seeds as possible to germinate and to produce seedlings. When a sufficient number of adult plants are reached to ensure progenies, it is important to let the populations constitute their own soil seed bank. If sufficiently large, this seed reserve will buffer the poor seed production that could occur some years and enhance the demographic stability of the population. After successful germination, the possible fates of a seedling within the first year are either to die or to reach the juvenile or the lignified stage. Reducing the mortality of seedlings is an important challenge for *in situ* conservation of wild lima beans in the Central Valley, as most individuals die at this stage in natural populations (Degreef, 1998). In synthetic populations, death rates can be decreased by appropriate management practices, in particular using mulch to maintain soil moisture during the dry season.

23.8 Conclusions and Recommendations

The project has developed a uniquely detailed and comprehensive data set of ecogeographic, genetic and demographic information on a wild crop relative. It has then used this data set to develop an *in situ* conservation strategy for *P. lunatus* in the study region based on the selection of key natural sites, the establishment of complementary synthetic populations and targeted management interventions.

In order to build on these scientific achievements, it is important to improve the strategy of conservation and the methodology used to reach it. Some further research activities could be highlighted.

- Monitoring (through genetic characterization as well as phenological and demographic observations) and refinement of the management strategy should be carried out in the two types of populations (existing and synthetic) to ensure long-term conservation. In particular, plant growth and development, extent of gene flow among patches within microconservation sites and appearance of novel multilocus genotypes through genetic hybridization should be analysed.
- GIS tools should be used to examine in more depth any relationship between the distribution of genetic variation in the Central Valley and ecogeographic factors. For the conservation objective, it is essential to identify those populations that best represent ecological and genetic diversity. Data from various genetic markers (biochemical and molecular) could be tested against microscale passport data to highlight the combinations of factors that best orient the choice of individuals or populations for inclusion in conservation programmes. Such data will also be relevant for the determination of the minimum sample size required for maintaining a given level of allelic diversity.
- A 'carrying capacity' component should be added to the demographic model which has been developed, in order to determine the effects of plant density on mortality and growth rates of lima bean individuals. A study should be carried out to evaluate the impact of the 'extinction–recolonization' process on the genetic structure of populations. As the soil seed bank plays a significant role in population survival, it should be interesting to compare the genetic diversity between populations established from the soil seed bank after a local extinction and populations established with previously collected seeds from the same original populations.

As gene flow is a key element in determining the genetic structure of the wild populations, it is essential to follow-up their study in selected regions of the Central Valley by using microsatellite markers. A specific objective will be to measure the impact of gene flow on a very large scale through the sampling of individuals located at large distances from a central population.

References

Baudoin, J.-P., Degreef, J., Hardy, O., Janart, F. and Zoro Bi, I. (1998) Development of an *in situ* conservation strategy for wild Lima bean (*Phaseolus lunatus* L.) populations in the Central Valley of Costa Rica. In: Owens, S.J. and Rudall, P.J. (eds). *Reproductive Biology*. Royal Botanic Gardens, Kew, UK, pp. 417–426.

Baudoin, J.-P., Rocha, O., Degreef, J., Maquet, A. and Guarino, L. (2004) *Ecogeography, Demography, Diversity and Conservation of* Phaseolus lunatus *L. in the Central Valley of Costa Rica*. Systematic and Ecogeographic Studies on Crop Genepools 12. International Plant Genetic Resources Institute, Rome, Italy.

Bierzychudek, P. (1982) Life histories and demography of shade-tolerant temperate forest herbs: a review. *New Phytologist* 90, 757–776.

Caswell, H. (1989) *Matrix Population Models: Construction, Analysis and Interpretation*. Sinauer Associates, Sunderland, Massachusetts.

Charron, D. and Gagnon, D. (1991) The demography of northern populations of *Panax quinquefolium* (American Ginseng). *Journal of Ecology* 79, 431–445.

Crawford, T.J. (1984) The estimation of neighbourhood parameters for plant populations. *Heredity* 52(2), 273–283.

Cruden, R.W. (1977) Pollen-ovule ratios: a conservative indicator of breeding systems in flowering plants. *Evolution* 31(1), 32–46.

Damman, H. and Cain, M.L. (1998) Population growth and viability analyses of the clonal woodland herb, *Asarum canadense*. *Journal of Ecology* 86, 13–26.

De Kroon, H., Plaisier, A., Van Groenendael, J. and Caswell, H. (1986) Elasticity: the relative contribution of demographic parameters to population growth rate. *Ecology* 67, 1427–1431.

Degreef, J. (1998) Développement d'un Modèle Démographique et Application à la Conservation In Situ des Populations Sauvages de Haricot de Lima (*Phaseolus lunatus* L.) dans la Vallée Centrale du Costa Rica. PhD thesis, Gembloux Agricultural University, Gembloux, Belgium.

Degreef, J., Baudoin, J.-P. and Rocha, O.J. (1997) Case studies on breeding systems and its consequences for germplasm conservation. 2. Demography of wild Lima bean populations in the Central Valley of Costa Rica. *Genetic Resources and Crop Evolution* 44, 429–438.

Degreef, J., Rocha, O.J., Vanderborght, T. and Baudoin, J.-P. (2002) Soil seed bank and seed dormancy in wild populations of Lima bean (*Fabaceae*): considerations for *in situ* and *ex situ* conservation. *American Journal of Botany* 89, 1644–1650.

Ehrlen, J. and Van Groenendael, J. (1998) Direct perturbation analysis for better conservation. *Conservation Biology* 12, 470–474.

Given, D.R. (1993) *Principles and Practice of Plant Conservation*. Chapman & Hall, London.

Gliddon, C., Belhassen, E. and Gouyon, P.H. (1987) Genetic neighbourhoods in plants with diverse systems of mating and different datterns of growth. *Heredity* 59, 29–32.

Gómez, L.D. (1986) *Vegetación de Costa Rica*. Editorial Universidad Estatal A Distancia, San José, Costa Rica.

Hardy, O., Dubois, S., Zoro Bi, I. and Baudoin, J.-P. (1997) Gene dispersal and its consequences on the genetic structure of wild populations of Lima bean (*Phaseolus lunatus*) in Costa Rica. *Plant Genetic Resources Newsletter* 109, 1–6.

Holdridge, L.R. (1966) The life zone system. *Adansonia* 6, 199–203.

Lefkovitch, L.P. (1965) The study of population growth in organisms grouped by stages. *Biometrics* 21, 1–18.

Legendre, S. and Clobert, J. (1995) ULM, a software for conservation and evolutionary biologists. *Journal of Applied Statistics* 22, 817–834.

Maquet, A., Zoro Bi, I., Rocha, O.J. and Baudoin, J.-P. (1996) Case studies on breeding systems and its consequences for germplasm conservation. 1. Isozyme diversity in wild Lima bean populations in the Central Valley of Costa Rica. *Genetic Resources and Crop Evolution* 43, 309–318.

Maxted, N., Ford-Lloyd, B.V. and Hawkes, J.G. (1997) *Plant Genetic Conservation: The* In Situ *Approach*. Chapman & Hall, London.

Menges, E.S. and Dolan, R.W. (1998) Demographic viability of populations of *Silene regia* in midwestern prairies: relationships with fire management, genetic variation, geographic location, population size and isolation. *Journal of Ecology* 86, 63–78.

Mesterton-Gibbons, M. (1993) Why demographic elasticities sum to one? A postscript to De Kroon *et al. Ecology* 74, 2467–2468.

Meurrens, F., Degreef, J., Rocha, O.J. and Baudoin, J.-P. (2001) Demographic study in microconservation sites with a view to maintaining *In Situ* wild Lima beans (*Phaseolus lunatus* L.) in the Central Valley of Costa Rica. *Plant Genetic Resources Newsletter* 128, 45–50.

Oostermeijer, J.G., Brugman, M.L., De Boer, E.R. and Den Nijs, H.C. (1996) Temporal and spatial variation in the demography of *Gentiana pneumonanthe*, a rare perennial herb. *Journal of Ecology* 84, 153–166.

Ouédraogo, M. (2003) Etude de la Variabilité Génétique Et Du Flux De Gènes Chez Des Populations Sauvages de *Phaseolus lunatus* L. dans la Vallée Centrale du Costa Rica à l'aide de Marqueurs Enzymatiques et Microsatellites. PhD thesis, Gembloux Agricultural University, Gembloux, Belgium.

Rocha, O.J., Macaya, G. and Baudoin, J.-P. (1997) Causes of local extinction and recolonization determined by 3 years of monitoring wild populations of *Phaseolus lunatus* L. in the Central Valley of Costa Rica. *Plant Genetic Resources Newsletter* 112, 44–48.

Rocha, O.J., Degreef, J., Barrantes, D., Castro, E., Macaya, G. and Guarino, L. (2002) Metapopulation dynamics of Lima bean (*Phaseolus Lunatus* L.) in the Central Valley of Costa Rica. In: Engels, J.M.M., Ramanatha, V., Brown, A.H.D. and Jackson, M.T. (eds) *Managing Plant Genetic Diversity*. CAB International, Wallingford, UK, pp. 205–215.

Silvertown, J., Franco, M. and Menges, E. (1996) Interpretation of elasticity matrices as an aid to the management of plant populations for conservation. *Conservation Biology* 10, 591–597.

Slatkin, M. (1985) Rare alleles as indicators of gene flow. *Evolution* 39, 53–65.

Slatkin, M. (1993) Isolation by distance in equilibrium and non-equilibrium populations. *Evolution* 47, 264–279.

Standley, P. (1937) *Flora of Costa Rica*. Field Museum of Natural History, Chicago, Illinois.

Tiebout, H.M. and Anderson, R.A. (1997) A comparison of corridors and intrinsic connectivity to promote dispersal in transient successional landscapes. *Conservation Biology* 11, 620–627.

Vargas, E.M., Macaya, G., Baudoin, J.-P. and Rocha, O.J. (2000) Variation in the content of phaseolin in 37 wild populations of Lima beans (*Phaseolus lunatus* L.) in the Central Valley of Costa Rica. *Plant Genetic Resources Newsletter* 121, 53–58.

Vargas, E.M., Macaya, G., Baudoin, J.-P. and Rocha, O.J. (2001) Case studies on breeding systems and their consequences for germplasm conservation. 3. Electrophoretic mobility of phaseolins in wild populations of Lima beans (*Phaseolus lunatus* L.) in the Central Valley of Costa Rica. *Genetic Resources and Crop Evolution* 48, 109–120.

Wright, S. (1946) Isolation by distance under diverse mating systems. *Genetics* 31, 39–59.

Wright, S. (1951) The genetical structure of populations. *Annals of Eugenetics* 15, 323–354.

Yonezawa, K. (2000) *In situ* conservation strategies for plant species: some comments based on the recent advances in population genetic theories. In: *In Situ* Conservation Research, *Proceedings of MAFF International Workshop on Genetic Resources*, 13–15 October 1999. National Institute of Agrobiological Resources, Tsukuba, Japan, pp. 43–81.

Zoro Bi, I. (1999) Variabilité Génétique des Populations Sauvages de *Phaseolus lunatus* L. dans la Vallée Centrale du Costa Rica et ses Implications dans la Mise au Point d'une Stratégie de Conservation *In Situ*. PhD thesis, Gembloux Agricultural University, Gembloux, Belgium.

Zoro Bi, I., Maquet, A., Degreef, J., Wathelet, B. and Baudoin, J.-P. (1998) Sample size for collecting seeds in germplasm conservation: the case of the Lima bean (*Phaseolus lunatus* L). *Theoretical and Applied Genetics* 97, 187–194.

Zoro Bi, I., Maquet, A. and Baudoin, J.-P. (2005) Mating system of wild *Phaseolus lunatus* L. and its relationship to population size. *Heredity* 94, 153–158.

24 Population Performance of *Arnica montana* L. in Different Habitats

J. RADUŠIENĖ AND J. LABOKAS

24.1 Introduction

An important aspect of conservation of endangered species is the monitoring of the populations' structure over time. The structure of populations may be described by classifying the plants by age, size or life stage (Rabotnov, 1985). It is often impossible to establish the age of individuals as both size and reproductive capacity are poorly correlated with age. The best way of describing populations is through the determination of the relative proportions of individuals in the different ontogenetic stages of their life cycle. This method has been used in various plant species studies (Rabotnov, 1985; Oostermeijer *et al.*, 1994; Hegland *et al.*, 2001; Aguraiuja *et al.*, 2004). On the basis of population stage structure, it is possible to make an assessment of the status of populations, classifying them as 'dynamic', 'normal' or 'regressive' (Hegland *et al.*, 2001). The population is called 'dynamic' when young stages are represented to a significantly greater extent. A 'normal' population is a population with a relatively high proportion of adults and considerable number of young plants. Finally, over-representation of the mature stage indicates a 'regressive' status of the population. The information of population stage structure can be used as an indicator of present vitality and as a way to predict its dynamics (Hanski, 1998).

Arnica montana L. is regarded as a critically endangered species in Belgium, Bosnia, Croatia and Luxemburg; endangered in Belarus and the Netherlands; vulnerable in Estonia, Germany, Hungary, Latvia, Lithuania, Portugal and Romania; and near threatened in Denmark and Norway (Lange, 1998). The decrease of populations is mostly considered to be taking place as a result of environmental changes (land reclamation, use of fertilizers and atmospheric acidification), habitat destruction and over-exploitation for medicinal raw material which is picked mainly from the wild (Dueck and Elderson, 1992).

The more detailed analysis of the status of arnica populations has just begun in Lithuania. The aim of this chapter is to understand the present status of *A. montana* populations in different habitats and to give a forecast for further performance of populations, which is of primary importance for the creation of a proper conservation strategy.

24.2 Materials and Methods

Over the period of 2002–2004 field surveys were conducted in 42 subpopulations of *A. montana* in south-eastern and eastern parts of Lithuania (Fig. 24.1). Data of site characteristics and distance between subpopulations are given in Table 24.1. The distance between the northernmost and southernmost points of arnica distribution is 121.4 km, and between the easternmost and westernmost points 139 km. Vegetation relevées were made in the sites where arnica was studied. The vegetation was described using the Braun-Blanquet scale of plant abundance. A more detailed syntaxonomical interpretation of the vegetation data was made by the computer package program BIODIVERSITY PROFESSIONAL (Natural History Museum, London, and Scottish Association for Marine Sciences, Oban, Scotland).

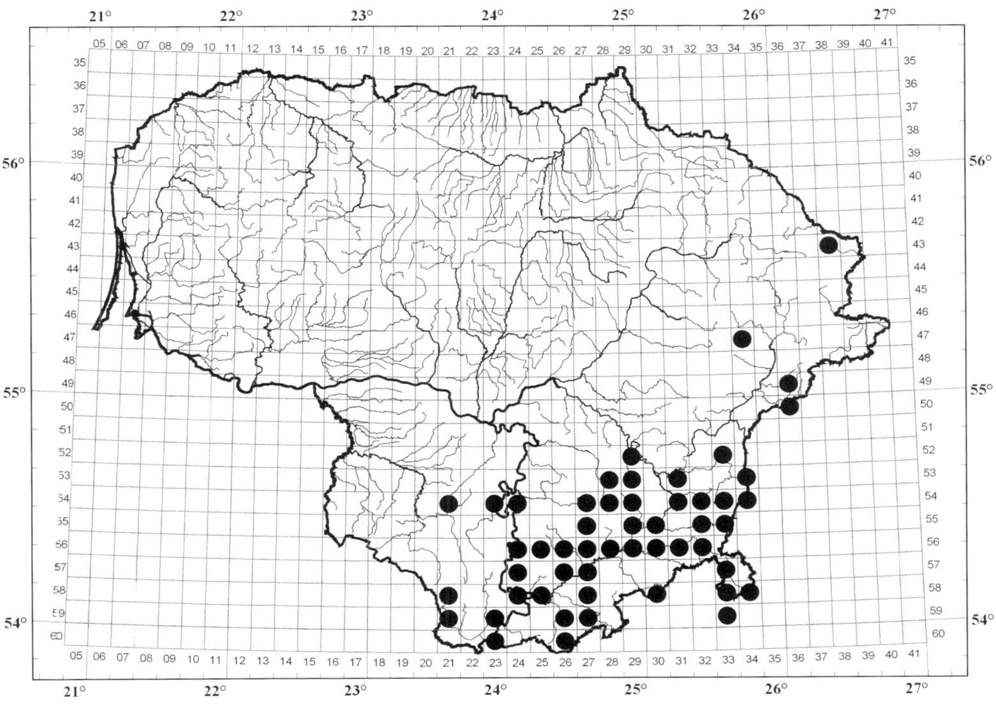

Fig. 24.1. Distribution of *Arnica montana* in Lithuania. The map of territory of Lithuania is divided into squares 6' N–S by 10' E–W (~10 × 10 km^2). All subpopulations in one square were marked as one dot.

Table 24.1. Site characteristics of *Arnica montana* subpopulations.

No.	Latitude N	Longitude E	Forest district	Sub-population size in m^2	Distances between sub-populations (km), as subsequent N–S points	Plant community[a]
1	54°33'38.5''[b]	25°19'42.2''	Juodšiliai	1		1
2	54°31'33''	25°39'25''*	Medininkai	1500	21.5688	3
3	54°22'23.3''	25°15'14.8''	Gudeliai	600	31.1406	3
4	54°22'13.5''	25°15'40.8''	Gudeliai	600	0.5612	3
5	54°08'30.0''	23°43'44.2''	Veisiejai	10	102.805	1
6	54°07'47.8''	24°20'03.7''	Puvočiai	100	65.9646	1
7	54°07'33.8''	24°20'15.5''	Puvočiai	30	0.4844	1
8	54°07'33.8''	24°20'44.8''	Puvočiai	225	0.5255	2
9	54°06'51.5''	24°23'13.1''	Puvočiai	140	3.4775	2
10	54°06'41.4''	24°23'16.2''	Puvočiai	25	0.3140	1
11	54°06'38.1''	24°22'37.1''	Puvočiai	1000	0.7130	2
12	54°06'36.9''	24°23'20.3''	Puvočiai	6000	0.7819	2
13	54°06'36.0''	24°21'30.6''	Puvočiai	100	1.9940	1
14	54°06'35.2''	24°22'35.8''	Puvočiai	100	1.1787	1
15	54°06'34.4''	24°22'23.8''	Puvočiai	0.3	0.2197	2
16	54°06'32.2''	24°22'13.8''	Puvočiai	100	0.1915	2
17	54°06'28.2''	24°19'07.1''	Puvočiai	50	3.3740	2
18	54°06'25.5''	24°19'21.7''	Puvočiai	75	0.2702	2
19	54°06'25.2''	24°19'23.6''	Puvočiai	0.05	0.0363	2
20	54°06'18.8''	24°23'53.5''	Puvočiai	100	4.8996	1
21	54°06'18.6''	24°19'07.4''	Puvočiai	600	5.1850	1
22	54°06'18.0''	24°20'44.0''	Puvočiai	1800	1.7585	1
23	54°06'15.8''	24°19'00.8''	Puvočiai	200	0.1572	2
24	54°06'13.3''	24°20'15.5''	Puvočiai	25	1.3611	1
25	54°06'09.3''	24°19'33.0''	Puvočiai	100	0.7714	1
26	54°06'07.8''	24°21'19.1''	Puvočiai	100	1.9228	1
27	54°06'02.9''	24°23'39.3''	Puvočiai	100	2.5430	1
28	54°06'01.3''	24°19'57.6''	Puvočiai	50	4.0253	2
29	54°06'00.3''	24°21'40.8''	Puvočiai	0.04	1.8678	1
30	54°05'50.8''	24°19'06.2''	Puvočiai	0.16	2.8094	1
31	54°05'41.2''	24°19'06.7''	Puvočiai	9	0.2783	2
32	54°05'38.7''	24°18'26.1''	Puvočiai	300	0.7313	1
33	54°05'33.2''	24°18'33.2''	Puvočiai	2500	0.2000	2
34	54°04'54.3''	24°18'19.0''	Margionys	4	1.2324	2
35	54°04'52.8''	24°18'12.8''	Margionys	150	0.1412	2
36	54°03'44.0''	24°18'07.0''	Margionys	0.06	2.1046	1
37	54°03'35.4''	24°18'19.2''	Margionys	144	0.3534	2
38	54°04'44''	24°18'25''	Margionys	150	0.2283	2
39	54°04'50.9''	24°18'21.4''	Margionys	100	2.1046	2
40	54°03'24.5''	23°42'45.4''	Kapčiamiestis	10	38.7329	1
41	54°02'48.4''	23°41'18.1''	Kapčiamiestis	200	1.9322	1
42	54°02'20.9''[b]	23°41'13.8''[b]	Kapčiamiestis	2000	0.8380	1

[a]Plant communities: 1 – *Peucedano-Pinetum*; 2 – *Cladonio-Pinetum*; 3 – *Polygalo-Nardetum strictae*.
[b]Indicates central points of subpopulations.

Plant communities were estimated according to the Lithuanian vegetation classification system (Balevičienė, 1991; Balevičienė *et al.*, 1998). Subpopulations of *A. montana* were grouped according to their dependence on the appropriate communities for further analysis.

Assessment of population structure was derived through direct counting of rosettes in fragments of population (subpopulations) and expressed as proportions of plant life stages. Following Rabotnov (1969), six developmental stages were distinguished to describe the stage structure of the population: seedlings and juveniles; immatures; vegetative adults; generative; subsenile; and senile. Seedlings are individuals developed from seeds, whereas juveniles are small ramets originated from development of buds of rhizomes. As the morphological differentiation of seedlings and juveniles is difficult, they were considered in one group. Immatures were defined as ramets with brighter and distinctly smaller leaves than adults. Vegetative adults were characterized for having ramets with leaves of maximum size. Generative plants were formed by rosettes with 1–3 flowering stems and leaves identical to those of vegetative adults. Subsenile plants were adults that never reached the flowering state and moved directly to the senile stage. Finally, senile plants were formed by old rosettes with much reduced leaves. The Student's *t*-statistic was performed for testing the significance of the representation developmental stages difference between habitats. A test for homogeneity of variance (Levene's test) was applied and either the separate-variance or pooled-variance *t*-test was used.

24.3 Results and Discussion

Two different habitat types, forest and meadow, were distinguished within the range of *A. montana*. The species occurs in the forest habitats of *Cladonio-Pinetum* Juraszek 1927 and *Peucedano-Pinetum* W. Matuszkiewicz (1982) 1986 plant communities and in a meadow of *Polygalo-Nardetum strictae* Oberdorfer 1957 community. A short description of the characteristic species of plant communities with *A. montana* is given in Table 24.2. The forest habitats of arnica are different from those described in other European areas where the species is distributed in mountain and submountain areas and in heathlands and grasslands of lowland areas (Luijten *et al.*, 1996). Specific habitats typical for Lithuania may be a result of the species performance on the north-eastern border of its distribution range.

Over 80% of all forest subpopulations were described in open stands near forest block lines or pathways where the woodland layer was very thin and did not exceed 30% coverage. The rest of the subpopulations were found in more shaded forest stands. Growth habitats varied in moisture and nutrient conditions. Forest subpopulations occurred in dry areas, while meadow subpopulations were in wet sites. The soils were characterized by high acid-to-acid reaction (pH 3.02–4.56), low content of total nitrogen (N, 0.024–0.377%) and varying concentrations of humus (0.76–8.12 mg/kg), available phosphorus (P_2O_5, 34.2–341.5 mg/kg) and available potassium (K_2O, 11.7–71.4 mg/kg).

Arnica is a polycarpic plant where new rosettes mainly originate from a renewal bud in the joint of the generative plant. As a result of this vegetative

Table 24.2. Plant communities of *Arnica montana* with their characteristic species.

Characteristic species	*Cladonio-Pinetum*	*Peucedano-Pinetum*	*Polygalo-Nardetum strictae*
Herbs	*Festuca ovina* *Corynephorus canescens* *Calamagrostis epigejos* *Calamagrostis arundinacea* *Danthonia decumbens* *Arctostaphylos uva-ursi* *Vaccinium vitis-idaea* *Vaccinium myrtillus* *Pilosella officinarum* *Hieracium umbellatum* *Hieracium murorum* *Thymus serpyllum*	*Picea abies* *Chimaphila umbellata* *Peucedanum oreoselinum* *Vaccinium vitis-idaea* *Vaccinium myrtillus* *Pilosella officinarum* *Agrostis capillaris* *Dryopteris carthusiana* *Lycopodium clavatum* *Peucedanum oreoselinum* *Scorzonera humilis* *Diphasiastrum complanatum*	*Festuca ovina* *Nardus stricta* *Holcus lanatus* *Deschampsia flexuosa* *Carex pallescens* *Bistorta major* *Lychnis flos-cuculi* *Polygala vulgaris* *Viola canina* *Potentilla erecta* *Hypericum maculatum*
Mosses	*Pleurozium schreberi* *Dicranum polysetum* *Polytrichum juniperinum*	*Pleurozium schreberi* *Dicranum polysetum* *Hylocomium splendens*	*Pleurozium schreberi* *Dicranum bonjeanii* *Plagiomnium affine* *Aulacomnium palustre*
Lichens	*Cladonia rangiferina* *Cladonia arbuscula* *Cladonia gracilis*	*Cladonia rangiferina* *Cladonia arbuscula*	Not found
Total number of species	80	108	46

reproduction, which is dominant in the life performance of arnica, the subpopulations were found to be fragmented into small clusters of intermingled ramets (<0.1 m^2) belonging only to a few genets. The high proportion of new discrete genets pointed to the spread of the species. The size of subpopulations ranged from 0.04 to 6000 m^2 (Table 24.1). The largest subpopulations occurred in the meadow habitat. Important differences between forest and meadow subpopulations were found in the relative density of individuals of arnica (Fig. 24.2). Density of arnica varied from 1 to 10 plants/m^2 in 35% of all studied cases in the forest habitat, while, in meadow, the highest frequency (50% of all cases) corresponded to subpopulations with a density of 20–30 plants/m^2. Similarly, subpopulations of arnica from the two habitat types differed significantly in mean total plant density with 17.78 and 38.82 plants/m^2 in forest and meadow, respectively.

Population structure was assessed by the density and proportion of individuals in the different ontogenetic stages of the life cycle (Fig. 24.3). The *t*-test comparisons of plant densities of life stages in forest and meadow habitats

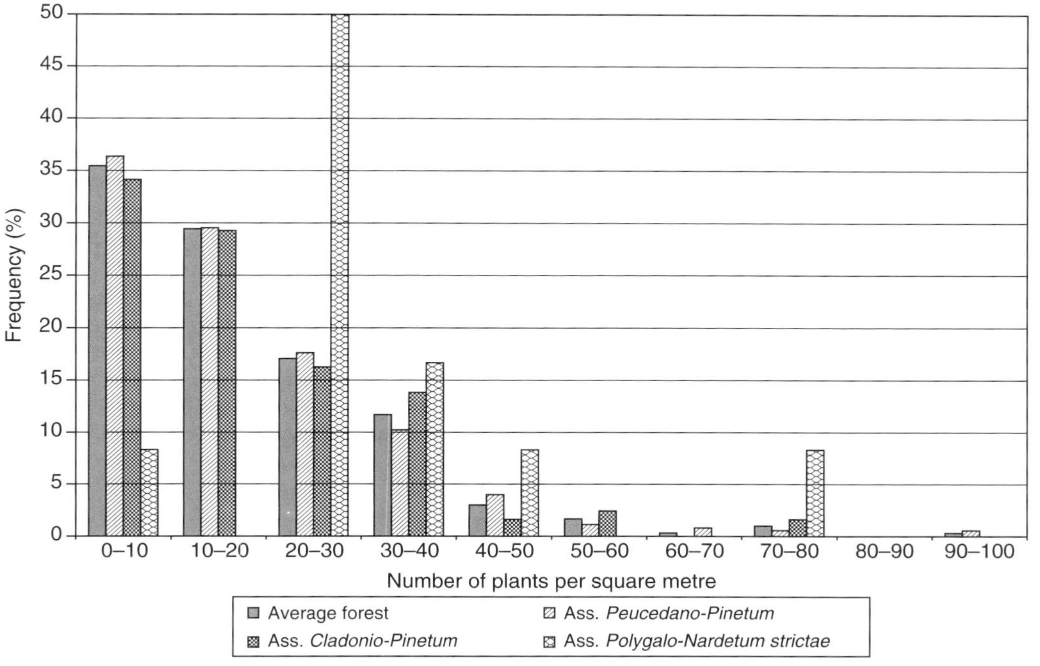

Fig. 24.2. Density of *Arnica montana* individuals in forest (*Cladonio-Pinetum* and *Peucedano-Pinetum*) and meadow (*Polygalo-Nardetum strictae*) plant associations.

showed significant differences in the number of immature (6.22 and 17.08, respectively; $t = 3.43$, $p < 0.05$, df = 357) and generative (1.77 and 9.25, respectively; $t = 2.51$, $p < 0.03$, df = 357) individuals, which were considerably higher in meadow than in forest (Table 24.3). The comparison of life-stage structure of subpopulations in those habitats showed highly significant differences

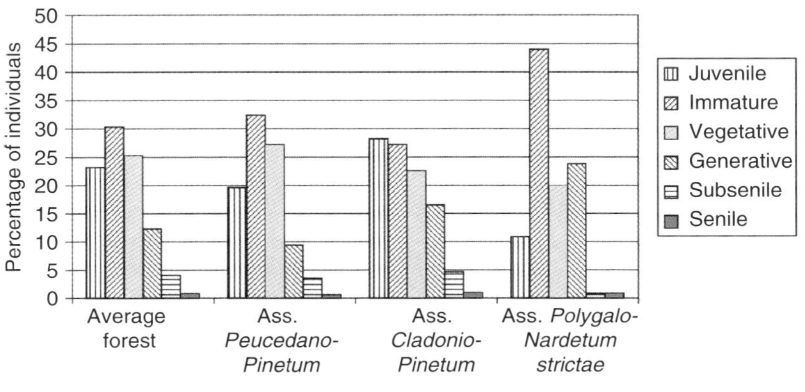

Fig. 24.3. Life-stage structure of *Arnica montana* subpopulations in different plant associations.

Table 24.3. *t*-Test comparison of mean values of plant density and proportions of life-stage structure variables of *Arnica montana* between forest and meadow subpopulations.

Stage structure variables	Density (number of plants per m²)				Proportion (%)			
	M				M			
	Forest	Meadow	*t*	*p*	Forest	Meadow	*t*	*p*
Juveniles	3.76	4.25	−0.40	0.69	19.13	9.55	4.38	0.01
Immatures	6.22	17.08	−3.43	0.05	32.88	48.68	−3.13	0.01
Vegetatives	5.21	7.58	−0.94	0.37	32.50	15.11	3.25	0.01
Generatives	1.77	9.25	−2.51	0.03	12.20	24.65	−2.79	0.01
Subseniles	0.695	0.33	0.47	0.64	2.90	0.81	0.78	0.44
Seniles	0.12	0.33	−1.43	0.15	0.44	1.19	−0.63	0.54

Degrees of freedom: I 357. M: mean; *t*: Student's statistic; *p*: significance level.

in the proportions of plants in all stages with exception of subseniles and seniles, which are of low importance in the stage structure of populations. The proportions of immature and generative individuals (48.68% and 24.65%, respectively) in meadow habitat were larger than those in forest (32.88% and 12.20%, respectively) (t = 3.13 and t = 2.79, respectively; $p < 0.05$, df = 357), while juveniles and vegetatives were less represented in meadow (9.55% and 15.11%, respectively) subpopulations than in forest (19.13% and 32.5%, respectively) (t = 4.38 and t = 3.25, respectively; $p < 0.05$, df = 357). The differences in plant density and proportions of stage structure were not statistically significant ($p > 0.05$) between two forest communities of *Cladonio-Pinetum* and *Peucedano-Pinetum* (Table 24.4). There appeared to be no association between plant community and stage structure of population.

Table 24.4. *t*-Test comparison of mean values of plant density and proportions of life-stage structure variables of *Arnica montana* between subpopulations of *Cladonio-Pinetum* (1) and *Peucedano-Pinetum* (2) communities.

Stage structure variables	Density (number of plants per m²)				Proportion (%)			
	M				M			
	1	2	*t*	*p*	1	2	*t*	*p*
Juveniles	3.73	3.78	−0.12	0.91	19.10	19.16	−0.32	0.98
Immatures	6.20	6.24	−0.07	0.94	32.91	32.85	0.31	0.98
Vegetatives	5.18	5.24	−0.13	0.89	32.61	32.39	0.11	0.91
Generatives	1.74	1.81	−0.29	0.77	12.16	12.25	−0.06	0.96
Subseniles	0.68	0.7	0.72	0.94	2.79	2.94	−0.04	0.97
Seniles	0.123	0.12	0.69	0.95	0.44	0.43	0.06	0.95

Degrees of freedom: I 345. M: mean; *t*: Student's statistic; *p*: significance level.

Studied meadow subpopulations were significantly over-represented by immature stage individuals. Proportion of generative plants was half that of immature plants. Generative stage is very important in age structure of arnica population as this stage indicates a relatively high turnover of plants in population. On this basis, following previous reports (Hegland *et al.*, 2001; Aguraiuja *et al.*, 2004) one may claim that the status of meadow subpopulations could be considered as a dynamic one. The forest subpopulations were in worse condition, since their plant density was considerably lower. These subpopulations were dominated by young plants, while the generative stage was characterized by low proportion and by very low density of plants. The forest subpopulations condition can be classified as normal and has the potential for the establishment of dynamic subpopulations due to the high proportion of young stages. The whole population of *A. montana* seemed to be in relatively good condition, since the stage structure was currently indicated to be dynamic or normal. At the same time, subpopulations of arnica in two habitats had differences in stage structure. The meadow population being in dynamic stage represents optimal development, while the forest population corresponds to less vitality and indicates worse conditions for the species on the border of its distribution area. All investigated subpopulations of arnica could be considered as local populations. To determine actual population size in terms of genets, elaborated genetic methods need to be applied.

Acknowledgements

The study was conducted under the Programme 'Scientific Research of Plant Genetic Resources' supported by the Ministry of Education and Science of the Republic of Lithuania.

References

Aguraiuja, R., Moora, M. and Zobel, M. (2004) Population stage structure of Hawaiian endemic fern taxa of *Diellia* (*Aspleniaceae*): implications for monitoring and regional dynamics. *Canadian Journal of Botany* 82, 1438–1445.

Balevičienė, J. (1991) *Sintaksonomo-fitogeografičeskaja Struktura Rastitel'nosti Litvy.* Vilnius.

Balevičienė, J., Kizienė, B., Lazdauskaitė, Ž., Patalauskaitė, D., Rašomavičius, V., Sinkevičienė, Z., Tučienė, A. and Venckus, Z. (1998) *Vegetation of Lithuania. Meadows* (in Lithuanian). Kaunas, Vilnius.

Dueck, T.A. and Elderson, J. (1992) Influence of NH_3 and SO_2 on the growth and competitive ability of *Arnica montana* L. and *Viola canina* L. *New Phytologist* 122, 507–514.

Hanski, I. (1998) Metapopulation dynamics. *Nature* 369, 41–49.

Hegland, S.J., van Leeuwen, M. and Oostermeijer, J.G.B. (2001) Population structure of *Salvia pratensis* in relation to vegetation and management of dutch dry floodplain grasslands. *Journal of Applied Ecology* 38, 1277–1289.

Lange, D. (1998) *Europe's Medicinal and Aromatic Plants: Their Use, Trade and Conservation.* TRAFFIC International, Cambridge.

Luijten, S.H., Oostermeijer, J.G., van Leeuwen, N.C. and den Nijs, C.H.M. (1996) Reproductive success and clonal genetic structure of the rare *Arnica montana* (*Compositae*) in the Netherlands. *Plant Systematics and Evolution* 201, 15–30.

Oostermeijer, J.G.B., Van't Veer, R. and Den Nijs, J.C.M. (1994) Population structure of the rare, long-lived perennial *Gentiana pneumonanthe* in relation to vegetation and management in the Netherlands. *Journal of Applied Ecology* 31, 428–438.

Rabotnov, T.A. (1969) On coenopopulations of perennial herbaceous plants in natural coenoses. *Vegetatio* 19, 87–95.

Rabotnov, T.A. (1985) Dynamics of plant coenotic populations. In: Withe, J. (ed.) *Handbook of Vegetation Science*. Dr. W. Junk Publishers, Dordrecht/Boston/Lancaster, pp. 121–142.

25 A Designated Nature Reserve for *In Situ* Conservation of Wild Emmer Wheat (*Triticum dicoccoides* (Körn.) Aaronsohn) in Northern Israel

D. Kaplan

The discovery of the wild emmer wheat (*Triticum dicoccoides* (Körn.) Aaronsohn) in the north of Israel by Aaronsohn in 1906 (Aaronsohn, 1909; Openheimer and Even-Ari, 1940), after *c.* 9000 years, was a revolution in improvement of one of the three most important energy sources for the human diet. It is now well established that Neolithic agriculture developed in the Middle East and depended primarily on the domestication and subsequent cultivation of wild emmer wheat. Natural populations of wild emmer wheat are confined to the Fertile Crescent. The species has its centre of origin and diversity, and attains its widest morophological variation and ecological range in Israel, Jordan, southern Syria and Lebanon (Horovitz and Feldman, 1991).

The rediscovery of the wild emmer wheat evolved worldwide research and development of new varieties. Continuous investigations have been carried out in Israel since 1908. With the remarkable revolution in biology and plant science, wild emmer wheat research is undergoing a revival, which has attracted the attention of the scientific community, public institutions and government to the natural resources and indigenous plants in Israel and has raised the awareness of *in situ* conservation of food progenitors.

The research carried out in Israel over the last two decades of the 20th century resulted in mapping of the wild emmer wheat populations as well as their genetic characteristics. A wide range of habitats with high affinity to grasslands was shown.

Two international workshops carried out in Israel in 1990 (Horovitz and Feldman, 1991) and in 1999 (Eyal and Hadas, 1999) summarized the research and supported the idea that wild emmer wheat harbours rich genetic resource polymorphism appropriate for wheat improvement. It was strongly recommended in these workshops to establish nature reserves dedicated to *in situ* conservation of the wild emmer wheat.

An analysis of grasslands and wild emmer wheat in nature reserves in Israel was carried out based on Israel Nature and Parks database and on the BIOGIS database (BIOGIS, 2005; The Hebrew University of Jerusalem, 2005). It was found that nature reserves in Israel cover 12% of the country area. In the desert ecosystems, their coverage is 20%, while in the Mediterranean ecosystems it is only 6%. Grasslands encompass 20,800 ha, i.e. 22% of the nature reserves in the Mediterranean ecosystems of Israel.

The distribution map of the wild emmer wheat is based on non-systematic observations collected by the Israel Flora Database of the Hebrew University of Jerusalem (2005), available on their web site and on the Israel Nature and Parks Authority database (Israel Nature and Parks, 2005). Although the information is non-systematic and incomplete, it supplies quite a good assessment of the spatial distribution of the species in Israel. Analysis of the *T. dicoccoides* distribution map shows that 43% (n = 77) of the locations are in nature reserves (Fig. 25.1).

A long-range transect through different populations of the wild emmer wheat from the Golan in the east to Mount Meron in the centre of Galilee has shown the highest morphological and genetic differentiations over very short distances in the Kibbutz Ammiad area (eastern Galilee), on the karst limestone

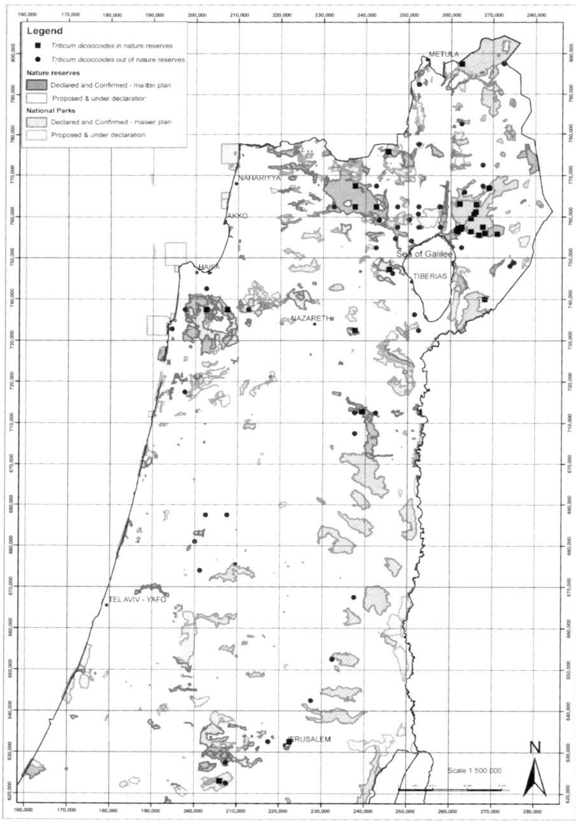

Fig. 25.1. *Triticum dicoccoides* distribution map in Israel. (From Israel Nature and Parks database and The Hebrew University BIOGIS database.)

formation of the Eocene era, providing diverse microhabitats (Anikster and Noy-Meir. 1991). Therefore, the Israel Nature and Park Authority chose the Ammiad region as a nature reserve dedicated to *in situ* conservation of wild emmer wheat. An area of 380 ha was allocated (Fig. 25.2) and the very long process of declaration of the nature reserve is ongoing. The process of nature reserve's declaration in Israel is long and encumbered with three levels of building and planning committees of the Interior Ministry being involved, as well as the governmental land bureau and private stakeholders.

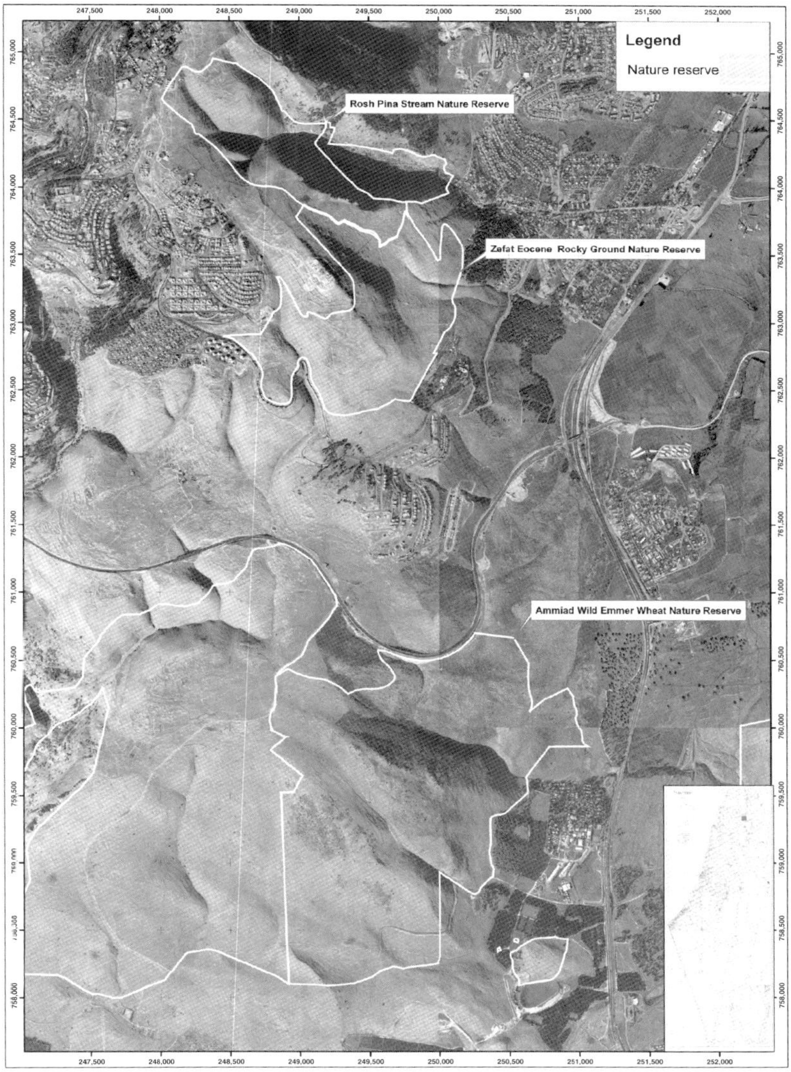

Fig. 25.2. Grasslands nature reserves in eastern Galilee, Israel, designated for wild emmer wheat.

The first stage was empowered by the local building and planning committees, and the local private stakeholders. Kibbutz Ammiad, whose cattle rangeland is in the proposed reserve, opposed the reserve declaration, anxious that their grazing rights and the grazing regime would be hurt. The declaration documents ensure the grazing regime and define the grazing pressure. The growing town of Zefat, neighbouring the reserve to the north, opposes the reserve's declaration, concerned for the growth limitations of the town to the south. The land bureau opposed in order to reserve land for future unexpected development of the area.

The Kibbutz that was primarily opposed to declaration of the nature reserve, had to rescind the decision when realizing that development of the nearby town Zefat was a greater threat than a nature reserve with grazing rights for their own herd ensured in the nature reserve regulations.

This gave way to the local planning committee to recomend the regional planning committee to approve the reserve. The land bureau agreed to rescind their objection, and in spite of the opposition of Zefat's town planning committee, the declaration process went ahead to its final stages.

The nature reserve is dedicated to preserve not only wild emmer wheat, but also some other progenitors of crops, like *Hordeum spontaneum* (barley), *Beta vulgaris* (beet) and *Olea europaea* (olive), of which few wild specimens are located at the very top of the hill, on upper Eocene limestone, as well as the rich grassland as a whole, emcompassing more than 400 vascular species.

The nature reserve management is based on deferred grazing by cattle from March through November and a grazing pressure of ~2 ha/cow/year. Several exclosures were established for wild emmer wheat research and will also serve as grazing monitoring control. The research in the reserve is carried out by a research consortium among The Hebrew University of Jerusalem, The Weizmann Institute in Rehovot and Tel Aviv University, and covers genetics and ecology of the species.

The nature reserve will not be open to the public. The nature reserve regulations in Israel prevent access except on designated and marked trails, unexisting in this reserve. There is also little expected interest for the public to visit, though researchers and specially interested groups will be welcomed.

The combination of interests of farmers to keep their rangeland and the research, carried out in Israel over the last three decades, resulted in the Ammiad wild emmer wheat nature reserve establishment, as the first 'scientific' reserve in Israel, now in the final stages of declaration.

Three other nearby nature reserves, with similar habitats, light cattle grazing and with populations of wild emmer wheat, have been located, and are in the first stages of the long way for declaration. These reserves encompass an area of 1173 ha, including the Rosh-Pina stream (Fig. 25.2), where wild emmer wheat was first discovered by Aaronsohn in 1906.

Acknowledgements

The author is grateful for the help of Sussana Kauffman for the GIS mapping and editing.

References

Aaronsohn, A. (1909) Contribution à l'Histoire des Céréales. Le Blé, l'orge et le Seigle à l'état Sauvage. *Bulletin of the Society of Botany of France* 56, 196–203, 237–245, 251–258.

Anikster, Y. and Noy-Meir, I. (1991) The wild-wheat field laboratory at Ammiad. *Israel Journal of Botany* 40, 351–362.

BIOGIS (2005) *The Hebrew University BIOGIS Database*. The Hebrew University, Jerusalem, Israel.

Eyal, Z. and Hadas, Y. (eds) (1999) *The Aaronsohn Lectures on Wild Emmer Wheat. An 80th Anniversary (1919–1999) Memorial Symposium*. The Tel-Aviv University Institute for Cereal Crops Improvement and the Aaronsohn Foundation LTD, Zichron Yaakov, Israel.

Hebrew University of Jerusalem (2005) *BIOGIS, Israel Biodiversity Information System*. Available at: http://www.biogis.huji.ac.il (accessed October 2005)

Horovitz, A. and Feldman, M. (eds) (1991) International workshop on the dynamic *in-situ* conservation of wild relatives of major cultivated plants. *Israel Journal of Botany* 40, 509–519.

Openheimer, H.R. and Even-Ari, M. (eds) (1940) *The Flora of the West Bank of the Jordan. Dairy of A. Aaronsohn*. The Aaronsohn Foundation LTD, Zichron Yaakov, Israel.

26 Integrating Wild Plants and Landrace Conservation in Farming Systems: a Perspective from Italy

V. Negri, F. Branca and G. Castellini

26.1 Use of Wild-harvested Plants and Landraces in Italy

Wild-harvested plants are commonly collected and used for food, medicinal and/or ornamental purposes in Italy. Their diffusion and uses are linked to the habitat and cultural diversity which characterize Italy. Over 40 wild plant species are reported to be still commonly collected and used for human consumption in the peninsula, but many others were surely used in the past (Branca, 1992; Hammer *et al.*, 1999; Manzi, 1999; Ranfa, 2004). Many of them are crop wild relatives (CWR), i.e. *Borago officinalis* L., *Beta* spp., *Capparis* spp., *Cichorium intybus* L., *Cynara cardunculus* L., *Lactuca perennis* L., *L. serriola* L., *L. viminea* (L.) Presl., *Brassica nigra* (L.) Koch., *B. fruticulosa* Cyr., *Diplotaxis erucoides* (L.) DC, *D. muralis*, *D. tenuifolia* (L.) DC, *Eruca sativa* Miller, *Nasturtium officinale* R. Br., *Raphanus raphanistrum* L., *Sinapis alba* L., *S. arvensis* L., *Melissa officinalis* L., *Mentha acquatica* L., *M. longifolia* (L.) Huds., *M. pulegium* L., *Origanum* spp., *Rosmarinus officinalis* L., *Thymus* spp., *Glycyrrhiza glabra* L., *Allium ampeloprasum* L., *A. orsinum* L., *A. vineale* L., *Asparagus acutifolius* L., *A. albus* L., *Anethum graveolens* L., *Apium nodiflorum* (L.) Lag., *Daucus carota* L., *Foeniculum vulgare* Miller and *Valerianella carinata* Loisel. The list would be much longer if medicinal and ornamental plants, plants used for particular preparations (i.e. liquors), trees and forages were included (cf. Hammer *et al.*, 1999).

Some of the above-mentioned species (*A. acutifolius, A. albus, B. nigra, B. fruticulosa, C. cardunculus* and *Origanum* spp.), which were once only collected by local residents, are now actively collected by non-residents and are often sold in local markets, at the greengrocers as well as in restaurants because they are greatly appreciated in the preparation of typical regional dishes. The consumer is usually willing to pay a very high price for them. For example, the price of wild asparagus in Perugia is 3–4 times higher than that of cultivated asparagus. However, it is difficult to give accurate figures relative to the marketing of these species, because it does not follow the normal commercial circuits. It is also difficult to estimate the impact of intensive collection on the viability of natural populations, since no detailed study has ever been carried out on the subject. Personal observations in Umbria indicate that some asparagus populations are in decline because of overexploitation which hampers the rejuvenation of plants through young sprouts. In addition, habitat changes and destruction continuously lead to the loss of entire populations.

All the same, landraces (LRs) are still found and used throughout Italy. Over 400 LRs belonging to different species of forages, cereals, pulses, garden crops and fruit trees were found in central Italy over a 10-year collection period (Negri, 2003; Albertini *et al.*, 2005). In the context of conserving and using CWR, LRs which are cultivated in the vicinity where their relatives grow are of particular interest. In Italy, it is possible to find LRs of *Apium graveolens* L., *Beta vulgaris* L., *Brassica oleracea* L., *B. rapa* L., *C. intybus* L., *Cynara scolimus* L., *Lactuca sativa* L., *Lathyrus cicera* L., *L. sativus* L., *Pisum sativum* L. and *Salsola soda* L. where their wild relatives are also present, or were present until recently.

The above-mentioned study in central Italy (Negri, 2003) showed that, of the LRs used for food, 67.1% (161 out of a total of 240 recorded cases) were used within the family, 9.2% were prevalently sold on the local market and 23.7% on a broader market.

As a matter of fact, many LRs are still a consistent part of horticultural production in Italy. The most consistent examples of native crops where LRs are used for wide-market production include *B. oleracea* in Sicily and *C. scolimus* throughout Italy. Several forms and types related to cauliflower (*B. oleracea* var. *botrytis*), Italian broccoli (*B. oleracea* var. *italica*), kale (*B. oleracea* var. *acephala*) and kohlrabi (*B. oleracea* var. *gongylodes*) are grown on peri-urban farms and in home gardens in response to an active demand for such products on local markets to prepare traditional dishes. Sicily is the cradle from where cauliflower and broccoli originated and diversified (Gray, 1989). The LR seeds are still produced on farms and are named according to the month of harvest (from September to April), e.g. 'Agostino', 'Settembrino', 'Ottobrino', 'Sanmartinaro', 'Natalino', 'Iennarotu', 'Marzuddu' and 'Aprilotu' (Branca and Iapichino, 1997).

As for artichokes, Italy has approximately 90 different varieties which is more than in any other country in the world. Some varieties never cross regional boundaries, while others are sold nationwide as well as exported. Local Italian markets usually carry the regional varieties as well as a few of the country's most popular varieties. Artichokes can be prickly or smooth, fat or slim, and vary from an intense green colour to violet-green. The slim Catanese variety from Sicily and the big, round Romanesco variety, renowned for its softness, from the area surrounding Rome, are very widespread on the market. Artichokes with prickly spines are primarily grown in Liguria, Sardinia and Sicily; among them the Albenga variety is particularly delicious. They are so tender that they are best when eaten raw. Gastronomical specialists prize the Violetto cultivated in southern Italy, the Masedu from Sardinia, the Veneto from Chioggia and the Empoli from Tuscany. The LRs used for wide-scale production are generally cultivated using high input agronomic techniques.

Apart from the LRs sold on the local and wider-ranging markets, most LRs are used directly by farm families. Garden crops are mostly grown by elderly farmers (average age = 63.6 years) who operate small farms or have home gardens and use traditional farming systems which, none the less, always include the use of mechanical tools for soil preparation and the occasional use of pesticides and chemical fertilizers. There is no one situation in on-farm conservation and management in the studied area: the main factors involved appear to be a fragmented habitat and the presence of relatively elderly farmers. The main reasons why these LRs have been maintained on-farm are: their resistance or good productivity under difficult or harsh climatic conditions, local traditions and organoleptic peculiarities, which make them highly prized and expensive on the local markets and/or simply because the farm families like them.

LRs are threatened due to changes in the socio-economic context that have occurred since World War II. There has been 73% genetic erosion in southern Italy over a period of 30 years (Hammer *et al*., 1996). Few data are available regarding the present level of genetic erosion in central Italy. Repeated collections in 1995 in the Mount Amiata area (Tuscany) after 4 years and in the Trasimeno Lake area (Umbria) after 10 years revealed that only one-third of the farmers who had maintained germplasm were still alive.

Few people still live in the country and many of them are old and unable to carry on their activity any longer. So the extinction of LRs is still in progress and is mostly associated with the ageing of the older population in the country. Other reasons for the loss of LRs are: the younger people remaining in the country are often employed part-time in agriculture and often find it more convenient to buy the seeds rather than produce them. Seed harvesting, cleaning and conditioning require time and sometimes appropriate equipment which is not always present on the farm. Lack of skill is another constraint in reproducing seed or propagating plants, since the younger farmers often do not know techniques that were usually applied in the past (such as grafting). This makes it difficult to increase cultivation or, what is even more troubling for plant genetic resource conservation, to continue cultivation. Action to rescue and preserve this wealth of biological diversity is urgently needed.

In the end, both CWR and LRs are threatened as a consequence of overexploitation, habitat alteration and loss and socio-economic changes.

26.2 Some Examples of Successful (and Unsuccessful) Rescues of Neglected Landraces from Central Italy

26.2.1 The black celery ('sedano nero') from Trevi (Umbria)

A 'black' LR of celery, *A. graveolens* L. var. *dulce* (Mill.) Pers., is grown in Umbria (Italy) near the little town of Trevi. The term 'black' refers to the wild physiological characteristic of maintaining green petioles (not self-blanching) if not subjected to the agronomic whitening treatment. The cultivated area is small (~2 ha) so production is limited and mainly destined for the local market during the local 'black celery fair' (October). The survival of this LR could be helped by the increased request for local traditional products to which consumers attribute superior quality and which are sold on the regional gastronomic circuit. For this purpose this LR has been registered in the list of typical local products of the region which was recently prepared by the region of Umbria.

A. graveolens was domesticated in Italy and some CWR of this species exist in the investigated area (*A. nodiflorum, Levisticum officinale* W. D. J. Koch, L., *Smyrnium olusatrum* L.). Wild forms of *A. graveolens* L. were identified in the *Flora* of *Umbria* (Batelli, 1886, 1887, 1888; Cicioni, 1895; Barsali, 1929, 1931, 1932) until the beginning of the 20th century. However, it is difficult to say if they were truly wild forms or cultivated forms that had escaped or are still present. In contrast, *A. nodiflorum*, a perennial rooted hydrophytic plant (water celery) is widespread throughout Umbria, like in the entire national territory, although its habitat is continuing to decrease because of insufficient ecological protection. This species is a representative plant of the flora of the springs of the river Clitumnus that are located at a few kilometres from Trevi. The taxonomic relationships of the above-mentioned species to cultivated celery are largely undetermined (Muminović *et al.*, 2004). *A. nodiflorum* has been hybridized with celery by Pink *et al.* (1983) to introduce resistance to *Septoria apiicola* Speg. and leaf miners (*Liriomyza* spp.), but no information is available on the characteristics of the hybrids. Later attempts to hybridize these species have failed (Quiros, 1993). As for intergeneric crosses, there are at least two independent reports on the successful hybridization between celery and parsley (*Petroselinum sativum* Hoffm.) which also belongs to the *A. graveolens* gene pool (Madjarova and Bubarova, 1978; Honma and Lacy, 1980). Farmers in Trevi state that hybridization between celery and parsley commonly occurs.

27.2.2 The Sicilian cauliflowers

LRs of violet and green cauliflowers are grown in Sicily, where white curded types are rarely grown. Production is characterized by a wide variation of harvesting dates and of curd morphology and colour. The violet cauliflower is of interest because of its aroma, sugar and glucosinolate content in comparison with the white ones. About 50 LRs of this crop, differing in harvesting period, curd size, colour and glucosinolate content were studied in an attempt to qualify their production for national and European markets (Branca, 1998, 2000, 2002a; Acciarri *et al.*, 2004). In the violet cauliflower, LRs contain glucoraphanin (precursor of the sulphorafane of great interest for its antioxidant and antitumoral actions) and lack sinigrin (glucosinolate which has biotoxic effects in humans and has recently been utilized for biofumigation of agricultural fields), whereas the opposite occurs in the commercial white cauliflower cultivars (Branca *et al.*, 2002). This aspect and the typical colour and taste could foster the diversification and qualification of vegetable production requested by several markets oriented to functional foods. More than 100 selective lines and 40 androgenic lines were bred by DOFATA for this purpose. DOFATA and CONVASE (the consortium of the Italian seed companies) are currently collaborating to register two new commercial cultivars which will support high-quality production and open new markets

26.2.3 The cowpea ('Fagiolina') from the Trasimeno Lake area, Perugia

Cowpea (*Vigna unguiculata* subsp. *unguiculata* cv gr. *unguiculata* (L.) Walp.) is the *phaseolus* that was grown by the Romans (Pliny). Traditionally, the fresh pods and seeds of this species are used for human consumption in Italy. The area around Trasimeno Lake is densely inhabited and has many services, the land is fertile and the climate is mild. The intensive agriculture in the area is based on maize and horticultural crops. None the less, clearly distinguishable cowpea LRs were still present during a germplasm exploration and collection mission carried out in 1994 (Negri *et al.*, 2000; Tosti and Negri, 2002). At that time, most farms produced cowpea for their own use; only one farm produced a small white-seeded LR for the town market in Perugia where the demand largely exceeded production. The demand was due to the peculiar shape and colour of the product and the consumers' opinion that the LR has a better taste than the common cowpea. The presumed better quality of cowpea LRs from Trasimeno Lake was tested in an ad hoc experiment. The results showed significant differences between the LRs and a variety commonly found on the market with respect to organoleptic characteristics, such as taste and visual appeal, crude protein content and total carbohydrate percentages of dry matter (Negri *et al.*, 2001). Financial support for research and the cultivation of these LRs was initially given by local authorities. Cultivation using the traditional system requires a great deal of labour especially for the cropping and cleaning of the product, but one of the farmers developed an almost completely mechanized cultivation method and was able to sell the crop to a wider market. As a consequence of improved agricultural techniques, the steady increase in demand and the support given by local authorities, the area in which the small, white-seeded LR is cultivated has greatly increased while the price in the regional capital (Perugia) has also steadily increased from 6 euros/kg in 1994 to the present 19 euros/kg. The Fagiolina of Trasimeno Lake has become a must in many top restaurants even outside Umbria. In addition, a commission is presently evaluating the possibility and methods for applying for the protected designation of origin (PDO), which covers foodstuffs that are produced, processed and prepared in a given geographical area using recognized know-how. Quality labels, such as PDO, can encourage diverse agricultural production in a rural development context by adding value to the products. In this respect, the great morphological and agronomic variation among LRs poses a problem from a wider commercial perspective, because consumers usually expect standard products when they are protected by a quality mark.

26.2.4 The 'fagiolo A pisello' at Colle di Tora, Rieti

In Colle di Tora, located in the Appennine mountains of the Rieti Province (Lazio Region), on the steep hillsides of Turano Lake, at altitudes ranging from 750 to

900 m asl, five farmers grow a *Phaseolus vulgaris* L. LR called 'A pisello'. Because of the particular characteristics of the seed (smooth, white and round with a doughy taste) and its very limited production, it has a rich local niche market. In fact, in the town of Rieti the cost is 16 euros/kg. This LR is quite delicate during the first period of growth and has very particular thermal, humidity and edaphic requirements. Consequently, it cannot be grown successfully under different conditions. Although the market has the perspective of being expanded, this LR, contrary to the examples mentioned above, appears to be in danger of extinction. In fact, it is only cultivated by elderly farmers who are not interested in the potential business. The few younger people in the village do not seem to be interested in continuing its cultivation due to the heavy work involved and the inconvenient place where it can be grown. In addition, gene introgression from other LRs cultivated in the area is suspected. Because the 'A pisello' LR runs a severe risk of extinction in the next few years, several efforts have been funded by local (regional) authorities in an attempt to save it. In particular, the Agency for Rural Development of Lazio Region is giving economic support for its cultivation. A bushy growth habit is being introduced in a few genotypes with the hope of saving, at least partially, this germplasm and promoting wider cultivation by local people. Farmer-assisted selection for anthracnose resistance and trueness to type is in progress to help overcome problems related to disease susceptibility and gene introgression from other LRs and cultivars.

26.3 Genetic Diversity

Knowledge of the LR levels of diversity is fundamental for LR use in breeding, as well as for planning *in situ* (on-farm) conservation activities. If genetically similar LRs exist in a certain area, a single farm could carry out the conservation activity. If, however, the LRs are different, several farms would need to be involved in their preservation. The level of variation within a population is also important because it affects the persistence of the population over time (Soulé, 1987; Nunney and Campbell, 1993). Finally, changes can be monitored by assessing variation and monitoring is essential if the effectiveness of conservation is to be verified.

Nine amplified fragment length polymorphism (AFLP) markers were used to detect variation and relationships among six populations of 'black celery from Trevi', four celery cultivars, one *A. nodiflorum* population, collected near the area where the LR is cultivated and one population of parsley. Five bulks of 16 plants for each celery accession were used. A total of 568 bands were detected, of which 305 (53.7%) were polymorphic among celery accessions. Figure 26.1 reports relationships among the investigated materials. The six LR populations were separated from the cultivars and belonged to the same cluster and each of them was distinguished from the other LR populations. The germplasm of *A. nodiflorum* and *P. sativum* is distinct from the cultivated celery forms. Cluster analysis of genetic similarity shows a closer relationship of celery to *P. sativum* (GS = 0.187) than to *A. nodiflorum* (GS = 0.15), in spite of the fact that it belongs to the same *genus*.

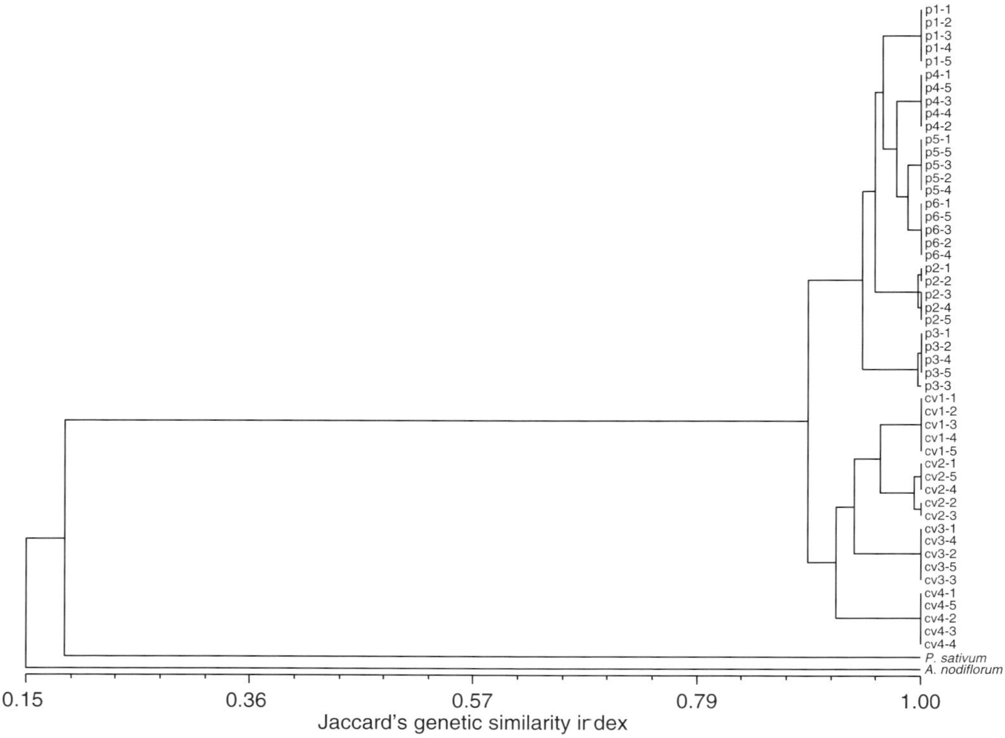

Fig. 26.1. Association among *Apium graveolens* var. *dulce* (six accessions of 'Sedano Nero di Trevi' LR and four cultivars) and related species *A. nodiflorum* and *Petroselinum sativum*, as revealed by average linkage (unweighted pair group method with arithmetic mean (UPGMA)) cluster analysis of Jaccard's genetic similarity (GS) coefficients calculated from AFLP data of nine primer combinations.

AFLP and selective amplification of microsatellite polymorphic loci (SAMPL) molecular markers appear also to be useful in the analysis of the limited genetic diversity among the cowpea LRs found around Trasimeno Lake (Tosti and Negri, 2002). Studies carried out with AFLP and SAMPL showed that these LRs are more closely related between themselves than to other LRs from different parts of Italy (V. Negri, 2003, unpublished data). In addition, the entire Trasimeno cowpea population appears to be a structured population in which a substantial differentiation is maintained at the subpopulation level (Tosti and Negri, 2005).

Similar to what was observed in cowpea from Trasimeno Lake, molecular markers were able to discriminate 'A pisello' LR from other Italian LRs and commercial cultivars (Negri and Tosti, 2002a,b) and the LR is a structured population (Negri and Tiranti, 2005).

Studies carried out with molecular markers (SSR, AFLP and random amplified polymorphic DNA (RAPD)) on wild cardoon (*C. cardunculus*) populations and cultivated LRs of *C. scolymus* from Sardinia and Sicily (Lanteri *et al.*, 2004; Lanteri and Barcaccia, 2005; Portis *et al.*, 2005a,b) showed that: (i) there is wide genetic variation among Italian artichoke LRs; (ii) genetic variation among artichokes belonging to the same varietal type is sometimes greater than that found among differently named accessions that come from different areas; and (iii) most of the genetic variation was present within, rather than between populations, which is consistent with previously reported data regarding outbreeding species (Bussel, 1999; Gaudel *et al.*, 2000).

Despite the high level of within-population diversity, most of the wild and cultivated populations were differentiated from one another.

A complex interaction of factors (drift, LR isolation, farmer selection, migration within LRs) explains the observed pattern of diversity in both inbreeding and outbreeding species. Italian germplasm appears to be a useful source of variation for breeding purposes with a remarkable level of variation both within and among wild populations and LRs. This variation deserves to be protected and maintained for the future.

In this respect, only *in situ* and on-farm conservation can safeguard genetic resources by maintaining their ability to evolve in the face of biotic and abiotic pressures, social and cultural changes in order to meet unpredictable future demands (Brown, 2000; Brush, 2000).

26.4 Some Examples of Exploitation of Wild Plants (Cultivating Wild Plants)

To safeguard wild crops in Sicily from the threat caused by overexploitation and habitat destruction and to promote product innovation in the vegetable crop industry, the feasibility of growing five wild species (beet, chicory, leafy cabbage, fennel and borage) was assessed in different agronomic trials. The effects of sowing date, growing environment, plant density and nitrate nutrition were also studied. The experimental trials showed a good adaptation to growing techniques. Autumn sowing was much more favourable than winter sowing for borage, but the latter was most effective for obtaining good productions in the other species. Cultivation in a cold greenhouse caused a significant yield increase, due to the improved environmental conditions in the protected growing environment (Branca, 2001, 2002b). The productive level of the wild species, in comparison with similar growing species and forms strictly related to them, was modest. These differences could be reduced if more appropriate growing protocols were found for each of the studied species. Wild species products had traits that were similar to those of the corresponding crops except for ascorbic acid and calcium content. High nitrate levels did not cause significant differences in terms of yield, but reduced radiation caused a slower growth rate in wild chicory and borage. All experimental variants caused more or less evident effects on yield and product quality (Tribulato *et al.*, in press). More knowledge about their requirements for setting up better growing protocols and for identifying their growth limits is needed.

26.5 Perspective for the Conservation of Wild Plants and Landrace Conservation within Farming Systems

In light of what has been detected with population genetic studies, the best strategy for maintaining CWR and LR diversity is to preserve each population *in situ* or on the farm from which it came. In particular, the investigation carried out on LRs to date, suggests that variation can be maintained and increased if their cultivation is continued over time (i.e. in a dynamic situation).

Because of the value which consumers attribute to CWR and LRs for the preparation of particular traditional Italian dishes, one perspective for *in situ* and on-farm conservation would be to promote their use to obtain typical products. This can also be done in 'modern' farming systems. The genetic and organoleptic differences detected among LRs and the varieties commonly found on the market could be the basis for requesting quality labels, which would promote *in situ* (on-farm) conservation of LRs.

However, most of the LRs are grown only for private consumption. These LRs are at great risk of extinction due to the advanced age of the farmers who grow them. Reinforcement of the links between the rural communities, their plant genetic resources and pride in their heirloom are needed to stimulate and convince the younger generations to continue growing LRs. As for wild species, *in situ* conservation activities should be based mostly on recovering the traditional use of these plants in farm families as well as on recovering the traditional management of the territory. In addition, promotion of their use in local tourist circuits or markets could also serve the purpose, but great attention should be paid to the risk of overexploitation, for which the possibility of cultivating them should be better explored.

Acknowledgements

Provincia di Perugia and Ente Parco del Trasimeno funded the work on the 'Fagiolina' LR (project responsible person: Prof. V. Negri), Agenzia Regionale per lo Sviluppo e l'Innovazione dell'Agricoltura del Lazio and the Italian Ministry of Research funded the work on the 'A pisello' LR (responsible persons of the projects: Prof. M. Falcinelli, DBVBAZ director and Prof. V. Negri, respectively) and Regione Umbria funded the work carried out on 'Sedano nero di Trevi' LR (project responsible person: Prof. M. Falcinelli).

References

Acciarri, N., Branca, F., Sabatini, E., Argento, S. and Magnifico, V. (2004) Miglioramento Genetico dei Cavolfiori a Corimbo Bianco e Colorato. *L'Informatore Agrario*, LX, 25, 33–36.

Albertini, E., Buonumori, G., Castellini, G., Concezzi, L., Dalla Ragione, I., Desantis, F., Falcinelli, M., Falistocco, E., Fatichenti, F., Ferranti, F., Giampiccolo, C., Mauceri, S., Micheli, M., Pagiotti, R., Paladin, C., Porfiri, O., Prosperi, F., Rossi, F., Scerna, G.C., Standardi, A., Reale, L. and Torricelli, R. (2005) *La Biodiversità Vegetale in Umbria e la Sua Conservazione*. Edizioni 3A-PTA, Pantalla Di Todi (Perugia), Italy.

Barsali, E. (1929) Prodromo della Flora Umbra. *Nuovo Giornale Botanico Italiano* 36, 548–623.
Barsali, E. (1931) Prodromo della Flora Umbra. *Nuovo Giornale Botanico Italiano* 38, 624–689.
Barsali, E. (1932) Prodromo della Flora Umbra. *Nuovo Giornale Botanico Italiano* 39, 346–415, 459–602.
Batelli, A. (1886) Prima Contribuzione alla Flora Umbra. *Annali Dell'Università Libera Di Perugia* (Anno I) 1, 2–56.
Batelli, A. (1887) Seconda Contribuzione alla Flora Umbra. *Annali Dell'Università Libera Di Perugia* (Anno II) 2, 136–248.
Batelli, A. (1888) Terza Contribuzione alla Flora Umbra. *Annali Dell'Università Libera Di Perugia* (Anno III) 2, 29–46.
Branca, F. (1992) Studi su Specie Erbacee della Flora Spontanea di Interesse Alimentare. Doctoral thesis, Università Degli Studi Di Catania, Italy.
Branca, F. (1998) Caratterizzazione di Tipi di Cavolfiore Violetto. *Proceedings of IV Giornate Scientifiche Società Orticola Italiana*. SOI, Firenze, Italy, pp. 79–80.
Branca, F. (2000) Valutazione di Cultivar Locali di Cavolfiore Violetto. *Proceedings of the Workshop Risultati Del Primo Anno Di Attività Del Piano Nazionale Di Ricerca Per l'Orticoltura Del MIPA*, Sirmione, 29 Marzo 2000, SOI, Firenze, Italy, pp. 114–115.
Branca, F. (2001) Prove di Coltivazione di Specie Spontanee Utilizzate. In Sicilia a Scopi Alimentari. *Italus Hortus*, 8, 4, 22–26.
Branca, F. (2002a) Linee di Attività per il Miglioramento del Cavolfiore Violetto. *Proceedings of VI Giornate Scientifiche Società Orticola Italiana*. Spoleto, 23 Aprile 2002, SOI, Firenze, Italy, pp. 461–462.
Branca, F. (2002b) Indirizzi per la Coltivazione di Specie Spontanee di Interesse Alimentare In Sicilia. *Etnobotanica Nella Provincia di Catania*. Nuova Zangara Stampa, Siracusa, pp. 55–63.
Branca, F. and Iapichino, G. (1997) Some wild and cultivated *Brassicaceae* exploited in Sicily as vegetables. *FAO/IPGRI Plant Genetic Resources Newsletter* 110, 22–28.
Branca, F., Li, G., Goyal, S. and Quiros, C. (2002). Survey of Aliphatic Glucosinolates in Sicilian Wild and Cultivated Brassicaceae. *Phytochemistry* 59, 717–724.
Brown, A.H.D. (2000) The genetic structure of crop landraces and the challenge to conserve them *in situ* on farms. In: Brush, S.B. (ed.) *Genes in the Field*. IPGRI, Rome, Italy/IDRC, Ottawa, Canada/Lewis, Boca Raton, Florida, pp. 29–48.
Brush, S.B. (2000) The issues of *in situ* conservation of crop genetic resources. In: Brush, S.B. (ed.) *Genes in the Field*. IPGRI, Rome, Italy/IDRC, Ottawa, Canada/Lewis, Boca Raton, Florida, pp. 3–26.
Bussel, J.D. (1999) The distribution of random amplified polymorphic DNA (RAPD) diversity amongst populations of *Isotoma petraea* (*Lobeliaceae*). *Molecular Ecology* 8, 775–789.
Cicioni, C. (1895) *La Flora del Trasimeno: Osservazioni Generali*. Tipografia Santucci, Perugia, Italy.
Gaudel, M., Taberlet, P. and Till Bottraud, I. (2000) Genetic diversity in an endangered alpine plant, *Eryngium alpinum* (*Apiaceae*), inferred from amplified fragment length polymorphism markers. *Molecular Ecology* 9, 1625–1637.
Gray, A.R. (1989) Taxonomy and evolution of broccolis and cauliflowers. *Baileya* 23, 28–46.
Hammer, K., Knupffer, H., Xhuveli, L. and Perrino, P. (1996) Estimating genetic erosion in landraces. Two case studies. *Genetic Resources and Crop Evolution* 43, 329–336.
Hammer, K., Knupffer, H., Laghetti, G. and Perrino, P. (1999) *Seed From the Past. A Catalogue of Crop Germplasm in Central and North Italy*. Istituto del Germoplasma del Consiglio Nazionale delle Ricerche, Bari, Italy.
Honma, S. and Lacy, M. (1980) Hybridization between Pascal Celery and Parsley. *Euphytica* 29, 801–805.
Lanteri, S. and Barcaccia, G. (2005) Molecular Based Analysis for Crop Germplasm Preservation. *Proceedings of the International Workshop 'The Role of Biotechnology for the*

Characterization and Conservation of Crop, Forestry, Animal and Fishery Genetic Resources', Turin, 5–7 March 2005, Fondazione Per Le Biotecnologie, pp. 55–66.

Lanteri, S., Saba, E., Cardinu, M., Mallica, G.M., Baghino, L. and Portis, E. (2004) Amplified fragment length polymorphism for genetic diversity assessment in globe artichoke. *Theoretical and Applied Genetics* 108, 1534–1544.

Madjarova, D.J. and Bubarova, M.G. (1978) New forms obtained by hybridization of *Apium graveolens* and *Petroselinum hortense. Acta Horticulturae* 73, 65–72.

Manzi, A. (1999) *Le Piante Alimentari In Abruzzo*. Casa Editrice Tinari, Chieti, Italy.

Muminović, J., Melchinger, A.E. and Lübberstedt, T. (2004) Prospects for celeriac *(Apium Graveolens* Var. *Rapaceum)* improvement by using genetic resources of *Apium*, as determined by AFLP markers and morphological characterization. *Plant Genetic Resources* 2(3), 189–198.

Negri, V. (2003) Landraces in Central Italy: where and why they are conserved and perspectives for their on farm conservation. *Genetic Resources and Crop Evolution* 50(8), 871–885.

Negri, V. and Tiranti, B. (2005) Molecular Analysis for *Ex Situ* and On-Farm Conservation of Common Bean (*Phaseolus vulgaris* L.) Italian Germplasm. *Proceedings of the International Workshop on the Role of Biotechnology for the Characterization and Conservation of Crop, Forestry, Animal and Fishery Genetic Resource*. Turin, 5–7 March 2005, pp. 221–222.

Negri, V. and Tosti, N. (2002a) Genetic diversity within a common bean landrace of potential economic value: its relevance for on-farm conservation and product certification. *Journal of Genetics and Breeding* 56, 113–118.

Negri, V. and Tosti, N. (2002b) *Phaseolus* genetic diversity maintained on farm in Central Italy. *Genetic Resources and Crop Evolution* 49, 511–520.

Negri, V., Tosti, N., Falcinelli, M. and Veronesi, F. (2000) Characterization of thirteen cowpea landraces from Umbria (Italy). *Genetic Resources and Crop Evolution* 47, 141–146.

Negri, V., Floridi, S. and Montanari, L. (2001) Organoleptic and chemical evaluation of Italian cowpea landraces from a restricted area. *Italian Journal of Food Science* 13, 383–390.

Nunney, L. and Campbell, K.A. (1993) Assessing minimum viable population size: demography meets population genetics. *Trends in Ecology and Evolution* 8, 234–239.

Pink, D.A., Walkey, D.G., Stanley, A.R., Carter, P.J., Smith, B.M. Mee, C. and Bolland, C.J. (1983) Genetics of Disease Resistance. *National Vegetable Research Station. Annual Report*, Wellesbourne, Warwick, UK, pp. 58–59.

Portis, E., Mauromicale, G., Barchi, L., Mauro, R. and Lanteri, S. (2005a) Population structure and genetic variation in an autochthonous globe artichoke germplasm from Sicily Island. *Plant Science* 168, 1592–1598.

Portis, E., Acquadro, A., Comino, C., Mauromicale, G., Saba, E. and Lanteri, S. (2005b) Genetic structure of island populations of wild cardoons [*Cynara cardunculus* L. var. *sylvestris* (Lamk) Fiori] detected by AFLPS and SSRS. *Plant Science* 169, 199–210.

Quiros, C.F. (1993) Celery – *Apium graveolens* L. In: Kalloo, G. and Bergh, B.O. (eds.) *Genetic Improvement of Vegetable Crops*. Pergamon Press, Oxford, UK, pp. 523–534.

Ranfa, A. (2004) *Piante Amiche e Nemiche dell'uomo*. Ali & No Editrice, Perugia, Italy.

Soulé, M.E. (1987) *Viable Population for Conservation*. Cambridge University Press, Cambridge.

Tosti, N. and Negri, V. (2002) Efficiency of three pcr-based markers in assessing genetic variation among cowpea landraces. *Genome* 45, 268–275.

Tosti, N. and Negri, V. (2005) On-going on-farm microevolutionary processes in neighboring cowpea landraces revealed by molecular markers. *Theoretical and Applied Genetics* 110, 1275–1283.

Tribulato, A., La Malfa, G. and Branca, F. *Il Contributo delle Piante Mediterranee allo Sviluppo dell'orticoltura. Proceedings of 2° Convegno Nazionale Piante Mediterranee*, Agrigento, 7–8 Ottobre 2004 (in press).

VI *Ex Situ* Conservation

27 *Ex Situ* Conservation of Wild Species: Services Provided by Botanic Gardens

P.P. SMITH

27.1 Introduction

Botanic gardens have traditionally played a central role in educating the public and raising awareness about the need for plant conservation. However, they also have an important practical role to play in the conservation of wild plants, including crop wild relatives (CWR). Despite their major importance to *ex situ* conservation of wild species, botanic gardens have not always been fast to recognize their own potential or to engage fully with *in situ* and *ex situ* conservation practitioners (Lowry and Smith, 2003; Smith *et al.*, 2004).

This chapter discusses a range of conservation services routinely provided by botanic gardens, and gives examples from the Royal Botanic Garden (RBG) Kew's Millennium Seed Bank Project (MSBP) of conservation services that RBG Kew and partner botanic gardens have been able to provide to applied conservation projects worldwide. The MSBP is a 10-year international programme (2000–2010) with two main aims:

- Collect and conserve seeds from 24,200 species, principally from the drylands, by 2010;
- Develop bilateral research, training and capacity-building relationships worldwide in order to support and advance the seed conservation effort.

The MSBP currently comprises partnerships in 18 countries with 45 institutions. Four areas of conservation action are described: (i) provision of botanical diversity information; (ii) seed conservation services; (iii) conservation genetics research; and (iv) horticultural services.

27.2 Provision of Botanical Diversity Information

Botanic gardens, particularly those with associated herbaria or other botanical reference collections, are the repositories of vast amounts of taxonomic and botanical diversity information. For example, RBG Kew maintains a herbarium of more than seven million specimens from all over the world. Each specimen has label information associated with it, detailing the plant name, locality, phenology, date of collection, habitat information, uses and so on. Much of this information is useful to conservationists. In addition, the curators and taxonomists who work on this reference material are encouraged to develop plant identification skills which have obvious applications in the field. Of course, not all botanic gardens have associated herbaria, but they all have living collections, and the horticulturalists who create and maintain these collections often have excellent field knowledge.

The MSBP and its partner institutions recognized early on that plant diversity information and plant recognition skills are in short supply and, as such, represent a major constraint to targeted seed collecting. If the aim is to collect seed from a threatened species, the field teams need to know where to find it, when it will be in seed and then be able to recognize it when they see it in the field. To this end, the MSBP currently employs a team of 12 people in Kew's herbarium to digitize herbarium specimens and incorporate them into a geographical information system (GIS). Querying the GIS produces phenology maps for species, detailing both localities and timing of seed set. This information is then incorporated into a targeted collection guide, and combined with species descriptions and images of herbarium specimens to create a field tool for local seed collectors. Targeted collection guides are particularly useful for rare or obscure species, which may not be familiar to the local field team.

This approach is presently being employed for ten MSBP partner countries. For example, in Botswana, the plant Red Data List (RDL) (Golding, 2002) includes 43 taxa, 22 of which are data deficient. The MSBP Botswana field team has been using the specimen data for the taxa on this list to locate these populations and collect seeds from them. In the first year of the Botswana project (2003/2004), 37 populations of 18 taxa on the RDL were found using this approach. Seed was collected for ten of these taxa. However, the conservation outcomes of this methodology are far greater than seed banking alone, because locating these populations allows the seed-collecting teams to carry out full field assessments of status and threats for data-deficient taxa. A specific example of this is the case of *Erlangea remifolia* Wild & G.V. Pope (*Asteraceae*), which is designated data deficient on the Botswana RDL, but which field survey now suggests is highly habitat-specific, and so far only one population of four individuals has been found. On the other hand, *Hoodia currori* (Hook.) Decne subsp. *lugardii* (N.E.Br.) Bruyns, currently designated as vulnerable, has been found to be common and widespread.

The MSBP has fully utilized people with good field identification skills. An example of this is the expedition mounted to the Northern Cape (South Africa) in February 2001 (Smith *et al.*, 2001). A major focus of this seed-collecting expedition was the family *Mesembryanthemaceae*. With around 1800 taxa in South Africa, this is a difficult group for non-specialists to master. For this reason, Priscilla Burgoyne from the National Botanic Institute (now South African National Biodiversity Institute) in Pretoria joined the expedition. One of the rarest species that Priscilla was able to identify was *Cylindrophyllum hallii* (L.) Bolus, a species not collected since 1960. Seed was collected from this very rare species (currently down to only six individuals) and has since been grown to produce *ex situ* living collections from which further seed is being harvested. Although Priscilla is a specialist in one taxonomic group, she also has local knowledge, and this is critical to the success of any field expedition. In general, people with local knowledge of plants across the taxonomic spectrum are most useful for conservation work because they are able to carry out a range of botanic survey and inventory activities in specific areas of conservation importance.

Good plant identification skills are not just required in the field. Plant taxonomists have an important role to play in identifying accessions held in *ex situ* collections. A quick scan of the World Information and Early Warning System (FAO, 2006) reveals that 8.3% and 18.2% of *Acacia* and *Trifolium* accessions, respectively, held in WIEWS gene banks are not identified to the species level. Even where an accession has a binomial assigned to it, it is not always correctly identified. Taylor *et al*. (1982) report that of the 13 *Trifolium* seed lots collected from seed banks for cytological study, seven (54%) were misidentified.

27.3 Seed Conservation Services

Botanic gardens with seed banks and/or seed conservation expertise have a significant contribution to make to the conservation of CWR or other wild species. Seed banks will often hold data on seed morphology, seed storage behaviour (orthodox or recalcitrant), germination, dormancy or longevity. Any of these information may be critical to ensure effective *ex situ* conservation of target plant species. For example, if seed banking is being considered as a conservation strategy, it is important to know whether seed is orthodox, i.e. likely to survive drying and freezing. If, on the other hand, the seed is recalcitrant, conventional seed banking will have to be discounted and another methodology employed (e.g. cryopreservation or maintenance as a living collection in a field gene bank or botanic garden). Likewise, if a species is to be propagated for regeneration, reintroduction or sustainable use, it will usually need to be germinated from the seed. This may involve breaking dormancy through an appropriate physical or chemical treatment.

Kew's Millennium Seed Bank (MSB) is currently the largest wild seed bank in the world (Smith *et al*., 1998). It has more than 40 seed scientists working on seed conservation-related problems. The MSB holds species from 131 countries, and the scientists who work there carry out around 10,000 germination tests each year. For the majority of species tested at the MSB, this is the first time that germination in a laboratory has been attempted. Many of the more difficult species to germinate exhibit dormancy mechanisms, which must be broken before germination can take place. The MSB freely publishes all of its germination and other data on the Kew web site in its Seed Information Database (SID) (Royal Botanic Garden, Kew, 2006). SID currently holds data on ~20,000 plant species.

Apart from seed conservation data, the seed-related skills found in seed banks and botanic gardens can be invaluable to conservation practitioners.

Seed collecting and handling are skilled activities that may take many years to develop, and will have an important bearing on seed quality. In a recent case study in Madagascar, staff from the MSB provided training and advice in seed collection and processing to Quebec Madagascar Minerals (QMMs), a subsidiary of Rio Tinto. QMMs are mining ilmenite (titanium dioxide) in the southeast of the country, and have set up a restoration programme aimed at restoring 200 ha of littoral forest on the mining path. Some of the plant species in this area are down to a few dozen individuals, and the restoration team was unable to achieve good germination rates for some of these. Visiting seed scientists from the MSB quickly discovered that the problem lay with the conditions under which the harvested seed was being kept. Under ambient conditions of high temperature and humidity, the seed was dying in a matter of weeks. The purchase of a drier and a refrigerator has solved these problems.

27.4 Conservation Genetics Research

Apart from revolutionizing plant systematics, molecular biology has found application in wild plant conservation through the use of genetic fingerprinting techniques. Genetic variation in wild or cultivated plant populations can be characterized using amplified fragment length polymorphism (AFLP) and/or plastid microsatellites. This information can be used in a number of conservation applications, including the development of sampling strategies for *ex situ* conservation (e.g. seed banking) and in the reintroduction of the wild plants of appropriate genotypic diversity and provenance. This kind of approach can be used to avoid founder effects, determine effective population size and avoid outbreeding or inbreeding depression, any one of which can influence a reintroduction programme.

RBG Kew has a well-established conservation genetics unit, which has been involved in a number of reintroduction programmes. One of these is that of *Cypripedium calceolus* L., the lady's slipper orchid (Fay *et al.*, 2003). This species is reduced to only one surviving plant in the wild in the United Kingdom. In addition to this single wild plant there are several plants of wild origin in living collections and two introduced specimens growing in semi-natural habitat. Microsatellite characterization of all these plants revealed that the one surviving wild plant and those of wild origin in living collections all possessed one or other of the two western European fingerprints. However, the two introduced plants were found to be of different origin – one matching material from Austria and the other shown to be of a different species altogether, the morphologically similar North American *C. parviflorum* Salisb. Both of these plants have been excluded from further propagation exercises.

27.5 Horticultural Services

Growing plants outside their natural habitat is a skilled endeavour, and the gardening staff in botanic gardens must acquire those skills if the garden is to succeed. These same skills have applications in conservation, particularly in species reintroduction, habitat restoration and sustainable use programmes. Appropriate propagation techniques, media and conditions are critical, as are introduction sites, timing of introduction and management of competitors, diseases and pests. Most horticulturalists deal with these kinds of problems on a daily basis.

RBG Kew has over 200 horticultural staff, and a number of them have worked with partner institutions through the MSBP. A recent example of this is in Botswana, where one of the MSBP partners, Veld Products Research and Development (VPRD) are running propagation trials on 40 wild plant species. All of these species are valued by local rural communities, and they include food plants, bush teas, medicinal plants and ornamentals. Depending on the results of the trials, some of these plants may have potential in rural industry, e.g. *Rhigozum brevispinosum* Kuntze (*Bignoniaceae*), which is used as an ornamental plant in rural gardens. In the spring, it produces profuse yellow flowers, rather like *Forsythia* in the Northern Hemisphere. This species has been propagated by VPRD with the help of a Kew horticulturalist. Further advice has come from a gardener from the National Botanic Research Institute in Namibia, who has suggested that watering the *Rhigozum* discourages it from flowering.

27.6 Training

While all of the conservation services listed above will not be available from any one botanic garden, botanic garden networks are actively exchanging skills through project collaborations, joint fieldwork and training programmes. RBG Kew runs four diploma courses – herbarium techniques; plant conservation strategies; seed conservation techniques; and botanic garden management – aimed at personnel from botanic gardens or similar institutions. RBG Kew's MSBP trained more than 570 people in seed conservation techniques over a total of more than 5600 training days between 2000 and 2004. This training ranged from informal instruction in the field to exchange visits to research laboratories to regional and United Kingdom-based formal courses. This kind of effort means that conservation-oriented skills are on the increase in botanic gardens. Despite this, there is much to do to change perceptions from inside and outside.

27.7 Conclusions

Although there is a lot of conservation work going on in botanic gardens, conservation is rarely seen as the main priority. This is because most botanic gardens are primarily visitor attractions, and it is through visitors that they raise their revenue. Most botanic gardens are able to manage the balance between producing a spectacle that will attract visitors and providing educational facilities or interpretation. This is because these two areas of activity are largely compatible. Indeed, many botanic gardens would say that their conservation effort stops there – they provide a valuable service by educating the public about conservation issues. Balancing the role of visitor attraction with that of scientific and conservation organization is much more difficult, but, in an increasingly anthropogenic world where wild species and 'natural' habitats are being squeezed into smaller and smaller corners, the role of botanic gardens will become ever more important (Blackmore and Paterson, 2006). Many botanic gardens are already the last resort for species that are extinct in the wild. RBG Kew holds around 2000 threatened species from all over the world in its living collections – more than the entire native flora of the United Kingdom. What is to be done with these species? Maintaining living collections is expensive and risky – disease, disaster or even old age can mean the end of the line for a species. Clearly, banking of seed is a good idea, but seed banking is not an end in itself. If we are ever to make use of these species, we need to cultivate them – this means reintroduction of species or species assemblages, but often it will just mean looking after them, preferably *in situ*.

Botanic gardens need to recognize the roles that they can play. Above all they need to engage with conservation practitioners, including local communities, government agencies, conservation NGOs and the private sector. It is only by broadening their range of users that they can find real social relevance.

References

Blackmore, S. and Paterson, D.S. (2006) Gardening the Earth: the contribution of botanic gardens to plant conservation and habitat restoration. In: Leadlay, E. and Jury, S. (eds) *Taxonomy and Plant Conservation*. Cambridge University Press, Cambridge, pp. 266–273.

FAO (2006) WIEWS. Available at: http://apps3.fao.org/wiews/wiews.jsp

Fay, M.F., Bone, R., Cook, P. and Maurin, O. (2003) Genetic characterization of *Cypripedium calceolus* (Orchidaceae) using plastid DNA length polymorphisms: biogeographic patterns and identification of introduced plants. Report to English Nature. Royal Botanic Gardens, Kew, UK.

Golding, J.S. (2002) (ed.) *Southern African Plant Red Data List*. Southern African Botanic Diversity Network Report Series No. 14. National Botanic Institute, Pretoria, South Africa, pp. 135–156.

Lowry, P.P. II and Smith, P.P. (2003) Closing the gulf between botanists and conservationists. *Conservation Biology* 17(4), 1175–1176.

Royal Botanic Garden Kew (2006) *Seed Information Database*. Available at: http://www.kew.org/data/sid/

Smith, P.P., Burgoyne, P. and van Wyk, E. (2001) Rare plants rediscovered in the Northern Cape. *SABONET News* 6(1), 51–52.

Smith, P.P., Lowry, P.P. II, Timberlake, J. and Golding, J.S. (2004) The work of taxonomy. *Conservation Biology* 18(1), 8–9.

Smith, R.D., Linington, S.H. and Wechsberg, G.E. (1998) The Millennium Seed Bank, the Convention on Biological Diversity and the Dry Tropics. In: Prendergast, H.D.V., Etkin, N.L., Harris, D.R. and Houghton P.J. (eds) *Plants for Food and Medicine: Proceedings of the Joint Conference of the Society for Economic Botany and the International Society for Ethnopharmacology*. The Royal Botanic Gardens, Kew, UK, pp. 251–261.

Taylor, N.L., Gillett, J.M. and Giri, N. (1982) Morphological observations and chromosome numbers in *Trifolium* L. section *Chronosemium* Ser. *Cytologia* 48, 671–677.

28 Conservation of Spanish Wild Oats: *Avena canariensis*, *A. prostrata* and *A. murphyi*

P. García, L.E. Sáenz de Miera, F.J. Vences, M. Benchacho and M. Pérez de la Vega

28.1 Introduction

The genus *Avena* L. is widely distributed throughout the world and comprises species with three ploidy levels: diploids ($2n = 14$); tetraploids ($2n = 28$); and hexaploids ($2n = 42$). The centres of origin of these species are unknown, but the western Mediterranean has been suggested for some tetraploid and hexaploid oats as the majority of biological species coexist in the area between southern Spain, western Algeria and the Canary Islands (Rajhathy and Thomas, 1974).

Most of the species described in Spain are abundantly present throughout Europe (i.e. *A. barbata* Pott. ex Link, *A. sterilis* L., *A. fatua* L.); however, three of them are restricted to a few scarce populations in a limited area (*A. canariensis*, *A. prostrata* and *A. murphyi*).

Avena canariensis Baum Rajhathy et Sampson ($2n = 14$), described by Baum *et al.* (1973), is an endemic species to the Canary Islands, especially in Lanzarote and Fuerteventura. This species is catalogued in the endangered (EN) category under the IUCN criteria (VV. AA., 2000).

Avena prostrata Ladizinsky was identified as a different species by Ladizinsky (1971a). Morphologically, it is very similar to other diploid and tetraploid oats, such as *A. hirtula*, *A. wiestii* or *A. barbata*, and usually grows in mixed stands with them in a restricted area of south-east Spain (Murcia and Almeria provinces), although one population has been described in Morocco. Ladizinsky (1971a) considered *A. prostrata* as apparently common in this area and it seems unlikely that populations were under threat of extinction (Leggett *et al.*, 1992). However, the *Avena* Working Group of the European Cooperative Programme for Crop Resources Network (ECP/GR) consider that *A. prostrata* should possibly be a Red List species based on the small number of described populations and because the primary habitat where it grows has been progressively taken over by the glasshouse industry (Leggett, 1992).

Avena murphyi Ladizinsky is a tetraploid species of oat ($2n = 4x = 28$) described by Ladizinsky (1971b) after he had discovered it in a collecting mission. This author considered *A. murphyi* a native of southern Spain, being a common species in the valley of Laguna de la Janda (Cádiz), but with a narrow distribution. It does not occur in other parts of Andalusia. In order to protect this species, the Andalusian Government included *A. murphyi* in the Red List (Boletín Oficial de la Junta de Andalucía, 1994, 2003) and the taxon appeared in the Libro Rojo de la Flora Silvestre Amenazada de Andalucía (Blanca *et al.*, 2000). *A. murphyi* is also included in the Red List of Spanish Vascular Flora (VV. AA., 2000) in the vulnerable category. *A. murphyi* was also described in Morocco, limited to some areas in Tangier.

Despite the fact that Spain is particularly rich in wild oat species, they were poorly represented in the Centro Nacional de Recursos Fitogenéticos (Spanish Plant Genetic Resources Center). Several projects have been funded to collect and establish a representative collection of these materials. In 1999, a programme started in which five samples of *A. murphyi* were obtained with the inestimable collaboration of Prof. Valdés (Valdés *et al.*, 2000). Populations of *A. canariensis* were collected in 2001. In the case of *A. prostrata*, a collection of 17 populations sampled in 1996 in south-east Spain and without a definitive taxonomical classification was studied in order to see if they included individuals of *A. prostrata*. Only two out of the 17 populations contained a significant proportion of plants of this species.

The level of genetic variation and the way it is structured in this collection has been studied by using two different genetic markers (isozymes and inter-simple sequence repeats (ISSRs)). These results can be directly applied to design sampling strategies for *ex situ* conservation of genetic resources and may help to define which populations would be of most interest in order to preserve *in situ* the maximum genetic variability in these species.

28.2 Materials and Methods

The plants in this study were collected in eight populations of *Avena canariensis*, two of *A. prostrata* and five of *A. murphyi*, located at the sites shown in Fig. 28.1. The number of individuals analysed and the geographical characteristics of the sampling sites are shown in Table 28.1. The number of plants analysed for isozymes reflects the size of the population.

In all sampled populations, a single panicle of each plant was collected and stored separately. When multiplication was required, one seed per plant was sown and the descendant seeds were stored separately. In this way, and due to the high levels of self-fertilization, the genetic structure of the original sample has remained unchanged.

28.2.1 Isozyme electrophoresis

A single seed was sown from each panicle and crude extracts from the leaves of 15-day-old seedlings were electrophoresed following the Hutchinson *et al.* (1983) procedure. The enzymes assayed were glutamate oxalacetate transaminase

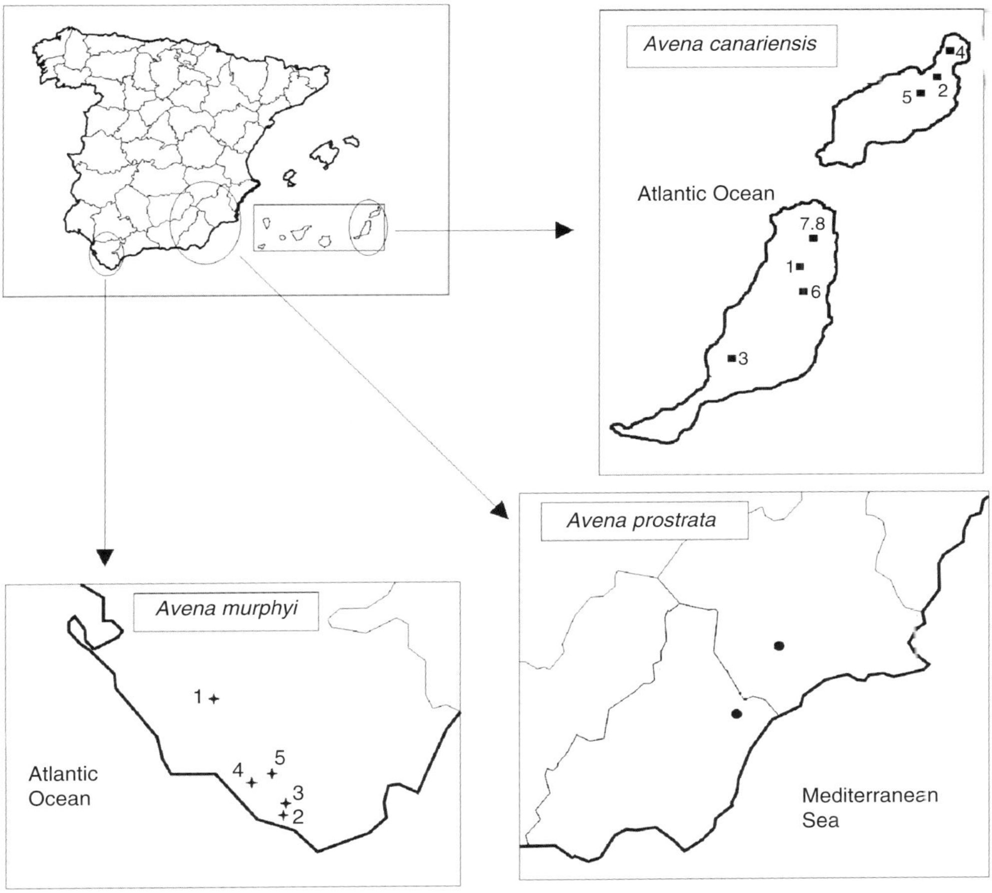

Fig. 28.1. Collection localities for three *Avena* spp.

(GOT, EC 2.6.1.1), leucin aminopeptidase (LAP, EC 3.4.11.1), malate dehydrogenase (MDH, EC 1.1.1.37), phosphoglucoisomerase (PGI, EC 5.3.1.9), phosphoglucomutase (PGM, EC 2.7.5.1), 6-phosphogluconate dehydrogenase (6-PGD, EC 1.1.1.44), esterase (EST, EC 3.1.1.1.2), peroxidase (PRX, EC1.11.1.7) and acid phosphatase (ACP, EC 3.1.1.2). Genetic interpretation of band patterns and nomenclature for loci follow those previously applied to *A. barbata* by Hutchinson *et al.* (1983) and García *et al.* (1989).

28.2.2 DNA extraction

DNA was extracted from fresh leaves using the Dneasy Plant Mini kit (QIAGEN) following the manufacturer's instructions, and it was quantified in agarose gels by comparison with a size marker by means of the TOTALLAB v1.11 software (Phoretix).

Table 28.1. Location and sample size of the populations of *Avena* studied.

Populations	No. of plants (isozymes)	No. of plants (ISSRs)	Latitude	Longitude	Altitude (m)	Spanish province/island
Avena canariensis						
La Matilla	5	3	28° 33' N	13° 57' W	381	Fuerteventura
Los Valles	62	10	29° 06' N	13° 30' W	517	Lanzarote
Maguez	30	10	29° 10' N	13° 27' W	188	Fuerteventura
Pajara	66	10	28° 18' N	14° 08' W	221	Lanzarote
Teguise	64	9	39° 03' N	13° 33' W	295	Lanzarote
Triquivijate	5	2	28° 28' N	13° 56' W	217	Fuerteventura
Villaverde A	14	4	28° 37' N	13° 54' W	250	Fuerteventura
Villaverde B	55	9	28° 37' N	13° 54' W	250	Fuerteventura
A. prostrata						
San Francisco	58	10	37° 24' N	1° 50' W	350	Almería
San Julian	35	9	37° 42' N	1° 37' W	315	Murcia
A. murphyi						
A323	25	8	36° 22' N	5° 55' W	25	Cadiz
Bolonia	81	13	36° 06' N	5° 44' W	70	Cadiz
Facinas	18	7	37° 08' N	5° 43' W	110	Cadiz
La Negra	25	9	36° 10' N	5° 49' W	80	Cadiz
Tahivilla	108	10	36° 12' N	5° 45' W	20	Cadiz

28.2.3 ISSR procedure

The ISSR-polymerase chain reaction (PCR) technique was carried out using nine primers (807, 808, 809, 810, 812, 815, 816, 817 and 818, University of British Columbia Biotechnology Laboratory, Vancouver, Canada; primer set no. 9). The reaction mixture (25 µl) contained between 10 and 50 ng of DNA, 2.5 µl of 10x *Taq* DNA polymerase buffer (50 mM Tris-HCl, pH 8.0, 100 mM NaCl, 0.1 mM EDTA, 5 mM DTT, 50% glycerol and 1.0% Triton X-100), 12.5 pmol of primer, 2 mM $MgCl_2$, 0.2 mM each of the four dNTP, 2% formamide and 1 U *Taq* DNA polymerase (Promega). The reactions were performed in a thermal cycler Perkin-Elmer mod 9700 under the following conditions: 7 min at 94°C, followed by 45 cycles of 30 s at 94°C, 45 s at 52°C and 2 min at 72°C, with a final step of 7 min at 72°C. The amplification products were separated on 1.8% agarose gels and the bands visualized with ethidium bromide.

Each band position was considered a different locus with two alternatives (presence versus absence), and all plants were considered as homozygous since these three species have very high levels of self-fertilization (higher than 95%) (P. García, 2003, unpublished data; Morikawa and Leggett, 1990).

28.2.4 Statistical analysis

The level of polymorphism was estimated by means of the following parameters: mean number of alleles per locus (A); percentage of polymorphic loci (P); and gene diversity index (H_j) for each locus and population calculated as:

$$H_j = 1 - \sum_{i=1}^{n} p_i^2$$

In which p_i is the frequency of the *i*th allele and *n* is the number of alleles observed for the *j*th locus and population (Nei, 1973). In *A. murphyi*, due to its allotetraploid condition, p_i is the frequency of the *i*th genotype and *n* is the number of genotypes observed in the *j*th locus and population, for the reasons which are explained in Section 3.

In order to analyse the distribution of variation within and among populations, total genetic diversity (H_T), diversity within populations (H_S), genetic diversity among populations (D_{ST}) and the coefficient of genetic differentiation (G_{ST}) were calculated following the statistics of Nei (1973). The G_{ST} value was used to estimate the level of gene flow (Nm) based on the relation $F_{ST} = 1/(4\,Nm + 1)$ (Wright, 1951) considering F_{ST} as equivalent to G_{ST}. In addition, the gene flow between populations was estimated from the frequencies of 'private alleles' (alleles found in only one population) by following the Barton and Slatkin (1986) procedure.

Nei's genetic identity (Nei, 1972) was calculated for each of the possible pairwise combinations among the same species populations and a phenogram was generated using the unweighted pair group method with arithmetic mean (UPGMA) method (Sneath and Sokal, 1973).

Correlations between the matrices of genetic distances obtained with the two different types of markers and between genetic and geographical distances were tested with a Mantel test (Mantel, 1974).

The analyses were performed using the computer programs GARRET, TULKAS and MANTEL written by L. Saénz de Miera, University of León, Spain (1999, unpublished data).

28.3 Results

28.3.1 Isozymes

The isozyme analysis of the seven systems described in materials and methods allowed for the study of 11 loci (*Got1, Got2, Pgd1, Lap1, Prx1, Est1, Pgm1, Pgi1, Mdh1, Mdh2* and *Mdh3*) in the three species. An additional locus in the system 6-PGD (*Pgd2*) which was not well resolved in *A. murphyi* could be scored in *A. canariensis* and *A. prostrata*, and in *A. murphyi* the system ACP allowed for the study of the locus *Acp2* which could not be analysed in the other two species.

All plants showed homozygous genotypes, with the exception of one plant in Los Valles (*A. canariensis*) and one plant in San Julian (*A. prostrata*), both heterozygous for the locus *Est1*. *A. murphyi* is an allotetraploid species with two different genomes (A and C). This fact was reflected in a high percentage of heterozygous phenotypes, although the plants breed true, which means that most of isozymatic loci scored were duplicated in the two homoeologous genomes and they were homozygous for different alleles. Thus, in *A. murphyi* there exist populations with monomorphic loci in which all plants showed the same heterozygous phenotype. For instance, for *Mdh1* all plants presented two alleles named *3* and *4*, and the subsequent heterozygous phenotype, although they had a homozygous genotype (the genotype for the locus *Mdh1* in one of its genomes would be *33* and for the other would be *44*). For this reason, in the estimation of the gene diversity index, the allelic frequencies were substituted by genotypic frequencies.

All analysed populations were polymorphic, but the number of variable loci ranged from one in the populations of *A. prostrata* to seven in *A. canariensis* (La Matilla). At specific level, *A. canariensis* showed the highest number of polymorphic loci (9), *A. murphyi* presented eight variable loci and the lowest value was for *A. prostrata* (2). Polymorphism values for the analysed loci are given in Table 28.2, and the variability parameters within populations and species in Table 28.3.

The analysis of genetic diversity within and among populations was carried out using Nei's method, in which the total diversity in the species (H_T) is the result of average diversity in the populations (H_S) plus the average gene diversity among populations (D_{ST}). The relationship $D_{ST}/H_T = G_{ST}$ measures the relative magnitude of differentiation among populations (Table 28.4). The three species showed a similar G_{ST} value, around 0.4, indicating a significant differentiation

Table 28.2. Genetic diversity parameters per locus.

Species/locus	No. of alleles per locus	Mean H_j values (range)	Species H_j values
Avena canariensis			
Got1	2	0.056 (0.00–0.44)	0.080
Got2	4	0.240 (0.00–0.34)	0.569
6Pgd1	2	0.120 (0.00–0.44)	0.168
6Pgd2	4	0.222 (0.00–0.60)	0.418
Lap1	3	0.297 (0.00–0.64)	0.482
Prx1	1	0.000	0.000
Est1	3	0.216 (0.00–0.51)	0.489
Pgm1	1	0.000	0.000
Pg11	1	0.056 (0.00–0.44)	0.080
Mdh1	2	0.048 (0.00–0.32)	0.056
Mdh2	3	0.040 (0.00–0.32)	0.049
Mdh3	2	0.000	0.000
A. prostrata			
Got1	1	0.000	0.000
Got2	1	0.000	0.000
6Pgd1	1	0.000	0.000
6Pgd2	2	0.000	0.361
Lap1	1	0.000	0.000
Prx1	1	0.000	0.000
Est1	3	0.550 (0.51–0.60)	0.599
Pgm1	1	0.000	0.000
Pgi1	1	0.000	0.000

Continued

Table 28.2. *Continued*

Species/locus	No. of alleles per locus	Mean H_j values (range)	Species H_t values
Mdh1	1	0.000	0.000
Mdh2	1	0.000	0.000
Mdh3	1	0.000	0.000
A. murphyi			
Got1	1	0.000	0.000
Got2	2	0.079 (0.00–0.35)	0.182
6Pgd1	2	0.232 (0.00–0.57)	0.308
Lap1	6	0.179 (0.00–0.50)	0.272
Prx1	4	0.278 (0.00–0.51)	0.475
Acp2	3	0.132 (0.00–0.10)	0.097
Est1	9	0.548 (0.00–0.75)	0.732
Pgm1	3	0.175 (0.00–0.49)	0.445
Pgi1	4	0.241 (0.00–0.67)	0.539
Mdh1	2	0.000	0.000
Mdh2	1	0.000	0.000
Mdh3	1	0.000	0.000

Table 28.3. Genetic diversity parameters within populations and in the species.

	Isozymes				ISSRs			
Population	A	npa	% P	H	A	npa	% P	H
		12 loci				66 loci		
Avena canariensis	2.33	–	75.0	0.181	1.89	–	89.4	0.226
La Matilla	1.67	5	58.3	0.155	1.03	2	3.0	0.013
Los Valles	1.67	2	50.0	0.170	1.39	3	39.4	0.105
Maguez	1.17	0	16.7	0.021	1.23	2	22.7	0.060
Pajara	1.58	0	41.7	0.145	1.62	4	62.1	0.218
Teguise	1.25	0	16.7	0.010	1.12	1	12.1	0.033
Triquivijate	1.33	1	33.3	0.105	1.17	0	16.7	0.074
Villaverde A	1.33	0	33.3	0.042	1.05	0	4.5	0.010
Villaverde B	1.33	0	33.3	0.115	1.11	0	10.6	0.053
		12 loci				81 loci		
A. prostrata	1.33	–	16.7	0.080	1.68	–	67.9	0.211
San Francisco	1.25	2	8.3	0.049	1.41	7	40.7	0.148
San Julian	1.17	1	8.3	0.045	1.57	9	56.8	0.176
		12 loci				59 loci		
A. murphyi	3.17	–	66.7	0.254	1.86	–	87.4	0.266
A323	2.00	0	25.0	0.166	1.22	1	22.0	0.065
Bolonia	2.08	3	58.3	0.154	1.46	7	45.8	0.166
Facinas	1.83	2	33.3	0.148	1.02	0	1.7	0.004
La Negra	2.00	1	16.7	0.093	1.27	4	27.1	0.083
Tahivilla	2.58	2	58.3	0.217	1.47	5	47.5	0.151

A: number of alleles per locus; npa: number of private alleles; % P: percentage of polymorphic loci; H: gene diversity (heterozygosity).

Table 28.4. Summary of genetic variability distribution in three *Avena* spp.

Species	Isoenzymes				ISSRs			
	H_T	H_S	D_{ST}	G_{ST}	H_T	H_S	D_{ST}	G_{ST}
Avena canariensis	0.181	0.108	0.073	0.402	0.226	0.071	0.155	0.686
Fuerteventura	0.173	0.107	0.066	0.374	0.213	0.042	0.171	0.802
Lanzarote	0.176	0.115	0.061	0.344	0.201	0.119	0.082	0.407
A. prostrata	0.080	0.046	0.034	0.428	0.211	0.162	0.049	0.232
A. murphyi	0.254	0.155	0.099	0.389	0.266	0.094	0.172	0.647

among populations. The Nm values estimated from G_{ST} were 0.372 for *A. canariensis*, 0.334 for *A. prostrata* and 0.393 for *A. murphyi*. When using the private alleles method (Barton and Slatkin, 1986), the values of Nm were 0.135, 0.008 and 0.205, respectively.

Because the populations of *A. canariensis* were collected from two islands, two different analyses were carried out. The results are shown in Table 28.4, and as it can be seen, the parameters are very similar in both groups and to the total values.

The relationships between populations were studied by calculating their pairwise Nei's genetic identities and UPGMA clustering (Fig. 28.2). Genetic identities ranged between 0.799 and 0.959 for *A. canariensis*, between 0.703 and 0.974 for *A. murphyi*, and it was 0.905 for the two populations of *A. prostrata*.

28.3.2 ISSR markers

In the ISSR analysis, nine primers were used in the analysis of *A. canariensis* and *A. prostrata*, and seven in *A. murphyi* (primers 807 and 812 did not yield good amplifications in a large number of plants). Band patterns obtained with a primer were quite different for different species, henceforth, each one was scored independently. The bands scored had sizes between 200 and 1500 bp, and the number of bands per primer ranged from three (primer 809 in *A. canariensis*) to 17 (primer 807 in *A. prostrata*), with an average number of bands per primer of 7.3 for *A. canariensis*, 9.0 for *A. prostrata* and 8.4 for *A. murphyi*. As for isozymes, all populations showed polymorphism in a number of loci. A summary of the heterozygosities for each locus within populations and in the species is shown in Fig. 28.3, and the variability parameters within populations and species in Table 28.3.

For *A. canariensis* and *A. murphyi*, the heterozygosities within populations were usually of the same magnitude or lower than those obtained with isozymes, while *A. prostrata* showed a higher heterozygosity value with ISSRs. However, the heterozygosities for the species were higher for ISSRs than for isozymes, although in *A. canariensis* and *A. murphyi* the differences were not statistically significant. With this distribution of the genetic variability, and

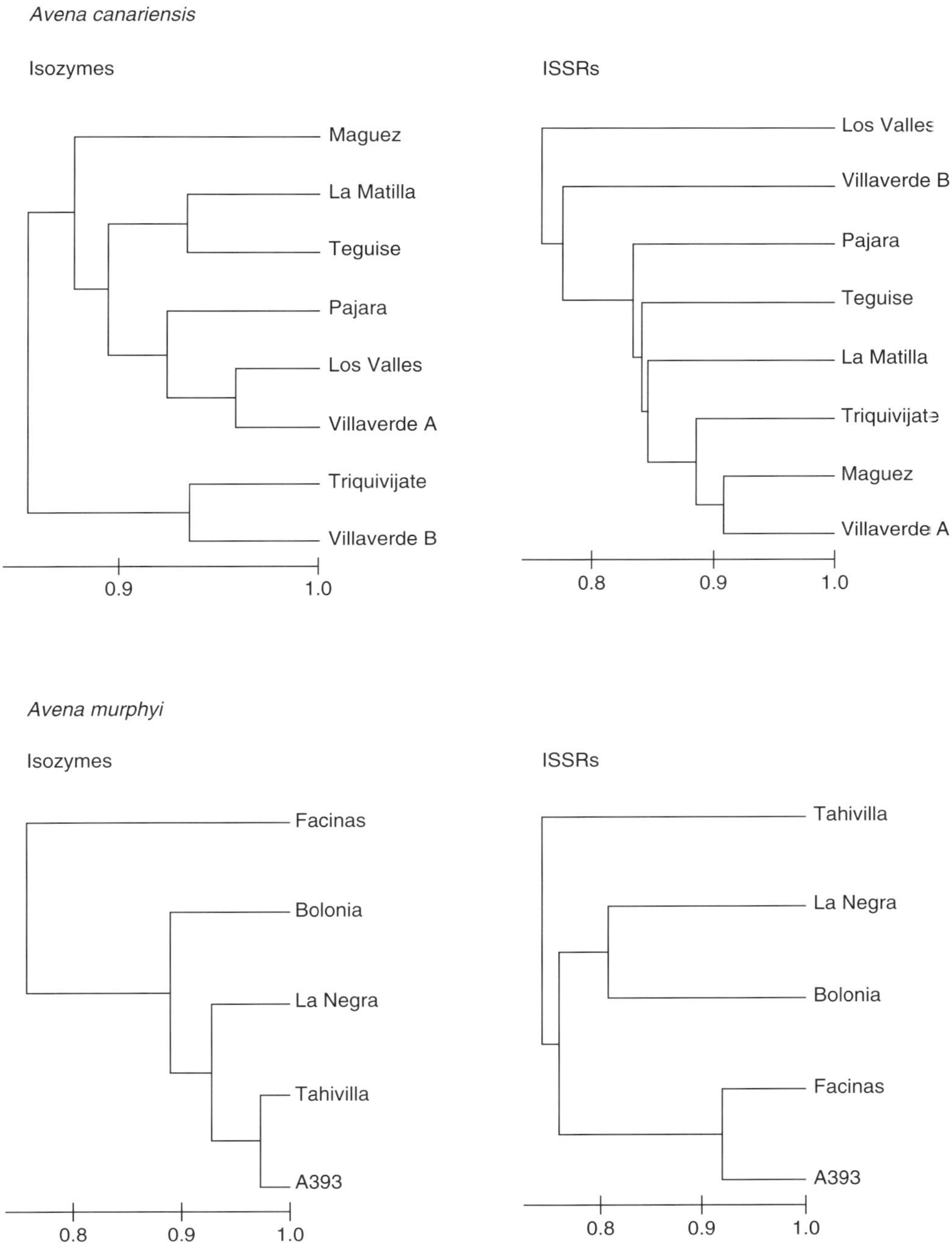

Fig. 28.2. UPGMA dendrograms of *Avena canariensis* and *A. murphyi* populations based on Nei's genetic identities obtained for isozymes and ISSRs.

(a)

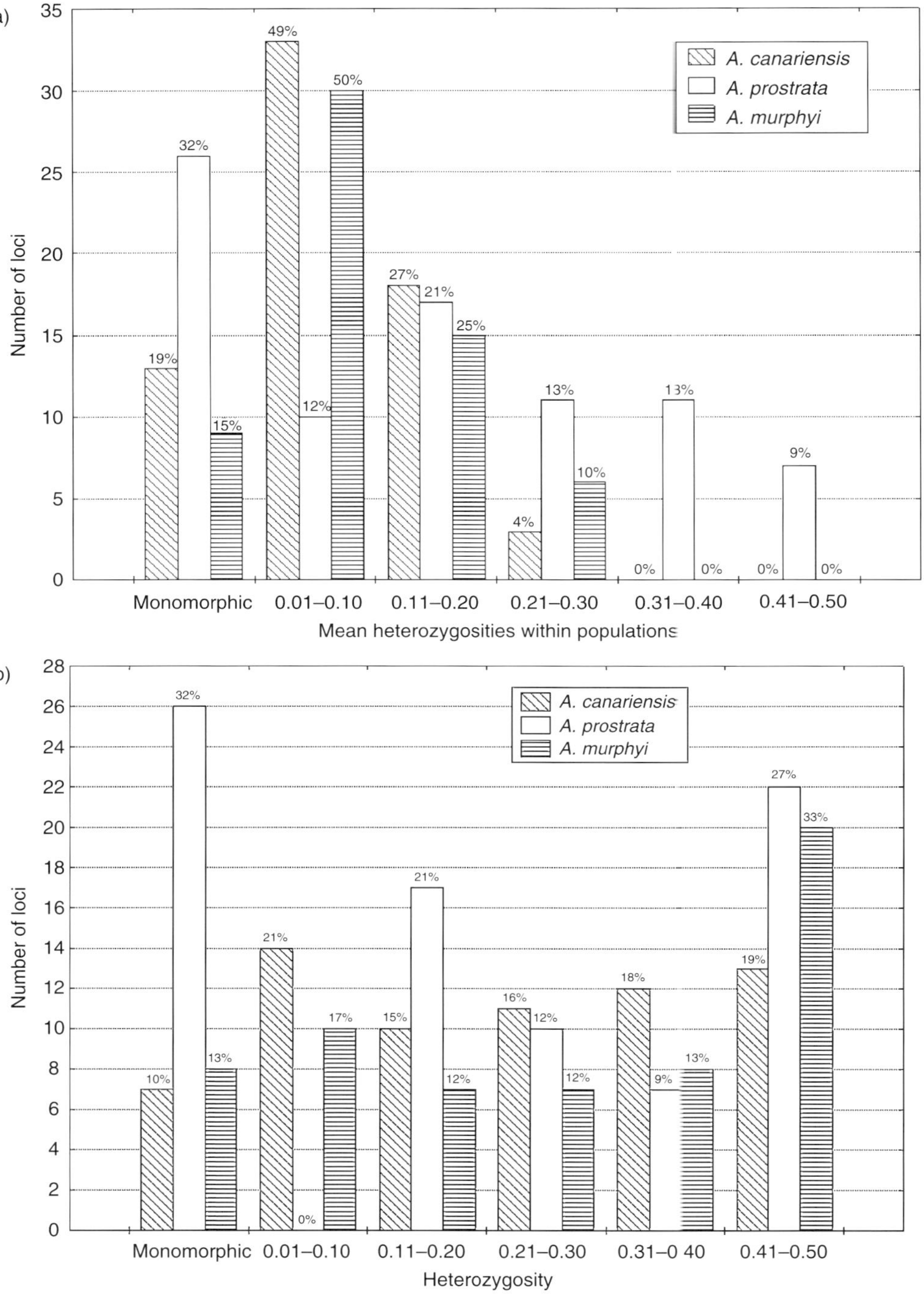

(b)

Fig. 28.3. Distribution of heterozygosities in ISSR loci within populations (a) and in species (b). Percentage of loci is given above bars.

comparing with isozymes, a higher coefficient of genetic differentiation between populations has been obtained in *A. canariensis* and *A. murphyi*, and it has been lower in *A. prostrata*. The Nm values for the G_{ST} values estimated were 0.114 for *A. canariensis*, 0.828 for *A. prostrata* and 0.136 for *A. murphyi*. The method of the private alleles yielded the values 0.110, 0.177 and 0.143, respectively.

In a similar way to isozymes, the populations of *A. canariensis* of each island were studied independently, with the results shown in Table 28.4. Although the total variability was similar for both the islands, the heterozygosity within populations was higher in Lanzarote, which results in a lesser differentiation between the populations from this island.

The UPGMA phenograms from Nei's genetic identities are shown in Fig. 28.2, and as it can be inferred from them, the identities are usually slightly lower with ISSRs than with isozymes. The values ranged from 0.653 to 0.910 for *A. canariensis*, from 0.730 to 0.919 in *A. murphyi*, and were 0.881 for the two populations of *A. prostrata*.

The different grouping of populations shown for the different markers in the dendrograms could be a by-product of the clustering method. Thus, Mantel tests were carried out to calculate the correspondence between isozymes and ISSRs genetic distances for *A. canariensis* and *A murphyi*. In both species, correlations were not statistically significant. In *A. prostrata*, this analysis was not performed since only two populations were studied.

In addition, Mantel tests were carried out between genetic and geographic distances, and were not significant in any case.

28.4 Discussion

Isozymatic analyses of *A. canariensis* were previously carried out by Morikawa and Leggett (1990) who detected a high level of variability in the 19 populations studied, higher than other species of oats, such as *A. barbata* and *A. fatua*. They also showed that the geographic distribution of the variability was not homogenous, where Fuerteventura populations were more polymorphic than the populations of Lanzarote. Finally, they detected the existence of ecotypes, already reported by Craig *et al.* (1974), which were isoenzymatically and morphologically distinguishable. In this work, the populations studied showed similar heterozygosity values for both the islands, and they were similar to that found in Lanzarote populations by Morikawa and Leggett (1990). Thus, the total variability found was lower than those described previously. The lack of habitat data does not allow us to establish whether these results are due to a bias in the habitats sampled or simply reflects the lower number of populations and/or the differences in the isozymatic systems analysed.

Genetic data for a rare or endemic species with a restricted distribution are more informative when compared with data for widespread congeners with similar life histories, since it allows a better understanding of the causes of rarity (Gitzendanner and Soltis, 2000). Morikawa and Leggett (1990) compared the results for *A. canariensis* to those for Californian populations of *A. barbata* and *A. fatua*. However, these two species are polyploid, and the American populations show low levels of genetic variability because they were introduced to California during European exploration and colonization. The species *A. hirtula* seems more appropriate for doing comparisons since it is diploid and grows in a wide geographical range. Our populations of *A. canariensis* have only shown a slightly lower level of gene diversity than *A. hirtula* in which the values of H_T, H_S and G_{ST} found in ten populations collected in the Iberian Peninsula were 0.251, 0.091 and 0.633, respectively (Ceballos and García, 1997, unpublished data). Although endemics and rare species have generally less genetic variation than their widespread congeners, it is not uncommon to find similar or even higher genetic variability in the rare species (Gitzendanner and Soltis, 2000). In a review on the level of genetic variability of 69 species, Canarian endemics have been reported to show a relatively high diversity when compared to endemics of Pacific islands (Francisco-Ortega *et al.*, 2000). The high variability levels in the endemic *A. canariensis* could be explained due to its relative abundance in its distribution range; thus, the number of plants is not necessarily reduced. In addition, the populations of *A. canariensis* may maintain the genetic variability due to microhabitat selection, since they seem to be highly adapted to different soil conditions (Morikawa and Leggett, 1990).

The isozymatic data for *A. murphyi* show that a substantial amount of genetic variation is contained in its populations in spite of their small number and size. This fact is even more remarkable when comparing results with *A. barbata*, another tetraploid species which is very common and maintains large populations. In a study carried out with 42 Spanish populations of *A. barbata* (García, 1988), the values of H_T, H_S and G_{ST} were 0.368, 0.174 and 0.527. The fact that *A. murphyi* populations contain the same level of genetic variability as the large and abundant *A. barbata* shows that *A. murphyi* still presents a remarkable genetic richness. A possible explanation would be a recent reduction in the number of individuals and populations, since most of the described populations have been considered as small residuals of larger populations confined to marginal sites (Valdés *et al.*, 2000). In the case of *A. murphyi* the progressive agricultural use of the habitat in which it grows has been clear, which points out that human activities are the main factor for the low number of populations. A similar picture arises from Moroccan populations; thus, the collection expeditions carried out by Saidi and Ladizinsky (2003) from 1998 to 2001 showed that *A. murphyi* was not present in most of the previously reported sites due to urbanization and cultivation, and they called for measures to protect this species.

The use of isozymatic loci has produced important information on the evolutionary biology and adaptive strategies in *Avena*, but the low number of loci available and the lack of polymorphism for some systems have driven to the use of DNA markers. This has led to the widespread use of PCR methodologies with arbitrary primers, whose main advantages are a higher level of polymorphism and more comprehensive coverage of the genome. Among these, ISSRs are very easy to develop and present a good level of reproducibility.

The polymorphism values obtained with ISSR have been slightly higher at species level, however, the average heterozygosities within populations were lower. This is in contrast with the general tendency obtained with ISSRs to produce higher estimates of within-population variation than isozymes and other dominant DNA markers such as RAPDs and AFLPs (Nybom, 2004). Nevertheless, heterozygosities with both markers have shown a positive correlation for *A. canariensis* and *A. murphyi*, though not significant.

The genetic distances obtained with the two markers in *A. canariensis* and *A. murphyi* were remarkably different, and this is evident in the dendrograms (Fig. 28.2). Similar results have been found in *A. barbata* using isozymes and ISSRs (Guma, 2001), or in *A. sterilis* for isozymes and RFLPs (Beer *et al.*, 1993) or isozymes and RAPDs (Heun *et al.*, 1994). In our dendrograms, the only point in common was that an association between genetic and geographical distances did not exist.

The incongruity among markers in our study might be due to the limited number of plants analysed for ISSRs and also to the fact that different markers reveal the genetic characteristics of different regions of DNA (protein-coding genes for isozymes and mainly non-coding areas when using ISSR loci), which are under particular evolutionary forces. It is usually thought that estimates using ISSRs should give a more accurate picture of the relationships between populations than those of isozymes since a higher number of loci can be studied and henceforth a wider coverage of the genome is achieved. This idea is probably true when the SSR sequences (the annealing sites for the primers used in the ISSR-PCR) are randomly distributed in the genome; and the availability of the DNA sequence for several plants allows us to know the location of SSRs. The data show that in some species such as tomato and *Arabidopsis*, SSRs are preferentially located in some specific regions, thus the data of genetic variability apply to a limited part of the genome (Barth *et al.*, 2002). The data of the distribution of SSRs in *Avena* are limited, but Li *et al.* (2000) found that a high proportion of *Avena* SSRs was associated with known repetitive elements that show a low level of polymorphism.

Despite their differences, both markers have shown a significant differentiation among populations and to a similar degree for the three species, with the exception of *A. prostrata* for ISSRs. In order to reach this genetic structure, gene flow must be restricted and/or there is natural selection favouring particular genotypes in different sites. As indicated previously, in *A. canariensis* the occurrence of differential selection is supported by the results of Morikawa and Leggett (1990). These authors stated that the populations appeared to be adapted to distinct microhabitats, especially soil conditions, and they found that distinctive ecotypes, with different morphological traits, were marked for the presence of characteristic isozyme genotypes. For *A. murphyi*, further studies would be needed, but the high level of variability and the similar edaphic and climatic characteristics of the sampled sites seem to point to the limited gene flow as the main agent to account for the population differentiation.

The G_{ST} values showed that at least 40% of the total genetic variation is due to population differences. This fact implies that if the *in situ* conservation is focused on preserving only one population, even if it is the most variable, it would affect the genetic richness to a limited degree. For instance, Valdés *et al.* (2000) proposed the conservation of one of the larger and more variable populations of *A. murphyi* (Bolonia), and although this would be an excellent measure, only about 61% of the total variation would be protected. If the conservation was directed to the two larger and variable populations (Bolonia and Tahivilla), at least 95% of the variability would be under protection.

A. canariensis is a different case since some populations are under protection because they are grown in national parks and it is a relatively common species in Fuerteventura and Lanzarote. However, because of the high value of genetic differentiation among populations and its adaptive nature, an intensive sampling of populations for *ex situ* conservation should be undertaken. In addition, specific populations outside the national parks with high levels of variability or that are genetically unique should be protected. Using as an example of the populations analysed in this study, a complex picture arises due to different levels of variability obtained with the two markers but taking into account the genetic and demographic characteristics, Los Valles and Pajara should be a prior consideration.

A. prostrata results are so limited that no conclusion can be inferred and the need for further collections, especially since the primary habitat where it grows, has been progressively taken over by the glasshouse industry (Leggett, 1992).

When focused on the conservation of specific populations, it is necessary to take into account that the private alleles of the other populations would not be included. Although the adaptive significance of these private alleles is a controversial issue, especially because they are usually in low frequencies in the populations, they might be an indicator of the presence of other exclusive characteristics in the populations (Neel and Ellstrand, 2003).

Finally, it is necessary to bear in mind that the relationship between the variation in molecular markers and the quantitative characters implied in adaptation is not clear, and although Mckay and Latta (2002) give theoretical arguments for the validity of molecular markers to predict the adaptive potential, the establishment of priorities based exclusively on markers must only be taken as a first step towards the genetic conservation of a species.

The need to undertake further collections for *A. canariensis* and *A. prostrata* for *ex situ* conservation is evident. For *A. murphyi*, the search has been quite exhaustive, and although new populations could be found, the data available calls for the necessity of an urgent *in situ* conservation programme.

Acknowledgements

This work has been supported by the Instituto Nacional de Investigacion y Tenologia Agraria y Alimentaria (INIA) of Spain (projects No. RF00–001 and RF02–002), and a doctoral fellowship to M. Benchacho. The authors thank Eshan Dulloo and Brian Ford-Lloyd for their valuable comments on the manuscript.

References

Barth, S., Melchinger, A.E. and Lübberstedt, Th. (2002) Genetic diversity in *Arabidopsis thaliana* L. Heynh. investigated by cleaved amplified polymorphic sequence (CAPS) and inter-simple sequence repeat (ISSR) markers. *Molecular Ecology* 11, 495–505.

Barton, N.H. and Slatkin, M. (1986) A quasi-equilibrium theory of the distribution of rare alleles in a subdivided population. *Heredity* 56, 409–415.

Baum, B.R., Rajhathy, T. and Sampson, D.R. (1973) An important new diploid *avena* species discovery on the Canary Islands. *Canadian Journal of Botany* 51, 759–762.

Beer, S.C., Goffreda, J., Phillips, T.D., Murphy, J.P. and Sorrells, M.E. (1993) assessment of genetic variation in *Avena sterilis* using morphological traits, isozyme and RFLPs. *Crop Science* 33, 1386–1393.

Blanca, G., Cabezudo, B., Hernández-Bermejo, J.E., Herrera, C.M., Muñoz, J. and Valdés, B. (eds) (2000) *Libro Rojo de la Flora Silvestre Amenazada de Andalucía*. Consejería de Medio Ambiente, Junta de Andalucía. Sevilla, Spain.

Boletín Oficial de la Junta de Andalucía (1994) Decreto 104/1994, de 10 de Mayo, por el Que se Establece el Catálogo Andaluz de Especies de la Flora Silvestre Amenazada. *Boletín Oficial De La Junta de Andalucía* 107/1994.

Boletín Oficial de la Junta de Andalucía (2003) Ley 8/2003, de 28 de Octubre, de la Flora y la Fauna Silvestres. *Boletín Oficial de la Junta de Andalucía* 218, 23790–23810.

Craig, I.L., Murray, B.E. and Rajhathy, T. (1974) *Avena canariensis*: morphological and electrophoretic polymorphism and relationship to the *A. magna – A. murphyi* complex and *A. sterilis*. *Canadian Journal of Genetics and Cytology* 16, 677–689.

Francisco-Ortega, J., Santos Guerra, A., Kim, S.-C. and Crawford, D.J. (2000) Plant genetic diversity in the Canary Islands: a conservation perspective. *American Journal of Botany* 87, 909–919.

García, P. (1988) Análisis de la Estructura Genética de Poblaciones Españolas de *Avena barbata* Pott ex Link. PhD thesis. Universidad de León. Spain.

García, P., Vences, F.J., Sáenz De Miera, L.E., Allard, R.W. and Pérez De La Vega, M. (1989) Allelic and genotypic composition of ancestral Spanish and colonial Californian gene pools of *Avena barbata*: evolutionary implications. *Genetics* 122, 687–694.

Gitzendanner, M.A. and Soltis, P.S. (2000) Patterns of genetic variation in rare and widespread plant congeners. *American Journal of Botany* 87, 783–792.

Guma, I.R. (2001) Estudio de la Variabilidad y la Estructura Genética en Poblaciones de *Avena barbata* de Argentina. PhD thesis. Universidad de León. Spain.

Heun, M., Murphy, J.P. and Phillips, T.D. (1994) A comparison of RAPD and isozyme analyse for determining the genetic relationships among *Avena sterilis* L. accessions. *Theoretical and Applied Genetics* 87, 689–696.

Hutchinson, E.S., Hakim-Elahi, A., Miller, R.D. and Allard, R.W. (1983) The genetics of the diploidized tetraploid *Avena barbata*: acid phosphatase, esterase, leucine aminopeptidase, peroxidase and 6-phosphogluconate dehydrogenase loci. *Journal of Heredity* 74, 325–330.

Ladizinsky, G. (1971a) *Avena prostrata*: a new diploid species of oat. *Israel Journal of Botany* 20, 297–301.

Ladizinsky, G. (1971b) *Avena murphyi*: a new tetraploid species of oat from southern Spain. *Israel Journal of Botany* 20, 24–27.

Leggett, J.M. (1992) The conservation and exploitation of wild oat species. In: Barr, A.R. and Medd, R.W. (eds) *Proceedings of the Fourth International Oat Conference. Volume II. Wild Oats in World Agriculture*. Organizing Committee, Fourth International Oat Conference, South Australia, pp. 57–60.

Leggett, J.M., Ladizinsky, G., Hagberg, P. and Obanni, M. (1992) The distribution of nine *Avena* species in Spain and Morocco. *Canadian Journal of Botany* 70, 240–244.

Li, C.D., Rossnagel, B.G. and Scoles, G.J. (2000) The development of oat microsatellite markers and their use in identifying relationships among *Avena* species and oat cultivars. *Theoretical and Applied Genetics* 101, 1259–1268.

Mantel, N. (1974) The detection of disease clustering and a generalized regression approach. *Cancer Research* 27, 209–220.

Mckay, J.K. and Latta, R.G. (2002) Adaptive population divergence: markers, QTL and traits. *Trends in Ecology and Evolution* 17, 285–291.

Morikawa, T. and Leggett, J.M. (1990) Isozyme polymorphism in natural populations of *Avena canariensis* from the Canary Islands. *Heredity* 64, 403–411.

Neel, N.C. and Ellstrand, N.C. (2003) Conservation of genetic diversity in the endangered plant *Eriogorium ovalifolium* var. *vineum* (Polygonaceae). *Conservation Genetics* 4, 337–352.

Nei, M. (1972) Genetic distance between populations. *The American Naturalist* 106, 283–292.

Nei, M. (1973) Analysis of gene diversity in subdivided populations. *Proceedings of the National Academy of Sciences of the United States of America* 70, 3321–3323.

Nybom, H. (2004) Comparison of different nuclear DNA markers for estimating intraspecific genetic diversity in plants. *Molecular Ecology* 13, 1143–1155.

Rajhathy, T. and Thomas, H. (1974) *Cytogenetics of Oats*. The Genetics Society of Canada. Ottawa, Ontario, Canada.

Saidi, N. and Ladizinsky, G. (2003) Distribution and ecology of the wild tetraploid oat species *Avena Magna* and *A. Murphyi* in Morocco. In: Lipman, E. Maggioni, L., Knüppfer, H., Veteläinen, M., Ellis, R., Leggett, J.M., Germeier, C., Kleijer, G. and Podyma, W. (Compilers) *Reports of a Cereal Network Meeting, Yerevan, Armenia, 3–5 July 2003*. International Plant Genetic Resources Institute, Rome, Italy.

Sneath, P.H.A. and Sokal, H.H. (1973) *Numerical Taxonomy*. Freeman, San Francisco, California.

Valdés, B., Ocaña, M.E., Parra, R. and Pina, F.J. (2000) *Avena Murphyi* Ladizinsky. In: Blanca, G., Cabezudo, B., Hernández-Bermejo, J.E., Herrera, C.M., Muñoz, J. and Valdés, B. (eds) *Libro Rojo de la Flora Silvestre Amenazada de Andalucía. Tomo II: Especies Vulnerables*. Consejería de Medio Ambiente, Junta de Andalucía. Sevilla, Spain, pp. 67–69.

VV. AA. (2000) Lista Roja de Flora Vascular Española (Valoración Según Categorías UICN). *Conservación Vegetal* 6, 11–38.

Wright, S. (1951) The genetical structure of populations. *Annals of Eugenics* 15, 323–354.

29 Analysis of Wild *Lactuca* Gene Bank Accessions and Implications for Wild Species Conservation

T.S. RAJICIC AND K.J. DEHMER

29.1 Introduction

Cultivated lettuce (*Lactuca sativa* L.) is the best-known member of *Lactuca* L. genus. It is a crop species widely used in human nutrition. Its wild progenitor *L. serriola* L. and other wild *Lactuca* species are considered to be an important source of disease resistance genes (Sicard *et al.*, 1999; Jeuken and Lindhout, 2002). This was one of the major reasons why the interest in collecting wild *Lactuca* species emerged. Nowadays *Lactuca* accessions are common in the gene bank collections of many countries (Lebeda *et al.*, 2004). However, for the appropriate use of such collections in many areas (from botanical and taxonomical research to breeding purposes), it is necessary to investigate them and have the data about genetic diversity and population structure of the accessions preserved in the gene banks. Such information, together with passport data and phenotypic data about accessions provide a solid basis for the potential use of the collection for different studies that aim at breeding purposes. There is the possibility of crosses between some *Lactuca* species and in order to increase the benefit in the use of *Lactuca* collections there is a need for research on the phylogenetic relationships within this genus.

The European Union (EU)-funded GENE-MINE project was dealing with improved use of germplasm collections with the aid of novel methodologies for integration, analysis and presentation of genetic data and genetic diversity of wild crop relatives with the emphasis on exploitation and efficiency of *ex situ* conservation of *Lactuca* in gene banks. Within this project we were involved in the research on *Lactuca* collections with the use of molecular markers. In this chapter we present data on our research on 12 wild *Lactuca* species preserved in the gene banks. This is only part of a larger study regarding accessibility and redundancy of wild *Lactucas* in *ex situ* collection in order to promote a more effective use of the material preserved *ex situ*. Initially, accessions were chosen in order to study redundancies within collections. It turned out that within this potentially redundant material, the whole collections of some of the wild *Lactuca* species are contained. With the use of molecular AFLP markers (Vos *et al.*, 1995) we have studied diversity preserved within 12 *Lactuca* species and have drawn some conclusions regarding conservation of rarely collected wild species.

29.2 Materials and Methods

Seeds for the studies were provided from six gene banks: CGN, Wageningen (The Netherlands), HRIGRU, Wellesbourne (United Kingdom), IPK, Gatersleben (Germany), RICP, Olomouc (Czech Republic), USDA-ARS, Salinas and USDA-WG, Pullman (USA). Investigated species included: *L. aculeata* Boiss. & Kotschy ex Boiss., *L. altaica* Fisch. et C. A. Mey., *L. dentata* (Thunb.) C. B. Rob., *L. dregeana* DC., *L. indica* L., *L. livida* Boiss. et Reut., *L. perennis* L., *L. quercina* L., *L. saligna* L., *L. serriola* L., *L. tatarica* (L.) C. A. Mey. and *L. virosa* L.

AFLP analyses were performed with three primer combinations (Table 29.1) that were kindly provided to us by Keygene NV, Wageningen, the Netherlands. As many as 78 accessions with 20 individual plants each were grown in the greenhouse. Leaves were harvested individually and genomic DNA was isolated according to Doyle and Doyle (1990), modified for a robotic liquid handling system-based extraction in a 96-well plate. The AFLP procedure was performed according to Dehmer (2001) Pre-selective amplification (preamplification) was performed in two steps, first with primers with a total of two, then with three selective bases, respectively (Table 29.1), with identical protocol. Diluted, restricted and ligated DNA of 3.5 µl was added to 10 pmol each of *Eco*RI and *Mse*I primers, 200 pmol dNTPs, 2.25 nmol $Mg(OAc)_2$, 1 × PCR buffer, 0.3 U *Taq* polymerase (Eppendorf, Hamburg, Germany) in a total of 15 µl. After every preamplification, DNA was again diluted 1:50 in TE. For selective amplifications, three primer combinations were used with a total of seven selective nucleotides (Table 29.1; *Eco*RI primers fluorescence-labelled). The products of the three differently labelled selective amplifications were pooled, and fragment analysis was performed on the MegaBACE 1000 sequencer (Amersham Biosciences Europe, Freiburg, Germany), genotyping protocol.

Data were analysed with FRAGMENT PROFILER 1.2 software (Amersham Biosciences). Peaks in the size range from 70 to 415 base pairs were scored as present or absent, and a binary matrix was created. Similarity was determined according to Nei's original measures of genetic identity and genetic distance (Nei, 1972). Population parameters (G_{st}, H_t, H_s, D_{st}; Nei, 1987) and percentage of polymorphic loci were calculated with the POPGENE software (v1.32). Genetic distance matrices were loaded into NTSYS 2.1 software and principal coordinate analysis was conducted using DCENTER and EIGENVEC procedures. Analysis of molecular variance (AMOVA) was performed with WINAMOVA 1.55 software package (Excoffier *et al.*, 1992). Variance components were tested for significance by a non-parametric resampling approach using 1000 permutated data sets.

Table 29.1. List of primers and adaptors used. (Primer information provided by Keygene NV.)

Adaptors	Sequences
*Eco*RI	*5' CTCGTAGACTGCGTACC 3'*
	3' AATTGGTACGCAGTC 5'
*Mse*I	*5' GACGATGAGTCCTGAG 3'*
	3' TACTCAGGACTCAT 5'
Preamplification primers	
*Eco*RI + 1 primer	*E00 + A*
*Mse*I + 1 primer	*M00 + C*
*Mse*I + 2 primer	*M00 + CT*
Amplification primers	
*Eco*RI + 3 primer	*E00 + ACA*
*Mse*I + 4 primer	*M00 + CTAT*
*Mse*I + 4 primer	*M00 + CTTC*
*Mse*I + 4 primer	*M00 + CTTT*

29.3 Results

After processing the AFLP genotyping data by calculating pairwise genetic distances according to Nei (1972), a principal component analysis (PCO) plot was obtained (Fig. 29.1) that shows several species-specific clusters: *L. indica*, *L. perennis*, *L. saligna* and *L. tatarica* are each forming distinct groups, while the largest group is consisting of a mixture of *L. aculeata*, *L. altaica*, *L. dentata*, *L. dregeana* and *L. serriola* accessions (further referred to as '*serriola*-like' group). Adjacent to this group is a joint cluster of one accession of *L. livida* and *L. virosa* each (Fig. 29.1). Parameters on genetic diversity of the species involved in our studies are presented in Table 29.2. The largest number of polymorphic loci per species has been found in *L. serriola* accessions (38.7%), whereas only 15.7% loci have been polymorphic in *L. perennis*. The proportion of genetic differentiation between accessions of one species varies from 0.91 for *L. livida* to 0.26 for *L. indica*. Gene diversity within the species has the highest value for *L. indica* and *L. tatarica* (H_t = 0.12) and the lowest value for *L. dentata* (H_t = 0.03; Table 29.2). Considering gene diversity within accessions, very low diversity was present in accessions of *L. livida* and *L. saligna* (Table 29.2), which coincides with the latter findings of duplicates within these species (data not presented here).

There are accessions that are not clustering according to their taxonomic designation in the ILDB (International *Lactuca* Database, www.plant.wageningen-ur.nl/cgn/ildb). Most of them are placed within the *serriola*-like group. For these accessions, morphological analyses also suggest that a taxonomical redetermination should be done (data from colleagues in RICP, not presented here). These accessions are found among the following species: *L. dentata*, *L. dregeana*, *L. livida* and *L. quercina*. From *L. livida*, only one accession clusters with a *L. virosa* accession and not within *serriola*-like group. On a molecular basis,

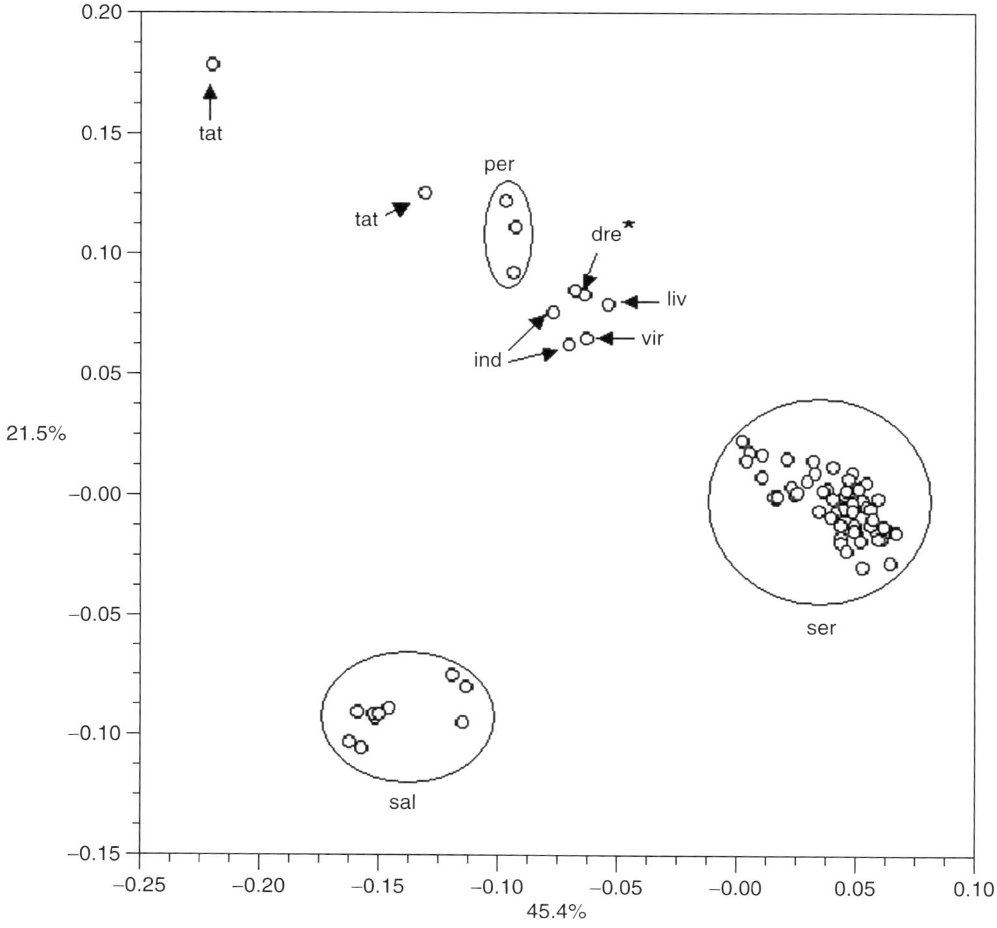

Fig. 29.1. PCO plot of different *Lactuca* species: dre = *L. dregeana*; ind = *L. indica*; liv = *L. livida*; per = *L. perennis*; sal = *L. saligna*; ser = *serriola*-like group; tat = *L. tatarica*; vir = *L. virosa*; * = incorrect species designation also confirmed morphologically.

three of the four accessions labelled as *L. virosa* in the database are related within some *serriola*-like entries; only one is clustering apart from the *serriola*-like group, together with the *L. livida* entry mentioned above.

From the species *L. aculeata*, *L. dentata*, *L. dregeana*, *L. livida* and *L. quercina* all available gene bank accessions were included in the analyses. In order to further assess genetic diversity of these species preserved *ex situ*, analysis of molecular variance was performed. In Table 29.3, components of molecular variance are compiled. Variance between accessions of *L. livida* and *L. dregeana* is much higher (90.4% and 84.8%, respectively) than variance within accessions, which is in accordance with the high G_{st} values for these species (0.91 and 0.90, respectively, Table 29.2). Within *L. aculeata* and *L. dentata*, variance within accessions is higher than variance between them (Table 29.3).

Table 29.2. AFLP-based genetic variability of wild *Lactuca* gene bank accessions.

Species	% of polymorphic loci	G_{st}	H_t ± SD	H_s ± SD	D_{st}
L. aculeata	19.1	0.41	0.06 ± 0.02	0.03 ± 0.006	0.03
L. altaica	20.2	0.61	0.05 ± 0.01	0.02 ± 0.004	0.03
L. dentata	12.6	0.27	0.03 ± 0.01	0.02 ± 0.005	0.01
L. dregeana	27.4	0.90	0.09 ± 0.02	0.009 ± 0.0009	0.08
L. indica	37.5	0.26	0.12 ± 0.03	0.08 ± 0.002	0.04
L. livida	27.7	0.91	0.07 ± 0.01	0.006 ± 0.0007	0.06
L. perennis	15.7	0.38	0.05 ± 0.02	0.03 ± 0.008	0.02
L. quercina	23.8	0.50	0.08 ± 0.03	0.04 ± 0.008	0.04
L. saligna	18.5	0.87	0.05 ± 0.01	0.006 ± 0.0004	0.04
L. serriola	38.7	0.82	0.06 ± 0.02	0.01 ± 0.006	0.05
L. tatarica	34.7	0.41	0.12 ± 0.03	0.07 ± 0.001	0.05
L. virosa	28.3	0.76	0.09 ± 0.02	0.02 ± 0.002	0.07

Note: SD = standard deviation.

Table 29.3. Molecular variance components within rare *Lactuca* species. All available accessions from the gene banks are included in the research. Analyses have been performed on two hierarchical levels: between accessions and within accessions belonging to a certain species. *P* values are derived from permutation tests and present probability of observing larger variance components at random.

Species	Variance component	
L. aculeata	Variance between accessions	5.93 (44.3%)
	Variance within accessions	7.45 (55.7%)
		$P < 0.001$
L. dentata	Variance between accessions	1.15 (18.51%)
	Variance within accessions	5.05 (81.49%)
		$P = 0.19$
L. dregeana	Variance between accessions	13.13 (84.8%)
	Variance within accessions	2.35 (15.19%)
		$P < 0.001$
L. livida	Variance between accessions	8.95 (90.4%)
	Variance within accessions	0.95 (9.6%)
		$P < 0.001$
L. quercina	Variance between accessions	12.83 (58.78%)
	Variance within accessions	9.00 (41.22%)
		$P < 0.001$

PCO (details not shown) demonstrate that within *L. livida*, three accessions seem to be differing from the others. The first two accessions are already described as *L. dregeana*, and the third is the only *L. livida* clustering with *L. virosa* outside of the *serriola*-like group (Fig. 29.1). Between the rest of the *L. livida* accessions there seems to be only a small genetic distance, and this is further confirmed by the existence of duplicates within these samples. In *L. dregeana*, there is a higher diversity between accessions. The species *L. aculeata*, *L. dentata* and *L. quercina* consist of 'more dispersed' accessions. In all cases, the first and second principal component already describe most of the diversity (at least 80%).

29.4 Discussion

In our studies on the relationships among *Lactuca* species, we have found that the examined accessions are clustered in a few groups. Species like *L. indica*, *L. perennis*, *L. saligna* and *L. tatarica* are forming separate clusters. One group consists of accessions of *L. aculeata*, *L. altaica*, *L. dentata*, *L. dregeana* and *L. serriola*. Koopman *et al.* (2001) referred to this group as *serriola*-like species, a term also used here. *L. serriola* is in our studies 'dispersed' among other *serriola*-like species. Two accessions representing *L. livida* and *L. virosa* are clustering adjacent to the *serriola*-like group, but not within it. Similar findings have been reported with the use of different marker systems by other authors (Kesseli and Michelmore, 1986; Kesseli *et al.*, 1991; De Vries and Van Raamsdonk, 1994; Hill *et al.*, 1996; Koopman *et al.*, 1998; Koopman *et al.*, 2001).

The clustering of putative *L. livida*, *L. quercina* and *L. virosa* accessions among the *serriola*-like group is mostly due to the wrong/incomplete taxonomical determination, confirmed by both the morphological and the molecular findings. The species status of *L. livida* is not well defined, accessions have been found to be *L. dregeana* or could not be completely determined on the species level. The position of two *L. quercina* accessions is not quite clear and not well supported; this group was not properly resolved. A reason for this might be the fact that we had problems with germination of one of the two accessions, so not enough plants were obtained and included in the analyses. Besides the mentioned species, incorrect determinations were found within *L. dentata*, *L. dregeana* and *L. serriola*.

Another aspect of our research is looking at some of the mentioned species as the whole germplasm resources available in *ex situ* collections – this applies to the five species: *L. aculeata*, *L. dentata*, *L. dregeana*, *L. livida* and *L. quercina*. These are rarely collected species with the small number of accessions maintained within the gene banks. This might lead to their incorrect determination, since they might be difficult to recognize due to the scarcity in collections. Ideally, genetic diversity within collections is distributed mainly between accessions, rather than within (Van Treuren *et al.*, 2001). *L. aculeata* accessions display similar variance distribution within and between accessions. The larger proportion of variability in *L. dentata* accessions is distributed within accessions, rather than between them. In *L. dregeana*, *L. livida* and *L. quercina*, the major part of variance exists between accessions, with a low variance within accessions, especially in *L. livida* and *L. dregeana*. This leads to the conclusion that accessions from these species are more homogeneous than *L. dentata* accessions. A more detailed analysis with PCO shows that *L. livida* collection forms three gene pools, one consisting of the accessions among which duplication has been confirmed, and two more distant pools (data not presented here). Considering the highly similar *L. livida* accessions, duplication within it probably occurred because of a frequent exchange of this rare species among the gene banks. We were expecting a similar situation with the other rarely collected *Lactuca* species, but it does not seem to be the case. Although the major proportion of variation detected in *L. dentata* is due to the variance within accessions rather than between them, PCO plots are showing that the two accessions from this group are distant enough not to be considered as duplicates. Variance within accessions of *L. dregeana* is showing one group of related accessions (with the exception of one accession), but still distinct from each other, so it probably is not the case that most of them resulted from the exchange of the same material.

Based on previous results on *Lactuca* species relations, as well as on suggestions from morphology and molecular findings on some of our accessions, we believe that some recommendations on wild *Lactuca* conservation in the gene banks can be given. The reliability of the taxonomic classification of these wild species proved to be low. If time and space allow it, the taxonomical status of accessions from collections of wild and rare species should be checked. Detailed analysis of accessions that represent the entire *ex situ* gene pool of one species provides gene bank curators with information on the amount of diversity within such material, and how unique the material in their gene bank is. Rarely collected species are likely to be more frequently exchanged among the gene banks, so the probability of redundancies within these accessions is high. The amount of diversity residing within and between accessions of inbreeding species is also important to know. When variability within an accession is higher than variability between different accessions, this might hint at paying a higher attention to the storage and reproduction protocols. It could also give indications for future collecting activities and here, together with the data on diversity within the collected species, the knowledge on the natural distribution of the species would be of high importance. Decisions on increasing or decreasing the number of wild accessions obviously depend on a large number of criteria, and will largely depend on the funding status of a gene bank and the priority of the species concerned. Knowledge about the true taxonomic identity of the accessions and the structure of the gene pool, such as generated in this study, can support the difficult decisions the gene bank curator has to make in the attempt to better serve the end-users.

Acknowledgements

This work is a part of the EU-funded GENE-MINE project (QLK5-CT-2000-00722). The authors would like to thank Keygene NV, Wageningen, the Netherlands, for providing information about the primer combinations and to all gene banks that were mentioned in the chapter, for providing us with the seed material. We would also like to thank to Dr Ales Lebeda, Dr Eva Kristkova from Palacky University, Czech Republic for providing us data from their morphological analyses and Dr Theo van Hintum, CGN, Wageningen, the

Netherlands, Dr Elena Potokina, School of Biosciences, University of Birmingham, United Kingdom, Dr Frank Blattner and Dr Andreas Graner, IPK, Gatersleben, Germany, for fruitful discussions.

References

De Vries, I.M. and Van Raamsdonk, L.W.D. (1994) Numerical morphological analysis of lettuce cultivars and species (*Lactuca* sect. *Lactuca*, Asteraceae). *Plant Systematics and Evolution* 193, 125–141.

Dehmer, K.J. (2001) Conclusions on the taxonomy of the *Solanum nigrum* complex by molecular analyses of IPK germplasm accessions. In: Van den Berg, R.G., Barendse, G.W.M., Van der Weerden, G.M. and Mariani, C. (eds) *Solanaceae V: Advances in taxonomy and utilization*. University Press, Nijmegen, The Netherlands, pp. 85–96.

Doyle, J.J. and Doyle, J.L. (1990) Isolation of plant DNA from fresh tissue. *Focus* 12, 13–15.

Excoffier, L., Smouse, P.E. and Quattro, J.M. (1992) Analysis of molecular variance inferred from metric distances among DNA haplotypes: application to human mitochondrial DNA restriction data. *Genetics* 131, 479–491.

Hill, M., Witsenboer, H., Zabeau, M., Vos, P., Kesseli, R. and Michelmore, R. (1996) PCR-based fingerprinting using AFLPs as a tool for studying genetic relationships in *Lactuca* spp. *Theoretical and Applied Genetics* 93, 1202–1210.

Jeuken, M. and Lindhout, P. (2002) *Lactuca saligna*, a non-host for lettuce downy mildew (*Bremia lactucae*), harbors a new race-specific *Dm* gene and three QTLs for resistance. *Theoretical and Applied Genetics* 105, 384–391.

Kesseli, R.V. and Michelmore, R.W. (1986) Genetic variation and phylogenies detected from isozyme markers in species of *Lactuca*. *Journal of Heredity* 77, 324–331.

Kesseli, R., Ochoa, O. and Michelmore, R. (1991) Variation at RFLP loci in *Lactuca* spp. and origin of cultivated lettuce (*L. sativa*). *Genome* 34, 430–436.

Koopman, W.J.M., Guetta, E., Van de Wiel, C.C.M., Vosman, B. and Van den Berg, R. G. (1998) Phylogenetic relationships among *Lactuca* (Asteraceae) species and related genera based on ITS-1 DNA sequences. *American Journal of Botany* 85, 1517–1530.

Koopman, W.J.M., Zevenbergen, M.J. and Van den Berg, R.G. (2001) Species relationships in *Lactucas*. I. (Lactucaceae, Asteraceae) Inferred from AFLP fingerprints. *American Journal of Botany* 88, 1881–1887.

Lebeda, A., Doležalová, I. and Astley, D. (2004) Representation of wild *Lactuca* wpp. (Asteraceae, Lactuceae) in world Genebank collections. *Genetic Resources and Crop Evolution* 51, 167–174.

Nei, M. (1972) Genetic distance between populations. *American Naturalist* 106, 283–292.

Nei, M. (1987) *Molecular Evolutionary Genetics*. Columbia University Press, New York.

Sicard, D., Woo, S.-S., Arroyo-Garcia, R., Ochoa, O., Nguyen, D., Korol, A., Nevo, E. and Michelmore, R. (1999) Molecular diversity at the major cluster of disease resistance Genes in cultivated and wild *Lactuca* spp. *Theoretical and Applied Genetics* 99, 405–418.

Van Treuren, R., Van Soest, L.J.M. and Van Hintum, Th.J.L. (2001) Marker-assisted rationalisation of genetic resource collections: a case study in Flax Using AFLPs. *Theoretical and Applied Genetics* 103, 144–152.

Vos, P., Hogers, R., Bleeker, M., Reijans, M., Van de Lee, T., Hornes, M., Frijters, A., Pot, J., Peleman, J., Kuiper, M. and Zabeau, M. (1995) AFLP: a new technique for DNA fingerprinting. *Nucleic Acids Research* 23, 4407–4414.

30 The Role of Botanic Gardens in the Conservation of Crop Wild Relatives

S. SHARROCK AND D. WYSE-JACKSON

30.1 Introduction

Crop wild relatives (CWR) include taxa that are closely related to the species of direct socio-economic importance as well as to the ancestors of modern crops. Genes from CWR make a direct contribution to increasing the quantity and quality of our food supply and the species themselves form a vital part of both natural and agricultural ecosystems. Promoting the conservation of wild crop relatives constitutes one of the 20 agreed activities of the Food and Agriculture Organization's *Global Plan of Action for the Conservation and Sustainable Utilization of Plant Genetic Resources for Food and Agriculture* (FAO, 1996), is an important component of the implementation of the *International Agenda for Botanic Gardens in Conservation* (Wyse Jackson and Sutherland, 2000) and contributes to several targets of the *Global Strategy for Plant Conservation* (Targets 8, 9 and 13) (CBD, 2002).

Botanic gardens play a major role in the conservation of plant genetic resources. There are over 2500 botanic gardens in existence worldwide and collectively they contain over six million plant accessions and an estimated 80,000 plant species (Wyse Jackson, 1999). Many botanic gardens are playing an active role in both the *in situ* and *ex situ* conservation of CWR.

Botanic Gardens Conservation International (BGCI) has a particular focus on the conservation of wild plant species. As a partner in a GEF-funded project on the *in situ* conservation of CWR entitled '*In situ* conservation of crop wild relatives through enhanced information management and field application', BGCI is already playing a role in the conservation of CWR (Lane, 2005). BGCI's role in this project focuses on providing information on CWR that are conserved in botanic garden collections through the information portal that is being developed by the project. Through its extensive network of botanic garden partners, BGCI also aims to raise awareness about the importance of CWR and the need for the long-term conservation of such valuable germplasm.

30.2 Survey of CWR in Botanic Garden Collections

Socio-economically important plant species include food, fodder and forage crops, medicinal plants, spices, ornamental and forestry species, as well as plants used for industrial purposes, such as oils and fibres. Many of these species, especially medicinal and ornamental plants, are widely grown in botanic gardens and form an important part of the *ex situ* conservation collections of such gardens. The role that botanic gardens are playing in the conservation of wild relatives of major food crops however is less clear This chapter provides the results of an initial investigation into the conservation of wild relatives of food crops by botanic gardens.

For this survey, only those crops included in Annex 1 of the *International Treaty on Plant Genetic Resources for Food and Agriculture* were considered (CGRFA, 2001) (Table 30.1). The Treaty, which came into force in June 2004, aims to ensure that plant genetic resources for food and agriculture, which are vital for human survival, are conserved and sustainably used and that benefits from their use are equitably and fairly distributed. The Treaty represents a multilateral system of facilitated access and benefit sharing for the crops and forages most important for food security. The crops listed in Annex 1 are those considered not only to be of highest value for food security but are also those for which there is a high degree of interdependence among countries with respect to their genetic diversity.

In order to carry out the survey, two main databases were consulted: BGCI's PlantSearch database (http://www.bgci.org/conservation/plant_search.html), which includes approximately 130,000 taxa from over 600 botanic gardens worldwide, and the System-wide Information Network for Genetic Resources (SINGER) database maintained by the International Plant Genetic Resources Institute (IPGRI) (http://singer.grinfo.net/).

30.3 CWR Species in Botanic Garden Collections

The survey revealed that species of all 50 genera included in Annex 1 of the *International Treaty* are present in botanic garden collections, and in some cases large numbers of species are recorded. For example, 107 species of breadfruit (*Artocarpus* Forst.), 82 species of *Lathyrus* L. and 122 species of the *Brassica* L. complex are listed in the database (Table 30.1).

A comparison was made with the number of species recorded in the SINGER database for the same set of species. It can be seen that in many cases the two databases are complementary, in that a number of genera with large numbers of species recorded in botanic garden collections have few species recorded in the SINGER database. Taking the examples listed above, it can be seen that SINGER includes only five species of *Artocarpus* Forst., 46 species of *Lathyrus* and 33 *Brassica* complex species. In other cases, many more species are recorded in SINGER than in BGCI's database. For example BGCI records only 85 species of *Ipomoea*, while SINGER has 340 and records for *Vigna* Savi. are 12 and 88 species respectively.

Table 30.1. CWR in botanic garden collections (PlantSearch) and the SINGER database based on the crops included in Annex 1 of the International Treaty on Plant Genetic Resources for Food and Agriculture compared with the number of species in each genera given in the Plant Book (Mabberley, 1997) (see text for further information). (*Brassica* genera include *Brassica* L., *Armoracia* Gilib., *Barbarea* R.Br., *Camelina* Crantz., *Crambe* L., *Diplotaxis* D.C., *Eruca* Mill., *Isatis* L., *Lepidium* L., *Raphanobrassica, Raphanus* L., *Rorippa* Scop. and *Sinapis* L.)

		Number of species			Number of rare and threatened species				Number of gardens with special collections
Crop	Genus	Mabberley (1997)	Plant-Search	SINGER	IUCN, 2004	IUCN 1997 (Walter and Gillett, 1998)	PlantSearch/ IUCN 2004	PlantSearch/ IUCN 1997	
Breadfruit	*Artocarpus* Forst.	50	107	5	5	5	1	1	5
Asparagus	*Asparagus* L.	140	86	2	1	13	0	6	2
Oat	*Avena* L.	25	14	19	0	3	0	0	2
Beet	*Beta* L.	13	5	10	0	5	0	2	1
Brassica	13 genera	370	122	33	2	118	0	31	7
Pigeon pea	*Cajanus* DC.	37	2	19	0	2	0	0	0
Chickpea	*Cicer* L.	40	16	43	0	6	0	2	1
Citrus	*Citrus* L.	16	18	21	1	3	0	0	20
Coconut	*Cocos* L.	1	9	1	0	0	0	0	3
Major aroids	*Colocasia* Schott.	8	11	2	0	0	0	0	3
	Xanthosoma Schott.	57	25	3	2	1	0	0	
Carrot	*Daucus* L.	22	7	1	0	3	0	0	0
Yams	*Dioscorea* L.	850	60	72	4	68	0	4	9
Finger millet	*Eleusine* Gaertn.	9	4	6	0	0	0	0	0
Strawberry	*Fragaria* L.	12	16	1	0	1	0	0	4
Sunflower	*Helianthus* L.	50	36	18	0	18	0	1	1
Barley	*Hordeum* L.	20	17	26	0	2	0	0	1
Sweet potato	*Ipomoea* L.	650	85	340	1	45	0	0	0
Grass pea	*Lathyrus* L.	160	82	46	0	24	0	2	2
Lentil	*Lens* Mill.	4	2	6	0	0	0	0	0

Continued

Table 30.1. *Continued*

Crop	Genus	Number of species			Number of rare and threatened species				Number of gardens with special collections
		Mabberley (1997)	Plant-Search	SINGER	IUCN, 2004	IUCN 1997 (Walter and Gillett, 1998)	PlantSearch/ IUCN 2004	PlantSearch/ IUCN 1997	
Apple	*Malus* Mill.	55	62	20	3	5	1	3	48
Cassava	*Manihot* Mill.	98	15	104	0	69	0	1	2
Banana	*Musa* L.	35	51	22	0	3	0	1	13
Rice	*Oryza* L.	18	5	27	0	3	0	0	0
Pearl Millet	*Pennisetum* Rich.	130	23	32	0	5	0	0	0
Beans	*Phaseolus* L.	36	28	50	3	2	0	1	0
Pea	*Pisum* L.	2	2	4	0	0	0	0	1
Rye	*Secale* L.	3	6	4	0	5	0	1	1
Potato	*Solanum* L.	1700	190	250	43	129	1	9	1
Aubergine	*Solanum* L.								1
Sorghum	*Sorghum* Moench.	24	15	19	0	2	0	0	0
Triticale	*Triticosecale* Wittmack.		0	0	0		0	0	0
Wheat	*Triticum* L.	4	23	17	0	3	0	2	3
	Agropyron Gaertn.	15	9	17	0	9	0	0	
	Elymus L.	150	36	43	0	11	0	0	
Faba bean/ vetch	*Vicia* L.	140	77	75	0	23	0	5	0
Cowpea	*Vigna* Savi.	150	12	88	0	4	0	0	0
Maize	*Zea* L.	4	5	4	0	3	0	1	0
Total		5098	1283	1453	65	593	3	73	131

It can be seen from Table 30.1 that botanic garden collections hold a total of 1283 species of selected crop plants – this compares with 1453 species listed in SINGER – a database that contains only crop data. Given that the BGCI PlantSearch database presently holds records for only 600 or so gardens, out of the over 2400 gardens that exist in the world, it is clear that botanic gardens are an important source of CWR germplasm.

30.4 Rare and Threatened CWR

The direct link between the BGCI PlantSearch database and the IUCN Red Lists from 1997 (Walter and Gillett, 1998) and 2004 (IUCN, 2004) allowed an analysis to be made of how many rare and threatened CWR species are included in botanic garden collections. As shown in Table 30.1, according to the 1997 data, a total of 73 rare and threatened species can be identified in botanic garden collections out of a total of 593, whereas using the 2004 data (based on changed IUCN Red Listing criteria), this falls to only three species out of 65. This reflects the relatively limited capture of data on the global conservation status of plant species post 1997. IUCN is currently addressing the need to increase the rate of plant Red Listing and BGCI is becoming increasingly involved in this activity. It will be important to prioritize useful plant species for Red Listing as recognized by IUCN and other partners in the Global Partnership for Plant Conservation (GPPC).

30.5 Other Roles of Botanic Gardens in the Conservation of CWR

Botanic garden collections can be a useful source of plants that are of local importance, even if not listed as rare and threatened. It can be seen from Table 30.1 that nine botanic gardens have yams included in special collections. These include species such as *Dioscorea dumetorum* Kunth., *D. hispida* Dennst. and *D. pentaphylla* L., species that are used in times of famine. Other yam species found in botanic garden collections include *D. floribunda* Mart.& Gal. and *D. balcanica* Kosanin. (a European species) that are useful sources of the steroid diosgenin – a source material for oral contraceptives.

A number of botanic gardens around the world are involved in extensive research and conservation on crop species. These include, for example, the Fairchild Botanical Garden in Florida, USA, which maintains an extensive collection of mango germplasm. As well as conserving mango diversity, Fairchild works to raise public awareness about this diversity through its annual mango festival and is working on the commercial development of the crop.

Other botanic gardens with special crop-based programmes include:

- The National Tropical Botanic Garden, Hawaii, USA – maintains a breadfruit germplasm collection.
- Wuhan Botanic Garden, China – kiwi conservation (*Actinidia* Lindl.) (62 of 66 species are in China), grape conservation and breeding.

- Jardín Botánico de Chacras de Coria, Mendoza, Argentina – wild populations of tomatoes and potatoes, *Solanum ruiz-lealii* Brucher., *Solanum kurtzianum* Bitter.
- Proyecto Jardín Botánico de la Ciudad Universitaria, Argentina – collections of *Phaseolus vulgaris* var. *vulgaris* L. and its wild relative *P. vulgaris* var. *aborigineus* (Burk.) C.Baudet.
- Jardín Agrobotánico, Universidad Nacional de la Plata, Buenos Aires, Argentina – research and breeding of maize using its wild relatives.
- The National Botanic Garden (NBG) of Belgium – a seed bank of wild *Phaseolineae* recognized as a base collection by IPGRI. The NBG has also extracted ethnobotanical knowledge from the herbarium specimens of Central Africa belonging to two economically important families (*Cucurbitaceae*, *Leguminosae*) as a prototype project.
- Portuguese botanic gardens have developed best practices for the conservation and sustainable use of the herbs *Mentha cervina* L., *Mentha pulegium* L. and *Thymbra capitata* (L.) Cav.

6 Conclusions

It is clear that botanic gardens are playing an important role in the conservation of a wide range of CWR. This includes not only the conservation of diversity, but also research and breeding to provide new crops and raising public awareness about the importance of CWR. Botanic gardens are also important players in the overall task of conserving CWR through the horticultural and taxonomic expertise they can provide and in many cases as repositories of indigenous knowledge – especially about the crops and their relatives that grow in the locality of the garden.

References

CBD (2002) *Global Strategy for Plant Conservation (GSPC)*. Secretariat of the Convention on Biological Diversity, Montreal, Canada.

Commission on Genetic Resources for Food and Agriculture (CGRFA) (2001) *International Treaty on Plant Genetic Resources for Food and Agriculture*. Available at: http://www.fao.org/ag/cgrfa/itpgr.htm (accessed 17 November 2005)

FAO (1996) *Global Plan of Action for the Conservation and Sustainable Utilisation of Plant Genetic Resources for Food and Agriculture*. Food and Agriculture Organization of the United Nations, Rome, Italy.

IUCN (2004) *2004 IUCN Red List of Threatened Species*. Available at: http://www.redlist.org/ (accessed 17 November 2005)

Lane, A. (2005) A global initiative to conserve crop wild relatives *in situ*. *BG Journal* 2(2), 13–15.

Mabberley, D.J. (1997) *The Plant-Book*, 2nd edn. Cambridge University Press, Cambridge.

Walter, K.S. and Gillett, H.J. (eds) (1998) *1997 IUCN Red List of Threatened Plants*. Compiled by the World Conservation Monitoring Centre. IUCN – The World Conservation Union, Gland, Switzerland and Cambridge.

Wyse Jackson, P.S. (1999) Experimentation on a large scale – an analysis of the holdings and resources of botanic gardens. *BGCNews* 3(3), 27–30.

Wyse Jackson, P.S. and Sutherland, L.A. (2000) *International Agenda for Botanic Gardens in Conservation*. Botanic Gardens Conservation International, London.

31 A National Italian Network to Improve Seed Conservation of Wild Native Species ('RIBES')

C. Bonomi, G. Rossi and G. Bedini

31.1 Introduction

The conservation of biological diversity is a major challenge which many nations have committed to solve by signing and ratifying the Convention on Biological Diversity (CBD) adopted in Rio in 1992. Italy ratified it in 1994 (ref. L.124 dated 14 February 1994). The CBD Conference of Parties (COP) at its meeting in April 2002 adopted the 2010 biodiversity target (Decision VII/26) of: '*to achieve by 2010 a significant reduction in the current and continuing rate of biodiversity loss at the global, regional, and national level as a contribution to poverty alleviation and to the benefit to all life on earth*'. This target was endorsed by ministers responsible for CBD implementation during a ministerial round-table discussion in April 2002 (Hague Ministerial Declaration), and was endorsed by world leaders during the World Summit on Sustainable Development in September 2002 (WSSD Plan of Implementation). The COP also, at its 6th meeting in The Hague, adopted a Global Strategy for Plant Conservation (GSPC) (Decision VI/9). This strategy calls for concerted actions to halt the current and continuing loss of plant diversity, by providing a framework to facilitate harmony between existing initiative aimed at plant conservation in line with targets set by WSSD agreed in Johannesburg in 2002. The GSPC recommends 16 targets that the parties to the CBD should endeavour to attain by 2010. These targets are divided into five groups dealing with understanding and documenting plant diversity, conserving plant diversity, sustainable uses of plant diversity, promoting education and awareness and building capacity. *Ex situ* conservation is considered as an important complementary approach to *in situ* conservation and target 8 of the GSPC urges all contracting parties to conserve 60% of threatened plant species in accessible *ex situ* collections preferably in the country of origin and include 10% of them in recovery and restoration programmes by 2010. At the European level, the European Plant Conservation Strategy (EPCS), also adopted in 2002 by the

Council of Europe, also targets *ex situ* conservation and raises this target to 80% of the threatened plant species.

As party to the CBD, Italy developed measures to meet its obligations to these initiatives, by preparing a National Action Plan for Biodiversity Conservation, which is still at its preliminary planning phase and has not yet been approved or implemented. The only document so far approved is a 'Strategic Guidelines and preliminary plan for the implementation of the CBD in Italy' (Linee strategiche e programma preliminare per l'attuazione della Convenzione della Biodiversità in Italia – Del. CIPE 16/04/1994). However, other actions are in place to conserve endangered species and natural habitats at national level, including the management of a national network of protected areas, covering approximately 10% of the national territory (L. n. 394/1991). A network of SCI (Sites of Communitarian Interest) has been developed on a regional basis as part of the implementation of the EU Habitat Directive (92/43/CEE) in the Natura 2000 framework. Moreover the Ministry of the Environment is currently selecting and appointing National Centres for Forestry Biodiversity Conservation. Still, additional actions are needed to attain an effective national plant conservation strategy for the endangered native flora whose extinction risk is still severe (Conti *et al.*, 1992, 1997; Walter and Gillet, 1998; APAT, 2003; Scoppola *et al.*, 2003; Scoppola and Caporali, 2005). As far as *ex situ* plant conservation is concerned, the second report on the implementation of the CBD produced in 2001 by the Italian CBD focal point, the Ministry for the Environment, Directorate for Nature Protection (MATT, 2001), made reference to individual actions for species protection undertaken by some Italian Botanic Gardens such as conservation programmes for the endangered *Zelkova sicula* Di Pasquale, Garfi et Quezel and *Abies nebrodensis* (Lojac.) Mattei in Sicily and *Marsilea quadrifolia* L. in Emilia-Romagna. These activities are part of the statutory duties of botanic gardens as stated in the Action Plan for Botanic Gardens in the European Union, which recommends a correct management of *ex situ* collections for conservation purposes (Target C2). Although these actions are commendable, there is a lack of coordination at the national level to ensure that the resources are being utilized efficiently and the experiences being generated by individuals are being shared among the plant conservation community and the stakeholders involved in the conservation of biological diversity at the national level. Consequently networks at national, regional and international levels should be promoted to coordinate plant conservation activities as recommended by target 16 of the GSPC.

In Europe, there are already national networks especially dedicated to *ex situ* plant conservation. In Spain REDBAG (RED de BAncos de Germoplasma) is a network that coordinates *ex situ* conservation activities of Spanish Botanic Gardens. The CBN (Reseau de Conservatoires Botaniques Nationaux) in France is a governmental network which is in charge of both *ex situ* and *in situ* plant conservation. At regional level in Europe two EU-funded network projects are specifically focused on *ex situ* conservation and seed banking of native threatened plants. The first project is called ENSCONET and is funded in the VI Framework Programme as an Integrated Infrastructure Initiative (the European Native Seed CONservation NETwork, 6th FP III-CA,

RICA-CT-2004-506109). It includes 24 partners from 17 EU countries and the Italian representatives are the universities of Pavia and Pisa and Trento Natural History Museum. Another European project (GEN-MEDOC) is funded in the Interreg IIIB funding framework (Création d'un réseau de centres de conservation du matériel génétique de la flore des régions méditerranéennes de l'espace MEDOCC, FESR-Interreg III B, 4.3.20031-E-060) active for the western Mediterranean where Italy is represented by the universities of Catania in Sicily and Cagliari in Sardinia.

31.2 The Italian Perspective: a National Network for Seed Conservation

In Italy, in the absence of a national coordination mechanism, many local administrations created centres for *ex situ* plant and seed conservation. Among the others are Piedmont (act LR 22/83); Lombardy (act DGR VII/16038, 16.1.2004); Trentino (act del. PAT 1159, 24.5.2002); Tuscany (act LR 56/2000 del 1175, 22.11.2004); Livorno (act dec.106 del, 12.10.2004); Cagliari district in Sardinia (act LP n. 2037, 16.05.1997); Sicily (act POR Sicilia 2000–2006 dec. C(2000) 2348, 8.08.2000); and Palermo district (act del. 0207/24, 27.2.1998). In this context, various institutions created local gene banks, receiving substantial financial supports that enable them to create state-of-the-art conservation facilities.

In order to create a better coordination of *ex situ* conservation activities in Italy, an 'Italian Seed Bank Network for the *ex situ* conservation of the Italian native flora' was created and named RIBES (Rete Italiana Banche del germoplasma per la conservazione *ex situ* della flora spontanea italiana).

The idea to create a national seed bank network originated during a national workshop on seed banking techniques (Seed banks as a conservation tool for native species. Building a national network in an European perspective) held in Trento, north-east Italy, in April 2004 (Bonomi *et al.*, in press). This meeting provided an opportunity for a good discussion among 14 Italian groups working on seed conservation as well as guest speakers who presented case studies from other European institutions including the Royal Botanic Gardens Kew (United Kingdom), the botanic garden of Cordoba and Gran Canaria (Spain), the Conservatoire Botanique National Alpin, Gap (France), the Botanic Gardens of Warsaw (Poland) and IPGRI (Rome). The seed bank network was further developed during other meetings held in Rome (October 2004), Milan (December 2004) and Pavia (February 2005).

31.3 Purpose and Structure of RIBES

The founder group agreed on the purpose of the network, and its formal organization. It was decided that RIBES would focus on native plant conservation and would promote all necessary collaborative actions needed for the *ex situ* conservation of two different but correlated groups of plants:

1. Native species threatened with extinction at international, national and local level according to international, national and local legislation and scientific documents;
2. Native species particularly important from a biogeographical and ecological point of view that might be utilized for land stabilization and habitat restoration projects.

From a formal point of view RIBES will be a scientific association based on a participative approach and a democratic decision-making process and will be registered as a non-profit organization. Its activities will be regulated by a charter and various internal regulations.

The potential members of the network committed themselves to implement these decisions by signing a Memorandum of Understanding aiming to formally constitute RIBES in a year's time. The MoU was signed in Pavia on 9 February 2005 at the grand opening of the Lombardy Seed Bank in Pavia Botanic Garden in the presence of the representatives of the CBD national focal point, the National Environmental Protection Agency, the Italian Botanical Society, Italian National Parks Federation, Royal Botanic Gardens Kew, regional council of Lombardy and Tuscany. RIBES MoU is structured in three parts: an introduction specifying the scientific and legal justification of the proposed action (reference to inventories of threatened species, reference to local, national and EU legislation, international conventions and agreements); a short description of the network focus; a synoptic reference to its aims and objectives; and a time limit and implementation plan to bind all contracting parties to the formal constitution of the association in front of a public officer within a year of the memorandum signature. The MoU also contains the full contact details of participating institutions and their legal representatives.

RIBES was formally constituted on 3 December 2005 in Trento, with the official signature of the Constitutive Act in front of a public officer by the legal representatives of 18 founding members.

31.4 Objectives of RIBES

The specific objectives of RIBES as included in its statute and approved by all founding members are the following:

1. Promote the dissemination at local and national level of the knowledge on critical issues and state-of-the-art facilities and operating procedures for *ex situ* native plant conservation by means of newsletters, congresses, workshops and various other events;
2. Set and update minimum standards to be adopted for the proper management of *ex situ* conservation programmes.
3. Make sure, as far as possible, that *ex situ* collections are managed and conserved according to internationally approved standards.
4. Disseminate, as far as possible, information on EU and national programmes centred on *ex situ* plant conservation.
5. Create a national register of the species that are currently conserved *ex situ*.

6. Contribute to other programmes and initiatives for the conservation of biological diversity.
7. Develop research activities to gain a better understanding of *ex situ* conservation and plant propagation techniques that might be used in reintroduction projects.
8. Set up and put in operation specific information systems to document native germplasm collections in order to certify their origin.
9. Develop at local and national level specific education programmes aimed at schools and at the wider pubic to raise awareness on the importance of *ex situ* conservation of biological diversity.
10. Promote training activities on *ex situ* conservation of native species.
11. Cooperate with other institutions having similar aims.

These objectives will be implemented through an action plan and specific working groups that will address specific issues such as seed collection, seed curation and germination, data management and dissemination. The working groups will be run with a participative approach and will adopt a national perspective setting priorities for conservation at national level. They will discuss best practice and operating protocols and set minimum and recommended standards for *ex situ* conservation of wild species. Although the network will not focus primarily on crop wild relatives, those included in the Italian native flora will be included. Nevertheless it aims to establish active links with the crop wild relative conservation community in order to integrate possible overlapping areas of interest, producing mutual benefits.

31.5 Members of RIBES

Members of the network are mainly university botanic gardens but also include local governmental agencies, national parks, non-profit organizations and private companies. They represent most Italian regions and include key members that are already involved in other EU networks such as ENSCONET and GENMEDOC, providing in this way an active connection with the European context. The full member list is in Appendix 1.

31.6 Conclusions

RIBES will liaise with relevant stakeholders that hold key information on the conservation status of the threatened species such as the Italian Botanical Society and the Botanic Gardens Community (Scoppola *et al.*, 2003; Scoppola and Caporali, 2005). RIBES will also bridge the gap between scholars of plant and seed science and plant conservation managers based in natural parks and protected areas providing them with means, techniques and opportunities to experiment plant reintroductions and populations reinforcements. RIBES will seek to operate in close connection with the CBD Focal Point, the Ministry of the Environment, offering it a powerful means to implement the CBD and the GSPC. RIBES therefore plans to contribute effectively to the national implementation of several GSPC targets. It is hoped that it will also activate transnational cooperation with other nations in biogeographical units, e.g. Alpine and Mediterranean regions.

Acknowledgements

The authors wish to thank the Italian National Biodiversity Focal Point Ministry of the Environment (MATT-DCN, Roma); Ms B. Piotto (APAT), Region Lombardy; Trento Natural History Museum; Region Tuscany; Professor F. Garbari (Pisa), Professor P. Grossoni (Firenze), Professor M. Mariotti (Genova), Mr S. Linington (Millennium Seed Bank, Royal Botanic Gardens, Kew), Ms V. Dominione (Pavia).

References

APAT (2003) *Environmental Data Yearbook*. APAT, Agency for Environmental Protection and Technical Services, Rome. Available at: http://www.sinanet.apat.it

Bonomi, C., Rossi, G. and Bedini, G. (eds) (in press) Atti del Convegno 'Banche del Germoplasma: uno Strumento per la Conservazione. Verso una Rete Nazionale in Prospettiva Europea'. Trento, 1–2 Aprile 2004. *Studi Trent. Sci. Nat., Acta Biol.* Suppl. 81 (in stampa).

Conti, F., Manzi, A. and Pedrotti, F. (1992) *Libro Rosso delle Piante d'Italia*. Tipar, Rome, Italy.

Conti, F., Manzi, A. and Pedrotti, F. (1997) *Liste Rosse Regionali delle Piante d'Italia*. Tipar, Rome, Italy.

MATT (2001) *National Report to the Convention on Biological Diversity*. Rome, Italy. www.minambiente.it

Scoppola, A. and Caporali,C. (2005) Le Specie Vulnerabili, Endemiche e Rare della Flora Vascolare Italiana. In: Blasi C. (ed.) *Sistema Biodiversità Italia*. SBI – Commissione per la Promozione della Ricerca Botanica, MATT, Direzione per la Conservazione della Natura.- DCN. (In stampa).

Scoppola, A., Caporali, C., Gallozzi, M.R. and Blasi, C. (2003) Aggiornamento delle Conoscenze Floristiche a Scala Nazionale: Commenti e Primi Risultati. *Inform. Bot. Ital* 35(1), 178–197.

Walters, K.S. and Gillet, H.J. (eds) (1998) *1997 IUCN Red List of Threatened Plants*. IUCN, Gland, Switzerland and Cambridge.

Appendix

RIBES members, listed in geographical order from north to south, the contact person is in brackets.

1. Germplasm Bank of South-western Alps Cuneo natural parks management agency (B. Gallino)
2. Lombardy Seed Bank, Lombardy centre for the Native Flora (G. Rossi)
3. Trentino Seed Bank, Trento Natural History Museum (C. Bonomi)
4. Germplasm Bank of Padua Botanic Garden, University of Padua (G. Cassina)

5. Laboratory for the conservation of Liguria plant diversity, Hanbury Botanic Gardens, University of Genoa (S. Giammarino)
6. Germplasm Bank of Pisa Botanic Garden, University of Pisa (G. Bedini)
7. Livorno Germplasm Bank, Livorno District Council (M. Lupi)
8. Germplasm Bank for the conservation of anfi-adriatic species, Polytechnic University of Marche (E. Biondi)
9. Germplasm Bank of Viterbo Botanic Garden, Tuscia University (A. Scoppola)
10. Germplasm Bank of Rome Botanic Garden, University of Rome La Sapienza (A. Scoppola)
11. Germplasm Bank of the Central Appennine, National Park Gran Sasso and Laga (D. Di Santo)
12. Germplasm Bank of Majella National Park (M. Di Cecco)
13. Germplasm Bank of Molise, University of Molise (A. Stanisci)
14. Germplasm Bank of CODRA Mediterranea s.r.l. (E. Lanzillotti)
15. Germplasm Bank of Sardinia, University of Cagliari (G. Bacchetta)
16. Germplasm Bank of Palermo Botanic Garden, University of Palermo (A. Scialabba)
17. Germplasm Bank of Catania Botanic Garden, University of Catania (P. Pavone)
18. Germplasm Bank of the Mediterranean ONLUS (I. Li Vigni)

32 Linking *In Situ* and *Ex Situ* Conservation with Use of Crop Wild Relatives

N. Maxted and S.P. Kell

32.1 Introduction

The goal of plant genetic conservation is primarily direct use through exploitation for crop improvement. We expend resources on the maintenance of genetic, species and ecosystem diversity because of their immediate or potential utilization value to humankind. This intimate link between plant genetic diversity, conservation and utilization is acknowledged in the objectives of the Convention on Biological Diversity (CBD, 1992). There are two basic strategies for conservation of plants: *in situ* and *ex situ*. *In situ* conservation focuses on conserving plant diversity where it is currently found, while *ex situ* conservation involves the movement of the plant diversity to another location where it is conserved. Historically, at least for plant genetic conservation, much greater priority has been placed on *ex situ* conservation; but recently, encouraged by the emphasis of the CBD, *in situ* conservation activities have been given higher priority and *ex situ* conservation has begun to be considered more as a back-up for the more desirable maintenance of biodiversity where it is currently found. Ford-Lloyd and Maxted (1993) argued that the two strategies should not be seen as being in opposition to one another, rather that they should be seen as complementing each other to provide maximum preservation of plant diversity. This chapter investigates the potential linkages between *in situ* and *ex situ* conservation, how *ex situ* conserved germplasm, particularly of crop wild relatives (CWR), might be better used in support of *in situ* and ecosystem conservation, and how *in situ* conserved germplasm might be better utilized.

Both the complementary nature of *in situ* and *ex situ* conservation and the desire to link conserved diversity to use are emphasized in the methodology for plant genetic conservation (Fig. 32.1) proposed by Maxted *et al.* (1997a). As can be seen, the raw materials of genetic conservation are plant genetic diversity, and the product of conservation is utilized genetic diversity. Conservation is the process that links plant genetic diversity to utilization and

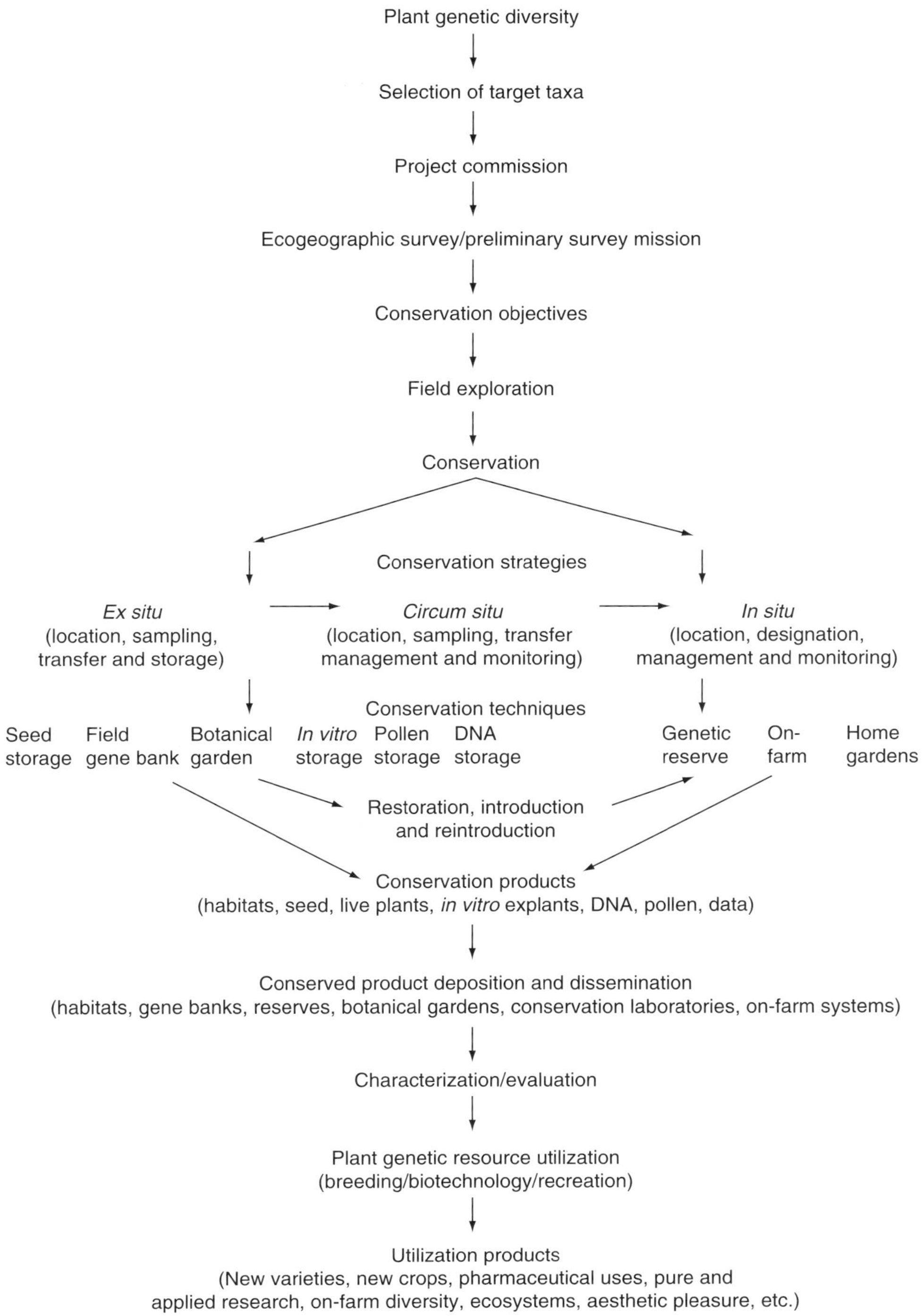

Fig. 32.1. Model of plant genetic conservation. (Adapted from Maxted *et al.*, 1997a.)

actively retains and protects the diversity of the gene pool, habitat or ecosystem, with a view to actual or potential human exploitation.

It is too often overlooked that conservation has a real economic cost; it would be difficult if not irresponsible to persuade society to meet this cost unless the conserved material could be shown to be of some 'value' to society. It is relatively easy to argue the economic benefit (value) that might accrue from conservation and subsequent utilization and exploitation of crop varieties or landraces in breeding programmes, but it is more difficult to ascribe economic value to wild species closely related to crops or more distantly related CWR species. However, Maxted *et al.* (1997b) argued that virtually all plants are likely to be of some value to society, whether in terms of immediate crop breeding potential (particularly with the application of biotechnological techniques) or for pharmaceutical use, but also for far less overt forms of utilization, such as recreation, ecotourism, educational use or simply making people feel 'good' to think that nature is 'safe'. Like all biodiversity, genetic diversity is part of any nation's heritage, alongside its art and culture, but pragmatically it can be argued that conservation can only be truly sustainable if plant diversity is effectively linked to some form of utilization. Therefore, it is important when applying a conservation technique for CWR, such as genetic reserve (where a protected area is managed to conserve the genetic diversity of a species; Maxted *et al.*, 1997c) or seed conservation in gene banks, that an explicit link to actual or potential utilization is made.

The exploitation of CWR genetic diversity for crop improvement has historically been the major driving force for its exploration and *ex situ* gene bank conservation (Hodgkin and Arora, 1999; Hajjar and Hodgkin, 2007; Hodgkin and Hajjar, Chapter 38, this volume), but the same opportunities are likely to apply equally to *in situ* conserved CWR diversity. However, acting as a gene donor is not the only use of CWR species. For instance, Cook (1995), when developing the International Working Group on Taxonomic Databases for Plant Sciences (TDWG) standard, groups plant use into 13 broad categories, as shown in Table 32.1. Therefore, even if CWR are literally defined by their use as gene donors, examples could be found of CWR use in each of these 13 categories.

However, whatever use is to be made of conserved CWR germplasm, the material must be accompanied by adequate information that allows the user to distinguish which accessions they should access and ultimately utilize. This ancillary germplasm information can be passport, characterization or evaluation data, although FAO (1998) concludes that lack of characterization and evaluation data is a major limitation to germplasm use. Although it is not stated, we presume that the FAO conclusion relates to *ex situ* conserved germplasm alone. It is likely that the characterization and evaluation data associated with *in situ* conserved germplasm are even less readily available; therefore, much use of CWR is likely to be based solely on passport data associated with the original collection locality.

As a minimum requirement, passport data should be gathered on each accession at the time of collection, and made available to potential users. Passport descriptors are defined by IPGRI (IPGRI, 1997) as follows:

Table 32.1. TDWG plant use standards. (From Cook, 1995.)

Category	Detail
Food	Food, including beverages, for humans only
Food additives	Processing agents and other additive ingredients used in food preparation
Animal food	Forage and fodder for vertebrate animals only
Bee plants	Pollen or nectar sources for honey production
Invertebrate food	Plants eaten by invertebrates that are useful to humans, such as silkworms, lac insects and edible grubs, are covered here
Materials	Woods, fibres, cork, cane, tannins, latex, resins, gums, waxes, oils, lipids, etc. and their derived products
Fuels	Fuels – wood, charcoal, petroleum substitutes, fuel alcohols, etc.
Social uses	Plants used for social purposes, rather than food or medicines, e.g. masticatories, smoking, narcotics, drugs, contraceptives and abortifacients and plants with ritual or religious significance
Vertebrate poisons	Plants which are poisonous to vertebrates, both accidentally and usefully, e.g. for hunting and fishing
Non-vertebrate poisons	Accidental or useful poisons (e.g. molluscicides, herbicides, insecticides) to non-vertebrate animals, plants, bacteria and fungi are included
Medicines	Both human and veterinary
Environmental uses	Plants used as intercrops and nurse crops, ornamentals, barrier hedges, shade plants, windbreaks, soil improvers or for revegetation, erosion control, waste purifiers, and indicators of metals, pollution or underground water
Gene sources	CWR that may possess traits or qualities, such as disease resistance, cold hardiness, etc. of value in breeding programmes

> These provide the basic information used for the general management of the accession (including the registration at the gene bank and other identification information) and describe parameters that should be observed when the accession is originally collected.

Extensive guidelines for gathering and recording passport data in the field have recently been published by Moss and Guarino (1995).

Where resources are available to further enhance the use value of conserved germplasm, detailed characterization and thorough evaluation are essential. This enables, for instance, the identification of valuable pest and disease resistance or stress tolerance characteristics, which can be used to improve crop varieties, or chemical assays for pharmacologically active constituents. Characterization and evaluation data describe the highly heritable phenotypic and genotypic characteristics of an accession, and are both defined by IPGRI (1997):

> Characterization descriptors: These enable an easy and quick discrimination between phenotypes. They are generally highly heritable, can be easily seen by the eye and are equally expressed in all environments. In addition, these may include a limited number of additional traits desirable by a consensus of users of the particular crop.

Evaluation descriptors: Many of these descriptors in this category are susceptible to environmental differences but are generally useful in crop improvement and others may involve complex biochemical or molecular characterization. They include yield, agronomic performance, stress susceptibilities and biochemical and cytological traits.

Characterization data are historically largely morphological characters that can be seen easily by the eye but they are increasingly being replaced by direct assessment of the genome using various molecular techniques. To facilitate and standardize the characterization of variants of different crop species, IPGRI has published extensive descriptor lists for many crop species (see http://www.ipgri.cgiar.org/). Many traits required by users are too genetically complex to be screened for in the preliminary characterization of germplasm accessions. These data are revealed at the stage of evaluation of germplasm for advanced traits, many of which may be subject to strong genotype by environment (G × E) interactions and hence be site-specific (FAO, 1998). However, the evaluation of germplasm is costly and tends to be very specific; for example, a breeder will evaluate a collection for drought resistance, having first screened the passport data for accessions with provenances from comparatively dry locations. Therefore, not only do the majority of gene bank accessions have incomplete passport data, but their characterization and evaluation information is even more limited, and may be very specific, according to the priorities of local users.

There is therefore a need for greater linkage between *in situ* and *ex situ* conservation of CWR. The end point of conservation is utilization, whether the germplasm is conserved *ex situ* or *in situ* and whatever form of utilization is proposed, there is a need for (at least) minimum passport data, ideally supported by characterization and evaluation data that are related to foreseen potential use. For example, if a conserved CWR accession is potentially to be used as a gene donor, then it should be adequately characterized and evaluated for the trait of interest, e.g. disease or pest resistance, while if a wild species is to be used as a sand dune stabilizer, then the potential wild species accession must have been evaluated for this use.

32.2 Linking *In Situ* to *Ex Situ* Conservation and Use

32.2.1 Safety *ex situ* back-up

Ex situ safety duplication of germplasm conserved *in situ* is needed more now than ever before, not least because of the impact of climate change on natural plant diversity (Thuiller *et al.*, 2005; Van Vuuren *et al.*, 2006). Human-induced climatic changes have accelerated global warming over the last 30 years (Osborn and Briffa, 2005). Temperature increases are predicted to be in the range of 1.1–6.4°C by 2100 (IPPC, 2007), which is likely to result in large-scale extinctions (Thomas *et al.*, 2004). It would be foolish to focus conservation action on *in situ* activities that may be prone to fires, flooding, plagues, vandalism, etc. when it is relatively easy, once the reserve is established, to sample and collect a representative sample of diversity for *ex situ* storage. Then, if for some reason the original population declines or goes extinct, the manager can always obtain seed from the gene bank to attempt to restore the natural population using samples of the original population. Seeds collected from populations at the site are likely to stand the best chance of population re-establishment because they are genetically adapted to grow at the site.

The point should also be made that germplasm users are always likely to find access to germplasm easier via a gene bank that routinely deals with potential user enquiries than a genetic reserve or on-farm conservation project. The seasonality of the availability of seed (the most common form of germplasm dispatched for use) means that it is only available for relatively short periods of the year (*in situ*), whereas seed from gene banks is available throughout the year. Therefore, if the *in situ* germplasm from the reserve or on-farm project is duplicated and available via the gene bank, the gene bank may be seen to act as a staging post for those wishing to utilize the germplasm originally conserved *in situ*.

32.2.2 *In situ* characterization

As already emphasized, one of the major limitations to germplasm use is lack of characterization or evaluation data – how is the germplasm user to decide which germplasm to use if there is no way of distinguishing which is most fit for their purpose? Possibly because of the potential magnitude of the task of *in situ* characterization or possibly because the protected area manager does not routinely undertake characterization or evaluation trials, it seems unlikely that actual *in situ* characterization or evaluation is feasible without significant additional resources being made available. However, one way of circumventing this problem is to undertake 'virtual' or 'predictive' characterization. That is the remote characterization of the conditions under which natural populations exist, using population passport data that is much more readily available. Simply knowing the location of CWR populations and the conditions under which they grow means a significant amount of characterization data can be deduced. For example, if a plant breeder works in a country that is likely to suffer lower rainfall following climate change, then he or she can search for adaptation to drought by sourcing germplasm from slightly drier conditions that currently occur in the breeders target country. Also, the breeder might look for germplasm from areas that are suffering other environmental stresses, e.g. disease hot spots, as germplasm from these areas may have adapted to the local conditions and these useful traits may prove important to further development of cultivated plants. Geographic information system (GIS) analysis to characterize populations by overlaying distribution maps of CWR with GIS layers on environmental data, including climate and soil or pest and disease occurrence, is becoming routine. Thus, statistical analysis techniques can be used to classify and predict the distribution of certain characteristics. Although this form of 'virtual' or 'predictive' *in situ* characterization is likely to be conducted centrally, rather than within individual reserves or on-farm projects, the process should greatly enhance the use of *in situ* conserved germplasm.

32.3 Linking *Ex Situ* to *In Situ* Conservation and Use

32.3.1 Genetic reserve enrichment

Given that the goal of genetic reserve conservation is the conservation of genetic diversity of natural wild populations, effectively in a semi-closed system, the introduction of any non-indigenous germplasm would present a competitive threat to native genetic diversity and therefore would be deleterious. The introduction of alien germplasm into a protected area is likely to impact on the genetic integrity of the original population and may even lead to outbreeding depression, where hybrids between the native and alien population are less fit for local environmental conditions than native populations. Therefore, ideally, no *ex situ* conserved germplasm should be introduced into a genetic reserve unless it was originally collected from that site or is genetically matched to the host population.

As a precursor for the establishment of a genetic reserve, the conservationist should know the minimum viable population (MVP) number for the target taxon and whether that population size exists at the reserve site. In practice, however, the conservationist rarely has *carte blanche* to establish a genetic reserve on purely scientific criteria. In practice, it is more common for the conservationist to be provided with access to an area of state-owned land, thus avoiding the need to purchase new privately owned land, which has no immediate competing land use, in which case, the conservationist is often required to establish the genetic reserve with a suboptimal population size. In these cases, the initial conservation goal will be to increase the target taxon population to the sustainable MVP number. An approach that has been used is to 'enrich' (sometimes referred to as enhancement/reinforcement/augmentation) the target taxon population at the reserve site. This has been experimented with in the project financed by the Global Environment Facility (GEF), through UNDP, on the 'Conservation and Sustainable Use of Dryland Agrobiodiversity in Jordan, Lebanon, Palestine Authority and Syria' (http://www.icarda.org/Gef.html).

On occasions when it is necessary to enrich genetic reserve populations with alien genetic diversity (i.e. no local provenance material is available), germplasm should only be used if it meets three basic criteria: (i) the taxonomic identification of the germplasm has been verified; (ii) the material is sourced from a provenance that is relatively close to the target site; and (iii) the material is sourced from populations that have a homoclinal and ecogeographic match with the populations being enriched. In the case of the 'Conservation and Sustainable Use of Dryland Agrobiodiversity in Jordan, Lebanon, Palestine Authority and Syria' project, no germplasm was found that met these criteria in gene banks; therefore, fresh germplasm was collected from neighbouring populations to enrich the target taxa in the genetic reserves.

Historically, little attention has been paid to the genetic implications of the introduction of alien germplasm but as stressed in the previous section, the closer the introduced alien germplasm is to the native, in terms of a genetic and homoclinic match, the more likely the reintroduction or recovery is to be successful. The *IUCN/SSC Guidelines for Reintroductions* (IUCN, 1998) were prepared as a response to growing occurrence of reintroduction programmes worldwide.

32.3.2 Ecosystem enrichment

Outside of the specific context of genetic reserve conservation, the more general goal of ecosystem conservation is to maintain the diversity of living organisms, their habitats and the interrelationships between organisms and their environment. This involves maintaining the various elements of biodiversity and their biotic and abiotic relationships, without the introduction of exotic germplasm from *ex situ* gene banks. However, in reality this is not always possible – where biological diversity is seriously threatened or has been lost within an ecosystem, there may be a need for action at the population or habitat level within that ecosystem or the creation of entirely new habitats. There are a number of related techniques that have the objective of maintaining plant populations that have either gone extinct or are severely threatened within an ecosystem. These may be categorized as: (i) reintroduction (the introduction of an organism into a habitat in which it was known to exist at some time in the past but is not currently present); or (ii) recovery (the management of a threatened species to increase population numbers to a self-sustaining level).

Whether the species goes extinct or is reduced below its MVP number, there is likely to be some deleterious factor causing the change in population numbers; therefore, attention needs to be paid to resolving this problem before the species is reintroduced, or the outcome is unlikely to be successful. It is also essential to ensure that sufficient numbers of individuals are introduced to the site to establish a self-sustaining population, and the clarification of this number is likely to require species-level research. Whatever the reason for reintroduction or recovery, there will be a need for close management and regular monitoring of the site to ensure the long-term success of the project.

Habitat restoration extends conservation beyond intact ecosystems to encompass the rehabilitation of degraded habitats (e.g. mining spoil tips, former quarry sites, motorway roadsides, brown field urban sites, derelict intensive agricultural land, logged forest) and opportunities for the creation of new areas of habitat on previously degraded or intensively used land. Ecological restoration is the process of assisting the recovery and management of ecological integrity. Ecological integrity includes a critical range of variability in biodiversity, ecological processes and structures, regional and historical context and sustainable cultural practices (Society for Ecological Restoration, 2006). Article 8 of the CBD (CBD, 1992) includes a commitment to rehabilitate and restore degraded ecosystems; therefore, *ex situ* conserved germplasm is likely to be of increasing use in habitat restoration projects.

Needless to say, it is impossible to restore an entire habitat from *ex situ* germplasm accessions – a habitat may contain thousands of plant, animal and microbial species, all of which interact with one another. However, it may be possible to plant keystone species and encourage other species to immigrate to the restoration site from the surrounding ecosystem, from the remaining habitat fragments not completely destroyed when the site was degraded or from the soil seed bank. Following reintroduction of keystone species, restoration effort may be concentrated on reinstating processes or establishing management regimes which favour the target species and eradicate or control problem species which currently dominate the site. Where the physical conditions of the site are harsh or rapid vegetation establishment is required, a 'nurse' species may be used to promote vegetation establishment. These are typically fast growing, short-lived species which provide a favourable micro-environment for the germination and establishment of the target species and then senesce, or in the case of trees, are removed, once the target vegetation becomes established.

Increasingly, consideration of the genetic provenance of the material used in ecological restoration has developed because of the problems of using germplasm from non-homologous ecogeographic locations that fail in new locations, and has undoubtedly resulted in homogenization of the ecosystem. In some cases, particular strains or cultivars have been selected for their rapid growth or large growth forms, often radically different to the native strains they are intended to represent. For example, in the United Kingdom, much of the planting stock for trees and shrubs is of European continental origin, often collected many hundreds of miles from where it is used, while many grassland mixtures contain agricultural cultivars. Effective restoration initiatives demand locally sourced material. For example, the burgeoning interest in the creation of 'wildflower meadows' in the United Kingdom has spawned a growing industry of specialist seed companies ready to supply the market with carefully designed mixtures suited to particular soil and management conditions. However, the high level of demand for locally sourced germplasm has frequently led to plant materials being sourced at great distances from the restoration site, coupled with commercial production of seeds and plants under agricultural conditions, to supply the restoration industry. While it is desirable to use locally sourced material, excessive local sourcing or local harvesting of germplasm could lead to further threats to local populations. It is also questionable whether gene banks with local germplasm would be able to supply sufficient quantities of seed for the restoration activity, so there is likely to be a bulking phase to generate sufficient material for the restoration.

Passport, characterization and evaluation data are even more critical for restoration projects because the degraded sites may have remnant problems for the reintroduced species, such as high levels of minerals, salt or trampling. The data could be used to match germplasm to the characteristics of the degraded site, improving the chances of restoration success.

32.3.3 Plant introduction, refuges, species substitutions and mitigation

Many degraded habitats that contain CWR species may have undergone extreme, and in many cases, rapid changes, and major ecological processes are likely to have been disrupted or destroyed, e.g. flood, road building, urban development projects, this habitat disruption may be ongoing, therefore, restoration may not be a viable option at the site. In these cases, where the original habitat cannot effectively be restored the most sustainable option is to rescue or salvage the CWR population and introduction or translocation of the habitat components by human agency outside its natural range. It involves the collection of wild CWR plants and deliberate transportation to an alternative *in situ* or *ex situ* site which is not facing the same threat of habitat destruction. Another option may be to establish a refuge, which is an artificial, highly modified, 'virtual' island, free from predators and herbivores. As with the other conservation techniques summarized, passport, characterization and evaluation data could be used to match germplasm to the characteristics of the host site to better ensure the success of the technique.

Biodiversity restoration is most controversial when it is carried out in the context of compensatory mitigation (Race, 1985; Race and Fonseca, 1996; Breaux and Serefiddin, 1999; Zedler and Callaway, 1999; Crooks and Ledoux, 2000; Bauer *et al.*, 2004). The word 'mitigate' may be defined as 'to lessen the severity' or 'to appease' or 'mollify'. It is usually used in the context of restoration when a site is earmarked for development, and the planners offer compensation for the destruction of the habitat by 'constructing' a replacement habitat elsewhere, or by translocation of a population from the site to an alternative location. As already discussed, the creation of habitats is never a substitute for retaining the original habitat; in fact, Howard (1996) estimates that only 12.5% of mitigation translocations are successful. Therefore, while this option is not favoured by conservationists, when there is no other choice, the CWR diversity used to create a replacement habitat could be taken from *ex situ* conserved germplasm.

32.4 Improving the *Ex Situ* Conservation/Utilization Link

As already noted, there is an explicit link between the expenditure of resources on conservation and the utilization of the conserved material for human benefit. We must therefore attempt to ensure that we make maximum use of *ex situ* conserved germplasm. Simmonds (1962) was the first to point out that mismanaged or underutilized germplasm collections could be regarded as mere museum exhibits gathering dust. Also, by implication, the resources expended on collecting and conserving *ex situ* conserved germplasm – estimated to be US$5.3 billion by Hawkes *et al.* (2000) for the 6.1 million accessions in world gene banks (FAO, 1998) – would have been misspent. The onus is equally placed on those managing *in situ* CWR as for those managing *ex situ* conserved diversity to ensure maximum use of the conserved material. How can we expect the public, who primarily fund such collections and protected areas, to continue their support if they are not being used?

Both Given (1995) and FAO (1998) estimated that approximately two-thirds of globally conserved *ex situ* germplasm lacks basic passport data, 80% lacks characterization data and 95% lacks evaluation data. Given (1995) estimated that only approximately 1% of gene bank accessions are appropriately catalogued and ready for use. The figures are unavailable for *in situ* conserved germplasm, but it seems likely that there are currently no genetic reserves where the conserved target species are fully characterized, evaluated, and ready for utilization. Unless the professionals involved with CWR conservation and use can ensure that conserved germplasm is held in a form better suited for breeders and other user groups and that there is a more seamless gradation of conservation into utilization, then the situation is unlikely to change.

A major step forward in improving the accessibility of *ex situ* conserved CWR germplasm has been taken by the EC-funded project, European Plant Genetic Resources Information Infra-Structure (EPGRIS). The project established the web-based, European Internet Search Catalogue of *Ex Situ* PGR Accessions (EURISCO) (ECPGR, undated). The catalogue are automatically updated from national inventories and is an effective way of promoting the use of *ex situ* conserved CWR. At present the passport data included are limited but this is likely to be improved with each updating iteration and therefore will further aid germplasm selection by the PGR user community in Europe. EURISCO will provide a means of sourcing local material for *in situ* introduction or reintroduction by matching germplasm passport data with the conservation site details. Through gap analysis of the accession contained in the system it is also likely to promote a more systematic duplication of *in situ* genetic diversity.

An equally important step forward was taken in improving access to *in situ* conserved diversity through the development of the Crop Wild Relative Information System (CWRIS) (PGR Forum, 2005) and CWR Catalogue for Europe and the Mediterranean (Kell *et al.*, 2005; Kell *et al.*, Chapter 5, this volume). While EURISCO and CWRIS are significant steps forward, there is still much that needs to be done to improve the recording and management of data associated with CWR germplasm samples conserved *ex situ* and *in situ*.

Better conservation planning, field work, genetic reserve and gene bank conservation will also enhance usage of *in situ* and *ex situ* conserved CWR germplasm, as outlined below.

32.4.1 Conservation planning

There are various factors that should be considered when selecting which CWR taxa to prioritize and conserve (Maxted *et al.*, 1997b; Ford-Lloyd *et al.*, Chapter 6, this volume). The more efficient these choices, the more likely it is that the material collected will be utilized. If the goal is to improve usage of *ex situ* conserved germplasm within *in situ* or ecosystem conservation projects, then the breadth of the species conserved needs to be widened. Currently, only 4% of government-funded gene banks and 14% of CGIAR gene bank accessions are of wild species (FAO, 1998) – the vast majority of collections being devoted to advanced breeders material and landraces. If *in situ* or ecosystem CWR conservation project use is to be a priority, then a much broader representation of total plant diversity is required. However, it would not be feasible or efficient to establish an *in situ* genetic reserve for each CWR species; therefore, there is a need within each country and region to establish priority sites where CWR species can be conserved.

Full ecogeographic representation is also required in the conserved material, whether *in situ* or *ex situ* – having a single accession of a CWR species present in a gene bank or a population present in one genetic reserve does not mean that species is conserved. The full range of ecogeographically diverse sites where the species is found needs to be sampled or represented. This involves careful analysis of the ecogeography of the species across its full range.

32.4.2 Field conservation

The better the quality of the sample transferred *ex situ*, the more likely it is to be utilized. Therefore, the collector must try to ensure that the sample is of a sufficient size to avoid the need for regeneration and the likely genetic selection of the original sample for the cultivation conditions under which it is regenerated. The sample should represent the full range of genetic variation found in the population sampled (Brown and Marshall, 1995; Hawkes *et al.*, 2000).

The more complete the associated passport data collected, the more useful the germplasm accessions are to the users. Although the specific data recorded by each *ex situ* collection mission or *in situ* population manager is likely to vary depending on the target taxon or target area, a 'good collector' should always record certain information: sample labelling (expedition identifier, collector name and number, date, type of material), sample identification (scientific name, vernacular name), sampling information (population estimate – number of plants in population, covering what area, sampling method), collecting site location (country, province, precise location, latitude, longitude, altitude, farmer's name), collecting site description and context (site disturbance, physiography, soil, biotic factors), population information (phenology, pests and diseases, uses).

32.4.3 Gene bank or genetic reserve conservation

Improving usage of *in situ* or *ex situ* conserved germplasm can be improved by publishing collecting reports and publicizing reserve holdings. Following the collection expedition, the existence of novel diversity can be signalled to potential users by publishing collecting reports. Once the genetic reserve is established the reserve manager can equally publish a review of the material found in the reserve. The precise format of the publication will vary, but it may include details of the species ecogeographic characteristics, material availability, and initial characterization and pre-evaluation data. For example, knowing that an accession is native to an area with high heavy lead concentrations would be very useful to the conservationist trying to locate material to use in the rehabilitation of lead mine spoil tips.

Once the collection is safely conserved, the collection manager must be proactive and advertise the material in the collection. He or she can draw the existence of the collection to potential users by various means, including published catalogues, and accessible collection databases (either making the database directly available on the Internet or by a link to the collection via a meta-database, such as Bioversity International's collection holdings database). The SINGER database of CG institute germplasm holdings can be queried (SINGER, no date) and a large proportion of European gene bank holdings are available via

EURISCO (ECPGR, undated). CWRIS (PGR Forum, 2005) also provides a user-friendly window into *in situ* conserved diversity, which will be enhanced as the system is developed and improved over the coming years.

Samples from mismanaged reserves or collections are less likely to be of use and therefore used. Gene bank samples held *ex situ* must be kept in optimum storage conditions to avoid differential genetic erosion. Human management errors through mislabelling or storing the same accession under different accession number in the same or different institutes must be avoided, and if the sample has to be regenerated it must be done under suitable conditions (i.e. if it is an outbreeder cages must be used to avoid outcrossing) (Smith *et al*., 2003). While for native CWR diversity held *in situ* in genetic reserves the population should be appropriately managed and monitored to ensure that the genetic integrity is maintained and available for exploitation (Iriondo *et al*., 2007).

As has already been stated the more detailed the passport, characterization and evaluation the more likely the conserved material is to be utilized. This will assist the user select accessions that match their requirements (e.g. an accession with a high drought or salt tolerance is likely to be useful in sand dune stabilization). More detailed evaluation for drought or salt tolerance, or the deliberate infection of the material with diseases or pests to screen for particular biotic resistance will obviously enhance the utilization potential of accessions found to have desirable traits.

However, in the future there is likely to be an increasing evaluation of the germplasm collections for biochemical and molecular markers to assess the genetic diversity of these species (Kang, 2002). It is also often advisable to involve potential germplasm users in characterization and evaluation process. Any characterization and evaluation results should be published and made fully available (e.g. via EURISCO and CWRIS), to ensure the potential users are aware of the diversity of material available.

As mentioned above, it is not feasible or efficient to establish *in situ* genetic reserves for each CWR species. Therefore, there is a need within each country and region to identify important plant areas for CWR species (Important CWR Genetic Reserves) and in each, establish reserve sites that contain a high concentration of CWR species. Within the draft Global Strategy for CWR Conservation and Use (see Heywood *et al*., Chapter 49, this volume), it is recommended that to consolidate *in situ* CWR conservation, an interconnected network of such sites is established within each country, region and globally, with five CWR genetic reserves per country, 25 per continent and 100 globally.

With the increasing size of many collections and the ever-limited funds available for characterization and evaluation, it may be necessary to develop a core collection of representative, well-characterized and evaluated CWR accessions, to assist the potential user to select the desired traits and the accessions that will prove most useful. As pointed out by Hodgkin *et al*. (1995) the core collection should be seen as a point of entry to the entire collection and not a separate entity in itself.

The quality of the service provided to germplasm users by germplasm curators affects the potential for utilization. If e-mails remain unanswered, the user is supplied with dead or misidentified seed, or inadequate passport data is available, the user is unlikely to repeat their request and the material will remain unnecessarily underutilized. The user community needs to be identified and surveyed to ensure the service provided meets their requirements.

Unfortunately, the people who conserve and utilize germplasm are seen as people of distinct professions, often located in two distinct and remote locations. However, utilization can be improved by bringing conservationists and germplasm users physically as well as professionally together. The mixing of the communities was a goal at the First International Conference on Crop Wild Relative Conservation and Use and the need to link conservation to use is at the foundation of the newly established CWR Specialist Group.

32.5 Specific Issues Related to *In Situ* and Ecosystem Conservation of CWR

Sourcing the right material for restoration and reintroduction can present practical dilemmas – establishing an appropriate provenance is essential. For example:

- *How local is local*: The genetic structure of populations and metapopulations should be thoroughly understood in order to establish the spatial extent of adaptation.
- *Population extinctions*: In some circumstances, a population may have gone extinct from an area and the only source material that is available for reintroduction comes from a different population. Should the reintroduction go ahead with the non-local material?
- *Lack of choice*: Many species have been reduced to extremely low numbers of populations and numbers of individuals with populations; therefore, in order to reinstate populations with sufficient genetic variation, founder individuals for reintroduction stock may have to be sampled from several different sites. Combined with this, historical data are often lacking and conservationists may be starting with little or no knowledge of the species' original locations. In such extreme cases, decisions to carry out reintroduction are based on a great deal of uncertainty and expediency.
- *Genetic contamination*: On severely degraded sites, it could be argued that the reintroduction of non-locally sourced material can do little more harm than has already been done to the environment. However, the suggestion that an area may be genetically contaminated in this way also raises ethical questions.

Putting all these issues aside, the success of a reintroduction project may actually depend upon using locally sourced material. Non-local material may simply not survive in its new environment, and worse still, by using it to supplement existing populations, it may even lead to the genetic degradation or, in a worst-case scenario, to the complete collapse of the population as a whole.

- *Genetically variable reintroduction stock*: As noted by Knapp and Dyer (1998), it is widely agreed that the presence of genetic variation is a critical factor in the success of species reintroductions and restoration programmes.

In order for reintroduced populations to persist, they need to be large enough to avoid the loss of alleles due to random genetic drift, and increased inbreeding associated with population genetic bottlenecks. This is obviously related to the establishment of the MVP size for the species, and may require specific research to resolve.

- Ex situ *raised reintroduction material: loss of genetic diversity*: In addition to the need for adequate sampling to ensure that the accession is genetically diverse, a further critical consideration is the potential for loss of genetic diversity in *ex situ* cultivated or regenerated stock. Most reintroduction projects rely on the use of *ex situ* facilities before the reintroduction can take place. This is particularly true of rare and threatened species management. For species that are presumed extinct in the wild, the only known material may be held *ex situ*, or, even if wild populations remain extant, they are frequently very small. In this case, the conservationist faces particular challenges, as they are likely to be working from a very limited genetic base. In the worst cases, the only material that is known to exist, whether in one or more gene banks, may have been collected from a very limited genetic base in the first place. The stock will then have been subjected to storage and/or multiplication in an artificial environment. Therefore, the material is likely to lose genetic variation and suffer secondary selection pressures, which may affect its fitness to survive when reintroduced to the wild.
- *Use of non-native species*: Many biodiversity restoration projects attempt to improve the structure and functioning of a habitat. A first step in rehabilitation of a severely degraded site may be amelioration, where the initial requirement is for land stabilization to avoid further species and genetic erosion. In such circumstances, non-native species are often selected for replanting. The benefits of using non-native species may include the fact that they are fast-growing, resistant to the harsh conditions prevailing at the degraded site (e.g. drought, salinity, shallow soils and/or resistance to pests and disease), easy to propagate and grow in large quantities and can act as a nurse crop for the reintroduction of native species. However, using non-native species can result in problems because of their possible invasive nature, the fact that they are often planted in monocultures which may make them vulnerable to pest and disease, the possibility that they may hybridize with the native species, and are unlikely to support the same range of native fauna and microfauna as native species.

32.6 Conclusions

If we are to ensure efficient conservation and utilization, we must follow the example of 'good' collection managers and ensure that CWR diversity is effectively sampled, efficiently conserved, well characterized and evaluated, advertised thoroughly and disseminated widely. No matter how large the collections held, unless they have appropriate passport data, and are adequately characterized and evaluated, the potential user will be unable to identify desirable genotypes for use, and the collections will remain unused. Conservationists have a product, plant genetic diversity, and like 'good shop keepers' they must present their product in an appropriate manner, and advertise their 'goods' to ensure the continued high level of support from public funds. No one will want to visit a 'poorly organized shop' where the 'goods' are covered in dust. Therefore, the conservationist, like the 'good shop keeper', must ensure their collections are available to meet the varied needs of the broad user community. This will ensure that the general public will continue their long-term support for active *in situ* and *ex situ* CWR conservation, as well as underwriting conservation and utilization sustainability for the long-term benefit of all humankind.

The majority of worldwide and European PGR conservationists have in the past focused their activities on *ex situ* conservation of crops. Post CBD (CBD, 1992) this situation has changed, not only because the CBD explicitly gives emphasis to *in situ* conservation, but also because it is now generally agreed that the world's gene banks already contain a good representation of global crop (if not CWR) genetic diversity. The focus has now turned to *in situ* conservation of CWR diversity, and this, linked to the growing public awareness of environmental issues, has established new challenges for the PGR community. As these conservation activities are almost exclusively state-funded we must ensure we can maintain the conservation/utilization link and give priority to meeting the germplasm user communities' needs – conservation of CWR is not an end in itself or an academic exercise – it is a means of ensuring food security and the sustainability of humankind itself.

Acknowledgements

We would like to acknowledge Jose Iriondo, Devendra Gauchan and Luigi Guarino for their comments on drafts of this chapter.

References

Bauer, D.M., Cyr, N.E. and Swallow, S.K. (2004) Public preferences for compensatory mitigation of salt marsh losses: a contingent choice of alternatives. *Conservation Biology* 18(2), 401–411.

Breaux, A. and Serefiddin, F. (1999) Validity of performance criteria and a tentative model for regulatory use in compensatory wetland mitigation permitting. *Environmental Management* 24(3), 327–336.

Brown, A.D.H. and Marshall, D.R. (1995) A basic sampling strategy: theory and practice. In: Guarino, L., Ramanatha Rao, V. and Reid, R. (eds) *Collecting Plant Genetic Diversity: Technical Guidelines*. CAB International, Wallingford, UK, pp. 75–92.

Convention on Biological Diversity (1992) *Convention on Biological Diversity: Text and Annexes*. Secretariat of the Convention on Biological Diversity, Montreal.

Cook, F.E.M. (1995) *Economic Botany Data Collection Standards*. Royal Botanic Gardens, Kew, UK.

Crooks, S. and Ledoux, L. (2000) Mitigation banking: potential applications in the UK. *Environmental and Waste Management* 3(4), 215–222.

ECPGR (undated) *European Internet Search Catalogue of Ex Situ PGR Accessions*. Bioversity International. Available at: http://eurisco.ecpgr.org/ (accessed 5 April 2005)
FAO (1998) *The State of the World's Plant Genetic Resources for Food and Agriculture*. FAO, Rome, Italy.
Ford-Lloyd, B. and Maxted, N. (1993) Preserving diversity. *Nature* 361, 579.
Given, D. (1995) *Principles and Practice of Plant Conservation*. Chapman & Hall, London.
Hajjar, R. and Hodgkin, T. (2007) The use of wild relatives in crop improvement: a survey of developments over the last 20 years. *Euphytica* 10.1007/s10681-007-9363-0.
Hawkes, J.G., Maxted, N. and Ford-Lloyd, B.V. (2000) *The Ex Situ Conservation of Plant Genetic Resources*. Kluwer Academic, Dordrecht, The Netherlands.
Hodgkin, T. and Arora, R. (1999) Developing conservation strategies and activities for crop wild relatives. In: Balakrishna, P. (ed.) *In Situ Conservation: an Indian Perspective*. IARC, New Delhi.
Hodgkin, T., Brown, A.D.H., van Hintum, Th.J.L. and Morales, E.A.V. (1995) *Core Collections of Plant Genetic Resources*. Wiley, Chichester, UK.
Howard, J.L. (1996) *Populus tremuloides*. In: *USDA Forest Service, Rocky Mountain Research Station, Fire Sciences Laboratory. Fire Effects Information System*. Available at: http://www.fs.fed.us/database/feis/ (accessed 14 April 2007)
IPGRI (1997) *Annual Report*. IPGRI, Rome.
IPPC (2007) *Climate Change 2007: Fourth Assessment Report*. Intergovernmental Panel on Climate Change Secretariat, Geneva, Switzerland.
Iriondo, J.M., Dulloo, E. and Maxted, N. (eds) (2007) *Plant Genetic Population Management*. CAB International, Wallingford, UK.
IUCN (1998) *Guidelines for Re-introduction*. Prepared by IUCN/SSC Re-introduction Specialist Group, IUCN, Gland, Switzerland and Cambridge.
Kang, M.S. (2002) *Quantitative Genetics, Genomics and Plant Breeding*. CAB International, Wallingford, UK.
Kell, S.P., Knüpffer, H., Jury, S.L., Maxted, N. and Ford-Lloyd, B.V. (2005) *Catalogue of Crop Wild Relatives for Europe and the Mediterranean*. Available at: The Crop Wild Relative Information System (CWRIS – Available at: http://cwris.ecpgr.org/) and on CD-ROM. University of Birmingham, UK.
Knapp, E.E. and Dyer, A.R. (1998) When do genetic considerations require special approaches to ecological restoration? In: Fielder, P.L. and Kareiva, P.M. (eds) *Conservation Biology*, 2nd edn. Chapman & Hill, New York, pp. 345–363.
Maxted, N., Ford-Lloyd, B.V. and Hawkes, J.G. (1997a) Complementary conservation strategies. In: Maxted, N., Ford-Lloyd, B.V. and Hawkes, J.G. (eds) *Plant Genetic Conservation: the* In Situ *Approach*. Chapman & Hall, London pp. 20–55.
Maxted, N., Hawkes, J.G., Guarino, L. and Sawkins, M. (1997b) The selection of taxa for plant genetic conservation. *Genetic Resources and Crop Evolution* 44, 337–348.
Maxted, N., Hawkes, J.G., Ford-Lloyd, B.V. and Williams, J.T. (1997c) A practical model for in situ genetic conservation. In: Maxted, N., Ford-Lloyd, B.V. and Hawkes, J.G. (eds) *Plant Genetic Conservation: the* In Situ *Approach*. Chapman & Hall, London, pp. 545–592.
Moss, H. and Guarino, L. (1995) Gathering and recording data in the field. In: Guarino, L., Ramanatha Rao, V. and Reid, R. (eds) *Collecting Plant Genetic Diversity: Technical Guidelines*. CAB International, Wallingford, UK, pp. 367–417.
Osborn, T.J. and Briffa, K.R. (2005) The spatial extent of 20th-century warmth in the context of the past 1200 years. *Science* 311, 841–844.
PGR Forum (2005) *Crop Wild Relative Information System (CWRIS)*. University of Birmingham, UK. Available at: http://cwris.ecpgr.org/ (accessed 3 April 2007)
Race, M.S. (1985) Critique of present wetlands mitigation policies in the United States based on an analysis of past restoration projects in San Francisco bay. *Environmental Management* 9(1), 71–81.

Race, M.S. and Fonseca, M.S. (1996) Fixing compensatory mitigation: what will it take? *Ecological Applications* 6(1), 94–101.

Simmonds, N.W. (1962) Variability in crop plants, its use and conservation. *Biological Reviews* 37, 422–465.

SINGER (no date) *System-wide Information Network for Genetic Resources*. Bioversity International. Available at: http://singer.grinfo.net/ (accessed 5 April 2007)

Smith, R.D., Dickie, J.B., Linington, S.H., Pritchard, H.W. and Probert, R.J. (eds) (2003) *Seed Conservation: Turning Science into Practice*. Royal Botanic Gardens, Kew, UK.

Society for Ecological Restoration (2006) *Society for Ecological Restoration Online*. Avaliable at: http://www.ser.org/about.asp (accessed 14 April 2007).

Thomas, C.D., Cameron, A., Green, R.E., Bakkenes, M., Beaumont, L.J., Collingham, Y.C., Erasmus, B.F.N., Ferreira De Siqeira, M., Grainger, A., Hannah, L., Hughes, L., Huntley, B., Van Jaarsveld, A.S., Midgley, G.F., Miles, L., Ortega-Huertas, M.A., Peterson, A.T., Phillips, O.L. and Williams, S.E. (2004) Extinction risk from climate change. *Nature* 427, 145–148.

Thuiller, W., Lavorel, S., Arau' jo, M.B., Sykes, M.T. and Prentice, I.C. (2005) Climate change threats to plant diversity in Europe. *Proceedings of the National Academy of Sciences of the United States of America* 102(23), 8245–8250.

Van Vuuren, D.P., Sala, O.E. and Pereira, H.M. (2006) The future of vascular plant diversity under four global scenarios. *Ecology and Society* 11, 25.

Zedler, J.B. and Callaway, J.C. (1999) Tracking wetland restoration: do mitigation sites follow desired trajectories? *Restoration Ecology* 7(1), 69–73.

VII Information Management

33 CWRIS: an Information Management System to Aid Crop Wild Relative Conservation and Sustainable Use

S.P. Kell, J.D. Moore, J.M. Iriondo, M.A. Scholten, B.V. Ford-Lloyd and N. Maxted

33.1 Introduction

It is widely accepted within the plant genetic resource (PGR) conservation and user community that a major factor hindering effective conservation is lack of easy access to data, as well as obstacles to information exchange due to the many different approaches in managing data. To conserve and sustainably utilize crop wild relatives (CWR), we first need to identify their names, locations, biology, ecological requirements, habitat specificity, threats and current conservation status. There is a myriad of information about CWR and their habitats available, but much of this information is dispersed; for example, information may be found in journal articles, textbooks, unpublished reports and other grey literature, web sites, databases belonging to individuals or institutions, herbaria, gene banks and botanic gardens. Information from these disparate sources has to be collated and analysed to formulate and implement comprehensive conservation strategies. The data sources are particularly disperse for CWR because of the broad taxonomic range of species and the fact that much data are held by those outside of the PGR community. Accessing such information is not only time-consuming, but comparing data sets is often difficult due to the diversity of information management models used. If CWR are to be conserved and sustainably utilized, a means of bringing together this information into an accessible and standard format is required. It may also be necessary for those involved in CWR and *in situ* conservation to adopt data collection and information management standards. This has been achieved, to a large degree, for the collection and *ex situ* conservation of PGR using standard data descriptors, such as the FAO/IPGRI Multi-Crop Passport Descriptors (FAO/IPGRI, 2001); however, data collection and recording standards for *in situ* CWR populations were not available prior to the development of the information management system presented in this chapter.

Recent advances in information technology, including programs that make use of the Internet to connect information and present it in one platform, e.g. GIS Open Source software, have opened the door to greater possibilities in terms of providing common access to information via a single data portal. There is a strong move away from creating static databases that hold information in one place, towards linking existing sources of information together in a public arena. However, in order to do this, a common language for CWR data management needs to be agreed among the PGR community.

The European Community (EC) recognizes the need to tackle the known loss of CWR genetic diversity across Europe. The EC-funded project, PGR Forum, was established to address the issues of how these resources can be conserved and sustainably utilized (Maxted *et al.*, Chapter 1, this volume; PGR Forum, 2003–2005). One of PGR Forum's aims was to address the critical need for effective management of CWR information by developing methodologies for data handling. To this end, the project developed a CWR information management model, which is implemented in the Crop Wild Relative Information System (CWRIS) (PGR Forum, 2005), the first information management system specifically designed to facilitate CWR conservation and sustainable use.

33.2 The Crop Wild Relative Information System (CWRIS)

CWRIS comprises two main dimensions: access to taxon information – currently, access is available to the Catalogue of Crop Wild Relatives for Europe and the Mediterranean (Kell *et al.*, 2005; Kell *et al.*, Chapter 5, this volume) – and CWR descriptors for conservation and use for individual CWR taxa. The Catalogue contains more than 25,000 species records and in excess of 273,000 records of taxon occurrences in 130 geographical units across the region. Although CWRIS currently provides access to information on crops and CWR in the Euro-Mediterranean region, the information management model can be applied to the crop and CWR flora of any country or region of the world.

The CWR descriptors provide a comprehensive set of data standards that can be used to effectively manage genetic conservation of CWR taxa and their component populations. The descriptors provide the structure within which existing data can be accessed, mapped on to the data model, or communicated, and in which novel data can be collated.

CWRIS is available as an online information management system (PGR Forum, 2005), and on CD-ROM (Moore and Kell, 2005). CWRIS has been designed to facilitate access to CWR data for a diverse range of user communities, including plant breeders, protected area managers, policy makers, conservationists, taxonomists and the wider public. Workshops and an online survey were conducted to establish specific user requirements. CWRIS provides the opportunity for users to carry out taxon searches and to select and extract a list of CWR taxa for geographical areas of interest. CWRIS also provides access to ancillary information on the taxa of interest, via links to external online resources such as Mansfeld's World Database of Agricultural and Horticultural Crops (Hanelt and IPK Gaterslaben, 2001; IPK Gaterslaben, 2003), GRIN Taxonomy

(USDA, ARS, National Genetic Resources Programme, 2006), European Nature Information System (EUNIS) (EEA, 2007a), the IUCN Red List (IUCN, 2006), Electronic Plant Information Centre (EPIC) (Royal Botanic Gardens, Kew, 2002) and key publication search engines. The opportunity exists to link to any number of useful and relevant online information resources that will benefit the CWR conservation and use community.

This chapter explains how the components of CWRIS were developed and puts forward recommendations for future expansion and use of the system.

33.3 The CWR Information Management Model

33.3.1 Developing the information management model

The CWR information management model has been designed with provision of a core taxonomic data set to which detailed taxon information can be attached – the basic information needed, for example, to monitor change in CWR populations over time. At present, the CWR Catalogue for Europe and the Mediterranean (Kell *et al*., 2005; Kell *et al*., Chapter 5, this volume) provides the baseline data accessible via CWRIS, which defines the number of CWR taxa in the region, their taxonomic status, nomenclature and occurrence within defined geographical units. However, the information management model can be used to manage crop and CWR information in any country or region of the world.

Building on models for data management proposed by Ford-Lloyd and Maxted (1997), Hawkes *et al*. (2000) and Kell and Maxted (2003), PGR Forum undertook to create a data management model specifically for the genetic conservation of CWR. The model defines the minimum data that are required in order to develop comprehensive conservation strategies and provides a standard which can be utilized by the PGR conservation community.

The CWR information management model is hierarchical, starting with the root (level one) descriptors: taxon information, and site and population information (Appendix 1). Within each of the two root descriptors are several categories of more detailed information (level two). For example, site and population information contains the descriptors: site/population location, geomorphology, geology and soil, habitat, population size, population structure, population dynamics, population management, biotic interactions, ethnography, characterization and evaluation, local threats and conservation measures. In turn, each category leads to a further set of detailed descriptors (level three). For example, population structure contains the descriptors: number of subpopulations or microhabitats, spatial pattern of individuals, population density, isolation/fragmentation, location of individuals, reproductive/vegetative ratio and sex ratio. Each level three descriptor is assigned data standards to enforce consistency in data recording, storage and retrieval. Existing data standards were utilized in the model where possible and appropriate; however, there were a number of data types for which new data standards were required, or existing ones adapted. For example, there were no standards available for breeding system, pollination mechanism or plant habit, and novel standards were proposed for these categories.

For usage, the Taxonomic Databases Working Group (TDWG) standards for economic botany (Cook, 1995) were reviewed and compared with data standards of other databases (Scholten *et al.*, 2004). Scholten *et al.* found that while the original TDWG standards for economic botany databases provide a very detailed set of use standards, they do not provide sufficient detail for the major category 'gene sources' (which provides information about the specific use of a taxon, e.g. cold tolerance, drought resistance, high yield), which is particularly relevant to ensure CWR are able to be fully utilized in breeding programmes. Other economic botany databases, such as GRIN (Wiersema and Leon, 1999; USDA, ARS, National Genetic Resources Programme, 2006), applying a modified TDWG standard, provide a more detailed classification within gene sources, i.e. the degree of relationship between a CWR and its domesticated relative and the method followed to use a CWR as a gene source e.g. 'used as a hybrid parent'. The CWR database managed at the N.I. Vavilov Research Institute (VIR) in Russia (T. Smekalova, Russia, 2003, personal communication) gives an even further detailed degree of relatedness between crops and wild relatives, while the IUCN Utilization Authority File (IUCN, 1995–2006a) provides detailed classification of the scale of usage. Scholten *et al.* (2004) recommend adaptation of Cook's original TDWG standard in order to accommodate CWR as a distinct category in economic botany databases.

For the classification of habitats, two data standards were used in the information management model: EUNIS Habitat Types classification (EEA, 2007b) and the IUCN Habitats Authority File (IUCN, 1995–2006b). The EUNIS Habitat Types classification is a comprehensive pan-European system to facilitate the harmonized description and collection of data across Europe for use in environmental reporting and for assistance to the Natura 2000 process (EU Birds and Habitats Directives). The IUCN Habitat Types Authority File is the standard used by the IUCN Red List Programme when describing the habitat in which a taxon occurs. Inclusion of both standards means that in the information management model, habitat codes are selected, stored and retrievable from both classification systems, making the CWR data of greater use in a wide range of applications.

The IUCN Conservation Actions Authority File (IUCN, 1995–2006c) is an example of a data standard that requires further investigation regarding its applicability to the CWR information management model. While this and other data standards have been carefully debated and reviewed before release, many have not previously been considered in the context of CWR taxa.

Enforcing the use of data standards, in other words, forcing the use of a set of predefined criteria in order to define an object attribute within the model is extremely important, otherwise data retrieval and analysis become complex and cumbersome. However, in a few cases, where no suitable standard exists, or enforcing a limited choice of attributes is not appropriate, the model allows for the inclusion of free text elements.

An Extensible Mark-up Language (XML) schema corresponding to the data model was written as part of PGR Forum's commitment to making the information management concept available for access and sharing within the PGR community; the full schema can be viewed online (see Moore *et al.*, 2005).

33.3.2 Testing the information management model: the CWR case studies

The CWR information management model underwent an iterative process of testing by CWR user groups and as part of this process, a number of case study data sets were collated for selected CWR taxa. An online data entry module was created for this purpose. The objective of collating the case study data sets was not only to test the CWR information management model but also to illustrate its functionality via presentation of the case studies online (PGR Forum, 2005). The use and testing of the model in this way was important to steer further adjustments and development; feedback from each testing cycle was collated and the model was refined accordingly. Perhaps not surprisingly, the case studies reveal that there is a particular lack of detailed information available at site and population level for many taxa, and that if data are available, they are not in a format that can easily be translated into a standard data model. This emphasizes further the need for a standard model and data exchange formats to ease data sharing between CWR conservation practitioners.

33.3.3 Referencing system

In addition to the detailed taxon, site and population descriptors, a full referencing system that allows all object attributes to be linked to resources – which may include a literature source, a web site, a database, a map, an illustration or a person – is fundamental to the CWR information management model. Any data item can be linked to any number of references and a number of different reference classes are available; for instance, records can be linked to the original data source, the name and contact details of the person entering the data and additional information of interest. In this way, the referencing system ensures that the provenance of each data item can be tracked and further details of interest can be provided; where appropriate, a link can be made directly to an online source or to the contact details of the originator. The basic details of each person, literature item, web site, etc., are stored within the CWRIS database and any data item can be linked to any reference. The referencing system is fully extensible, so that new kinds of references can be added, and linked to, if necessary (e.g. multimedia, or other types of additional data), and new classes of referencing could be added (e.g. information about primary and secondary sources of data). This component is essential for data quality assessment. In the past, many biological data systems have accumulated impressive amounts of information, but the lack of tools for data quality assessment have greatly diminished their value.

33.4 Translating the CWR Data Management Model into an Online System

33.4.1 Establishing user requirements

Knowing and understanding the requirements of the CWR stakeholder community was fundamental to the development of CWRIS. The PGR Forum project itself brought together a CWR stakeholder group representing a broad cross section of the PGR community; including plant breeders, conservation ecologists, population biologists and geneticists, gene bank managers, taxonomists and information management specialists. Discussions held during workshops devoted to the development of the information management system and via an online discussion forum and e-mail contributed directly to establishing user requirements. In addition, a user requirement survey was conducted via an online questionnaire.

The aim of the user requirement survey was not to decide on the precise data types (or fields) needed in the information system, but to ascertain how the broad categories of data types might be queried in the system. For example, while user A may simply need a list of all *Lupinus* L. spp. occurring in Spain, user B may require distribution maps of *Lupinus* spp. occurring in protected areas in Spain and user C may demand a list of uses of *Lupinus* spp., contact details for researchers in *Lupinus*, and a record of accessions held in *ex situ* collections. These three queries are related to each other via the taxon of interest, but they clearly demand very different functionalities in the information system. This interrelationship between the broad data categories can only be translated into practical terms by analysing the types of information retrieval required by the user. Although the user requirement survey provides a good indication of the type of information retrieval required and gives some direction in system development, some requirements only emerge once users actually use an information system; therefore, an iterative 'develop and review' approach was required.

Results of the user requirement survey showed that potential users of CWRIS have a wide range of research interests and that users would demand very different capabilities of the system. Analysis of the responses indicated that ten broad information types were required in the system to cater for their needs (Table 33.1). An important result of the survey was that although the majority of respondents want to access CWR information via the Internet, some users want to access the information on CD-ROM. The survey also provided an opportunity to find out about the types of CWR information available in the respondents' institutes, in what format the data are stored and whether they are freely accessible. Around half of the reported CWR information sources are available in the public domain – primarily via searchable online databases; for those that are not, a number of respondents indicated that there are plans to make the data available in the public domain in the future. Results showed that a wide variety of data management systems are used to manage the data, that only a few are web-based Structured Query Language (SQL) databases and that some data are already available in XML format. Respondents also provided a number of online resources and contacts to further assist in development of CWRIS. This information forms a valuable list of potential information sources of interest to users of CWRIS, and helps establish whether such information may already be in a format that would be suitable for linking via the CWRIS XML schema. It also provides an indication of general trends in information management practices.

The user requirement survey underlined a number of key issues with regard to information about CWR conservation and sustainable use. In particular, the

Table 33.1. The ten main information categories required in CWRIS to cater for the user community.

1. Taxonomy and nomenclature
2. Degree of relationship between crop and CWR
3. Uses: historic, current and potential
4. Current, historical and potential distribution, including:
   - Country occurrence/extent of occurrence
   - Number of populations
   - Record of extinctions
   - Mapping function/GIS layers
5. Genetic diversity and biology
6. Ecology and habitat
7. Threat status
8. Conservation measures, including:
   - Occurrence in named protected areas and genetic reserves
   - Conservation management techniques
   - *Ex situ* holdings in gene banks
9. References to specific research projects
10. Contacts

information required to take many evidence-based conservation decisions is currently lacking or fragmented, the data required for such decisions are varied and complex, and the needs and capabilities of the user base are diverse. Where such data are available, integrating them is currently not easy. Most users stated that they wish to ask questions with a number of key dimensions – taxon-specific, geographic area-specific, conservation/threat-specific, and sometimes, time-specific questions, and questions combining any number of these dimensions. This diversity of both source data and requirement led the development team to adopt the flexible data warehouse architecture which was employed, because it suits such an application.

33.4.2 CWRIS technical specifications

CWRIS uses open standards for data storage and retrieval. The open database, MySQL, has been used to support the CWRIS data schema. MySQL is used as a standard in many public projects for managing scientific data, especially where the data are published online, and are available for all standard computing platforms (Unix/Linux, Windows, Apple OSX). The CWRIS database schema is a generic 'data warehouse' fact-table design, with multiple dimensions, so could equally be managed in other database management systems, if necessary. The object relationships, to support the various classes of data, are stored as separate classes in the same database table as the data themselves, and so can be queried and manipulated using a common set of modules. The CWRIS data management, data entry and data retrieval modules have been written in the open language Perl, and also, in parallel, in the proprietary Microsoft language VB Script. Because of this, CWRIS can be hosted on any web server, either open or Microsoft ASP-based. The warehouse object schema was implemented using the CWRIS data management module. The objects supporting the case study data sets were based on the XML schema, which the project developed (see Moore *et al.*, 2005). The objects supporting the taxonomic data were based on a modified version of the corresponding Euro+Med PlantBase object relationships (Euro+Med PlantBase, 2005) – these objects can be used for any taxonomic information, not just the Euro-Mediterranean data. The objects linking project data to references were bespoke to this project. A number of user-access levels were defined, including read only (for unregistered site visitors), editor (for contributors to case studies), manager (for bulk-loading taxonomic data) and designer (for defining warehouse object relationships). The database was populated using a combination of SQL scripts for loading bulk taxonomic data and the CWRIS front-end for populating individual case studies.

The system has been designed to cater for changes in taxonomic classification and nomenclature, by automatically retrieving updates to the core taxon data set (which can be implemented at regular time intervals, e.g. weekly). At present, CWRIS provides access to the taxon data provided by Euro+Med PlantBase (2005), which is undergoing a process of critical review and updating; therefore, the automatic retrieval function is critical to ensure taxonomic and nomenclatural consistency. The same will be the case for the flora of any country or region of the world – because taxonomic classification and nomenclature do not remain static, this is an important function of the system.

33.4.3 Accessing CWR inventories online

All information (facts or objects) is linked to a taxon name – this is an important feature of CWRIS as it enforces taxonomic and nomenclatural standards. Users can access taxon information either by browsing or searching for a taxon name, or access lists of taxa for each country or intra-national region. Taxa are organized from genus level and below, with the full taxonomy and nomenclature obtainable by following links from individual taxa to the Euro+Med PlantBase (2005) data which forms the taxonomic core of the Catalogue. Note that this feature can equally be applied to taxonomic information from other sources, whether national, regional or global. An example of a view of the taxon information is shown in Fig. 33.1 for the genus *Beta* L. The user can see a list of species contained in the genus; species highlighted in bold are cultivated taxa and the rest are the wild relatives. Selection of a taxon from the list reveals a list of synonyms, subspecific taxa, geographical distribution and a link to a case study if available. An example is shown in Fig. 33.2 for *B. patellaris* Moq. From here the user can obtain further information about the taxon's classification and nomenclature by following the link to the original source of taxonomic data. The user can also follow the link to the case study to learn more about the taxon and the in-depth CWR taxon information management model, find information about the taxon from external sources, or find out which other CWR taxa occur in the taxon's geographic range.

CWRIS has been designed to cater for different taxonomic classification systems as far as possible. Therefore, while an accepted taxonomy should ide-

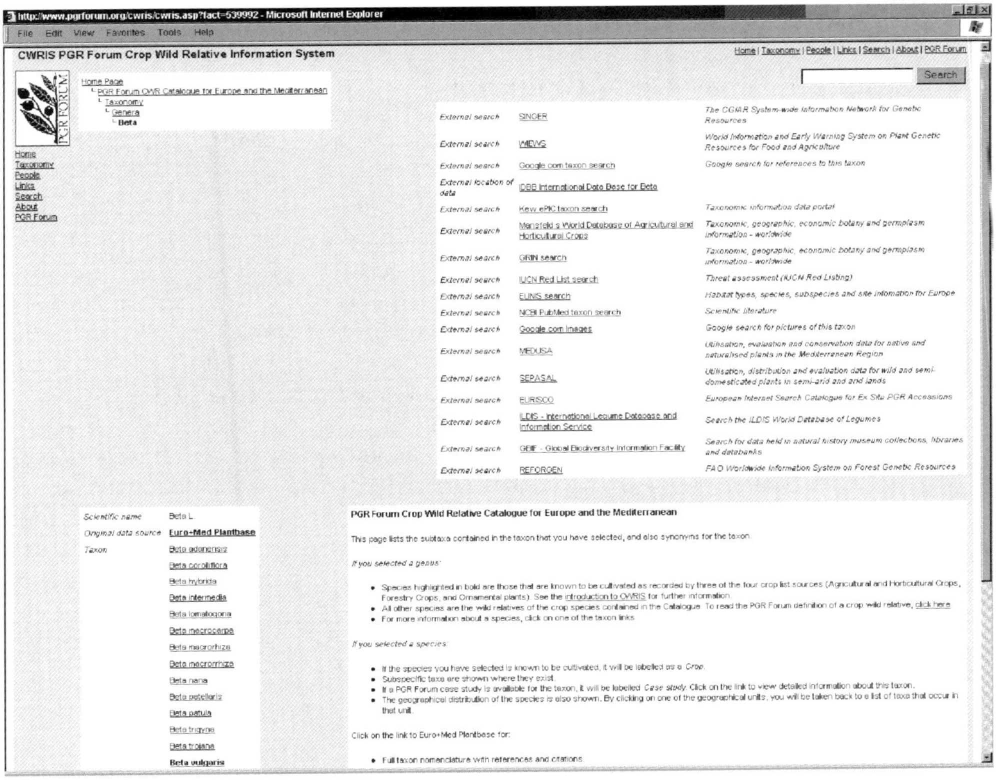

Fig. 33.1. View of CWRIS showing the CWR catalogue arranged at genus level and below. The example shows records for the genus *Beta* L.

ally be adopted (currently that of Euro+Med PlantBase), the database can be searched on synonyms as well as accepted names. For example, in the case of *Beta*, some scientists recognize *B. webbiana* Moq. as an accepted name, but Euro+Med PlantBase does not. Therefore, while *B. webbiana* does not appear in the list of species within *Beta* in CWRIS, a search for the taxon informs the user that the taxon is a synonym of *Patellifolia webbiana* (Moq.) A.J. Scott, Ford-Lloyd & J.T. Williams, and the system provides a link to the accepted taxon.

33.4.4 Accessing external information sources

CWRIS not only comprises a facility for accessing crop and CWR taxon inventories and a taxon information management model, it is also a portal providing access to sources of information on the taxa included. At present, CWRIS contains links to a selection of external online information sources that can be consulted to find information about taxa; however, to complete every field in

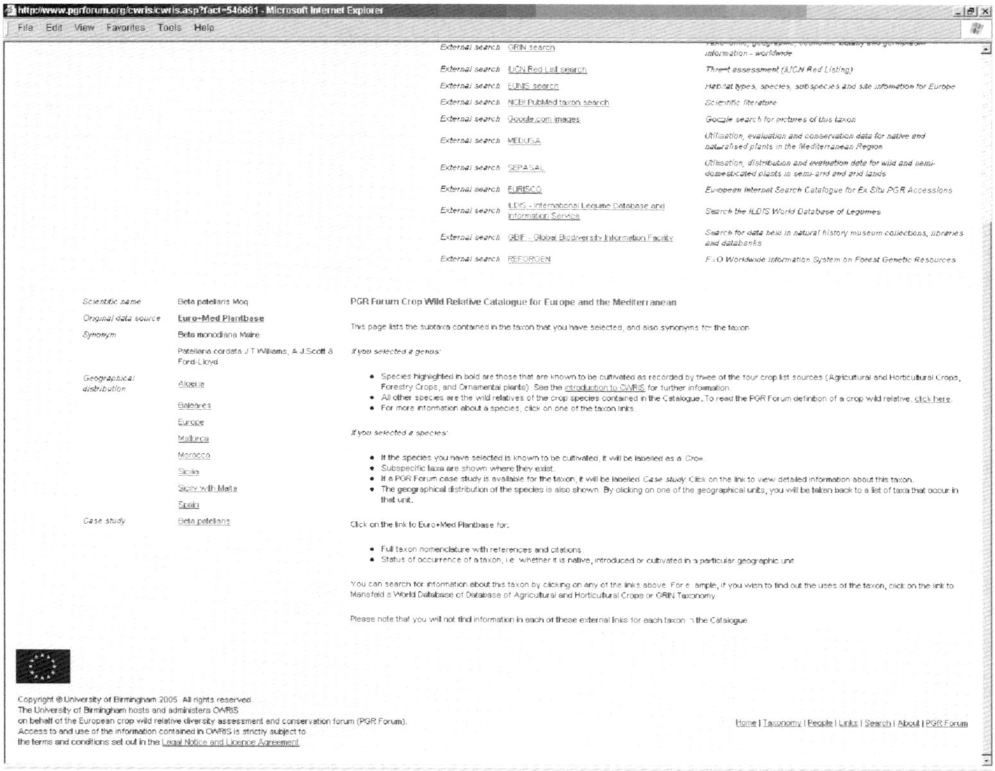

Fig. 33.2. View of CWRIS showing information for *Beta patellaris* Moq.

the information management model via external links is not possible because this information does not exist in a readily accessible format for a large number of taxa. The data model is designed for the effective genetic conservation of CWR, which means that detailed time-series information about populations of taxa is needed. We know that for the majority of CWR taxa, such data do not exist simply because there are insufficient resources to undertake detailed population monitoring and genetic sampling for a large number of taxa. In a limited number of cases, detailed information does exist, but most likely it is stored in a database on a researcher's hard drive or at best is published in journals.

Despite these limitations, it is possible to retrieve a great deal of information about CWR taxa by consulting external information sources. For example, by consulting ePIC (electronic Plant Information Centre) (Royal Botanic Gardens, Kew, 2002), a range of information can be accessed, including nomenclatural, bibliographic (including floras) and Kew collections data, and information on seed storage behaviour and physiology. Other examples of CWR information sources include: GRIN (USDA, ARS, National Genetic Resources Programme, 2006), which contains information on economic importance, distribution, references and accessions held in the USDA National Plant Germplasm System; Mansfeld's World Database of Agricultural and Horticultural Crops (Hanelt and IPK Gaterslaben, 2001; IPK Gaterslaben, 2003), which holds detailed information on uses, as well as bibliographic and distribution data; EUNIS (EEA, 2007a) which provides access to data on species, habitats and sites compiled within the framework of Natura 2000; and the IUCN Red List of Threatened Species (IUCN, 2006), which gives detailed information about Red Listed taxa, including the Red List Category and Criteria of the assessment, taxon distribution, habitat and ecology, threats, links to other sources of useful information and bibliographic data. These are just a few examples of external sources of data that will help in the establishment of CWR conservation strategies. Of course, the information is not available for every taxon, but where it exists, CWRIS provides a valuable tool for accessing it without the need to search each external source individually. Currently, CWRIS provides access to general sources of information linked via the taxon name; however, any number of further external links can be added, including links to national-level information via the geographic units listed in the crop and CWR inventory.

Because CWR have been accorded a higher profile in recent years, some database managers are taking steps to 'tag' CWR in their own information systems. Examples include the EUNIS Database web application (EEA, 2007a), which includes access to data on species, habitats and sites compiled within the framework of Natura 2000, and the Plant Search database managed by Botanic Gardens Conservation International (BGCI, 2007), which contains information on around 1600 institutions worldwide that maintain live plant collections. There are many other examples of existing information management systems that could help in providing access to CWR information by adding a CWR code to their database. This is particularly important in the context of ecological conservation of habitats and biomes. For example, national information portals, such as NBN Gateway (NBN Trust, 2007), which gives access to a comprehensive range of species, habitat and geographical data sets in the United Kingdom, could add a code to tag the CWR species. This is a relatively simple procedure that can be achieved by requesting a list of CWR taxa included in CWRIS and cross-checking which of these taxa are included in the respective database – quite a straightforward course of action, but with potentially far-reaching results.

33.4.5 Presentation of CWR case studies

As described earlier, a number of CWR case studies were compiled to test and illustrate the CWR taxon data model. CWRIS users can view the case studies by following a link from the home page or from the taxon name for which the case study has been compiled. The online case studies present a hierarchical view of the CWR descriptors, with hyperlinks from one level to the next. Information is stored in simple tables which can be viewed on screen. All data contained in the case studies are fully referenced to ensure that each data item is linked to an author. Figure 33.3 shows a view of part of the case study for *Erodium paularense* Fern. Gonz. & Izco. The view shows the taxon biological

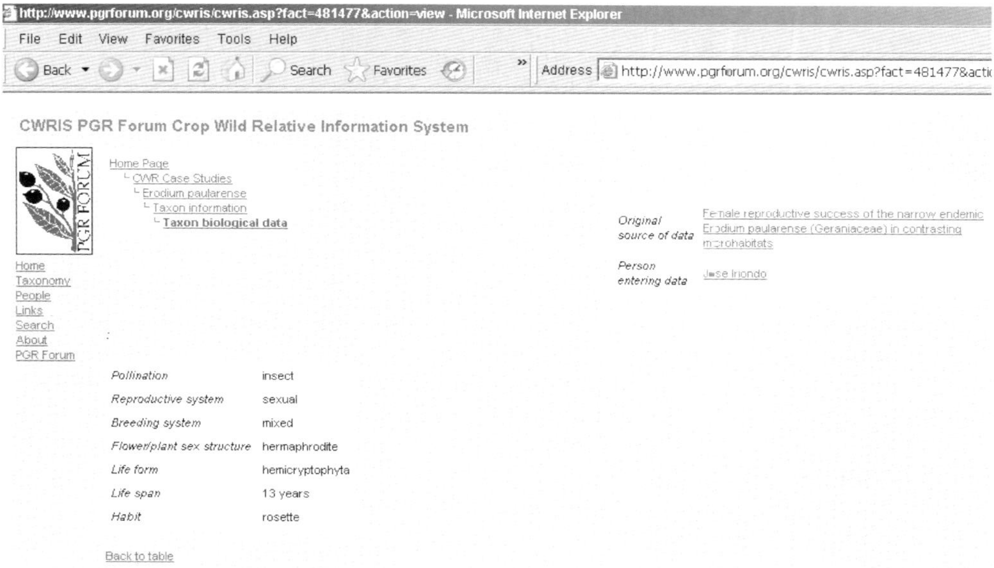

Fig. 33.3. CWRIS case studies, illustrating the schema for management of CWR taxon data. The example shows the taxon biological data table for *Erodium paularense* Fern. Gonz. & Izco. In the top right side of the screen are links to the original source of data contained in the table and the name of the person that entered the data.

data table, links to the original source of data contained in the table and the name of the person who entered the data. Clicking on the reference hyperlinks provides the user with full reference details and in the case of a person, their full contact details.

33.5 Examples of CWRIS Use Cases

33.5.1 Creating a national CWR inventory

An inventory of the resource to be conserved is an essential early step in any conservation programme (CBD, 2002; Lughadha, 2004; Mace, 2004). One of the uses of CWRIS, combined with the Crop Wild Relative Catalogue for Europe and the Mediterranean (Kell *et al.*, 2005), that has proved most valuable is the extraction of lists of CWR taxa to form the basis of a national CWR inventory; examples are provided by Codd (2005), Markkola (2005), Scholten *et al.* (Chapter 7, this volume), Maxted *et al.* (in press), Magos Brehm *et al.* (in press) and H. Fitzgerald and N. Maxted (unpublished data). Users gain access to the data by selecting the country or geographical units of interest, in order to produce a list of taxa that occur within the selected geographical area. These data can then be cross-checked against local floras, databases and other documentation, verified and edited as necessary. A methodology for creating national CWR inventories has been proposed by Maxted *et al.* (in press) – these inventories form the basis for the development of national CWR conservation strategies.

33.5.2 Obtaining taxon distribution data

Rather than focusing on a single geographic area, another fundamental use of CWRIS is to determine the distribution of a crop or CWR taxon. By selecting the taxon of interest, the user is provided with the countries or geographical units in which the taxon is known to occur. By utilizing the link from the taxon page to Euro+Med PlantBase (2005), the user can obtain the status of the taxon in each geographical unit of its range, i.e. whether native, introduced or cultivated. This feature can be used to link to any source providing the same information for taxa in other regions of the world. At present, there are no mapping facilities built in to CWRIS; however, by utilizing the links from the taxon page to the external sources of information, it is possible to retrieve distribution maps for some taxa.

33.5.3 Breeding potential and other uses

One of the main arguments for the conservation of CWR is to maintain the genetic diversity inherent in CWR populations, which is of potential use in plant breeding; therefore, one of the primary functions of the CWR data schema and of CWRIS is to provide information about the breeding potential of taxa. The CWR information model includes descriptors for the degree of relatedness of the CWR to the crop, expressed using the Gene Pool concept of Harlan and de Wet (1971) and the Taxon Group concept of Maxted *et al.* (2006). It is possible to obtain information about the genetic relationships within some taxa, but only for a relatively small percentage, because this level of data is only available for taxon groups that have been extensively researched – primarily the crops of major socio-economic importance. However, even for the major crop groups, information on genetic relatedness is only available for a relatively small number of taxa. Where this information is available, it is possible to find it via CWRIS using links to external information sources such as GRIN Taxonomy (USDA, ARS, National Genetic Resources Programme, 2006) and Mansfeld's World Database of Agricultural and Horticultural Crops (Hanelt and IPK Gaterslaben, 2001; IPK Gaterslaben, 2003). For example, GRIN Taxonomy provides details about the economic importance of taxa where available, including their potential as gene sources, while Mansfeld's World Database of Agricultural and Horticultural Crops provides general information on plant uses and links to National Centre for Biotechnology Information databases (NCBI, 2006) containing extensive molecular information. In all instances, the user is provided with reference details so that they can find the original source of data regarding the specified use.

Where genetic information is not available for taxa, the Taxon Group concept can be applied as a proxy indicator of breeding potential (see Maxted *et al.*,

2006). However, accessing this information electronically is only possible if the system containing the taxonomic information has incorporated the taxonomic hierarchy, thus showing which taxa are contained in which Taxon Group, e.g. subgenus, section or series. The two concepts used to define the degree of relationship between the CWR and its associated crop are integral to the CWR schema – a fundamental feature of the information management system.

Interest in uses of CWR is not limited to their breeding potential; some users may wish to find out what the general use of a taxon is, e.g. food, fodder, forage, forestry, ornamental, medicinal and aromatic uses. Data standards for taxon uses have already been developed and the standard adopted by TDWG (Taxonomic Databases Working Group) is that of Cook (1995). Cook (1995) provides a comprehensive system for categorizing these general plant uses; however, placing the approximately 25,000 species contained in the CWR Catalogue into a use category (not to mention the thousands of other crop and CWR species of the world) would require an enormous amount of work and is not practical. Information managers must not reinvent the wheel – where the data exist elsewhere, the system should provide access to the data, rather than replicating them. Currently, a wide range of online databases provide information on the uses of many crop and CWR taxa. Beyond those already mentioned, such as GRIN and Mansfeld's database, other examples include the MEDUSA Database (MEDUSA Network, 2002), which can be searched for information on taxon uses and utilizes an adapted version of the TDWG standard (C. Johnson, Reading, 2003, personal communication), SEPASAL (Survey of Economic Plants for Arid and Semi-Arid Lands) (Royal Botanic Gardens, Kew, 1999), which utilizes the full TDWG use standard and REFORGEN (the FAO Worldwide Information System on Forest Genetic Resources) (FAO, no date), which provides information on uses for some taxa. CWRIS provides links from the taxon records to these and other sources.

The CWR information management model includes descriptors for taxon uses adopting the TDWG standard (Cook, 1995) – thus, when the model is used in the future for collation of novel data sets, information on taxon uses can be accommodated. Further, the crop and CWR taxa are categorized according to the use groups: agricultural and horticultural crops (including food, fodder, forage and industrial species), forestry species, ornamental species, and medicinal and aromatic plants. This has been possible because the lists of taxa within a genus were originally derived from queries against the four crop genus sources. Currently, CWRIS does not present this view, but this function is included in future development plans.

33.5.4 Community curation of links to validated information sources

Prior to the PGR Forum project, there was no single accepted online reference source of links to information of particular relevance to conservation and sustainable use of CWR. The CWRIS database provides a first step in providing such a resource and currently maintains a basic set of validated links which are common to all taxon groups. CWRIS can be used, in an extended way, to become a community-curated resource of such links to validated data sources.

The CWRIS database can be used to host the activities of a group of researchers, each with responsibility for curating links within their own area of expertise in particular taxon groups, geographic areas, or biological, conservation or sustainable utilization practice. By dividing the task of maintaining such links among an online research community, in a similar way to the Open Directory Project (1998–2007), the database has the potential to host a definitive community resource, providing easy access to validated sources of online CWR research. Establishing such a resource might require additional support in the form of a small secretariat to administer the community's activities.

33.6 Future Needs and Development Potential of CWRIS

The development of CWRIS has required a multifaceted approach involving the user community, PGR specialists and information technology experts. CWRIS comprises two interlinked parts: (i) an information management model for CWR conservation and sustainable use with an emphasis on site and population data (and corresponding XML schema); and (ii) an online information management system and portal providing access to crop and CWR inventories, as well as access to information on taxonomic status and nomenclature, distribution, uses and other types of information of relevance to CWR conservation and sustainable use. A snapshot of the CWRIS database (incorporating the CWR Catalogue for Europe and the Mediterranean) has also been published on CD-ROM for fast access to the data without an internet connection (see Moore and Kell, 2005). There is however a need to continue to develop, refine and improve CWRIS.

For example, critical for future CWR conservation and sustainable use is the collation of detailed location and demographic data over time. Compilation of the CWRIS case studies revealed that there is generally a lack of detailed site and population data available. There are probably multiple factors contributing to this: lack of availability of a standard data collection model for site and population data, site and population descriptive data stored in inaccessible files and scarcity of resources for collecting the data. It may be argued that providing a detailed data schema for management of site and population data is pointless if the data do not exist. However, as has been noted, often the data do exist but are not currently in an easily accessible format. Where the data do not exist, it is important to encourage practitioners, and to lobby funding agencies to provide the resources needed to collect and make available these vital data. The model presents the minimum data types required for the effective conservation and sustainable use of CWR. The implication of the model is that without these data, the genetic diversity of our CWR heritage cannot effectively be conserved and utilized. Conservation must be pursued at the microlevel, that is, monitoring and management of populations; knowing the locations of populations of taxa is not enough.

Access to data on CWR populations *in situ* can arguably be collated most effectively at national level, starting with the priority taxa (see Ford-Lloyd *et al.*, Chapter 6, this volume). Researchers, including taxonomists, who frequently have detailed knowledge of populations, must be encouraged to collect demographic data. Data can be fed into a central repository via the CWRIS taxon information management model and made available via a common portal. Such a system will be complex and will require substantial resources to create and manage, but the technology is available and it is possible. The most difficult part of the process will be to secure regional and national resources to regularly monitor populations and ensure continuity over time and between researchers. Therefore, one of the next steps in the development of CWRIS is to establish the linkage methodology between national and regional data sets. The detailed data descriptors are already available within CWRIS and the technology is now required to convert the data management model into a workable web-based data gathering and data mining system. This could be achieved by following a similar protocol to that implemented in EURISCO (European Internet Search Catalogue of *Ex Situ* PGR Accessions) (ECPGR, no date), which acts as a repository and portal for information on *ex situ* PGR collections in Europe. A simple querying system to facilitate data mining of national inventory information also needs to be developed.

An alternative, and perhaps more desirable model to work toward, is the development of CWRIS as a portal and data mining system providing access to dispersed data sets, rather than collating the data in one place. This route is achievable with today's Internet technology and avoids the process of uploading and updating data.

Although CWRIS currently provides access to data associated with taxa within the Euro-Mediterranean region, this does not limit the scope of the system as a global information portal. The core taxonomic information in CWRIS can be expanded to include the floras of other regions and countries of the world. Where this information is not available, the system can still act as a portal and data mining system for global information by linking taxon information to a list of genera containing crops, which can be generated from sources such as Mansfeld's World Database of Agricultural Crops (Hanelt and IPK Gatersleben, 2001; IPK Gaterslaben, 2003) (Kell *et al.*, Chapter 5, this volume). However, it should be possible to create links between CWRIS and databases such as Species 2000 (2007) and GBIF (Global Biodiversity Information Facility) (GBIF, no date), which could provide the core taxonomic information needed as the foundations for access to global crop and CWR information. The CWR information management model can also be used independently of the online system itself to manage CWR information at any level, e.g. *in situ* conservation sites, *ex situ* collections and national or regional CWR data sets.

The CWRIS developers ended the project with an extensive 'wish-list' of additions/alterations to the system. For example, the addition of a greater number of links to information of relevance to CWR conservation and use and the organization of these information sources into different categories to aid ease of user access would be beneficial. Information sources can be added either at taxon or geographic level. At present, the external links are mainly at taxon level; however, an opportunity also exists to link external information sources at geographic level. For example, when the user follows a link to their country of interest, it would be useful for them to be able to view a list of links to detailed information on CWR of that country. There are many examples of information sources at national level and these could be collated along with taxon-level information sources and added to CWRIS as community-validated information sources, as described in the use cases. The addition of a list of nations, linked to the geographic units in the database would also improve the user interface.

Improving access to crop and CWR information according to use is also needed. The CWR Catalogue for Europe and the Mediterranean (Kell *et al.*, 2005), which is the taxonomic data set that CWRIS currently provides access to, was created by gathering data on four groups of PGR: agricultural and horticultural crops, forestry species, ornamental species and medicinal and aromatic plants (Kell *et al.*, Chapter 5, this volume). While it would clearly be useful to the user to be able to search the system on plant uses, to assign individual uses to all the taxa included is not practical nor desirable, since this would simply be duplicating information found in existing databases. What can be done immediately, however, is to group the taxa into the four use groups listed above, thus improving user access to information of interest. Further, it would be quite simple to tag the major and minor food crop genera – for example, as defined by Groombridge and Jenkins (2002) (Kell *et al.*, Chapter 5, this volume) – and provide a simple search facility via crop groups and vernacular crop names.

To put these and other developments into practice and to maintain a dynamic, secure and accessible system, the long-term sustainability of CWRIS is paramount – all too often the results of short-term projects fail due to lack of resources and infrastructure to maintain access to the products after the project's lifetime. To ensure the continued availability of CWRIS to the conservation and use communities, CWRIS (which was developed at and hosted by the University of Birmingham, United Kingdom) is now hosted by Bioversity International on behalf of the Secretariat of the European Cooperative Programme for Plant Genetic Resources (ECPGR). Further funding is also being sought to develop the system further, particularly with regard to collating CWR and landrace national inventory data across Europe. The vision is to develop a single PGR information portal that encompasses crop and CWR *in situ* and *ex situ* population data. Plans are afoot to develop such a system building on the EURISCO and CWRIS models (S. Gaiji, Italy, 2006, personal communication).

CWRIS has been developed with the goal of providing a model for collation and management of CWR conservation and sustainable use data and a system for accessing this information – this goal has been fully achieved. The creation of CWRIS is a major achievement of a cooperative EC-funded project and is a significant step forward for the PGR community, both within Europe and worldwide, but the model requires full implementation if the conservation of CWR is to be secured.

Acknowledgements

The concepts discussed in this chapter were stimulated by PGR Forum (the European crop wild relative diversity assessment and conservation forum – EVK2-2001-00192 – http://www.pgrforum.org/), funded by the EC Fifth Framework Programme for Energy, Environment and Sustainable Development. We are grateful to participants in the CWRIS user acceptance development panel for their time spent testing and providing feedback on the system.

References

BGCI (2007) *Plant Search*. Botanic Gardens Conservation International. Available at: http://www.bgci.org/plant_search.php/ (accessed 26 March 2007)

CBD (2002) *Global Strategy for Plant Conservation*. Secretariat of the Convention on Biological Diversity, Montreal. Available at: http://www.biodiv.org/programmes/cross-cutting/plant/default.asp (accessed 6 April 2007)

Codd, R.B. (2005) Conservation Action Planning for UK Crop Wild Relatives. MSc thesis, University of Birmingham, Birmingham, UK. Available at: http://www.pgrforum.org/Documents/UOB_Theses/Codd_R_UOB_MRes_Thesis.pdf (accessed 27 March 2007)

Cook, F.E.M. (1995) *Economic Botany Data Collection Standard*. Prepared for the International Working Group on Taxonomic Databases for Plant Sciences (TDWG). Royal Botanic Gardens, Kew, UK. Available at: http://www.rbgkew.org.uk/tdwguses/ (accessed 3 April 2007)

ECPGR (no date) *European Internet Search Catalogue of* Ex Situ *PGR Accessions*. Bioversity International. Available at: http://eurisco.ecpgr.org/ (accessed 5 April 2007)

EEA (2007a) *European Nature Information System (EUNIS)*. European Environment Agency. Available at: http://eunis.eea.europa.eu/index.jsp (accessed 13 April 2007)

EEA (2007b) *EUNIS Habitat Type Hierarchical View*. European Environment Agency. Available at: http://eunis.eea.europa.eu/habitats-code-browser.jsp (accessed 13 April 2007).

Euro+Med PlantBase (2005) Euro+Med PlantBase: The Information Resource for Euro-Mediterranean Plant Diversity. Dipartimento di Scienze botaniche ed Orto botanico. Universitá degli studi di Palermo. Available at: http://www.emplantbase.org/home.html

FAO (no date) *REFORGEN – the FAO Forestry Database on Forest Genetic Resources*. Available at: http://www.fao.org/forestry/site/39116/en/ (accessed 3 April 2007)

FAO/IPGRI (2001) *FAO/IPGRI Multi-Crop Passport Descriptors*. Food and Agriculture Organization of the United Nations, Rome and International Plant Genetic Resources Institute, Rome.

Ford-Lloyd, B.V. and Maxted, N. (1997) Genetic conservation information management. In: Maxted, N., Ford-Lloyd, B.V. and Hawkes, J.G. (eds) *Plant Genetic Conservation: The* In Situ *Approach*. Chapman & Hall, London, pp. 176–191.

GBIF (no date) *Global Biodiversity Information Facility*. Available at: http://www.gbif.org/ (accessed 6 April 2007)

Groombridge, B. and M.D. Jenkins (2002) *World Atlas of Biodiversity*, prepared by the UNEP World Conservation Monitoring Centre. University of California Press, Berkeley, California.

Hanelt, P. and IPK Gatersleben (eds) (2001) *Mansfeld's Encyclopedia of Agricultural and Horticultural Crops*. 6 vols. 1st Engl. ed. Springer, Berlin/Heidelberg/New York.

Harlan, J. and de Wet, J. (1971) Towards a rational classification of cultivated plants. *Taxon* 20, 509–517.

Hawkes, J.G., Maxted, N. and Ford-Lloyd, B.V. (2000) *The* Ex Situ *Conservation of Plant Genetic Resources*. Kluwer Academic, Dordrecht, The Netherlands.

IPK Gaterslaben (2003) *Mansfeld's World Database of Agricultural and Horticultural Crops*. Leibniz Institute of Plant Genetics and Crop Plant Research. Available at: http://mansfeld.ipk-gatersleben.de/ (accessed 3 April 2007)

IUCN (1995–2006a) *Utilization Authority File (Version 1.0)*. IUCN/SSC Authority Files for Habitats, Threats, Conservation Actions and Utilization of Species. Available at: http://www.iucn.org/themes/ssc/sis/authority.htm (accessed 3 April 2007)

IUCN (1995–2006b) *Habitats Authority File (Version 2.1)*. IUCN/SSC Authority Files for Habitats, Threats, Conservation Actions and Utilization of Species. Available at: http://www.iucn.org/themes/ssc/sis/authority.htm (accessed 3 April 2007)

IUCN (1995–2006c) *Conservation Actions Authority File (Version 1.0)*. IUCN/SSC Authority Files for Habitats, Threats, Conservation Actions and Utilization of Species. Available at: http://www.iucn.org/themes/ssc/sis/authority.htm (accessed 3 April 2007)

IUCN (2006) *The IUCN Species Survival Commission 2006 Red List of Threatened Species*. Available at: http://www.redlist.org (accessed 27 March 2007)

Kell, S.P. and Maxted, N. (2003) Orchid conservation data: management, access and use. In: Dixon, K.W., Kell, S.P., Barrett, R.L. and Cribb, P.J. (eds) *Orchid Conservation*. Natural History Publications (Borneo), Kota Kinabalu, Sabah, Malaysia, pp. 329–346.

Kell, S.P., Knüpffer, H., Jury, S.L., Maxted, N. and Ford-Lloyd, B.V. (2005) *Catalogue of Crop Wild Relatives for Europe and the Mediterranean*. Crop Wild Relative Information System (CWRIS – available at: http://cwris.ecpgr.org/) and on CD-ROM. University of Birmingham, UK.

Lughadha, E.N. (2004) Towards a working list of all known plant species. *Philosophical Transactions of the Royal Society of London Series B-Biological Sciences* 359 (1444), 681–687.

Mace, G.M. (2004) The role of taxonomy in species conservation. *Philosophical Transactions of the Royal Society of London Series B-Biological Sciences* 359, 711–719.

Magos Brehm, J., Maxted, N., Ford-Lloyd, B.V. and Martins-Loução, M.A. (in press) National inventories of crop wild relatives and wild harvested plants: case-study for Portugal. *Genetic Resources and Crop Evolution.*

Markkola, H. (2005) Regional Red List Assessment and Biodiversity Action Plans for Crop Wild Relatives in Ireland. MSc thesis, University of Birmingham, Birmingham, UK. Available at: http://www.pgrforum.org/Documents/UOB_Theses/Markkola_H_UOB_MSc_Thesis.pdf (accessed 27 March 2007)

Maxted, N., Ford-Lloyd, B.V., Jury, S.L., Kell, S.P. and Scholten, M.A. (2006) Towards a definition of a crop wild relative. *Biodiversity and Conservation* 15(8), 2673–2685.

Maxted, N., Scholten, M.A., Codd, R. and Ford-Lloyd, B.V. (in press) Creation and use of a national inventory of crop wild relatives. *Biological Conservation*.

MEDUSA Network (2002) *The MEDUSA Database*. Available at: http://medusa.maich.gr/database/ (accessed 3 April 2007)

Moore, J.D. and Kell, S.P. (eds) (2005) *PGR Forum CD-ROM*. University of Birmingham, UK.

Moore, J.D., Kell, S.P., Maxted, N., Iriondo, J.M., Ford-Lloyd, B.V. and contributors (2005) *International Crop Wild Relative Information Schema*. University of Birmingham, UK. Available at: http://www.pgrforum.org/cwris/cwr_20050823.xsd (accessed 6 April 2007)

NBN Trust (2007) *NBN Gateway*. National Biodiversity Trust. Available at: http://www.searchnbn.net/ (accessed 6 April 2007)

NCBI (2006) *National Centre for Biotechnology Information, U.S. National Library of Medicine*. Available at: http://www.ncbi.nlm.nih.gov/ (accessed 3 April 2007)

Open Directory Project (1998–2007) *Open Directory Project*. Netscape. Available at: http://www.dmoz.org/ (accessed 6 April 2007)

PGR Forum (2003–2005) *European Crop Wild Relative Diversity Assessment and Conservation Forum*. University of Birmingham, UK. Available at: http://www.pgrforum.org/ (accessed 3 April 2007)

PGR Forum (2005) *Crop Wild Relative Information System (CWRIS)*. University of Birmingham, UK. Available at: http://cwris.ecpgr.org/ (accessed 3 April 2007)

Royal Botanic Gardens, Kew (1999) Survey of Economic Plants for Arid and Semi-Arid Lands (SEPASAL) database. Available at: http://www.rbgkew.org.uk/ceb/sepasal/internet/ (accessed 3 April 2007)

Royal Botanic Gardens, Kew (2002) *Electronic Plant Information Centre*. Available at: http://www.kew.org/epic/ (accessed 7 April 2007)

Scholten, M.A., Kell, S.P., Maxted, N. and Ford-Lloyd, B.V. (2004) *Economic Botany Standards for Crop Wild Relatives.* Presentation given at PGR Forum Workshop 5, Genetic Erosion and Genetic Pollution Assessment Methodologies. Terçeira Island, Portugal, September 8–11, 2004.

Species 2000 (2007) *Species 2000*. Available at: http://www.sp2000.org/ (accessed 6 April 2007)

USDA, ARS, National Genetic Resources Programme (2006) *Germplasm Resources Information Network – (GRIN)* [Online Database]. National Germplasm Resources Laboratory, Beltsville, Maryland. Available at: http://www.ars-grin.gov/cgi-bin/npgs/html/index.pl (accessed 3 April 2007)

Wiersema, J.H. and Leon, D. (1999) *World Economic Plants: A Standard Reference*. CRC Press, Washington DC.

Appendix 1. The CWR information management model: basic descriptors. Each level three descriptor is assigned data standards to enforce consistency in data recording, storage and retrieval (not shown).

TAXON INFORMATION

Taxon nomenclature

Scientific name
Genus
First epithet
Second epithet
Rank
Hybrid flag
Author team parenthesis
Author team
Cultivated plant name elements
Principal synonyms
Botanical type
Vernacular name
Closest relative

Taxon biological data

Reproductive system
Breeding system
Flower/plant sex structure
Pollination
Life form
Life span
Generation time
Habit

Taxon ecogeographical data

Geographical distribution
Migration
Approximate extent of occurrence
Area of occupancy
Altitudinal zone
Altitudinal range
Soil description
Soil type
Soil texture
pH
Climatic preference
Habitat
Habitat status
Vegetation
Phenology
Status

Taxon population level data

Number of locations or populations in which the taxon is distributed
Overall population number
Population trends
Biotic interactions

Taxon utilization

Uses/ethnobotany
Target crop
Degree of relationship to crop

Taxon threats

Threats
Red List assessment

Taxon conservation actions

Legislation

In situ: protected area, *in situ* management plan, reintroduction, translocation etc.

Ex situ: accessions, time of collection, place of storage etc.

Recovery plans

Taxon references

Publications
Herbarium specimens
Photographs
Illustrations

SITE AND POPULATION INFORMATION

Site/population location
Country
Administrative unit
Named area
Nearest named place
Coordinates
WFS
WMS
Altitude

Site geomorphology
Position in relief/geomorphology
Aspect
Aspect text
Slope

Site geology and soil
Geology with effect on soil
Geology/stratigraphy
Soil type
Soil texture
Soil moisture regime
Soil depth
pH
Percentage organic matter, elements
Salinity
Derived soil evaluation

Site habitat
Habitat type
Microclimate
Vegetation
Vegetation stratification
Inventory of accompanying plant species
Land use type(s)
Land use intensity
Anthropogenic effects
Owner
Site detail
Location of population with regard to species distribution

Population size
Boundaries
Diameter
Approximate area of occupancy
Population size
Number of mature individuals

Population structure
Number of subpopulations or microhabitats
Spatial pattern of individuals
Population density
Isolation/fragmentation
Location of individuals
Reproductive/vegetative ratio
Sex ratio for dioecious species

Population dynamics
Survey date
Trend

Population management
Site management plan
Management interventions
Monitoring detail

Biotic interactions
Dominant vegetation
Associated vegetation
Keystone species
Dominant herbivores
Associated herbivores
Grazing pressure
Pollinators
Seed dispersers
Percentage tree cover
Type of tree cover

Ethnography
Local ethnic group
Language
Traditional use of site

Characterization and evaluation
Preliminary characterization

Local threat
Threat category
Threat reason

Conservation measures
Legislation
In situ
Ex situ

34 Crop Wild Relatives in the ECPGR Central Crop Databases: a Case Study in *Beta* L. and *Avena* L.

C.U. Germeier and L. Frese

34.1 Introduction

Many accessions in gene bank collections and a multitude of data on them relate to wild species and crop wild relatives (CWR). The information on genetic resources held in European collections has been collated in the European Central Crop Databases (ECCDBs), promoted within the framework of the European Cooperative Programme for Crop Genetic Resources Networks (ECP/GR) since the mid-1980s. Recently these have been supplemented by EURISCO, a central repository of passport data hosted at IPGRI to standardize, formalize and facilitate the regular acquisition of passport data from the gene banks (IPGRI, 2003). The scope of the ECCDBs goes beyond passport data towards more descriptive data, mainly from characterization and evaluation and it is targeted at the interests of users of genetic resources in food and agriculture, mainly breeders and agronomists. Collating these data crop-wise in crop-specific repositories reflects the crop-specific interests of this user community, the conservation biologists dealing with specific CWR being part of it.

This chapter suggests to the CWR community to use the ECCDBs wherever crop-specific interests prevail. Representation of CWR in the ECCDBs and assumed common interests of the breeding-oriented (*ex situ*) plant genetic resources (PGR) domain and the CWR domain are exemplified with the International Database for *Beta* (IDBB) and the European *Avena* Database (EADB). It is suggested to build on the ECCDBs for the crop-wise integration of data on *ex situ* and *in situ* collections, with a focus on their use for the mutual benefit of breeders, agronomists and conservation biologists. Geographic data on collecting sites of *ex situ* samples can be used to identify and recommend *in situ* management sites. In many cases data on *ex situ* and *in situ* collections serve similar objectives (e.g. characterization and evaluation) and a similar user community. In other cases, when synecologically targeted information systems are more appropriate for the CWR community, interfaces between those and the ECCDBs should be developed.

The IDBB was founded in 1987 in Wageningen (IBPGR, 1987). It has been maintained by the Braunschweig Genetic Resources Collection (BGRC) since late 1991 and since 1996 further developed by the Gene Bank of the Federal Centre for Breeding Research on Cultivated Plants (BAZ) (Germeier and Frese, 2001a). It serves as a central repository of passport, characterization and evaluation data within the World *Beta* Network (WBN) and recently took up characterization and evaluation data produced in the European Union-funded GENRES CT95–42 project (Frese *et al.*, 2000) with 11 European partners. Currently, the database contains passport data for 10,485 accessions from 29 collections in 25 countries located in the Northern Hemisphere with a focus on European contributors. The EADB was established at the BGRC, the former gene bank of the Federal Agricultural Research Centre (FAL) in 1984 (IBPGR, 1984). The database contains passport data of 34,146 accessions representing the *Avena* collections from 24 European contributors. It recently took up characterization and evaluation data produced in the GENRES CT99–106 project (Germeier and Frese, 2002) with five European partners. Both databases have been redesigned during the last 5 years, following identical modelling approaches (Germeier and Frese, 2001b; Germeier *et al.*, 2003) and made accessible on the Internet by an identical web application. Direct access is provided by BAZ to the development versions (BAZ, 2005a,b), which are the most up-to-date versions but may be temporarily affected by development activity. Stable releases are mirrored at the German Centre for Documentation and Information in Agriculture (BAZ and ZADI 2005a,b). Both databases can also be accessed from a central entry point to the ECCDBs hosted at IPGRI (ECPGR, 2005).

A brief introduction to the domain model, on which the databases and web applications are based, is given and possible interfaces to a CWR information system are highlighted. The representation of CWR in the databases, the method to find them and the results available for them are outlined. Some remarks are given on the taxonomic concepts used in the databases. Availability of geographic collecting information and characterization and evaluation data are especially considered.

34.2 Plant Genetic Resources in the View of Breeding Research – a Domain Model

A domain model describes a part of the real world to be covered in an information system. It identifies real world objects to be pictured by software design objects. PGR collections (*ex situ*) have often been affiliated with breeding research institutions. Their documentation has been designed to serve commercial breeders and breeding research on cultivated plants with their prevailing interests in two major data categories: passport data and characterization and evaluation data. The ECCDBs presented here reflect the view of breeding research on PGR (Fig. 34.1) and the crop-specific approach of conventional breeding, determined by the outer limits of the tertiary gene pool.

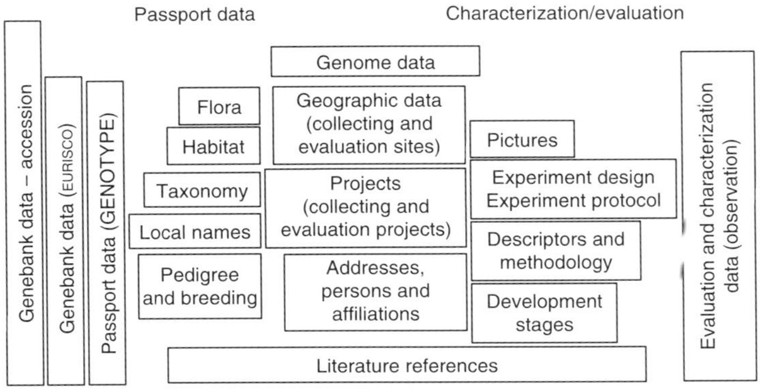

Fig. 34.1. Plant genetic resources in the domain of breeding research.

All ECCDBs started with a list of accessions in a so-called Multicrop Passport format (Hazekamp *et al.*, 1997), which now is provided by EURISCO as well (FAO/IPGRI, 2001). In the case of the IDBB and EADB additional data structures have been developed to cover the following tasks (Germeier and Frese, 2001a; Germeier *et al.*, 2003):

- Describe results of a duplicate search in a set of tables normalized according to the principles of relational database design.
- Provide a harmonized and corrected view to each duplicate group without touching originally provided data.
- Systematize originally provided botanical names according to multiple taxonomic views.
- Document characterization and evaluation data according to scientific standards.
- Normalize collecting information according to geographical and vegetation ecology standards.

It is suggested that a domain model as presented in Fig. 34.1 and implemented for the EADB and IDBB will be applicable to CWR and their *in situ* collections as well to meet the demands in breeding and breeding research. Description of the site, habitat and vegetation will be of particular importance for germplasm maintained *in situ* and they are demanded and increasingly recorded during the collection of *ex situ* materials as well.

In situ management, especially the management of protected areas, additionally touches the domain of nature conservation with a more synecological and less crop-specific approach (e.g. managing priority lists of taxa, managing groups of associated taxa in protected areas). Proper analysis of the various domain models and their interfaces will reveal common objects, which should be shared between the different domains, and specialized objects which should be added for a domain-specific view. Such an interdisciplinary analysis would make the reuse of already existing software or software to be developed in a collaborative effort possible and would make software development for genetic resources applications more synergistic by avoiding redundant implementations.

34.3 How to Search Data for Crop Wild Relatives in the IDBB and EADB

The web interface for both databases is an identical application written in PHP: Hypertext Preprocessor (The PHP Group, 2001–2005) and the way to search for CWR thus is identical for *Avena* and *Beta*. The web application provides a highly interactive user interface. It displays all contents, which can be searched for in drop-down lists and allows stepwise formulation of even complex queries combining conditions on passport, characterization and evaluation data using logical combinations with 'and', 'or' and parentheses. It controls and displays the number of hits at each step to avoid queries with no results.

According to the Multicrop Passport Descriptors a special attribute (sample status) should provide the development status of an accession. Sample status of 'wild' or 'weedy' could be a first approach to find accessions of wild material of the respective genus. As many as 3153 accessions have a sample status 'wild', 161 are designated 'weedy' in the IDBB, compared to 3473 accessions belonging to non-cultivated species. In the EADB only 1403 accessions are indicated as 'wild' or 'weedy' by their sample status compared to 4334 accessions of non-cultivated species.

Thus sample status information is not appropriate to find all accessions of wild species. Those that are affected by breeding research (crossing, selection) have got the sample status of a breeder's line. Additionally, sample status information, despite being demanded by the Multicrop Passport Descriptors, often cannot be made available by the holders of the accessions providing the original information. As the relation to taxonomy is ambiguous (cf. wild, weedy or breeder's line), it is not possible to reconstruct missing original information. In order to find all accessions of wild species it is more appropriate to search explicitly for those species.

34.4 Taxonomic Concepts Used in the EADB and IDBB

Searching wild species by scientific names is considered the preferred way of finding CWR in genetic resources databases. Yet taxonomy is a matter of dispute as well. Multiple taxonomic systems have been proposed for the genus *Avena* (Baum, 1977; Loskutov, 2002). Ladizinsky recommended the use of a biological species concept, which strictly uses a biological species definition based on barriers for intercrossing (Ladizinsky and Zohary, 1971; Ladizinsky, 1989). Criticism on this approach came from workers in genetic resources (Baum, 1977; Loskutov, 1998, 2002), who need to identify morphologically different forms, even if maintained by geographic isolation only or the influence of man in cultivated forms. Although a decision had been taken to implement Ladizinsky's system for the EADB (Schittenhelm and Seidewitz, 1993), this was not fully implemented. Also, the maintenance of a consistent taxonomic approach through database updates from sources using different concepts would be difficult. Given the progress in database technology and new concepts for taxonomic databases (Berendsohn, 1995; Pullan *et al.*, 2000), it finally seemed more appropriate not to abolish taxonomic information available with various systems of different granularity, but rather to give different taxonomic views a representation in the database. Additional data structures have been implemented, which annotate respective taxonomic systems (represented by major taxonomic monographs on the genus) probably used for an originally reported scientific name and provide a translation to the other systems, especially the biological species concept of Ladizinsky.

In the IDBB, an earlier decision was taken and the systems of Letschert *et al.* (1994) and Lange *et al.* (1999) have been consistently implemented throughout the database (Table 34.1) resulting in the abolishment of all scientific names not fitting into these concepts.

In Table 34.2 all scientific names for *Avena* spp. are listed for which a considerable number of accessions have been reported, irrespective of the taxonomic system used. Resulting from Ladizinsky's initiative some contributors began to reduce species designations to subspecies rank (e.g. *A. fatua* L. to *A. sativa* subsp. *fatua* (L.) Thell., *A. sterilis* L. to *A. sativa* subsp. *sterilis* (L.) De Wet). These cases are generally summarized in Table 34.2 with the former species names (*A. fatua, A. sterilis*).

Table 34.1. Representation of wild species and collecting information in the International Database for *Beta* (IDBB).

	Accessions		Collecting site information available (%)[a]				
Species	Total	%[b]	Country	District	Site	Coordinates	Elevation
			Primary gene pool				
B. vulgaris subsp. *maritima*	1975	18.8	94	83	83	78	73
B. vulgaris subsp. *adanensis*	78	0.7	100	96	96	78	91
B. macrocarpa	105	1.0	80	68	72	68	59
B. patula	21	0.2	71	71	10	62	5
			Secondary gene pool				
B. corolliflora	133	1.3	86	77	73	76	71
B. macrorhiza	95	0.9	75	59	53	48	45
B. lomatogona	277	2.6	83	71	69	68	61
B. intermedia	335	3.2	97	76	76	74	76
B. trigyna	118	1.1	51	19	16	14	10
B. nana[c]	59	0.6	100	49	88	88	88
			Tertiary gene pool				
B. patellaris	160	1.5	51	46	36	31	31
B. procumbens	76	0.7	61	54	50	47	49
B. webbiana	41	0.4	44	41	37	39	37
Total	3473	33.0					

[a]Percentage of accessions of the taxon with each type of passport data.
[b]Percentage of total accessions of *Beta* documented in the database.
[c]So far it was not possible to cross *B. nana* with any other species of the genus *Beta. B. nana* is classified in the secondary gene pool as the species is more closely related to section *Corollinae* than to section *Procumbentes* or *Beta* (Shen *et al.*, 1998).

Table 34.2. Representation of wild species in the European *Avena* Database (EADB).

	Accessions		Collecting site information available (%)[a]					
Species	Total	%[b]	Country	Region	District	Site	Coordinates	Elevation
			Hexaploid species					
A. sterili [c]	1968	5.76	97	7	3	20	7	9
A. ludoviciana[d]	463	1.36	99	26	8	26	0	23
A. fatua	547	1.60	97	26	10	38	5	14
A. occidentalis	7	0.02	100	57	57	43	0	0
			Tetraploid species					
A. barbata[e]	344	1.01	97	35	6	41	27	21
A. abyssinica	159	0.47	96	0	0	1	0	0
A. vaviloviana	64	0.19	97	16	0	16	0	6
A. magna[f]	63	0.18	87	3	35	67	19	13
A. murphyi	19	0.06	89	11	84	95	42	68
A. damascena	16	0.05	100	0	50	94	50	50
A. agadiriana	6	0.02	100	0	83	100	83	83
A. macrostachya	3	0.01	100	0	0	100	0	0
			Diploid species					
A. strigosa[g]	366	1.07	90	3	5	11	9	9
A. brevis	56	0.16	82	0	2	4	0	0
A. hirtula	44	0.13	95	27	14	84	45	18
A. canariensis	43	0.13	100	81	84	81	5	44
A. longiglumis	38	0.11	92	0	18	53	26	16
A. wiestii	32	0.09	100	25	16	66	0	22
A. clauda	30	0.09	100	17	10	63	3	27
A. pilosa[h]	29	0.08	100	28	21	52	3	31
A. prostrata	13	0.04	100	0	85	92	69	69
A. atlantica	12	0.04	100	0	83	100	83	83
A. ventricosa	8	0.02	100	0	0	50	0	0
A. bruhnsiana	4	0.01	100	100	0	0	0	0
Total	4334	12.70						

[a]Percentage of accessions of the taxon with each type of passport data.
[b]Percentage of total accessions of *Avena* documented in the database.
[c]Includes *A. persica* Steud., but not *A. sterilis* subsp. *ludoviciana* (Dur.) Gillet et Magne.
[d]Includes *A. trichophylla* C. Koch.
[e]Includes *A. hirsuta* Moench.
[f]Syn. *A. maroccana* Gand.
[g]Does not include *A. strigosa* subsp. *brevis* Husn.
[h]Includes *A. eriantha* Durieu.

34.5 Representation of Wild Species and Geographic Information for the Collecting Site in the IDBB and EADB

Table 34.1 shows the representation of wild species in the IDBB. As many as 3475 accessions belong to 12 wild species including two subspecies of *Beta vulgaris* L. These make up more than 30% of the total accessions documented in the database. Almost 20% are species belonging to the primary gene pool, with an emphasis on *B. vulgaris* subsp. *maritima* (L.) Arcang. represented by 1975 accessions and some of each of *B. vulgaris* subsp. *adanensis* (Pamuk.) Ford-Lloyd & Will. (78), *B. macrocarpa* Guss. (105) and *B. patula* Ait. (21 accessions). The secondary gene pool with *B. corolliflora* Zosimovich, *B. macrorhiza* Steven, *B. lomatogona* Fisch & Meyer, *B. intermedia* Bunge, *B. trigyna* Wald. & Kid. and *B. nana* Boiss. & Heldr. is represented by 1017, and the tertiary gene pool with *B. patellaris* Moq., *B. webbiana* Moq. and *B. procumbens* Smith by 277 accessions. In most cases collecting sites are well documented.

The representation of wild species in the EADB is shown in Table 34.2. The most important hexaploid wild oats (*A. sterilis* L., *A. ludoviciana* Durieu, *A. fatua* L.) are represented with 2978 accessions (almost 9%), tetraploid species (*A. barbata* Pott ex Link, *A. abyssinica* Hochst., *A. vaviloviana* (Malz.) Mordv, *A. magna* Murphy et Terr., *A. murphyi* Ladiz., *A. damascena* Raj. et Baum, *A. agadiriana* Baum et Fedak, *A. macrostachya* Bal. ex Coss. et Dur.) with 674, and diploid species (*A. strigosa* Schreber, *A. brevis* Roth, *A. hirtula* Lag., *A. canariensis* Baum, Raj. et Samp., *A. longiglumis* Durieu, *A. wiestii* Steud., *A. clauda* Durieu, *A. pilosa* (Roem et Schult.) M. Bieb., *A. prostrata* Ladiz., *A. atlantica* Baum et Fedak, *A. ventricosa* Bal. ex Coss., *A. bruhnsiana* Gruner) with 675 accessions. In total 4334 accessions of wild species make up almost 13% of the European collection represented in the database. Geographic documentation of collecting sites is poor compared to the IDBB.

34.6 Results Available from the IDBB and EADB

Table 34.3 lists the types of results which are displayed by the web application. Listings can be downloaded in a MS Excel compatible ASCII format. If Excel is available at the client computer, downloaded data are automatically displayed as an Excel worksheet.

Table 34.3. Results available from the IDBB and EADB.

Button	Item	Downloadable
Cross table	Rankings (1–9) for characterization and evaluation data listed as a spreadsheet to give a preliminary and easy to read first information	Yes
Observations	Comprehensive listing of characterization and evaluation data including (if available) date of observation, development stage of the observed plants, original data in specified SI units, percentages or scores, descriptive statistics and ranking	Yes
Standard observations	Comprehensive listing of characterization and evaluation data for material explicitly designated as standards in the respective experiments	Yes
Observation methodology	Descriptor definition, methods and protocols used for taking observations	–

Continued

Table 34.3. *Continued*

Button	Item	Downloadable
Experimental details	Responsible person and institute, experiment design, treatments and experiment site	–
Passport data	Taxonomy including synonyms, accession name, origin country, collecting information, accessions including probable duplicates	Yes
Map collecting sites	Under construction: will display collecting sites and characterization and evaluation data on a geographic map	
Gene banks	Addresses of gene banks holding the selected accessions	–
Alleles	Currently only in the EADB: identified alleles (Simons *et al.*, 1978) and accessions carrying these alleles	–/yes

34.7 Characterization and Evaluation Data Available for Crop Wild Relatives

The EADB currently contains almost 170,000 characterization and evaluation observations. A rather low proportion is dedicated to wild species, as GENRES 95–106 (Germeier and Frese, 2002) focused on landraces. A great part of the observations on wild species available in the EADB (Table 34.4), mainly for resistance to diseases (Barley yellow dwarf virus 1331, *Erysiphe graminis* DC. f. sp. *avenae* 11

Table 34.4. Characterization and evaluation data (mainly scores on diseases and pests) for accessions of wild species available in the EADB.

	Number of observations			
Species	Total	%[a]	Diseases	Pests
A. ludoviciana	1407	0.83	1197	21[illegible]
A. sterilis	1238	0.73	1055	183
A. fatua	731	0.43	623	108
A. strigosa	657	0.39	51	
A. barbata	579	0.34	528	51
A. vaviloviana	336	0.20	303	33
A. abyssinica	261	0.15	33	
A. wiestii	204	0.12	179	25
A. magna	133	0.08	123	10
A. longiglumis	121	0.07	109	12
A. hirtula	110	0.07	100	10
A. pilosa	109	0.06	94	15
A. clauda	104	0.06	92	12
A. canariensis	74	0.04	69	5
A. occidentalis	43	0.03	38	5
A. murphyi	27	0.02	25	2

Continued

Table 34.4. *Continued*

Species	Number of observations			
	Total	%[a]	Diseases	Pests
A. brevis	20	0.01	4	
A. bruhnsiana	15	0.01	12	3
A. ventricosa	15	0.01	13	2
A. damascena	14	0.01	11	3
A. atlantica	11	0.01	10	1
A. macrostachya	10	0.01	8	2
A. agadiriana	8	0.00	7	1
A. prostrata	4	0.00	4	
Total	6231	3.68	4688	693

[a]Percentage of total available characterization and evaluation data.

Helminthosporium leaf spot (*Drechslera avenae* (Eidam) Sharif, teleomorph = *Pyrenophora avenae* Ito & Kurib.) 877, *Myrothecium verrucaria* Ditm. ex Steudel 23, *Puccinia coronata* Corda var. *avenae* W.P. Fraser Ledingham 1134, *Puccinia graminis* Persoon f. sp. *avenae* 1126, *Septoria avenae* A.B. Frank (teleomorph: *Phaeosphaeria avenaria* G. F. Weber) 186 observations) and pests (*Oscinella frit* L., 693 observations) have been recently contributed by I. Loskutov, N.I. Vavilov Research Institute of Plant Industry, St Petersburg, Russia.

Currently, 36,513 observations on characterization and evaluation descriptors are available in the IDBB. For accessions of wild species 16,084 (44%) observations are available on 56 evaluation and characterization descriptors (Table 34.5). These include resistance to virus diseases (Beet mild yellow virus 323, Beet yellows virus 366, Rhizomania 336 observations) and fungus diseases (*Aphanomyces cochlioides* Drechs. 379, *Cercospora beticola* Sacc. 345, *Erysiphe betae* (Vanha) Weltzien 368, *Polymyxa betae* Keskin 15, *Pythium*

Table 34.5. Classes of characterization and evaluation data for accessions of wild species available in the IDBB.

Species	Number of observations								
	Total	%[a]	Fungus diseases	Virus diseases	Pests	Phenology	Breeding system	Yield	Morphology
B. vulgaris subsp. *maritima*	11,919	32.6	1321	781	27	1636	501	235	7418
B. vulgaris subsp. *adanensis*	1,259	3.4	80	64		165	69	11	870
B. macrocarpa	858	2.3	79	52	1	135	37	19	535
B. corolliflora	475	1.3	85	35		158	13	26	158
B. macrorhiza	461	1.3	32	21		94	37	25	252

Continued

Table 34.5. *Continued*

			Number of observations						
Species	Total	%[a]	Fungus diseases	Virus diseases	Pests	Phenology	Breeding system	Yield	Morpho-logy
B. lomatogona	287	0.8	40	32		72	3	14	126
B. patellaris	252	0.7	25	10		85	20	19	93
B. nana	198	0.5				19			179
B. procumbens	104	0.3	7	5		34	3	8	47
B. patula	104	0.3	6	3		14	9	1	71
B. intermedia	82	0.2	18	12		8	7	2	35
B. webbiana	56	0.2	5	5		23		4	19
B. trigyna	29	0.1	12	5		1	11		
Total	16,084	44	1710	1025	28	2444	710	364	9803

[a]Percentage of total available characterization and evaluation data

ultimum Trow 346, *Rhizoctonia solani* Kühn 257 observations). Most of these data were contributed from the EU-funded GENRES CT 95–42 project (Frese *et al.*, 2000).

34.8 Known Alleles for Resistance Against Pests and Diseases and Other Traits

By crossing experiments alleles have been identified and named for several traits, especially for genes relating to resistance against pests and diseases. Simons *et al.* (1978) provide a list of these results. Those referring to accessions within the collection held by the N.I. Vavilov Research Institute of Plant Industry have been digitized and provided to the EADB by I. Loskutov (Table 34.6). As

Table 34.6. Number of accessions of wild species with known alleles available in the EADB.

		Number of accessions						
Species	Total	Male sterility	Disease resistance	Leaf	Metabolism	Quality	Rachilla	Stem
A. strigosa	162	13	149					
A. sterilis	58		51	1	2	2		2
A. brevis	40	4	36					
A. fatua	31		15		6	6	4	
A. barbata	11		8	3				
A. ludoviciana	11		11					
A. ventricosa	4		4					
A. vaviloviana	2		2					
A. maroccana	2		2					
A. hirtula	1		1					
A. abyssinica	1		1					

duplicated accessions of the reference material probably exist in many other collections, this information most likely applies to these accessions as well.

Acknowledgements

The usefulness of the databases essentially depends on the willingness of partners to share their data with the central information system as well as on the amount and quality of the data. We acknowledge curators of national gene banks for providing passport data, members of the International Institute for Beet Research (IRRB) for input in kind evaluation work, the European Union for funding characterization and evaluation projects on *Beta* (GENRES CT95–42) and *Avena* (GENRES CT99–106) and the project groups for *Avena* (L. Bondo, A. Katsiotis, J. Koenig, M. Leggett, F. Ottosson, M. Veteläinen) and *Beta* (E. DeAmbrogio, M. Asher, B. Bentzer, W. Beyer, E. Biancardi, A. Börner, B. Ford-Lloyd, G. Koch, H. Loeffler, M. Luterbacher, G. Mandolino, W. Mechelke, O. Scholten, N. Stavropoulos) for providing data to the ECCDBs. Further we acknowledge the ECP/GR for funding the preparation of oat data at the N.I. Vavilov Research Institute of Plant Industry and I. Loskutov for providing the EADB with these data.

References

Baum, B.R. (1977) *Oats: Wild and Cultivated – A Monograph of the Genus Avena L. (Poaceae)*. Printing and Publishing Supply and Services Canada, Ottawa, Canada.

BAZ (2005a) EADB European *Avena* Database, Federal Centre on Breeding Research on Cultivated Plants. Available at: http://eadb.bafz.de (accessed 14 April 2007)

BAZ (2005b) IDBB International Database for *Beta*, Federal Centre on Breeding Research on Cultivated Plants. Available at: http://idbb.bafz.de (accessed 14 April 2007)

BAZ and ZADI (2005a) EADB European *Avena* Database, Federal Centre on Breeding Research on Cultivated Plants, German Centre for Documentation and Information in Agriculture. Available at: http://eadb.zadi.de/eadb/ (accessed 14 April 2007)

BAZ and ZADI (2005b) IDBB International Database for *Beta*, Federal Centre on Breeding Research on Cultivated Plants, German Centre for Documentation and Information in Agriculture. Available at: http://idbb.zadi.de/idbb/ (accessed 14 April 2007)

Berendsohn, W. (1995) The concept of potential taxa in databases. *Taxon* 44, 207–212.

ECPGR (2005) European Cooperative Programme for Crop Genetic Resources Networks ECP/GR European Central Crop Databases (ECCDBs). Available at: http://www.ecpgr.cgiar.org/Databases/databases.htm (accessed 14 April 2007)

FAO/IPGRI (2001) FAO/IPGRI Multi-Crop Passport Descriptors. Available at: http://www.ipgri.cgiar.org/publications/pdf/124.pdf (accessed 14 April 2007)

Frese, L., Ziegler, D., Bücken S., Kraus R., Germeier C.U. (2000) GEN RES #42. Evaluation and Enhancement of *Beta* Collections for Extensification of Agricultural Production. Available at: http://www.fal.de/bgrc/eu9542 (accessed 10 September 2005)

Germeier, C. and Frese, L. (2001a) The International Database for *Beta* – an Expert System for *Beta* Genetic Resources, *Proceedings of the 64th IIRB Congress, June 2001, Bruges (B)*, International Institute for Beet Research, Bruges, Belgium, pp. 123–132.

Germeier, C.U. and Frese, L. (2001b) A data model for the evaluation and characterization of plant genetic resources. In: Swiecicki, W., Naganowska, B. and Wolkon, B. (eds) *Broad Variation and Precise Characterization – Limitation for the Future*. Proceedings of the XVIth Eucarpia Section Genetic Resources Workshop, 16–20 May 2001, European Association for Research on Plant Breeding Section Genetic Resources, Poznan, Polen, Poland, pp. 174–177.

Germeier, C.U. and Frese, L. (2002) GENRES #CT99–106: Evaluation and Enhancement of *Avena* Collections for Extensification of the Genetic Basis of *Avena* for Quality and Resistance Breeding. Available at: http://www.fal.de/bgrc/eu99106 (accessed 10 September 2005)

Germeier, C.U., Frese, L. and Bücken, S. (2003) Concepts and data models for treatment of duplicate groups and sharing of responsibility in genetic resources information systems, *Genetic Resources and Crop Evolution* 50, 693–705.

Hazekamp, Th., Serwinski, J. and Alercia, A. (1997) Multicrop passport descriptors. In: Lipman, E., Jongen, M.W.M., Hintum Van, Th.J.L., Gass, T. and Maggioni, L. (Compilers) *Central Crop Databases: Tools for Plant Genetic Resources Management*. International Plant Genetic Resources Institute, Rome, Italy/CGN, Wageningen, The Netherlands, pp. 35–39 and pp. 75–78.

IBPGR (1984) *Report of a Working Group of Oat*. UNDP/IBPGR European Cooperative Programme for the Conservation and Exchange of Crop Genetic Resources. International Board for Plant Genetic Resources, Rome, Italy.

IBPGR (1987) *Report of a Beta Workshop*. European Cooperative Programme for the Conservation and Exchange of Crop Genetic Resources. International Board for Plant Genetic Resources, Rome, Italy.

IPGRI (2003) EURISCO – Finding Seeds for the Future. Available at: http://eurisco.ecpgr.org (accessed 10 September 2005)

Ladizinsky, G. (1989) Biological Species and Wild Genetic Resources in *Avena*. In: *IBPGR 1989 Report of a Working Group on Avena (Third Meeting)*. European Cooperative Programme for the Conservation and Exchange of Crop Genetic Resources. International Board for Plant Genetic Resources, Rome, Italy.

Ladizinsky, G. and Zohary, D. (1971) Notes on species delimination, species relationships and polyploidy in *Avena* L. *Euphytica* 20, 3, 380–395.

Lange, W., De Bock, Th.S.M. and Brandenburg, W.A (1999) Taxonomy and cultonomy of beet (*Beta vulgaris* L.). *Botanical Journal of the Linnean Society* 130, 81–96.

Letschert, J.P.W., Lange, W., Frese, L., and van den Berg, R.G. (1994) Taxonomy of *Beta* section *Beta*. *Journal of Sugar Beet Research* 31, 69–85.

Loskutov, I. (2002) *Systematical Approaches to the Genus* Avena *L*. Available at: http://www.vir.nw.ru/avena/sys_app.htm (accessed 14 April 2007)

Loskutov, I.G. (1998) Database and taxonomy of VIR's world collection of the genus *Avena* L. In: Maggioni L., Leggett, M., Bücken, S. and Lipman, E. (compilers) *Report of a Working Group on Avena. Fifth meeting, Vilnius, Lithuania, 7–9 May 1998*, IPGRI, Rome, Italy, pp. 26–31.

PHP Group (2001–2005) *php*. Available at: http://www.php.net (accessed 14 April 2007)

Pullan, M.R., Watson, M.F., Kennedy, J.B., Raguenaud, C. and Hyam, R. (2000) The Prometheus taxonomic model: a practical approach to representing multiple classifications. *Taxon* 49, 55–75.

Schittenhelm, S. and Seidewitz, L. (1993) Progress on the European *Avena* database. In: Frison, E.A., Koenig, J. and Schittenhelm, S. (eds) *Report of the Fourth Meeting of the ECP/GR Avena Working Group*. International Board for Plant Genetic Resources, Rome, Italy, pp. 37–49.

Shen, Y., Ford-Lloyd, B.V. and Newbury, H. (1998) Genetic relationships within the genus *Beta* determined using both PCR-based marker and DNA sequencing techniques. *Heredity* 80, 624–632.

Simons, M.D., Martens, J.W., McKenzie, R.I.H., Nishiyama, I., Sadanaga, K., Sebesta, J. and Thomas, H. (1978) *Oats: a Standardized System of Nomenclature for Genes and Chromosomes and Catalogue of Genes Governing Characters*, Agriculture Handbook 509, United States Department of Agriculture, Washington DC

35 Crop Wild Relative Information: Developing a Tool for its Management and Use

I. THORMANN, A. LANE, K. DURAH, M.E. DULLOO AND S. GAIJI

35.1 Introduction

The importance of the relatives, whether close or distant, of modern crops to continuing improvement in crop production cannot be underestimated. Indeed, modern cultivars of most crops now contain genes that are derived from a wild relative whether through natural gene flow or human-induced processes. Crop wild relatives (CWR) continue to play a critical and leading role in the search for new genes to improve both quantity and quality of agricultural production and to reduce risk of crop losses by providing resistance to biotic and abiotic disturbances.

'*In situ* conservation of crop wild relatives through enhanced information management and field application' (UNEP/GEF Crop Wild Relatives Project) is a UNEP/GEF-supported project that addresses national and global needs to improve food security through effective conservation and sustainable use of CWR. This large, multifaceted, 5-year project was launched in 2004 and involves five countries and six international organizations to focus on improved management and use of information on these species.

The five countries, Armenia, Bolivia, Madagascar, Sri Lanka and Uzbekistan, each have significant numbers of important and threatened CWR and all are among the world's biodiversity hot spots – places that have the highest concentrations of unique biodiversity on the planet (Mittermeier *et al.*, 2005). Bioversity International (formerly known as IPGRI) is the project manager, and five other international conservation agencies are formally partners in the project and are contributing experience, expertise and information to the project. These are the Food and Agriculture Organization of the United Nations (FAO), Botanic Gardens Conservation International (BGCI), the United Nations Environment Programme's World Conservation Monitoring Centre (UNEP-WCMC), the World Conservation Union (IUCN) and the Information and Coordination Centre for Biological Diversity of the German Federal Agency for Agriculture and Nutrition (BLE). In addition, the project is collaborating with other global and regional initiatives that involve CWR and/or information management and use, such as GBIF (Global Biodiversity Information Facility) and PGR Forum (European Crop Wild Relative Diversity Assessment and Conservation Forum) (see Maxted *et al.*, Chapter 1, this volume).

35.2 Crop Wild Relatives in the Project Countries

All five countries participating in the project possess a rich diversity of CWR. Armenia possesses many species of wild relatives of domesticated crops, including three of the four known wild species of wheat (*Triticum boeoticum* Boiss., *T. urartu* Tumanian ex Gandilyan and *T. araraticum* Jakubz.), many species belonging to the genus *Aegilops* (e.g. *Ae. tauschii* Coss., *Ae. cylindrical* Host, *Ae. triuncialis* L.) and wild relatives of rye and barley. Wild apple and pear species grow in most of Armenia's forests, together with wild forms of several other fruits and nuts (e.g. quince, apricot, sweet and sour cherry, walnut, pistachio and fig). The Caucasus Mountains form a significant feature of the country determining much of the character of the biodiversity.

Bolivia's location within the Andean region, where several important biomes are represented within a limited geographical area, and where mountain ecosystems form one of the major components, means that it is rich in natural biodiversity. It is within one of the world centres of crop domestication and within centres of diversity of important crops such as potato, sweet potato, maize, groundnut, cassava, cotton, tobacco, cocoa, bean and pepper, as well as several local Andean tubers (i.e. *Ullucus tuberosus* Caldas, *Oxalis* spp.), quinoa (*Chenopodium quinoa* Willd.), tarwi (*Lupinus mutabilis* Sweet) and others. Most of the CWR of these and other Bolivian species are characterized by environmental and soil stress tolerance, disease resistance and other adaptive traits that could be useful for crop improvement programmes, and are of particular interest to Bolivia.

Madagascar's rich environment and climates have generated a remarkable natural biodiversity and it is recognized as a world centre of plant diversity. Much of this diversity is associated with the island's mountain ecosystems. The numerous CWR in Madagascar include two wild relatives of rice (*Oryza longistaminata* A. Chev. & Roehr. and *O. punctata* Kotschy ex Steud.), which possess virus and pest resistance, one wild relative of sorghum (*Sorghum verticilliflorum* (Steud.) Stapf), two wild relatives of *Vigna* (*V. vexillata* (L.) A. Rich. and *V. angivensis* Baker) and a wild relative of banana (*Musa perrieri* Claverie). Most significantly, the country possesses almost 50 species of wild coffee (section *Mascarocoffea* Chev.) with interesting and unique traits.

In terms of species, genes and ecosystems, Sri Lanka has very high forest biodiversity and agrobiodiversity. The highland areas support montane subtropical broadleaf hill forest and wet temperate forest. Important wild relatives of crops include cereal relatives in the genera *Oryza* L., *Eleusine* Gaertn. and *Panicum* L. There are also important legumes (e.g. *Vigna* Savi, *Cajanus* DC.), vegetables (e.g. *Abelmoschus* Medik., *Solanum* L., *Ipomoea* L.), oilseeds (e.g. *Sesamum* L.), fruits (e.g. *Citrus* L., *Garcinia* L., *Mangifera* L., *Musa* L.) and other economic plants (e.g. *Cinnamomum* Schaeff., *Curcuma* L., *Tamarindus* L.).

Vavilov (1935) identified Uzbekistan as one of the centres of origin of many modern crop plants. It has some of the nearest wild relatives of cultivated onion (*Allium oschaninii* O. Fedtsch., *A. vavilovii* Popov & Vved., *A. praemixtum* Vved., *A. pskemense* B. Fedtsch.), as well as many wild fruit and nut species (*Vitis vinifera* L., *Pistacia vera* L., *Malus sieversii* (Ledeb.) M. Roem., *Pyrus turcomanica* Maleev and *Rubus caesius* L.). Uzbekistan forms part of the global centre of plant diversity in the mountains of middle Asia and is in a centre of crop plant diversity.

35.3 Threats to Crop Wild Relatives

The natural populations of many CWR in these and other countries are increasingly at risk and they are at present poorly conserved *in situ*. In the wild, they are under threat from habitat loss, degradation and fragmentation, invasive species, agricultural expansion and global climate change. The economic pressure on farmers to grow improved high yielding varieties in place of more traditional varieties is reducing on-farm diversity and increasing risk of crop losses from pests, diseases and unfavourable climatic conditions. There is an urgent need to address the effective conservation and strategic use of CWR.

In the project countries, there has been very little progress in conserving CWR, as is the case the world over. The reason for this situation is that there are major constraints that can be attributed to the challenges involved in developing conservation plans for a group of species which vary in their individual, biological, ecological and use characteristics. Also there are no collaborative frameworks between different agencies (agriculture, forestry and environmental) to work together to implement effective conservation and use activities for target priority species. Awareness among decision makers of the importance of CWR is also often lacking.

A major limitation on effective and sustainable *in situ* conservation of CWR is in the capacity to bring together and use information. A number of studies have shown that substantial amounts of information often exist on CWR (e.g. Thormann *et al*., 1999), but that it is dispersed among different institutions and agencies in different countries and international organizations. Due to the sometimes large geographic separation between cultivated crops and their wild relatives and the innate complexities of *in situ* conservation, conservation efforts need to be undertaken on a regional or even global scale and to adopt integrated multi-institutional and multidisciplinary approaches.

Thus, the UNEP/GEF Crop Wild Relatives Project was developed with a major focus on the systematic compilation, access and use of information related to CWR, at international and national levels. An analysis of this information is a first critical step prior to the development and implementation of *in situ* conservation and monitoring strategies for CWR at the national level.

35.4 Information Management

35.4.1 Information availability

Many international conservation organizations maintain online databases that contain data relevant for CWR. Some of the major information sources are held by the international partners of the UNEP/GEF project mentioned above. Bioversity maintains the System-wide Information Network for Genetic Resources (SINGER, 2005) of the Centres of the Consultative Group on International Agricultural Research (CGIAR). SINGER provides access to data on more than half a million samples, conserved in the CGIAR gene banks. Information about protected areas can be obtained from UNEP-WCMC's World Database on Protected Areas WDPA (UNEP-WCMC, 2006). FAO's World Information and Early Warning System WIEWS (FAO, 1990–2005) provides information about *ex situ* collections existing around the world. Data about endangered species and Red List information can be obtained from IUCN's Red List database (IUCN, 2006), while BGCI maintains a global database documenting botanic gardens all over the world (BGCI, 2006). Many other databases and information resources do exist and are available for searches at the respective web sites of the institutions and organizations that manage them.

In all partner countries, information on CWR exists in different sources such as herbaria and gene banks that can be used to determine the likely origin and sometimes specific location of populations of species of CWR. Information on the extent and distribution of protected areas is also available from responsible agencies in the various ministries that cover environmental management issues.

The results of the GEF PDF B phase of the project showed that information held by national institutions is usually dispersed and in a form that is not easily used. Not all of the information is available in computerized form and most of the location data (e.g. latitude and longitude) has still to be digitized. Armenia, for example, had no CWR-related database in electronic format. Where parts of the information have been computerized (e.g. Bolivia, Madagascar and Sri Lanka), the different agencies have developed independent information management systems with different data structures and formats. Nine institutions had been evaluated in Bolivia during the preparatory phase of the project for their institutional capacity for handling CWR information. Five of these institutions manage computerized data, one of them making them available on the Internet. The other four institutions do not yet manage computerized data. Noteworthy is the national inventory of CWR in Bolivia that was carried out during the GEF PDF B phase of the project by the Fundacion Amigos de la Naturaleza (FAN) and the Museo de Historia Natural Noel Kempff with support from the United States Department of Agriculture (USDA) and Bioversity. The inventory compiled data from main herbaria in Bolivia, Argentina and the USA, and two Bolivian gene banks. In Uzbekistan, three databases containing records of *ex situ* collections have been developed during the past years, but using different software programs.

In addition to the purely technical difficulties in sharing data due to different data structure and format or lack of digitized data, a major constraint in most countries is the absence of clear policies to share information. Project partners will develop collaborative agreements for information access and exchange.

35.4.2 Information systems

Given that information about CWR is widely dispersed at both national and international levels and that it is kept in many different formats, it is very difficult for the countries to access and make effective use of the existing data to develop conservation plans. As mentioned previously, relevant data from international databases is currently available to the global community. However, it can only be obtained by accessing databases individually.

One of the main objectives of the UNEP/GEF Crop Wild Relatives Project is therefore to develop a global information system that will provide easy access and search facilities for information on CWR. An international information portal dedicated to CWR will be developed to serve as a gateway to such information. Users will be able to search databases and information resources, starting with the above listed databases maintained by the international project partners, through the portal. Building on the most recent developments in approaches to information harvesting, such as GBIF/BIOCASE web service protocols, the international portal will provide access to distributed information sources while at the same time retaining the autonomous management and individual structure of the resource databases.

The international web portal will be linked to five national information systems, one in each partner country. The portal will achieve this through agreed scientific standards between the partners. The national information systems will bring together relevant existing national information sources and enrich them with new data that will be generated from field surveys during the course of the project. Making national data and information resources available through a national inventory and from the portal will enhance its use for analysis, conservation planning and decision-making purposes at country level. The information flow within the global system is represented in Fig. 35.1.

National data and information will be compared and analysed and this will be particularly interesting for those countries working on the wild relatives of the same crops, for example banana and rice, which are important for both Madagascar and Sri Lanka. However, integration and exchange of information from different national systems is potentially difficult in the UNEP/GEF Crop Wild Relatives Project. To achieve more coherency and consistency in data management and to simplify the processes of information collation, evaluation and exchange, Bioversity is developing a tool, 'Genetic Resources Information System', GRIS v4.0.

35.5 Genetic Resources Information System (GRIS)

GRIS is an open source software that was developed by Bioversity. GRIS was conceived as a modelling tool to document and manage plant genetic resources,

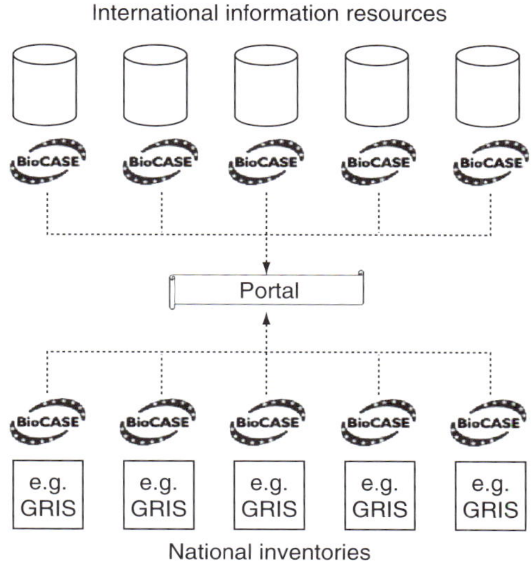

Fig. 35.1. Illustration of information flow between sources in the global information system being developed within the framework of the UNEP/GEF Crop Wild Relatives Project.

from both *in situ* and *ex situ* perspectives. An earlier version of GRIS (v3.1) is already being used at national programme levels in Morocco, Algeria, Tunisia, Libya, Egypt, Lebanon, Yemen, Oman and Vietnam. This integrated system is well suited to be used at national level to store and manage information related to conservation of CWR, which are the focus of both *in situ* and *ex situ* conservation actions. Bioversity, in collaboration with project partners, is now working on modifying and refining GRIS to ensure that it encompasses the needs of national partners and can be used in other projects such as the UNEP/GEF Crop Wild Relatives Project and other ongoing and proposed projects that address conservation of genetic resources. For example, the UNEP-GEF project '*In Situ*/On-Farm Conservation and Use of Agricultural Biodiversity (Horticultural Crops and Wild Fruit Species) in Central Asia' will be implemented in 2006 and will use GRIS v4.0 as the information system for the project. Project partners and countries that will use GRIS v4.0 will receive training and support from Bioversity in setting up GRIS and getting it to work in their countries. A key feature of the system is a help desk that will provide users with help and technical support or assistance as they may require.

GRIS consists of two types of databases: (i) static databases to document passport-related information; and (ii) dynamic user-defined databases to document the characterization and evaluation-related information. Information is entered into all databases according to clearly defined standards (e.g. taxonomic nomenclature, date formats, weights and measures, etc.) and using agreed descriptors. The inbuilt user-defined databases allow users to modify databases to suit their needs.

GRIS uses a conceptual modelling language which means that relationships between biological, environmental and other types of information can be described and the impacts of changes in one variable on another can be predicted through modelling. Conceptual models can be very helpful in applied studies of ecological species or systems, where the objective is to predict the response to a particular stressor or remedial action. GRIS also has a facility to visualize data on GIS maps and display the area where a particular species was sampled and whether it has been collected. These data can be used to identify actual and potential sampling sites for *ex situ* collections and *in situ* surveys.

35.5.1 *In situ* information

GRIS includes a wide range of data fields to cover information collected during field surveys of species and species habitat including:

- Ecogeographic surveys:
  - Site information such as point location, climate, soil, geological features and social factors such as land use and population pressure;
  - Taxon information including taxonomy, population dynamics, conservation status, threats and uses.
- Monitoring and evaluation:
  - Assessing change over time in, for example, species populations, conservation status, threats, habitat condition and land use.

GRIS can also capture and view indigenous knowledge through integrating multimedia features such as video, audio clips and photographic images into GRIS and linking these multimedia files to a particular sample or species.

35.5.2 *Ex situ* information

GRIS maximizes both operational and data management efficiency in *ex situ* collections by assigning each accession with a bar code that efficiently allows:

- Optical identification;
- Collection site mapping;
- Automatic data entry/retrieval;
- Import/export germplasm data.

To manage the storage location and status of each sample, GRIS keeps track of samples' locations and records their viability test results in order to:

- Monitor number of seeds per accession;
- Monitor seed viability;
- Estimate regeneration date.

To enhance the reliability of collecting survey information, remote and mobile technology is incorporated into GRIS to make use of optional information and communication technology tools like Personal Digital Assistance (PDA) to use electronic survey forms, digital cameras, MP3 player, Global Positioning System

(GPS) and a PDA-modem for remote transfer of survey information to the GRIS host computer. Using these tools will make data transfer more efficient but they are not necessary for operation of the system.

GRIS has a data exchange module that provides a powerful way of sharing and disseminating information between partners irrespective of other databases they might have. It allows users to provide passport data information in any possible format. Furthermore, the database tables were made in a structured and scalable manner to allow easy migration or conversion of existing national databases into the GRIS data model.

GRIS will support multi-user and client/server architecture which allows space and mobility to the user and more flexibility in implementation and accommodates different users' operational scenarios. Multilingual capacity, using dynamic language definition (DLD) technology, adds value to GRIS and facilitates international usage. In addition to the extended CWR/*in situ* module that will characterize GRIS v4.0, further developments are planned such as the integration of on-farm data, gap analysis, breeding and molecular tools.

35.5.3 Using the information system to support conservation actions

Information management and effective conservation management are inextricably linked. In this project, the collation and application of a range of information categories is central to the development of *in situ* conservation plans. To develop and implement conservation strategies that are needed to conserve priority CWR *in situ*, it is essential to understand where species occur and where they are at greatest risk. Obviously it is not possible to survey all areas in a country for the occurrence of CWR, or for that matter, any target species. Thus, conservation planners need to be able to predict species ecological and geographical distribution from analysis of data on species location and habitat, which is available from information databases held by various relevant institutions. Habitat information is particularly important here as it refers to the place where the species usually lives, often described in terms of physical features such as topography and soil moisture and by associated dominant forms and therefore can provide useful indicators to identify interspecific and intraspecific diversity of target taxa. For example, a species may predominantly occur in swampy valleys at low elevations and within a certain rainfall and temperature range.

Overlaying accurate information such as vegetation type, climate, terrain (elevation, slope, relief and aspect) and soils will allow modelling of species' distributions and their diversity patterns. Other information on existing and proposed developments such as human settlements, roads and dams is also essential both for rationalizing field surveys and for assessing ongoing and emerging pressures on populations of CWR through land development projects. Social and economic information such as local uses of certain species are important to determine economic value, cultural importance and also potential pressures due to exploitation.

Species distribution mapping in the early stages of the project using *ex situ* data will allow scientists to more reliably select potential sites for *in situ* surveys. Analysis of both sets of information combined with the data layers mentioned above will form the baseline against which the impact of targeted conservation strategies and of identified threats can be assessed over time.

One of the sustainable outcomes of the project will be the refinement of technologies to link information on species distribution and other spatial and temporal data through GIS analysis. The use of GIS allows better analysis of data and prediction of future conditions, which is essential for conservation management and planning. The biological and physical information collected by this project and managed in GRIS can be seamlessly transferred to GIS programs such as FLORAMAP and DIVA. The project will further develop and test procedures for using spatial information as a tool in conservation of CWR.

35.6 Conclusion

Effectively managed biological, physical, social and economic information is critical for planning and implementing conservation strategies, monitoring ongoing conservation status and ultimately minimizing loss of species populations as a result of new or escalating threats. However, *in situ* conservation of CWR is currently hampered by lack of capacity to bring together and use relevant information. GRIS v4.0 is being developed to address this major constraint at the national level. When fully developed it will integrate and manage information from both *in situ* and *ex situ* sources and will be an important tool to support more informed management decisions for species and their populations. It is envisaged that GRIS will eventually be widely used to manage *in situ*, *ex situ* and on-farm information relevant to conservation of genetic resources and will thereby optimize cost-effectiveness and efficiency in management and exchange of information between national and international users.

References

BGCI (2006) *Botanic Gardens Search*. Available at: http://www.bgci.org/garden_search.php (accessed 20 March 2006)

FAO (1990–2005) *World Information and Early Warning System*. Available at: http://apps3.fao.org/wiews/wiews.jsp (accessed 20 March 2006)

IUCN (2006) *The IUCN Red List of Threatened Species*. Available at: http://www.redlist.org/ (accessed 20 March 2006)

Mittermeier, R.A., Gil, P.R., Hoffman, M., Pilgrim, J., Brooks, T., Goettsch Mittermeier, C., Lamoreux, J. and da Fonseca, G.A.B. (2005) *Hotspots revisited*. Conservation International, Washington, DC.

SINGER (2005) *The System-wide Information Network for Genetic Resources*. Available at: http://singer.grinfo.net/ (accessed 20 March 2006)

Thormann, I., Jarvis, D.I., Dearing, J.A. and Hodgkin, T. (1999) Internationally available information sources for the development of *in situ* conservation strategies for wild species useful for food and agriculture. *Plant Genetic Resources Newsletter* 118, 38–50.

UNEP–WCMC (2006) *World Database of Protected Areas*. Available at: http://sea.unep-wcmc.org/wdbpa/ (accessed 20 March 2006)

Vavilov, N.I. (1935) *Teoreticeskie osnovy selekcii*. T.l. Moskva & Leningrad.

36 Managing Passport Data Associated with Seed Collections from Wild Populations: Increasing Potential for Conservation and Use of Crop Wild Relatives in Israel

R. HADAS, A. SIROTA, M. AGAMI AND A. HOROVITZ

36.1 Introduction

Israel's location in the Middle East heartland of genetic diversity for many major agricultural crops makes it a part of the 'Fertile Crescent', where, in Neolithic times, native useful plant species were first cultivated instead of being harvested from the wild. It is located in an area where the Mediterranean plant-geographical region borders on the tail ends of the Irano–Turanian, Saharo–Arabian and Sudanian regions (Zohary and Feinbrun-Dothan, 1966–1986; Danin and Plitman, 1987). Elements of all four regions contribute to the species richness of the area. Danin (2004) lists 2750 species in 138 families for the 'area of the Flora Palaestina', i.e. an area of less than 50,000 km^2 that comprises the present area of Israel and the area east of it up to the 36th meridian. The local species diversity is accentuated by a concentration of dissimilar ecological ranges: marine, mountainous, steppe and desert. About 10% of the species in this flora are related to Old World crop plants as wild-growing relatives, feral derivatives or direct ancestors of crop plants. Other wild species are in folk use as medicinal or spice plants. The region of Israel and its neighbours are thus a unique centre for a wealth of wild relatives of a large variety of important cultivated plants. The destruction of wild habitats due to spreading urbanization, intensive farming and other land use makes the preservation of wild genetic riches an imperative. Similarly, the residual landraces which local farmers have replaced by elite cultivars need to be preserved. To meet these needs, the Israeli Gene Bank for Agricultural Crops (IGB) was established in 1979 by Israel's National Council for Research and Development (now the Ministry of Science, Culture and Sport) and the Ministry of Agriculture and Rural Development. At present, these two governmental bodies, together with scientists from academic institutions and from Israel's seed industry, collaborate in efforts to conserve plant genetic resources in the area. The main effort is invested in a national seed bank. In addition, there are small living collections and clonal nurseries of some of the candidate species, and a few selected populations are being conserved *in situ*. A recent additional objective of the IGB is the conservation of other native plant species that are rare and in danger of extinction.

36.2 Seed Collections for *Ex Situ* Conservation

Since 1997 the IGB has collaborated with scientists at Tel Aviv University, the Hebrew University and the local plant information service 'Rotem' in rescue collections from endangered wild-growing populations of crop plant relatives and, in some cases, from abandoned orchards. Initially this project was supported by the United States Department of Agriculture (USDA) and the 'Yad Hanadiv' Foundation. So far, most of the collections were made in a dozen development areas where the vegetation is endangered. A list of species or species subtaxa in need of conservation was compiled. Table 36.1 shows the distribution of these taxa in different categories of breeding use and in different categories of countywide abundance. Over a fifth of the target species are potential breeding resources for numerous crop categories in addition to the one to which they have been assigned in Table 36.1. To characterize its holdings with regard to abundance in nature, the IGB uses the groupings of the Rotem Plant Information Service. Based on accumulated and ongoing plant search in many thousands of sites, this organization has divided the native flora into approximate abundance categories, and these are summarized for the assignments in Table 36.1. Since the objective in rescue collections is the conservation of genotypes and alleles, which may be lost when a given population of even the commonest species becomes extinct, no priority is given to conser-

Table 36.1. Number of target taxa in different categories of breeding use and countywide abundance.

Crop category for which the taxon is a genetic resource	No. of target taxa (species and subspecies)	Common	Rare	Very rare[a]
Cereals	16	6	5	5
Dye plants	4	3	–	1
Fibre plants	4	2	1	1
Forage and fodder plants	32	11	16	5
Fruits and nuts	11	4	6	1
Medicinal plants	73	46	20	7
Oil plants	2	–	2	–
Ornamentals	35	14	13	8
Pulses	15	4	10	1
Spice, perfume and tea plants	40	17	18	5
Vegetables and tubers	28	16	9	3
Total	260	123	100	37

[a]Observed in less than 31 sites.

vation of rare species in this approach. As can be seen from the table, nearly half of the taxa grow abundantly in our area. Among such common plants are crop wild relatives (CWR) species such as wild barley and wild oats. Others are medicinal plants, either species with domesticated counterparts such as holy thistle, *Silybum marianum* (L.) Gaertn. (*Compositae*), species on the fringe of domestication such as small caltrops, *Tribulus terrestris* L. (*Zygophyllaceae*), or invasive weeds that are at this stage harvested from the wild for culinary, medicinal or veterinary uses, such as mesquite, *Prosopis farcta* (Banks et Sol.) MacBride (*Mimosaceae*).

Where possible, 20 or more plants are sampled to represent a collection site population, and the total collection is treated as an accession. So far, some 3000 accessions from endangered habitats have been collected and deposited in the bank. However, in a future approach, selected species will be collected throughout their distribution area in the country rather than in areas where their habitats are under an immediate threat of extinction.

36.3 Documentation

The documentation supplied by the compiler has to be reliable and precise, so that it can supply empirical data for present and future users and analysts. It has to convey data that have the same unanimous meaning for the compiler and the user. Within a descriptor, different entries conveying the same datum have to be uniform.

To achieve all this, the data can be categorized and coded and entered into the database as fixed templates in a language that is internationally understood. The infrastructure of the database has to have a large holding capacity, a capacity to accommodate multiple users, and automatic back-up systems. Categorization by use of fixed templates has many advantages. Categorized data can be compared, queried and manipulated, but many of the innumerable nuances exhibited by the wild population from which the collection is made, and which one can still find in individual labels attached to herbarium specimens, are lost. However, in cases in which the template adopted for a variable embraces a further series of variation, it provides a useful starting point for more detailed studies of selected aspects.

Up to now our database has been maintained in Access format. It is now being transformed to MySQL to establish a better interface with the user. Where applicable, we follow the general guidelines of FAO/IPGRI Multicrop Passport Descriptors (MCPDs) proposed by Alercia *et al.* (2001). While this format fits accessions of cultivars and breeders' lines, it is inadequate for wild plant collections. For these latter, the database is augmented by a set of data that specify the methods by which a collection is sampled and another set with details on the provenance of the collection and the structure of its source population.

The current format of our database for individual accessions is shown in Table 36.2. Under 'Administrative status data' we include the nomenclature of the accession adopted by us, which is that of the *Flora Palaestina* (Zohary and Feinbrun-Dothan, 1966–1986) and its revisions. When a revised name is used,

Table 36.2. Descriptors in the database (MCPDs and their abbreviations are shown in capital letters).

Administrative status data	Collection data	Provenance data
ACCENUMB	COLLECTORS	Latitude N
INSTCODE	COLLDATE	Longitude E
ACQDATE	No. of plants sampled	COLLSITE
ORIGCTY	Sampling method: *all plants* *selected plants* *random*	District
DONORCODE	Sampling strategy: *transects* *diffuse* *single points, etc.*	Target area
SAMPLE TYPE (SAMPSTAT)	Distance between sampled plants	ELEVATION
FAMILY	Material collected: *single plant seed* *bulked seed* *herbarium vouchers*	Exposure/slope (degrees)
GENUS	Pests/diseases	COLLSRC = habitat
SPECIES	Parts stored	Associated species
SPECIES AUTHOR		Soil
SUBTAXON w author		Physical structure of population: *size, area, shape, density, plant distribution*
SynFloPal		Phenotypic structure of population *special features*
Common name		
Hebrew name		

the *Flora Palaestina* synonym is entered into the Status section of the database. Each taxon has a generic and specific Hebrew name. In addition, as far as available, an English common name is entered, and we are also compiling a list of Arabic common names for the Status section. An important additional entry in this section is the MCPD descriptor SAMPSTAT. Most of our collections belong to the sample status categories 'Wild natural' if collected from an undisturbed site and 'Wild disturbed' if the collection site is disturbed. Other categories are 'Replanted from the wild' to gardens or plots; 'Abandoned crop', mainly surviving orchard trees of almond, carob or olive on formerly cultivated hill terraces; and 'Presently growing crop'. In the last category, landraces and primitive varieties of unknown parentage are of interest.

In the second section of the database, entitled 'Collection data', entries for the descriptors 'Material collected', 'Sampling method', 'Sampling strategy' and 'Parts stored' are templates for which the collector encircles a Hebrew equivalent on the field collection form. The sampling method is not always a matter of free choice. In a small population, every plant has to be sampled. If plants are selected, the collector specifies the criteria for selection, such as earliest or latest fruiting plants which are the only available sources for collection, or plants with the largest fruit or single healthy plants among others that are disease infected, etc. There are many alternative sampling strategies, such as adjacent collections from the same sampling point, only a single collection from each of a series of patches, collections around the periphery of the sampling site, collections from different elevations on the site or from only shaded or only shade-free points. The collector's observation of infestation by pests and disease infection is also included in this section. The entry for 'Part stored' is made after the collection has been cleaned and specifies the type of propagule that is put into storage. The quantity of seeds, fruits or other propagules is also entered here.

In the third section, entitled 'Provenance data', entries for the descriptors 'District', 'Target area', 'Exposure', 'COLLSRC', 'Soil', 'Physical structure of population' and 'Phenotypic structure of population' are given in fixed templates. Up to now we have used the British Mandatory Palestine or Cassini-Soldner grid for latitude and longitude coordinates, which we give in four digits and hundreds of metres. Alternatively we use the corresponding Universal Transverse Mercator (UTM) units. In the descriptor COLLSITE, the location of the collection site is further specified by its relation to the nearest geographical landmark.

The area of the Flora Palaestina has been divided into a series of phytogeographical districts, and we assign each collection to its source District. The descriptor 'Target area' refers to one of the selected endangered areas in which most of the collections have been made so far. In the region under discussion, the descriptors 'Elevation' or 'Altitudinal range' include below sea-level elevations of the Jordan Rift Valley. The descriptor 'Exposure' indicates whether the sampling site is flat, undulating or sloped, and the compass direction of the slope and its angle. To provide data for the 'Collection source' or habitat column, the collector encircles one or more categories of a wide choice, given on the field questionnaire. Some of these can be considered Mediterranean subcategories of the broader MCPD COLLSRC variants. Examples are Maquis and Park Forest as subcategories of Woodland and Batha and Garrigue as special forms of Scrubland. A common semi-wild habitat for CWR material, specific to hill regions in the modern Middle East, is the abandoned terraced orchard. In the course of time, it reverts to scrubland. We ask our collectors to name up to five plant species that grow on the sampling site together with the taxon that is being sampled. This descriptor helps to characterize the habitat. The descriptor 'Soil' is a further aid to habitat characterization. Up to now we have asked collectors to encircle soil types or bedrock types on the field form. The physical structure of the collection site population (Fig. 36.1) is recorded by its size in estimated number of individuals, its area, shape, pattern of plant distribution and plant density. If time allows the collectors to study the surroundings of the

Population size (estimated number of individuals)

1-10 ; 10-100 ; 100-500 ; 500-1000 ; >1000

Population area m^2

Population shape

Rectangular	Core w branches	Fragmented	On different levels
Uneven	Roundish	Oblong	Strip
Other *specify*			

Plant distribution

Uneven	Mainly single plants	Large patches
Even	Small even patches	Other *specify*

Plant density

Sparse	Medium	Dense	Irregular

Fig. 36.1. Part of field collection form: physical structure of the collection site population.

collection site, the collection site population can be given further qualifications in terms of its relation to a larger population of which it is a part. At this stage, our database contains no entries regarding the genetic structure of source populations. Collections for which such information exists are held and documented by the respective researchers. However, we have entries for phenotypic features of the collection site population. So far, we have only recorded exceptional features that affect the entire population (plant size, earliness, lateness), or variations in flower or fruit colour and different sex forms in polygamous populations.

Each accession is also accompanied by a management record, which documents its storage history. This includes the long-term/short-term storage type, the storage regime, the dates and results of germination tests, the distribution of samples to applicants and regeneration records.

In parallel with the standard database and the management record, we are beginning to compile a reference table for background information on individual candidate species, comprising global information, local information and biological data. This open-ended table should summarize information as it becomes available. The descriptors for global information are 'World distribution', all 'Scientific synonyms and English common names' known for the respective taxon, 'Related wild species', 'Related crops', 'Uses in breeding', 'Literature (genetic and PGR-related studies) and 'Representation in other seed banks'. Other databases, including the Crop Wild Relative Information System (CWRIS) (see Kell *et al.*, Chapter 33, this volume) should help us in compiling global information. Detailed local studies can sometimes furnish data for the second and third sections of the reference table, and this is exemplified here for *Anemone coronaria* L. in Table 36.3. The biological data should be of aid in conservation-related activities, such as prediction of gene flow in the source population, and in

Table 36.3. Local information and biological data for *Anemone coronaria* L. (*Ranunculaceae*, 2n = 2x = 16 (confirmed in Israel)).

	Local information
Local distribution	Species recorded in Phytogeographic Districts 1–19 and 23–24
Abundance	CC = very common in many of the Districts
Weediness	None
Presence of related spp.	None
Range of habitats	Wide, with phenotypic variation in soil and temperature preferences
Common pests/diseases	Anemone rust, *Tranzschelia pruni-spinosae*
	Biological data
Life form	Geophyte with lifespan of less than 10 years. Winter pollen source for wide range of insects. Most of these are non-pollinating.
Phenology	December–April flowering with phenotypic variation. Summer dormancy induced by long photoperiod and high temperatures.
Gender	Hermaphrodite with short female and longer male stage, very rarely male sterile.
Mating system	Largely outcrossed through protogyny. Selfing causes inbreeding depression.
Pollination	Wind, as well as insects that visit flowers for night shelter and/or mating: male bees (e.g. *Eucera*) and scarab beetles. Flowers are thermonastic and close at night.
Pollen flow	Within 1.5 m by wind; longer distances by night visitors.
Dispersal	Wind, sometimes up to several hundred metres
Germination	Embryo immature at time of dispersal; seed has an after-ripening requirement of up to 5 months. Germinates in dark and light at 10–15°C. Germination lag of 1–2 weeks caused by water-repellent woolly coat of nutlet.
Juvenile period	Time from germination to flowering is 2–4 years when unaided.

manipulations and regeneration of the stored accessions. It must be borne in mind that data for biological descriptors are apt to vary in different geographical regions, and that the data in the reference table are only applicable to local conditions. In a seed bank which engages in worldwide exchange of plant material, geographical variation, especially that in phenological traits such as dormancy behaviour and day length response, may need wider study.

36.4 Hopes for the Future

As a gene bank in a fluid stage of modernization, we aim at benefiting from the experience of larger organizations and the worldwide knowledge that has been invested in the topic of CWR conservation. Since we find ourselves in a privileged area with respect to plant genetic resources, we hope for and anticipate collaborative *ex situ* conservation efforts within a regional network of complementary neighbouring seed banks and clonal nurseries. Documentation should provide links between the different conservation ventures in the region, disclose gaps that need further coverage and study, enable us to become acquainted with regional gene pools in their totality and promote public awareness of the asset contained in the regional gene pools of wild relatives of crop plants.

Acknowledgements

The authors would like to thank Dr Yehoshua Anikster, who has been the pioneer and establisher of CWR collection in Israel, and all our colleagues and collectors for their contribution to our activities.

References

Alercia, A., Diulgheroff, S. and Metz, T. (2001) *Multicrop Passport Descriptors*. IPGRI Publications, Rome, Italy.

Danin, A. (2004) *Distribution Atlas of Plants in the Flora Palaestina area*. Israel Academy of Sciences and Humanities, Jerusalem, Israel.

Danin, A. and Plitman, U. (1987) Revision of the plant geographical territories of Israel and Sinai. *Plant Systematics and Evolution* 150, 43–53.

Zohary, M. and Feinbrun-Dothan, N. (1966–1986) *Flora Palaestina*, Volumes I–IV. Israel Academy of Sciences and Humanities, Jerusalem, Israel.

37 Some Thoughts on Sources of News about Crop Wild Relatives

L. Guarino

Where can we find news about crop wild relatives (CWR)? The conventional answer of most researchers working in the field would probably be to refer to the scientific literature. But notice the title of this chapter refers to 'news', rather than 'information', or even 'data'. By the time information reaches the scientific literature, it can hardly be called news, meaning timely. Which begs the question: do we really need *news* about CWR, in the same way as we undoubtedly need the sort of information we find in the scientific literature? I think the answer is yes, in particular because the threats that many CWR face are urgent and immediate, and because the news media can be extremely powerful vehicles for public awareness.

So where *can* we find timely information about CWR? It should not surprise anyone that I would suggest the first place to look is Google (http://google.com). Using the search expression 'crop wild relative' generated 1,430,000 Google hits (using '"crop wild relative"', i.e. with quotation marks brought the number down to 529). Just to give an idea, these are the first few hits:

> PGR Forum Home
> European **Crop Wild Relative** Diversity Assessment and Conservation Forum. LAUNCHED IN 2005: CWRIS: The PGR Forum **Crop Wild Relative** Information System ...
> www.pgrforum.org/ – 9k – 18 Feb 2006 – Cached – Similar pages
>
> Press Releases – June 2004 – Every **crop** needs its **wild** relatives ...
> The project, called In Situ Conservation of **Crop Wild** Relatives Through ...One **wild relative** has made it possible to increase the solids content of the ...
> www.unep.org/Documents.Multilingual/Default.asp?DocumentID=399&ArticleID=4542&l=en – 27k – Cached – Similar pages

sandiego.indymedia.org | **Crop** diversity, **wild relative**, GE contaminate
Crop wild relative species have already made substantial contributions to improving food production through the useful genes they contribute to new **crop** ...
sandiego.indymedia.org/en/2004/11/106437.shtml – 25k – Cached – Similar pages

BGCI – **Crop Wild Relative** Database Launched
BGCI: Botanic Gardens Conservation International, working in partnership with botanic gardens around the world to promote sustainability and plant ...
www.bgci.org.uk/news/cwr_database.html – 10k – 19 Feb 2006 – Cached – Similar pages

[PDF] Microsoft PowerPoint – Brian_p
File Format: PDF/Adobe Acrobat – View as HTML
European **crop wild relative**. diversity and conservation ... A **crop wild relative** is a **wild** plant taxon that has an indirect use derived from its ...
www.nerium.net/plantaeuropa/ Download/Oral_Presentations/Brian_p.pdf – Similar pages

DuPont Biotechnology: Gene Flow – **Crop** to **Wild**
proximity of the **crop** to the **wild relative**; the **wild** plant's tendency to ... Cross-pollination between a **crop** and a **wild relative**, giving rise to fertile ...
www2.dupont.com/Biotechnology/en_US/science_knowledge/gene_flow/**crop**To**wild**.html – 44k – Cached – Similar pages

IUCN
CONSERVING EUROPEAN **CROP WILD RELATIVE** DIVERSITY ... The 'First International Conference on **Crop Wild Relative** Conservation and Use' will be held in ...
www.iucn.org/themes/ssc/news/ebulletin2005/ebulletinapr05.htm – 34k – Cached – Similar pages

REPORT
Confined field experiments in **crop-wild relative** systems, starting with reasonable ... The research establishing **crop**-to-**wild relative** gene flow in the past ...
www.isb.vt.edu/brarg/brarg_wshop/Summary_gene_flow.htm – 15k – Cached – Similar pages

Welcome to CWRIS: the PGR Forum **Crop Wild Relative** Information System
Welcome to CWRIS: the PGR Forum **Crop Wild Relative** Information System. CWRIS is an online information management system specifically designed to facilitate ...
cwris.ecpgr.org/ – 6k – Cached – Similar pages

Wild relatives for better **crop** performance – Geneflow
Tropical Manioc Selection (TMS) cassava varieties were developed using crosses with a **wild relative** of the **crop** and have been adopted by a number of African ...
ipgri-pa.grinfo.net/index.php?itemid=1178&catid=27 – 9k – Cached – Similar pages

Remember that this list might well be somewhat different if it was carried out on a different day, because the Internet is always changing, but probably not all that much. While some of these hits might be defined as news, this would definitely not be the case for others. Fortunately, there is Google News (http://news.google.com). When I did a search on 'crop wild relative' on 20 February 2006, I got the following seven (mostly) interesting hits going back a month (using quotations around the string gave no hits at all):

Molecular breakthrough will add grist to the mill for wheat...
Innovations-Report, Germany – Feb 9, 2006

... This would have implications for **crop** improvement far beyond just wheat... unfortunately prevents the pairing of wheat and **wild relative** chromosomes precluding...

PLANT OF THE WEEK
San Francisco Chronicle, USA – 28 January 2006
... the plain, white-stemmed kind, was the most successful **crop** in my... sea beet (*Beta vulgaris* subspecies *maritima*), a small, unobtrusive, **wild relative** of spinach...

Mitigating transgene flow from crops
Checkbiotech.org (press release), Switzerland – 7 February 2006
... **crop** from inflow from **wild** or weedy relatives.1 The proposals to integrate the transgene in the plastid or mitochondrial genomes do not preclude the **relative**...

High protein found in low places
Fauquier Times–Democrat, VA – 14 February 2006
... on their feet to get any share of the **crop**. ... Black walnuts, growing **wild** in the woods, are even... A **relative** of the walnut, the butternut (*Juglans cinerea*) is...

Effects of insect-resistance transgenes on fecundity in rice...
Am J Botany (subscription) – 20 January 2006
... is useful for evaluating **crop** performance and the fitness of **crop–wild** hybrid progeny... from three transgenic lines, Bt, CpTI and Bt/CpTI, **relative** to isogenic...

On her own
Providence Journal (subscription), RI – 21 January 2006
... Indeed, if Ms. Barker's **relative** had not reported the... was 1. unhealthy, 2. unsafe due to **wild** animals, 3... dust underneath the carpets to grow a **crop** of potatoes...

Blue-ribbon wishes and guava dreams
Bradenton Herald, USA – 22 January 2006

... They work in **relative** anonymity, but the Manatee County Fair... Mountains of Colorado to the lush, **wild** Florida Everglades... out on the table and you **crop** until you...

Doing the search on another day would almost certainly give new results, because Google News does not monitor largely static web pages, but rather 4500 online news sources around the world, which can change from minute to minute. The overlap between the databases that search engines scour is not complete by any means. Searching Yahoo News (http://news.yahoo.com) with the same search expression and on the same date as the Google News search summarized above gave the following nine hits, highlighting the importance of using more than one search engine:

Molecular breakthrough at the John Innes Centre will add grist to the mill for wheat breeders
SeedQuest – 17 February, 7:45 AM
A team of scientists at the John Innes Centre in Norwich, UK led by Dr Graham Moore (photo) have a completely new understanding of the structure of a gene complex in wheat that controls the pairing of its chromosomes, knowledge of which has the potential to revolutionise wheat breeding.

Striped herd makes its mark
Denver Post – 15 February, 12:23 AM
Today, in the latest report on strange ranches of the Colorado flatlands, we find plumber and rancher Paul Whipple going through his typical morning checklist, which looks like this: Organize plumbing work orders. Check plumbing supplies. Feed the zebras.

The role of non-GM biotechnology in developing world agriculture
SeedQuest – 10 February, 10:01 AM
Discussions about the role of agricultural science in boosting food production tend to be dominated by controversy over the characteristics of genetically modified (GM) **crops** and the implications of their use.

Crop, Animal Diseases Must Be Checked in Time
AllAfrica.com – 8 February, 9:06 AM
ZIMBABWE has so far experienced a favourable first part of the 2005/6 rainy season that gave most parts of the country above average rainfall.

Banning GM **crops** not enough to save wildlife
New Scientist – 8 February, 1:18 PM
Results from the UK's environmental tests of transgenic **crops** are set for release, but non-GM herbicide-resistant **crops** might pose just as great a threat

Waves of grain
The Philadelphia Inquirer – 28 January, 2:17 AM
The grains of ancient civilizations are so in vogue. Except now they are called heirloom grains. Farro, once a standard daily ration of the Roman legions, is showing up as a side dish at the iconic BYO Django.

Putting Aperture Through Its Paces: Part I
PDNonline – 23 January, 9:00 AM
The following article, the first in our two-part online review of Aperture, focuses on the design goals behind Apple's new photo management software and some of the hotly debated features of the new program.

S.C. short on blue chips in 2006 recruiting game
The State – 22 January, 12:14 AM
Normally, Richland Northeast coach Jay Frye is accommodating when the media calls. But when his phone rang Thursday morning, Frye had to beg off for at least another day.

Blue-ribbon wishes and guava dreams
Bradenton Herald – 22 January, 12:12 AM
In a wired world where everyone's connected to a computer, an iPod or a cell phone, the people of Manatee County still stop and take the time to do things with their hands, engaging in crafts that harken back to a simpler era.

Though it may not be relevant to this discussion about news, it is worth noting that Google also has a function for searching the scientific literature. Google Scholar (http://scholar.google.com), which covers peer-reviewed papers, theses, books, abstracts and articles, from academic publishers, professional societies, preprint repositories, universities and other scholarly organizations, gave 48,800 hits on 'crop wild relative', the top of which (again on 20 February 2006) are listed below:

[CITATION] ... initiative in the Valencian Community (Spain) and its use to conserve populations of **crop wild** ...
E Laguna – **Crop wild relative**, 2004
Cited by 2 – Web Search

Assessing the risks of transgene escape through time and **crop-wild** hybrid persistence
CR Linder, J Schmitt – *Molecular Ecology*, 1994 – csa.com
... We suggest methods that can be used in conjunction with evaluation of the **relative** fitness of **crop-wild** hybrids that will determine the likelihood of back...
Cited by 39 – Web Search

[CITATION] Rationale for in situ management of **wild** Beta species
L Frese – **Crop wild relative**, 2004
Cited by 2 – Web Search

... backcross QTL analysis in a cross between an elite processing line of tomato and its **wild relative** ...
SD Tanksley, S Grandillo, TM Fulton, D Zamir, Y ... – *TAG Theoretical and Applied Genetics*, 1996 – Springer

... line of tomato and its **wild relative**... The potential of exploiting unadapted and **wild** germplasm via...QTL analysis for the enhancement of elite **crop** varieties is...
Cited by 118 – Web Search – BL Direct

When transgenes wander, should we worry – group of 10 »
NC Ellstrand – *Plant Physiology*, 2001 – plantphysiol.org
... Radish is outcrossing and insect pollinated. Its **wild relative** is the same species. What about a more important **crop**? What about a more important weed?...
Cited by 95 – Web Search – BL Direct

Introgression and DNA marker analysis of Lycopersicon peruvianum, a **wild relative** of the cultivated ...
TM Fulton, JC Nelson, SD Tanksley – *TAG Theoretical and Applied Genetics*, 1997 – Springer
... ycopersicon peruvianum is a **wild relative** of the cultivated tomato,.... This makes it a likely source of new genes for **crop** improvement and, in fact,....
Cited by 31 – Web Search – BL Direct

Genetic modification: Transgene introgression from genetically modified crops to their **wild ...** – group of 4 »
CN Stewart, MD Halfhill, SI Warwick – *Nature Reviews Genetics*, 2003 – nature.com
... 107, 528–539 (2003). The first evidence of transgene escape to a **wild relative** from a commercially released GM **crop**. | Article | PubMed | ISI | ChemPort |. 61. ...
Cited by 35 – Web Search – BL Direct

Assessment of genetic relationships between Setaria italica and its **wild relative** S. viridis using ...
M Le Thierry d'Ennequin, O Panaud, B Toupance, A ... – *TAG Theoretical and Applied Genetics* 2000 – Springer
... replaced by other cereals, and in Europe it has become a minor **crop** in Europe. Green foxtail millet, *Setaria viridis*, is its closest **wild relative** (de Wet...
Cited by 20 – Web Search – BL Direct

[CITATION] Changes in grassland management and its effect on plant diversity
Å Asdal – **Crop wild relative**, 2005
Cited by 1 – Web Search

GENE FLOW AND INTROGRESSION FROM DOMESTICATED PLANTS INTO THEIR **WILD** RELATIVES – group of 7 »
NC Ellstrand, HC Prentice, JF Hancock – *Annual Review of Ecology and Systematics*, 1999 – ecolsys annualreviews.org
... cases, the only evidence for hybridization is the presence of morphologically intermed ate plants in localities where the **crop** and **wild relative** are sympatric....
Cited by 205 – Web Search – BL Direct

Identification of trait-improving quantitative trait loci alleles from a **wild** rice **relative**, Oryza... – group of 5 »
J Xiao, J Li, S Grandillo, SN Ahn, L Yuan, SD ... – *Genetics*, 1998 – genetics.org
... with other loci introgressed from the **wild relative**, O. rufipogon... that the world's reservoir of **wild** and unadapted... in rice and possibly other **crop** species....
Cited by 45 – Web Search – BL Direct

Engineered Genes in **Wild** Populations: Fitness of Weed-**Crop** Hybrids of Raphanus Sativus – group of 3 »
T Klinger, NC Ellstrand – Ecological Applications, 1994 – JSTOR
... Kareiva *et al*. (1991). Expectations regarding the **relative** fitness of weed-**crop** hybrids and **wild** plants are conflicting. On the one...
Cited by 42 – Web Search – BL Direct

Wild relatives and **crop** cultivars: detecting natural introgression and farmer selection of new ... – group of 2 »
DI Jarvis T Hodgkin – *Molecular Ecology*, 1999 – blackwell-synergy.com
... For high frequency out-crossing crops, where few detectable traits separate the **crop** from its **wild relative**, detecting the extent of introgression will be...
Cited by 30 – Web Search – BL Direct

Carbon and nitrogen economy of 24 **wild** species differing in **relative** growth rate – group of 3 »
H Poorter, C Remkes, H Lambers – *Plant Physiology*, 1990 – bio.uu.nl
... J, Mooney HA (1989) Responses of **wild** plants to... JH, Hitz WD, Giaquinta RT (1984)
Crop productivity and... Ingestad T (1982) **Relative** addition rate and external...
Cited by 109 – Web Search

Clearly, all of these searches could have been modified and refined by adding things like a genus or crop name or a geographical area (e.g. a country) to the search expression. For instance, searching on 'crop wild relative India' gave the following single hit on Google News:

> PLANT OF THE WEEK
> *San Francisco Chronicle*, USA – 28 January 2006
> ... stemmed kind, was the most successful **crop** in my ... sea beet (Beta vulgaris subspecies maritima), a small, unobtrusive, **wild relative** of spinach ... and eastward as far as **India**, often on ...

Searching on 'crop wild relative rice' gave the following:

> Molecular breakthrough will add grist to the mill for wheat ...
> Innovations-Report, Germany – 9 February 2006
> ... would have implications for **crop** improvement far ... pairing of wheat and **wild relative** chromosomes precluding ... smaller sequenced genomes of **rice** and Brachypodium ...
>
> Effects of insect-resistance transgenes on fecundity in **rice** ...
> *Am J Botany* (subscription) – 20 January 2006
> ... **crop** performance and the fitness of **crop-wild** hybrid progeny ... the growth and fecundity of potted **rice** plants from ... Bt, CpTI and Bt/CpTI, **relative** to isogenic ...

Making judicious use of brackets and the Boolean operators 'AND' and 'OR' allows considerable sophistication in the search expression. For example, one could use 'crop (wild OR weedy) (relative OR ancestor)' to broaden the search a bit.

To repeat, the results shown above are already out of date. Doing the same Google or Yahoo News search on another day will probably give new results. So does that mean that in order to keep up with the news on CWR you have to carry out the search repeatedly, day after day? Well, yes, but there are things that will help you save some time. For example, a search expression can be set up as a Google Alert (http://www.google.com/alerts?hl=en), which will deliver an e-mail to a specified account on a daily or weekly basis containing all relevant hits for the period in question. Also, an RSS or Atom feed can be generated.

RSS is an early application of Extensible Markup Language (XML). It is a method used to describe a series of annotated web links. When an RSS file contains news items, it is referred to as a 'news feed'. With appropriate software, called 'RSS (Really Simple Syndication) readers', 'feed readers', 'feed aggregators' or 'news readers' (see http://en.wikipedia.org/wiki/News_aggregator; http://blogspace.com/rss/readers), users can 'subscribe' to any number of feeds and rapidly scan headlines for interesting items. You do not have to search a web site repeatedly for new and relevant content as links to this content are conveniently packaged in one or more RSS feeds for rapid browsing. Not all web sites provide RSS feeds, but their use is spreading fast. In addition to custom-made feeds from searches, as described above, Table 37.1 indicates RSS feeds that may on occasion contain news and information relevant to CWR.

The breakthrough for news feeds came in 2003–2004 with a surge of interest that can be attributed to the phenomenal success of weblogs and their routine incorporation of RSS feeds. Also known as a 'blog', a weblog is an online journal or diary that can be updated regularly, with the entries displayed in chronological order, the most recent first. A weblog can be used to bring together news about a particular subject from web sites, other weblogs, RSS feeds, print publications, word of mouth, etc., and then making that news available in different forms, including a web page, e-mail alerts and an RSS feed. The IPGRI Public Awareness feed listed in the table above is generated from what is essentially a weblog on plant genetic resources maintained by Dr Jeremy Cherfas. Another example of a PGR weblog, this time with a geographic focus, is Plant Genetic Resources News from the Pacific (http://papgren.blogspot.com). Dr Roland Bourdeix of CIRAD has a blog with a crop focus (on coconut) at http://diversiflora.blogspot.com.

Blogs can be set up to accept comments, which makes possible a discussion on specific topics arising from a given entry, or 'posting'. Another way to bring together, disseminate and discuss news on a particular subject is to set up

Table 37.1. RSS feeds that may contain CWR news and information.

All Headline News – Agriculture	http://www.allheadlinenews.com/rss/agriculture.xml?cat=Agriculture
EarthWire RSS Newsfeed	http://www.earthwire.org/feeds/getnews_uk_rss.xml
Eldis Biodiversity newsfeed	http://www.eldis.org/newsfeeds/rss/2/biodiversity.xml
Eureka Alert! Agriculture	http://www.eurekalert.org/rss/agriculture.xml
FAO What's new	http://www.fao.org/waicent/whats_new/RSS_2/fao_EN.xml
Forest Conservation Newsfeed	http://forests.org/rss/forest.xml
Future Harvest – Electronic Clippings file	http://ipgri-pa.grinfo.net/xml-rss2.php?blogid=3
IPGRI – Public Awareness	http://ipgri-pa.grinfo.net/xml-rss2.php?blogid=1
Moreover – Agriculture news	http://p.moreover.com/cgi-local/page?o=rss&c=Agriculture%20news
Moreover – Environment news	http://p.moreover.com/cgi-local/page?o=rss&c=Environment%20news
News4Sites – Biodiversity	http://www.news4sites.com/service/newsfeed.php?tech=rss&id=2578
WWF International: Latest news	http://www.panda.org/rss/news.cfm

an e-mail discussion group, for example using Yahoo (http://groups.yahoo.com/), Google (http://groups.google.com/), D-groups (http://www.dgroups.org/) or other software, as proved successful for the PGR Forum project (see Maxted *et al.*, Chapter 1, this volume). E-mail discussions can be either open-ended or of the more focused, time-limited e-conference type. You can search Yahoo and Google Groups postings in the usual way.

A couple of examples of e-mail discussion groups with a biodiversity conservation interest are the Rare Fruit Society of the Philippines (http://groups.yahoo.com/group/rarefruit-ph/) and PestNet (http://groups.yahoo.com/group/pest-net/). As for crop-focused lists, COGENT (International Coconut Genetic Resources Network) recently surveyed Yahoo discussion groups on coconut and found the following (similar lists could no doubt be drawn up for other crops):

The Centre for Information on Coconut Lethal Yellowing (**CICLY**) is intended to act as a clearing house for information about lethal yellowing and similar diseases of coconuts and other palms.
http://groups.yahoo.com/group/CICLY/
316 members
Started 1999

If you have other Interesting Coconut Conundrums Requiring Answers just write to **ICCRA**.
http://groups.yahoo.com/group/ICCRA/
149
2000

The **Coconut Protocol** is an initiative of the Global Filipinos under its Economic Affairs Committee. Our objective is to create awareness for the potential of the coconut industry to become the pillar of development of our economy and to work towards making this into a reality by bringing together the private and public sectors and government in this endeavor.
http://finance.groups.yahoo.com/group/CoconutProtocol/
92
2004

Vitro2Vivo. Although coconut embryo culture has been possible for many years it is not always easy to achieve reliable success when transferring plantlets from the *in vitro* laboratory to the *in vivo* nursery and field. Other plants may also be recalcitrant in this respect. This eGroup invites participation from any phytobiologist who is interested in exchanging views and information on the subject. The group was initially set up for coconut embryo culture by Hugh Harries and it is also open to other crops, particularly palms, and covers related issues in tissue culture.
Http://groups.yahoo.com/group/vitro2vivo/
79
1999

IndonesiaCoconut. This group for manufacturer/producer, seller, buyer, exporter and importer of all kinds of product from coconut, such as: coconut shell powder/coconut shell flour, active carbon, coconut fibre, charcoal, coconut milk, coconut flour/powder, etc.
http://finance.groups.yahoo.com/group/indonesiacoconut/
60
2002

CocoPest is dedicated to the exchange of information, news and opinions about the pests and diseases of coconut palms. There are many other sources of information about plant pests and diseases on the Internet. CocoPest does not compete with these, it simply provides convenient e-mail exchange for specialists who have the latest information as well as for the general public who want to know where to find that information.
http://groups.yahoo.com/group/cocopest/
55
2000

The **Pacific Coconut Timber** group:

- shares information and regional expertise on coconut wood utilization – processing, product development, marketing including related research and developments
- supports integrated senile coconut tree utilization in the Pacific Island countries and territories to improve rural livelihoods
- promotes the sustainablility of senile coconut tree utilization, industry and resource supply based on the country's policies
- facilitates standardization of coconut timber products trading and targets a high-level quality product profile
- supports a regional Pacific marketing strategy
- promotes coconut timber products for local and export trade and works together to serve export markets.

http://finance.groups.yahoo.com/group/pacific_coconut_timber/
44
2004

Coconut Australia (CA) is an internet-based discussion group to promote coconut particularly in Australia and worldwide generally. The group is restricted to members only. Although the group is dedicated to coconut issues relevant to Australia, membership is open to others around the globe who share the mission of the group. The moderator of CA facilitates the discussion and circulation of relevant information, but the message is entirely the opinion of the sender. The moderator reserves the right to cease membership.
http://finance.groups.yahoo.com/group/CoconutAustralia/
20
2005

Cocowood. Information exchange on coconut wood and coconut shell products
http://groups.yahoo.com/group/cocowood/
16
2001

Cocotimeline. Key knowledge on the coconut palm and where to find the information. Search the entire database of over 8500 entries for words and phrases
(http://cocos.arecaceae.com/)
http://groups.yahoo.com/group/cocotimeline/
9
2001

CoconutWater. Yes, we love coconut water and people in the Tropics have been enjoying this gift from heaven for thousands of years. Have fun coconut water lovers!
http://health.groups.yahoo.com/group/coconutwater_group/
2
2005

Another way to exchange information on CWR is through a wiki, a type of web site that allows users to easily add and edit content and is especially suited for collaborative writing. Perhaps the best-known example is Wikipedia (http://en.wikipedia.org/wiki/Wiki). This has entries on many crop plants, often including entries to its wild relatives. See, for example, the entry on wheat (http://en.wikipedia.org/wiki/Wheat).

It would be easy to start a blog (e.g. using Blogger, http://blogger.com), e-mail discussion group or even wiki on CWR, whether as a class of plants or by gene pool. The question is: would it last? Cost would not be a problem. Blogspot and Yahoo and Google Groups are free services. Maintaining a blog or e-mail group can be time-consuming, but a small number of dedicated volunteer moderators can keep the service ticking over without too much trouble. The real key to sustainability is usefulness. If enough people find the news exchanged useful in their work, they will find a way to keep the service going. This has been clearly shown by PestNet, which is focused on solving the urgent, practical, day-to-day plant protection problems faced by researchers, extension officers and quarantine people in the Pacific. Researchers and others working on CWR also face difficult questions, sometimes requiring immediate answers. How can I distinguish species A from species B? Is my species found in country C? How about in protected area D? Will road building in area E affect any CWR? The tools are certainly out there to get news and information relevant to these kinds of queries to the people who need them in timely fashion. Are we making the most of them as a CWR research and development community?

VIII Gene Donors for Crop Improvement

38 Using Crop Wild Relatives for Crop Improvement: Trends and Perspectives

T. HODGKIN AND R. HAJJAR

38.1 Introduction

Crop wild relatives (CWR) – species or other taxa more or less closely related to crops – have made an extremely significant contribution to modern agricultural production through the characteristics that they have contributed to plant cultivars. Many of these taxa are also useful in their own right, as keystone species in some ecosystems, medicinal species (e.g. many *Solanum* spp., *Vigna luteola*), forages (*Pennisetum purpureum*, *Aegilops* spp.), wild harvested foods, building materials, living fences, fuelwood or as rootstocks (e.g. *Malus* spp. or *Pyrus* spp.) (Heywood, 1999). However, their role as relatives of crops, valuable by virtue of that relationship and their potential contribution to crop production, gives a specific perspective to their conservation and use and raises certain issues in respect of their optimum maintenance that are rather different from those they have in common with other useful species.

Generalizing about the conservation and use of CWR is difficult. Wild relatives of crops comprise very different kinds of plant species: annuals, biennials and perennials, from ephemeral weeds to large long-lived tree species, are found in every part of the globe, occupy all kinds of different environments and possess more or less every possible adaptation or potentially useful trait. The ways in which individual species are best conserved, so as to meet the potential needs of users, and the extent and manner of their use in crop improvement are, in many respects, equally varied. However, some generalizations can be made which may help towards their better conservation and use.

It is likely that some CWR have been sources of genes for cultivated materials more or less continuously since the beginning of agriculture and the separation of domesticated crops from their wild relatives. Mostly the process of introgression of new genes into the crop will have been casual and unconscious but it may also have been deliberate (see Jarvis and Hodgkin, 1999, for review). As examples of this continuing process, Doggett and Majisu (1968) recorded that women farmers identified hybrids between wild and cultivated sorghum as suitable for planting, and Richards (1986) described situations in which Sierra Leone rice farmers received seedlots containing weedy rice.

The use of wild relatives as part of commercial plant breeding can be traced to the late 19th century with the development of rootstock cultivars from wild *Vitis* species to combat *Phylloxera* aphids and *Meloidogyne* nematodes in grapes (Prescott-Allen and Prescott-Allen, 1988). Another early recorded use of CWR was that of *Saccharum spontaneum* to obtain virus resistance in sugarcane before World War I (Stalker, 1980). The hybrid stocks also gave substantially improved yield and vigour and became the basis for the modern sugarcane industry so that all modern sugarcane varieties are derived from interspecific hybrids and contain three to five species in their pedigrees (Stalker, 1980). In 1941 the first tomato variety was released that contained a deliberately introduced gene from a wild relative (*Fusarium* resistance from *Lycopersicon pimpinellifolium*). Since then, tomatoes have probably become the crop which has made use of the widest variety of genes from the largest numbers of different species (Rick and Chetelat,1995).

During the 1940s and 1950s the value of genes in CWR which conferred useful characters in crop cultivars was more widely recognized (Plucknett *et al.*, 1987). Breeding programmes in a number of crops began to explore the potential for using crop relatives and in some cases significant advances were made. Well-known examples include the introduction of late blight resistance into potato from *Solanum demissum* and *S. stoloniferum*, resistance to viruses in potatoes from these species together with *S. chacaoense* and *S. acaule* (Ross, 1986), and the introduction of *Septoria*, stem rust and other disease resistances into wheat from *Aegilops tauschii* (Lagudah and Appels, 1993).

Through the 1970s and 1980s the use of crop relatives became increasingly common and, by the mid-1980s, in a major review, Prescott-Allen and Prescott-Allen (1988) asserted that the achievements were substantial enough to demonstrate the enormous potential of the genetic diversity of wild plants to improve yields and quality of domesticated crops and provide for the more rapid domestication of new crops. In this review they also drew attention to importance of maintaining as great a range as possible of the variation within these wild species and noted that this in turn required a clearer understanding of the special nature of wild genetic resources; how they differed from wildlife and other wild resources and from domesticated genetic resources.

As the importance of using CWR was recognized, so was the importance of their conservation and, during the 1980s, work on conserving CWR increased substantially. Between 1986 and 1992, the International Board for Plant Genetic Resources (IBPGR), working with national programmes, supported a number of collecting missions that focused primarily on collecting CWR (IBPGR, 1991), and ecogeographic approaches to collecting relevant for CWR were developed (Maxted *et al.*, 1995). Recognizing the difficulties involved in conserving CWR *ex situ*, and the desirability of maintaining adaptability and evolutionary processes, there was increasing interest in the possibilities of effective *in situ* conservation. Heywood (1997) noted that the conservation of CWR was best carried out in their natural habitats or ecosystems whenever possible, although he argued for complementary strategies involving an appropriate combination of approaches.

Developments in molecular genetics, improvements in interspecific hybridization and embryo rescue techniques and, of course, increasingly successful gene transfer procedures have continued to stimulate interest in the use of CWR. There has been a steadily expanding literature on useful genes and traits detected in CWR. Tanksley and McCouch (1997) argued that the effective use of molecular genetics allowed for a much more effective use of CWR and called for a new paradigm in which the search for desirable traits (the traditional approach to plant breeding) was replaced by a search for specific genes based on the use of molecular techniques. They argued that new molecular methods would allow for the detection and use of a much wider range of useful genes including those not normally expressed in the crop relative and those associated with quantitative traits.

In this chapter we wish to explore the use of CWR in the production of new cultivars, since the reviews of Prescott-Allen and Prescott-Allen (1986, 1988). In particular, we wish to examine whether there has been a substantial increase in the use of these species and what form any increased use has taken. A significantly increased use might suggest that the proposed change of paradigm suggested by Tanksley and McCouch (1997) is indeed occurring and that we have entered a new era of use of CWR. This might lead, in turn, to the need for new or different approaches to the conservation of these species so as to optimize their availability to users. We therefore discuss the extent to which conservation, distribution and use practices for CWR are taking account (or need to take account) of the changes or opportunities presented by molecular genetics and new use opportunities.

38.2 Approach

In this chapter we summarize the findings from two different kinds of investigation. The first involved examining the distribution of samples of CWR from the gene banks of the Future Harvest Centres. The second reviewed available information on the release of new cultivars with genes from CWR.

For the first study we collated data on the holdings and distribution of wild and weedy species held in the gene banks of the Future Harvest Centres over the last 20 or so years. The data were obtained from the CGIAR System-wide Information Network for Genetic Resources (SINGER, http://singer.grinfo.net/) during July 2005 for rice, wheat, barley, sorghum, millet, common bean, chickpea, cowpea, lentils, pigeon peas and groundnut. For each crop we obtained data on the current (2004 holdings) numbers of accessions maintained and the numbers of these designated as wild and weedy. We also analysed the numbers of accessions and samples distributed over the period 1985–2000 and the proportion of these that were designated as wild and weedy. The crops chosen for this survey were selected from those included in the more complete review of use described below, on the basis of data availability for the required time period, and accurate and more or less complete designation of wild and weedy samples in the database.

Information was collected on the use of CWR in the development of new cultivars of over 20 different crops. Most of the selected crops were also mandate crops of the Future Harvest Centres of the Consultative Group on International Agricultural Research (CGIAR) – rice, wheat, maize, barley, sorghum, millet, cassava, potato, chickpea, cowpea, lentils, soybean, pigeon peas, groundnut and bananas. Others – lettuce, sunflower, tomato and brassica crops – were chosen because of the known importance of CWR in their breeding. The information was collected and analysed as part of the work of the project '*In situ* Conservation of Crop Wild Relatives through Enhanced Information Management and Field Application' currently supported by the Global Environment Fund undertaken by a number of countries and international agencies with the International Plant Genetic Resources Institute and the United Nations Environment Programme. The longer-term aim is to incorporate the data collected into the proposed global information system that will be developed under this project and, in this way, to support conservation and use decisions.

As noted above, the focus of the study was on use as demonstrated by the release of cultivars containing one or more genes from the relevant CWR. CWR are also used extensively in research, trait evaluation, pre-breeding and breeding work. Our concern was primarily with evidence that these different activities had been successful in terms of cultivar production. Very little of this type of data exists in publications and most of the information was obtained directly from breeders themselves. This was one reason for focusing on CGIAR breeding programmes and mandate crops, where the information is in the public domain. In the course of this research, many breeders themselves noted the lack of complete records and the difficulty of keeping track of the further use of released varieties in other breeding programmes. Thus, much of the information presented is qualitative in nature based on the consensus views of several informants and the available literature. Some examples of the use of CWR may well have been missed and it is hoped that these can be added to the information system as they are detected. In this chapter we focus on the major trends detected while a more detailed paper (Hajjar and Hodgkin, 2007) presents the data from the survey.

38.3 The Amount of Distribution

The total holdings described as wild and weedy by SINGER for a number of different crops conserved *ex situ* by Future Harvest Centre gene banks range from about 400 accessions in the case of sorghum to over 5000 in the case of wheat. Most commonly, about 5% of the accessions are listed as wild or weedy although the range is from 1.1% (sorghum) to 10% (cowpea). These figures certainly underestimate the holdings of wild relatives in the gene banks of the Future Harvest Centres. First, there are a number of situations in which accessions are not recorded in SINGER as wild or weedy even though they come from wild species. Second, in clonally propagated crops such as potato or cassava, the proportion of accessions that belong to wild species is much higher (apparently over 40%). Because of difficulties in precisely classifying all the accessions in such collections, these were omitted from the analysis.

Table 38.1. Number of accessions for selected crops and their wild and weedy relatives in the gene banks of the Future Harvest Centres in 2004 (data from SINGER in 2005) compared with percentage holdings in 1983. (From Plucknett *et al.*, 1987.)

Crop	Total number of accessions	Wild and weedy accessions	%	% reported by Plucknett *et al.* (1986) citing IBPGR (1983)
Rice	127,723	4,274	3.35	2.0
Wheat	112,132	5,555	4.95	0
Barley	37,329	1,966	5.27	0.001
Sorghum	36,804	431	1.17	1.4
Millet	21,789	1,071	4.92	0.1
Common bean	34,197	2,420	7.08	0.5
Chickpea	29,279	423	1.44	–
Lentil	9,997	498	4.98	–
Pigeon pea	13,935	692	4.97	0.4
Groundnut	15,396	703	4.57	0.2

The numbers have grown very significantly over the last 20 years. Plucknett *et al.* (1987) reported that wild species typically accounted for less than 2% of gene bank accessions and it is clear from Table 38.1 that the numbers of accessions conserved in 1983 in the gene banks of the Future Harvest Centres was very much lower than it is today. This continuing growth to quite substantial numbers (often several thousand accessions) is perhaps surprising given the commonly reported difficulty in maintaining wild relatives *ex situ* (Heywood, 1997). It may reflect either the increasing importance given to conserving these species by gene bank managers or the increasing need for *ex situ* conservation in the face of identified threats to existing populations. It may also reflect expectations by collectors and gene bank managers of the increasing usefulness of wild species as suggested by Tanksley and McCouch (1997).

The average annual numbers of samples distributed of a selected range of crops was calculated over three different 5-year periods (1986–1990, 1991–1995, 1996–2000) (Table 38.2). The numbers of samples were used rather than the numbers of accessions to give a general view of use and to avoid problems of comparisons from year to year where it was not known if the same or different accessions were distributed. The distribution data given in SINGER refer only to distribution outside the centre – they give no information on use by the centre itself in its own breeding or research programmes. Iwanaga (1993) noted that 75% of the *Phaseolus* samples distributed by the Centro Internacional de Agricultura Tropical (CIAT) over 15 years were in fact used in other CIAT programmes. Including distribution to within centre activities would certainly substantially increase numbers of samples distributed. However, it is not clear whether it would lead to any great change in the proportion of samples distributed that were wild and weedy.

The results suggest that substantial numbers of samples of CWR have been distributed to users throughout the world. Change in numbers over the three periods is very variable. In rice, there is a steady increase in numbers of accessions distributed while in sorghum and millet there is a steady decrease in numbers.

Table 38.2. Average annual number of samples of selected crops designated as wild or weedy by SINGER distributed from the gene banks of the Future Harvest Centres for three 5-year periods from 1986 to 2000, and proportion of total distribution for that crop.

	1986–1990		1991–1995		1995–2000	
Crop	Samples distributed per year	% of total annual distribution	Samples distributed per year	% of total annual distribution	Samples distributed per year	% of total annual distribution
Rice	444	4.6	1,022	5.5	1,663	19.7
Wheat	761	31.5	3,901	44.5	2,841	29.5
Barley	55	1.9	252	6.2	412	7.1
Sorghum	174	0.9	71	0.9	31	1.5
Millet	154	2.4	89	3.6	62	7.2
Common bean	1,075	7.0	791	12.2	558	14.8
Chickpea	301	1.3	1,361	5.8	1,261	4.3
Lentil	81[a]	7.9[a]	801	16.5	615	17.7
Pigeon pea	355	9.0	428	17.1	278	22.6
Groundnut	104	2.3	250	8.8	252	8.9

[a]Data for only 2 years, 1989 and 1990.

Changes in numbers of samples distributed over the period certainly reflect in part, changes in total numbers of all samples of the crop distributed over the period. By contrast, the percentage of wild relatives samples distributed shows an increase over the whole period for eight out of ten of the crops. The only major exception is wheat where there has always been a very significant distribution of CWR but this seems to have peaked in the mid-period examined.

It seems likely, therefore, that there has been a steadily increasing demand for CWR by users as a proportion of the materials that they request from gene banks. It is interesting in this respect that it seems likely that the proportion of CWR accessions distributed by these gene banks is rather higher than the proportion these constitute of their total holdings (data not shown).

38.4 The Extent of Use in Breeding

Prescott-Allen and Prescott-Allen (1988) listed about 31 crops where wild relatives had been used to the extent that cultivars with genes from wild relatives were available. They also noted about 35 crops in which this was not the case. They classified crops according to crop groups such as cereals, roots and tubers, oils, vegetables and pulses, fruit and nuts, sugar, commodity and fibre crops, ignoring crops contributing less than 1% to production in their category. There are some strange omissions, such as cauliflowers and lettuce (the latter included in Prescott-Allen and Prescott-Allen, 1986), and some difficulties in interpretation (as to what constitutes a crop) but, of the 20 crops considered in our survey, they identified nine (rice, wheat, maize, barley, cassava, potato, soybean, sunflower, tomato) for which cultivars had been developed containing genes from wild relatives. To this we should add lettuce, as noted in Prescott-Allen and Prescott-Allen (1986). Combining information from both sources it appears that some 31 wild species had been utilized to provide genes for these ten crops.

Our survey indicated that cultivars containing genes from wild relatives had been produced for 17 of the 20 crops surveyed, the exceptions being cowpea, lentil and pigeon pea. In addition to a marked increase in the numbers of crops for which cultivars containing genes from CWR had been developed, there had been an increase in the number of wild species used. At least 56 species were identified as having been used for ten crops mentioned by Prescott-Allen and Prescott-Allen, nearly twice as many as noted in the earlier studies.

The most common use of CWR continues to be as a source of disease or pest resistance genes. Of about 200 trait/wild relative combinations noted in our survey, over 80% of the uses described involved disease and pest resistance and, where wild genes have been used, all the crops except chickpea and barley now seem to have cultivars with disease resistance genes obtained from crop relatives. However, in maize and groundnut, disease resistance seemed to be the only trait obtained from wild relatives. Other important characteristics included abiotic stress resistance (barley, chickpea, rice and sunflower), the development of F1 hybrids (through the introduction of cytoplasmic male sterility systems in sunflower and rice), yield increases and improved quality (particularly tomato). The following paragraphs illustrate these different uses.

38.4.1 Multiple use

Tomato and wheat stand out as crops in which extensive use has been made of CWR for a wide range of different breeding objectives. Tomato continues to be the crop for which breeders seem to have used the widest variety of wild species and greatest range of characters. The last 20 years have seen no evidence of a reduced use of CWR. Rather, as illustrated in a review by Rick and Chetelat (1995), there has been a steady increase in the numbers of genes transferred. Table 38.3 provides a summary from which the importance of wild relatives as sources of disease resistance and other traits can easily be seen. Tomato is also the example most often quoted (with sunflower) when the value of CWR is being described. It has been estimated that the transfer of genes conferring high sugar content to tomato were worth US$5–8 million/year to the tomato industry in California (FAO, 1998). Over the period of our survey there has also been an increasing use of more distant relatives, so that even genes from gene pool 3 species (Harlan and De Wet, 1971) have been introgressed into modern cultivars. This raises significant issues for conservation as gene pool 3 species are often given very low priority in conservation planning.

In wheat, traditional programmes for introgressing useful traits from wild relatives have been complemented by the development of synthetic wheat breeding programmes by Centro Internacional de Mejoramiento de Maiz y Trigo (CIMMYT). A number of different wild relatives (e.g. *Triticum boeoticum, Aegilops tauschii, Ae. ventricosum, Ae. elongatum*) have been used as

Table 38.3. Examples of use of the different wild relatives of tomatoes to illustrate multiple usage of crop wild relatives. (Data from Rick and Chetelat, 1995.)

Gene pool	Taxa	Characters
1	*L. esculentum* var. *cerasiformae*	Resistance to bacterial spot, wilt, broomrape
	L. pimpinellifolium	Increased fruit size
	L. cheesmanii	Resistance to TYLV, CMV, black mould
2	*L. chmielewskii*	Increased fruit size improved colour, increased soluble solids content
	L. hirsutum	Resistance to flea beetle, leaf miner, aphids
	L. parviflorum	Resistance to powdery mildew, Fusarium wilt
	L. pennellii	Resistance to whitefly, aphids, leaf miners, drought tolerance
3	*L. chilense*	Resistance to Gemini viruses, powdery mildew, bacterial spot, eel nematode, drought and salinity tolerance
	L. peruvianum	Root knot nematode resistance
	Solanum lycopersoides	Resistance to cucumber mosaic virus
	S. ochranthum	Resistance to late blight

sources of useful genes, and breeders have continued to isolate and introgress genes from wheat wild relatives with resistance to a number of important diseases (leaf and stem rust, yellow dwarf virus, powdery mildew, wheat streak mosaic virus) and increased protein content (Hoisington *et al.*, 1999). CIMMYT have gone one stage further and have developed a series of synthetic hexaploids based on crosses between *Aegilops tauschii* and *Triticum turgidum* lines. This re-synthesis of the crop, almost a kind of re-domestication, seems to represent a qualitative leap in breeding using CWR. Releases so far include synthetic wheats with improved waterlogging tolerance (Villareal *et al.*, 2001) and disease resistance (Mujeeb-Kazi *et al.*, 2001) and the release of the new cultivar Carmona in Spain. These have been used in further breeding programmes in a number of countries (http://www.cimmyt.org/english/wps/news/wild_wht.htm). Future releases are expected with enhanced yield potential, resistance to drought and higher content of essential minerals such as iron or zinc.

38.4.2 Disease resistance

As noted above, the search for and introgression of new disease resistance genes or alleles continue to be the predominant use of CWR in breeding programmes. New disease resistance alleles have been reported from tomato wild relatives at the rate of about one per year since 1982 (Rick and Chatelet, 1995), with virtually all resistance genes currently in commercial tomatoes coming from wild sources.

Wild relatives have been equally important donors of disease resistance in potato. Resistance to late blight has been introduced from *Solanum demissum* and *S. stoloniferum*, resistance to viruses from these species together with *S. chacoense* and *S. acaule* and resistance to potato cyst nematodes has come from *S. vernei* and *S. spegazinii* (Bradshaw and Ramsey, 2005). These wild species had all been used extensively in the 1980s (Ross, 1986). Since then these species have continued to be used together as sources of new resistance genes and genes from other species have also been introduced such as *S. bulbocastanum* which provided blight resistance for the cv. Biogold released in 2004. Perhaps the most interesting feature of the use of wild relatives in potato breeding is the way in which it has become integrated with the use of other materials (advanced breeding lines, traditional landraces) to become part of a more general strategy of base broadening for the crop (Bradshaw and Ramsey, 2005).

Resistance to fungal, bacterial and viral diseases provides the largest number of examples of use of wild relatives but other resistances have also been transferred. Thus, a related development, designed to assist in the control of the parasitic plant species broomrape, has been the introduction of resistance to imidazolinone and sulfonylurea herbicides from wild sunflower, *H. annuus* (Seiler and Gulya, 2004). These resistance genes have been transferred into cultivated hybrids under the trade name 'Clearfield', and are expected to be worth millions of dollars globally (G.J. Seiler, 2005, personal communication).

38.4.3 Abiotic stress

Wild relatives are often cited as being desirable because of their tolerance to drought, cold or other types of abiotic stress. However, only a few cases seem to have reached the stage of release. Genetic control of the desired traits is often quite complex, as they are often associated with characteristics undesirable in the crop and difficult to screen for and maintain through backcrossing programmes. Recent advances in this field include the first chickpea variety to be released by the Indian Agricultural Research Institute with drought and temperature tolerance derived from *Cicer reticulatum*. In 2004 ICARDA released six barley cultivars with drought tolerance derived from *Hordeum spontaneum* for use in Syria.

38.4.4 Heterosis and hybrid production

The use of wild species as a basis for the development of F1 cultivars using cytoplasmic male sterility (CMS) has become important in a whole range of crops. Of the ones surveyed here, such approaches have been tried in brassicas, pearl millet and others. The most important, however, continue to be rice and sunflower. In rice, the CMS source was wild *Oryza sativa* f. *spontanea* L. Hybrid cultivars were first released in 1976 and are currently planted on approximately 45% of China's rice area. Hybrid cultivars of sunflower using CMS were developed during the 1970s using both wild *Helianthus annuus* and *H. petiolaris*. Today F1 hybrids using PET 1 CMS from *H. petiolaris* are estimated to be grown on 90% of sunflower production area and it has been suggested that hybrid cultivars confer a 20% yield advantage over open-pollinated ones. These uses of CWR pre-date this survey although, in rice, the use of hybrid cultivars has dramatically expanded over the last 20 years. However, over the period of the survey it is likely that new sources of CMS have been tested and introduced even in the crops where they are well established.

38.4.5 Yield increases

It is difficult to point to examples of cultivars developed to possess genes for increased yield per se from wild relatives. Of course, F1 hybrids provide yield improvements through heterosis, as do the introduction of many disease resistance or stress tolerance traits. The rice variety NSICRc112 released in the Philippines from a cross between *O. sativa* and *O. longistaminata* is known to be high yielding (BRAR, IRRI, 2005, personal communication), but positive results from targeted searches for yield enhancing genes are apparently yet to result in released cultivars. Xiao *et al.* (1998) identified yield enhancing QTLs affecting tiller number and other yield traits in *Oryza rupifogon*. These were introgressed into cultivated rice but we are not aware of any varieties released from this crossing programme. Maintaining such traits in backcrossing programmes with elite cultivars is clearly difficult and requires the use of molecular markers. At present it is also difficult to develop screening and introgression programmes to handle the numbers of crosses and progenies that would be desirable to identify yield enhancing genes, confirm their expression in appropriate genetic backgrounds, transfer them and select desirable progeny.

38.4.6 Quality improvements

A range of contributions to improved quality have come from wild relatives including increased protein content in wheat and cassava; wild cassava genes derived from *Manihot oligantha* were reported to double protein levels (N. Nassar, 2005, personal communication). There are also a number of examples from tomato where genes for increased content of soluble solids and fruit size have both been obtained from wild relatives. One of the most interesting examples 'in the pipeline' is the planned release of broccoli cultivars with increased amounts of anti-cancer compounds. These were obtained from *Brassica villosa,* a GP1 relative of *B. oleracea,* and are described (2005) as currently being commercialized (D. Pink, Wellesbourne, 2005, personal communication). This is an interesting example because it seems more or less fortuitous, in that the trait was not deliberately screened for or sought as a breeding objective.

38.5 Discussion

In this chapter we have considered only two aspects of use of CWR – distribution of materials and the production of cultivars containing genes from wild relatives.

In fact, it is important to consider utilization of genetic resources as a more or less interconnected process, with different components that are all necessary for optimum use. Thus, characterization and evaluation are important in creating interest and knowledge about materials and will both stimulate and result from distribution of materials. Information from such studies needs to be made widely available to stimulate use. Research on trait expression and its genetic control are also important for any effective use of material in breeding programmes and, in the case of CWR, there will also be a need to work on hybridization and introgression procedures to ensure that any identified trait is satisfactorily transferred.

The survey on which this chapter is based was necessarily limited in scope. It would be valuable to analyse the use of wild relatives in research (as shown by research publications) or in breeding programmes (from direct surveys of breeders). A survey of the importance of wild relatives in research and of the kind of research being undertaken would be particularly interesting since it would provide more direct evidence on the uptake of the ideas put forward by Tanksley and McCouch (1997). The work on wheat re-synthesis undertaken by CIMMYT and the work on rice undertaken by Xiao *et al.* seem particularly relevant in this regard. One of the areas that has certainly seen a significant development has been that of interspecific hybridization where tissue culture, embryo rescue and ways of overcoming interspecific compatibility have significantly improved the use of gene pools 2 and 3 of many crops over the last 20 years.

There was evidence both of an increase in the size of holdings and distribution of CWR from the gene banks of the CGIAR Centres. There was also clear evidence that use of wild relatives had continued to increase in the 20 years since the surveys of Prescott-Allen and Prescott-Allen (1986, 1988). This increase manifested itself as a rise in the number of crops with cultivars containing genes from wild relatives, an increase in the numbers of species used to provide genes and an expansion in the range of traits for which genes from wild relatives were used. Perhaps it is this latter aspect that is most striking. Although pest and disease resistance continue to dominate in the use of wild relatives, the increasing interest in abiotic stress resistance and yield enhancing genes seems to break new ground. While we are not convinced that the new paradigm proposed by Tanksley and McCouch (1997) has been fully realized, there has certainly been a steady quantitative increase in use of CWR over the last 20 years.

As the use of CWR expands and improves it becomes increasingly important to ensure their proper conservation. At the same time, it has to be recognized that they are becoming more vulnerable and subject to genetic erosion. Meilleur and Hodgkin (2004) have discussed conservation needs for CWR and other chapters of this book also bear on this important question. However, in developing conservation plans, it may also be useful to consider user perspectives.

At present the major need remains for disease resistance genes controlled by relatively few major genes although the increasing interest in traits controlled by more complex gene action (e.g. abiotic stress resistance and yield-related traits) should be noted. In both cases there is often significant interest in traits for which there is within species variation in the target species. Thus any conservation strategy for CWR needs to ensure that a range of different populations from different ecogeographic areas are conserved. The ecogeographic survey approaches are therefore an essential element of designing an effective conservation strategy and, it is likely that the emphasis should be on trying to include as many different populations with potentially useful traits as possible in any action. It is this emphasis on the importance of conserving both within species and within population genetic diversity to meet user needs that provides a major difference from many conservation strategies that may be developed for other wild species.

User needs may not be related to conservation priorities. One example of this is *Lactuca serriola* which is a common weed and would not necessarily be included in any conservation programme. However, the continuing detection and introgression of disease resistance genes from this species is essential to the lettuce industry in Europe. Breeders need access to a wide diversity of populations of *L. serriola*, preferably from gene banks, and large collections have therefore been established (Lebeda and Astley, 1999).

Since users hope to be able to access materials from gene banks rather than having to maintain their own collections or carry out their own collecting missions, there is a continuing need for *ex situ* conservation of wild relatives. Recognizing the desirability of *in situ* conservation as a key element of any overall strategy, the question then becomes how much *ex situ* conservation and of what species? It would be valuable for users and conservation workers to explore this question together so as to combine meeting user needs with carrying out essential *ex situ* conservation of species and populations threatened in the wild for specific crops and their relatives. A balance will be needed which takes account of available resources, the interest of users and the concern by conservationists with threatened species which may not yet have been identified as directly useful.

There is clearly an increasing interest in, and need for, useful genes from CWR. While *in situ* conservation may be most desirable as an effective conservation strategy, *ex situ* conservation also has a role to play in providing security for highly threatened materials and meeting users needs. To do this gene banks need to do more than just store wild relatives, they need to be actively engaged in characterization and evaluation of these materials and may even want to work with others on key research questions (gene action and gene control), on identification of potential parents for crossing programmes and on pre-breeding. Although this may seem like an expanded agenda for conservation workers, it is likely to be a necessary part of demonstrating the value of the work they do and mobilizing the resources they need to do it.

References

Bradshaw, J.E. and Ramsey, G. (2005) Utilization of the commonwealth potato collection in potato breeding. *Euphytica* 146, 9–19.

Doggett, H. and Majisu, B.N. (1968) Disruptive selection in crop development. *Heredity* 23, 1–23.

FAO (1998) *The State of the World's Plant Genetic Resources for Food and Agriculture*. FAO, Rome.

Hajjar, R. and Hodgkin, T. (2007) The use of wild relatives in crop improvement: a survey of developments over the last 20 years. *Euphytica* 10.1007/s10681-007-9363-0.

Harlan, J.R. and de Wet, J.M.J. (1971) Toward a rational classification of cultivated plants. *Taxon* 20, 509–517.

Heywood, V.H. (1997) Introduction. In: Valdes, B., Heywood, V.H., Raimondo, F.M. and Zohary, D. (eds) Proceedings of the Workshops on 'Conservation of Wild Relatives of European Cultivated Plants'. *Bocconea* 7, 15–18.

Heywood, V.H. (1999) *Use and Potential of Wild Plants in Farm Households*. FAO, Rome.

Hoisington, D., Khairallah, M., Reeves, T., Ribaut, J. M., Skovmand, B., Taba, S. and Warburton, M. (1999) Plant genetic resources: what can they contribute toward increased crop productivity. *Proceedings of the National Academy of Sciences of the United States of America* 96, 5937–5943.

IBPGR (1991) *International Board for Plant Genetic Resources, Annual Report for 1990*. IBPGR, Rome.

Iwanaga, M. (1993) Enhancing links between germplasm conservation and use in a changing world. *International Crop Science* 1, 407–413.

Jarvis, D.I. and Hodgkin, T. (1999) Wild relatives and crop cultivars: detecting natural introgression and farmer selection of new genetic combinations in agroecosystems. *Molecular Ecology* 8, 159–173.

Lagudah, E.S. and Appels, R. (1993) Wild relatives as a source of new germplasm for cereal improvement. In: Imrie B.C. and Hacker J.B. (eds) *Focused Plant Improvement: Towards Responsible and Sustainable Agriculture*. Proceedings of the Tenth Australian Plant Breeding Conference, Gold Coast, Australia. 18–23 April 1993. Organizing Committee, Australian Convention and Travel Service, Canberra, Australia. pp. 10–21.

Lebeda, A. and Astley, D. (1999) World genetic resources of *Lactuca* spp., their taxonomy and biodiversity. In: Lebeda, A and Kristkova, E. (eds) *Eucarpia Leafy Vegetables'99*. Proceedings of the Eucarpia meeting on Leafy Vegetables and Breeding, Olomouc, Czech Republic, 8–11 June 1999, pp. 81–94.

Maxted, N., van Slageren, M.W. and Rihan, J.R. (1995) Ecogeographic surveys. In: Guarino, L., Ramanatha Rao, V. and Reid, R. (eds) *Collecting Plant Genetic Diversity: Technical Guidelines*. CAB International, Wallingford, UK, pp. 255–286.

Meilleur, B.A. and Hodgkin, T. (2004) *In situ* conservation of crop wild relatives: status and trends. *Biodiversity and Conservation* 13, 663–684.

Mujeeb-Kazi, A., Fuentes-Davila, G., Villareal, R.L., Cortes, A., Roasas, V. and Delgado, R. (2001) Registration of 10 synthetic hexaploid wheat and six bread wheat germplasms resistant to karnal bunt. *Crop Science* 41, 1652–1653.

Plucknett, D.L., Smith, N., Williams, J.T. and Anishetty, N.M. (1987) *Gene Banks and the World's Food*. Princeton University Press, Princeton, New Jersey.

Prescott-Allen, C. and Prescott-Allen, R. (1986) *First Resource: Wild Species in the North American Economy*. Yale University Press, New Haven, Connecticut.

Prescott-Allen, R. and Prescott-Allen, C. (1988) *Genes from the Wild: Using Wild Genetic Resources for Food and Raw Materials*. Earthscan Publications, London.

Richards, P. (1986) *Coping with Hunger: Hazard and Experiment in an African Rice-Farming System*. Allen & Unwin, London.

Rick, C.M. and Chetelat, R.T. (1995) Utilization of related wild species for tomato improvement. *Acta Horticulturae* 412, 21–38.

Ross, H. (1986) Potato breeding – problems and perspectives. *Fortschritte der Pflanzenzüchtung* 13. Paul Parey, Berlin and Hamburg, Germany.

Seiler, G.J. and Gulya Jr., T.J. (2004) Exploration for wild *Helianthus* species in North America: challenges and opportunities in the search for global treasures. *International 16th Sunflower Conference Proceedings* 1, 43–68.

Stalker, H.T. (1980) Utilization of wild species for crop improvement. *Advances in Agronomy* 33, 111–145.

Tanksley S.D. and McCouch, S.R. (1997) Seed banks and molecular maps: unlocking genetic potential from the wild. *Science* 277, 1063–1066.

Villareal, R.L., Sayre, K., Banuelos, O. and Mujeeb-Kazi, A. (2001) Registration of four synthetic hexaploid wheat (*Triticum turgidum/Aegilops tauschii*) germplasm lines tolerant to waterlogging. *Crop Science* 41, 274.

Xiao, J., Li, J., Grandillo, S., Ahn, S.N., Yuan, L., Tanksley, S.D. and McCouch, S.R. (1998) Identification of trait-improving quantitative trait loci alleles from a wild rice relative. *Oryza rufipogon*. *Genetics* 150, 899–909.

39 The Secondary Gene Pool of Barley as Gene Donors for Crop Improvement

M. SCHOLZ, B. RUGE-WEHLING, A. HABEKUß, G. PENDINEN, O. SCHRADER, K. FLATH, E. GROßE AND P. WEHLING

39.1 Introduction

The wild species *Hordeum bulbosum* L. is a potential resource of novel genes for barley improvement by introgression breeding, which represents the secondary gene pool. It is distributed around the Mediterranean and into the Middle East. This wild species exists in both diploid ($2n = 2x = 14$) and tetraploid ($2n = 4x = 28$) cytotypes. It is an obligate outbreeder and is perennial.

Among 111 accessions originated from the Middle East, Europe, North and South America and Australia, Michel (1996) identified one tetraploid ($2n = 4x = 28$) *H. bulbosum* accession (BAZ-3, which confers a complete resistance (immunity) to barley yellow dwarf virus (BYDV-PAV 1 ASL)). In addition, this accession carries resistance to the soilborne mosaic virus complex (BaMMV, BaYMV-1 and BaYMV-2), powdery mildew (resistance was effective against 26 isolates), leaf rust, typhula blight and the cereal cyst nematodes *Heterodera avenae* Wollenweber and *H. filipjevi* (Madzhidov) Stelter. It may, thus, serve as an excellent source of resistance to many diseases.

A crossing programme was initiated in 2001 to develop new interspecific *H. vulgare* ×× *H. bulbosum* hybrids suitable to transfer desired characters into cultivated barley through introgression.

39.2 Materials and Methods

Interspecific hybridization was carried out using the three diploid ($2n = 2x = 14$) *H. vulgare* cvs. 'Igri' (VV-1), 'Nikel' (VV-2) and 'Borwina' (VV-3), one tetraploid ($2n = 4x = 28$) cultivar 'Borwina' (VVVV) and the selected *H. bulbosum* accession (Michel, 1996) BAZ-3 (BBBB). VV-1, VV-2 and VV-3 were used as female parents. Reciprocal crosses were also made between VV-1 and BBBB, as well as VVVV and BBBB. VV-1 was used to produce BC1 offspring, with the triploid ($2n = 3x = 21$), partially fertile hybrid H2 (BAZ-50.131) used as pollinator.

About 14–20 days after pollination, embryo rescue was performed. Using selfing, backcrossing with VV-1 and anther culture technique, progenies were obtained from a BC1 genotype (BAZ-60.001) with resistance to BYDV-PAV 1 ASL.

Chromosome counts were done on root tips using the Feulgen technique. The nuclear DNA content was analysed by flow cytometry. Resistance tests were performed as follows:

BYDV-PAV 1 ASL: Niks *et al.* (2004), BaMMV/BaYMV-1, BaMMV/BaYMV-2: Proeseler (1993), leaf rust: Ruge *et al.* (2004), powdery mildew: Flath (2000), cereal cyst nematodes: biotest following Große.

Sequence-tagged-site (STS) marker analysis was carried out as described by Ruge *et al.* (2003). Genomic *in situ* hybridization (GISH) was performed according to Schrader *et al.* (2000).

39.3 Results and Discussion

The seed sets per flower and the yields of embryos and plantlets from the caryopses and embryos, respectively, in *Hordeum vulgare* × *H. bulbosum* crosses highly depended on the *H. vulgare* cultivar that was used as a female parent (Table 39.1). Cv. 'Borwina' ($2x$, $4x$) led to the highest frequency of embryos (VV-3: 67.9%/VVVV: 44.3%) compared to cv. 'Igri' (VV-1: 5.9%) and cv. 'Nikel' (VV-2: 17.2%). When cv. 'Borwina' ($4x$) was used as pollinator, however, only 7% of the embryos could be obtained compared to 44.3% achieved with the opposite direction. The regeneration rate of embryos into plants was also influenced by the genotype and the cross direction.

When cv. 'Nikel' (VV-2) was used as female parent only 6 plants out of 69 embryos could be regenerated compared to 17 plants out of 20 embryos from cv. 'Igri' ×× *H. bulbosum* crosses. The regeneration rates of embryos into plants were much lower when *H. bulbosum* was used as female parent (36.4%: BBBB ×× VV-1; 22.1%: BBBB ×× VVVV; not shown in Table 39.1). Michel

Table 39.1. Yields of hybrid plants obtained from reciprocal crosses between *Hordeum vulgare* ($2n = 14$: VV-1, VV-2, VV-3; $2n = 28$: VVVV) and *H. bulbosum* ($2n = 28$, BBBB).

Combinations		Florets pollinated	Embryos cultured		F1 plants	
			No.	%	No.	%
VV-1	BBBB	339	20	5.9	17	5.0
VV-2	BBBB	401	69	17.2	6	1.5
VV-3	BBBB	78	53	67.9	50	64.0
VVVV	BBBB	463	205	44.3	200	43.2
BBBB	VV-1	1108	74	6.7	36	3.2
BBBB	VVVV	752	53	7.0	40	5.3

(1996) also reported a slower seed set in this cross direction. Apparently, there is an effect of the cytoplasm of *H. bulbosum* on the vigour of the embryos.

All offspring plants from VV/VVVV ×× BBBB crosses that were analysed proved to be triploid ($2n = 3x = 21$). Offspring from BBBB ×× VV crosses were triploid or tetraploid and all offspring from BBBB ×× VVVV crosses were tetraploid ($2n = 4x = 28$). Ruge *et al.* (2003) demonstrated that fertile tetraploid VVBB hybrids are useful for gene transfer into *H. vulgare*.

A differentiation of alleles at molecular markers from *H. bulbosum* (*b* alleles) and *H. vulgare* (*v* alleles) in the hybrids and their parents was attempted by using STS anchor markers located on the seven chromosomes of barley. The hybrid character of three interspecific F1 offspring from BBBB × VV (H1, H3) and VV × BBBB crosses (H2), respectively, could be demonstrated (Table 39.2).

In three instances, anchor markers failed to verify the presence of the respective orthologous *H. bulbosum* chromosome or chromosomal region (Table 39.2).

In hybrid H3, *H. bulbosum* chromosome 1 could not be detected since marker STS-*ABG373* displayed the *v* allele only. *ABG373* is located distally on the long arm of barley chromosome 1H (Qi *et al.*, 1996). STS primers for additional 1H anchor markers gave no polymorphisms between the two genomes (not shown). Absence of the *b* alleles of markers was also observed in hybrids H2 and H3 for chromosomes 7HS and 7HL, respectively, as demonstrated by markers *MWG530* and *MWG2031*, respectively (Table 39.2, Fig. 39.1).

Cytological analysis displayed 21 chromosomes for all three hybrids. However, the absence of *b*-marker alleles for the distally located markers on 1HL (hybrid H3) and 7HS (hybrid H2), respectively, as well as the proximally located marker MWG2031 (~85 cM) on 7HL (hybrid H3) may be indicative for partial elimination of *H. bulbosum* chromatin.

Table 39.2. Identification of interspecific hybrids (H1–H3) using barley STS markers derived from RFLP anchor markers.

Expected genome constitution	STS marker	Hybrid H1 BBBB × VV BBV	Hybrid H2 VV × BBBB VBB	Hybrid H3 BBBB × VV BBV
Chromosome 1H	**ABG373**	*bv*	*vb*	*v-*
Chromosome 2H	*MWG2133*	*bv*	*vb*	*bv*
	MWG2146	*bv*	*vb*	*bv*
	MWG520	*bv*	*vb*	*bv*
Chromosome 3H	*MWG549*	*bv*	*vb*	*bv*
Chromosome 4H	*WG622*	*bv*	*vb*	*bv*
Chromosome 5H	*MWG877*	*bv*	*vb*	*bv*
	MWG583	*bv*	*vb*	*bv*
Chromosome 6H	*MWG2218*	*bv*	*vb*	*bv*
	MWG934	*bv*	*vb*	*bv*
Chromosome 7H	*MWG530*	*bv*	*v-*	*bv*
	MWG2031	*bv*	*vb*	*v-*

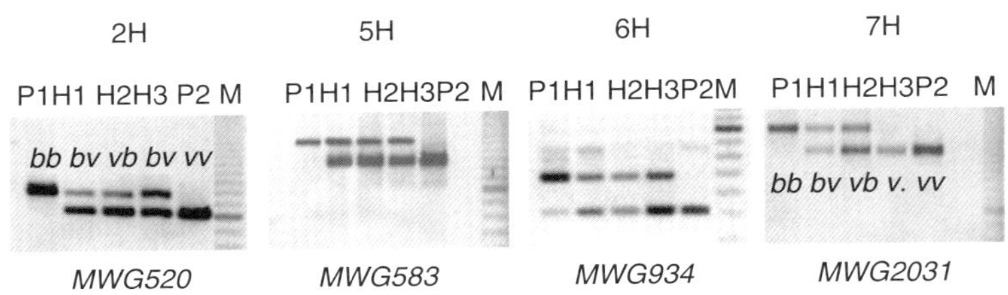

Fig. 39.1. Identification of interspecific hybrids by means of STS markers. P1, *Hordeum bulbosum* parent; P2, *H. vulgare* parent; H1, H2 and H3, interspecific hybrids.

Elimination of marker alleles of the *H. bulbosum* parent in the two hybrids appears to be independent of whether this parent was used as male or female (Table 39.2). The loss of *H. bulbosum* chromatin in the two interspecific hybrids will have to be analysed in its extension in more detail.

Meiotic metaphase I analysis in triploid and tetraploid interspecific hybrids revealed pairing between *H. bulbosum* and *H. vulgare* chromosomes. The mean frequency of trivalent pairing configurations of the H2 hybrid ranged from 0.63 to 0.9 with a maximum of four per pollen mother cell (PMC) (Table 39.3, Fig. 39.2a). One tetraploid hybrid that arose from BBBB × VVVV crosses (H4) showed a very high intergenomic pairing. Tetravalents occurred in 96% of the PMCs. The mean frequency for tetravalents was 3.1 with a maximum of 6 per cell.

The mean frequency of tetravalents ranged from 1.8 to 2.7 (Table 39.3, Fig. 39.2b). The high intergenomic pairing implicates that recombination can occur in this hybrid at a very high level. All other tetraploid interspecific hybrids which were analysed showed a lower intergenomic pairing.

The hybrids H1 and H2 verified with molecular markers proved to be non-infectable by BYDV-PAV 1 ASL. In addition, they also displayed resistance to leaf rust.

Table 39.3. Meiotic chromosome pairing at metaphase I in the interspecific hybrids H1–H4.

Hybrid AGC[a]	Ploidy level	Nos. of PMCs	I	II Rod	II Ring	II Total	III	IV	Chiasmata per cell
H1	21	20	6.7	1.3	4.65	5.95	0.8	0	11.4
BBV			(5–11)	(0–3)	(3–7)	(0–7)	(0–2)		
H2	21	75[b]	6.2–6.8	1.1–2.3	3.9–5.0	6.1–6.2	0.6–0.9	0	11.5–13.0
VBB			(3–8)	(0–4)	(1–7)	(0–7)	(0–4)		
H3	28	25	0.4	0.92	10.8	11.72	0.16	0.92	26.7
BBVV			(0–2)	(0–4)	(6–14)	(0–14)	(0–1)	(0–2)	
H4	28	34[c]	0	0.1–0.3	8.5–9.9	8.6–10.2	0	1.8–2.7	27.3–27.9
BBVV				(0–2)	(2–14)	(0–14)		(0–6)	

[a]Assumed genomic constitution.
[b]Mean of three plants.
[c]Mean of two plants.

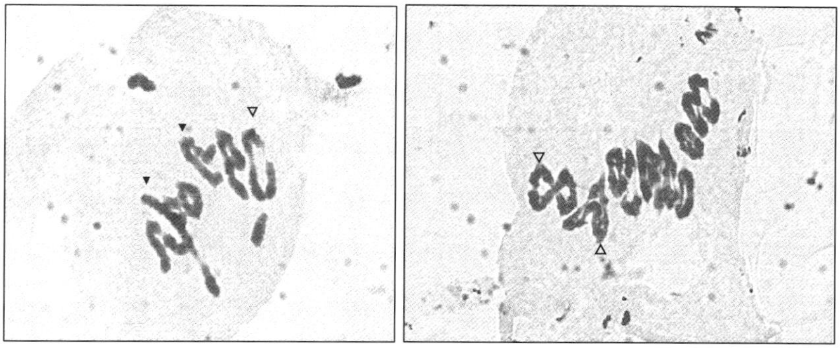

Fig. 39.2. Meiotic chromosome behaviour in triploid (a) and tetraploid (b) interspecific *Hordeum vulgare* ×× *H. bulbosum* hybrids. (a) Metaphase I showing four univalents, one rod and three ring bivalents, two-rod (closed triangles) and one-ring trivalent (open triangle). (b) Metaphase I showing ten-ring bivalents and two tetravalents (open triangles).

After backcrossing the partially fertile hybrid H2 to VV-1, one BC1 plant (BAZ-60.001) could be obtained that did not develop any symptoms and stayed free of BYDV virus according to enzyme-linked immunosorbent assay (ELISA) results. This BC1 plant is perennial and resembled *H. vulgare* morphologically except that the tillers were shorter and thinner (Fig. 39.3).

We identified four *H. bulbosum* signals of the BC1 plant at the terminal end of the *H. vulgare* chromosomes using GISH technique (Fig. 39.4a). To regenerate plants with stable gene introgression, selfing, backcrossing and

Fig. 39.3. Spikes of cross parents, P1, *Hordeum bulbosum*; P2, *H. vulgare*; H2, triploid *H. vulgare* ×× *H. bulbosum* hybrid; BC1, progeny from backcrossing H2 to P2.

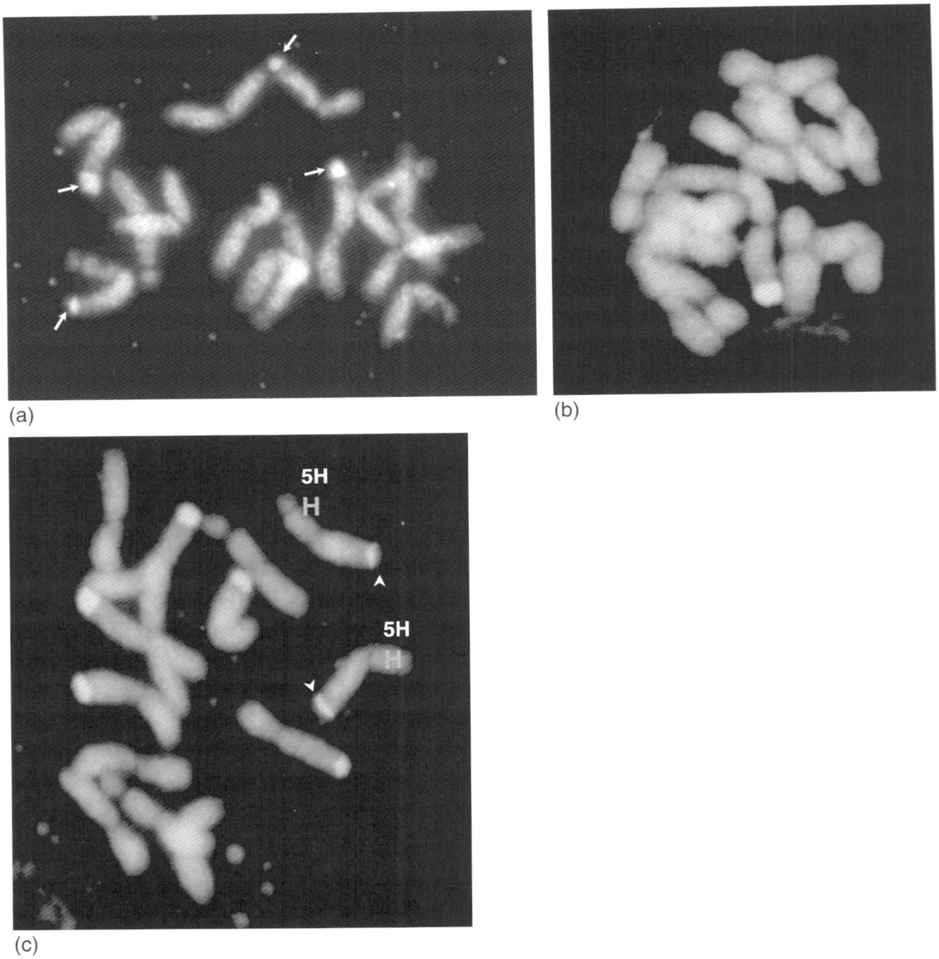

Fig. 39.4. (a) Genomic *in situ* hybridization of mitotic chromosomes from the diploid Bc1 plant (Baz-60.001) showing four highlighted signals (arrows) of *H. bulbosum* chromatin at the terminal end of the *H. vulgare* chromosomes, (b) from the diploid Bc2 plant showing only one signal of *H. bulbosum* and (c) from a Bc1f2 plant showing seven signals of *H. bulbosum.*

anther culture techniques have been used. Haploid, diploid and triploid progenies having one (Fig. 39.4b) to seven introgressions have been produced from the BC1 plant. Some of the BC2 and BC1F2 genotypes displayed resistance to BYDV-PAV 1 ASL, the soilborne virus complex (BaMMV, BaYMV-1 and BaYMV-1–2), the cereal cyst nematode *H. filipjevi* and nine isolates of powdery mildew. To utilize these novel genes in agriculture, mapping populations will need to be developed for genetic analysis together with molecular marker analysis.

References

Flath, K. (2000) Testing of resistance of field and horticultural crops to fungi, bacteria and viruses. In: Bartels, G. and Backhaus, G.F. (eds) *Mitteilungen aus der Biologischen Bundesanstalt für Land- und Forstwirtschaft, Berlin-Dahlem* 373, pp. 6–8.

Michel, M. (1996) Untersuchungen zur Übertragung von Resistenzgenen aus der Wildart *Hordeum bulbosum* L. in die Kulturgerste *Hordeum vulgare* L. PhD thesis, Technische Universität München, Munich, Germany.

Niks, R.E., Habekuß, A., Bekele, B. and Ordon, F. (2004) A novel major gene on chromosome 6H for resistance of barley against the barley yellow dwarf virus. *Theoretical and Applied Genetics* 109, 1536–1543.

Proeseler, G. (1993) *Triticum durum* Desf. a further host of barley mild mosaic virus (BaMMV)). *Journal of Phytopathology* 36, 262–264.

Qi, X., Stam, P. and Lindhout, P. (1996) Comparison and integration of four barley genetic maps. *Genome* 39, 379–394.

Ruge, B., Linz, A., Pickering, R., Proeseler, G., Greif, P. and Wehling, P. (2003) Mapping of *Rym14*^{Hb}, a gene introgressed from *Hordeum bulbosum* and conferring resistance to BaMMV and BaYMV in barley. *Theoretical and Applied Genetics* 107, 965–971.

Ruge, B., Linz, A., Habekuß, A., Flath, K. and Wehling, P. (2004) Introgression and mapping of novel resistance genes from the secondary gene pool of barley, *Hordeum bulbosum*. *Czech Journal of Genetics and Plant Breeding* 40, 138.

Schrader, O., Budahn, H. and Ahne, R. (2000) Detection of 5S and 25S rRNA Genes in *Sinapis alba, Raphanus sativus* and *Brassica napus* by double fluorescence *in situ* hybridization. *Theoretical and Applied Genetics* 100, 665–669.

40 Exploitation of Wild Cereals for Wheat Improvement in the Institute for Cereal Crops Improvement

E. Millet, J. Manisterski and P. Ben-Yehuda

40.1 Introduction

Due to the single or few plant origins of domesticated wheat (Zohary, 1999), cultivated wheat species feature a narrow genetic basis. Another portion of the genetic variation of cultivated wheat was lost during the intensive wheat breeding and replacement of traditional landraces by modern cultivars in recent years. The Institute for Cereal Crops Improvement (ICCI) at Tel Aviv University is aware of the reduced diversity of wheat and invests considerable efforts in collection of different wild wheat relatives that are native to Israel and evaluation of their breeding value as potential donors of desirable traits for wheat improvement. In this chapter, which is focused on the exploitation of wild cereals for wheat improvement at the ICCI, we will describe traits from the wild that are desirable in cultivated wheat, with examples of those that are currently studied at the ICCI. In addition, we will describe three different current projects of gene transfer from wild cereals to wheat.

40.2 Origin of Cultivated Wheat and Genetic Relationships Among *Triticeae* Species

The genetic constitution of the different *Aegilops* and *Triticum* species is derived from their evolution (Fig. 40.1; Feldman, 2001). It is believed that they all have a common diploid ancestor ($2n = 14$) which diverged about 4.5 million years ago to different *Aegilops* and *Triticum* species (Huang *et al.*, 2002). Among them are *T. urartu* (genome A), *Ae. speltoides* (genome S) and *Ae. tauschii* (genome D). By hybridization between diploid species having genome B and genome A, which occurred spontaneously some 500,000 years ago, a new tetraploid species, namely *T. turgidum* subsp. *dicoccoides* (wild emmer wheat) evolved, having BA genome and $2n = 4x = 28$ chromosomes. The

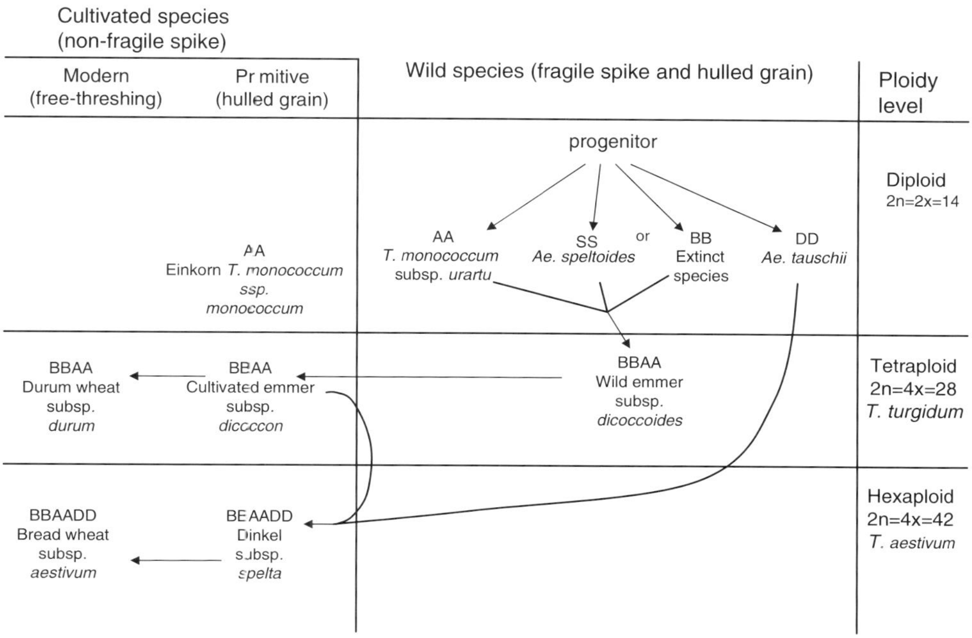

Fig. 40.1. The origin of cultivated wheat. A, B, D and S denote species genomes. (Adapted from Feldman, 2001.)

donor of the B genome and the cytoplasm of wheat have not been found so far and may be extinct. Yet, high similarity between genome B of wheat and genome S of *Ae. speltoides* suggests that they have a common progenitor.

Wild emmer was domesticated by the early farmers in the Middle East about 10,000 years ago by selection for a non-brittle spiked prototype to generate the cultivated emmer wheat, *T. turgidum* subsp. *dicoccon*. From this primitive crop, by further selection to free-threshing spike type, the current subsp. *durum* was derived. Comparisons of molecular and phenotypic diversity in wild and domesticated *T. turgidum* species (Huang *et al.*, 1999) suggest monophyletic origin of cultivated wheat (Zohary, 1999).

As a result of this domestication event, and the spread of domesticated tetraploid wheat to eastern Turkey and western Iran, where *Ae. tauschii* ($2n = 14$; genome DD) is found, D genome was spontaneously combined with BA genomes of cultivated emmer wheat about 9500 years ago to generate the new hexaploid species *T. aestivum* subsp. *spelta* (spelt wheat) having BAD genomes and $2n = 6x = 42$ chromosomes. Spelt wheat has tough glumes and bread wheat was derived from it by further selection to free-threshing spiked types.

Bread wheat and durum wheat are the most widespread cultivated wheat species. As described above, both are allopolyploids but exhibit diploid-like chromosome behaviour, i.e. pairing between only homoeologous chromosomes. This behaviour is governed by *Ph* genes that suppress homoeologous pairing. Each genome of tetraploid and hexaploid wheat consists of seven chromosome pairs that are classified into seven homoeologous groups according to the relatedness of any chromosome with its homoeologues in the other genomes. Genes on homoeologous chromosomes usually maintain sinteny except in chromosome regions that underwent rearrangement that occurred after the evolution of the polyploids. Yet, there may be considerable differences in the allelic composition of the homoeologous chromosomes and in the expression of the three homoeoalleles (Adams *et al.*, 2003).

The wild wheat relatives carry homoeologous chromosomes to those of wheat and consequently, homoeoalleles can be transferred into their wheat homoeologues. The homoeologous relationships between the different genomes of wheat, namely B, A and D, will be similar to their relationships with another alien *Triticeae* genome that was introduced into wheat. The best-known example of such a case is the man-made species triticale which combines the two durum or three bread wheat genomes with the rye genome (R) in hexaploid or octoploid triticale, respectively. In both ploidy levels the chromosomes maintain diploid behaviour and no recombinations usually occur between the wheat and the rye genomes.

40.3 Gene Transfer Principles

Exploitation of wild species for crop improvement confronts different obstacles that are exerted by the nature of wild plants (undesirable and non-agronomic traits), by biological barriers (crossability, endosperm development), by unequal ploidy level and by non-homology between the candidate chromosomes. However, in wheat, a polyploid species that can bear aneuploidy and genome manipulations, special genetic lines were developed to facilitate gene transfer. The need to use different techniques and the ease of gene transfer depend, to a great deal, on the degree of relatedness between wheat and the donor of the gene controlling the desirable trait.

Transfer of genes between homoeologous genomes is straightforward and employs simple plant breeding methods. However, gene transfer between homoeologous genomes requires more sophisticated methodologies which are generally termed as chromosome engineering.

40.4 Systemic Evaluation of Alien Species and Gene Transfer

The systemic procedure to evaluate the breeding value of an alien wheat relative species and to transfer genes controlling desirable traits is described elsewhere (Feldman and Sears, 1981; Feldman, 1988) and follows this pathway:

1. Generation of an amphiploid between wheat and a line of alien *Triticeae* species exhibiting the trait. This step is accomplished by pollinating the wheat parent by the alien species, deriving haploid F_1 offspring and doubling its chromosomes by using colchicine treatment. This amphiploid contains both the wheat and the alien species genomes and is expected to show the desirable trait provided that no epistasis is exerted by the wheat homoeoalleles.

2. Generation of a disomic addition line. This step is required in order to exclude the chance for whole genome interaction, to allocate the gene to a specific chromosome and to facilitate next steps. The amphiploid is crossed with the wheat parent and individual plants with the wheat chromosome complement plus one chromosome are selected. They are allowed to self and plants with the wheat chromosome complement plus one pair are selected from the progenies. Plants showing the desired trait carry an extra chromosome which could be identified by cytogenetic or marker analysis.
3. Generation of a disomic substitution line. This step best reflects the gene transfer since the alien allele substitute for its cultivated homoeoallele. The portion of the alien genetic material is reduced to only one chromosome pair; yet, this chromosome pair may introduce some undesirable alien alleles that substitute for their wheat homoeologues. This step is achieved by crossing the addition line with a line which is monosomic for the selected chromosome (of the same homoeologous group of the alien chromosome) and selection of F_1 offspring lacking the selected wheat chromosome (and having a single alien chromosome). Plants with a pair of substituting chromosomes are selected from the progenies of the selfed F_1.
4. Generation of a recombinant line of wheat with an alien segment carrying the desirable gene. This is straightforward in the case of homologous genomes. However, with homoeologous genomes, the disomic alien substitution line has to be crossed with wheat line deficient for *Ph* allele and then backcrossed again to this line for obtaining progenies that are homozygous deficient for *Ph*. The production of an alien recombinant line, that eliminates undesirable alien genes while it retains on a selected wheat chromosome a small homoeologous alien segment with the desirable gene, is termed chromosome engineering.

Chromosome engineering enables gene transfer between homoeologous chromosomes which naturally do not pair and recombine. In practice, to transfer a selected alien allele to a selected wheat homoeologous chromosome there is no need to follow all the steps described above. Two versions for chromosome engineering are proposed. In the first one the diploid donor of the gene is used to pollinate a wheat plant which is mutant for *Ph* allele (*ph1ph1*) and an amphiploid wheat-alien species is generated. The haploid F_1 having the wheat and the alien genomes has no *Ph* allele and therefore homoeologous pairing will occur between wheat and alien chromosomes as well as between wheat and wheat chromosomes. The wheat genome is recovered by a series of backcrosses to a wheat cultivar while maintaining the desired recombinant chromosome by selection for the desirable trait. Unfortunately, the F_1 has very low fertility and only few seeds are obtained which will reduce the chance to select a BC_1 offspring with the desirable trait. The alternative procedure is to double first the chromosomes of the F_1 hybrid between the *Ph* mutant line and the alien species and then to cross it to the recurrent wheat parent. In this case many seeds are obtained but because homologues are present for each chromosome, the rate of pairing between homoeologous wheat and alien chromosomes is reduced to a degree that renders this procedure inefficient.

40.5 Traits with Potentially Economical Value

Most traits to be exploited from wild cereals are ascribed to one of the following categories:

1. *Resistance to biotic stresses*. Wild cereals are an invaluable resource for resistance to biotic stresses and considerable success with alien gene transfer is ascribed to this class of resistance including resistance to various diseases and pests. Most of the resistance genes of this type are monogenic and easy to handle.
2. *Resistance to abiotic stresses*. Wild cereals are not as susceptible as cultivated wheat to growing conditions and have a high intrinsic degree of resistance or tolerance to abiotic stresses. So wild cereals show tolerance to salt, heat and drought, as well as to mineral toxicity (e.g. aluminum) or deficiency (zinc).
3. *Technological quality*. Most wheat is processed into well-defined products such as bread, pasta and biscuits. Today much is known about the grain properties that are required for a good product. The inclusion, for example, in a wheat cultivar, of specific high molecular weight glutenin subunits that are found in *T. turgidum* subsp. *dicoccoides* or in *Ae. longissima* but not in cultivated wheat, may improve the making of such products.
4. *Nutritional quality*. During many years of breeding high grain yield was the major criterion for success. Nowadays there is more awareness of the nutritional value of the grains and recent technologies allow fast and accurate grain analyses so it is possible to exploit wild cereals for a better food. Wild emmer was found to be a good resource both for higher protein and higher zinc content.

40.6 Alien Gene Transfer Projects at the ICCI

40.6.1 Transfer of a yellow rust resistance gene from wild emmer to bread wheat

Many of the wild emmer accessions show good resistance to yellow rust. However, so far only two yellow rust resistance genes, namely *Yr15* and *H52*, have been identified in wild emmer and transferred to wheat, both located on the short arm of chromosome 1B about 10 cM apart (McIntosh *et al.*, 1966; Peng *et al.*, 1999). Ten yellow rust resistant lines from our collection were crossed with two tester lines carrying *Yr15* or *H52*, and F_1 and F_2 progenies were obtained and inoculated with the yellow rust pathogen. Only the F_2 progenies derived from crosses with the wild emmer line TD2000 segregated for yellow rust resistant and susceptible plants, while in all the other combinations no susceptible plant was obtained, indicating identity of the Yr genes between the tested lines and the testers. Line TD2000 was crossed with a leading Israeli spring cultivar as well as US winter cultivars and pentaploid F_1 plants were obtained. These plants are partially sterile but it is possible to obtain some seeds from backcrosses to the wheat cultivar as a recurrent parent. Five backcrosses accompanied with selection for resistant plants are planned to recover the cultivated wheat genetic background while retaining the *Yr* gene. This gene is currently being mapped by molecular markers. In addition, attempts are being made to generate a breeder friendly PCR marker to facilitate the transfer of this gene to cultivars in breeding programmes.

40.6.2 Transfer of a leaf rust resistance gene from *Ae. speltoides* to bread wheat

Ae. speltoides exhibits excellent resistance to wheat leaf rust caused by *Puccinia triticina* fs. *tritici*. The very low frequency of leaf rust susceptible *Ae. speltoides* accessions found by us and by Niks (1987) and others cited by him suggests that this species is only an occasional host for this fungus. Hence, our expectation is that resistance transferred from *Ae. speltoides* to wheat will be more durable. In fact, five major genes from *Ae. speltoides* namely, *Lr28, Lr35, Lr36, Lr47* and *Lr51*, have been already used to improve wheat resistance to leaf rust (McIntosh *et al.*, 2003). However, they all show a selective resistance to specific races of the fungus and would not account for the complete immunity of *Ae. speltoides* to leaf rust. Our approach to reveal this resistance is to generate wheat recombinant lines, each with a chromosome segment from *Ae. speltoides* that confer resistance to the susceptible wheat parent. The spectrum of pathogen races that are virulent to each line will be determined and if none of the recombinant lines will maintain the immunity of the alien species then different lines will be intercrossed to broaden the range of resistance to various races of the pathogen.

The procedure for generation of wheat-*Ae. speltoides* introgression lines is depicted in Fig. 40.2. We take advantage of the fact that some *Ae. speltoides* lines are of a high pairing type, i.e. they allow homoeologous pairing between the wheat and the alien homoeologous chromosomes when combined in an hybrid. Despite the haploid hybrid being considerably infertile, it still set a few seeds per spike when used as a female parent and pollinated by a wheat cultivar. Therefore, to ensure the inclusion of the desirable alien chromosome segment in the recombinant lines many florets (in a number of spikes) were pollinated. The resultant BC_1 plants were inoculated by leaf rust and only resistant offspring will be backcrossed to the wheat cultivar as a recurrent parent. Five backcross cycles accompanied with selection for resistance are supposed to recover the wheat genome and sweep away the alien chromosomes and chromosome segments except those carrying the desirable genes.

40.6.3 Transfer of a yellow rust resistance gene from *Ae. sharonensis* to bread wheat

Sharon goat grass (*Ae. sharonensis*) is a diploid relative of cultivated wheat and is endemic to the coastal plain of Israel. It possesses the S^l genome, which is closely related to the B genome of wheat. We have found that *Ae. sharonensis*

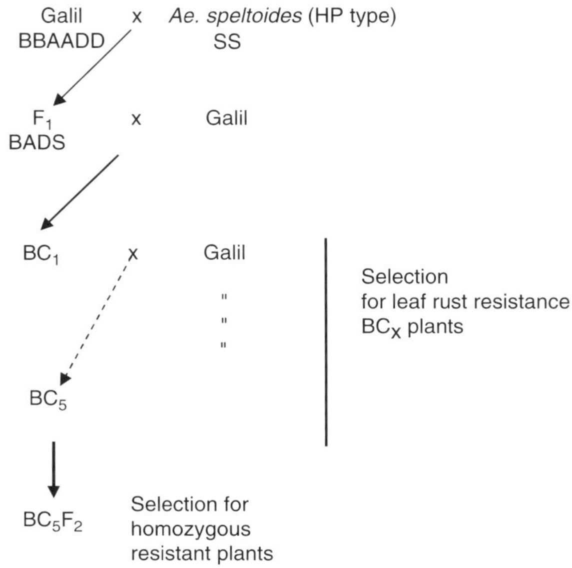

Fig. 40.2. Transfer of leaf rust resistance from *Aegilops speltoides* to wheat. BAD and S denote bread wheat and *Ae. speltoides* genomes, respectively.

is resistant to many important diseases of wheat including stem rust, leaf rust, yellow rust and powdery mildew. Despite *Ae. sharonensis* being a potentially valuable donor of resistance genes, it has rarely been used for wheat improvement, presumably because it possesses a gametocidal gene that prevents successful hybridization with wheat. The recent development of a genetic line that suppresses this gametocidal gene creates an exciting opportunity for introgressing economically important genes from this underutilized member of the secondary gene pool into cultivated wheat.

Yellow rust is a serious foliar disease of wheat causing tremendous yield loss in many parts of the world in which genetic resistance to the fungus has been broken. Introgression of resistance from an exotic resource may confer wheat with a more durable resistance.

Unlike in *Ae. speltoides*, the presence of a homoeologous pairing suppression genetic system in *Ae. sharonensis* is not known to us. Hence the procedure for gene transfer to wheat from this alien species (Fig. 40.3) requires the use of *ph1* mutant allele from wheat. A special genetic line, namely, an amphiploid between genome A of *T. monococcum* and genome D of *Ae. tauschii*, was prepared to facilitate the transfer into genome B of wheat. This amphiploid was pollinated by a leaf rust resistant line of *Ae. sharonensis* and the haploid hybrid is undergoing colchicine treatment for chromosome doubling. The resultant AADDS^lS^l amphiploid will be used to pollinate a wheat *ph1* mutant to yield a hybrid with a double dose of A and D genomes and a single dose of B and S^l genomes with no normal *Ph1* on chromosome 5B. Therefore homoeologous pairing will occur mainly between chromosomes of B and S^l genomes. Gametes of this hybrid will contain mainly the whole A and

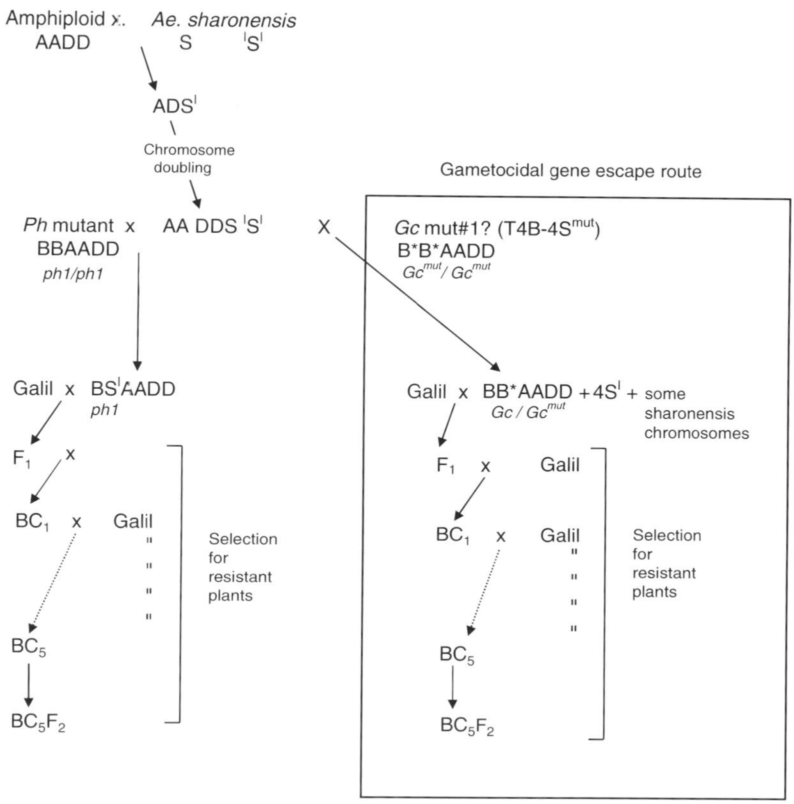

Fig. 40.3. Transfer of yellow rust resistance genes from *Aegilops sharonensis* to wheat. BAD and S^l denote for bread wheat and *Ae. sharonensis* genomes, respectively; *ph1* is a recessive mutation at the *Ph1* locus that allows for homoeologous pairing; *Gc* and Gc^{mut} are gametocidal and 'anti gametocidal' mutant allele, respectively.

D genomes and some chromosomes from B and S^l genomes including recombinant chromosomes from these two genomes.

Gametocidal gene is present at least in part of the *Ae. sharonensis* lines. This gene is a selfish gene that results in the transmission of only gametes containing it and abortion of the rest of the gametes by inducing chromosome breaks in them. It is located on chromosome $4S^l$ and therefore all viable gametes of a wheat-*Ae. sharonensis* hybrid and offspring thereof will contain this chromosome. Therefore, it is required in our procedure to allow selection against this chromosome. If no gametocidal gene is present in the *Ae. sharonensis* line that is used as is evident from the lack of chromosome breaks in the gametes of the $AABS^lDD$ hybrid, then backcrosses to a wheat cultivar as a recurrent parent accompanied with a selection for resistant backcross offspring will select against *Ae. sharonensis* chromosomes and chromosome segments except those carrying the resistance genes, and recover the wheat genetic background. However, if chromosome breakage will be found at meiosis of the hybrid, an indication for the presence of a gametocidal gene, then the *Gc2* mutant (Friebe *et al.*, 2003) will be used to pollinate the hybrid (Fig. 40.3: gametocidal gene escape route). It is known that in plants heterozygous for the gametocidal gene $Gc2/Gc2^{mut}$ there is no preferential selection for the selfish gene and therefore, in this generation, there is a possibility to get rid of chromosome $4S^l$. This step will be followed by backcrosses to the wheat cultivar and selection for resistant plants as before.

40.7 Conclusions

Wild wheat relatives are invaluable genetic resources of various desirable traits for wheat improvement. The genetic relationships of wheat with its wild relatives which are the outcome of wheat evolution and the allopolyploid nature of wheat allow gene transfer from wheat wild relatives into wheat. The selected procedure for gene transfer and the ease of this task depend on the degree of relatedness between wheat and its wild relatives. Different genetic lines are available and various methodologies were developed to facilitate gene transfer from wild cereals into wheat.

References

Adams, K.L., Cronn, R., Percifield, R. and Wendel, J.F. (2003) Genes duplicated by polyploidy show unequal contributions to the transcriptome and organ-specific reciprocal silencing. *Proceedings of the National Academy of Sciences of the United States of America* 100, 4649–4654.

Feldman, M. (1988) Cytogenetic and molecular approaches to alien gene transfer in wheat. In: Miller, T.E. and Koebner, R.M.D. (eds) *Proceedings of the 7th International Wheat Genetics Symposium*. The Institute of Plant Sciences Research, Cambridge, UK, pp. 23–32.

Feldman, M. (2001) Origin of cultivated wheat. In: Bonjean, A.P. and Angus, W.J. (eds) *The World Wheat Book*. Lavoisier, Paris, pp. 3–56.

Feldman, M. and Sears, E.R. (1981) The wild gene resources of wheat. *Scientific American* 244, 102–112.

Friebe, B., Zhang, P. and Nasuda, S. (2003) Characterization of a knock-out mutation at the *Gc2* locus in wheat. *Chromosoma* 111, 509–517.

Huang, L., Millet, E., Rong, J.K., Wendel, J.F., Anikster, Y. and Feldman, M. (1999) Restriction fragment length polymorphism in wild and cultivated tetraploid wheat. *Israel Journal of Plant Science* 47, 213–224.

Huang, S., Sirikhachornkit, A., Su, X.J., Faris, J., Gill, B., Haselkorn, R. and Gornicki, P. (2002) Genes encoding plastid acetyl-CoA carboxylase and 3-phosphoglycerate kinase of the *Triticum/Aegilops* complex and the evolutionary history of polyploidy wheat. *Proceedings of the National Academy of Sciences of the United States of America* 99, 8133–8138.

McIntosh, R.A., Silk, J. and The, T.T. (1996) Cytogenetic studies in wheat.17. Monosomic analysis and linkage relationships of gene *Yr15* for resistance to stripe rust. *Euphytica* 89, 395–399.

McIntosh, R.A., Yamazaki, Y., Devos, K.M., Dubcovsky, J., Rogers, W.J. and Appels, R. (2003) Catalogue of gene symbols for wheat. In: Pogna, N.E., Romano, M., Pogna, E.A., and Galterio, G. (eds) *Proceedings of the 10th International Wheat Genetics Symposium*, Vol. 4. and MacGene2003 database on CD, S.I.M.I., Rome.

Niks, R.E. (1987) Nonhost plant species as donors for resistance to pathogens with narrow host range. 1. Determination of nonhost status. *Euphytica* 36, 841–852.

Peng, J.H., Fahima, T., RÖder, M.S., Li, Y.C., Dahan, A., Grama, A., Ronin, Y.I., Korol, A.B. and Nevo, E. (1999) Microsattelite tagging of the stripe-rust resistance gene *yrh52* derived from wild emmer wheat, *Triticum dicoccoides*, and suggestive negative crossover interference on chromosome 1B. *Theoretical and Applied Genetics* 98, 862–872.

Zohary, D. (1999) Monophyletic vs. polyphyletic origin of the crops on which agriculture was founded in the near east. *Genetic Resources and Crop Evolution* 46, 133–142.

41 Using Crop Wild Relatives as Sources of Useful Genes

G. Sonnante and D. Pignone

41.1 Introduction

Crop wild relatives (CWR) are a source of strategic germplasm for the present and future generations. In the past, the majority of the information on plant genetic resources (PGR) was based on phenotype, geographic origin, social history and parentage. Useful characters were mostly searched in the field, but this could pose problems of phenotypic expression and environmental influence. Most of the research on wild relatives has regarded the search for resistance to biotic stress, especially resistance to the pests with highest economic impact. Successful examples of this strategy are found in tomato, cereals, sunflower, etc. Most of the work done in the past was based on traditional breeding techniques using wide hybridization to introduce wild genes into a crop followed by selection, based on visual examination of the offspring. The procedure was long and often only partially successful due to constraints in the selection of the elite lines to be used in further breeding (Lenné and Wood, 1991; Hawkes, 1977).

In the last decades, advances in plant science and technology have provided new insights into the study and use of CWR genetic resources, dramatically increasing the effectiveness of the actions; in fact, techniques derived from plant genomics can address the weaknesses inherent to non-molecular methods. The development of DNA markers and sequencing has provided new powerful ways of assessing genetic relationships and diversity, performing comparative linkage analysis, isolating useful genes from CWR, providing tools for marker assisted selection, etc. (Tanksley and McCouch, 1997; De Vienne, 2003). Within this frame, some programmes were started at the Institute of Plant Genetics, in order to introduce wild genes from *Aegilops* species into durum wheat, using molecular cytogenetics to assist selection of the recombinant lines possessing the target gene and the least proportion of the *Aegilops* genome (Blanco *et al*., 2002). Recently, studies have evolved towards the investigation of gene variants in some crop gene pools, also in the view of isolating possible alleles that can be useful in plant breeding and/or in the development of new molecules.

Three of these examples concern MSI proteinase inhibitors in wild rocket and other *Brassicaceae*, Bowman-Birk inhibitors in wild lentil, and genes coding for phenylalanine ammonia lyase (PAL) in artichoke and its wild relatives.

41.2 MSI Proteinase Inhibitors in Wild Rocket and Other *Brassicaceae*

In 1994 the Rocket Genetic Resources Network, an international cooperative group, was established under the aegis of IPGRI, with the aim of promoting the conservation and utilization of underutilized Mediterranean species belonging to the genera *Eruca* and *Diplotaxis* (Padulosi, 1995; Pignone, 1997). A follow-up of this initiative was the search, within the genome of rocket species, for genes coding for potentially useful characters. A candidate gene was the trypsin inhibitor gene *mti-2* initially isolated in *Sinapis alba*. Trypsin inhibitor molecules, in fact, besides a role in controlling endogenous proteinases during seed dormancy, have proven very useful in controlling sucking insect attacks owing to their role against proteolytic enzymes present in the pest digestive apparatus (Ryan, 1990). Due to the rapid adaptation of insects to this class of natural pesticides, there is a strong interest in searching for new sources of natural trypsin inhibitors in order to better study their structure and mechanism of action on insect digestive enzymes.

The MTI-2 protein is a serine protease inhibitor belonging to a novel, recently described class of trypsin inhibitors (Ceciliani *et al.*, 1994; Ceci *et al.*, 1995; De Leo *et al.*, 2002). The gene coding for this protein, *mti-2*, is a discontinuous gene formed by two exons and one intron. The mature active protein, comprising the reactive loop, is entirely coded by the second exon.

In order to identify new sources of inhibitors within the *Brassicaceae* family, primers were designed based on the sequence of the mustard *mti-2* gene, in the region of the second exon coding for the mature protein. Genomic DNA from members of the *Brassicaceae* listed in Table 41.1, counting cultivated, low-domesticated and wild species, was amplified using a primer pair designed to surround a region of the gene including the sequence coding for the mature protein. These primers produced amplifications only in *Diplotaxis muralis* and *D. tenuifolia*, whereas no amplification was detected in the other species. The cloning of these amplicons allowed the detection of nine different sequences, whose translation product showed a high degree of similarity to the MTI-2 protein. These sequences showed limited variation in some important residues of the active site. DNA from all the remaining taxa was amplified using another primer pair which was designed in highly conserved regions of the sequence coding for the mature protein. Amplifications were obtained in *Brassica oleracea, B. rapa, B. tournefortii, Raphanus sativus* and *Crambe hispanica*. These amplifications led to the detection of 15 new amplicons. The corresponding proteins, translated *in silico*, showed some variation in length

Table 41.1. List of plant material analysed.

Plant species	Family
Brassica oleracea	*Brassicaceae*
B. rapa	*Brassicaceae*
B. tourneforti	*Brassicaceae*
Crambe hispanica	*Brassicaceae*
Diplotaxis muralis	*Brassicaceae*
D. tenuifolia	*Brassicaceae*
Raphanus sativus	*Brassicaceae*
Lens culinaris subsp. *culinaris*	*Fabaceae*
L. culinaris subsp. *orientalis*	*Fabaceae*
L. tomentosus	*Fabaceae*
L. odemensis	*Fabaceae*
L. lamottei	*Fabaceae*
L. ervoides	*Fabaceae*
L. nigricans	*Fabaceae*
Cynara cardunculus var. *scolymus*	*Asteraceae*
C. cardunculus var. *sylvestris*	*Asteraceae*
C. cornigera	*Asteraceae*
C. humilis	*Asteraceae*
C. baetica	*Asteraceae*

and in amino acid composition, including the region of the active site. While most of the inhibitors showed amino acid residues basically corresponding to the trypsin inhibition activity, others exhibited different residues, suggesting that their targets might be different enzymes (Volpicella *et al.*, 2006).

The obtained results demonstrate that putative genes orthologous to *mti-2* are widely present in several species of the *Brassicaceae* and that possibly they are very active against insect gut proteases. Experiments are in progress in order to express some of these newly identified sequences in appropriate vectors and to test the expressed proteins on insect gut proteinases. Moreover, the target activity against proteases other than trypsin may suggest that in different species, patterns of expression differ from those observed in mustard. In addition, the detection in the same species of more than one variant sequence might indicate the presence of paralogous genes possibly having different organ expression and target of action in plant defence. Altogether these results demonstrate that the approach to the study of this family of inhibitors through DNA sequence homology is able to detect new, non-described protein variants in a group of species showing high genetic cohesiveness to important crops (cole crops, rapeseed, radish, etc.) possibly useful in their genetic improvement.

41.3 BBI Proteinase Inhibitors in the *Lens* Gene Pool

Bowman-Birk protease inhibitors (BBIs) are a family of small polypeptides with two active sites against proteases, typically found in pulses (Odani and Ikenaka,

1976; Prakash *et al.*, 1996). Like MTI inhibitors, these molecules seem to be involved in plant defence from attacks by insects and pathogens (Ryan, 1990). Recently, a possible involvement of BBIs in the prevention of tumorigenesis *in vitro* has been hypothesized (Maki *et al.*, 1994; Clemente *et al.*, 2004). In pea, three classes of BBIs have been described: one (*TI1*) showing one active site for trypsin and the other active site for chymotrypsin; the other two (*TI6* and *TI9*) possessing two sites for trypsin. These BBIs differ in their sequence, active site sequence, inhibitor function and patterns of expression (Domoney *et al.*, 1995, 2002).

In order to assess the presence of BBI coding genes in lentil and its wild relatives, the whole *Lens* gene pool, including the crop and six wild taxa, was analysed. Primers were designed on the pea sequences. Two primer pairs, namely F1R1 and F1R4, produced amplificates in all taxa, which were checked for similarity against sequence databases; the highest similarity values were observed for F1R1 and pea *TI1* on one side, and F1R4 and *TI6* on the other, while the similarity between F1R1 and F1R4 was much lower, indicating that two different paralogous gene classes coding for BBIs in *Lens* had been isolated (Sonnante *et al.*, 2005a).

A similarity analysis of all the *Lens* sequences, compared to the ones from pea, showed that all F1R1 amplificates clustered together with the pea *TI1* sequence; on the other hand, all F1R4 grouped together with pea *TI6* and *TI9*. These results demonstrate that each taxon of *Lens* possesses at least two members of the BBI family, paralogous to each other, and orthologous to the corresponding pea genes. Although *Lens* sequences were, within each class, rather similar to each other, those from *L. nigricans* always showed a higher level of differentiation. The translation *in silico* allowed prediction of the amino acid composition of the deduced protein. The *Lens* BBIs obtained from F1R1 sequences were similar to pea major seed TI1, with two active sites, one for trypsin and the other for chymotypsin. On the other hand, the amino acid composition of F1R4 deduced proteins was similar to pea TI6 inhibitor, with two active sites for trypsin (Fig. 41.1).

Although protein sequences were similar within the genus *Lens*, some variations were observed in relevant sites. In particular, *Lens* F1R1 proteins (TI1-like) showed two differences in trypsin active region as compared to pea: one in P1–P1' position (RS and YS, respectively), and another in P2' position (Q instead of N); in the chymotrypsin active site, all *Lens* taxa showed the same residues as pea, but in the P5' position there was a Gln (Q) residue instead of a Lys (K), except in *L. nigricans*, which showed the same residue as pea (K). *Lens nigricans* also showed three additional amino acid substitutions within the mature protein, compared to the cultivated lentil. A higher variation was observed in the F1R4 derived protein (TI6-like): several amino acid substitutions were observed in the precursor of the protein and at the beginning of the mature protein, not only between lentil and pea (a total of three substitutions in the two trypsin active sites), but also within the genus *Lens*. Also in this case, *L. nigricans* displayed the most variable amino acid sequence, with three amino acid substitutions as compared to cultivated lentil. The substitutions observed in important sites of the mature BBI proteins in lentil and its wild relatives might account for the different specificity and biological role of these inhibitors.

A

Orientalis 1

MVLMNKKAIMKLALMLFLLGFTANVVNARFDSTSFITQVLSNGDDVKSACCDTCLCTRSQPPTCRCVDVRESCHSACDKCV CAYSNPPQCQCYDTHNFCYKTCH----------

Orientalis 3

MVLMNKKAIMKLALMLFLLGFTANVVNARFDSTSFITQVLSNGDDVKSACCDTCLCTRSQPPTCRCVDVRESCHSACDKCV CAYSNPPQCQCYDTHNFCYKTCH----------

Odemensis 2

MVLMNKKAIMKLALMLFLLGFTANVVNARFDSTSFITQVLSNGDDVKSACCDTCLCTRSQPPTCRCVDVRESCHSACDKCV CAYSNPPQCQCYDTHNFCYKTCH----------

Odemensis 1

MVLMNKKAIMKLALMLFLLGFTANVVNARFDSTSFITQVLSNGDDVKSACCDTCLCTRSQPPTCRCVDVRESCHSACDKCV CAYSNPPQCQCYDTHNFCYKTCH----------

Lamottei

MVLMNKKAIMKLALMLFLLGFTANVVNARFDSTSFITQVLSNGDDVKSACCDTCLCTRSQPPTCRCVDVRESCHSACDKCV CAYSNPPQCQCYDTHNFCYKTCH----------

Culinaris micro

MVLMNKKAIMKLALMLFLLGFTANVVNARFDSTSFITQVLSNGDDVKSACCDTCLCTRSQPPTCRCVDVRESCHSACDKCV CAYSNPPQCQCYDTHNFCYKTCH----------

Orientalis 2

MVLMNKKAIMKLVLMLFLLGFTANVVNARFDSTSFITQVLSNGDDVKSACCDTCLCTRSQPPTCRCVDVRESCHSACDKCV CAYSNPPQCQCYDTHNFCYKTCH----------

Tomentosum

MVLMNKKAIMKLVLMLFLLGFTANVVNARFDSTSFITQVLSNGDDVKSACCDTCLCTRSQPPTCRCVDVRESCHSACDKCV CAYSNPPQCQCYDTHNFCYKTCH----------

Culinaris macro

MVLMNKKAIMKLALMLFLLGFTANVVDARFDSTSFITQVLSNGDDVKSACCDTCLCTRSQPPTCRCVDVRESCHSACDKCV CAYSNPPQCQCYDTHNFCYKTCH----------

Ervoides

MVLMNKKAIMKLALMVFLLGFTANVVNARFDSTSFITQVLSNGDDVKSACCDTCLCTRSQPPTCRCVDVRESCHSACDKCV CAYSNPPQCQCYDTHNFCYKTCH----------

Nigricans --

GDDVKSACCDTCLCTRSQPPTCRCVDVGETCHSACNKCVCAYSNPPKCQCYDTHNFCYKTCH----------

Pisum TI1

MELMNKKAMMKLALMVFLLSFAANVVNARFDSTSFITQVLSNGDDVKSACCDTCLCTKSNPPTCRCVDVRETCHSACDSCI CAYSNPPKCQCFDTHKFCYKACHNSEVEEVIKN

B

Culinaris macro MVLMNKKTMMKLVLMLFLLGFTATVADARFDSTFFITQLFANVDA------
SNKACCNSCPCTRSIPPKCSCSDIGETCHSACKSCLCTRSIPPQCRCTDVTNFCYKNCN

Orientalis 3 MVLMNKKTMMKLVLMLFLLGFTATVADARFDSTFFITQLFANVDA------
SNKACCNSCPCTRSIPPKCSCSDIGETCHSACKSCLCTRSIPPQCRCTDVTNFCYKNCN

Orientalis 1 MVLMNKKTMMKLALMLFLLGFTATVADARFDSTFFITQLFANGDA------
SNKACCNSCPCTRSIPPKCSCSDIGETCHSACKSCLCTRSIPPQCRCTDVTNFCYKNCN

Odemensis 1 MVLMNKKTMMKLALMLFLLGFTATVVDARFDSTFFITQLFANVDA------
SNKACCNSCPCTRSIPPKCSCSDIGETCHSACKSCLCTRSIPPQCRCTDVTNFCYKNCN

Odemensis 2 MVLMNKKTMMKLALMLFLLGFTATVVDARFDSTFFITQLFANVDA------
SNKACCNSCPCTRSIPPKCSCSDIGETCHSACKSCLCTRSIPPQCRCTDVTNFCYKNCN

Lamottei MVLMNKKTMMKLALMLFLLGFTATVVDARFDSTFFITQLFANGDA------
SNKACCNSCPCTRSIPPKCSCSDIGETCHSACKSCLCTRSIPPQCRCTDVTNFCYKNCN

Tomentosum MVLMNKKTMMKLALMLFLLGFTATVVDARFDSTFFITQLFSNDDA------
SNKACCNSCPCTRSIPPKCSCSDIGETCHSACKSCLCTRSIPPQCRCTDVTNFCYKNCN

Fig. 41.1. Multiple sequence alignment for A) F1R1 class and B) F1R4 class. Light grey box: trypsin active site, dark grey box: chymotrypsin active site; polymorphisms with *Pisum* (in the active site region) and within *Lens* are boxed in black. (From Sonnante *et al.*, 2005a.) *Continued.*

Orientalis 2 MVLMNKKTMMKLALMLFLLGFTATVVDARFDSTFFITQLFSNGDA------

SNKACCNSCPCTRSIPPKCSCSDIGETCHSACKSCLCTRSIPPQCRCTDVTNFCYKNCN

Ervoides MVLMNKKTMMKLALMLFLLGFTATVVDARFDSTFFITQLFSNGDA------

SNKACCNSCPCTRSIPPKCSCSDIGETCHSACKSCLCTRSIPPQCRCTDVTNFCYKKCN

Nigricans MVLMNKKTMMKLALMLFLLGFTATVVDARFDSTFFITQLFSNGDA------

SNKACCNSCPCTRSIPPKCRCTDIGETCHSACKSCLCTRSIPPQCRCTDVTNFCYKNCN

Pisum TI6 MVLMNKKAMMKLALMLFLLGFTATVVDARFDSDSFIIQLLSKGDA------

SNKACCDSCLCTKSIPPRCRCNDTGETCHSACKTCICTRSLPPQCRCIDITDFCYEKRN

Pisum TI9

MELINTKKMMKLALMVFLLGFTATVVDARFDSTSFITQLLSNGDAGYSIKSTTTACCDSCICTKSIPPQCHCTDVGKTCHSGC

NLCLCTRSFPPQCHCTDTNDFCYQKCN

Legend: the names are those of the species and subspecies of the genus *Lens* reported in Table 41.1. *L. culinaris* micro and macro refer to small- and large-seeded cultivated lentil respectively.

Fig. 41.1. *Continued*

41.4 Phenylalanine Ammonia Lyase Genes in *Cynara*

Cultivated and wild artichokes represent a good source of natural antioxidant compounds mostly derived from the metabolic pathway of phenylpropanoids. Leaf extracts from these plants are rich in phenolic compounds which protect cell membranes, show antibacterial, antioxidative, anti-HIV, bile expelling, hepatoprotective, urinative and choleretic activities as well as the ability to inhibit cholesterol biosynthesis and LDL oxidation (Kraft, 1997; Fritsche *et al*., 2002; Schutz *et al*., 2004).

Phenylalanine ammonia-lyase (PAL) is an enzyme which catalyses the first reaction in the general phenylpropanoid pathway leading to the production of phenolic compounds with a significant range of biological functions, being involved in plant defence against pests and predators (Wink, 1988), as UV protectants (Hahlbrock and Grisebach, 1979) and as signal molecules both internally and for communication with other organisms (Lynn and Chang, 1990).

In order to isolate and characterize *pal* genes from artichoke and its wild relatives, genomic DNA and cDNA were amplified with primers designed on other PAL sequences present in the databases. Starting from this approach and screening a genomic library, it was possible to isolate three *pal* genes in artichoke, differing from each other for nucleotide sequence and expression pattern; moreover, they contain one intron each of variable length and sequence (De Paolis *et al*., 2004; Sonnante *et al*., 2005b). Highly conserved regions of the nucleotide sequence of one of these genes, namely *pal1*, were considered reliable spots for designing primers to specifically amplify the intron and part of the second exon in the artichoke wild progenitor (*C. cardunculus* var. *sylvestris*) and other wild *Cynara* species (see Table 41.1). Studies are still in progress, but preliminary results indicate that the intron of the *pal1* gene, although not too variable, might provide useful information on the level of sequence divergence in the different taxonomical entities used in this study. This approach was preferred due to the length of the *pal* gene: it allows the research to be concentrated on those taxa showing the highest divergence from artichoke. In fact, the artichoke direct wild progenitor did not show enough variation to justify expectations of finding new gene variants to be employed in artichoke breeding.

41.5 Discussion

Since the early 1920s plant geneticists have become aware of the potential of plant genetic resources as a source of genetic resistance to major pests of crop species. Since then, the importance of CWR as donors of resistance genes has become widely accepted by plant breeders. Recently, the identification of genes that impact the amounts and types of natural products produced by plants has become a novel interest for research focused on the use of plant genetic resources. Scientists have developed methods to search for these new genes at the plant, cell or tissue level, and transfer them to production systems.

Advanced methods are available for gene transfer without the use of GM technology, which has little consumer's acceptance, especially in EU countries. Many important crop species are genetically cohesive with at least some CWR, e.g. *Brassica*, wheat, tomato, etc. (Stevens and Rick, 1986; Jørgensen *et al.*, 1996; Jarvis and Hodgkin, 1999), while in other cases biotechnological tools have been developed to assist in obtaining wide hybrids. Moreover, the approaches of Marker Assisted Selection enormously increase the efficiency of selection of recombinant offspring thus reducing time and cost of selection (Jennings and Iglesias, 2002). Nevertheless, the obstacle of the acceptance by GM technology mostly applies to field crops both for human consumption and for forage or technical use. Consumers are not concerned by GM technology when it applies to cell cultures.

The most promising field of the use of genetic resources from CWR appears to be the development of new pharmaceutically active compounds in plant bioreactors. In this framework, the relative ease of transforming some organisms, like tobacco, and the possibility of *in vitro* maintaining transformed cells in culture where they grow at high rates and produce the desired compounds, appears to be at present the edge of applied research on the utilization of CWR (Stoger *et al.*, 2002).

To fully exploit the potential of this new frontier of application, it is essential to explore plants at the genomic or proteomic level. In this sense the studies on evolution of crops' gene pool or the search, isolation and *in vitro* expression of orthologous genes in related plant genera appear to be the necessary support to achieve the full potential of this new field of research.

References

Blanco, A., Cenci, A., Simeone, R., Pignone, D. and Galasso, I. (2002) Characterization and transfer of a recombinant chromosome 1AS.1AL-1DL with the glu-D1 gene. *Cellular and Molecular Biology Letters* 7, 559–567.

Ceci, L.R., Spoto, N., De Virgilio, M. and Gallerani, R. (1995) The gene coding for the mustard trypsin inhibitor-2 is discontinuous and wound-inducible. *FEBS Letters* 364, 179–181.

Ceciliani, F., Bortolotti, F., Menegatti, E., Ronchi, S., Ascenzi, P. and Palmieri, S. (1994) Purification, inhibitory properties, amino acid sequence and identification of the reactive site of a new serine proteinase inhibitor from oil-rape (*Brassica napus*) seed. *FEBS Letter* 4, 221–224.

Clemente, A., Mackenzie, D.A., Jeenes, D.J. and Domoney, C. (2004) The effect of variation within inhibitory domains on the activity of pea protease inhibitors from the Bowman-Birk class. *Protein Expression and Purification* 36, 106–114.

De Leo, F., Volpicella, M., Licciulli, F. and Liuni, S. (2002) PLANT-Pis: a database for plant protease inhibitors and their genes. *Nucleic Acids Research* 30, 347–348.

De Paolis, A., Pignone, D. and Sonnante, G. (2004) *Characterization of the Phenylalanine Ammonia-Lyase Genes in Artichoke*. Proceedings of the XLVIII Congress of the Italian Society for Agricultural Genetics (SIGA), Lecce, Italy.

De Vienne, D. (2003) *Molecular Markers in Plant Genetics and Biotechnology*. Institut National de la Recherche Agronomique, Versailles, France.

Domoney, C., Welham, T., Sidebottom, C. and Firmin, J.L. (1995) Multiple isoforms of *Pisum* trypsin inhibitors result from modification of two primary gene products. *FEBS Letters* 360, 15–20.

Domoney, C., Welham, T., Ellis, N., Mozzanega, P. and Turner, L. (2002) Three classes of proteinase inhibitor gene have distinct but overlapping patterns of expression in *Pisum sativum* plants. *Plant Molecular Biology* 48, 319–329.

Fritsche, J., Beindorff, C.M., Dachtler, M., Zhang, H. and Lammers, J.G. (2002) Isolation, characterization and determination of minor artichoke (*Cynara scolymus* L.) leaf extract compounds. *European Food Research and Technology* 215, 149–157.

Hahlbrock, K. and Grisebach, H. (1979) Enzymic controls in biosynthesis of lignin and flavonoids. *Annual Review of Plant Physiology* 30, 105–130.

Hawkes, J.G. (1977) The importance of wild germplasm in plant breeding. *Euphytica* 26, 615–621.

Jarvis, D.I. and Hodgkin, T. (1999) Wild relatives and crop cultivars: detecting natural introgression and farmer selection of new genetic combinations in agro-ecosystems. *Molecular Ecology* 8, 159–173.

Jennings, D.L. and Iglesias, C. (2002) Breeding for crop improvement. In: Hillocks, R.J., Thresh, J.M. and Bellotti, A.C. (eds) *Cassava – Biology, Production and Utilization*. CAB International, Wallingford, UK, pp. 149–158.

Jørgensen, R.B., Andersen, B., Landbo, L. and Mikkelsen, T.R. (1996) Spontaneous hybridization between oilseed rape (*Brassica napus*) and weedy relatives. *Acta Horticulturae (ISHS)* 407, 193–200.

Kraft, K. (1997) Artichoke leaf extract – recent findings reflecting effects on lipid metabolism, liver and gastrointestinal tract. *Phyto-Medicine* 4, 369–378.

Lenné, J.M. and Wood, D. (1991) Plant diseases and the use of wild germplasm. *Annual Review of Phytopathology* 29, 35–63.

Lynn, D.G. and Chang, M. (1990) Phenolic signals in cohabitation: implications for plant development. *Annual Review of Plant Physiology and Plant Molecular Biology* 41, 497–526.

Maki, P.A., Paterson, Y. and Kennedy, A.R. (1994) Studies related to the potential antigenicity of the Bowman-Birk inhibitor, an anticarcinogenic protease inhibitor isolated from soybeans. *Nutrition and Cancer* 22, 185–193.

Odani, S. and Ikenaka, T. (1976) The aminoacid sequence of two soybean double headed proteinase inhibitors and evolutionary considerations on the legume proteinase inhibitors. *Journal of Biochemistry* 80, 641–643.

Padulosi, S. (1995) *Rocket Genetic Resources Network*. International Plant Genetic Resouces Institute, Rome, Italy.

Pignone, D. (1997). Present status of rocket genetic resources and conservation activities. In: Padulosi, S. and Pignone, D. (eds) *Rocket: a Mediterranean Crop for the World*. Report of Project on Underutilized Mediterranean Species, International Plant Genetic Resources Institute, pp. 2–12.

Prakash, B., Selvaraj, S., Murthy, M.R.N., Sreerama, Y.N., Rao, D.R. and Gowda, L.R. (1996) Analysis of the aminoacid sequences of the plant Bowman-Birk inhibitors. *Journal of Molecular Evolution* 42, 560–569.

Ryan, C.A. (1990) Protease inhibitors in plants: genes for improving defences against insects and pathogens. *Annual Review of Phytopathology* 28, 425–449.

Schutz, K., Kammerer, D., Carle, R. and Schieber, A. (2004) Identification and quantification of caffeoylquinic acids and flavonoids from artichoke (*Cynara scolymus* L.) heads, juice, and pomace. *Journal of Agricultural and Food Chemistry* 30, 4090–4096.

Sonnante, G., De Paolis, A. and Pignone, D. (2005a) Bowman–Birk inhibitors in *Lens*: identification and characterization of two paralogous gene classes in cultivated lentil and wild relatives. *Theoretical and Applied Genetics* 110, 596–604.

Sonnante, G., Orlando, M., Pignone, D. and De Paolis, A. (2005b) Expression of *pal* genes in artichoke organs. *Proceedings of the XLIX Congress of the Italian Society for Agricultural Genetics (SIGA)*. Potenza, Italy.

Stevens, M. and Rick, C.M. (1986) Genetics and breeding. In: Atherton, J. and Rudich, G. (eds) *The Tomato Crop. A Scientific Basis for Improvement*. Chapman & Hall, New York, pp. 35–109.

Stoger, E., Sack, M., Perrin, Y., Vaquero, C., Torres, E., Twyman, R.M., Christou, P. and Fischer, R. (2002) Practical considerations for pharmaceutical antibody production in different crop systems. *Molecular Breeding* 9, 149–158.

Tanksley, S.D. and McCouch, S.R. (1997) Seed banks and molecular maps: unlocking genetic potential from the wild. *Science* 277, 1063–1066.

Volpicella, M., De Leo, F., Sonnante, G., Pignone, D., Gallerani, R. and Ceci, L.R. (2006) Identification and characterisation of novel protease inhibitors from Cruciferae. 1. *Proceedings of* III *SOI Plant Breeding Workshop*. IAM, Valenzano (BA), pp. 569–573.

Wink, M. (1988) Plant breeding: importance of plant secondary metabolites for protection against pathogens and herbivores. *Theoretical and Applied Genetics* 75, 225–233.

42 Genetic Systems and the Conservation of Wild Relatives of Crops

D. Zohary

42.1 Introduction

Genetic systems are the ways by which the genetic material in each biological species is organized and transmitted from one generation to the next. In other words, the genetic systems represent the various reproductive strategies that have evolved in plants and animals. Prominent among the elements that build up the various genetic systems is chromosome organization, namely chromosome number, ploidy level, chromosomal sex determination and the amount and patterns of genetic recombination. Equally important is the mode of reproduction that each species has evolved – whether it is cross-pollination or self-pollination; or whether it is based on sexual or on asexual reproduction. As stressed by Darlington (1958), in each biological species the genetic system determines its capability to undergo evolutionary change. Moreover, genetic systems are themselves genetically controlled, and are being moulded by selection. In each population, and in each species, selection acts to maintain – or to modify – its genetic system, just as it acts to maintain or to change morphological or physiological traits.

This chapter aims to stress the fact that different and contrasting genetic systems have evolved in the flowering plants. It focuses on some of the main components of these systems, namely the various modes of reproduction that evolved in plants, and on the massive occurrence of polyploidy present in them. Since each genetic system moulds the gene pools in different and contrasting ways, the reproduction and the chromosomal set-up of each crop wild relative requires individual attention – when conservation and/or breeding work are being considered. The following components should be of particular concern when one tries to formulate strategies for the conservation of wild relatives of cultivated plants.

42.2 Cross-pollination

Cross-pollination constitutes the basic and most common breeding system in higher plants and animals. It is also the most extensively examined system – both theoretically and experimentally. In fact, the classical formulation of Hardy-Weinberg equilibrium, which is so central in population genetics, deals with this kind of a reproductive system (panmixis resulting from cross-fertilization). Cross-pollination characterizes most of the wild relatives of the fruit crops (such as apple, pear, almond, pistachio, date palm, fig) and forest trees; as well as many forage plants and ornamentals, several vegetables (e.g. cabbage, turnip, asparagus, spinach) and a few of the grain crops (e.g. rye, pearl millet, maize).

In the large majority of animals, cross-fertilization is brought about by chromosomal sex determination that leads to sexual dimorphism (50% males: 50% females). Under this mating system, self-pollination is automatically blocked since each individual produces only a single type of gametes. In contrast, such bisexuality (= dioecy) is rather rare in the higher plants. It occurs only in a small fraction of the plant species. Most plant species are hermaphrodites or monoecious. Each individual produces both types of gametes. However, they have evolved a whole battery of devices to prevent self-pollination and to safeguard cross-fertilization.

In the flowering plants, the most widespread mechanism to assure cross-fertilization is self-incompatibility. Darlington and Mather (1949) estimated that about half of the angiosperms use this genetic device to safeguard cross-fertilization. It is the main mechanism that assures outcrossing in many of the members of the families *Cruciferae, Leguminosae, Rosaceae, Compositae, Liliaceae* and *Gramineae,* i.e. the families that rank among the richest in crops' representation and the information available on their wild relatives. Several systems of self-incompatibility have evolved in the flowering plants (De Nettancourt, 1993). Each seems to have originated relatively early in the evolution of the angiosperms, and each characterizes a large taxonomic group (a whole family, or even a series of families). The majority of the self-incompatibility systems are governed by a single, multi-allelic gene locus (S), and in some families (such as *Gramineae* and *Liliaceae*) by two gene loci. Rarely, one finds also systems governed by several self-incompatibility genes. Finally, in several groups of plants, self-incompatibility is not poly-allellic, and it is reinforced by floral polymorphism such as distyly or tristyly. Another rather common device to prevent self-fertilization is temporal separation by protandry or protogyny (hermaphroditic flowers but the anthers and the stigma mature at different times).

The efficiency of the poly-allelic self-incompatibility systems depends on the number of the S alleles present in the populations. The larger the number of S alleles, the better are the chances that the seeds will be produced after crossing between individuals possessing different self-incompatibility genotypes. In species where self-incompatibility was extensively studied (such as some rosaceous fruit trees and some clovers) dozens of S alleles were detected in the tested populations.

42.3 Self-pollination

Self-pollination (more correctly, almost full self-pollination) is a common breeding system in plants. This type of reproduction is widespread among annual species. About 50% of the native annual plants growing in the Mediterranean basin and in temperate Europe seem to be selfers. Among others, this type of pollination characterizes all the eight 'founder crops' that initiated Neolithic agriculture in the Fertile Crescent belt (hard wheat, einkorn wheat, barley, lentil, pea, chickpea, bitter vetch and flax) as well as their closely related (members of gene pool 1) wild relatives. In addition, scores of other Mediterranean and European annual species that are of great interest to conservationists and plant breeders are selfers.

Predominantly self-pollinated species show several reproductive features that are very different from those present in cross-pollinated plants. The following traits are closely associated with the reproduction and speciation in these plants:

1. Self-pollination is a powerful reproductive isolation barrier (Zohary, 1999) that effectively blocks gene exchange: (i) between different inbred lines, and thus maintains the identity of the various inbred lines both in wild populations and in the cultivated crops; and (ii) between the evolving crops and their wild progenitors, thus preventing massive contamination and genetic swamping. In other words, from the very start, self-pollination sets the various inbred lines (both the wild inbred forms and the domestic inbred cultivars) on independent, evolutionary tracts. In contrast, such an isolation barrier does not exist in the cross-pollinated plants. Therefore, speciation patterns under these two breeding systems are quite different. In cross-pollinated plants, speciation is almost totally allopatric. In self-pollinated plants, it is frequently sympatric and apparently relatively speedier.

2. The structuring of genetic variation in populations in selfers differs markedly from that present in cross-pollinated, largely panmictic populations. So is the operation of selection under the two mating systems. Cross-pollinated populations usually contain a considerable amount of genetic variation, and in each sexual cycle their genes are recombined. As a result of this repeated reshuffling, the units of selection in outcrossers are mostly the individual genes, and the alleles are selected on the basis of their individual general performance. In contrast, self-fertilization brings about drastic reduction of recombination and the splitting of the variation present in the population into homozygous, true breeding gene combinations (Pérez de la Vega and Garcia, 1997). Unlike the situation in outcrossers, the units of selection, in selfers, are not the individual genes but mainly the various homozygous inbreed combinations ('genotypes') that selfing keeps reproductively apart. Competition is therefore between different inbred lines, and genetic polymorphism is maintained by coexistence in populations of different homozygous lines. In addition, one should take into account the fact that in almost all selfers, self-pollination is not absolute. Rare crosses between lines do take place. Genetic flexibility is also maintained.

3. Some of the main domestication traits that characterize crops and distinguish them from their wild progenitors are controlled by *recessive mutations* in one or two major genes. Well-studied examples of such mutations are those controlling the breakdown of the wild mode of seed dispersal, or the inhibition of germination in the Near Eastern cereals and pulses. Again, selection for such recessive mutations is more effective in selfers than in outcrossers. Under a system of cross-pollination heterozygous individuals will be common, and in each such individual the recessive allele will be phenotypically masked by its dominant 'wild-type' allele. Under self-pollination practically only homozygous mutant and homozygous wild-type individuals will be present, masking the recessive mutation. For this reason masking will be minimal. Selection for domestication traits will be more effective in selfers, and the emergence of cultigens will be faster.

4. Molecular genetic tests performed in stands of predominantly self-pollinated wild relatives of cultivated plants such as wild barley *Hordeum spontaneum* (Nevo *et al.*, 1979), wild wheat *Triticum dicoccoides* (Nevo *et al.*, 1982), and wild oats *Avena barbata* and *A. hirtula* (Pérez de la Vega and Garcia, 1997) showed how varied the populations of these wild cereals can be; and how tightly associated are the given multi-locus genotypes (inbred lines) with definable micro-niches, soil types and climatic conditions.

42.4 Asexual Reproduction

Sexual reproduction is not the only way by which the flowering plants are propagating themselves. Many of these plants also multiply asexually. In a considerable number of plant taxa, sexual reproduction has been strongly suppressed and largely replaced by various types of vegetative propagation. Populations of such plant species contain – partly or almost fully – clusters of vegetative clones. Reproducing asexually, such plants are able to maintain highly heterozygous, superior genotypes (frequently of hybrid origin), evade unbalance and breakdowns in meiosis, and produce numerous identical progeny with 'fixed' genotypes able to exploit their immediate fitness. The residual sexual reproduction that persists in many of these plants provides them with some genetic flexibility for coping with future changes.

Apomixis, namely the production of viable seed without fertilization (Gustafsson, 1946–1947; Stebbins, 1950; Grant, 1981), has been, evolutionally, the most successful type of vegetative reproduction in the flowering plants. Here the shift from sexual to asexual reproduction did not affect seed production. However, while in sexually reproducing plants the seed's embryo is formed by fertilization, in apomicts the embryo develops from a maternal cell. Thus, asexual reproduction hitch-hikes on seed dispersal. The seeds (the principal means of dispersal in plants) keep their vital function as effective propagules, ensuring effective dissemination. Yet instead of dispersing sexual embryos, seeds produced by apomicts are mostly agamospermous and contain copies of the genetic constitution of their mother plants.

Because asexual plants evade meiosis and gamete formation, they are able to tolerate chromosome combinations that in sexually reproducing plants would result in meiotic irregularities and seed sterility, and will be promptly weeded out. In other words, the shift from sexual reproduction to vegetative propagation is often associated with the build-up of unbalanced polyploid and even aneuploid chromosome complements. Some apomictic groups contain not only tetraploid and/or hexaploid relatively stable clones – but also a wide range of greatly unbalanced chromosome combinations like triploids, pentaploids or aneuploids. Such meiotically unbalanced chromosomal types can maintain themselves only by avoiding sexual reproduction.

Most apomictic plants are able to keep genetic flexibility in their populations by retention of some ability for sexual reproduction. In some genera (e.g. *Crepis*, *Rubus*, *Crataegus*, *Poa*, *Taraxacum*) this genetic system has had a considerable evolutionary success; and the adoption of apomixis led to the build-up of enormous agamic complexes (Grant, 1981), comprising hundreds of morphologically distinct apomictic forms or agamospecies. Again, in conservation of apomicts one has to consider: (i) their unique genetic structure (numerous and frequently highly heterozygous clones); and (ii) their origin (derived from sexual forms, and frequently products of interspecific hybridization). Understanding how a given agamic complex evolves, and knowledge of the distribution and the ecological specificities of its main sexual and asexual forms are essential for the planning of effective conservation of these complex groups of plants.

Relatively few crops are apomicts and/or have apomictic wild relatives. Prominent among them are the blackberry (*Rubus fruticosus*), numerous hawthorn (*Crataegus*) species and several of the *Citrus* crops such as oranges and mandarins.

42.5 Polyploidy

While animals are overwhelmingly diploid, a large proportion of the flowering plants have had a polyploid origin. Several attempts have been made to estimate the frequency of polyploid species among the world's flowering plants (for reviews see Stebbins, 1971; Grant, 1981). The estimates vary, depending on the criteria used by different authors for identification of polyploids. In the angiosperms they range between 35% and 50%. This is a large slice indeed. Very likely the proportion of polyploid taxa among the wild relatives of crop plants falls into this range too.

As already stressed by Stebbins (1971), the diploid and polyploid chromosomal systems are different in their responses to selection, and in their ability to accumulate and maintain genetic variation. Directional selection operates more effectively in diploids. In contrast, a polyploid chromosome set-up is better adjusted for fusion of genetic variation. Through polyploidy, populations can combine adaptive traits that have evolved in two or several diploid species. Polyploid species also have a better capacity to absorb – through interspecific hybridization – alien genetic variation. Furthermore, polyploidy is less associated with adaptive radiation and rapid speciation, but excels in genetic fusion. It often results in the build-up of polyploid complexes comprising of: (i) diploid pillars; and (ii) a large, inter-connecting polyploid superstructure. Frequently, polyploid forms in such complexes are widely distributed and successfully colonize disturbed or newly opened habitats, while the diploids are ecologically more specific and restricted to smaller areas. This is the case: (i) in the cross-pollinated cocksfoot grass *Dactylis glomerata* (Humphreys, 1997; Lumaret, 1997) and lucerne (*Medicago sativa*); (ii) in the predominantly self-pollinated wheat group (*Triticum–Aegilops*); and (iii) in the largely asexual blackberry (*Rubus fruticosus*). The latter is a huge agamic complex containing diploid 2*x* sexual forms, and a large number of 2*x*, 3*x*, 4*x*, 5*x*, 6*x* apomict forms. In such polyploid agamic complexes the whole polyploid complex (both the sexual forms and a representative collection of agamic lines) ought to be the conservation target. As shown by Lumaret (1997) the rarer and ecologically more restricted diploid 'pillars' are often more endangered. They should be of special concern to conservationists.

References

Darlington, C.D. (1958) *The Evolution of Genetic Systems*, 2nd edn. Basic Books, New York.

Darlington, C.D. and Mather, K. (1949) *The Elements of Genetics*. Allen & Unwin, London.

De Nettancourt, D. (1993) Self- and cross-incompatibility systems. In: Hayward, M.D., Bosemark, N.O. and Romagosa, I. (eds) *Plant Breeding: Principles and Prospects*. Chapman & Hall, London, pp. 203–212.

Grant, V. (1981) *Plant Speciation*, 2nd edn. Columbia University Press, New York.

Gustafsson, Å. (1946–1947) Apomixis in higher plants. *Lunds Universitets Årsskrift* 42–43, 1–370.

Humphreys, M.O. (1997) The use of extended gene pools comprising related species to improve the environmental adaptability of temperate forage grasses. *Bocconea* 7, 177–185.

Lumaret, R. (1997) Polyploidy and the critical size of natural populations: the case of cocksfoot (*Dactylis glomerata* L.) a grass used as fodder plant. *Bocconea* 7, 130–133.

Nevo, E., Zohary, D., Brown, A.H.D. and Haber, M. (1979) Genetic diversity and environmental associations of wild barley, *Hordeum spontaneum*, in Israel. *Evolution* 33, 815–833.

Nevo, E., Golenberg, E., Beiles, A., Brown, A.H.D. and Zohary, D. (1982) Genetic diversity and environmental associations of wild wheat *Triticum dicoccoides* in Israel. *Theoretical and Applied Genetics* 62, 241–254.

Pérez de la Vega, M. and Garcia, P. (1997) Genetic structure of self-pollinating species: the case of wild *Avena*. *Bocconea* 7, 141–152.

Stebbins, G.L. (1950) *Variation and Evolution in Plants*. Columbia University Press, New York.

Stebbins, G.L. (1971) *Chromosomal Evolution in Higher Plants*. Arnold, London.

Zohary, D. (1999) Speciation under self-pollination. In: Wasser, S.P. (ed.) *Evolutionary Theory and Processes: Modern Perspectives*. Kluwer Academic, Dodrecht, The Netherlands, pp. 301–307.

IX Use of Crop Wild Relatives and Underutilized Species

43 The Use and Economic Potential of Wild Species: an Overview

V.H. HEYWOOD

43.1 Introduction

It is widely stated that humans are dependent on only 30 or so species that comprise the staples which supply most of human nutrition and that three of these – rice, wheat and maize – provide more than half of the planet's food. Only 103 species of food plants contribute 90% of national per-capita supplies (Prescott-Allen and Prescott-Allen, 1990) although about 7000 plant species are or have been cultivated to some degree. The agricultural revolution that began some 10,000 years ago gradually led to a simplification of agriculture as people came to rely more and more on domesticated and improved varieties (cultivars) and paid less attention to wild species, thus leading to a significant reduction in our dietary diversity (Grivetti, 1980; Ogle and Grivetti, 1985). On the other hand, even today, in communities across the world, a combination of wild, semi-domesticated species and minor crops provide variety to the table in the form of vegetables, fruits, herbs and spices as well as vitamins and micronutrients while others are a source of oils, fibres, fuels, intoxicants, ornaments and medicines. These range from locally consumed species such as leaf greens and wild fruits to economically important non-timber forest products obtained by extractivism such as palm hearts, Brazil nuts and rubber and the trade, most of it uncontrolled and much of it illegal, in ornamentals including cycads, orchids, cacti and succulents and bulbs. The use of wild plants in most societies forms part of indigenous knowledge systems and practices that have been developed over many generations and which play an important part in decision making in local agriculture, food production, human and animal health and management of natural resources (Slikkerveer, 1994). Often women play a key role in these activities.

Although humans will continue to be dependent to a very large degree on continuing development of the major staple crops for most of their basic nutrition needs, traditional cropping systems are said to provide as much as 20% of the world's food supply. In fact it has been estimated that more than 3 million ha survive under traditional agriculture as raised fields, terraces, swidden fallows, polycultures, home gardens and other agroforestry systems (Altieri, 1999). While these traditional systems seldom have the potential to produce marketable surpluses, they do make a major contribution to food security and may contribute marginally or substantially to farm household income (Heywood, 1999). As Frison (2005) observes, stronger demand for the products of agricultural biodiversity builds a market that rural farmers – usually women – can supply and thus boosts incomes and fights poverty. There are also earning opportunities in non-food products, such as biofuels, medicinal plants and the like.

As the recent report *The Wealth of the Poor* (WRI, 2005) notes: 'Natural resources are the fundamental building block of most rural livelihoods in developing nations, and not just during lean times' and wild plants and their ecosystems are the main components of these resources.

The role of wild species in these agroecosystems and in natural ecosystems, in supplying nutrition as well as a range of other products has been widely neglected by the agricultural, scientific and development communities. This chapter will review the extent to which and the way in which wild species are used today and the contributions they make to meeting nutritional needs, household economies, health and trade, as well as exploring the reasons for their comparative neglect and possible solutions.

43.2 The Wild Species/Domestication Interface

It is surprisingly difficult in practice to define wild species as opposed to domesticated plants as there is a complete spectrum between completely wild and completely domesticated species, depending on the degree of human intervention or management involved. When applied to species or plants the term 'wild' refers to those that grow spontaneously in self-maintaining populations in natural or semi-natural ecosystems and can exist, independent of direct human action. It is contrasted with 'cultivated' or 'domesticated' species/plants that have arisen through human action, such as selection or breeding, and depend on management for their continued existence (Heywood, 1999).

In many parts of the world, plants may be found in various stages of domestication as a result of human selection, however slight it may have been. This is especially true for trees that have been widely planted, although genetically and culturally in a nearly wild state. In Borneo, for example, mangoes (*Mangifera* spp.) have a long history of semi-cultivation along the rivers and *M. bompardii* is common in semi-cultivation (Kostermans and Bompard, 1993).

Many species that are domesticated and widely cultivated, such as *Pistacia vera*, the pistachio, are also still extensively collected from the wild. For example, much of the market demand for rocket (*Eruca sativa* and *Diplotaxis* spp.) is met by harvesting the plant from the wild and wild types can be easily spotted in Italian vegetable markets (Bianco, 1995).

Domestication grew out of gathering food and this almost imperceptibly led to cultivation. It is a long and complex process and many plants are found in various stages of domestication as a result of human selection, however slight. Many species, especially trees, are widely planted, although genetically and culturally in a nearly wild state. Domestication/cultivation is contrasted with wild harvesting but some species are gathered from the wild in habitats that are managed to some degree. An example is the lowbush blueberry (*Vaccinium angustifolium*) industry, which produces somewhere over 50 million kilos of blueberries per year, based on intensively managed wild stands in the Maritime Provinces in Maine (USA). According to Finn (personal communication, 1999), human management of *Vaccinium* is an 'ancient' practice: it has been documented that Native Americans periodically burned areas in a regular rotation to keep production high.

A common situation is where plants grow wild in ecosystems that have themselves been 'domesticated' (Box 43.1). For example, many of the plant species of the Mayan home gardens are native wild species of the Yucatán peninsula, mainly tall trees that are left to stand when the forest is cut down to establish a new home garden. Forests have been intensively used for many thousands of years and their management of forest resources has resulted in the 'domestication of the landscape'. The term 'domiculture' was introduced to describe this kind of domestication as opposed to the conventional genetic modification of plants through selection and breeding.

Some wild fruits in temperate regions are wild harvested in natural habitats that are quite closely managed, a tradition that goes back to the times of the American Indians in the case of some species of wild blueberry (*Vaccinium* spp.) (Finn, 1999; personal communication).

Many species that have been domesticated, escaped from cultivation and have become naturalized. Weedy forms, cultivated, semi-cultivated side by side,

Box 43.1. Domestication of ecosystems. (From: Heywood, 1999.)

A common situation is where plants grow wild in ecosystems that have themselves been 'domesticated': for example, many of the plant species of the Mayan home gardens are native wild species of the Yucatán peninsula, mainly tall trees that are left to stand when the forest is cut down to establish a new home garden. Moreover, these plants may grow spontaneously from seeds or other propagules that are either already present in the home garden or naturally dispersed in the home garden from the nearby forest or from neighbouring home gardens. Likewise in the highlands of Irian Jaya, occupied by people of the Mek group, a pygmaean folk with a Neolithic culture, when new garden land was required, an area of secondary forest was cleared, and also sometimes primary forest, for shifting cultivation. All the young trees were removed with the exception of small groups of *Pandanus* spp. whose fruits and leaves were used. *Pandanus* is not planted but always collected from the forest. Leaving behind forms with particular characteristics (e.g. large fruits) can be regarded as a first stage of domestication.

A similar situation is found in many parts of Africa where 'farm trees' can be found scattered throughout areas of cultivated land within and near farm fields. These trees are actively managed, protected and harvested by the farmers to provide fuelwood, fodder, poles for construction, and a range of edible fruits and nuts. In the Sahel with its sandy soils of low fertility, the presence of scattered *Acacia albida* trees in millet or sorghum fields increases crop yields up to two and a half times over that obtained in open fields.

so that we have a whole range of different situations ranging from completely wild to semi-domesticated through selection, to fully domesticated through selection and breeding to escape from cultivation to become naturalized.

Some social anthropologists insist that ideas such as 'wild' and 'domesticated' are culturally specific and would, for example, have quite different meanings to a European farmer and a Kayapó Indian, as in the case just mentioned (Posey, 1992).

43.3 Diversity of Wild Species Used by Humans

Some 400,000 species or more of flowering plants are believed to exist today (Govaerts, 2001; Bramwell, 2002),[1] and despite the widespread belief that most of this great diversity is ignored by humans, it can be shown that thousands of plant species are used in human activities and can be considered resources. Many of the species that are grown locally are scarcely or only partially domesticated and many thousands more are gathered from the wild. Not surprisingly, most of the partially domesticated or wild-collected species are found in the tropics. For example, the Plant Resources Project of South-East Asia (PROSEA) records nearly 7000 species in its Basic List of species (some of them exotic) used by humankind in that area (Jansen *et al.*, 1991) and the Plant Resources of Tropical Africa (PROTA) project estimates a similar number for the region. No comparable figures are available for Latin America, and the database and information system on wild plant species with economic potential of the Latin American member countries of the Andrés Bello Agreement (Bolivia, Chile, Colombia, Cuba, Ecuador, Panamá, Perú and Venezuela) and the series of 11 volumes of the series *Especies vegetales Promisorias* so far published gives far from complete coverage. But assuming similar levels in the neotropics to those of Africa and Asia, we can extrapolate to a figure of 24,000 species for the tropics as a whole. In addition several thousand plant species are used in human activities in Mediterranean and temperate regions of the world.

To these must be added two major groups – plants used in traditional medicine systems which may be estimated at between 65,000–118,000 species (see below) and those used as ornamental species in the horticultural and nursery trade and in general garden collections (excluding botanic gardens) which possibly total some 30,000 species.

43.4 Main Usage Groups

The contribution that wild plants make to local communities and household economies can be considered under 14 main headings according to their usage (Table 43.1) or about 13 widely recognized commodity groups (Table 43.2) although many species are multi-purpose and belong to more than one of these groupings, notable examples being mangroves and palms.

[1] But see Scotland and Wortley (2003) who give an estimate of 223,300, (Thorne 2002) who indicates 260,000 and (Prance *et al.*, 2000) who suggest 300,000–320,000 species.

Table 43.1. Classification of the main usage categories of wild plants. (From Cook, 1995.)

- Food plants (including beverages for humans) – seeds, fruits, leaves, stems, petioles, roots, tubers, etc.
- Food additives (including processing agents and additive ingredients used in food preparations)
- Animal food (including forage and fodder for vertebrates)
- Bee plants (including pollen or nectar sources for honey production)
- Invertebrate foods (including plants eaten by invertebrates useful to humans, e.g. silkworms)
- Materials (including woods, fibres, tannins, latex, resins, essential oils, waxes, oils)
- Fuels (including fuelwood, charcoal, fuel alcohol)
- Social uses (including masticatories, smoking, hallucinogens, psychoactive drugs, contraceptives, abortifacients, plants used for ritual or religious purposes).
- Vertebrate poisons (including both accidental or useful poisonous plants e.g. hunting, fishing)
- Non-vertebrate poisons (including accidental and useful poisons e.g. mollusci-, herbi-, insecti-, bacteria- and fungicides)
- Medicines (including human and veterinary uses)
- Environmental uses (including ornamentals, barrier hedges, wind-breaks, soil improvers, erosion control, indicators of heavy metals, pollution or underground water)
- Cosmetic and perfumery plants
- Genetic resources (including wild relatives of crops)

Table 43.2. Wild species commodity groups.

- Cereals and pseudo-cereals
- Leaf and root vegetables
- Grain legumes
- Fruits and nuts
- Medicinal and aromatic plants (MAPs)
- Herbs and spices
- Energy crops
- Fibres
- Ornamentals
- Feed and forages
- Beverage and stimulants
- Timber
- Non-wood forest products

43.5 Dietary Diversity from Wild Species

There is now considerable evidence that a diversity of wild species provides considerable benefit to human health and productivity. Swaminathan (2005) calls for food-based nutritional literacy to be promoted as part of the educational curriculum worldwide, so that young people learn the value of a diverse diet in promoting good nutrition and better health. Just as traditional or underutilized crops may be more nutritious than exotic imported species, wild species play a similar role. Various kinds of food plant are involved but leafy green vegetables are of special importance in many regions and some examples are given below.

43.6 Wild Leafy Vegetables

A remarkable diversity of leafy vegetables or 'wild greens' is found in different parts of the world, not only in the tropics, but also in Mediterranean and temperate zones (Heywood, 1999). In Africa, for example, traditional vegetables represent a valuable resource, and are extremely important for human nutrition and as a source of income throughout the continent (Chweya and Eyzaguirre, 1999). Often they supply most of the necessary daily requirements for vitamins A, B complex and C of the rural poor (Guarino, 1997) and supply minerals and trace elements and may sometimes be better nutritionally than introduced cultivated vegetables (Jansen van Rensburg *et al.*, 2004). For example, an evaluation of the significance of dietary folate from wild vegetables in Vietnam indicates that a majority of the women (87%) got some dietary folate from wild vegetables and nearly one-third had mean daily folate intakes of >50 μg from such sources (Ogle *et al.*, 2001). Wild greens represent an undervalued reservoir of diversity and are of major importance for food security, nutrition and poverty alleviation throughout Africa.

Hundreds of species are involved and they are either actively cultivated in small patches in home gardens or grow as weeds in marginal areas or wild in forest areas. Leafy vegetables are important in the diet of many African countries. According to Jansen van Rensburg *et al.* (2004), indigenous leafy vegetables can play an important part in alleviating hunger and malnutrition in sub-Saharan Africa and are important sources of micronutrients including vitamins A and C, iron and other nutrients and are sometimes better nutritional sources than the modern vegetables. In Kenya, about 200 species growing naturally are used as leafy vegetables (Maundu, 1997) and 1000 in sub-Saharan Africa as a whole.

In Nigeria, for example, where the diet is dominated by starchy staple foods, traditional vegetables are essential sources of proteins, vitamins, minerals and amino acids. The majority of these vegetables are still being harvested from the wild (Okafor, 1997) although an IPGRI project is challenging conventional beliefs about these underutilized species. Development specialists did not think that leafy vegetables were cultivated very widely, but rather that they were gathered from the wild. Socio-economic research sponsored by IPGRI found the reverse. Farmers actively cultivated leafy vegetables and managed them according to the diversity they knew was within the species. For example, bitter leaf (*Vernonia amygdalina*) has several distinct genotypes with different degrees of bitterness that different cultural groups prefer. Farmers would select the material they planted depending on who would be buying and eating the leaves.[2] Another widely used group of leafy vegetables is the black nightshades of the *Solanum nigrum* complex, particularly in the West African forest zones. They are also a source of fruit and medicinal herbs. *S. nigrum* grows easily and

[2] http://www.ipgri.cgiar.org/system/page.asp?theme=4

wild in the forest but seeds of the preferred types are prized in markets and form a good source of income for the forest farmers.

In Asia, leafy vegetables are also an important component of many rural diets. In a study of the ethnobotany of the Iban and Kelabati tribes of Sarawak, it was found that wild-growing vegetables are gathered from a variety of vegetation types. Of a total of 315 species that are considered edible vegetables by them, 120 were used for their leaves and green shoots and 86 for their white shoots, although curiously only a few were used by both communities (Christensen, 2002). The criteria for their popularity are that they must be easy to find, easy to collect, easy to prepare and good tasting.

Many of these species are poorly known and have been neglected by both the scientific and agronomic and development communities (Heywood, 1999; Jansen van Rensburg *et al.*, 2004) so that little research on them is undertaken. Significant work has, however, been undertaken recently by IPGRI in a project on traditional African leafy vegetables that focuses on neglected vegetable species where women are the principal experts and users of genetic diversity, linking the conservation of PGR to raising incomes and nutrition of both rural and urban poor. It is yielding important insights into the diversity, uses and farmer management of germplasm.

43.7 Medicinal Plants

According to the recently published WHO Traditional Medicine Strategy (WHO, 2002), the use of traditional medicines (which are largely plant-based) remains widespread in developing countries with up to 80% of the population using them in Africa. In China, traditional medicine accounts for 40% of all health care and more than 80% of the material of the 700,000 t/year of medicinal plants that are reportedly used for direct decoction in traditional medicine and as ingredients in officinal medicine comes from wild sources (Xiao, 1991; He and Cheng, 1991). It reports that in many Asian countries traditional medicine continues to be widely used, even though allopathic medicine is often readily available. In Japan, for example, 60–70% of allopathic doctors prescribe kampo medicines for their patients. In Latin America, it is reported 71% of the population in Chile and 40% of the population in Colombia use traditional medicine.

At the same time, we are witnessing an ever increasing tendency in the developed world to employ wild plants, as part of complementary and alternative medicines or therapies, as herbal remedies, or as sources of neutraceuticals.

At present, we cannot estimate the total number of species that are used as medicinal plants or in ethnomedicine with a high degree of accuracy. This partly depends on how medicinal plants are defined.[3] A list prepared by WHO

[3] If a narrow definition that the term medicinal applied to a plant indicates that it contains a substance or substances which modulate beneficially the physiology of sick mammals, and that has been used by humans for that purpose, is adopted, the number of species will be much smaller than if the more widely applied definition of say, Srivavasta *et al.* (1996) 'those that are commonly used treating and preventing specific ailments and diseases, and that are generally considered to play a beneficial role in health care' is applied.

contained 21,000 names and Farnsworth and Soejarto suggested that 28% of the world's plant species have been used ethnomedically, an estimate based on an extrapolation from the NAPRALERT database. This would indicate a figure of between 65,000 and 118,000 depending on the total number of plant species believed to exist (see above). In Europe, at least 2000 medicinal and aromatic species are used on a commercial basis, of which two-thirds, 1200–1300 species are native to Europe.

Although the literature on medicinal plant use is very extensive, the information for individual countries is often incomplete: catalogues or compendia for many countries have been published but not complete inventories. For example, the excellent compilation *The Medicinal Plants of India* (Jain and DeFilipps, 1991) covers some 1813 species but more than 7000 wild species are used medicinally by communities and healthcare systems in India (Lambert, 1996).

In West Africa, the vast majority of drugs are obtained from the wild (Cole, 1996). A review of the use, trade and conservation of medicinal plants and aromatic plants in Europe was undertaken by TRAFFIC Europe (Lange, 1998) and for south-east Europe this included information on Albania, Bulgaria, Hungary and Turkey. The report highlighted the fact that wild collection is the major source of material of medicinal and aromatic plants, 30–50%, reaching as high as 75–80% in Bulgaria and almost 100% in Albania and Turkey. A focus in the report was on conservation needs and those species that are threatened by trade but for the majority of species that are used as medicinals or aromatics, we have little information on the ways they are gathered, traded or used. As Schippmann (1999) points out:

> Knowledge about the resource is also very poor amongst the stakeholders. Far too many importers, despite their good intention, are content to leave issues of environmental[ly] responsible sourcing to local exporters and harvesters and are unaware of the destructive effects their trade is having on some wild plant populations and habitats. The resource is often utilized and over-harvested without understanding its biology.

43.8 Harvesting too Much, Growing and Protecting too Little

Open access to medicinal plants in the wild is perhaps one of the main reasons for the current unsustainable levels of harvesting. Other factors contributing to unsustainability include lack of sufficient data on wild plant populations, marketing and trading; inadequate regulations and legal protection (including intellectual property rights for local practitioners with local knowledge); and poor access to appropriate technology for sound harvesting and plantation development.

Government support for and supervision of medicinal plant development are often weak. In some countries, public sector agencies exercise monopoly control over the purchasing and processing of such plants and other forest products, fostering inefficiencies, thwarting commercial development and preventing fair pricing for collectors. But even when they maintain such controls, exporting nations generally reap low returns, since royalty payments and permit requirements are usually not enforced. In Nepal, for instance, less than 25% of the total medicinal plant trade is actually registered.

Nevertheless, attempts are now under way to cultivate some easily grown species and protect important natural habitats in order to reduce the pressure on these vital resources. Indeed, cultivation offers the best hope for conserving many medicinal plants found in the wild while maintaining harvested supplies at today's levels. Cultivation also permits better species identification, improved quality control and increased prospects for genetic improvements. According to Alexander McCalla, director of the Agriculture and Natural Resources Department of the World Bank:

> What looks like a problem actually provides numerous opportunities for developing nations to advance rural well-being. After all, medicinal plants are one of the few (legal) developing country natural products that sell at premium prices. Thus, the global clamor for more herbal ingredients creates possibilities for the local cultivation of medicinal crops as well as for the regulated and sustainable harvest of wild stands. Such endeavors could help raise rural employment in the developing countries, boost commerce around the world, and perhaps contribute to the health of millions.
>
> (McCalla in Nickel and Sennhauser, 2001)

If existing supplies of medicinal plants are to keep up with demand, they will need adequate protection through development of appropriate institutions, policies and legislation. Local communities need support and encouragement to protect these resources. To complement cultivation of adaptable species, harvesting from the wild must be guided by accurate inventories and knowledge about the species concerned. Above all, overexploitation of rare and endangered species must be avoided.

43.9 Wild Harvesting and the Dangers of Over-harvesting

The collection of local plant material from the wild for food, fuel, medicinal (and other) purposes dates back to earliest times when humans learned to distinguish which plants had properties that helped treat ailments. It is still extensively practised by local communities that are dependent on medicinal plants as important components of their healthcare system and becomes a matter of concern when demand exceeds supply because of habitat loss, growing populations and other factors.

The type of wild harvesting or gathering of medicinal and aromatic plants that is a cause of concern in the context of biodiversity conservation is where parts of the target species – leaves, stems, bark, roots, flowers, fruits, seeds or whole plants – are collected in some quantity for medicinal purposes, either as part of traditional medicine systems, or for commercial exploitation by national or international pharmaceutical companies. The concern stems from the possible effects of wild harvesting on the regeneration or even survival of the populations of the species that are sampled.

As the collectors are usually paid very low prices for this wild-harvested material, prices that are maintained artificially low by small groups of traders acting as a monopoly may have the effect of encouraging over-harvesting so that the collectors get sufficient reward for their work. Of course, such low prices do not allow for the costs involved in managing or replacing the resources, and little attention is paid to these aspects by importing companies that require the collection of large quantities of wild material.

Most of the evidence regarding the extent and significance of wild harvesting is anecdotal and relatively few detailed case studies have been undertaken. While most countries are of course aware that wild harvesting of some of their biological resources is being carried out, few are aware of the scale. The data are not gathered and often the consequences are not obvious until the damage has been caused and particular species threatened with imminent local extinction.

In South Africa, it has been estimated that native medicinal plants are used by more than 60% of the population in their health care needs or cultural practices and *c.*3000 species are used by an estimated 200,000 indigenous traditional healers (Coetzee *et al.*, 1999). The indigenous medicinal plant industry is large, but almost entirely based on collecting from the wild. The current demand for the numerous plant species used in indigenous medicines exceeds supply. As a consequence, several plant species have been put at risk. For example in KwaZulu Natal, wild ginger (*Siphonochilus aethiopicus*) and the pepper-bark tree (*Warburgia salutaris*), have become extinct outside of protected areas (Mander, 1999).

The amount of material that is required by companies investigating how natural products work, varies considerably. In most cases, only small quantities of wild-harvested plant material are used, as when pharmaceutical companies are looking for specific active compounds that may eventually lead to the development of a new drug; on the other hand, in some cases, quite large quantities may be needed and when dealing with rare endemic species that only occur in small populations and are not cultivated, serious problems arise, as in the recent case of the Pacific yew, *Taxus brevifolia*, where the first attempts at the synthesis of the anti-cancer agent taxol, derived from the bark posed a problem to the survival of the species because of the quantities of material required, especially as to obtain the bark the tree had to be killed (Hamburger *et al.,* 1991). The restricted availability of the yew bark severely restricted the work. Since a large specimen of Pacific yew 2 feet in diameter requires 200 years to grow and yields about 5 pounds of bark and the yield of taxol obtained was very low – 0.004% – the extinction and destruction of the old-growth forests of the Pacific yew was a real threat.

In some cases, even when it is possible to synthesize a drug, it may prove less costly to extract active ingredients from wild plants (WHO, IUCN and WWF, 1993). The quantities involved can be astonishingly great in some cases and there can be little doubt as to the damage that their collection has caused to the wild populations of the plants concerned. For example, Cunningham and Mbenkum (1993) reported that 900 t of *Voacanga africana* seed, used for the industrial production of the alkaloid tabersonine, a depressor of central nervous system activity in geriatric patients, were exported from Cameroon to France between 1985 and 1991, and 11,537 t of the bark of *Prunus africana* (red stinkwood), used to treat prostatitis, in the same period. In some cases, although synthesis or cultivation is possible, harvesting material from the wild is cheaper.

Another group of plants harvested commercially from the wild is for the cut-flower trade. This occurs in many parts of the world, for example in the fynbos in South Africa. According to Turpie *et al.* (1998):

> [T]he wild flower industry has two components: fresh flowers and dried flowers, both of which continue to be harvested from the wild. Products comprise flowers (Proteaceae) and greens (comprising many taxa including *Leucadendron* foliage, ericas, etc. for use as filler material) for the fresh industry, and flowers, including *Leucadendron* cones and other products, for the dried flower industry.

At least 100 species are used in the wild flower industry (Cowling and Richardson, 1995). However, the numbers of species used, and indeed, which species are used, change subject to fluctuating market demands created by local and overseas fashions.

It has been calculated that as a whole the fynbos flower industry currently generates a gross income of R149.3 million/year (1997 prices), of which R91.5 million/year and R37.8 million/year is from the export of fresh and dried flowers, respectively, and R20 million/year is from local sales (*SAPPEX News*, July 1999, cited in Turpie *et al.*, 1998). And Turpie *et al.* (1998) estimate that natural vegetation (veld) is responsible for 57.6%, or R86 million, of this turnover.

In Australia also, a large number of native cut-flower species are picked from the wild (bush-picked), and are generally of poor quality and available in only small quantities (Jones, 1995, cited in Trupie *et al.*, 1998). A remarkable example from there reported by Smith (2000) is *Doryanthes excelsa*, one of the most outstanding monocots found in the Australian bushland:

> It carries massive flower spikes (scapes) that may attain 8 metres in height. These impressive flowering stems are highly sought after to provide floral designers with dramatic feature flowers for large imposing hotel foyer arrangements. At present the commercial appeal for this unique Australian plant is escalating with an increase in demand both on local and overseas markets. At present very few stems come from commercial row production with the vast majority of supply coming from bushland to the north of Sydney. The high returns for cut flowers has created a situation where stems are being removed illegally from the roadside, private properties and national parks.
>
> (Smith, 2000)

43.10 Ornamental Plants

A major group of plants that are often overlooked or at least underestimated in considerations of the uses of wild plants are the many thousands of species grown as ornamentals in parks and in public and private gardens and in the horticultural trade.

As summarized by Heywood (2003):

> While only about 1–200 species are used intensively in commercial floriculture (e.g. carnations, chrysanthemums, gerbera, narcissus, orchids, tulips, lilies, roses,

pansies and violas, saintpaulias, etc.) and 4–500 as house plants, several thousand species of herbs, shrubs and trees are traded commercially by nurseries and garden centres as ornamentals or amenity species. Most of these have been introduced from the wild with little selection or breeding. In Europe alone, 12,000 species are found in cultivation in general garden collections (i.e. excluding specialist collections and botanic gardens). In addition, specialist collections (often very large) of many other species and/or cultivars of groups such as orchids, bromeliads, cacti and succulents, primulas, rhododendrons, conifers and cycads are maintained in several centres such as botanic gardens and specialist nurseries, as are 'national collections' of cultivated species and cultivars in some countries.

The size of the trade in ornamental and amenity horticulture is difficult to estimate but probably runs into many billions of dollars annually. In the USA alone, landscaping plants, cut flowers and houseplants are part of an industry worth more than $11 billion in annual sales and the total world exports of bulbs and tubers in 1998 was worth $675,589,000 while the global floricultural exports in the same year amounted to $ 8,394,750,000 (Pertwee, 1999). There is considerable potential for further development and in recent years much effort has been directed in some countries such as Australia and South Africa at introducing many new native wild species into commercial cultivation.

43.11 Extractivism

A particular case of the use of wild plants is that known as extractivism, a term applied to the systematic exploitation of forest products that are intended for sale on local, national or international markets (Lescure *et al.*, 1994). It is best studied in tropical forest in Brazil where examples of the native species exploited include Brazil nut (*Bertholletia excelsa*) and various palm species (Table 43.3).

Table 43.3. Principal extractive products derived from native plant species in the Brazilian Amazon. (From Lescure *et al.*, 1994.)

Species	Family	Part used	Practice	Product
Euterpe precatoria	*Palmae*	Fruits	Picking	Fruits
Euterpe oleracea	*Palmae*	Fruits	Picking	Fruits
		Buds	Pruning	Palm hearts
Carapa guianensis	*Meliaceae*	Seeds	Gathering	Oil
Carapa procera	*Meliaceae*	Seeds	Gathering	Oil
Orbignya cf *phalerata*	*Palmae*	Leaves	Pruning	Thatching
Manilkara bidentata	*Sapotaceae*	Latex	Felling	Gum
Bertholettia excelsa	*Lecythidaceae*	Seeds	Gathering	Seeds
Castilloa ulei	*Moraceae*	Latex	Tapping	Gum
Astrocaryum chambira	*Palmae*	Leaves	Pruning	Fibres
Heteropsis spp.	*Araceae*	Aerial roots	Pruning	Fibres
Copaifera spp.	*Leguminosae*	Oleo-resin	Tapping	Oleo-resin
Dipteryx odorata	*Leguminosae*	Seeds	Gathering	Cumarin

Continued

Table 43.3. *Continued*

Species	Family	Part used	Practice	Product
Astrocarym jauari	*Palmae*	Bud	Pruning	Palm hearts
Manilkara spp.	*Sapotaceae*	Latex	Felling	Gum
Ptychopetalum olacoides	*Olaceae*	Root	Lifting	Medicinal
Jessenia bataua	*Palmae*	Fruits	Gathering	Fruits
Aniba rosaeodora	*Lauraceae*	Stem	Felling	Linalol
Myrcia citrifolia	*Myrtaceae*	Leaves	Pruning	Medicinal
Leopoldina piassaba	*Palmae*	Leaves, sheaths	Pruning	Fibres
Licaria pucherii	*Lauraceae*	Seeds	Gathering	Medicinal
Hevea spp.	*Euphorbiaceae*	Latex	Tapping	Gum
Couma macrocarpa	*Apocynaceae*	Latex	Felling	Gum
Couma utilis	*Apocynaceae*	Latex	Tapping	Gum
Derris spp.	*Leguminosae*	Roots	Lifting	Rotenone
Astrocaryum aculeatum	*Palmae*	Fruits	Gathering	Fruits
Virola sirinamensis	*Myristicaceae*	Seeds	Gathering	Oil

43.12 Major International or Regional Initiatives

While much of the work on the assessment and use of wild species used by humans is undertaken at a local level or on selected groups of species, some major international or regional initiatives exist.

PROSEA (Plant Resources of South-east Asia), an international programme that focuses on the documentation of information on plant resources of South-east Asia, covers the fields of agriculture, forestry, horticulture and botany. It is a research programme, making the knowledge available for education and extension, and is ecologically focused on promoting plant resources for sustainable tropical land use systems and is committed to conservation of biodiversity and rural development through diversification of resources and application of farmers' knowledge.

As many as 19 volumes of the 20-volume *PROSEA Handbook* originally planned are now published, covering nearly 7000 species of South-east Asia divided into 19 commodity groups (Table 43.4).

PROTEA (Plant Resources of Tropical Africa) is an international, not-for-profit foundation whose aim is to synthesize the dispersed information on the approximately 7000 useful plants of tropical Africa and to provide wide access to the information in the form of web databases, books, CD-Roms and special products.

The MEDUSA Network of Useful Plants of the Mediterranean Region was established by CIHEAM-MAICh, with the support of the European Union Directorate General I, for the identification, conservation and sustainable use of the wild plants of the Mediterranean Region. The Network comprises National Focal Point Coordinators from the countries of the region and also includes representatives of international organizations. The objectives of the network are (Heywood and Skoula, 1999):

Table 43.4. PROSEA commodity groups.

• Pulses	• Cereals
• Edible fruits and nuts	• Auxiliary plants
• Dye and tannin-producing plants	• Medicinal and poisonous plants
• Forages	• Spices
• Timber trees	• Vegetable oils and fats
• Rattans	• Cryptogams
• Bamboos	• Stimulants
• Vegetables	• Fibre plants
• Plants yielding non-seed carbohydrates	• Exudates
	• Essential-oil plants

1. The identification of native and naturalized plants of the Mediterranean region, according to use categories such as food, food additives, animal food, bee plants, invertebrate foods, materials, fuels, social uses, vertebrate poisons, non-vertebrate poisons, medicines, perfumery and cosmetics, environmental uses and gene sources.
2. The creation of a regional information system that will include: scientific plant name and authority, vernacular names, plant description, chemical data, distribution, habitat description, uses, conservation status, present and past ways of trading, marketing and dispensing and indigenous knowledge (ethnobiology and ethnopharmacology), including references to literature sources.
3. Preliminary evaluation of the conservation status and potential utilization in agriculture of these plants as alternative minor crops.

43.13 Medicinal Plants

Databases for medicinal plants include:

- NAPRALERT, an acronym of NAtural PRoducts ALERT, is a dynamic database containing information from more than 137,722 bibliographic records on 118,730 natural products. It contains ethnomedical data on plants, biological effects reported for extracts of living organisms and the occurrence of secondary chemical metabolites in living organisms. It includes information on pharmacological and taxonomic distribution and is from the laboratory of Dr Norman Farnsworth, a pioneer in plant drug research. Search results delivered to customers in computer-readable form remain the property of the database producer. http://info.cas.org/ONLINE/DBSS/napralertss.html
- The Natural Medicines Comprehensive Database (www.naturaldatabase.com);
- The Asian Pacific Information Network on Medicinal and Aromatic Plants (APINMAP) (http://www.pchrd.dost.gov.ph/) is a UNESCO-sponsored voluntary network of organizations in 14 Asian and Pacific countries, namely Australia, People's Republic of China, India, Indonesia, Republic of Korea, Malaysia, Nepal, Pakistan, Papua New Guinea, The Philippines, Sri Lanka, Thailand, Turkey and Vietnam. Its objective is to promote exchange of

information relating to medicinal and aromatic plants between its member organizations. Databases and other resources held by each organization are shared with others. APINMAP resources include an integrated APINMAP database containing bibliographic and factual information on medicinal plants, lists of research projects, institutions and personnel.

43.14 Reasons for Neglect of Wild Plants as Resources and Obstacles to their Development

Attention is often drawn to the neglect on the part of agriculture and development agencies and by governments of the enormous resources that wild species represent. Lamb's (1993) comments for non-woody forest products that: '[T]here is almost everywhere a lack of hard facts, figures and published science-based information about the extraction, use, profitability and potential of non-wood forest products. This makes it still harder to integrate their use into development schemes. . .' are generally applicable to all categories of wild species used by humans. The reasons for this neglect are many and they are summarized in Table 43.5.

On the positive side, however, there have been signs in the past few years, that the importance of the biodiversity represented by wild plants that are used by humans in various ways is receiving wider recognition, in particular their contribution to farm household security and other human needs. For example,

Table 43.5. Reasons for neglect of wild plants as resources.

- A lack of information about the extent of their use and importance to rural economies
- A lack of information, especially statistics, concerning the economic value of wild plants
- A lack of information and reliable methods for measuring their contribution to farm households and the rural economy
- A lack of world markets, except for a small number of products
- A lack of market research and commercial information
- The irregularity of supply of wild plant products
- The lack of quality standards
- The lack of storage and processing technology for many of the products
- The availability of substitutes
- The bias in favour of large-scale agriculture
- Changing fashions
- Flawed perceptions on the part of national authorities and resource managers regarding the value and potential of wild species
- Prejudice on the part of some planners and development experts against the very concept of 'wild' products because they do not all fit into conventional categories or formal markets, or because they seem to have a retrograde or archaic 'back-to-nature' aspect
- Distrust on the part of some technologists because the use of wild plants involves societal factors that are difficult to assess and evaluate and which are often only be usefully discussed in non-technical terms
- Unwillingness of many scientists, development experts and agencies to get involved in issues of access and IPR

the FAO Leipzig Global Plan of Action, which sets out a global strategy for the conservation and sustainable use of plant genetic resources for food and agriculture with an emphasis on productivity, sustainability and equity (Cooper *et al.*, 1998), includes among its recommendations for priority activities:

- Promotion of the *in situ* conservation of wild crop relatives and wild plants for food production;
- Promotion of the development and commercialization of underdeveloped crops and species;
- Promotion of sustainable agriculture through diversification of crop production and broader diversity in crops;
- Expansion of *ex situ* conservation activities, including collecting material of many local plants that are important for food and agriculture.

The GPA complements the action being taken in fulfilment of the requirements of the Convention on Biological Diversity. There is now 'official recognition' by governments of the need to widen our approach to the conservation and sustainable use of species so as to take into account the very wide range that are involved or associated with human activities.

The role of indigenous communities is now recognized as an essential component of any strategy for the conservation and sustainable use of biodiversity as stressed in the Global Biodiversity Assessment (Heywood, 1995). Although the latter included a whole section on the all-pervading influence of human action, the complexity of the social, cultural, ethical, religious and other human interactions with biodiversity and agroecological systems led UNEP to commission a separate and complementary volume on cultural and spiritual values of biodiversity (Posey, 1999). As the STAP Expert Group on Sustainable Use of Biodiversity (UNEP, 1998) notes: 'Development of sustainable use projects requires a paradigm shift from a focus on protection and the development of protected areas to considering also such skills as dealing with the interaction of socio-economic and ecological systems.'

43.15 Actions Needed

It would not be appropriate here to propose a strategy for addressing the above issues but any such plan would probably have to include the following action points:

- Establishment of a priority list of wild species under exploitation in the country/region;
- Assessing their detailed conservation status;
- Assessing the extent of exploitation of the populations of these species;
- Ensuring that the recently adopted Bonn Guidelines on Access to Genetic Resources and Fair and Equitable Sharing of the Benefit are properly implemented;
- Protecting and promoting indigenous knowledge;
- Identifying and documenting the contribution of local use and management of these plant resources to improved livelihood and biocultural conservation efforts;

- Reviewing current practice and methodologies for domestication and cultivation of intensively used wild species within the region;
- Determining the state of knowledge and identifying gaps in the knowledge-base concerning the chemical composition of plants used in traditional health care systems in the region;
- Development of a programme to involve local communities through their participation in 'conservation through use' programmes;
- Promoting public awareness of the monetary and social value of the wild species;
- Development of existing databases to provide the necessary informatics background and structure and information flow as well as other measures necessary for the dissemination of information to interested parties and to the public.

References

Altieri, M.A. (1999) The agroecological dimensions of biodiversity in traditional farming systems. In: Posey, D.A. (ed.) *Cultural and Spiritual Values of Biodiversity*. For UNEP, Intermediate Technology Publications, London, pp. 291–297.

Bianco, V.V. (1995) Rocket, an ancient underutilized vegetable crop and its potential. In: Padulosi, S. (compiler) *Rocket Genetic Resources Network. Report of the First Meeting, 13–15 November 1994, Lisbon, Portugal*. International Plant Genetic Resources Institute, Rome, Italy.

Bramwell, D. (2002) How many plant species are there? *Plant Talk* 28, 32–34.

Christensen, H. (2002) *Ethnobotany of the Iban and the Kelabit*. Forest Department Sarawak, Malaysia, NEPCon, Denmark and University of Aarhus, Denmark.

Chweya, J.A. and Eyzaguirre, P.B. (eds) (1999) *The Biodiversity of Traditional Leafy Vegetables*. International Plant Genetic Resources Institute, Rome, Italy.

Coetzee, C., Jefthas, E. and Reinten, E. (1999) Indigenous plant genetic resources of South Africa. In: Janick, J. (ed.) *Perspectives on New Crops and New Uses*. ASHS Press, Alexandria, Virginia, pp. 160–163.

Cole, N.H.A. (1996) Diversity of medicinal plants in West African habitats. In: van der Maesen, L.G., van der Burgt, X.M. and van Medenbach de Rooy, J.M. (eds) *The Biodiversity of African Plants*, Kluwer Academic, Dordrecht, The Netherlands, pp. 704–713.

Cook, F.E.M. (1995) *Economic Botany – Data Collection Standard*. Prepared for the International Working Group on Taxonomic Databases for Plant Sciences (TDWG) Royal Botanic Gardens, Kew, UK.

Cooper, H.D., Spillane, C., Kermali, I. and Anishetty, N.M. (1998) Harnessing plant genetic resources for sustainable agriculture. *Plant Genetic Resources Newsletter* 114, 1–8.

Cowling, R.M. and Richardson, D.M. (1995) *Fynbos: South Africa's Unique Floral Kingdom*. Fernwood Press, Cape Town, South Africa.

Cunningham, A.B. and Mbenkum, F.T. (1993) *Medicinal Bark in International Trade: A Case Study of the Afromontane Tree* Prunus africana. Report to WWF International.

Finn, C. (1999) Temperate berry crops. In: Janick, J. (ed.) *Perspectives on New Crops and New Uses*. ASHS Press, Alexandria, Virginia, pp. 324–334.

Frison, E. (2005) *Agricultural Biodiversity and Livelihoods. The Role of Biodiversity in Achieving the United Nations Millennium Development Goal of Freedom from Hunger and Poverty*. MSSRF, Chennai.

Govaerts, R. (2001) How many species of seed plant are there? *Taxon* 50, 1085–1090.

Grivetti, L.E. (1980) Goat kraal gardens and plant domestication. Thoughts on ancient and modern food production. *Ecology of Food and Nutrition* 10, 5–7.

Guarino, L. (ed.) (1997) *Traditional African Vegetables. Promoting the Conservation and Use of Underutilized and Neglected Crops.* 16. Proceedings of the IPGRI International Workshop on Genetic Resources of Traditional Vegetables in Africa: Conservation and Use, 29–31 August 1995, ICRAF-HQ, Nairobi, Kenya. Institute of Plant Genetics and Crop Plant Research, Gatersleben/International Plant Genetic Resources Institute, Rome, Italy.

Hamburger, M., Marston, A. and Hostettmann, K. (1991) Search for new drugs of plant origin. *Advances in Drug Research* 20, 167–215.

He, S. and Chweng, Z. (1991) The role of Chinese botanical gardens in conservation of medicinal plants. In: Akerele, O., Heywood, V. and Synge, H. (eds) *The Conservation of Medicinal Plants*. Cambridge University Press, Cambridge, pp. 229–237.

Heywood, V.H. (1995) (ed.) *Global Biodiversity Assessment*. UNEP and Cambridge University Press, Cambridge.

Heywood, V.H. (1999) *Use and Potential of Wild Plants in Farm Households*. Farm Systems Management Series 15, FAO, Rome.

Heywood, V.H. (2003) Conservation and sustainable use of wild species as sources of new ornamentals. In: Düzyaman, E. and Tüzel, Y. (eds) *Proceedings of the International Symposium on Sustainable Use of Plant Biodiversity to Promote New Opportunities for Horticultural Production Development. Acta Horticulturae* 598, 43–53.

Heywood, V. and Skoula, M. (1999) The MEDUSA network: conservation and sustainable use of wild plants of the Mediterranean region. In: Janick, J. (ed.) *Perspectives on New Crops and New Uses*. ASHS Press, Alexandria, Virginia, pp. 148–151.

Jain, S.K. and DeFilipps, A.A. (1991) *Medicinal Plants of India*. 2 vols. Reference Publications, Inc., Algonac, Michigan.

Jansen, P.C.M., Lemmens, R.H.M.J., Oyen, L.P.A., Siemonsma, J.S., Stavast, F.M. and van Valkenburg, J.L.C.H. (eds) (1991) *Basic List of Species and Commodity Grouping. Final version* PROSEA Foundation, Bogor.

Jansen van Rensburg, W.S.,Venter, S.I., Netshiluvhi, T.R., van den Heever, E., Vorster, H.J. and de Ronde J.A. (2004) Role of indigenous leafy vegetables in combating hunger and malnutrition. *South African Journal of Botany* 70, 52–59.

Jones, R.B. (1995) New ornamental crops in Australia. *Acta Horticulturae* 397, 59–70.

Kostermans, A.J.G.H. and Bompard, J.-M. (1993) *The Mangoes*. Academic Press, London.

Lamb, R. (1993) *More than Wood – Special Options on Multiple Use of Forests*. Forestry Topics Reports No. 4. FAO, Rome.

Lambert, J.H.D. (1996) *Towards an Agenda to Conserve and Enhance India's Medicinal Plant Heritage*. AGRAF, AGRAF, The World Bank, Washington, DC.

Lange, D. (1998) *Europe's Medicinal and Aromatic Plants: Their Use, Trade and Conservation*. TRAFFIC International, Cambridge.

Lescure, J.-P., Pinton, F. and Emperaire, L. (1994) People and forest products in central Amazonia: the multidisciplinary approach of extractivism. In: Clusener-Godt & Sachs (eds) *Extractivism in the Brazilian Amazon: Perspectives on Regional Development. MAB Digest* 18, 14–33.

Mander, M. (1999) *Marketing of Indigenous Medicinal Plants in South Africa: A Case Study in KwaZulu-Natal: Summary of Findings*. Food and Agricultural Organisation of the United Nations, Forest Products Division, Rome.

Maundu, P.M. (1997) The status of traditional vegetable utilization in Kenya. In: Guarino, L. (ed.) *Traditional African Vegetables. Promoting the Conservation and Use of Underutilized and Neglected Crops.* 16. Proceedings of the IPGRI International Workshop on Genetic Resources of Traditional Vegetables in Africa: Conservation and Use, 29–31 August 1995,

ICRAF-HQ, Nairobi, Kenya. Institute of Plant Genetics and Crop Plant Research, Gatersleben/International Plant Genetic Resources Institute, Rome, Italy, pp. 92–95.

Nickel, W. and Sennhauser, E. (2001) *Medicinal Plants: Local Heritage with Global Importance*. Available at: http://lnweb18.worldbank.org/sar/sa.nsf/0/fae63d87e2bd14038525687f0057e0d1?OpenDocument

Ogle, B.M. and Grivetti, M.E. (1985) Legacy of the chameleon. Edible wild plants in the kingdom of Swaziland, Southern Africa. A cultural, ecological, nutritional study. Part 2. Demographics, species availability and dietary use, analysis by ecological zone. *Ecology of Food and Nutrition* 17, 1–30.

Ogle, B.M., Johansson, M., Tuyet, H.T. and Johannesson, L. (2001) Evaluation of the significance of dietary foliate from wild vegetables in Vietnam. *Asia Pacific Journal of Clinical Nutrition* 10, 216.

Okafor, J.C. (1997) Conservation and use of traditional vegetables from woody forest species in Southeastern Nigeria. In: Guarino, L. (ed.) *Traditional African Vegetables. Promoting the Conservation and Use of Underutilized and Neglected Crops*. 16. Proceedings of the IPGRI International Workshop on Genetic Resources of Traditional Vegetables in Africa: Conservation and Use, 29–31 August 1995, ICRAF-HQ, Nairobi, Kenya. International Plant Genetics Resources Institute, Rome, pp. 5–11.

Pertwee, J. (1999) *International Floriculture Trade Statistics 1999*. Pathfast Publishing, Frinton-on-Sea, UK.

Posey, D.A. (1992) Interpreting and applying the 'Reality' of indigenous concepts: what is necessary to learn from the natives? In: Redford, K.H. and Padoch, C. (eds) *Conservation of Neotropical Forests. Working from Traditional Resource Use*. Columbia University Press, New York, pp. 21–34.

Posey, D.A. (1999) *Cultural and Spiritual Values of Biodiversity UNEP*. Intermediate Technology Publications, London.

Prance, G.T., Beentje, H., Dransfield, J. and Johns, R. (2000) The tropical flora remains undercollected. *Annals of the Missouri Botanical Garden* 87, 67–71.

Prescott-Allen, R. and Prescott-Allen, C. (1990) How many plants feed the world? *Conservation Biology* 4, 365–374.

Schippmann, U. (1999) Summary remarks and conclusions. Medicinal plant trade in Europe: conservation and supply. *Proceedings, First International Symposium on the Conservation of Medicinal Plants in Trade in Europe*. TRAFFIC Europe.

Scotland, R.W. and Wortley, A. (2003) How many species of seed plants are there? *Taxon* 52, 101–104.

Slikkerveer, L. (1994) *Indigenous Agricultural Knowledge Systems in Developing Countries: A Bibliography*. Indigenous Knowledge Systems Research and Development Studies No. 1. Special Issue: INDAKS Project Report 1 in collaboration with the European Commission DG XII. Leiden, the Netherlands, Leiden Ethnosystems and Development Programme (LEAD).

Smith, J. (2000) *Micro-propagation of the Gymea Lily. A Report for the Rural Industries Research and Development Corporation*. RIRDC Publication No 00/36: Barton, ACT, Australia.

Srivavasta, J., Lambert, J. and Vietmeyer, N. (1996) *Medicinal Plants: An Expanding Role for Development*. World Bank Technical Paper 320, Washington, DC.

Swaminathan, M.S. (2005) *Presidential Address*. International Consultation on the Role of Biodiversity in Achieving the United Nations Millennium Development Goal of Freedom from Hunger and Poverty, MSSRF, Chennai.

Thorne, R.F. (2002) How many species of seed plants are there? *Taxon* 51, 511–522.

Turpie, J., Heydenrych, B. and Hassan, R. (1998) Accounting for stock and flow values of woody land resources: methods and results from South Africa. In: Heydenrych, B. and Turpie, J. (eds) *Accounting for Natural Resources in South Africa: A Preliminary Assessment of Fynbos*

Stocks, Yields and Values in the Western Cape. Report submitted to South African National Parks. Available at: http://www.ranesa.co.za/fynbossa_een.htm

UNEP (1998) *STAP Expert Group Workshop on Sustainable Use of Biodiversity*. Malaysia, 24–28 November 1997. Reported to 11th STAP meeting, 21–23 January 1998, Agenda Item 6.

WHO (2002) *WHO Traditional Medicine Strategy 2002–2005*. World Health Organization, Geneva.

WHO, IUCN and WWF (1993) *Guidelines on the Conservation of Medicinal Plants*. Gland, Switzerland.

WRI (2005) *World Resources 2005 – The Wealth of the Poor: Managing Ecosystems to Fight Poverty*. United Nations Development Programme, United Nations Environment Programme, The World Bank, World Resources Institute.

Xiao, P. (1991) The Chinese approach to medicinal plants – Their utilization and conservation. In: Akerele, O., Heywood, V.H. and Synge, H. (eds) *The Conservation of Medicinal Plants*. Cambridge University Press, Cambridge, pp. 305–319.

44 Minor Crops and Underutilized Species: Lessons and Prospects

S. Padulosi, I. Hoeschle-Zeledon and P. Bordoni

44.1 Introduction: What Are Minor and Underutilized Crops?

These are words loaded with cultural meaning and hence they are far from being understood in the same way by all (Padulosi *et al.*, 2002, Padulosi and Hoeschle-Zeledon, 2004). Yet, agreement among workers on the meaning of these terms facilitates better communication and collaboration among stakeholders engaged in the promotion of such species.

Between the two, minor crop is perhaps the most ambiguous term: minor to what? to whom? and where? Take the example of saffron (*Crocus sativus* L.). This species is often considered a minor crop because of the limited extensions of its cultivations. It is though not minor in terms of income generation and hence it cannot be assimilated to thousands of other 'really' minor crops whose income generation is very marginal for their growers. Defining what is a minor crop is however possible within a specific context, for example when setting regulatory framework in the use of agrochemicals for plant protection. In this case, the American Pesticide Association (FQPA) has defined minor crops as those 'cultivated over fewer than 300,000 acres of land' (http://www.pmac.net/minorcp.htm).

Minor and underutilized as referred to crops are terms often used interchangeably. In reality, this is also the case for many other terms used for their appeal to a specific audience, being that of policy makers, development specialists, funding agencies, farmers or the public at large.

With the view of enhancing plant biodiversity for improving livelihood of people, these – variously described species – are receiving greater attention by the international community and the issue of defining, as much as possible, what we mean by them is strategic to canvassing political and possibly financial support to their cause. This chapter will be focusing only on underutilized species in view of the fact that out of all terms mentioned above, this is by far the most common in literature. The term generally carries a positive message and is preferred over others because of the social and economic intrinsic opportunities that it alludes to. However, it must be also added that in poor countries, the term underutilized often carries the connotation of unpopular and old fashioned. (For further speculations over the use of terms such as neglected crop, see Eyzaguirre *et al.*, 1999.)

In Table 44.1 we have listed those traits that make underutilized crops distinct from commodities or other so-called major crops. The relevance of their

Table 44.1. Describing underutilized crops, their relevance to people and for R&D interventions.

Main trait of underutilized crops[a]	Why they are important to people?	Why they are underutilized?	Why they require attention from research and/ or development?
Important in local consumption and production systems	They are sources of nutrition, income, risk mitigation and better livelihood for local people	Underutilized in the context of greater/better organized markets	R&D is needed in order to both fully realize currently poorly tapped benefits and strengthen their contribution to livelihood. Particularly relevance for nutrition is not acknowledged
Highly adapted to agroecological niches and marginal areas	Strategic asset particularly for the poor and the marginalized	R&D has been focusing traditionally on commodity crops	Research is needed to develop enhanced cultivation systems in such marginal lands
Ignored by policy makers and excluded from R&D agendas	Victim of biased policies in support of major crops (e.g. during the Green Revolution)	Gradually left aside because of less competitiveness in the markets	As point 1
Represented by wild species, ecotypes, landraces	High adaptability to marginal/less favourable environments; high genetic variability that can be used in different contexts	As point 2	Research needed to domesticate them, select better varieties and provide germplasm to users; also to allow development of sufficiently uniform products for the market
Cultivated and utilized drawing on IK	They form an important genetic and cultural asset in the hands of the poor; they are part of the identity of local communities	As point 2	R&D is needed to enhance cultivation and ultimate use through enhancement of IK practices
Hardly represented in *ex situ* gene banks	As point 2	As point 2	Research is needed to sample, conserve and characterize their diversity
Characterized by fragile or non-existent seed supply systems	As point 2	As point 2	R&D is critical in order to establish regular and quality provision of seed to users

[a]We refer to both cultivated and wild species.

main traits to people, reasons for the status of underuse and justifications on why attention by R&D is being voiced are also offered. Complementary description of these species and their distinction with major crops on the basis of cultural, commercial, ecological and farm management criteria are also provided in Table 44.2.

Notwithstanding the descriptions provided in Tables 44.1 and 44.2, the ultimate identification of underutilized species is not a 'black and white' matter and disagreements among workers are common. Time (underutilized crops may change their status over a short period of time, hence their definition today may not be applicable tomorrow) and space (same species may have different status across a country, a region or even a continent) are, for instance, factors that need to be taken always into consideration. In fact, differences in the status of a species within a country and/or within a region are the norm and not an exception. Take for example the case of rocket (*Eruca vesicaria* (L.) Cav. subsp. *sativa* (Miller) Thell.), which is a booming crop in Italy, but a highly underutilized species in Syria, Egypt and other Mediterranean countries. Furthermore, different stakeholders often have different views and perceptions on what is underutilized, when a species is no more underutilized and in some cases is even threatened by genetic erosion due to an upsurge of excessive unsustainable utilization (e.g. the case of devil's claw in Namibia and Botswana; Grote, 2003). Lack of data on existing uses and socio-economic importance of certain species is an additional difficulty. As a way to reconcile different opinions, particularly when deciding which species should be included in a novel R&D project, we would like to recommend the use of what we call the 'user's lens approach' based on the principles of social equity and fairness. Based on this approach, we should be asking whether or not our intended efforts are going to ultimately lift up the use of the selected species and bring real benefits to communities (particularly those poor and marginalized) who have used and

Table 44.2. Discriminatory features between major and underutilized crops. (From Thies, 2000.)

Major crops	Underutilized crops
Farm management criteria	
High production – high risk	Low production – low risk
Few products	Many products
Homogeneous produce	Heterogeneous produce
Certified seed	Local seed
Commercialization criteria	
Satisfy modern nutritional habits	Satisfy local nutritional habits
Regional, national, international markets	Mainly local markets
Dependency on world market prices	Mainly local markets
Subsidies and incentives	No subsidies or incentives
Ecological criteria	
Highly sensitive to climate	Locally adapted
Cultural criteria	
Internationally promoted using global standards	Relevant to local identities/values and selected using local criteria

safeguarded the species over generations in spite of the continued lack of R&D attention. Such an approach should be truly participatory to share the opinion of various stakeholder groups (from farmers to processors, women groups, traders, etc.). The outcome of such consultations will ensure that limited resources will be channelled to species with highest opportunity for improving people's livelihoods and in so doing bringing greatest benefits from lesser-used agrobiodiversity (Padulosi, 1999).

44.2 Underutilized Species and CWR

In the context of crop wild relatives (CWR), underutilized species can be grouped into three distinct groups each characterized by a different status with regard to the level of attention received by R&D. The first group includes underutilized wild relatives of commodity crops, i.e. species belonging to the primary or secondary gene pool of major crops which have not been fully exploited for the improvement of their cultivated relatives. Reasons behind their poor exploitation may be related to lack of demand for particular traits that they may possess, lack of scientific evidence of their potential, lack of readily available germplasm in *ex situ* collections or limited access to these due to IPR issues and technical and/or institutional difficulties experienced in carrying out breeding programmes. The second group refers to wild relatives of cultivated underutilized crops that have not been tapped at all (or if so very marginally) for the improvement of their cultivated relatives. It may well be that the status of underuse of their cultivated relatives is partly due to little investments in breeding programmes using such wild resources. In this group we find species such as *Artocarpus mariannensis* Trecul. the wild relative of breadfruit, *Coriandrum testiculatum* L. the wild relative of coriander (*C. sativum* L.), *Chenopodium berlandieri* Moq. the wild relative of quinoa (*C. quinoa* Willd.), *Vigna subterranea* (L.) Verdc. var. *spontanea* (Harms) Hepper the wild relative of bambara groundnut (*V. subterranea* (L.) Verdc.), *Annona diversifolia* Staff. the wild relative of cherimoya (*A. cherimola*) Miller and *Punica protopunica* Balf. f. the wild relative of pomegranate (*P. granatum* L.). The third group is made up of those species which have never been domesticated or cultivated and henceforth are used directly as such from the wild. Many wild leafy vegetables, e.g. *Solanum nigrum* L. and *Cleome gynandra* L., belong to this group, which plays an important role in nutritional security as sources of vitamins, minerals and micronutrients in the diets of local communities in sub-Saharan Africa. Their use enhancement is being promoted in a number of international projects currently being carried out in Africa (Johns and Eyzaguirre, 2006). Other species of great market potential, but still uncultivated and harvested in the wild by local populations, are the artichoke-like vegetable *Gundelia tournefortii* L. ('akoub' in Arabic – used for its tender shoots – a real delicacy in Lebanese restaurants), growing in Lebanon and other Middle Eastern countries and the small tree grewia (*Grewia asiatica* L.) found in arid and semi-arid regions of Africa (well appreciated particularly in Sudan) where its sweet berries are used for making jams and juices. Out of these three categories, the latter group of species is the least addressed by R&D and the survival of such resources (along with the maintenance of associated knowledge on how to use the species) is almost entirely in the hands of local communities, whose role of stewardship will never be acknowledged enough. This group comprises also the majority of underutilized medicinal and aromatic plants which are prone to over-harvest and unsustainable management practices in view of increasing requests by the market for natural bioremedies and natural flavours. The unsustainability of such harvests should be of great concern to all, because by threatening the very existence of such local resources, they jeopardize income-generating opportunities of many landless local people who rely on them for their own survival.

Which is the most strategic group for R&D interventions? For the reasons provided earlier, we believe that the scientific community should be focusing particularly on groups 2 and 3 in view of the need to safeguard important resources for local populations.

44.3 Evolving Interest and Milestones

One of the objectives of this chapter is to provide an overview of evolution of the attention paid to underutilized species over the last 30 years. We are not able to be fully comprehensive in this assessment, but we intend to highlight those milestones that underscore the increasing recognition of the values of underutilized species by the international community.

44.3.1 Period: 1970–1980

This period is characterized by a predominant R&D focus on major commodities (staples and industrial crops in particular). The work of the Agricultural Research Institutes (including CGIAR centres) on staple crops is yielding important impacts in terms of hunger and poverty reduction. However, it also leads to a narrowing of the range of species being cultivated and a dramatic loss of intra-varietal diversity of staple crops (Hawkes, 1983; Brush, 1995). Deployment of a broader basket of species to mitigate the impact of crop failures and hence to fight periodical food insecurity is not perceived as an issue yet. A milestone is the document of the US National Academy of Science drawing the attention to underutilized fruit trees (NAS, 1975), as promising plants for improving quality of life in the tropics through providing food, better nutrition and income.

44.3.2 Period: 1981–1990

Attention on underutilized species starts building up. This is also due to an increased recognition of the importance of CWR by the CGIAR, manifested in enhanced germplasm collecting expeditions to gather wild progenitors of cultivated commodities in search of resistance traits. An example is provided by the International Institute of Tropical Agriculture (IITA), starting a thorough programme financed by the Italian Government in 1990 in support of wild cowpea (Laghetti *et al.*, 1990). This project contributes to scientific research on wild *Vigna* species, mostly underutilized (e.g. *Vigna marina* Merr., *Vigna vexillata* (L.) A. Rich., etc.), used by local populations in Africa as sources of forage or leafy vegetables for human consumption (Maxted *et al.*, 2005). In 1987 the International Conference on 'New Crops for Food and Industry' organized by the University of Southampton was held, and as a result of one of its recommendations, the International Center for Underutilized Crops (ICUC) was established. In 2005, ICUC moved to IWMI Headquarters in Colombo, Sri Lanka (http://www.icuc-iwmi.org/). In 1988, the University of Purdue (USA) organized the first of a series of symposia on new crops, mainly looking for alternatives to major crops for US farmers (Janick and Simon, 1990, 1993; Janick, 1996, 1999; Janick and Whipkey, 2002).

As one of the first countries, India recognized the importance of underutilized species as a means to attain sustainable agricultural production, improve the nutritional value of food for large sections of the population and reduce the country's dependence on food imports. In 1982, the All Indian Coordinated Research Project on Underutilized Plants was launched with a list of priority species to be addressed. International collaboration on underutilized species was promoted through newly funded biodiversity projects such as those supported by the Overseas Development Agencies (ODAs), which mobilize funds to IITA (e.g. GTZ – Germany to assist survey, collection and study of bambara groundnut (*Vigna subterranea*) (Begemann, 1988)) or to ICUC (DFID – United Kingdom support to underutilized tropical fruit trees through DFID's Forestry Research Programme (FRP) – Global Programme on fruits for the future; further details can be found on the Internet at: http://www.frp.uk.com/project_dissemination_details.cfm/projectID/5562/projectCode/R7187/disID/1965).

The drawbacks of the Green Revolution (e.g. loss of interspecific and intraspecific biodiversity in farmers' fields as a result of the focus on fewer crops and replacement of landraces by high yielding varieties) start to be acknowledged in literature (Smale, 1997). The concept of sustainable agriculture makes its first appearance in scientific papers (Dimitri and Richman, 2000) so are innovative approaches based on the deployment of greater diversity in farmers fields using – inter alia species so far considered underutilized (case of alley cropping introduced in sub-Saharan Africa, revolving around the use of leguminous crops, such as *Leucaena* species in order to maintain and/or restore fertility in farmers' fields (Kang *et al.*, 1995)).

44.3.3 Period: 1991–2000

In 1992, the Convention of Biological Diversity (CBD) stressed the concept of sustainability (http://www.biodiv.org/convention/articles.asp) rooted in agricultural and cultural diversity supportive of nutritional needs, incomes and greater protection from biotic and abiotic stresses.

The CBD has a tremendous impact in raising awareness of people at the highest level on the value of biodiversity, including underutilized species. The CBD also introduces new values such as the environmental services provided by a vast biodiversity which will have a profound impact in the years to come in influencing countries' strategies in agricultural activities, more conducive than in the past to safeguarding less commercialized crops and species (e.g. underutilized species). In the early 1990s, the AVRDC launched a number of projects focusing specifically on traditional African vegetables. The CGIAR revised its mission statement, no more limited to food security, but broadened so as to include more explicitly poverty reduction and protection of the environment: opportunities to work on species not necessarily used in food production were highlighted. An IDRC-supported study recommended IPGRI's greater involvement in medicinal plants in view of the fact that many of these species are underutilized and neglected by R&D despite their high income-generation potential (Leaman *et al.*, 1999).

The decade is marked by a remarkable increase in ODA's support to underutilized species. Italy, IDRC, ADB (Asian Development Bank), the European Commission, the Netherlands and other donors join Germany and the United Kingdom in financing ad hoc projects and networks (Table 44.3). Nevertheless, funding still remains limited against the background of large R&D gaps. Therefore, international cooperation is advocated by stakeholders as the only way to achieve a visible impact in this domain. Other symposia on new crops are organized by the University of Purdue. Such meetings provide an important platform to the scientific community for sharing experiences and lessons directly related to underutilized species and their development into new crops.

The FAO process of the International Conference and Programme for Plant Genetic Resources (ICPPGR) leading to the IV Technical Conference, which was held in Leipzig in 1996, represented a unique opportunity for scientists to raise the visibility of underutilized species. Preparatory national and regional meetings (e.g. the European meeting held in Nitra, Slovakia) contributed through country-driven bottom-up approaches to the development of the Global Plan of Action (GPA) for Plant Genetic Resources for Food and Agriculture (PGRFA). The GPA, listing 20 activities, covered an array of themes and from conservation to sustainable use (FAO, 1996) and provided unprecedented visibility to underutilized species, dedicating a specific activity to their promotion: activity 12, 'Promoting development and commercialization of underutilized crops and species' (http://www.fao.org/ag/agp/agps/GpaEN/gpaact12.htm). This certainly represented the first and most important milestone in the process of recognition by governments of underutilized species, paving the way to subsequent endorsements by other UN organizations and donor agencies. The 1996 FAO State of World Report (SWR) depicted a worrying situation with regards to conservation of non-commodity crops (including underutilized species), very poorly represented in the surveyed 300 *ex situ* gene banks and collections. According to a study made by IPGRI (Padulosi *et al.*, 2002), less than 22% of the estimated 6 million samples held in gene banks around the world were made of non-commodity crops and of this portfolio (inclusive of underutilized species), species were extremely scarcely represented in terms of intraspecific diversity

Table 44.3. Examples of projects focusing on underutilized species during 1991–2000.

Title	Funding agency	Objective
1. Conservation and use of underutilized Mediterranean species (UMS Project)	Italian Ministry of Foreign Affairs	Promote the utilization of underutilized Mediterranean species (focus on rocket, hulled wheat, oregano and pistachio) through collaborative research networks.
2. Monographs on promoting the conservation and use of underutilized and neglected crops	Deutsche Gesellschaft für Technische Zusammenarbeit (GTZ) GmbH, Germany	Sharing of information and awareness creation
3. Underutilized Mediterranean Fruit Crops Centre International des Hautes Etudes Agronomiques Méditerranéennes (CIHEAM)	CIHEAM/EU	Identification of new alternatives in the agricultural sector in the Mediterranean region
4. CIHEAM–MAICH MEDUSA project	CIHEAM–MAICH/EU	Identification of native and naturalized plants of the Mediterranean region
5. ICUC – fruits for the future	DFID and others	Collate, publish and distribute information to assist in research, development and promotion of prioritized underutilized tropical fruit trees
6. GENRES Project 'Conservation, evaluation, exploitation and collection of minor fruit tree species' EC Project GENRES 29	EU	Enhancement of conservation and utilization of minor fruit tree species
7. BAMFOOD	EU	Increasing the productivity of Bambara groundnut (*Vigna subterranea*) for sustainable food production in semi-arid Africa
8. PROSEA	The Netherlands Government, DFID, EU, FINNIDA, Tropenbos, Yayasan Sarana Wanajaya, GLAXO, IDRC and others	Documentation of information on plant resources of South-east Asia

(more than 80% of these – on average – are made of less than ten accessions). It is anticipated that the next SWR for PGRFA (to be released in 2008) will provide the global community with an update on the situation regarding *ex situ* and *in situ* maintenance of underutilized species along with a review of progress made towards their conservation since 1996.

In 1998, in the framework of Italy's campaign in support of the development of the FAO International Treaty on PGR, a panel of experts gathered in Florence, Italy to discuss the development of an alternative List of Species to Annex I of the International Treaty. Specific discussions are held on the possibility to include underutilized species in the Treaty based on suggestions provided by IPGRI (Padulosi, 2000). It is interesting to note that Annex I of the approved Treaty contains today a total of 80 genera of which only 15 include underutilized species.

An important endorsement of the value of underutilized horticultural crops (particularly fruit trees, vegetables, medicinal and aromatic plants) was also recorded at the International ISHS Congress held in Rome in 1998, during which calls for their better conservation and use were reiterated in several scientific contributions.

In 1999, the IFAD-supported workshop organized in Chennai, India, by the CGIAR PGR Policy Committee, covered specifically underutilized species (Enlarging the Basis of Food Security: Role of Underutilized Species). The meeting yielded large support from attending CGIAR centres and donors. It is interesting to note that this meeting represented the first time ever that the CGIAR discussed underutilized species in a formal way and its deliberations will be important instruments in support of subsequent promotional campaigns for these species.

44.3.4 Period: 2001 to date

The last 5 years witnessed a dramatic increase of support by the international community. The Global Forum on Agricultural Research (GFAR) endorsed the need for assisting NARS in the promotion of these species at its meeting in Dresden (GFAR, 2001 Conference[1]). As an answer to that, the German Government agreed to mobilize financial resources towards the establishment of a Global Facilitation Unit for Underutilized Species (GFU), which will be based at IPGRI Headquarters in Maccarese (Rome), Italy. GFU is a major effort that aims at increasing the contribution of underutilized species to food security and poverty alleviation of the rural and urban poor through facilitating access to information on underutilized species, performing policy analysis and providing advice to policy makers on how to create an enabling policy environment for underutilized species and enhancing public awareness of these species (http://www.underutilized-species.org).

In addition, Germany is also funding a GTZ Mutliregional Project 'People and Biodiversity in Rural Areas' that supports national partners in improving existing value chains of underutilized crops and breeds and analyses the economic potential of other underutilized species and breeds in selected regions. Several publications on the topic have been published and workshops organized, such as the publication by GTZ on 'Promising and underutilized crops and breeds' (Thies, 2000).

The PROTA Network (Plant Resources of Tropical Africa) was founded in order to provide access to information on 7000 tropical African plants most of them little known or underutilized.

[1] http://www.fao.org/documents/show_cdr.asp?url_file=//docrep/004/y0554e/y0554e00.htm

The AVRDC Strategy for the 2001–2010 decade was launched: one objective is to increase diversity of indigenous and underutilized vegetables for better nutrition and health and income (AVRDC, 2002).

The first global UN Project on neglected and underutilized species (NUS) (Enhancing the Contribution of Neglected and Underutilized Species to Food Security and to Incomes of the Rural Poor) was successfully launched in 2001 and ended in 2005. This IFAD-supported effort represented a unique opportunity to test the hypothesis that NUS are strategic crops in support of poverty reduction and empowerment of the poor. Its results have been extremely encouraging particularly in India and Latin America (Padulosi *et al*., 2003) and a follow-up phase is being launched in 2007 (Bioversity International, 2007).

IPGRI published its Strategy on Neglected and Underutilized Species (NUS) in 2002 (IPGRI, 2002) recommending interventions in eight main strategic areas, namely: (i) gathering and sharing information; (ii) priority setting; (iii) promoting production and use; (iv) maintaining diversity; (v) marketing; (vi) strengthening partnerships and capacities; (vii) developing effective policies; and (viii) improving public awareness (IPGRI, 2002 – http://www.ipgri.cgiar.org/nus/strategy.htm). A joint ICUC–IPGRI analysis of the status of underutilized species is also published in 2002 (Williams and Haq, 2000).

A BMZ-funded workshop on underutilized species was organized by the GFU, GTZ and InWEnt in Leipzig, Germany during which the need for mainstreaming underutilized crops in R&D agendas was stressed with the aim of fully exploiting the potential of these species. A follow-up technical consultation on marketing strategies and capacity building for underutilized species was also organized by GFU and IPGRI in Macerata, Italy, in 2004.

An issue of the LEISA magazine entirely dedicated to underutilized species was published in 2004 (LEISA, 2004) receiving a large interest among stakeholders.

At the 7th Meeting of the Conference of Parties to the CBD in 2004 a recommendation of the Subsidiary Body on Scientific, Technical and Technological Advice (SBSTTA) was endorsed. This recommendation suggests activities that contribute to improved food security and human nutrition through enhanced use of crop and livestock diversity, and conservation and sustainable use of underutilized species. SBSTTA underlines that identification of constraints and success factors in marketing underutilized species is a key aspect for their promotion and that capacity building at different levels is highly needed.

In 2005, the International Horticultural Assessment commissioned by USAID was also published. This work (engaging 750 participants, 60 countries and 3 regional workshops and involving as well a major survey) is a further strong endorsement of the value of underutilized crops to revitalize the agricultural sector in crisis. More than a third of promising horticultural species are underutilized; grouping fruits, vegetable crops, herbs, spices and ornamentals, 79 out of 226 belong to the latter category.

Excerpts of the report related to horticultural perspectives in sub-Sahara and Latin America are also very supportive towards underutilized species. For instance in sub-Saharan Africa the report stresses that despite diverse biophysical constraints such as drought and low soil fertility, the region calls for expanded cultivations of its underutilized and indigenous crops adapted to harsh conditions (e.g. leafy vegetables (*Cleome gynandra* L., *Solanum macrocarpon* L., *Moringa oleifera* Lam., *Hibiscus sabdariffa* L.), fruits (*Ziziphus mauritania* Lam.), medicinal plants and vegetables). In Latin America and the Caribbean underutilized fruit trees represent opportunities to generate new markets. It is interesting to note that one specific recommendation of the report calls for more efforts in the area of documenting regional knowledge on cultivation and traditional use of these species.

In June 2005, at the end of a broadly participative process, the CGIAR published its research priorities for the period 2005–2015 (CGIAR, 2005). Underutilized species (featured as Underutilized Plant Genetic Resources/ UPGR) were given high visibility under the System Priority 1b ('Promotion, conservation and characterization of underutilized plant genetic resources to increase the income of the poor') and Priority 3a ('Increasing income from fruit and vegetables' in consideration that many of the latter are considered in fact underutilized species). In addition, underutilized species are also considered indirectly through Priority 3d ('Sustainable income generation from forests and trees') and Priority 4d ('Sustainable agroecological intensification in low- and high-potential environments'). The emergence of niche and high-value markets for underutilized crops in developed countries provides a potential pathway out of poverty for farmers in developing countries, and hence UPGR are relevant also to Priority 5b ('Making international and domestic markets work for the poor').

In April 2005, 100 R&D experts and policy makers with varied backgrounds from 25 countries took part in an International Consultation at the M S. Swaminathan Research Foundation in Chennai, India. This meeting represented a major milestone in support of agricultural biodiversity, including underutilized species. The Consultation, jointly organized by IPGRI, GFU and MSSRF, was called to discuss how biodiversity can help the world to achieve the Millennium Development Goals, and in particular the goal of freedom from hunger and poverty (Bala Ravi *et al.*, 2006). The 'Chennai Platform for Action' resulting from this Consultation, in its ten recommendations, emphasizes the importance of underutilized species and calls upon policy makers to promote specific interventions in support of these species (http://www.underutilizedspecies.org/ documents/PUBLICATIONS/ chennai_declaration_en.pdf).

During this period, increased visibility on underutilized species was provided by dedicated web sites on the Internet from both international and national agencies (see Table 44.4). For the first time, in 2004, a major donor (EU) made a specific call within its 6th Framework for 'Research to increase the sustainable use and productivity of annual and perennial underutilized tropical and subtropical crops and species important for the livelihoods of local populations'. EU recognizes that these crops have potential for wider use and could significantly contribute to food security, agricultural diversification and income generation.

During the course of 2006, ICUC–IPGRI–GFU carried out an electronic consultation to design a strategic framework for R&D on underutilized species. This was followed by two regional strategy workshops held in Colombo, Sri Lanka (16–17 March) and Nairobi, Kenya (24–25 May). An international workshop on *Moringa* species and other leafy vegetables with high nutritional value has also successfully taken place in Accra, Ghana in November 2006.

Table 44.4. Web sites dedicated to underutilized species.

Name	URL
Purdue New Crop Resource Online Program	http://www.hort.purdue.edu/newcrop/default.html
Purdue Famine Foods	http://www.hort.purdue.edu/newcrop/famine foods/ff_home.html
Orphan Commodities on ECOPORT	http://ecoport.org/perl/ecoport15.pl?C13 =Y& searchType=searchKeyWordProfile&kwpId= PL****&subjectType=E&maxCheckbox=144
IPGRI Neglected and Underutilized Crop Species (NUS)	http://www.ipgri.cgiar.org/nus/
Andean roots and tubers at the International Potato Centre (CIP)	http://www.cipotato.org/artc/artc.htm
Cultivos Andinos	http://www.rlc.fao.org/prior/segalim/prodalim/ prodveg/cdrom/indice_gral.htm
AVRDC indigenous vegetables	http://www.avrdc.org.tw/closerlook/indig enous_veg.html
Australian New Crops Web Site	http://www.newcrops.uq.edu.au/
Famine Food Field Guide	http://www.africa.upenn.edu/faminefood/index. htm
Jefferson Institute	http://www.jeffersoninstitute.org/
SAFIRE – The Southern Alliance for Indigenous Resources	http://www.safireweb.org/
CIKS (Centre for Indian Knowledge Systems)	http://www.ciks.org/
Green Foundation	http://www.greenconserve.com/
Indigenous Knowledge for Development Programme –The World Bank	http://www.worldbank.org/afr/ik/index.htm
Winners & Losers – Investigating the human and ecological impacts of the commercialisation of non-timber forest products (NTFPs)	http://www.ceh-wallingford.ac.uk/research/ winners/index.html
ECO-SEA – The Ethnobotanical Conservation Organization for South-East Asia	http://www.ecosea.org/index.html
PhytoTrade Africa	http://www.phytotradeafrica.com/
Plant Resources of Tropical Africa (PROTA)	http://www.prota.org/
Plant Resources of South-east Asia (PROSEA)	http://www.proseanet.org/index.htm
Underutilized Tropical Fruits of Asia Network (UTFANET)	http://www.civil.soton.ac.uk/icuc/utfanet/
Useful Plants of the Mediterranean Region Network (MEDUSA)	http://medusa.maich.gr/
The Tree Against Hunger, Enset-Based Agricultural Systems in Ethiopia	http://www.aaas.org/international/africa/enset/ index.shtml
Bambara Groundnut Network (BAMNET)	http://www.genres.de/bambara/
International Centre of Research and Training on Sea buckthorn (ICRTS)	http://www.icrts.org/
Moringanews	http://www.moringanews.org/
Taro Network for South-east Asia and Oceania (TANSAO)	http://www.nari.org.pg/research/wlmp/tansao. htm
International Network for Bamboo and Rattan (INBAR)	http://www.inbar.int/
Fonio	http://fonio.cirad.fr/en/index.html

An international symposium on Contribution of African Botanica to Humanity coorganized by ISHS, UDECOM and Lyceum INAABD took place between 3 and 7 October 2006, in Guinea. Participants from just over 40 countries gathered to focus on two major themes: food and nutrition; and medicinal and other uses of plants. The symposium also covered areas such as the clinical uses and the protection of unique plant species that are becoming more and more rare. Other aspects related to the cultural dimension of use of plants for the well-being of people were also discussed. For further information visit the web site: www.botaniqueafricaine.com. Another important event was the AVRDC International Conference on Indigenous Vegetables, held in Hyderabad, India in December 2006. ISHS has approved and will establish a working group on underutilized species within the Commission on Plant Genetic Resources. GFU and ICUC have been invited to chair it.

44.4 Characterizing Stakeholders Working Today on Underutilized Species

Currently major actors at the international level engaged in the promotion of underutilized species include IPGRI, GFU, ICUC, CIAT, CIP and AVRDC. At the national level, the following agencies dedicate substantial resources in support of these species: the M.S. Swaminathan Research Foundation (MSSRF), India, Centro de Investigación de Recursos Naturales y Medio Ambiente (CIRNMA), Peru; Promoción e Investigación de Productos Andinos – (PROINPA) – Bolivia; the Green Foundation, India.

A survey carried out by GFU between 2003 and 2005 shows that currently about 170 organizations and institutions globally are working on themes related in one way or another to underutilized species. The experts involved in the work are mainly located in Europe (36%) followed by sub-Saharan Africa (22%), the Americas and Asia/Pacific (18% each) and Central/West Asia (7%). Their background of expertise is predominantly genetic resources conservation (40%). Only a limited number has expertise in marketing (11%), socio-economics (11%) or policy and legal issues (8%). Therefore, the projects that such organizations and experts carry out deal mainly with applied research on conservation and characterization followed by information/documentation and public awareness. Projects aiming at improved nutrition, food security, income generation or conservation of cultural traditions are of lower priority in these efforts. This situation is also explained by the type of institutions that work on underutilized species. National and international research centres, universities and a relatively small number of NGOs are the main actors. Development and farmers' organizations, as well as the business and industrial sector, are only marginally involved. Greater participation of non-academia, private sector and farmers' associations is needed to consistently exploit the income potential of underutilized species.

44.5 Lessons Learned and Perspectives

44.5.1 Lessons from ethnobotanic surveys

In addition to genetic erosion faced by some species (particularly medicinal and aromatic plants) due to over-regulated or unregulated harvests (e.g. *Orchis* L., *Ophrys* L., *Serapias* L. genera collected in the wild in Turkey to prepare the famous drink 'salep'), there is – in our opinion – another 'less visible' drama unfolding that is putting in jeopardy the future of underutilized species, or rather their successful and meaningful use in our lives. This is the loss of indigenous knowledge (IK) which has been safeguarded by generations of users and is today under threat of being lost within a few decades. Surveys have revealed that IK is no longer transmitted from one generation to the next because of social changes and less attention given by society members and their leaders to its proper safeguard. For instance, in Lebanon, Bioversity International-supported research studies have documented that the wealth of knowledge is maintained only by people over 60 years old who are very marginally transmitting it to new generations, because of the scarce interest of these in acquiring it or because of splitting of large family clans in which elders are no longer living close to the younger ones. Another case is that of '*Pistic*' a traditional soup made out of 56 species (all of them underutilized) gathered in the wild in Italy's Friuli Venezia Giulia region. Only very few elderly people still know where to find these plants and how to prepare the dish. The question that comes to our mind is: how many *Pistic* we will be soon loosing in our regions, countries and continents around the world? Interventions to promote and safeguard traditions associated with underutilized species should have an imperative emergency for all of us.

44.5.2 Conservation outlook

Considering the dwindling budget for conservation of PGR, we believe that it is unlikely that adequate support will be mobilized to consistently conserve underutilized species through *ex situ* conservation measures. This position is supported by the following considerations:

- Sheer number of underutilized species;
- Limited knowledge about their conservation requirements;
- Limited capacities of gene banks (physical and human);
- *Ex situ* conservation does not ensure proper safeguarding of IK associated to underutilized species;
- Difficulties in replicating required environmental conditions for *ex situ* maintenance and multiplication of germplasm;
- Increasing concerns by countries and local communities over IPR issues;
- Greater demands from communities to share the responsibilities for PGR conservation;
- Increased attention of donor agencies towards *in situ*/on-farm conservation.

In the light of these observations, it is reasonable to believe that *in situ*/on-farm conservation will be the most sustainable way to maintaining diversity of

underutilized species. This fact would also underscore the guardian role of communities over such resources and ensure continued access to these resources in support of their livelihood options.

44.5.3 Lessons from community-based work

The following example is drawn from Bioversity's global project on neglected and underutilized species which demonstrated the livelihood benefits derived from highly participatory approaches to promote use enhancement of target species in India and Nepal.

In India and Nepal, productivity improvement for minor millets (*Eleusine coracana*, *Setaria italica*, *Panicum sumatrense*) was successfully obtained thanks to a number of measures that actively involved farmers and other community members and included variety selection, capacity building in quality seed production, seed bank establishment, improvement of agronomic practices, etc. These approaches were socially engineered by establishing self-help groups of farm women and men, building entrepreneurship in production, utilization and marketing. Judicious millet variety choice, need-based intercrops and their planting ratios, optimum seed rate, planting methods and production management constituted components of agronomic refinement. Many farmer participatory demonstrations using improved production practices were conducted along with organization of field days and field walking workshops. As a result of these interventions, yields increased by 18–64% in finger millet, 31–63% in foxtail millet and 19–43% in little millet. Furthermore, some traditional varieties although low in yield possessed remarkable adaptive and quality superiority, which farmers in some regions preferred over varieties with increased productivity.

The project introduced simple grain mills for small householders, organized in Women Self Help Groups (SHGs), reducing considerably the drudgery of processing the harvest. This approach benefited particularly women as main food processors. It directly enhanced consumption at household level and opened opportunities for developing diversified value-added products fetching a far higher market price than the primary produce, the grain. The Home Science Departments of the Universities of Agricultural Sciences of Bangalore and Darward offered strategic support in product development and training farm women and other stakeholders. SHGs empowered with training in value addition and product development and support from microcredit were able to establish nutritious millet-based entrepreneurship. The economic profit margins increased by about 44% by SHGs from value-addition activities.

Overall, 67 trainings/workshops/exhibitions were organized by the Indian and Nepalese project partners during the 3-year period of the project. Among farmers adopting improved seed and agronomic practices, yield enhancement generated an income in the range of Rs 750–2250 (US$18–52)/ha. Grains of these crops normally fetch a farm gate price of Rs 5000 and Rs 6000 (US$116 and US$140)/t of finger and little/foxtail millet, respectively. On value addition of finger millet to flour, semolina and malt, the additional incomes generated by the SHGs were Rs 1100 (US$25), Rs 4300 (US$100), Rs 18,550 (US$430)/t, respectively. Similarly, milling and polishing, little and foxtail millet to rice has the potential to fetch an income of Rs 12,980 (US$300)/t. These different processing activities also generated additional employment of 15–45 man days/t of grain.

44.6 Market Prospects

The growing demand in both developed and developing countries for more diversity in food and other products, more healthy, nutritious and at the same time sustainably produced food offers opportunities for farmers and other actors of the value chains in developing countries to participate in newly emerging markets. The expansion of supermarkets in many countries of the South provides modern sales outlets targeting a specific consumer segment. There is good awareness about the requirements that these markets pose on producers and processors with regard to quality standards and reliable supply. Best opportunities for income generation are often seen in export markets, which are much more demanding in terms of quality and safety. One of the potentially biggest markets, the EU, however, is extremely difficult to access with food products which have not been marketed within its territory to a significant degree before May 1997 in view of existing policies. Most products developed from underutilized species do not fulfil these criteria. Anyone who wants to market them is challenged by the Novel Food Regulation (EC) 258/97 that subjects all novel foods to a stringent scientific safety assessment. The Regulation fails to distinguish between genuinely novel foods which may have unknown properties and foods that are exotic within the EU, but have a long history of consumption in their countries of origin. The provision of the required scientific data is far beyond the means of the small producers and traders in developing countries and the predominantly small importers in Europe. The Regulation clearly represents a barrier to trade in these products because of the costs involved and the lengthy process to obtain approval. Despite the fact that trade is recognized and considered for development objectives, the EU Novel Food Regulation is both stifling economic progress and conflicting with the development agenda of the EU member countries and working against other global agreements such as the UN Millennium Development Goals. The European Commission has recognized inadequacies of the Regulation in its treatment of exotic foods and is currently preparing to revise it. Organizations such as the UNCTAD (United Nations Conference on Trade and Development BioTrade Facilitation Programme), CBI (Centre for the Promotion of imports from developing countries based in the Netherlands), IPGRI, GFU and GTZ in a joint effort are working with the European Commission to correct the before-mentioned shortcomings (http://www.underutilized-species.org/documents/Publications/cbi_unctad_paper_on_eu_nfr.pdf).

44.7 Lessons from Networking

Networking remains crucial in order to achieve successfully the sustainable conservation and use of underutilized species. It is seen as the most effective mechanism to address the complexity of issues related to the promotion of these species and to integrate the diverse range of partners that are involved in this process. Collaboration should be pursued among all actors of the production-to-use chain as well as among international and national bodies which play a strategic role in creating the 'enabling external environment' for the successful performance of these chains. For instance, interministerial collaboration (e.g. among the Ministries of Agriculture, Commerce and Education) is conducive towards favourable policies to support upscaling and wider adoption of good practices. The participation of the private sector cannot be overemphasized in view of their leading role in translating small scale initiatives into viable business opportunities.

44.8 Concluding Remarks

Poor communities around the world depend on the vast array of underutilized species and their wild relatives for their livelihood. Such species are part of a strategic, culturally important, asset in support of nutrition, health and income generation.

Among the reasons for today's increased attention on underutilized species are the following:

- Alternative source of income;
- Collapse of commodity prices;
- Greater appreciation of biodiversity in enhancing livelihood;
- Participation of communities in setting research agendas;
- Stronger NARS willing and now able to invest beyond commodity crops;
- Search for cultural identities in an increased globalized and 'mobile' world;
- Increased multi-ethnicity in cities and demands for traditional food by immigrant communities;
- Better understanding of the limits of the 'Green Revolution' and the importance of a diversified food production and diet;
- Recognition of the role of underutilized species in empowering marginalized members of society, particularly women.

With regard to priorities that should be guiding the work on underutilized species, the following areas have been pointed out in numerous meetings and fora:

- Economics (especially identification of novel markets and innovative marketing strategies);
- Nutrition (especially validation of nutritional potentials of species and nutrient-saving processing technology);
- Empowerment of communities (including enhancing negotiating capacities);
- Policies (especially with regard to the inclusion of target species in national and international R&D agendas);
- Conservation (targeting both biodiversity and associated knowledge);
- Provision of quality seed (development of seed systems).

The challenges that we will however continue to be confronted with are still numerous. They include the need to change the often negative image associated with underutilized species, building long-lasting partnerships between the public and the private sector, convincing the private sector to invest in areas that albeit promising are still relatively unexplored, maintaining intraspecific diversity in markets while promoting produce with specific traits, bioprospecting and benefit sharing and intersectorial cooperation among various ministries and agencies.

With reference particularly to underutilized wild species we should consider that their use enhancement may not necessarily imply their domestication. In fact, it is a common observation that species brought to cultivation often tend to loose some of their valuable organoleptic qualities. This is the case for instance of sage (*Salvia officinalis* L.) growing wild in Lebanon which has been observed to reduce its marked fragrance in cultivation trials (Noun, 2003, unpublished data). Furthermore, an increasing number of consumers prefer products that originate in the wild to those cultivated in view of their naturalness and the fact that they have not undergone agronomic practices at all (which further underlines the message of naturalness used in their own marketing strategy). In connection to this aspect, adequate attention should be directed towards the development of sustainable management practices for species that will continue to be harvested in the wild, in order to ensure on the one hand the safeguarding of genetic diversity and on the other the fair share of benefit for local people who will be encouraged to continue to play their role of custodian of such resources.

The GFU and ICUC are providing neutral platforms for all stakeholders to address these challenges. The conferences and consultations organized by the two entities have contributed to greater awareness of policy makers, donors and the public about the importance of underutilized species.

GFU and ICUC have been facilitating the development of a strategic framework for research and development of underutilized species in Asia and Africa (Jaenicke and Hoeschle-Zeledon, 2006). If widely adopted by stakeholders, this framework is expected to yield increased collaboration of national and international players in the public and private sector making use of their unique strengths and comparative advantages. It will also allow addressing priority issues of individual stakeholders without losing sight of the existing gaps and eventually lead to greater impact of undertaken research efforts aiming at promoting underutilized species to enhance people's livelihoods.

References

AVRDC (2002) *Empowering Small-Scale Farmers for Knowledge-Based Agriculture: AVRDC Strategy 2010.* Asian Vegetable Research and Development Center. Shanhua, Taiwan. Available at: http://www.avrdc.org.tw/pdf/strategy2010.pdf

Bala Ravi, S., Hoeschle-Zeledon, I., Swaminathan, M.S. and Frison, E. (eds) (2006) Hunger and poverty: the role of biodiversity. Report of an International Consultation on the Role of Biodiversity in Achieving the UN Millennium Development Goal of Freedom from Hunger and Poverty, Chennai, India, 18–19 April 2005. M.S. Swaminathan Research Foundation, International Plant Genetic Resources Institute, Global facilitation Unit for Underutilized species. Bioversity International, Maccarese, Rome, Italy.

Begemann, F. (1988) *Ecogeographic Differentiation of Bambara Groundnut (*Vigna subterranea*) in the Collection of the International Institute of Tropical Agriculture (IITA).* Wissenschaftlicher Fachverlag Dr Fleck, Niederkleen, Germany.

Bioversity International (2007) Neglected no more. Achievements of the IFAD-NUS Project (2001–2005) and framework for its follow up initiative (2007–2009). Bioversity International, Maccarese, Rome, Italy.

Brush, S.B. (1995) *In situ* conservation of landraces in centers of crop diversity. *Crop Science* 35, 346–354.

CGIAR (2005) System Priorities for CGIAR Research 2005–2015. Consultative Group on International Agricultural Research. Science Council Secretariat. December 2005. Available at: http://www.sciencecouncil.cgiar.org/activities/spps/pubs/Priorities%20Dec%2005.pdf

Dimitri, C. and Richman, N. (2000) *Food, Agriculture, Conservation, and Trade Act of 1990 (FACTA), Public Law 101–624, Title XVI, Subtitle A, Section 1603*. DCL Government Printing Office, 1990, Washington, DC.

Eyzaguirre, P., Padulosi, S. and Hodgkin, T. (1999) IPGRI's strategy for neglected and underutilized species and the human dimension of agrobiodiversity. In Padulosi, S. (ed.) *Priority Setting for Underutilized and Neglected Plant Species of the Mediterranean Region*. International Plant Genetic Resources Institute, Rome, Italy.

FAO (1996) *Global Plan of Action for the Conservation and Sustainable Utilization of Plant Genetic Resources for Food and Agriculture and the Leipzig Declaration*. Adopted by the International Technical Conference on Plant Genetic Resources, Leipzig, Germany, 17–23 June 1996. Food and Agriculture Organization of the United Nations, Rome, Italy.

Fruits for the future. Available at: http://www.odi.org.uk/tropics/projects/3293.htm

GFAR (2001) Proceedings of the first GFAR Conference, Dresden, Germany – Short report of the proceedings of '*GFAR-2000 Conference: Highlights and Follow-up Action*'. Available at: http://www.fao.org/documents/show_cdr.asp?url_file=//docrep/004/y0554ey0554 e00.htm

Grote, K. (2003) *The Increased Harvest and Trade of Devil's Claw (*Harpagophytum procumbens*) and Its Impacts on the Peoples and Environment of Namibia, Botswana and South Africa – Report for the GFU*. Available at: http://www.underutilized-species.org/documents/devils_claw.pdf

Hawkes, J.G. (1983) *The Diversity of Crop Plants*. Harvard University Press, Cambridge, Massachusetts.

IPGRI (2002) *Neglected and Underutilized Plant Species: Strategic Action Plan of the International Plant Genetic Resources Institute*. International Plant Genetic Resources Institute, Rome, Italy. Available at: http://www.bioversityinternational.org/publications/pubsurvey.asp?id_publication=837

Jaenicke, H. and Hoeschle-Zeledon, I. (eds) (2006) *Strategic Framework for Underutilized Species Research and Development, with Special Reference to Asia and the Pacific, and Sub-Saharan Africa*. International Centre for Underutilized Crops, Colombo, Sri Lanka and Global Facilitation Unit for Underutilized Species, Rome, 33 pp.

Janick, J. (ed.) (1996) *Progress in New Crops*. ASHS Press, Alexandria, Virginia.

Janick, J. (ed.) (1999) *Perspectives on New Crops and New Uses*. ASHS Press, Alexandria, Virginia.

Janick, J. and Simon, J.E. (eds) (1990) *Advances in New Crops*. Timber Press, Portland, Oregon.

Janick, J. and Simon, J.E. (eds) (1993) *New Crops*. Wiley, New York.

Janick, J. and Whipkey, A. (eds) (2002) *Trends in New Crops and New Uses*. ASHS Press, Alexandria, Virginia.

Johns, T. and Eyzaguirre, P. (2006) Linking biodiversity, diet and health in policy and practice. *Proceedings of the Nutrition Society* 65(2), 182–189.

Kang, B.T., Osiname, A.O., Larbi, A. (eds) (1995) *Alley farming research and development. Proceedings of an International Conference on Alley Farming*. IITA/ICRAF/ILRI, Ibadan, Nigeria.

Laghetti, G., Padulosi, S., Hammer, K., Cifarelli, S. and Perrino, P. (1990) Cowpea (*Vigna unguiculata* (L.) Walp.) germplasm collection in southern Italy and preliminary evaluations. In: Ng, N.Q. and Monti, L.M. (eds) *Cowpea Genetic Resources: Contributions in Cowpea Exploration, Evaluation and Research from Italy*. International Institute of Tropical Agriculture, Ibadan, Nigeria. pp. 45–67.

Leaman, D.J., Fassil, H. and Thorman, I. (1999) *Conserving Medicinal and Aromatic Plant Species: Identifying the Contribution of the International Plant Genetic Resources Institute (IPGRI)*. Unpublished report. International Plant Genetic Resources Institute, Rome.

LEISA (2004) Magazine on Low External Input and Sustainable Agriculture. Available at: http://www.leisa.info/index.php?url=magazine-details.tpl&p[readOnly]=0&p[_id]=64433

Maxted, N., Mabuza-Diamini, P., Moss, H., Padulosi, S., Jarvis, A. and Guarino, L. (2005) *An Ecogeographic Study African* Vigna. Systematic and Ecogeographic Studies on Crop Genepools 11. International Plant Genetic Resources Institute, Italy.

National Academy of Science (NAS) (1975) *Underexploited tropical plants with promising economic value*. NAS, Washington, DC.

Padulosi, S. (ed.) (1999) *Priority Setting for underutilized and neglected plant species of the Mediterranean region*. Report of the IPGRI Conference, 9–11 February 1998, ICARDA, Aleppo. Syria. International Plant Genetic Resources Institute, Rome, Italy.

Padulosi, S. (2000) A comprehensive vs. limited list of crops: the role of underutilized crops and opportunities for international centres, donor communities and recipient countries. In: Broggio, M. (ed.) *Exploring Options for the List Approach – Proceeding International Workshop: Inter-dependence and Food Security which List of PGRFA for the Future Multilateral System?* Instituto Agronomico per l'Oltremare, 1–2 October 1998, Firenze, Italy.

Padulosi, S. and Hoeschle-Zeledon, I. (2004) Underutilized species: what are they? *LEISA Magazine* 20(1), 5–6. Available at: http://www.leisa.info/index.php?url=magazine-details.tpl&p[readOnly]=0&p[_id]=64433

Padulosi, S., Hodgkin, T., Williams, J.T. and Haq, N. (2002) Underutilized crops: trends, challenges and opportunities in the 21st century. In: Engels, J.M.M., Ramanatha Rao, V., Brown, A.H.D. and Jackson, M.T. (eds) *Managing Plant Genetic Diversity*. CAB International, Wallingford, UK, pp. 323–338.

Padulosi, S., Noun, J., Giuliani, A., Shuman, F., Rojas, W. and Ravi, B. (2003) Realizing the benefits in neglected and underutilized plant species through technology transfer and Human Resources Development. In: Schei, P.J. Sandlund, O.T. and Strand, R. (eds) *Proceedings of the Norway/UN Conference on Technology Transfer and Capacity Building*, 23–27 June, 2003, Trondheim, Norway.

Smale, M. (1997) The green revolution and wheat genetic diversity: some unfounded assumptions. *World Development* 25(8), 1257–1269.

Thies, E. (2000) *Promising and underutilized crops and breeds*. Deutsche Gesellschaft für Technische Zusammenarbeit (GTZ) GmbH, Eschborn Germany. Available at: http://www2.gtz.de/agrobiodiv/download/thies3.pdf

Williams, J.T. and Haq, N. (2000) *Global research on underutilized crops; an assessment of current activities and proposals for enhanced cooperation*. International Centre for Underutilized Crops, Southampton, UK. Available at: www.bioversityinternational.org/Publications/pubfile.asp?ID_PUB=792.

45 Conservation and Use of Wild-harvested Medicinal Plants in Sri Lanka

R.S.S. Ratnayake and C.S. Kariyawasam

45.1 Present Status of Medicinal Plants in Sri Lanka

Medicinal plants are a global asset to be safeguarded for the benefit of people all over the world (World Bank, 2004). Sri Lanka is home to a wealth of medicinal plants which have been used by Sri Lankans for millennia as the primary source of health care. According to recent surveys, there are 1432 medicinal plants in Sri Lanka of which 148 are endemic and 110 are threatened (BMARI, 2002). It is noteworthy that out of these endemic medicinal plants 61 species are threatened. Of all the medicinal plants in the island about 50% occur in natural forests (MENR, 2002) and some of them are found in particularly unique habitats. According to these investigations, taxonomic families such as the *Fabaceae*, *Euphorbiaceae*, *Poaceae*, *Rubiaceae* and *Asteraceae* contain the highest number of medicinal plant species.

Medicinal plants provide primary health care needs of about 80% of the people in the country even today (NASTEC, 2002). Approximately 269 species of medicinal plants are commonly used, of which 87 species are heavily used (>10,000 kg/year) by Ayurvedic[1] drug manufacturers in the country. In 2005, the most heavily used medicinal plant species in the country were *Justisia adhatoda*, *Benincasa hispida*, *Sida alnifolia*, *Terminalia belerica* and *Solanum melongena*. The total imported herbal material for Ayurvedic preparations in the year 2000 was 1.5 million kg valued at US$1.8 million and the five most frequently imported medicinal plant species were *Solanum virginianum*, *Mollugo cerviana*, *Zingiber officinale*, *Anethem graveolens* and *Cedrus dedora* (IUCN, 2001).

Medicinal plants conservation in Sri Lanka goes back in the history of the period of the kings in Anuradhapura era (6th century BC), the first royal domain. King Buddhadasa of the Anuradhapura kingdom paid special attention to

[1] Ayurveda is the medicinal plants-based herbal treatment system in the country.

planting and using medicinal herbs for healing. Ruins speak of a medicinal trough (medi-bath) used at that time. The development of systematic pharmacopoeias in Sri Lanka dates back as far as 3000 BC (Pilapitiya, 1999). Use of medicinal plants in Sri Lanka is closely associated with the cultural diversity and the traditional knowledge system in the country and substantially contributes to rural economies and health security (IUCN 2004). Rural communities, particularly elderly people, have fairly good knowledge on medicinal properties of plants and the various ailments for which they are used.

Traditional varieties and wild relatives of cultivated crops and a number of fruit species found in Sri Lanka also have very distinct medicinal properties. Most of them are wild, semi-wild or undomesticated, existing naturally in home gardens, jungles and reservations awaiting exploitation (DOA, 2002). So far, diversity, potential and opportunities to make use of these species and measures to conserve against erosion have not been implemented systematically.

Medicinal plants in Sri Lanka are not systematically cultivated for commercial purposes, except ginger and turmeric which are used more as condiments than medicine (Pilapitiya, 1999). Usually the local requirement of medicinal plants is extracted from the wild or gathered from home gardens. Nevertheless, *ex situ* conservation is carried out at several locations at a limited level. Several medicinal plant gardens have been established in different climatic zones of the country, all of which contain a considerable number of medicinal plant species. Usually practitioners as well as people maintain their own collections in home gardens for their own requirement. The Plant Genetic Resources Centre which is mandated to conserve agriculturally important germplasm of crops has paid very meagre attention to gene pool conservation of medicinal plants.

45.2 Threats to Medicinal Plants in Sri Lanka

Since large proportions of pharmaceutical drugs are derived from medicinal plants, the demand for these raw materials is gradually increasing. Therefore, commercial users unsustainably exploit the natural populations of medicinal plants to fulfil this high demand. The Sri Lanka Ayurveda Drugs Cooperation alone needs 57 native medicines every year (Perera, 2004). There were 104 Ayurveda drugs manufacturers registered with the Department of Ayurveda in Sri Lanka by the year 2000 (IUCN 2001). The national demand for herbal materials in the year 2000 was 3.86 million kg (valued US$3.87 million) and 68% of this was met from local supply while the remaining 32% was through imports (IUCN, 2001). At present, Sri Lanka imports herbal materials mainly from India. Therefore, it is imperative to promote the systematic commercial cultivation of medicinal plants in the country. Most of the wild plants in the country are freely accessible, which triggers the unsustainable harvesting and depletion of plant biodiversity in natural ecosystems that endangered many medicinal plant species. Destruction of important habitats of medicinal plants for unplanned development and agricultural activities is a serious problem that affects the conservation of medicinal plants in the country. It is well evident that the forest cover of the country has been reduced drastically from 65% in 1900 to 22% in the year 2000 (Bandarathilleke, 2001). Since forests are the storehouses of medicinal plants the rate of loss is evident. In agriculture, some of the important medicinal plants, such as *Cyperus rotundus,* are considered to be serious weeds and removed. This is a major threat for the survival of these species. But fortunately, most of the medicinal plants that grow as weeds cannot be easily eradicated and sometimes grow better when there are land disturbances (Rao and Arora, 2004). Information on medicinal plants is not easily available, and is scattered, fragmentary and a bit contradictory (Wijesundara, 2004). This hampers the implementation of effective conservation strategies, plans and programmes. Lack of a proper regulatory system to protect and safeguard medicinal plants and associated knowledge, low economic value and lack of market, lack of technology and technical skills to develop value-added products are also serious problems affecting conservation and use of medicinal plants in the country.

45.3 Medicinal Plant-based Research in Sri Lanka

Several institutions are conducting research studies on conservation and use of medicinal plants. Nevertheless, the development of research on herbal medicines in Sri Lanka is not satisfactory. An integrated approach is needed involving all concerned parties at all levels for the development of herbal products industry based on medicinal plants instead of conducting independent research studies (NASTEC, 2002). Considerable numbers of studies have been conducted on plant identification, propagation and conservation of selected species of medicinal plants. However, studies conducted at genetic level are quite inadequate. Several attempts have been made to identify and prioritize the research needs for conservation and use of medicinal plants in the country.

45.4 Herbal Medical System in Sri Lanka

The native medical system prevailing in Sri Lanka was '*Desiya Chikitsya*' which is a corpus of community-based knowledge. With time, '*Desiya Chikitsya*' became largely integrated with the Ayurveda system. As a common practice in this traditional medicine physicians prepare their own medicine, such as decoctions, herbal oils, herbal tea, herbal wines, herbal powders and herbal pills. Other than herbal medicine, herbs are used in Sri Lanka for cultural practices, customs, traditions, religious and spiritual beliefs, rituals, ceremonies, etc. Use of herbal medicines over centuries has proven that there are no adverse after-effects, side effects or danger of drug poisoning. Furthermore, herbal medicine is more popular in rural areas in Sri Lanka due to relatively low cost and availability of raw materials.

There are over 11,766 traditional physicians and 7740 Ayurveda physicians scattered all over the country (IUCN, 2001). Traditional healers have a rich stock of indigenous knowledge on medicinal plants and associated herbal treatment in their own secret prescriptions. Generally the knowledge associated with this traditional medical system transfers from generation to generation, usually from father to son. But little effort has been made to document this knowledge for future generations and therefore it is at a high risk of extinction with the demise of the holders. In fact, it is obvious that some of this knowledge has already been lost. The younger generation shows a lack of interest and acceptance of this knowledge is yet another problem. Historically traditional practitioners visit and treat patients voluntarily. Such practitioners are highly respected. This category of practitioners is now on the decline and is seen only in remote rural communities.

Although we have an old tradition of herbal medicine, it has not expanded in terms of co-modification and commercialization to compete with global trends. Herbal medicines in Sri Lanka do not have any regulatory status and they are sold based on medicinal properties (WHO, 2005). In this arena safety and efficacy of herbal medicines is sometimes in question. Except for a few attempts, the Ayurvedic products have remained non-standardized medicines (NASTEC, 2002). In traditional medicine, there is a built-in system of standards. This is very often not adhered to by the practitioners, due to various reasons such as misidentification and non-availability of specific material in required quantities.

Until a few years ago traditional medicine had not received its due recognition. But today there is a state ministry dealing with the subject of Indigenous Medicine which is mandated to implement policies, plans and programmes for the development of an indigenous medical system in the country. This Ministry basically deals with the promotion of Ayurvedic medical system and very often does not cover other traditional medical practices and knowledge systems. Currently, the subject of indigenous medicine is decentralized, similar to the western medical system, and has provincial-level ministries and administrative offices. But the most popular and widely spread medical system in the country, the Western medical system, comes under the Ministry of Health. It has however, little or no link with the traditional medical system. All Ayurvedic physicians trained in government or private medical colleges are obliged to register with the Ayurvedic Medical Council before they begin to practice. However, traditional healers do not come under this licensing system. The criteria adopted by the Department of Ayurveda for registration of traditional practitioners should be revised.

45.5 Present Institutional Legal and Policy Framework for Medicinal Plants

The Government of Sri Lanka has already prepared several policy documents which have some emphasis on conservation and sustainable utilization of medicinal plants and associated knowledge, the current focus of attention on intellectual property and benefit sharing is not broad enough to overcome the threats posed to that (IUCN, 2001). There is no comprehensive law to conserve traditional knowledge associated with medicinal plants in Sri Lanka (J. Gunawardena, Colombo, 2005, personal communication). But the Forest

Ordinance (1907) and the Fauna and Flora Protection Ordinance (1937) have provisions to protect medicinal plant species. The Ayurveda Act (1961) has very little significance in protecting them. The Intellectual Property Rights Act (2003) covers innovations and does not cover plant varieties, animal varieties and traditional knowledge.

Convention on Biological Diversity (CBD) provides the basis for the conservation and sustainable use of medicinal plants. Sri Lanka has signed and ratified the CBD and the country is obliged to implement the provisions of the convention. Even though 13 years have elapsed from the signature of the CBD, at the moment there is no law to cover the aspects of Access to Genetic Resources and benefit sharing holistically. Due to lack of legal, economical and other relevant instruments, in the country the bioprospecting process is immovable. In addition to the above inappropriateness of legal and other instruments related to bioprospecting, it adversely affects the connected technology transfer and capacity building. This situation has created underutilization of biological resources and increased biological piracy. About 100 of our important medicinal plants and their varieties such as *Salacia reticulate*, *Aloe vera*, *Momordica charantia*, *Gymnema sylvestre* and *Coscinium fenestratum* have already been taken out of the country and products have been patented by several multilateral companies. It is high time for the country to develop proper legal and other instruments to prevent biological piracy and facilitate access to genetic resources and benefit sharing in a sustainable manner. Sri Lanka needs a national policy on access to genetic resources and benefit sharing when considering the pharmaceutical prospecting of medicinal plants. In the Sri Lankan context, formulating national policy and strategic framework for conservation and sustainable use of medicinal plants is very important.

Preparation of species conservation profiles and recovery plans for selected threatened species and preparation of a national strategy for species conservation are important recent initiatives taken by the government to safeguard the available populations of threatened plant species that includes a considerable number of medicinal plants.

45.6 Conservation and Sustainable Use of Medicinal Plants Project (1998–2002)

A 5-year project on conservation and sustainable use of medicinal plants, which was initiated in 1998, is considered as the first successful participatory project carried out in Sri Lanka. The objectives of the project were to conserve globally and nationally significant medicinal plants, their important habitats and genetic stock while promoting their sustainable use. The project attempted to achieve these objectives by *in situ* conservation, *ex situ* cultivation and through the provision of information and institutional support together with a proper policy and legal framework. Major activities conducted by this project were:

1. *In situ* conservation – five medicinal plant conservation areas were established in different climatic zones of the country and these sites were used as a

centre for different project activities (World Bank, 2004). Enrichment planting, stream bank conservation planting and establishment of fire lines were successfully carried out. Studies were conducted to find out the sustainable levels of harvesting for five medicinal plant species.

2. *Ex situ* conservation – improvement of existing nurseries and establishment of new nurseries. Two new nurseries were also set up to serve the needs of other biogeographic zones, each one of the new nurseries maintains a collection of over 500 species (Silva and Wettasinghe, 2004). The research on propagation met with a creditable degree of success. Protocols for raising nursery plants were developed for 22 species and techno-guides were prepared to provide easy-to-follow steps on propagation, nursery establishment and cultivation. The education, awareness and extension activities on conservation and sustainable use conducted by the project achieved a notable success.

3. Information and institutional support with a proper policy and legal framework – three surveys namely, socio-economic survey, ethnobotanical survey and resource inventory survey were carried out with the participation of the community for baseline data collection. The current status of legislation on intellectual property rights was reviewed and subsequently legislation was formulated to establish an adequate legal regime to safeguard traditional knowledge relating to the use of medicinal plants.

4. It is commonly accepted that the conservation of medicinal plants in Sri Lanka gained a relatively high significance after implementation of this project.

45.7 Discussion

Medicinal plants occupied a pre-eminent place in indigenous medicine in Sri Lanka over thousands of years. Out of the recorded 1432 of medicinal plants in the country over 10% are endemic and about 8% are threatened. Around 80% of the population in Sri Lanka continues to depend on medicinal plants for primary healthcare. Usually the local requirements of medicinal plants are either extracted from the wild or collected from home gardens. The demand for raw materials of medicinal plants is gradually increasing. However, the development of research on conservation and use of medicinal plants in the country is not satisfactory. Major issues confronted with the conservation of medicinal plants in Sri Lanka are habitat destruction, unsustainable harvesting from the wild and pollution of habitats. Information generated to implement effective conservation strategies is limited. Systematic commercial cultivation of medicinal plants is not prevalent in the country but *ex situ* conservation is carried out at several locations to a limited extent. About 20,000 physicians use herbal medical system in the country. There is a rich traditional knowledge base associated with medicinal plants which is also at a high risk of extinction. Therefore a successful national policy and a strategic framework for the conservation and sustainable use of medicinal plants and associated knowledge are important. Sri Lanka needs an effective national policy on access to genetic resources when considering the pharmaceutical prospecting of medicinal plants and intellectual property rights.

References

Bandarathillake, H.M. (2001) *Administrative Report of the Conservator General of Forests Sri Lanka – 2001*. Forest Department, Ministry of Forestry and Environment, Colombo, Sri Lanka, pp. 59–60.

BMARI (2002) *The Checklist of Medicinal Plants in Sri Lanka*. Bandaranayake Memorial Ayurvedic Research Institute, Ministry of Indigenous Medicine, Sri Lanka.

DOA (2002) *Preliminary Survey of the Situation Analysis of Under-utilized Fruit Crops in Dry and Intermediate Zones of Sri Lanka*. Second Report on Under Utilized Fruit Development Project, Department of Agriculture, Sri Lanka, pp. 4–16.

IUCN (2001) *Statistics on the National Demand for Medicinal Plants*. Report on MPP/R/21, Sri Lanka Conservation and Sustainable Use of Medicinal Plants Project, IUCN – World Conservation Union, Sri Lanka.

IUCN (2004) *Three Case Studies on Legislations for Safeguarding Traditional Knowledge Relating to the Use of Medicinal Plants*. Report no MPP/R/39, Sri Lanka. Conservation and Sustainable Use of Medicinal Plants Project, IUCN – The World Conservation Union, Sri Lanka.

MENR (2002) *State of the Environment in Sri Lanka*. A national report prepared for the South Asian Association for Regional Cooperation, Ministry of Environment and Natural Resources, Colombo, Sri Lanka, pp. 245.

NASTEC (2002) *A National Program for Herbal Health Care*. NASTEC Monograph series 01. National Science and Technology Commission, Colombo, Sri Lanka, pp. 33–34.

Perera, D.L. (2004) *Truth and Myth of Green Piracy*. Biodiversity conservation and sustainable use of medicinal plants project in Sri Lanka, pp. 177.

Pilapitiya, U. (1999) The role of indigenous medicine in biodiversity and sustainable development. *Vidurawa,* Science Magazine of the National Science Foundation, Colombo, Sri Lanka, pp.16–19.

Rao, V.R. and Arora, R.K. (2004) Rationale for conservation of medicinal plants. In: Batugal, P., Jayashree Kanniah, A., Young, L.S. and Oliver, J.T. (eds) *Medicinal Plant Research in Asia*, Vol. 1: *The Framework and Project Work Plans*. International Plant Genetic Resource Institute (Regional office for Asia and Oceania), Serdang, Selangor DE, Malaysia, pp. 7–17.

Silva, M.A.T. and Wettasinghe, D.T. (2004) *Sri Lanka Conservation and Sustainable Use of Medicinal Plants*. Final evaluation report, IUCN – The World Conservation Union Sri Lanka, pp. 45.

Wijesundara, D.S.A. (2004) Inventory documentation and status of medicinal plants research in Sri Lanka. In: Batugal, P., Jayashree Kanniah, A., Young, L.S. and Oliver, J.T. (eds) *Medicinal Plant Research in Asia*, Vol. 1: *The Framework and Project Work Plans*. International Plant Genetic Resource Institute (Regional office for Asia and Oceania), Serdang, Selangor DE, Malaysia, pp. 184–195.

World Bank (2004) Medicinal plants: conservation and sustainable use in Sri Lanka. *IK Notes* 66.

World Health Organisation (2005) *Traditional Medicine and Regulations of Herbal Medicines*. Country summaries, Report of a WHO global survey, World Health Organization, Geneva, pp. 124–125.

46 Use of Wild Plant Species: the Market Perspective

S. Curtis

46.1 Introduction

Consumers are increasingly interested in the provenance of the goods they are purchasing. Wild plant species are wild harvested and used as crops then sold as products, especially for food and medicines, throughout the world. In the developed world attention is increasingly being focused on the sustainability of wild harvesting such species, aspects of which must be considered when developing national conservation programmes.

Neal's Yard Remedies may be used as an example of how one company ethically purchases and markets a number of wild plant species. One important area of difficulty that companies can experience is how to establish accurate information about the conservation status of certain plants. There are pressures from some organizations for companies to stop selling a number of plants, for example golden seal (*Hydrastis canadensis* L.), rosewood (*Aniba rosaeodora* Duke) and Atlas cedar (*Cedrus atlantica* (Endl.) G. Manetti ex Carriere), which needs to be balanced with encouraging more environmentally sensitive projects that support local economies. Atlas cedar and spikenard illustrate how, as a company, we have responded to a variety of pressures from campaigning organizations, commercial demands, media perceptions and customer requirements.

46.2 Ethical Foundation of Business

Neal's Yard Remedies was founded in 1981 by natural medicine enthusiasts as a shop where a comprehensive range of natural remedies such as medicinal herbs, essential oils and homoeopathic remedies could be bought, as well as plant-based cosmetics in distinctive blue glass bottles. They now have 25 shops in the United Kingdom and five in Japan. They sell over 200 dried medicinal herbs and 40 essential oils. Neal's Yard Remedies has established a reputation for being an 'ethical' company, their policies are made readily available on the web site. The commitments of Neal's Yard Remedies are:

- GMO free. We do not use genetically modified raw materials in our products. We require a GM-free statement form with supporting evidence for all ingredients that we purchase.
- Promote organic. We are a Soil Association-certified manufacturer and packer. We actively support organic methods of farming and use organically certified materials where available and practical.
- Responsibility. We consider our impact on the environment in everything we do. We aim only to buy raw materials where the source is sustainable for both the ecology and the local community.
- Fair trade. We actively support many projects, which promote fairly traded products. We believe in paying a sustainable price for goods, which reflects the real cost of the production of these goods.
- No animal compromise. Following consultation with British Union for the Abolition of Vivisection (BUAV) we only buy materials that have never been tested on animals or have not been tested on animals after the year 2000. Our range is vegetarian.
- Close to nature. Our products are formulated as naturally and as close to plant sources as possible. We never use petrochemical derivatives such as mineral oil and petrolatum, nor do our products contain synthetic aromas.

These are policies that the directors of Neal's Yard Remedies are personally committed to: they are also areas of concern for a growing number of the population. For example, throughout the United Kingdom, the organic market has grown rapidly over the last decade. Sales of organic food have increased tenfold to £1.12 billion in 2003/04 and during 2003–2004, organic sales grew by a further 10.2% to almost £2 million a week (Taylor Nelson Sofres, 2005).

46.3 Meeting the Ethical Challenge

The United Kingdom Soil Association believe that taste and food safety concerns are the most important factors in persuading people to try organic food for the first time and in encouraging consumers to increase spending on organic products. However, shoppers only become serious organic consumers when they are also persuaded of the health, environmental and animal welfare benefits of eating organic (Soil Association, 2004).

The Soil Association has introduced wild harvesting standards for certified organic products (Soil Association, 2005). To qualify for organic certification the grower has to complete an annual audit supplied by the Soil Association. This is helpful to a company such as Neal's Yard Remedies when trying to ensure that their products are harvested sustainably. However it only applies to those crops that are organically certified, and not to non-organic wild harvested products.

There is a similar story of growth with the fair trade market. One in every two adults in the United Kingdom now recognizes the Fairtrade Mark according to figures from Market and Opinion Research International (MORI, 2005). The survey shows that 50% of the adult population in 2005 could identify the certification mark, up from 25% in 2003 and 39% in 2004. The highest recognition of the Fairtrade Mark is among the 25–34 age group, which is now recognized by 55% of the adults in the United Kingdom. People in this group are now just as likely as older age groups to buy fair trade products regularly. The poll also shows that the majority of those buying fair trade are recent converts – an indication of future promise for fair trade sales. More than half of fair trade buyers (53%) first bought a fair trade product in the past year, including 7% who first bought fair trade in the last 3 months. MORI says this figure equates to 3% of all adults in the United Kingdom buying fair trade over the last 3 months. 'These figures hold great promise for the future of Fairtrade', says Harriet Lamb, Director of the Fairtrade Foundation (MORI, 2005). 'Reaching 50% of the population is a hugely significant marker for us. Companies should take note that the public are more canny and caring than they are often given credit for. Price is emphatically not their only concern when they go shopping – they do want the reassurance that farmers in developing countries receive a better deal.'

Neal's Yard Remedies has worked with a community in Ecuador to develop a market for a product previously unknown in Europe. A tribal community in the rainforest area of Ecuador produces oil from the fruit of the Seje tree (*Jessenia bataua*). This oil is used traditionally both internally and externally for its medicinal properties. Each year there is a surplus of this oil produced that Neal's Yard Remedies has contracted to purchase on a fair trade basis, i.e. we commit to purchasing a minimum quantity at a minimum price. A representative of the community has worked with Neal's Yard Remedies to inform them of the therapeutic properties of the oil so that we can develop suitable products utilizing it.

This arrangement with the tribal community in Ecuador enables them to earn currency while maintaining and developing value in a local wild-harvested species. We think that fair trade policies are an essential part of a conservation policy: a valuable and valued resource will be well looked after and managed sustainably if the local community benefits on a long-term basis.

Is Neal's Yard Remedies unusual in having an ethical and sustainable basis for its business? We like to think we are at the leading edge of companies demonstrating a commitment to issues of environmental and social sensitivity and sustainability but other often much larger companies have to be seen to be improving in these areas and are taking steps to catch up. McDonald's is now supplying only organic milk in the United Kingdom and in 2005 was given an award by the Royal Society for the Prevention of Cruelty to Animals for its abattoir policy. The Hong Kong and Shanghai Banking Corporation are sponsoring a massive botanic garden conservation project based at the Botanic Gardens Conservation International (BGCI, 2006) and Nestle has recently launched a fair trade coffee (BBC, 2006).

Neal's Yard Remedies often works directly with growers, suppliers and manufacturers to find ways of developing products suitable to our requirements. We visit as many as possible of them. This means that we can work in partnership and have more say in developing strategies; we grow with our suppliers and invest in them; we develop a relationship with them that has 'human' benefits to our staff and customers; we get better information about such things as harvest and availability; and we can invest in sustainable growing and fair trade schemes.

An example of a group of companies benefiting from working directly with growers can be seen from the area of sustainable forestry. The Forest Stewardship Council (FSC) is an independent international organization established to promote responsible management of the world's forests through standards setting, certification and labelling of forest products. Certain companies, such as B&Q in the United Kingdom, have worked hard to get all their suppliers to adopt FSC methods and be able to use the FSC mark on their products (FSC, 2006). This has transformed the retail timber market in the United Kingdom and has also given companies like B&Q a significant marketing advantage. The United Kingdom Ethical Purchasing Index that tracks the size of the market for ethical products using sales data from nine major retailers estimated that last year the sales of FSC certified products in the United Kingdom exceeded US$1.7 billion (FSC, 2005). The estimated size of the FSC global market in 2005 was estimated at US$5 billion.

The market in medicinal herbs is another fast-growing market. More than 80% of the world's population depends on herbal medicine for primary healthcare and more than 25% of the population of the United Kingdom uses it regularly. Mintel (2005) estimated that the United Kingdom herbal market in 2005 was worth £75 million and has grown 16% a year for the past 5 years. The new European Herbal Medicinal Products Directive (DTHMP) is likely to further stimulate growth for certain herbal remedies to become mass-market products.

On a global level, most material for herbal remedies is wild harvested. Neal's Yard Remedies sells over 200 dried medicinal herbs approximately 60% of which are wild harvested rather than cultivated. The rapidly growing demand for medicinal plants combined with habitat loss is putting pressure on many species. Some medicinal plants that were sold in the United Kingdom 10–15 years ago have now been placed on the IUCN Red List of Threatened Species (http://www.redlist.org/) and are no longer sold, for example lady's slipper root (*Cypripedium pubescens* Willd.). While other species are plentiful, such as juniper (*Juniperus communis* L.) that grows abundantly in southern France, and for these sustainable wild harvesting presents no threat. Golden seal (*Hydrastis canadensis* L.) is a medicinal herb that is wild harvested in North America. The increasing popularity of this useful antimicrobial herb put the wild populations under increased threat and as a result Neal's Yard Remedies will now only stock golden seal from sustainable cultivated sources.

The need for action in relation to medicinal herbs is recognized in the Global Strategy for Plant Conservation targets agreed by the Parties to the Convention on Biological Diversity (CBD, 2002). Plantlife International surveyed the global herbal industry (Plantlife International, 2004) and proposed the introduction of certification schemes with an appropriate chain of custody mechanisms, the development of a code of practice for industry, the incorporation of sustainability practices in law, more support for cultivation and a new programme of research and education.

The media, certain campaigning organizations and some customers do put pressure on companies such as Neal's Yard Remedies to prove their environmental credentials. *The Ecologist*, for example, has highlighted the dangers of overexploiting endangered medicinal plants, as has the *Daily Mail* and the *Daily Telegraph*. The campaigning organization 'Cropwatch' has a web site (http://www.cropwatch.org/) and newsletter that raises awareness of endangered plants that are distilled and used for their essential oil. Neal's Yard Remedies, perhaps particularly as a result of its ethical stance, regularly receives enquires concerning the source and sustainability of specific medicinal plants we market. All of which indicates the growing public interest in sustainability and ethical production.

46.4 How the Conservationist Might Assist Business

Accurate information about the environmental status of specific plants can be very difficult to establish. There is no internationally recognized list of wild harvested species or wild harvesting audit or certification scheme for all wild plants; there is a great need for this. Neal's Yard Remedies uses organically certified plants wherever possible, and as mentioned above these will include a wild harvesting audit; we are also in the process of developing our own audit of all our growers but this would be far more effective if it was a statutory requirement. A first step has been taken by the FSC which last year started to certify plants growing on the forest floor and not just trees. This is another well-controlled certifying body that indicates a plant has been sustainably wild harvested. Neal's Yard Remedies has never previously sold the essential oil of spikenard (*Nardostachys grandiflora* DC) because for many years it was unsustainably harvested, however, the FSC has just accredited a source of this oil from Nepal and Neal's Yard Remedies are for the first time considering supplying this oil. There is a need for global and national lists of wild-harvested species, wild harvesting audits and a certification scheme for all wild plants to ensure those that are exploited are exploited sustainably.

There are plants that Neal's Yard Remedies has chosen to stop selling because of concerns about their sustainability even though they are not on the IUCN red list. Rosewood (*Aniba rosaeodora* Duke), for example, was one of our most popular essential oils but rosewood trees are being cut down as part of the deforestation in Brazil. We stopped selling this oil 10 years ago so as not to support this deforestation; although we are hopeful that a plantation will provide a fully sustainable source as basis for future sales. Sometimes finding a different source of a plant is possible. Atlas cedar (*Cedrus atlantica* (Endl.) Carr.) is not an endangered species but is being decimated as part of the deforestation occurring in the Atlas Mountains of North Africa. We have managed to find a supply of Atlas cedar that is sustainably grown in France and we now only buy this rather than that previously sourced from Morocco. More research and entrepreneurial commitment needs to be focused on cultivation of those wild plants that have been harvested from the wild in significant quantities threatening natural populations.

46.5 Conclusions

There are many opportunities for conservation projects to work in partnership with companies. Questions scientists and conservationists should ask when looking at the feasibility of approaching a company to invest in a project should include: What are the traditional uses of the crop and how does this relate to the modern world? Is there a surplus? Does the crop or the local community make an interesting story? Is there someone involved in the project who can liaise/be a spokesperson for it? Good photographs and a well-written summary of the crop or project can go a long way to enrolling interest. Other factors that can add value to a wild crop are if it can be certified organic, be part of a fair trade project or if it can be FSC certified.

Few companies these days, especially smaller ones, are cash rich and are unlikely to simply give cash donations, but an increasing number will take the opportunity to work with a conservation project if there is perceived to be a mutual benefit.

References

BBC (2006) *Nestle Launches Fair Trade Coffee*. Available at: http://news.bbc.co.uk/1/hi/business/4318882.stm

BGCI (2006) *Botanic Gardens Conservation International*. Available at: http://www.bgci.org.uk/

CBD (2002) *Global Strategy for Plant Conservation*. Secretariat of the Convention on Biological Diversity, Montreal. Available at: http://www.biodiv.org/decisions/?lg=0&dec=VI/9

Forest Stewardship Council (2005) FSC Press Release. Available at: http://www.fsc-uk.info/

Forest Stewardship Council (2006) Forest Stewardship Council: UK Working Group. Available at: http://www.fsc-uk.info/

Mintel (2005) Complementary Medicines – UK. Market Research Com. Available at: http://www.marketresearch.com/

MORI (2005) Research for the Fairtrade Foundation. Published online 25 May 2005. Available at: http://www.ipsos-mori.com/polls/2005/fairtrade.shtml

Plantlife International (2004) Herbal Harvests for the Future. Plantlife International, Salisbury, UK.

Soil Association (2004) *Organic Food and Farming – Some Common Questions Answered*. Soil Association, Bristol, UK.

Soil Association (2005) *Soil Association Standards*. Soil Association, Bristol, UK.

Taylor Nelson Sofres (2005) *Organic Food: Understanding the Consumer and Increasing Sales*. Taylor Nelson Sofres, London.

47 Linking Conservation with Sustainable Use: *Quercus ilex* subsp. *rotundifolia* (Lam) O. Schwarz in Traditional Agro-sylvo-pastoral Systems in Southern Portugal

C.M. Sousa-Correia, J.M. Abreu, S. Ferreira-Dias, J.C. Rodrigues, A. Alves, N. Maxted, B.V. Ford-Lloyd

47.1 Introduction

Portuguese *montados* or Spanish *dehesas* are traditional Mediterranean wood pasture ecosystems where the vegetation is organized in a dispersed evergreen oak stratum (*Quercus suber* L. or *Quercus ilex* subsp. *rotundifolia* (Lam.) O. Schwarz or consociation of both) and an herbaceous substratum over which dispersed maquis communities are sometimes present. At present, evergreen oak density varies from 20 to 90 trees/ha (Alés, 1999) and as a typical landscape of the southern Iberian Peninsula. They cover about 3.1 million ha (Díaz *et al.*, 1997) from which 1.1 million are in Portugal (Direcção Geral de Florestas, 2001), occupying critical areas in terms of soil and water resources. *Montados* and *dehesas* have resulted from directed and progressively intensified human intervention over primitive Mediterranean evergreen oaks forests since Neolithic times, which involved the thinning of oaks, pines and shrubs to obtain a regularly spaced distribution of the retained trees (savannah-like).

These agricultural systems are primarily devoted to extensive livestock rearing, with both the acorn production and grass growth between trees maximized due to optimization of light dispersion in the plant stratum. Most of the present *montados* and *dehesas* were created during the last half of the 19th century and the beginning of the 20th century as a result of the increase in human population (Pulido *et al.*, 2001). They combine several interlocking land uses: forestry, agriculture and extensive acorns, oak foliage and herbage grazing, as well as wood, cork (in the case of *Q. suber* L.) and cereal production. The relative proportion of each will vary according to the soil productivity and historical circumstances (Pereira and Fonseca, 2003). They also have high biological significance in terms of biodiversity wealth (Díaz *et al.*, 1997; Alés, 1999) (see Table 47.1) and play a

Portugal. The aim of this work is to establish: (i) conservation strategies (*ex situ* and *in situ*) for *Q. ilex* subsp. *rotundifolia* in Portugal; and (ii) the application of a selection programme towards the genetic improvement of important agricultural quantitative traits, in this case the 'oil content' of the acorn.

The present study has focused on only one of the potential traits likely to affect yield in *montados* of Q. *ilex* subsp. *rotundifolia*: the 'oil content' in the acorn. This trait was targeted not only due to its economic value, but also to give continuity to complementary preceding studies developed in Portugal concerning characterization of this phenotypic trait as an aid to its genetic improvement. Changes in the Portuguese agricultural context during the 1990s, specifically the decay of cereal prices in Europe, created a situation in which the southern Portuguese cereal production (due to uncertainty of rainfall) was no longer viable and as a result the area of cereal cultivation was reduced to one-third of its original size in the last 10–15 years (Abreu and Mansinho, 2000). The intensive agricultural regimes based on crop cultivation are no longer attractive to farmers who are now looking at viable economic alternative uses of farmland in *montados*. Free-roaming livestock is presently seen as an alternative that conciliates the need for keeping cost (inputs) down with obtaining reasonable economic advantage from the land, especially when the high-quality products obtained have good acceptance in the market (e.g. the autochthonous black pig breed 'Porco Alentejano'). Abreu and Mansinho (2000) estimated that 200 million tonnes of acorns are used directly by animals annually.

In addition to the traditional use of holm oak acorns for animal feeding, the potential use of holm oak oil for human consumption has also been considered in the past. The interest in holm oak oil arises from the increasing tendency towards the consumption of vegetable oils in Europe, and of olive oil in particular, due to its neutraceutical properties, which has drawn attention to edible oils with similar composition to olive oil, such as holm oak oil. The new agricultural programme for acorn oil edible use is not well accepted yet within the olive oil farmer associations and industry (Goes, 1991). The extraction of the oil from *Q. ilex* subsp. *rotundifolia* acorns in a profitable way could reduce the importation of edible oils (up to 30%) and would give a great margin for differentiation of a national competitive product in the international market, thus greatly assisting with *montados* conservation. In terms of chemical structure, oak acorn oil and olive oil are very similar in fatty acids composition and Table 47.3 compares the fatty acid profiles and respective variation ranges of oak acorn oil and olive oil (Portaria no. 928/98; Commission of the Codex Alimentarius, 1993). Although oak acorn oil presents a fatty acid profile within the variation range exhibited by olive oil, the former may be richer in oleic acid, while the latter may present a slightly higher content in linoleic acid, an essential fatty acid to the human metabolism.

47.2 Materials and Methods

An overview of the methodology employed to conserve and sustainable use of *Q. ilex* subsp. *rotundifolia* (Lam.) O. Schwarz in their traditional agro-sylvo-pastoral systems in Portugal is presented in Fig. 47.1.

Table 47.3. Fatty acid composition comparison for Holm oak acorn oil and olive oil.

	Myristic acid (C 14:0)	Palmitic acid (C 16:0)	Palmitoleic acid (C 16:1)	Heptadecanoic acid (C 17:1)	Stearic acid (C 18:0)	Oleic acid (C18:1)	Linoleic acid (C 18:2)	Linolenic acid (C 18:3)	Arachidic acid (C 20:0)	Behenic acid (C 22:0)
Acorn[a]	Max 0.5	10.0–19.0	Max 1.0	Max 0.3	0.5–5.0	50.0–3.0	11.0–27.0	0.5–3.0	Max 1.0	0.2–0.8
Olive[b]	Max 0.5	7.5–20.0	0.3–3.5	Max 0.6	0.5–5.0	56.0–83.0	3.0–20.0	0.1–1.5	Max 0.8	Max 0.2

[a]Portaria no. 928/98, Diário da República-I Série-B, Portugal (1988).
[b]Codex Alimentarius (1993).

STAGE 1 Ecogeographic survey
1a Project design
1b Data collection and analysis
1c Data synthesis

STAGE 2 Genetic variation study
2a Collecting strategy A
2b Collecting mission
2c DNA extraction
2d SSRs analysis
2e Genetic data analysis

STAGE 3 Phenotypic variation study of 'oil content'
3a Collecting strategy B
3b Collecting mission
3c Calibration of the FT-NIR spectroscopy equipment
3d FT-NIR spectra of individual acorns
3e Phenotypic data analysis

STAGE 4 Molecular marker association study
4a Collecting strategy A,B
4b Collecting mission
4c DNA extractions
4d AFLP's
4e Association analysis between 'oil content' and AFLP's profile
4f Statistical model to predict potential 'oil content' in young trees

---

STAGE 5 Metabolomic marker association study
5a Collecting strategy C
5b Collecting mission
5c FT-IR spectra of maternal canopy leaves
5d Association analysis between 'oil content' and 'canopy leaves' metabolomic profiles
5e FT-IR spectra of maternal leaves from lateral shoot
5f Association analysis between 'canopy leaves' and 'lateral shoot' metabolomic profiles
5g Statistical model to predict potential 'oil content' in young trees

Fig. 47.1. Methodological scheme developed to sustain conservation and sustainable use of *Q. ilex* subsp. *rotundifolia* Portugal.

47.2.1 Stage 1: Ecogeographic survey

An ecogeographic survey of *Q. ilex* subsp. *rotundifolia* across its distribution area in Portugal was undertaken. The methodology followed was that defined by Maxted *et al.* (1995) and involved the collation, synthesis and analysis of ecogeographic data, and resulted in the development of a collecting strategy for fresh young leaf material (collecting strategy A), fully mature acorns (collecting strategy B), as well as the identification of the target species in the field. The geographic information was used to select diverse locations for collecting sites that represented different ecogeographic regions throughout Portugal. In addition, the ecogeographic survey provided information on the species conservation status (e.g. threat assessment, present location of *on-farm* and genetic reserves, and *in situ* and *ex situ* gap analysis) to be used in the establishment of complementary conservation strategies.

47.2.2 Stage 2: Genetic variation study

This stage provides information on the distribution of genetic variation across the *Q. ilex* subsp. *rotundifolia* distribution range in Portugal, the genetic structure of populations and total variation within species. As part of this stage vegetative material (canopy leaves) was collected from trees located in different ecogeographic sites (collecting strategy A) (Fig. 47.1). This population level information will assist in the establishment of a complementary conservation strategy for *Q. ilex* subsp. *rotundifolia* in Portugal and in establishing a baseline for monitoring schemes (e.g. frequency, methods and priority sites). In relation to sustainable use, it will contribute for the marker-assisted genetic improvement programme by aiding selection of propagation sources (e.g. location of source sites for propagation, number of mother trees used for propagation and monitoring for the decay of genetic diversity).

47.2.3 Stage 3: Phenotypic or metabolomic variation study of the trait 'oil content'

Here, the aim is to review the total range and distribution of phenotypic variation on the trait 'oil content' of the acorns of *Q. ilex* subsp. *rotundifolia* in Portugal and to assess variation between years, trees, places and 'year × trees' and 'year × place' interactions. This involves the collection of target acorns from trees located in different ecogeographic sites (collecting strategy B) (Fig. 47.1). It also involves the phenotypic screening of 'plus' trees among the phenotypic range of the trait 'oil content'.

Plant improvement for quality purposes frequently depends on evaluations over large numbers of individuals and these measurements were traditionally undertaken by chemical and physical methods of analysis, so a rapid, non-destructive method of evaluating seed quality traits is important. Fourier-transformed near-infrared spectroscopy (FT-NIR) is a multitrait technique of large application in the analysis of quality traits in food and agriculture, which is very fast and non-destructive, and requires no sample preparation (Pasquini, 2003). This screening will allow the effective identification of additional 'plus' trees to be used as mother trees during the first cycle of mass selection. Furthermore, the phenotypic variation study will provide guidelines on the establishment of field trials for evaluation purposes including the estimation of the 'tree × place' interaction in order to obtain partitioning between the environmental and genetic components of the total variation.

It is also proposed to undertake studies of the association of molecular marker with the trait 'oil content' in the acorns of *Q. ilex* subsp. *rotundifolia* to screen the offspring of the 'plus' trees, previously identified as high 'oil content' phenotypes (first cycle of selection), using the method developed in stage 2. Work is also underway to investigate the associations between metabolomic markers and the 'oil content' in the acorns of *Q. ilex* subsp. *rotundifolia* as a further means of selection of potential 'plus' trees among the offspring (second cycle of selection) of previously selected 'plus' trees (first selection cycle), during their early stages of development. However, this work is yet to be completed.

47.3 Results

47.3.1 Stage 1: Ecogeographic survey

A detailed distribution map for *Q. ilex* subsp. *rotundifolia* in Portugal was already available (Direcção Geral de Florestas, 2001), but additional geographic information was obtained from the most important Portuguese herbaria and the Herbarium of the Royal Botanic Gardens of Kew, London. The software ArcGIS 8.0 was used for overlaying the distribution map and a number of shape files (temperature, soil type, altitude, rainfall, pH, etc.) representing different ecological conditions for Portugal. Distinct ecogeographic regions were then defined as targets for collecting sites for the genetic, phenotypic and molecular or metabolomic marker association studies. The conservation status of *Q. ilex* subsp. *rotundifolia* was investigated and despite the existence of extensive legislation for the protection of the species *in situ*, the high rate of traditional agro-sylvo-pastoral ecosystems abandonment or conversion into other land uses suggests that legislation is proving ineffective to conserve the species and associated ecosystems and landscapes.

A threat assessment using the IUCN criteria was undertaken and it indicated 13.5% decline in the occupied area (720,000 ha) within one generation time at an average of 15 oaks per hectare per 30 years (1964–1994), which represents an expected reduction of 40% after three generations. However, latter predictions (J.M. Abreu, Porto, 2005, personal communication) estimate the decline to have intensified and over three generations it is expected to rise to 50%. This places *Q. ilex* subsp. *rotundifolia* in the endangered – A 4 (C) category of IUCN criteria at the regional level for Portugal. Furthermore, the assessment remains unchanged when neighbour conspecific populations are taken into account as the populations in Spain are also known to be in decline.

47.3.2 Stage 2: Genetic variation study

Genetic variation was studied using SSR markers for 19 populations sampled across the species distribution in Portugal and deliberately selecting distinct ecogeographic regions. F_{ST} values indicate small but significant differentiation between populations ($F_{ST} = 0.018$, $P < 0.05$). Small F_{ST} values are expected for *Q. ilex* subsp. *rotundifolia* due to its outcrossing mating system and long generation time. Overall, F_{IS} indicates small but significant levels of inbreeding ($F_{IS} = 0.164$, $P < 0.05$). Further studies are being developed to understand if the heterozygote deficiency observed results from the presence of null alleles.

Other factors that could potentially contribute for a deficiency in heterozygotes would be a Wahlund effect, where different phenology would favour the mating of individuals with overlapping flowering periods. Other reasons could be the selection process undertaken during the clearing of the original Mediterranean forest and up to the present time, which could favour the fixation of homozygotes. Furthermore, the very low rates of regeneration could contribute to low heterozygote numbers, despite gene flow being potentially high then and now. Populations show high and homogenous values of genetic diversity (He = 0.691 ± 0.026). Most of the diversity in holm oak is found within populations rather than between populations. Unweighted pair group method with arithmetic mean (UPGMA) phenogram based on pairwise genetic distances (Nei, 1978) suggests that genetic distances do not relate to geographical distances.

In terms of complementary conservation, the genetic variation study indicates that a small number of large populations will provide genetic diversity to most of the species, and their location is not necessarily related to ecogeographic conditions or geographic distances. While in terms of use, it appears that there is enough variation within each site so that 'plus' trees for propagation can be selected from within the same farm or ecogeographic region, in such a way that conservation of adaptive traits would be ensured. If the tree density on a farm or region was low so that 'plus' trees within a site could not be selected, other sites with similar ecological conditions could be utilized.

47.3.3 Stage 3: Phenotypic or metabolomic variation study of percent oil in the acorns

The phenotypic study was undertaken using FT-NIR spectroscopy which could be used on effective (high throughput of 1 sample/min and small sample preparation) large-scale screening and identification of 'plus' trees. This high phenotype identification consists of the first cycle of selection of the present genetic improvement programme for 'oil content'. For the phenotypic study, two consecutive years, 2003 (two populations) and 2004 (five populations) acorns were used. Each population is represented by 10–30 trees and five acorns per tree were individually analysed for 'oil content'. The estimation found that 'oil content' varied between 3.5% and 16.2%, but there were no significant differences between years or places except between trees ($P < 0.05$) as well as for the interaction 'years × trees'. These significant differences account for 27.7% of total variation, which may be allocated to both genetic and environmental effects. The great variation range indicates the importance of selecting 'plus' mother trees whose offspring would be used to increase the density of *montados* of *Q. ilex* subsp. *rotundifolia* in southern Portugal.

Correlation coefficients between 'oil content' and morphological characters of the acorn (e.g. weight, volume and weight of peel) were estimated; the highest correlation was found between the density of the (peeled) acorn and 'oil content', so that a trend was found, which showed that the more dense the acorns, the more oil these tended to yield (percentage oil).

In terms of conservation and use strategies, the phenotypic variation study supports the results obtained from the genetic variation study by suggesting that variation is larger between trees within populations than between populations. It also indicates that each region (ecogeographic area) is likely to contain the necessary variation, which will allow the identification of 'plus' trees for 'oil content', while conserving other adaptive traits. On the other hand, the high correlation found between density and 'oil content' of the acorn suggests that there are other production factors such as 'irrigation' that can favourably affect the oil content of the acorn.

47.4 Conclusions

The study shows clearly that both *montados*, as an agro-silvicultural ecosystem, and *Q. ilex* subsp. *rotundifolia* are declining in distributional range in southern Portugal and there is an urgent need for a coherent conservation strategy. One approach to halting decline is to increase the productivity and thus the economical viability of the *montados* production systems, so facilitating conservation of both *montados* and *Q. ilex* subsp. *rotundifolia*. Allied with this, increasing the density of evergreen oaks, improved tree management (irrigation, fertilization and pruning) and genetic improvement programmes targeting acorn-related quantitative traits will each benefit local farmers and land owners.

The study shows that sustainable use can be promoted by employing a selection programme for *Q. ilex* subsp. *rotundifolia* to identify 'plus' trees during the first cycle of mass selection for the increase in 'oil content' of the acorns. Subsequentially, a second selection cycle will be applied among the offspring of the 'plus' trees, based on models built upon statistical associations between molecular and/or metabolomic marker of vegetative material and the 'oil content' of the acorns. These statistical association models aim at predicting the potential for 'oil content' of very young trees and can accelerate the selection process in one generation time. The finding that genetic variation is larger between trees within populations than between populations indicates that each ecogeographic region has candidate 'plus' trees with high 'oil content' that can be used as sources of improving trees to benefit the overall productivity of the *montados* and therefore hopefully prevent its decline as a production system and also as refugia for broader biodiversity.

References

Abreu, J.M. and Mansinho, I.N. (2000) *Gleditsia triacanthos*: Planos a Médio Prazo Para a Valorização das Zonas de *Montado* de Azinho. *Sulco* 115, 19–20.

Alés, F.R. (1999) Dehesas y *Montados*. Bases ecológicas para su Gestión. *Revista Biologica* 17, 147–157.

Capelo, J.H. (1996) Origem e Diferenciação das Paisagens Florestais do Baixo Alentejo. Apontamentos Sobre a Prespectiva Geobotânica no Planeamento Florestal. *Revista Florestal* IX(3), 72–81.

David, T.S., Ferreira, M.I., Cohen, S., Pereira, J.S. and David, J.S. (2004) Constraints on transpiration from an evergreen oak tree in southern Portugal. *Agricultural and Forest Meteorology* 122, 193–205.

Díaz, M., Campos, P. and Pulido, F.J. (1997) The Spanish dehesa: a diversity in land use and wildlife. In: Pain, D.J. and Pienkowski, M.W. (eds) *Farming and Birds in Europe: The Common Agricultural Policy and Its Implications for Bird Conservation*. Academic Press, London, pp.178–209.

Direcção Geral de Florestas (2001) *National Forest Inventory*, Ministério da Agricultura e Pescas, Portugal (in Portuguese).

Direcção Geral de Florestas (2006) *Preliminary Results of the National Forest Inventory*, Ministério da Agricultura e Pescas, Portugal (in Portuguese).

Goes, B. (1991). *A floresta, sua importância e descrição das espécies de maior interesse em Portugal*, Portucel (in Portuguese).

Maxted, N., van Slageren, M.W. and Rihan, J. (1995) Ecogeographic surveys. In: Guarino, L., Ramanatha Rao, V. and Reid, R. (eds) *Collecting Plant Genetic Diversity: Technical Guidelines*. CAB International, Wallingford, UK, pp. 255–286.

Nei, M. (1978) Estimation of average heterozygosity and genetic distance from a small number of individuals. *Genetics* 89, 583–590.

Pasquini, C. (2003) Near infrared spectroscopy: fundamentals: practical aspects and analytical applications. *Journal of the Brazilian Chemical Society* 2, 198–219.

Pereira, P.M. and Fonseca, M.P. (2003) Nature vs. Nurture: the Making of the *Montado* Ecosystem. *Conservation Ecology* [Online] 7(3). Available at: http://www.consecol.org/vol7/iss3/art7

Pinto-Correia, T. (1993) Threatened Landscape in Alentejo, Portugal: the '*Montado*' and other 'Agro-Silvi-Pastoral' systems. *Landscape and Urban Planning* 24, 43–48.

Pulido, F.J., Díaz, M. and Trucios, S.J.H. (2001) Size structure and regeneration of spanish holm oak *Quercus ilex* Forests and Dehesas: effects of agroforestry use on their long term sustainability. *Forest Ecology and Management* 146, 1–13.

Varela, M.C. and Eriksson, G. (1995) Multipurpose gene conservation in *Quercus suber*: a Portuguese example. *Silvae Genetica* 44(1), 28–37.

X

Global Issues in Crop Wild Relative Conservation and Use

48 The Crop Wild Relative Specialist Group of the IUCN Species Survival Commission

M.E. Dulloo and N. Maxted

48.1 Introduction

The plant kingdom is highly diverse and assessments of global plant diversity vary from 250,000 (Mabberley, 1997) to 300,000–320,000 species (Prance *et al.*, 2000) or even to more than 400,000 (Bramwell, 2002). While the greater proportion of the Earth's plant diversity plays a crucial role in the proper functioning of the world's ecosystem, there is a smaller, but significant, group of plants which are of particular socio-economic importance because they contribute directly and indirectly to people's livelihood and global food security. Many of these are the close or more distant wild relatives of crop plants. The number of crop wild relatives (CWR) globally is unknown, but is substantial. Their conservation is often neglected due to falling between the responsibilities of the traditional agricultural and conservation communities. It is a truism to state that CWR have been neglected and their systematic conservation should receive much more attention. If the European and the Mediterranean floras are taken as an example, 25,687 species out of an estimated 30,983 species in the European and the Mediterranean region are regarded as CWR (see Kell *et al.*, Chapter 5, this volume). These include major crops such as oats (*Avena sativa* L.), sugarbeet (*Beta vulgaris* L.), apple (*Malus domestica* Borkh.), annual meadow grass (*Festuca pratensis* Hudson) and white clover (*Trifolium repens* L.), each of which have wild relatives in Europe. Many minor crops such as arnica (*Arnica montana* L.), asparagus (*Asparagus officinalis* L.), lettuce (*Lactuca sativa* L.) and sage (*Salvia officinalis* L.) have also been developed and domesticated in the region. So there is a need to raise the profile of CWR conservation and because these are socio-economically important species to ensure their conservation is linked to sustainable exploitation.

The importance of CWR has been discussed in detail by Prescott-Allen and Prescott-Allen (1983), Hoyt (1988), Maxted *et al.* (1997), Meilleur and Hodgkin (2004), Heywood and Dulloo (2006), Stolton *et al.* (2006) and elsewhere in this volume. Their conservation is receiving increasing attention by many organizations, such as Bioversity International (formerly International Plant Genetic Resources Institute, IPGRI) and other Consultative Group Centres, Botanical Gardens Conservation International (BGCI), Food and Agriculture Organization of the United Nations (FAO), United Nations Environment Programme (UNEP), as well as numerous national agricultural research programmes, and specific activities to ensure their conservation are being implemented. Recently, the World Conservation Union (International Union for the Conservation of Nature (IUCN)) decided to establish a Specialist Group within the Species Survival Commission. This chapter introduces the IUCN Species Survival Commission for the plant genetic resource (PGR) community, provides some information about the new IUCN Specialist Group on CWR and describes its scope, aims and objectives, and discusses the activities it intends to carry out.

48.2 IUCN Species Survival Commission

The Species Survival Commission (SSC) is one of the six Commissions of the IUCN – the World Conservation Union and was established in 1948. Since then it has grown into a global science-based network of 1000 volunteer experts. The Commission strategic plan 2001–2010 provides the basic building blocks for the activities of the Commission. The volunteer scientists are working together to achieve the vision of 'a world that values and conserves the present levels of biodiversity'. The Goal of the SSC is that 'The extinction crisis and massive loss in biodiversity are universally adopted as a shared responsibility, resulting in action to reduce this loss of diversity within species, among species and of ecosystems'. The major role of the SSC is to provide information to IUCN on the conservation species and their inherent values and their roles in ecosystem health and functioning, the provision of ecosystem services and the provision of support to human livelihoods (see Box 48.1).

Box 48.1. Vision, goal and objective of the IUCN/SSC.

Vision – 'a world that values and conserves the present levels of biodiversity.'

Goal – 'The extinction crisis and massive loss in biodiversity are universally adopted as a shared responsibility, resulting in action to reduce the loss of diversity within species, among species and of ecosystems.'

Objectives

1. Decisions and policies affecting biodiversity influence by sound interdisciplinary scientific information;
2. Modes of productions and consumption that promote the conservation of biodiversity adopted by users of natural resources;
3. Capacity increased to provide timely, innovative and practical solutions to conservation problems.

The SSC works through a large number of Specialist Groups which are constituted by IUCN SSC under its species programme. These Specialist Groups provide the breadth of expertise across the wide diversity of species of the planet. The majority of the Specialist Groups cover the conservation of particular taxa or groups of taxa from orchids to amphibians and crocodiles to tigers. Others have a regional focus such as the Mediterranean Plant Specialist Group and Indian Ocean Plant Specialist Group. Still others are concerned with interdisciplinary issues such as wildlife health, reintroduction, sustainable use, invasive species and conservation breeding.

48.3 Crop Wild Relative Specialist Group

The concept of establishing a Crop Wild Relative Specialist Group (CWRSG) has been brewing for a number of years within the scientific community of IUCN, IPGRI (now Bioversity International) and the University of Birmingham. Increasing concern over the rampant loss of agrobiodiversity and its negative impact undermining the foundations of sustainable agriculture, wealth creation and human well-being was and remains a global problem that required a global approach to resolution. With the new structuring of the SSC and the development of the new SSC programme cycle, a request was made in 2004 to the IUCN Plants Committee for the consideration of setting up a CWRSG to help address these global issues. This was endorsed by the Plant Committee and recommended SSC to establish the CWRSG. This was formally done on 26 July 2006, when the Chair of IUCN/SSC, Dr Jane Smart invited Ehsan Dulloo, Senior Scientist at IPGRI and Nigel Maxted, Senior Lecturer at University of Birmingham, to co-chair this group.

48.3.1 Objectives of the CWR Specialist Group

The objectives of the CWRSG are as follows:

- Help ensure that wild plant species of socio-economic value are adequately conserved and sustainably utilized;
- Promote integrated conservation and provide exemplar case studies, involving *in situ* and *ex situ* techniques, for wild plant species of socio-economic value;
- Develop effective strategies for gathering, documenting and disseminating information on wild plant species of socio-economic value diversity and conservation;
- Establish and maintain a global inventory and threat assessment of wild plant species of socio-economic value;
- Provision of advice, expertise and access to appropriate contacts to enhance the actions of individuals or organizations working on wild plant species of socio-economic value conservation;
- Increase awareness of the importance to agriculture and the environment of wild plant species of socio-economic value among governments, institutions, decision makers and the general public.

48.3.2 Activities of CWR Specialist Group

The first task of this new Specialist Group is to define the scope and working structure for their operation. The scope of the SSC is a global one and consequently it is envisaged that members would be drawn from all five continents representing real geographic breadth, as well as a having taxon-specific expertise focused on individual crops and their associated wild relatives. The scope of the CWR would also need to be defined. It is though desirable that the CWRSG will cover both CWR and wild-harvested species (WHS), both of which have important socio-economic importance for human well-being and value, particularly for the majority of the world engaged in subsistence agriculture. The strict distinction between CWR and WHS is often ill-defined and so it makes sense for both groups of species to be included in the proposed Specialist Group remit.

A strategic action plan would need to be developed for the conservation and sustainable utilization of CWR to help identify priority actions to be carried out. Work on developing such a plan began at the First International Conference on Crop Wild Relative Conservation and Use and a first draft of such a strategic action plan was agreed (see Heywood *et al.*, Chapter 49, this volume). One of the immediate activities is to start an inventory of CWR and the group will attempt to establish a global database on CWR. Specific information on biology, conservation status and extent of uses will be documented. One of the major responsibilities of IUCN/SSC Specialist Groups is to help to assess the category of threats for IUCN Red List. The group will collate information that can be used to assess the IUCN categories of threat for CWR and WHS species, working closely in collaboration with other taxonomic Specialist Groups and regional Specialist Groups.

The group will produce an annual newsletter to communicate the activities of the group among the group members and other people working on CWR and also as a strategy to enhance communication with other Specialist Groups within and the SSC secretariat and other IUCN bodies. The group will also promote exchange of information and scientific findings by organizing international conferences to discuss issues related to CWR conservation and use. Already, the First International Conference on Crop Wild Relative Conservation and Use was organized by the University of Birmingham and IPGRI, in collaboration with CRA-Istituto Sperimentale per la Frutticoltura di Roma (Italy) and the Agricultural Extension Service of the Regional Administration of Agrigento (Italy) in September 2005.

The work of the Specialist Group is supported by a team of experts who voluntarily donate their time for achieving the aims and objectives of the other. However, for the continued success of the group, it would be necessary to raise some core funding to enable the group to fulfil its diverse functions. In the initial stages, the group will interact mostly via e-mail and ad hoc working groups through ongoing CWR projects. In the mean time, both Bioversity International and the University of Birmingham are providing part-time administrative support for the Specialist Group. The CWRSG would need a sound financial footing and a fund-raising strategy would need to be developed.

Who should become a member of the CWRSG? You! If you have read this with interest and have an interest in CWR or WHS, you are the sort of person we would encourage to join the CWRSG. Admission to the group is free but by becoming a member the individual has to make a genuine commitment to work to enable the CWRSG to meet its objectives and promote CWR conservation and use. The CWRSG is still in the early stages of establishment, currently the web site is under development but by the time you read this it should be available, so we encourage you to do a web search for the IUCN SSC CWRSG and find out more about CWR conservation and use and the Specialist Group.

48.3.3 Strategic Linkages with Other Specialist Groups

This Specialist Group will interact closely with and endeavour to build upon the work of other relevant taxonomic and geographic Specialist Groups (Eastern and Southern Africa, China, India, the Mediterranean and Europe), as well as with other disciplinary Specialist Groups such as the Sustainable Use, Medicinal plant, Reintroduction and Conservation Breeding Specialist Groups. The CWRSG will also form strategic links with international stakeholders (e.g. Bioversity International, FAO, BGCI and GEF projects), and others along with regional and national institutions with white spot syndrome virus (WSSV) interests.

48.3.4 Projected Impact of CWR Specialist Group

The major focus and anticipated impact of the CWRSG will be on facilitating national, regional and global CWR and WHS conservation and use. The CWRSG will raise the profile of and stimulate both CWR and WHS conservation and use by improving public awareness of the importance of these species, both among the public in general and the scientific community. As responsibility for conservation of CWR has traditionally fallen between the agricultural and conservation communities there is a need to raise awareness even among the two communities that might be expected to have direct responsibility for their maintenance. The clarification of the importance to agriculture and the environment of wild plant species of socio-economic value among governments, institutions, decision makers and the general public cannot be overemphasized. A key element of the policies that results in adequately conserved and sustainably utilized CWR and WHS diversity will be a need for systematic analysis of the gaps in current conservation and utilization strategies, then when gaps are identified undertake remedial actions.

Experts within the Specialist Group, as well as engaging in IUCN Red List assessment of threat to CWR and WHS species, will develop and promote *in situ* and *ex situ* techniques for conservation that are applicable at the national, regional and global scale using diverse case studies. The CWRSG will help establish effective strategies for CWR data gathering, analysis and disseminating information through enhancement of the Crop Wild Relative Information System already established for European CWR taxa. As such the information systems developed will act as a global inventory of CWR and WHS diversity that will facilitate threat assessment and sustainable utilization.

Finally, experts from the CWRSG will provide an information source and 'help desk' for those seeking advice, expertise and access to appropriate contacts to enhance the actions of individuals or organizations working on all wild plant species of socio-economic value conservation. Thus, the impact of the CWRSG will be far reaching, helping ensure that our utilized plant options are maintained for the future in a changing environment.

References

Bramwell, D. (2002) How many plant species are there? *Plant Talk* 28, 32–34.

Heywood, V.H. and Dulloo, M.E. (2006) In Situ *Conservation of Wild Plant Species – A Critical Global Review of Good Practices*. IPGRI/FAO, Rome, Italy.

Hoyt, E. (1988) *Conserving the Wild Relatives of Crops*. IBPGR/IUCN/WWF, Rome, Italy.

Mabberley, D.J. (1997) *The Plant Book. A Portable Dictionary of Vascular Plants*, 2nd edn. Cambridge University Press, Cambridge.

Maxted, N., Ford-Lloyd, B.V. and Hawkes, J.G. (1997) *Plant Genetic Conservation: the* In Situ *Approach*. Chapman & Hall, London.

Meilleur, B.A. and Hodgkin, T. (2004) *In situ* conservation of crop wild relatives. *Biodiversity and Conservation* 13, 663–684

Prance, G.T., Beentje, H., Dransfield, H. and Johns, R. (2000) The tropical flora remain uncollected. *Annals of Missouri Botanic Garden* 87, 67–71.

Prescott-Allen, R. and Prescott-Allen, C. (1983) *Genes from the Wild: Using Wild Genetic Resources for Food and Raw Materials*. Earthscan Publications, London.

Stolton, S., Maxted, N., Ford-Lloyd, B., Kell, S.P. and Dudley, N. (2006) *Food Stores: Using Protected Areas to Secure Crop Genetic Diversity*. WWF Arguments for protection series. Gland, Switzerland.

49 Towards a Global Strategy for the Conservation and Use of Crop Wild Relatives

V.H. Heywood, S.P. Kell and N. Maxted

49.1 Introduction

Increasing recognition of the importance of crop wild relatives (CWR) as a source of genetic material for the continued genetic enhancement of our crops into the future has led to an increase in activities aimed at their conservation and sustainable use in various parts of the world. Such action has been made all the more urgent by the inexorable loss of biodiversity caused by habitat loss, fragmentation and simplification, the impact of invasive species, overexploitation and global change (demographic and climatic).

The conservation and sustainable use of CWR is, however, a complex and multidisciplinary process, as the work undertaken by PGR Forum has clearly shown (see Maxted *et al.*, Chapter 1, this volume; PGR Forum, 2003–2005a). It involves different agencies of government, non-governmental organizations (NGOs), universities and other institutions at national level, and UN agencies, intergovernmental organizations (IGOs) and NGOs at regional and global levels. It is clear that there is an urgent need for some level of coordination of the activities of these various bodies, both at local and global levels. Therefore, it was felt by the participants in the PGR Forum project and organizers of the First International Conference on Crop Wild Relative Conservation and Use that it would be valuable to propose a strategy that would bring the different strands together, and provide guidance for all those engaged in activities concerning CWR. A draft working version of the strategy was prepared and presented to the delegates at the Conference, which took place in Agrigento, Sicily, Italy, 14–17 September 2005 (see Kell *et al.*, 2005a; PGR Forum, 2003–2005b).

Conference participants were invited to join one of the six discussion groups which were set up to help formulate a draft for approval by the plenary session. The groups were led by Wiesław Podyma (Ministry of Agriculture and Rural Development, Poland), Nigel Maxted and Brian Ford-Lloyd (University of

Birmingham, United Kingdom), Sabine Roscher (German Documentation Centre for Information in Agriculture, Germany), José Iriondo (Universidad Politécnica de Madrid, Spain) and Jozef Turok (International Plant Genetic Resources Institute, Rome, Italy). Each group focused on a set of targets (later termed objectives) in the draft strategy, and presented their conclusions in a session at the end of the Conference.

49.2 The Proposed Strategy

We were conscious of the need to present a strategy that had clear objectives, with realistic, well-focused, time-bound, measurable targets, rather than open-ended aims, if it was to be of real use to the various user communities. The ultimate goal is effective conservation and sustainable use of CWR, including all wild plant species of socio-economic value, at national, regional and global levels.

We have called the strategy 'global', although many of the proposed components can be undertaken at different geographic levels and it is likely and indeed proper that national actions will be given priority, and in turn contribute to the regional and global dimensions being developed. But at national or local level, it will be helpful if those involved in implementing the strategy are informed of similar activities being undertaken in other countries in the region or in similar situations elsewhere. Although PGR Forum primarily involved European partners, the methodologies developed by the project participants are generic and equally applicable in other regions of the world. One of the major products of PGR Forum is the Catalogue of Crop Wild Relatives for Europe and the Mediterranean (Kell *et al.*, 2005b; Chapter 5, this volume); therefore, a logical extension would be to undertake regional projects for other adjacent regions (e.g. North Africa and Central Asia), as many CWR species in Europe and the Mediterranean are also found in these regions or are closely related to species that occur there. Regional action in other parts of the world should also be possible, although lack of a sound taxonomic base may be a limiting factor in some countries, especially those with rich and poorly documented floras.

The main objectives in the strategy (Heywood *et al.*, 2007; Appendix) are:

1. Prepare national CWR strategic action plans;
2. Prepare national CWR inventories;
3. Establish a global mechanism or clearing house for CWR conservation and use;
4. Create national priority CWR lists and identify priority CWR sites;
5. Create regional and global CWR priority lists and identify priority CWR sites;
6. Establish protocols for CWR information management and dissemination and provide national and global CWR information management systems;
7. Develop effective means of conserving and using CWR *in situ*;
8. Develop effective means of conserving and using CWR *ex situ*;
9. Assess CWR conservation and threat status;

10. Ensure effective security and legislation for CWR;
11. Promote sustainable utilization of CWR;
12. Initiate education and public awareness programmes on the importance of CWR.

Within each of these objectives we have suggested a series of targets. Targets can be defined as measurable or quantifiable estimates of the amount of particular elements or features of biodiversity to be included in strategies or action plans. They may be global, regional, national, intranational or local in coverage. Targets can be regarded as a means, by which the success or failure of previously established conservation actions or strategies can be measured, monitored and reported, and provide a guide for our actions and a means of measuring progress towards achieving agreed outcomes or goals (Heywood and Dulloo, 2006). The main focus today is on time-bound, quantitative targets for biodiversity conservation.

For the CWR Strategy, as many goals as possible should be achieved by the year 2010, in line with the Millennium Development Goals of the CBD, while for others, such a short time scale would be unrealistic and we have suggested a target of 2015. For each target, a lead organization is identified, contributing organizations suggested and a time frame given, together with references as appropriate.

The limited experience in applying biodiversity and conservation targets suggests that care should be taken to ensure that the targets are clear and unambiguous, especially bearing in mind the difficulties of defining biodiversity in a precise and measurable manner. If the goals are ambiguous or susceptible to different interpretations, there is a serious risk of debate as to whether in due course they have been met or not. They should be based on the best available scientific knowledge and there should be sufficient information about them to allow the baseline status of the target to be properly determined and for meaningful goals set. There should also be a reasonable expectation of the goals being met, although equally, they should not be set at such a level so as not to represent a challenge.

49.3 Future Action

Since the end of the PGR Forum project, which was marked by the First International Conference on CWR Conservation and Use, the authors have maintained close liaison with the UN Food and Agriculture Organization (FAO), Bioversity International and International Union for the Conservation of Nature (IUCN) (The World Conservation Union).

The draft strategy (see Heywood *et al.*, 2007; Appendix) is currently under review, and will be presented to the Eleventh Regular Session of the Commission on Genetic Resources for Food and Agriculture in June 2007, with a recommendation that it is implemented within the context of the International Treaty for Plant Genetic Resources for Food and Agriculture (ITPGRFA). Further, it is anticipated that the final version of the document will also be delivered to the CBD Secretariat as a paper for consideration by Subsidiary Body on Scientific, Technical and Technological Advice (SBSTTA), and will subsequently be passed by the ITPGRFA and SBSTTA to individual governments for implementation. It is also anticipated that the strategy will be sent to appropriate agencies or organizations, such as United Nations Development Programme (UNDP), United Nations Environment Programme (UNEP), the European Union (EU), World Bank, World Resources Institute (WRI), Conservation International (CI), Consultative Group on International Agricultural Research (CGIAR) and leading aid and agricultural agencies.

We feel that in view of their importance, CWR deserve to be given a much higher profile in the work programmes of the ITPGRFA and CBD and that information on CWR should be gathered as part the of ITPGRFA and CBD national reporting system. We also suggest that greater emphasis is placed on CWR in relevant national, regional and global legislative and policy instruments. Actions for the conservation and sustainable use of CWR should be included as a matter of course in national biodiversity strategies and action plans. Much more needs to be done to increase awareness of the value and importance of CWR by governments, the public and the agricultural, forestry, development, extension, conservation and biological sectors. We hope that the Global Strategy for CWR Conservation and Use will make a significant contribution towards achieving these objectives. If adopted, it will in effect provide an action plan for nations and regions to refer to and apply as appropriate when addressing the critical issue of effective CWR conservation and use.

We are fully aware of the plethora of strategies, action plans and declarations concerning different aspects of biodiversity that already exist and the demands that these make on limited human and financial resources; in drawing up the draft strategy we have tried to take these into account as far as possible. Some of the actions proposed are already partially being undertaken for other groups of plants, or in a more general context, and may only need adapting to the specific needs of CWR. For example, in many countries, species management, action or recovery plans have been prepared for rare or endangered wild species irrespective of their economic or social value, and it should therefore be possible to apply similar actions to CWR (indeed, some of the threatened species that are the subject of such plans may be CWR). It means different agencies learning to work much more closely than has hitherto been the norm.

Therefore, it is hoped that those charged with the task of taking forward the various actions will view the strategy in the context of existing policy, legislation and conservation initiatives where possible, rather than viewing it as yet another obligation in the seemingly endless demands of biodiversity conservation. The strategy can also provide the backdrop for the development of specific national and regional policy and legislative instruments.

Acknowledgements

The concepts discussed in this chapter were stimulated by PGR Forum (the European Crop Wild Relative Diversity Assessment and Conservation Forum – EVK2-2001-00192 – available at: http://www.pgrforum.org/), funded by the

EC Fifth Framework Programme for Energy, Environment and Sustainable Development. We also wish to acknowledge the contributions of delegates to the First International Conference on Crop Wild Relative Conservation and Use held in Agrigento, Sicily, Italy, 14–17 September 2005.

References

Heywood, V.H. and Dulloo, M.E. (2006) In Situ *Conservation of Wild Plant Species – A Critical Global Review of Good Practices*. IPGRI and FAO, Rome, Italy.

Heywood, V.H., Kell, S.P. and Maxted, N. (eds) (2007) *Draft Global Strategy for Crop Wild Relative Conservation and Use*. University of Birmingham, Birmingham, UK. Available at: http://www.pgrforum.org/Documents/Conference/Global_CWR_Strategy_DRAFT_11-04-07.pdf

Kell, S.P., Scholten, M. and Avanzato, D. (2005a) The First International Conference on Crop Wild Relative Conservation and Use: setting the scene for global action on conservation and sustainable use of CWR. *Crop Wild Relative* 5, 4–10.

Kell, S.P., Knüpffer, H., Jury, S.L., Maxted, N. and Ford-Lloyd, B.V. (2005b) *Catalogue of Crop Wild Relatives for Europe and the Mediterranean*. Available online via the Crop Wild Relative Information System (CWRIS – http://cwris.ecpgr.org/) and on CD-ROM. University of Birmingham, Birmingham, UK.

PGR Forum (2003–2005a) *European Crop Wild Relative Diversity Assessment and Conservation Forum*. University of Birmingham, Birmingham, UK. Available at: http://www.pgrforum.org/

PGR Forum (2003–2005b) *First International Conference on Crop Wild Relative Conservation and Use*. University of Birmingham, Birmingham, UK. Available at: http://www.pgrforum.org/Conference.htm

Appendix

Framework of the draft Global Strategy for Crop Wild Relative Conservation and Use. (To read the full draft working document, see Heywood *et al.*, 2007.)

Goal

Effective conservation and sustainable use of crop wild relatives, including all wild plant species of socio-economic value, at national, regional and global levels.

Objective 1: Prepare national CWR strategic action plans

Each country to prepare a national action plan for the inventory, survey, conservation (*in situ* and *ex situ*) and sustainable use of CWR by 2010.

Targets

1. Each country to develop a National Strategic Action Plan (NSAP) for CWR conservation and use;
2. Review existing biodiversity strategies and action plans in relation to CWR conservation and use;
3. NSAP for CWR conservation and use to be fully integrated into and complement existing national and regional biodiversity strategies and action plans;
4. Define responsibilities for preparation and implementation of NSAP;
5. Designate a National Focal Point for CWR;
6. Select political or social targets associated with CWR conservation, e.g. increase of awareness.

Objective 2: Prepare national CWR inventories

Each country to:

1. Prepare an inventory of the CWR growing in its territory by 2015;
2. Explore the possibility of cooperating with neighbouring countries in the production of regional catalogues by 2010.

Targets

1. Produce checklist of CWR;
2. Identify data sources, e.g. GIS and climate data;
3. Collate and manage information.

Objective 3: Establish a global mechanism or clearing house

FAO, CGIAR/IPGRI, IUCN and other relevant bodies, in cooperation with parties to the CBD, to put in place a global mechanism or clearing house for CWR conservation and use by 2010.

Targets
1. Report CWR diversity to FAO as part of Global Plan of Action for the Conservation and Sustainable Utilization of PGRFA commitments;
2. Establish global mechanism for CWR information.

Objective 4: Create national priority CWR lists and identify priority CWR sites

Each country to prepare a national priority list of CWR in need of urgent conservation action using existing priority-determining criteria by 2015.

Targets
1. Develop national priority CWR lists;
2. Develop conservation action plans for priority taxa;
3. Develop individual solutions to national priorities;
4. Identify within each country, five priority sites for the establishment of active CWR genetic reserves. These reserves should form an interrelated network of internationally, regionally and nationally important CWR genetic reserve sites for *in situ* conservation.

Objective 5: Create regional and global CWR priority lists and identify priority CWR sites

FAO, IPGRI, the CBD and other relevant bodies, in cooperation with parties to the CBD, to put in place a system for preparing regional and global CWR priority lists by 2015.

Targets
1. Develop regional and global priority CWR lists;
2. Develop conservation action plans for priority taxa;
3. Develop individual solutions to certain regional and global priorities;
4. Identify globally, and within each region, a small number of priority sites (global = 100, regional = 25) for the establishment of active CWR genetic reserves. These reserves should form an interrelated network of internationally and regionally important CWR genetic reserve sites for *in situ* conservation; that could then be further integrated with a minimum of five national priority sites.

Objective 6: Establish protocols for CWR information management and dissemination and provide national and global CWR information management systems

Existing information management systems for CWR to be harmonized and applied by national, regional and global organizations by 2015.

Targets
1. Integrate Crop Wild Relative Information System (CWRIS) into European Plant Genetic Resources Internet search catalogue (EURISCO);

2. Agree data standards and mechanisms for collation of National CWR Inventory (NI) data;
3. Harmonize CWRIS/EURISCO with the international GEF CWR project;
4. Improve data entry tools like GRIS.

Objective 7: Develop effective means of conserving and using CWR *in situ*

Targets

1. Establish the international, regional and national active genetic reserves identified under Objective 5, Target d;
2. National action to be taken to record the presence of CWR in each country's protected areas system;
3. Each country to assess whether the existing network of protected areas adequately represents the full range of national CWR diversity, and suggest additional reserve locations where required;
4. Link CWR *in situ* reserve sites with other current initiatives, such as the important plant area initiative and Natura 2000 network, and where appropriate establish genetic reserves linked to these initiatives;
5. Encourage UNESCO Man and the Biosphere Programme (MAB) to complete its floristic inventories in MAB Reserves and highlight which CWRs are known to occur in each;
6. Raise awareness among protected area managers of the importance of CWR and request them to take into account the maintenance and conservation needs of CWR when drawing up or revising management plans;
7. Involve local communities in planning community conservation of CWR and encourage them to participate in the management of reserves and other protected or non-protected areas in which CWR are known to occur;
8. Examine the potential role of microreserves in CWR conservation;
9. Countries and agencies to review the possibilities of conservation of CWR outside protected areas, including agroecosystems;
10. Countries and agencies to review possibilities for conservation of CWR outside protected areas through policy decisions (easements, set-aside and other appropriate mechanisms);
11. Promote traditional farming systems for both landrace and CWR conservation;
12. Establish protocols for the management and monitoring of genetic diversity in CWR populations;
13. Publish case studies for the complete genetic reserve location, establishment and routine maintenance process to act as templates for subsequent projects;
14. Publish protocols and examples of the integration of CWR *in situ* conservation and use as a means of promoting CWR use.

Objective 8: Develop effective means of conserving and using CWR *ex situ*

Targets
1. Undertake gap analysis of CWR representation in national, regional and global gene banks, including field gene banks (clonal collections);
2. Systematic collection of CWR diversity identified as being underrepresented in national, regional and global gene banks, including field gene banks (clonal collections);
3. In cooperation with International Association of Botanic Gardens (IABG), Botanic Gardens Conservation International (BGCI) and regional and national botanic garden networks, review the presence and status of CWR accessions in botanic gardens, universities and other holders of *ex situ* living collections;
4. Promote the establishment of local community seed banks for wild-harvested CWR and those cultivated for local usage.
5. Publish protocols and examples of the integration of CWR *ex situ* conservation and use as a means of promoting CWR use.

Objective 9: Assess CWR conservation and threat status

1. Red List all priority CWR taxa by 2015;
2. Assess the likely impact of global change on CWR diversity and survival by 2015;
3. Assess CWR diversity as gene sources to assist the likely impact of global change on crops;
4. Establish common protocols for assessing genetic erosion of CWR, including proxy indicators;
5. Establish common protocols for assessing crop to CWR gene flow and its consequences by 2010.

Targets
1. Assume responsibility to lead global Red Listing of all priority CWR taxa;
2. Assume responsibility to lead national Red Listing of all priority CWR taxa;
3. Establish functional expert committees at national level, comprising all key people;
4. Include a section on progress in national Red Listing in national reports on the state of the world PGR;
5. Undertake Red List threat assessments of national endemic CWR taxa;
6. Produce a first draft of a global Red List of priority CWR taxa based on national endemics;
7. Produce a report based on existing documentation on global change identifying areas at greater risk and priority CWR that are more vulnerable;
8. Develop and apply protocols for conservation assessment of *in situ*- and *ex situ*-conserved CWR diversity;

9. Conduct population-level research on selected CWR to aid threat and conservation assessment;
10. Produce national systematic conservation assessments;
11. Organize an international conference on protocols for assessing genetic erosion;
12. Submit research project proposal on protocols for assessing genetic erosion;
13. Organize an international conference on protocols for assessing crop to CWR gene flow and its consequences;
14. Submit research project proposal on protocols for assessing crop to CWR gene flow and its consequences;
15. Submit research project proposal on impact of climate change on longer term sustainability of CWR in genetic reserves;
16. Review impact of wild harvesting on CWR population sustainability.

Objective 10: Ensure effective security and legislation for CWR

Targets
1. Undertake a review of which CWR species are included in existing national, regional and global policy and legislative instruments, and afford protection to priority CWR taxa that are not already included;
2. Review the effectiveness of existing policy and legislation for the conservation and sustainable use of CWR;
3. Undertake a review of available funding opportunities for CWR conservation;
4. Encourage and promote close intersectoral coordination at national level with broad involvement of stakeholders;
5. Incorporate CWR into relevant national strategies.

Objective 11: Promote sustainable utilization of CWR

Targets
1. Promote sustainable use of CWR;
2. Encourage the use of CWR in breeding programmes.

Objective 12: Initiate education and public awareness programmes on the importance of CWR

Targets
1. Identify a means of integrating awareness of the importance of CWR into existing education and public awareness programmes at national, regional and global levels;
2. Identify and develop public awareness campaigns through existing groups that reach wide audiences, such as NGOs, the private sector, national associations of producers, farmers and industry;
3. Promote awareness of the importance of CWR conservation and use for professional protected area managers, plant breeders, biotechnologists and other potential CWR stakeholders.

Index

Table 47.1. Species richness, biomass and production of main Mediterranean succession ecological stages. (Adapted from Alés, 1999.)

Species richness	Forest 119/0.1 ha	Montado 135/0.1 ha	Matorral 10–25/0.1 ha	Pasture 135/0.1 ha
	Biomass (t/ha)			
Leaves	6–8	3.3	0.6–9.9	1–12
Wood	100–330	21.7	1.2–125	–
Roots	67–127	?	13.3	3–4
	Production (t/ha/year)			
Leaves, flowers, fruits	2.0–7.0	3.4–5.8	1.6–3.8	1–12
Wood	3.6–6.4	?	0.3–2	–

role in preventing desertification, decreasing soil erosion, retaining water and providing shade for wildlife (Pulido *et al.*, 2001; David *et al.*, 2004).

Changes in Portuguese agriculture during the 20th century led to modifications in these traditional agro-sylvo-pastoral systems to compensate for the loss of economic viability of their products (Goes, 1991). The traditional management was either replaced by intensification or extensification practices that resulted in different patterns of agricultural production and landscape (Pinto-Correia, 1993). Intensification in management occurred by additional deforestation of evergreen oaks and intensive cropping of fast-growing species, primarily wheat. Other changes, associated with intensification of agricultural practices in *montados*, included higher livestock densities, abandonment of forestry practice and conversion of agroforest or even exclusively forest land (clines and erosionable soils) into agricultural land (Pereira and Fonseca, 2003). Recently, the fall in cereal prices has led to conversion of *montados* into intensively irrigated monoculture of tomato, sunflower and maize, as well as conversion to olive orchards, vineyards or even reforestation using fast-growing species such as *Eucalyptus* and *Pinus* (Díaz *et al.*, 1997). Intensification has often resulted in increased susceptibility of evergreen oaks to pests and diseases, reduction of water reservoirs, wild habitat destruction, biodiversity decrease, soil erosion and fertility decrease and even to stages close to desertification (Pereira and Fonseca, 2003). In addition, it has been suggested that the intensification in management has had a negative impact upon the survival of acorns, seedlings and young trees so that the replacement of the old trees is unlikely (Pulido *et al.*, 2001). Alternatively, extensification in management resulted in changes in vegetation characterized by the appearance of different plant communities representative of the different successional stages of the Mediterranean ecosystems including *matorral* (represented by tall-shrub communities) and *phrygana* (represented by the under-shrub communities), as well as pasture land (represented by the annual, biennial and perennial herbaceous communities) (Capelo, 1996). Both intensification and extensification have a negative impact on biodiversity and the agricultural or forest use potential of the *montados,* is highly reduced (Goes, 1991).

The former National Forest Inventory (IFN) covering 1995–98 (Direcção Geral de Florestas, 2001) showed that the total area occupied by *Q. suber* increased slightly by 4.9% (33.000 ha) between 1990–92 and 1995–98, mainly due to the economic growth of the cork industry in Portugal. However, the later IFN survey covering 2005–06 (Direcção Geral de Florestas, 2006) showed the area has decreased by 10% from 1998 to 2006 (Table 47.2). For *Q. ilex* subsp. *rotundifolia*, although its occupation area suffered a severe 20% decrease between 1963 and 1989, it had stabilized until 2005–06 (Table 47.2). None the less, these numbers represent alterations in the occupied area only and do not represent tree mortality within each *montado*. The IFN (1995–98) survey concluded that the modal class for tree densities of *Q. ilex* subsp. *rotundifolia* is the smallest (>40 trees/ha), which represents 60–70% of the total occupied area while the other extreme density class (>200 tree/ha) represents only 1.6% of the total occupied area (DGF, 2001). Furthermore, as the density class <40 trees/ha is broad, Portuguese forest experts argue that a significant proportion of the occupied area within this class will average between 15 to 20 trees/ha and that the decay in tree density since the 1950s is likely to exceed the 60% in tree density on average per ha of *Q. ilex* subsp. *rotundifolia* (J.M. Abreu, Porto, 2003, personal communication).

Recently, these traditional ecosystems have become valued at national and international policy-making levels for their agronomic and ecological value, their wealth of biodiversity and potential for tourism. Consequently, *montados* are listed in the Appendix I of the European Community Habitats Directive (92/43/CEE). At the national level, extensive legislation has led to the implementation of measures towards the protection of *montados* of *Q. suber* and *Q. ilex* subsp. *rotundifolia*, the last being in 2001 (Decreto-Lei no. 169/2001). However, despite the existing legislation context for conservation and sustainable use of *montados*, an actual implementation strategy is still not well defined.

Long-term conservation of agricultural ecosystems must be linked to their productive value. A recent multipurpose conservation strategy was proposed by Varela and Eriksson (1995) for *Q. suber* in Portugal with the joint objectives of creating ideal conditions for conservation and future species evolution, as well as improvement of the production of the target species. They comment that the development of selection programmes for improvement of agricultural traits of *Q. suber* is necessary for the long-term conservation of its *montados*, as the increase in their economic viability is required for obtaining farmers' and landowners' commitment to conservation aims. The same argument by Varela and Eriksson (1995) for *Q. suber* can be applied to the conservation and sustainable use of *Q. ilex* subsp. *rotundifolia* in Portugal, which alongside *Q. suber* forms the basis of *montados* traditional Mediterranean wood pasture ecosystems in

Table 47.2. Changes in area occupied by the dominant oak species in Portuguese Montados. (From Direcção Geral de Florestas, 2005–2006.)

	1000 ha					
Forest species	1963–1966	1968–1980	1980–1989	1990–1992	1995–1998	2005–2006
Quercus suber	637	657	664	687	713	643
Quercus ilex subsp. *rotundifolia*	579	536	465	–	462	460